Concentrated Acids and Bases

NAME	APPROXIMATE WEIGHT PERCENT	APPROXIMATE MOLARITY	ML OF REAGENT NEEDED TO PREPARE 1 L OF ~1.0M SOLUTION
Acid			
Acetic	99.8	17.4	57.5
Hydrochloric	37.2	12.1	82.6
Hydrofluoric	49.0	28.9	34.6
Nitric	70.4	15.9	62.9
Perchloric	70.5	11.7	85.5
Phosphoric	85.5	14.8	67.6
Sulfuric	96.0	18.0	55.6
Base			
Ammonia[†]	28.0	14.5	69.0
Sodium hydroxide	50.5	19.4	51.5
Potassium hydroxide	45.0	11.7	85.5

[†] 28.0% ammonia is the same as 56.6% ammonium hydroxide.

Quantitative Chemical Analysis

Quantitative Chemical Analysis

DANIEL C. HARRIS
FRANKLIN AND MARSHALL COLLEGE

W. H. FREEMAN AND COMPANY
San Francisco

Project Editor: Pearl C. Vapnek
Copy Editor: David J. Cross
Designer: Robert Ishi
Production Coordinator: Linda Jupiter
Illustration Coordinator: Richard Quiñones
Artist: Eric Hieber
Compositor: Syntax International
Printer and Binder: Arcata Book Group

Cover illustration: Adapted from D. C. Harris and M. H. Gelb, *Biochim. Biophys. Acta,* **623,** 1 (1980).

Library of Congress Cataloging in Publication Data

Harris, Daniel C., 1948–
   Quantitative chemical analysis.

   Includes index.
   1. Chemistry, Analytic—Quantitative.   I. Title.
QD101.2.H37   545   82-7421
ISBN 0-7167-1347-0   AACR2

Copyright © 1982 by W. H. Freeman and Company

No part of this book may be reproduced by any mechanical, photographic, or electronic process, or in the form of a phonographic recording, nor may it be stored in a retrieval system, transmitted, or otherwise copied for public or private use, without written permission from the publisher.

Printed in the United States of America

1 2 3 4 5 6 7 8 9 0   KP   0 8 9 8 7 6 5 4 3 2

Contents

Preface   xiii

1   Units and Concentrations   1

1-1   SI UNITS   1

1-2   EXPRESSIONS OF CONCENTRATION   5

    Molarity   5
    Formality   6
    Percent Composition   6
    Parts per Million and Its Relatives   8
    Other Concentration Units   8

1-3   PREPARING SOLUTIONS   9

2   Tools of the Trade   13

2-1   LAB NOTEBOOK   13

2-2   ANALYTICAL BALANCE   14

    Principle of Operation   15
    Effect of Buoyancy   16
    Errors in Weighing   17

2-3   BURETS   17

2-4   VOLUMETRIC FLASKS   19

2-5   PIPETS AND SYRINGES   20

    Using a Transfer Pipet   21
    Delivering Small Volumes   21

2-6   FILTRATION   22

2-7   DRYING   23

3   Experimental Error   27

3-1   SIGNIFICANT FIGURES   27

    Arithmetic Operations   28

3-2   SIGNIFICANT FIGURES AND GRAPHS   31

3-3   TYPES OF ERRORS   32

    Systematic Error   32
    Random Error   32
    Precision and Accuracy   33
    Absolute and Relative Error   33

3-4   PROPAGATION OF ERROR   34

    Addition and Subtraction   34
    Multiplication and Division   35
    Mixed Operations   36
    Comment on Significant Figures   37

4   Statistics   41

4-1   GAUSSIAN ERROR CURVE   42

    Mean Value and Standard Deviation   43
    Standard Deviation and Probability   44

4-2   STUDENT'S t 46

    Confidence Intervals   46
    Comparison of Means   48

4-3   DEALING WITH BAD DATA   51

4-4   FINDING THE "BEST" STRAIGHT LINE   53

    Method of Least Squares   53
    How Reliable Are Least-Squares Parameters?   55
    A Practical Example   57

5   Solubility of Ionic Compounds   61

5-1   REVIEW OF CHEMICAL EQUILIBRIUM   61

    Manipulation of Equilibrium Constants   62
    Equilibrium and Thermodynamics   63
    Le Châtelier's Principle   65

5-2   SOLUBILITY PRODUCT   67

    Box 5-1   The mercurous ion   68

5-3   COMMON ION EFFECT   68

    Box 5-2   The logic of approximations   70

vi

CONTENTS

5-4   SEPARATION BY PRECIPITATION   70

5-5   COMPLEX FORMATION   72

Box 5-3   Notation for formation constants   73

Comment on Equilibrium Constants   74

6   Activity   81

6-1   ROLE OF IONIC STRENGTH IN IONIC EQUILIBRIA   81

Effect of Ionic Strength on Solubility of Salts   82
Explanation of Increased Solubility   82

6-2   ACTIVITY COEFFICIENTS   83

Demonstration 6-1   Effect of ionic strength on ion dissociation   84

Activity Coefficients of Ions   85
Activity Coefficients of Nonionic Compounds   88
High Ionic Strengths   88
Mean Activity Coefficient   89

6-3   USING ACTIVITY COEFFICIENTS   89

7   Systematic Treatment of Equilibrium   93

7-1   CHARGE BALANCE   93

7-2   MASS BALANCE   95

7-3   SYSTEMATIC TREATMENT OF EQUILIBRIUM   96

General Prescription   96
Ionization of Water   97
Solubility of Hg_2Cl_2 99

7-4   DEPENDENCE OF SOLUBILITY ON pH   100

Solubility of CaF_2 100

Box 7-1   All right, Dan, how would you really solve the CaF_2 problem?   102

Solubility of HgS   102

Box 7-2   pH and tooth decay   103

Recapitulation   105
Comments   106

8   Gravimetric Analysis   109

8-1   EXAMPLES OF GRAVIMETRIC ANALYSIS   109

Combustion Analysis   110

8-2   PRECIPITATION PROCESS   111

Solubility   111

Demonstration 8-1   Colloids and dialysis   112

Filterability   112
Purity   117
Composition of Product   118

8-3   SCOPE OF GRAVIMETRIC ANALYSIS   119

8-4   CALCULATIONS OF GRAVIMETRIC ANALYSIS   119

9   Precipitation Titrations   129

9-1   PRINCIPLES OF VOLUMETRIC ANALYSIS   129

9-2   CALCULATIONS OF VOLUMETRIC ANALYSIS   131

9-3   EXAMPLE OF A PRECIPITATION TITRATION   136

9-4   THE SHAPE OF A PRECIPITATION TITRATION CURVE   136

Box 9-1   Turbidimetry and nephelometry   137

Before the Equivalence Point   139
At the Equivalence Point   140
After the Equivalence Point   141
The Shape of the Titration Curve   141

9-5   TITRATION OF A MIXTURE   144

9-6   ENDPOINT DETECTION   146

Mohr Titration   146
Volhard Titration   147
Fajans Titration   148

Demonstration 9-1   Fajans titration   149

Scope of Indicator Methods   150

10   Introduction to Acids and Bases   155

10-1   WHAT ARE ACIDS AND BASES?   155

Two Classifications of Acid–Base Behavior   155
Conjugate Acids and Bases   157
Nature of H^+ and OH^- 157

10-2   pH   159

The pH Scale   160

10-3   STRENGTHS OF ACIDS AND BASES   161

Strong Acids and Bases   161

Demonstration 10-1   The HCl fountain   162

vii

CONTENTS

Box 10-1 The strange behavior
of hydrofluoric acid 164

Weak Acids and Bases 164
pK 167

Box 10-2 Carbonic acid 168

Relation Between K_A and K_B 169
Weak Is Conjugate to Weak 170
Using Appendix G 170

11 Acid–Base Equilibria 175

11-1 STRONG ACIDS AND BASES 175
The Dilemma 176
The Cure 176
Water Almost Never Produces 10^{-7} M H^+
and 10^{-7} M OH^- 178

11-2 WEAK ACIDS 178
A Typical Problem 178
A Better Way 180
Return to Chemistry 181
Essence of Weak-Acid Problems 182

Demonstration 11-1 Conductivity
of weak electrolytes 183

11-3 WEAK BASES 183
Standard Weak-Base Problem 184
Conjugate Acids and Bases, Revisited 185

11-4 BUFFERS 186
Mixing a Weak Acid and Its Conjugate Base 187
Henderson–Hasselbalch Equation 188

A Buffer in Action 190

Box 11-1 Strong plus weak react completely 191

Preparing a Buffer in Real Life! 193
Buffer Capacity 193

Demonstration 11-2 How buffers work 194

Limitations of Buffers 196
Summary 199

11-5 DIPROTIC ACIDS AND BASES 199
Acidic Form, H_2L^+ 203
Basic Form, L^- 204
Intermediate Form, HL 205

Box 11-2 More on successive approximations 208

Summary 210
Diprotic Buffers 211

11-6 POLYPROTIC ACIDS AND BASES 213

12 Acid–Base Titrations 221

12-1 TITRATION OF STRONG ACID
WITH STRONG BASE 221
Region 1: Before the Equivalence Point 224
Region 2: At the Equivalence Point 224
Region 3: After the Equivalence Point 225
Titration Curve 225

12-2 TITRATION OF WEAK ACID WITH STRONG BASE 225
Region 1: Before Base Is Added 226
Region 2: Before the Equivalence Point 226

Box 12-1 The answer to a nagging question 228

Region 3: At the Equivalence Point 228
Region 4: After the Equivalence Point 229
Titration Curve 230

12-3 TITRATION OF WEAK BASE WITH STRONG ACID 231

12-4 TITRATIONS IN DIPROTIC SYSTEMS 233
A Typical Case 233
Blurred Endpoints 236

12-5 FINDING THE ENDPOINT 237
Indicators 238

Demonstration 12-1 Indicators
and the acidity of CO_2 239

Potentiometric Endpoint Detection 240

Box 12-2 What does a negative pH mean? 242

12-6 PRACTICAL NOTES 246

12-7 TITRATIONS IN NONAQUEOUS SOLVENTS 248

13 Advanced Topics in Acid–Base Chemistry 255

13-1 WHICH IS THE PRINCIPAL SPECIES? 255

13-2 FRACTIONAL COMPOSITION EQUATIONS 257
Monoprotic Systems 257
Diprotic Systems 258

13-3 ISOELECTRIC AND ISOIONIC pH 260

13-4 REACTIONS OF WEAK ACIDS WITH WEAK BASES 261
Box 13-1 Isoelectric focussing 262
Case 1: K Is Large ($K \gg 1$) 264
Case 2: K Is Not Large 265
Case 3: Equimolar Mixture of HA and B 266

viii

CONTENTS

14 EDTA Titrations 271

14-1 METAL CHELATE COMPLEXES 271

Chelate Effect 272

14-2 EDTA 273

Acid–Base Properties 273

Box 14-1 Chelation therapy and thalassemia 274

EDTA Complexes 276
Conditional Formation Constant 277

14-3 EDTA TITRATION CURVES 280

Titration Calculations 280
Effect of Auxiliary Complexing Agents 283

14-4 METAL ION INDICATORS 283

Demonstration 14-1 Metal ion indicator color changes 284

14-5 EDTA TITRATION TECHNIQUES 286

Direct Titration 286
Back Titration 287
Displacement Titrations 288
Indirect Titrations 288
Masking 288

15 Fundamentals of Electrochemistry 293

15-1 BASIC CONCEPTS 293

Relation Between Chemistry and Electricity 293
Electrical Measurements 294

15-2 GALVANIC CELLS 297

A Cell in Action 298
Cell Conventions 299
Salt Bridge 300

Demonstration 15-1 The human salt bridge 302

Line Notation 302

15-3 STANDARD POTENTIALS 303

Table of Standard Potentials 305
Using E^0 Values 305

15-4 NERNST EQUATION 307

Box 15-1 Latimer diagrams 308

Using the Nernst Equation 309

Box 15-2 The cell potential does not depend on how you write the cell reaction 310

Box 15-3 Concentrations in the operating cell 312

Single-Electrode Potentials 313

15-5 RELATION OF E^0 AND THE EQUILIBRIUM CONSTANT 314

15-6 USING CELLS AS CHEMICAL PROBES 316

15-7 BIOCHEMISTS USE $E^{0\prime}$ 320

Relation Between E^0 and $E^{0\prime}$ 321

16 Electrodes and Potentiometry 333

16-1 REFERENCE ELECTRODES 333

How Analytical Chemists Write the Nernst Equation 335
Common Reference Electrodes 336

16-2 METALLIC INDICATOR ELECTRODES 336

Demonstration 16-1 Potentiometry with an oscillating reaction 337

16-3 WHAT IS A JUNCTION POTENTIAL? 342

16-4 pH MEASUREMENT WITH A GLASS ELECTRODE 343

Calibrating a Glass Electrode 346
Errors in pH Measurement 348

16-5 ION-SELECTIVE ELECTRODES 349

Solid-State Electrodes 350
Liquid-Based Ion-Selective Electrodes 351

Box 16-1 Microelectrodes inside living cells 353

Compound Electrodes 354
Use and Abuse of Ion-Selective Electrodes 355

17 Redox Titrations 363

17-1 THE SHAPE OF A REDOX TITRATION CURVE 364

A Slightly More Complicated Redox Calculation 370

Demonstration 17-1 Potentiometric titration of Fe^{2+} with MnO_4^- 373

17-2 TITRATION OF A MIXTURE 373

17-3 REDOX INDICATORS 374

The Starch–Iodine Complex 376

17-4 COMMON REDOX REAGENTS 376

Adjustment of Analyte Oxidation State 376
Oxidation with Potassium Permanganate 379
Oxidation with Cerium(IV) 382
Oxidation with Potassium Dichromate 383

Methods Involving Iodine 383
Analysis of Organic Compounds with Periodic Acid 388
Titrations with Reducing Agents 389

18 Electrogravimetric and Coulometric Analysis 395

18-1 ELECTROLYSIS: PUTTING ELECTRONS TO WORK 395

18-2 WHY THE VOLTAGE CHANGES
WHEN CURRENT FLOWS THROUGH A CELL 397

Demonstration 18-1 Electrochemical writing 398

Ohmic Potential 399

Box 18-1 Photo-assisted electrolysis 400

Concentration Polarization 402
Overpotential 403

18-3 ELECTROGRAVIMETRIC ANALYSIS 405

Current–Voltage Behavior During Electrolysis 408
Electrolysis at Constant Applied Potential 410
Constant-Current Electrolysis 412
Controlled-Potential Electrolysis 412

18-4 COULOMETRIC ANALYSIS 413

Some Details of Coulometry 416

Box-18-2 Measuring the Faraday constant 418

Counting Coulombs 421

19 Voltammetry 427

19-1 POLAROGRAPHY 427

Why We Use the Dropping-Mercury Electrode 429

19-2 SHAPE OF THE POLAROGRAM 430

Diffusion Current 431
Residual Current 433
Shape of the Polarographic Wave 434
Relation Between $E_{1/2}$ and E^0 435
Other Factors Affecting the Shape
of the Polarogram 437

19-3 APPLICATIONS OF POLAROGRAPHY 438

Quantitative Analysis 439
Polarographic Study of Chemical Equilibrium 443
Polarographic Study of Chemical Kinetics 445

19-4 POWERFUL VARIATIONS OF POLAROGRAPHY 446

Differential Pulse Polarography 447
Stripping Analysis 451
Cyclic Voltammetry 452

Box 19-1 Thin-layer differential pulse
polarography 453

Box 19-2 An optically transparent
thin-layer electrode 455

19-5 AMPEROMETRIC TITRATIONS 456

Systems with One Polarizable Electrode 456
Systems with Two Polarizable Electrodes 457

Box 19-3 The Clark oxygen electrode 458

Demonstration 19-1 The Karl Fischer jacks
of a pH meter 460

20 Spectrophotometry 467

20-1 PROPERTIES OF LIGHT 467

20-2 ABSORPTION OF LIGHT 468

Box 20-1 Why is there a logarithmic relation
between transmittance and concentration? 471

Demonstration 20-1 Absorption spectra 472

20-3 WHAT HAPPENS WHEN A MOLECULE
ABSORBS LIGHT? 473

Excited States of Molecules 473
What Happens to Absorbed Energy? 476

Box 20-2 Fluorescent lamps
and little-known fluorescent objects 478

20-4 THE SPECTROPHOTOMETER 479

Double-Beam Strategy 480
Major Components 481

20-5 ERRORS IN SPECTROPHOTOMETRY 485

Another Look at Beer's Law 486
Instrument Errors 487

20-6 TYPICAL ANALYTICAL PROCEDURES 488

Serum Iron Determination 489
Analysis of a Mixture 491

Box 20-3 Continuous-flow spectrophotometric
analysis 492

Spectrophotometric Titrations 495
Measuring an Equilibrium Constant:
The Scatchard Plot 496

20-7 LUMINESCENCE 499

Relation Between Absorption
and Emission Spectra 499
Emission Intensity 500

Box 20-4 Absorption-quenching
of emission spectra 503

Luminescence in Analytical Chemistry 504

CONTENTS

21 Atomic Spectroscopy 513

21-1 ABSORPTION, EMISSION, AND FLUORESCENCE 513

21-2 ATOMIZATION: FLAMES, FURNACES, AND PLASMAS 516

Premix Burner 516
Furnaces 517
Inductively Coupled Plasma 518
Effect of Temperature in Atomic Spectroscopy 519

21-3 INSTRUMENTATION 521

Source of Radiation 521
The Spectrophotometer 523

21-4 ANALYTICAL METHODS 524

Standard Curve 525

Box 21-1 The flame photometer in clinical chemistry 525

Standard Addition Method 526
Internal Standard Method 527

21-5 INTERFERENCE 528

22 Introduction to Analytical Separations 533

22-1 SOLVENT EXTRACTION 534

pH Effects 535
Extraction with a Metal Chelator 537
Some Extraction Strategies 538

Demonstration 22-1 Extraction with dithizone 540
Box 22-1 Crown ethers and ionophores 540

22-2 COUNTERCURRENT DISTRIBUTION 542

Theoretical Distribution 545
Bandwidth and Resolution 548

22-3 CHROMATOGRAPHY 551

Some Terminology 553
Plate Theory of Chromatography 554
Rate Theory of Chromatography 556
Resolution 558
A Touch of Reality 560

23 Chromatographic Methods 567

23-1 GAS CHROMATOGRAPHY 567

A Gas Chromatograph 567
The Chromatogram 570
Columns 573

Box 23-1 Glass capillary columns 574

Detectors 578
Quantitative Analysis 580

23-2 LIQUID CHROMATOGRAPHY 583

Classical Liquid Chromatography 583
High Performance Liquid Chromatography (HPLC) 585

23-3 ION-EXCHANGE CHROMATOGRAPHY 587

Ion Exchangers 588
Ion-Exchange Equilibria 592
Conducting Ion-Exchange Chromatography 594
Applications 596

23-4 MOLECULAR-EXCLUSION CHROMATOGRAPHY 597

Principles 597
Types of Gels 598
Applications 600

23-5 AFFINITY CHROMATOGRAPHY 601

23-6 PRACTICAL NOTES 602

Preparing the Stationary Phase 603
Pouring the Column 603
Applying the Sample 604
Running the Column 604
Gradients 605
Accessories 605

24 Experiments 611

24-1 CALIBRATION OF VOLUMETRIC GLASSWARE 611

Calibration of 50 mL Buret 613
Other Calibrations 614

24-2 GRAVIMETRIC DETERMINATION OF CALCIUM AS $CaC_2O_4 \cdot H_2O$ 615

24-3 GRAVIMETRIC DETERMINATION OF IRON AS Fe_2O_3 616

24-4 PREPARING STANDARD ACID AND BASE 617

Standard NaOH 617
Standard HCl 618

24-5 ANALYSIS OF A MIXTURE OF CARBONATE
AND BICARBONATE 619

24-6 KJELDAHL NITROGEN ANALYSIS 620

Digestion 620
Distillation 622

24-7 ANALYSIS OF AN ACID–BASE TITRATION CURVE:
THE GRAN PLOT 622

24-8 EDTA TITRATION OF Ca^{2+} AND Mg^{2+}
IN NATURAL WATERS 624

24-9 SYNTHESIS AND ANALYSIS
OF AMMONIUM DECAVANADATE 625

Synthesis 625
Analysis of Vanadium with $KMnO_4$ 626
Analysis of Ammonium Ion 626

24-10 IODIMETRIC TITRATION OF VITAMIN C 627

Preparation and Standardization
of Thiosulfate Solution 627
Analysis of Vitamin C 627

24-11 POTENTIOMETRIC HALIDE TITRATION WITH Ag^+ 627

24-12 ELECTROGRAVIMETRIC ANALYSIS OF COPPER 629

24-13 POLAROGRAPHIC MEASUREMENT
OF AN EQUILIBRIUM CONSTANT 629

24-14 COULOMETRIC TITRATION OF CYCLOHEXENE
WITH BROMINE 631

24-15 SPECTROPHOTOMETRIC DETERMINATION OF IRON
IN VITAMIN TABLETS 631

24-16 SPECTROPHOTOMETRIC MEASUREMENT
OF AN EQUILIBRIUM CONSTANT 633

24-17 PROPERTIES OF AN ION-EXCHANGE RESIN 634

24-18 QUANTITATIVE ANALYSIS
BY GAS CHROMATOGRAPHY OR HPLC 636

Appendixes 639

A Logarithms and Exponents 641
B Graphs of Straight Lines 644
C A Detailed Look at Propagation of Error 646
D Oxidation Numbers and Balancing
Redox Equations 649
E Normality 656
F Solubility Products 659
G Acid Dissociation Constants 663
H Standard Reduction Potentials 673

Glossary 683
Solutions to Exercises 701
Answers to Problems 731
Index 735

The Viking 2 landing site in Utopia Planitia on Mars in September 1976. Soil samples from near the base of the lander were scooped up for chemical analysis. [Jet Propulsion Laboratory.]

Preface

Almost every branch of science addresses fundamental questions to the analytical chemist. For example, the twin Viking spacecraft that set down on the surface of Mars in 1976 were actually analytical chemistry laboratories in flight. Each contained equipment to analyze the composition of the atmosphere and the soil, to search for telltale organic substances in the soil, and to detect life-related chemistry of organisms in the soil.

Life as we know it cannot exist without a plethora of organic compounds. Even in the sparsely populated Antarctic soil, there are enough organisms to yield myriad detectable organic vapors when the soil is heated. The gas chromatograms on the following page contrast the abundant vapors from Antarctic soil with the absence of any detectable vapors from Martian soil. The few small peaks in part b are from solvents used to clean the Viking equipment before it left Earth. Release of as few as 2 nanograms $(2 \times 10^{-9}$ g) of a compound per gram of soil would have been detectable. Some of Viking's biological experiments produced positive results that could be attributed to the inorganic chemistry of the superoxide ion, O_2^-, in the soil. Taken altogether, the weight of evidence from the Viking analytical experiments is that life does not exist at the two landing sites.

The questions addressed to the analytical chemist can be broadly classified into two categories: "What is in the sample?" (qualitative analysis) and "How much is in the sample?" (quantitative analysis). This text provides a foundation for understanding how the latter question is approached in the laboratory, although the two questions are usually entwined.

The need to answer these questions is so widespread that many university departments require students to take an introductory analytical chemistry course. This text emerged from just such a course taught at the University of California, Davis, from 1976 to 1980. Undergraduate chemistry

xiv

PREFACE

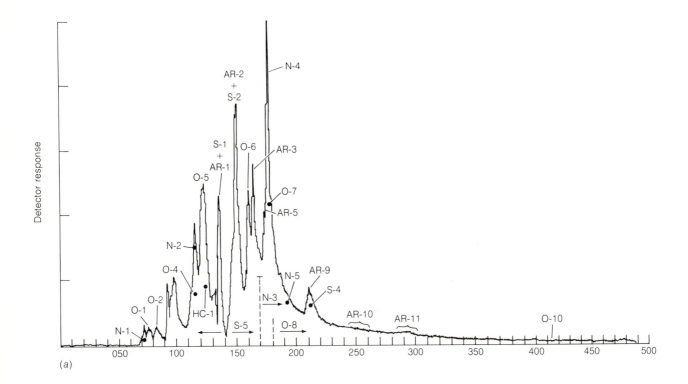

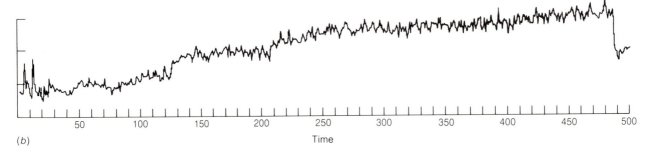

Gas chromatograms of Antarctic and Martian soil. Each peak in part a represents a compound that evaporated from the soil upon heating to 500°C. None of the spikes in part b could be attributed to the soil sample. The detector is a mass spectrograph that measures all molecules with a molecular weight greater than 47. The total time of analysis (500 units on the horizontal axis) corresponds to 85 minutes. [K. Biemann et al., *J. Geophys. Res.*, **82**, 4641 (1977).]

majors, all of whom took the course, constituted less than 10 percent of the class—the vast majority of whom were drawn from the life sciences. Thus, this text is designed to appeal to students with a broad range of interests. The manuscript has also been used for two years by chemistry majors at Franklin and Marshall College and has served their needs well.

The intent of the book is to present the subject matter of a traditional analytical chemistry course in a rigorous, readable, and interesting manner. The fundamental goals are to develop a sound physical understanding of the principles of analytical chemistry and to show how these principles are applied in chemistry and related disciplines. Topics include statistics, chemical equilibrium, acid–base chemistry, electrochemical equilibrium, electroanalytical methods, spectrophotometric techniques, methods of separation, and laboratory procedures. The last-named topic forms the basis of the concluding chapter, which consists of 18 experiments.

To help lighten the load of an admittedly dense subject, there are numerous examples, boxes, and lecture demonstrations interspersed at appropriate points. A full-color insert complements the demonstrations. Perhaps the most important feature of the book, however, is the collection of Exercises and Problems at the end of each chapter. To get the most from this course, you should work as many problems as time and energy permit. The Exercises tend to illustrate the most important but sometimes more complex problems. Detailed solutions to the Exercises, provided at the end of the book, present a wide variety of problem-solving techniques. Brief numerical answers to the Problems also appear at the end of the text. An extensive glossary is included as well.

I owe many thanks and much credit to the people who helped transform a set of lectures into a book. First and foremost are the many hundreds of students who used various predecessor manuscripts and who told me—in no uncertain terms—what was right and what was wrong. Professors Allen West (Lawrence University), Jacob H. Propp (University of Wisconsin, Oshkosh), Roger G. Bates (University of Florida), Arthur T. Hubbard (University of California, Santa Barbara), and Ted Haupert (Sacramento State University) provided many valuable comments on the manuscript. John Berg was a gentle and scholarly reviewer for some of the rougher parts of the manuscript, and my good friend David Ornitz helped ensure the accuracy of the solutions to the problems. Professor Marvin Goodrow, of Modesto Junior College, provided the spark that prompted me to transform a classroom tool into a publishable textbook. A great deal of my understanding of analytical chemistry, and approach to it, is derived from the excellent texts by D. A. Skoog and D. M. West. The form of the present book owes much to the guiding hands of Howard Boyer and Pearl C. Vapnek

xvi
PREFACE

at W. H. Freeman and Company. Some excellent ideas were also suggested by Jim Leisy, Jr. Most of the many versions of the manuscript were typed by Dee Kindelt, Doris Scarborough, Carol Strausser, and my wife, Sally. To all of these people, I express my deep appreciation.

A report issued in 1980 states that one out of seven medical laboratory test results in the United States is inaccurate. A person who understands what he or she is doing and who questions every procedure and every result is less likely to make a decision based on faulty data or faulty interpretations. The most important skill you can carry away from this course is that of independent, critical, and analytical reasoning.

July 1982 *Dan Harris*

Quantitative Chemical Analysis

1/ Units and Concentrations

The immediate goal of quantitative analysis is to answer the query "How much?" How much vanadium is contained in the ore? How much phosphate is bound to the enzyme? How much pesticide is contained in the groundwater? Some of the principles and methods that enable us to measure "how much" are the subject of this text. The ultimate goal of analytical chemistry is not just to measure "how much," but to use that knowledge for some greater purpose. This could be a scientific investigation, a policy decision, a cost analysis, philosophical satisfaction, or myriad other reasons. We begin with a brief discussion of units of measurement.

1-1 SI UNITS

Scientists are moving toward the worldwide adoption of a uniform system of measurement known as the *Système International d'Unités*—SI units. The fundamental units, from which all others are derived, are listed in Table 1-1. The standards of length, mass, and time are the familiar metric units of *meter* (m), *kilogram* (kg), and *second* (s). The other fundamental units of most concern to us measure electric current (*ampere*, A), temperature (*kelvin*, K), and amount of substance (*mole*, mol).

All of the remaining physical quantities, such as energy, force, and charge, can be expressed in terms of the fundamental units. Some of these derived quantities are shown in Table 1-2, along with their names and symbols. Conversion factors relating some common non-SI units to SI units are listed in Table 1-3.

Table 1-1

Fundamental SI units

Quantity	Unit	Symbol	Definition
Length	meter	m	The meter is the length of 1 650 763.73 wavelengths (in vacuum) of radiation corresponding to the transition $2p_{10}$-$5d_5$ of ^{86}Kr.
Mass	kilogram	kg	One kilogram is the mass of the prototype kilogram kept at Sevres, France. This is the only SI unit whose primary standard is not defined in terms of physical constants.
Time	second	s	The second is the duration of 9 192 631 770 periods of the radiation corresponding to the two hyperfine levels of the ground state of ^{113}Cs.
Electric current	ampere	A	One ampere is the amount of constant current that will produce a force of 2×10^{-7} N m^{-1} (newtons per meter of length) when maintained in two straight, parallel conductors of infinite length and negligible cross section, separated by one meter.
Temperature	kelvin	K	The thermodynamic temperature is defined such that the triple point of water (at which solid, liquid, and gaseous water are in equilibrium) is 273.16 K and the temperature of absolute zero is 0 K.
Luminous intensity	candela	cd	One candela is the luminous intensity emitted perpendicular to a surface of 1/600 000 m^2 of platinum at its melting point under a pressure of 101 325 N m^{-2} (1 atm).
Amount of substance	mole	mol	One mole of substance contains as many molecules (or atoms, if the substance is a monatomic element) as there are atoms of carbon in exactly 0.012 kg of ^{12}C. The number of particles in a mole is approximately $6.022\,045 \times 10^{23}$.

Table 1-1 (*continued*)

Quantity	Unit	Symbol	Definition
Plane angle	radian	rad	The radian is such that there are 2π radians in a complete circle.
Solid angle	steradian	sr	The steradian is defined such that there are 4π steradians in a complete sphere.

Table 1-2
Some SI-derived units with special names

Quantity	Units	Symbol	Expression in terms of other units	Expression in terms of SI base units
Frequency	hertz	Hz		s^{-1}
Force	newton	N		$m\ kg\ s^{-2}$
Pressure	pascal	Pa	N/m^2	$m^{-1}\ kg\ s^{-2}$
Energy, work, quantity of heat	joule	J	$N\cdot m$	$m^2\ kg\ s^{-2}$
Power, radiant flux	watt	W	J/s	$m^2\ kg\ s^{-3}$
Quantity of electricity, electric charge	coulomb	C	$A\cdot s$	$s\ A$
Electric potential, potential difference electromotive force	volt	V	W/A	$m^2\ kg\ s^{-3}\ A^{-1}$
Capacitance	farad	F	C/V	$m^{-2}\ kg^{-1}\ s^4\ A^2$
Electric resistance	ohm	Ω	V/A	$m^2\ kg\ s^{-3}\ A^{-2}$
Conductance	siemens	S	A/V	$m^{-2}\ kg^{-1}\ s^3\ A^2$
Magnetic flux	weber	Wb	$V\cdot s$	$m^2\ kg\ s^{-2}\ A^{-1}$
Magnetic flux density	tesla	T	Wb/m^2	$kg\ s^{-2}\ A^{-1}$
Inductance	henry	H	Wb/A	$m^2\ kg\ s^{-2}\ A^{-2}$
Luminous flux	lumen	lm		$cd\ sr$
Illuminance	lux	lx		$m^{-2}\ cd\ sr$

Table 1-3

Some conversion factors

Quantity	Unit	Symbol	SI equivalent
Volume	liter	L	$*10^{-3}$ m^3
	milliliter	mL	$*10^{-6}$ m^3
Length	Angstrom	Å	$*10^{-10}$ m
	inch	in.	$*0.025\ 4$ m
Mass	pound	lb	$*0.453\ 592\ 37$ kg
Force	dyne	dyn	$*10^{-5}$ N
Pressure	atmosphere	atm	$*101\ 325$ N m^{-2}
	torr	1 mm Hg	133.322 N m^{-2}
	pound/in^2	psi	$6\ 894.76$ N m^{-2}
Energy	erg	erg	$*10^{-7}$ J
	electron volt	eV	$1.602\ 189\ 2 \times 10^{-19}$ J
	calorie, thermochemical	cal	$*4.184$ J
	British thermal unit	Btu	$1\ 055.06$ J
Power	horsepower		745.700 W

Note: An asterisk (*) indicates that the conversion is exact (by definition).

The recommended way to write numbers is to leave a space between digits after every third digit on either side of the decimal point. An example is

$$1\ 032.971\ 35$$

Commas are *not* to be used for spacing. In Europe the decimal point is usually written as a comma, and the number above would appear as

$$1\ 032,971\ 34$$

EXAMPLE

The most common unit of pressure is the *atmosphere* (atm). The SI unit of pressure is the *pascal* (Pa), which equals a force of one newton per square meter (N m^{-2}). What pressure in pascals corresponds to a pressure of 0.268 atm?

Table 1-3 tells us that 1 atm is exactly 101 325 N m^{-2} = 101 325 Pa. We can write

$$(0.268\ \text{atm})\left(101\ 325\ \frac{\text{Pa}}{\text{atm}}\right) = 27\ 200\ \text{Pa}$$

The dimensions should be written beside each quantity, and the answer should be displayed with its units.

Various prefixes employed to indicate fractions or multiples of units are given in Table 1-4. It is inconvenient continually to write a number such as 3.2×10^{-11} s, so we write 32 ps instead. One *picosecond* (ps) is 10^{-12} s. To express 3.2×10^{-11} s in picoseconds, we perform the conversion as follows:

$$\frac{3.2 \times 10^{-11} \, s}{10^{-12} \, \dfrac{s}{ps}} = 32 \text{ ps}$$

The SI unit of volume (which has the dimensions length^3) is the *cubic meter* (m^3). The common unit of volume is the *liter* (L), which is defined as the volume of a cube 0.1 m on each edge. The *milliliter* (mL, 10^{-3} L) is exactly 1 cm^3. Small-scale work, especially in biochemistry, often employs *microliter* (μL, 10^{-6} L) volumes.

1-2 EXPRESSIONS OF CONCENTRATION

Concentration signifies how much of a substance is contained in a specified volume or mass. This section describes most of the common ways to express concentration. Normality and titer, which are not used in this text, are defined in the Glossary, and normality is further discussed in Appendix E.

Molarity

The most common unit of concentration is **molarity** (moles per liter), abbreviated M. Molarity can also be expressed as millimoles per milliliter, where one millimole (mmol) is 10^{-3} mol. A **mole** is defined as the number of atoms of ^{12}C in exactly 12 g of ^{12}C. This number of atoms is called *Avogadro's number* and its best value at this time is $6.022\,045 \times 10^{23}$. The term *gram-atom* is sometimes used for Avogadro's number of atoms, while *mole* is sometimes reserved for Avogadro's number of molecules. We will not make such a distinction. A mole is simply $6.022\,045 \times 10^{23}$ of anything.

The **molecular weight** (M.W.) of a substance is the number of grams that contain Avogadro's number of molecules. The molecular weight is simply the sum of the atomic weights of the constituent atoms. The terms "mass" and "weight" are usually used interchangeably. Weight actually refers to the force exerted by a mass in a gravitational field.

EXAMPLE

A solution is made by dissolving 12.00 g of benzene (C_6H_6) in enough hexane to give 250.0 mL of solution. Find the molarity of the benzene.

The molecular weight of benzene is 6 (atomic weight of carbon) + 6 (atomic weight of hydrogen) = 6 (12.011) + 6 (1.008) = 78.114 g/mol. The units of molecular weight,

---

Table 1-4
Prefixes

Prefix	Symbol	Factor
tera	T	10^{12}
giga	G	10^{9}
mega	M	10^{6}
kilo	k	10^{3}
hecto	h	10^{2}
deca	da	10^{1}
deci	d	10^{-1}
centi	c	10^{-2}
milli	m	10^{-3}
micro	μ	10^{-6}
nano	n	10^{-9}
pico	p	10^{-12}
femto	f	10^{-15}
atto	a	10^{-18}

$1 \, \mu L = 10^{-3} \text{ mL} = 10^{-6} \text{ L}$

$$\text{Molarity (M)} = \frac{\text{Moles of solute}}{\text{Liters of solution}}$$
$$= \frac{\text{Millimoles of solute}}{\text{Milliliters of solution}}$$

A mole is $6.022\,045 \times 10^{23}$ of *anything*.

Students have a propensity for confusing **moles** with **moles per liter,** especially on tests. When you want to write moles, the proper abbreviation is "mol." When you want to write moles per liter, use a capital M. Do not use the symbol "m," which means neither mol nor M. Writing the units beside all numbers is the best way to keep track of your calculations and to reduce the occurrence of silly errors.

Atomic weights appear on the inside cover of this text.

1 / UNITS AND CONCENTRATIONS

grams per mole, are often not written but simply understood. The number of moles in 12.00 g is

$$\frac{12.00 \text{ g}}{78.114 \text{ g/mol}} = 0.153\ 6 \text{ mol}$$

The molarity (moles per liter) is found by dividing the moles by the number of liters:

$$\frac{0.153\ 6 \text{ mol}}{0.250\ 0 \text{ L}} = 0.614\ 4 \text{ M}$$

Note that milliliters must first be converted to liters by dividing milliliters by 1000 mL/L:

$$\frac{250.0 \text{ mL}}{1\ 000 \text{ mL/L}} = 0.250\ 0 \text{ L}$$

Formality

HBr is a **strong electrolyte,** which means that it is virtually completely dissociated into H^+ and Br^- ions in aqueous solution. By contrast, acetic acid is a **weak electrolyte,** being only partially dissociated into $CH_3CO_2^-$ and H^+ in water.

If a solution is made by diluting 1.000 mol of HBr to 1.000 L with water, the **formal concentration** of HBr is 1.000 moles per liter. But the actual concentration of HBr molecules is nearly zero, because the HBr molecules have dissociated. The formal concentration refers to the amount of substance dissolved, without regard to its actual composition in solution. Rather than calling the HBr solution 1.000 M, it would be more correct to call it 1.000 F. The capital F is read "formal." Many texts use the terms "formality" and "molarity" interchangeably, and we will be guilty of the same simplification.

Unless you are fully aware of the chemistry of a particular compound, you rarely know its true molarity, but you can know its formal concentration from the amount weighed into a solution or measured by an analytical procedure. For this reason, the formal concentration is also called the **analytical concentration.**

The **formula weight** (F.W.) of a substance is the mass of one formula unit. For example, the formula weight of $AlCl_3$ is $[26.982 + 3(35.453)] = 133.341$. The true molecular weight in many nonpolar organic solvents is twice the formula weight, since the molecule exists as a dimer (pronounced **die**-mer) with the composition Al_2Cl_6. The formula weight refers to the species that has been written. We could refer to the formula weight of Al_2Cl_6, which is twice as much as the formula weight of $AlCl_3$.

$AlCl_3$ has the structure below in many solvents.

Cl behind page
↓
Cl, Cl, Cl
Al — Al
Cl, Cl, Cl
↑
Cl in front of page

Percent Composition

The percentage of a substance in a solution is most often expressed as a **weight percent,** which is defined as

$$\text{Weight percent (or wt/wt percent)} = \frac{\text{Mass of substance}}{\text{Mass of total solution}} \times 100$$

in which wt stands for weight. A solution labeled 40% (wt/wt) aqueous ethanol contains 40 g of ethanol per 100 g (not 100 mL) of solution. It is made by mixing 40 g of ethanol with 60 g of H_2O.

Other common percent units are

$$\begin{array}{c}\text{Volume percent} \\ \text{(or vol/vol percent)}\end{array} = \frac{\text{Volume of substance}}{\text{Volume of total solution}} \times 100$$

$$\begin{array}{c}\text{Weight/volume percent} \\ \text{(or wt/vol percent)}\end{array} = \frac{\text{Weight of substance (in grams)}}{\text{Volume of total solution (in milliliters)}} \times 100$$

in which vol stands for volume. Although the units of wt or vol should always be specified, wt/wt is usually implied when units are absent.

EXAMPLE

Commercial concentrated HCl is labeled 37.0%, which you may assume means weight percent. Its **density** (sometimes called **specific gravity**) is 1.18 g/mL. Find (a) the molarity of the HCl; (b) the mass of solution containing 0.100 mol of HCl; and (c) the volume of solution containing 0.100 mole of HCl.

(a) A 37.0% solution contains 37.0 g of HCl per 100 g of solution. The mass of one liter of solution is

$$(1\,000\ \text{mL})\left(1.18\ \frac{\text{g}}{\text{mL}}\right) = 1\,180\ \text{g}$$

The mass of HCl in 1 180 g of solution is

$$\left(0.370\ \frac{\text{g HCl}}{\text{g solution}}\right)(1\,180\ \text{g solution}) = 437\ \text{g HCl}$$

Since the molecular weight of HCl is 36.461, the molarity of HCl is

$$\frac{437\ \text{g/L}}{36.461\ \text{g/mol}} = 12.0\ \frac{\text{mol}}{\text{L}} = 12.0\ \text{M}$$

(b) Since 0.100 mol of HCl equals 3.65 g, the mass of solution containing 0.100 mol is

$$\frac{3.65\ \text{g HCl}}{0.370\ \text{g HCl/g solution}} = 9.85\ \text{g solution}$$

(c) The volume of solution containing 0.100 mol of HCl is

$$\frac{9.85\ \text{g solution}}{1.18\ \text{g solution/mL}} = 8.35\ \text{mL}$$

Parts per Million and Its Relatives

$$ppt = \frac{\text{g of substance}}{\text{g of sample}} \times 10^3$$

$$ppm = \frac{\text{g of substance}}{\text{g of sample}} \times 10^6$$

$$ppb = \frac{\text{g of substance}}{\text{g of sample}} \times 10^9$$

Composition is often expressed as **parts per million** (ppm), **parts per billion** (ppb), or **parts per thousand** (ppt). The term one part per million, for example, indicates one gram of the substance of interest per million grams of total solution or mixture. For an aqueous solution whose density is close to 1.00 g/mL, one ppm corresponds to 1 μg/mL or 1 mg/L.

EXAMPLE

A sample of salt water with a density of 1.02 g/mL contains 17.8 ppm NO_3^-. Calculate the molarity of nitrate in the water.

Molarity is moles per liter, and 17.8 ppm means that the water contains 17.8 μg of NO_3^- per gram of solution. One liter of solution weighs

$$\text{Mass of solution} = \text{Volume (mL)} \times \text{Density (g/mL)} = 1\,000 \times 1.02 = 1\,020 \text{ g}$$

One liter therefore contains

$$\text{grams of } NO_3^- = \frac{17.8 \times 10^{-6} \text{ g } NO_3^-}{\text{g solution}} \times 1\,020 \text{ g solution} = 0.018\,2 \text{ g of } NO_3^-$$

The molarity of nitrate, $[NO_3^-]$, is therefore

$$[NO_3^-] = \frac{\text{moles of } NO_3^-}{\text{liters of solution}} = \frac{0.018\,2 \text{ g } NO_3^-/(62.065 \text{ g } NO_3^-/\text{mol})}{\text{liters of solution}}$$

$$= 2.93 \times 10^{-4} \text{ M}$$

Other Concentration Units

Molality

$$\text{Molality (m)} = \frac{\text{Moles of solute}}{\text{Kilograms of solvent}}$$

The **molality,** m, defined as the number of moles of solute per kilogram of solvent, is useful for precise physical measurements. The reason is that molality is not temperature-dependent, whereas molarity is temperature-dependent. A dilute aqueous solution expands approximately 0.02% per degree Celsius when heated near 20°C. Therefore, the moles of solute per liter (molarity) decreases by this same percent.

Osmolarity

$$\text{Osmolarity} = \frac{\text{Moles of particles}}{\text{Liters of solution}}$$

Osmolarity, which is encountered in biochemical and clinical literature, is defined as the total moles of particles dissolved in one liter of solution. For nonelectrolytes such as glucose, the osmolarity is equal to the molarity. For the strong electrolyte $CaCl_2$, the osmolarity is three times the molarity, since each formula weight of $CaCl_2$ provides three moles of ions in solution ($Ca^{2+} + 2Cl^-$). Blood plasma is 0.308 osmolar.

1-3 PREPARING SOLUTIONS

If a pure solid or liquid reagent is to be used to prepare a solution of a given molarity, we simply weigh out the correct mass of reagent, dissolve it in the solvent, and dilute to the desired final volume. The dilution is usually done in a volumetric flask, as described in the next chapter. To prepare 1.00 M NaCl, we would not weigh out 1.00 mol of NaCl and mix it with 1.00 L of water, because the total volume of the mixture would not be 1.00 L.

EXAMPLE

What quantity of $H_2C_2O_4 \cdot 2H_2O$ (oxalic acid dihydrate) should be used to prepare 250 mL of 0.150 M aqueous oxalic acid?

The formula weight of oxalic acid dihydrate $(C_2H_6O_6)$ is 126.07. If we want to prepare 250 mL of 0.150 M oxalic acid, we will need

$$\left(\frac{250 \text{ mL}}{1\,000 \text{ mL/L}}\right)\left(0.150 \frac{\text{mol}}{\text{L}}\right) = 0.037\,5 \text{ mol}$$

This is equivalent to

$$(0.037\,5 \text{ mol oxalic acid})\left(126.07 \frac{\text{g of } H_2C_2O_4 \cdot 2H_2O}{\text{mol oxalic acid}}\right) = 4.73 \text{ g}$$

So 4.73 g of oxalic acid dihydrate should be dissolved in water and diluted to 250 mL.

It is frequently necessary to prepare a dilute solution of a reagent from a more concentrated solution. A useful equation for calculating the required volume of concentrated reagent is

$$M_{conc}V_{conc} = M_{dil}V_{dil} \qquad (1\text{-}1)$$

where conc refers to the concentrated solution and dil refers to the dilute solution.

> Since $M \cdot V = (\text{moles/L})(L) = \text{moles}$, Equation 1-1 simply states that the moles of solute in both solutions are equal. Dilution occurs because the volume has changed.

EXAMPLE

A solution of ammonia in water is also called "ammonium hydroxide" because of the equilibrium

$$\underset{\text{ammonia}}{NH_3} + H_2O \rightleftharpoons \underset{\text{ammonium}}{NH_4^+} + \underset{\text{hydroxide}}{OH^-}$$

The density of concentrated ammonium hydroxide, which contains 28.0% (wt/wt) NH_3, is 0.899 g/mL. What volume of this reagent should be diluted to 500 mL to make 0.100 M NH_3?

We begin by calculating the molarity of the concentrated reagent. Since the solution contains 0.899 g of solution per milliliter and there are 0.280 g of NH_3 per gram of solution (28.0% wt/wt), we can write

$$\text{Molarity of NH}_3 = \frac{\left(899\ \frac{\text{g of solution}}{\text{L}}\right)\left(0.280\ \frac{\text{g of NH}_3}{\text{g of solution}}\right)}{17.03\ \frac{\text{g of NH}_3}{\text{mol of NH}_3}} = 14.8\ \text{M}$$

To find the volume of 14.8 M NH_3 required to prepare 500 mL of 0.100 M NH_3, Equation 1-1 may be used:

$$M_{conc}V_{conc} = M_{dil}V_{dil}$$

$$\left(14.8\ \frac{\text{mol}}{\text{L}}\right)V_{conc}(\text{L}) = \left(0.10\ \frac{\text{mol}}{\text{L}}\right)(0.500\ \text{L}) \tag{1-2}$$

$$V_{conc} = 3.38 \times 10^{-3}\ \text{L} = 3.38\ \text{mL}$$

Note that both volumes in Equation 1-2 could be expressed in mL instead of L.

Summary

The SI base units are meter (m), kilogram (kg), second (s), ampere (A), kelvin (K), candela (cd) and mole (mol). Quantities such as force, pressure, and energy are measured in units derived from the base units. In calculations, the dimensions should be carried along with the numbers. Prefixes such as kilo and milli are used in the SI system to denote multiples of units. Common expressions of concentration are molarity (moles of solute per liter of solution), formality (formula units per liter), molality (moles of solute per kilogram of solution), percent composition, and parts per million. You should be able to calculate the quantities of reagents needed to prepare a given solution, and the equality $M_{conc}V_{conc} = M_{dil}V_{dil}$ is useful for this purpose.

Terms to Understand

analytical concentration
concentration
density
formal concentration
formula weight
molality
molarity
mole
molecular weight

osmolarity
parts per billion
parts per million
parts per thousand
specific gravity
strong electrolyte
volume percent
weak electrolyte
weight percent

Exercises[†]

1-A. A solution with a final volume of 500.0 mL was prepared by dissolving 25.00 mL of methanol (CH_3OH, density $= 0.791\,4$ g/mL) in chloroform.
   (a) Calculate the *molarity* of methanol in the solution.
   (b) If the solution has a density of 1.454 g/mL, find the *molality* of methanol.

1-B. A 48.0% (wt/wt) solution of HBr in water has a density of 1.50 g/mL.
   (a) What is the formal concentration (mol/L) of the solution?
   (b) What mass of solution contains 36.0 g of HBr?
   (c) What volume (mL) of solution contains 233 mmol of HBr?
   (d) How much solution is needed to prepare 0.250 L of 0.160 M HBr?

1-C. A solution contains 12.6 ppt of dissolved $MgCl_2$ (which is actually dissociated into $Mg^{2+} + 2Cl^-$). What is the concentration of Cl^- in parts per thousand?

Problems[‡]

1-1. State the fundamental quantities and their units in the SI system. Give one example of a derived quantity.

1-2. Write the name and number represented by each symbol. For example, for kW you should write kW = kilowatt = 10^3 watts.
   (a) mW   (c) KΩ   (e) TJ   (g) fg
   (b) pm   (d) μF   (f) ns   (h) dPa

1-3. Write each quantity using an appropriate prefix. For example, the quantity 1.01×10^5 Pa is written 101 kPa.
   (a) 10^{-13} J   (d) 10^{-10} m
   (b) $4.317\,28 \times 10^{-8}$ H   (e) 2.1×10^{13} W
   (c) $2.997\,9 \times 10^{14}$ Hz   (f) 48.3×10^{-20} mol

1-4. Newton's law tells us that Force = Mass × Acceleration. We also know that Energy = Force × Distance and Pressure = Force/Area. From these relations, derive the dimensions of newtons, joules, and pascals in terms of the fundamental SI quantities in Table 1-1.

1-5. What is the formal concentration of acetic acid when 2.67 g is dissolved in butanol to give 0.100 L of solution?

1-6. Find the molarity of pyridine (C_5H_5N) if 5.00 g is dissolved in water to give a total volume of 457 mL.

1-7. If 0.250 L of aqueous solution with a density of 1.00 g/mL contains 13.7 μg of pesticide, express the concentration of pesticide in (a) parts per million and (b) parts per billion.

1-8. A bottle of concentrated aqueous sulfuric acid, labeled 98.0% (wt/wt) H_2SO_4, has a concentration of 18.0 M.
   (a) How many milliliters of reagent should be diluted to 1.00 L to give 1.00 M H_2SO_4?
   (b) Calculate the density of 98.0% H_2SO_4.

1-9. Find the osmolarity of 1.00 L of solution containing 3.15 g $CaCl_2$, 0.153 g KCl, 1.57 g K_2SO_4, and 0.994 g of sucrose (table sugar, formula $= C_{12}H_{22}O_{11}$, a nonionic compound).

1-10. An aqueous solution containing 20.0% (wt/wt) KI has a density of 1.168 g/mL. Find the molality (not molarity) of the KI.

1-11. What is the density of 53.4% (wt/wt) aqueous NaOH if 16.7 mL of the solution produces 0.169 M NaOH when diluted to 2.00 L?

[†] Detailed solutions to Exercises are provided at the end of book.
[‡] Brief numerical answers to the Problems are given after the Solutions to Exercises.

2 / Tools of the Trade

Most of this text deals with fundamental "wet" chemical procedures, although elaborate instrumental techniques are discussed in later chapters. The principles developed in these early chapters are essential to the understanding of sophisticated techniques. In this chapter we describe some of the basic laboratory apparatus and manipulations associated with chemical measurements.

2-1 LAB NOTEBOOK

The critical functions of a lab notebook are to state *what was done* and *what you observed*. The greatest flaw, found even with experienced scientists, is that notebooks are difficult to understand. Hard as it is to believe, even the author of a notebook cannot understand his or her own notes after a few years. The problem is not usually one of legibility, but rather of poorly labeled entries and incomplete descriptions. The habit of writing in *complete sentences* is an excellent way to avoid incomplete descriptions.

Beginning students often find it useful to write a very complete description of an experiment, with formal sections dealing with purpose, methods, results, and conclusions. Arranging a notebook to accept numerical data prior to coming to the lab is an excellent way to prepare for an experiment.

The measure of scientific "truth" is the ability of different people to reproduce an experiment. Sometimes two scientists in different labs cannot reproduce each other's work, and neither has complete enough notebooks to understand why. Details that seem unimportant on the day of an experiment may become important months or years later. A good lab notebook will state everything that was done and will allow you or anyone else to repeat the experiment in exactly the same manner at any future date.

Keeping a complete and intelligible notebook should be second only to brushing your teeth at the start of the day.

2 / TOOLS OF THE TRADE

The notebook must:
1. State what was done.
2. State what was observed.
3. Be understandable to a stranger.

A complete notebook should also contain all of your observations. Long after you have forgotten what happened, you should be able to rely on your notes to tell what happened and perhaps provide a key observation needed to interpret an experiment. It could be that you do not understand what you observe during an experiment, but at some later time new knowledge will enable you to interpret old observations.

It is good practice to write a balanced chemical equation for every reaction you use. This helps you understand what you are doing and may point out what you do not understand about what you are doing.

2-2 ANALYTICAL BALANCE

The most common balance is the single-pan, semi-micro balance, with a total capacity of 100 to 200 g and a sensitivity of 0.01 or 0.1 mg. A typical balance is shown in Figure 2-1.

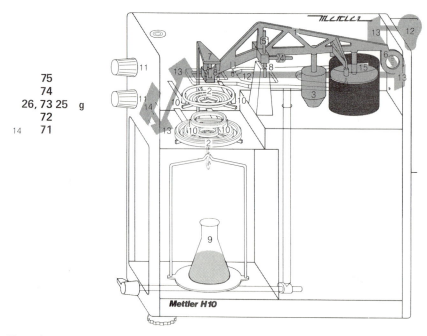

Figure 2-1
Cross section of a Mettler analytical balance, showing the projection of the optical scale from the back of the balance beam to the front of the instrument. (1) Balance beam. (2) Set of weights. (3) Fixed counterweight. (4) Zero-point adjustment weight. (5) Scale-deflection adjustment weight. (6) Graduation plate. (7) Parallelogram suspension. (8) Knife edges. (9) Sample to be weighed. (10) Weight-lifting mechanism. (11) Weight-control knobs. (12) Light bulb (light path). (13) Mirrors. (14) Readout panel. (15) Air damping. [Courtesy Mettler Instrument Corp., Hightstown, N.J.]

The usual weighing operation consists of first weighing a sheet of weighing paper or a receiving vessel on the balance pan. Then the substance to be weighed is added and a second reading is taken. The difference between the two masses corresponds to the mass of added substance. The mass of the empty weighing vessel is called the **tare.** Many balances can tare the receiving vessel. That is, with the receiver on the pan, the scale can be set to read zero. Then the substance to be weighed is added and its mass is read directly. No chemical should ever be placed directly on the balance pan. This protects the balance from corrosion and also permits you to recover all of the chemical being weighed.

Alternatively, weighing "by difference" is sometimes convenient. First a small vessel containing a reagent is weighed. Then some of the reagent is delivered to a receiver and the vessel is weighed again. The difference equals the mass of reagent delivered. Weighing by difference is especially useful for **hygroscopic reagents** (ones that rapidly absorb moisture from the air), since the weighing bottle can be kept closed during the weighing operations.

Principle of Operation

The principle of most balances is illustrated in Figure 2-2. The sample is

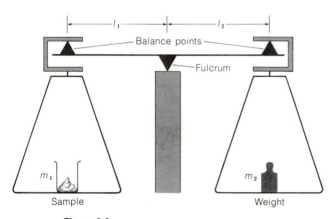

Figure 2-2
Principle of operation of a double-pan balance.

suspended on one side of a fulcrum, and a weight is placed on the opposite side. The two will be in perfect balance when

$$m_1 l_1 = m_2 l_2 \qquad (2\text{-}1)$$

where m_i is the mass on each side and l_i is the distance of the balance point from the fulcrum. Normally, $l_1 = l_2$, and m_2 is a known mass.

2 / TOOLS OF THE TRADE

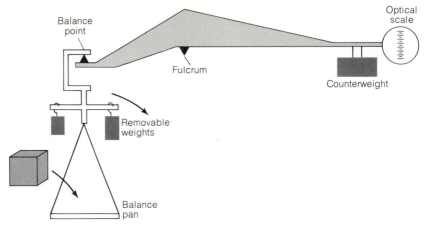

Figure 2-3
Principle of operation of a single-pan balance.
[Courtesy Mettler Instrument Corp., Hightstown, N.J.]

The schematic diagram of a single-pan balance in Figure 2-3 shows the empty balance pan and a set of *removable* weights suspended from the balance point to the left of the fulcrum. The counterweight at the right exactly balances the empty pan and the removable weights. When a sample is placed on the pan, the left side becomes heavier. The dials on the front of the balance are then used to *remove* some of the weights attached to the balance pan until the two sides are almost back in balance. The remaining deflection from horizontal of the balance beam is measured by an optical scale, located at the back of the balance, that is projected to the front of the instrument, as shown in Figure 2-1. The sum of the weights removed plus the reading on the optical scale equals the mass of the sample.

Effect of Buoyancy

For very accurate work, the buoyant effect of air must be considered. When a sample is placed on the balance pan, it displaces a certain amount of air. Removing this mass of air from the pan makes the object seem lighter than it actually is, because the pan was set to zero *with* the air on the pan. A similar effect applies to the removable weights on the balance. There will be a net effect of **buoyancy** whenever the density of the object being weighed is not equal to the density of the standard weights. For the most commonly encountered single-pan balances, which use stainless-steel weights, the true mass of the object is given by[†]

[†] J. E. Lewis and L. A. Woolf, *J. Chem. Ed.*, **48,** 639 (1971); F. F. Cantwell, B. Kratochvil, and W. E. Harris, *Anal. Chem.*, **50,** 1010 (1978). The notation *J. Chem. Ed.*, **48,** 639 (1971) means *Journal of Chemical Education*, volume 48, page 639, in the year 1971.

$$m = \frac{1.000\,011\,8\,m'\left(1 - \dfrac{d_a}{d_w}\right)}{1 - \dfrac{d_a}{d}} \approx \frac{0.999\,857\,m'}{1 - \dfrac{0.001\,2}{d}} \qquad (2\text{-}2)$$

where m = true mass of object weighed in vacuum

m' = mass read from balance

d_a = density of air (0.001 2 g/mL near 1 atm and 25°C)

d_w = density of weights (7.76 g/mL for stainless steel)

d = density of object being weighed

For extremely accurate work the correct air density at the particular values of temperature, pressure, and humidity should be used.

The magnitude of the buoyancy factor is illustrated by a few examples. When weighing water, with a density of 1.00 g/mL, the true mass is 1.001 1 g when the balance reads 1.000 0 g. The error is 0.11% in this case. For NaCl, with a density of 2.16 g/mL, the error would be 0.04%. For $AgNO_3$, whose density is 4.35 g/mL, the error is only 0.01%.

Errors in Weighing

Some care should be exercised to minimize weighing errors. A vessel being weighed should not be touched with your hands, since fingerprints will change its mass. A sample should always be at ambient temperature before weighing to avoid errors due to convective air currents. Cooling a sample that has been dried in an oven usually requires about half an hour in a desiccator at room temperature. The balance pan should be in its arrested position when adding a load and in its half-arrested position when dialing weights. This protects against abrupt forces that wear down the knife edges serving as fulcrum and balance point (Figure 2-3). Often a sensitive balance is placed on a heavy base (such as a large marble slab) to minimize the effect of room vibrations on the reading. Adjustable feet and a bubble meter at the top allow the balance to be maintained in a level position.

2-3 BURETS

A **buret** is a precisely bored glass tube with graduations enabling you to measure the volume of liquid delivered. This is done by reading the level before and after draining liquid from the buret. The typical burets in Figure 2-4 have Teflon stopcocks. A loosely fitting cap is used to keep dust out and

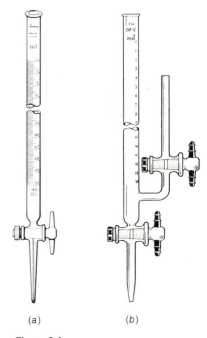

Figure 2-4
(a) A typical 50 mL buret. (b) Buret with sidearm for refilling. [Courtesy A. H. Thomas Co., Philadelphia, Pa.]

2 / TOOLS OF THE TRADE

Table 2-1
Tolerances of Class A burets

Buret volume (mL)	Smallest graduation (mL)	Tolerance (mL)
5	0.01	±0.01
10	0.05 *or* 0.02	±0.02
25	0.1	±0.03
50	0.1	±0.05
100	0.2	±0.10

Buret reading tips:
1. Read the bottom of the concave meniscus.
2. Avoid parallax.
3. Account for the thickness of the marking lines in your readings.

vapors in. The graduations of Class A burets are certified to meet the tolerances in Table 2-1.

When your read the height of liquid in a buret, it is important that your eye be at the same level as the top of the liquid. This minimizes the **parallax** error associated with reading the position of the liquid. If your eye is above this level, the liquid seems to be higher than it actually is. If your eye is too low, there appears to be less liquid than is actually present.

The surface of most liquids forms a concave **meniscus,** as shown in Figure 2-5. It is useful to use a piece of black tape on a white card as a background

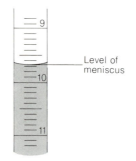

Figure 2-5
A portion of a buret with the level of the meniscus at 9.68 mL. You should always estimate the reading of any scale to the nearest tenth of a division. This corresponds to 0.01 mL for the buret in this figure.

for locating the precise position of the meniscus. To use this card, align the top of the black tape with the bottom of the meniscus and read the position on the buret. Some solutions, especially highly colored ones, appear to have two meniscuses. In such cases, either one may be used. The important point is to perform the readings reproducibly.

The markings of a buret normally have a thickness that is *not* negligible in comparison with the distance between markings. The thickness of the markings corresponds to about 0.02 mL, for a 50 mL buret. To most accurately use the buret, you should select one portion of the marking to be called zero. For example, say that the liquid level is at the mark when the bottom of the meniscus just touches the *top* of the mark on the glass. When the bottom of the meniscus is at the *bottom* of the mark, the reading is 0.02 mL greater.

Near the end point of a titration, it is desirable to deliver less than one drop at a time from the buret. This permits a finer location of the end point, since the volume of a drop is about 0.05 mL (from a 50 mL buret). To deliver a fraction of a drop, carefully open the stopcock until part of a drop is hanging from the buret tip. (Alternatively, you can allow just a fraction of a drop to emerge from the buret by very rapidly turning the stopcock through the open position.) Then touch the inside glass wall of the receiving flask to the buret tip to transfer the liquid to the wall of the flask. Carefully tip the flask so that the main body of liquid washes over the newly added liquid, and mix the contents. Near the end of a titration, the flask should be tipped and rotated to ensure that droplets on the walls containing unreacted analyte come in contact with the bulk solution. Liquid should drain evenly down the walls of a buret. If not, the buret should be cleaned with detergent and a buret brush. If this is insufficient, the buret should be soaked in chromic acid cleaning solution.† Volumetric glassware should never be soaked in an alkaline cleaning solution, since glass is slowly attacked by base. The tendency of liquid to stick to the walls of a buret can be diminished by draining the buret slowly. A slowly drained buret will also provide greater reproducibility of results. *A drainage rate of no more than 20 mL/min should be used.*

One of the most common errors in using a buret is caused by failure to expel the bubble of air that often forms directly beneath the stopcock (Figure 2-6). If an air bubble is present at the start of a titration, it may fill in during the titration, causing an error in the volume of liquid delivered from the buret. Usually the bubble can be dislodged by draining the buret for a second or two with the stopcock wide open. Sometimes a tenacious bubble must be expelled by carefully shaking the buret while draining liquid into a sink.

When you fill a buret with fresh solution, it is a good idea to rinse the buret several times with the new solution, discarding each wash. It is not necessary to fill the whole buret with the wash solution. Simply tilt the buret to allow the whole surface to come in contact with a small volume of liquid. This same washing technique can be applied to any vessel (such as a spectrophotometer cuvette or a pipet) that must be reused without opportunity for drying.

2-4 VOLUMETRIC FLASKS

A **volumetric flask** is calibrated to contain a particular volume of water at 20°C when the bottom of the meniscus is adjusted to the center of the line marked on the neck of the flask (Figure 2-7, Table 2-2). Most flasks bear the

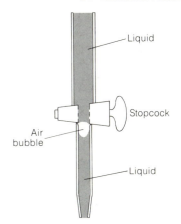

Figure 2-6
The air bubble often trapped beneath the stopcock of a buret should be expelled before using the apparatus.

Washing any glassware with a new solution is a good idea.

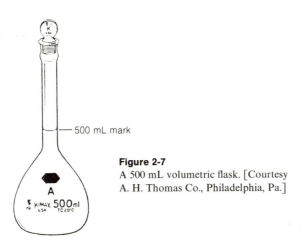

Figure 2-7
A 500 mL volumetric flask. [Courtesy A. H. Thomas Co., Philadelphia, Pa.]

Table 2-2
Tolerances of Class A volumetric flasks

Flask capacity (mL)	Tolerance (mL)
1	±0.02
2	±0.02
5	±0.02
10	±0.02
25	±0.03
50	±0.05
100	±0.08
200	±0.10
250	±0.12
500	±0.20
1000	±0.30
2000	±0.50

---

† Cleaning solution is prepared by dissolving 5 g of sodium dichromate in 5 mL of water and adding 100 mL of 98% sulfuric acid with stirring and cooling on ice. This mixture is an extremely powerful oxidant that eats clothing and people, as well as dirt and grease.

2 / TOOLS OF THE TRADE

Thermal expansion of water and glass is discussed in Section 24-1.

label "TC 20°C," which means that the flask is calibrated *to contain* the indicated volume at 20°C. (Other types of glassware may be calibrated *to deliver*, "TD," their indicated volume.) The temperature is important because both the liquid and the glass expand as they are heated.

To adjust the liquid level to the *center* of the mark of a volumetric flask (or of a pipet, which is calibrated in the same way), look at the mark from *above* or *below* the level of the mark. The front and back of the mark will not be aligned with each other and will describe an ellipse. Drain the liquid until the bottom of the meniscus is at the *center* of the ellipse (Figure 2-8).

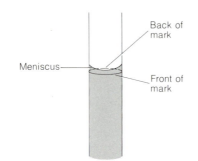

Figure 2-8
Proper position of the meniscus: at the center of the ellipse formed by the front and back of the calibration mark when viewed from above or below. Volumetric flasks and pipets are calibrated to this position.

This places the bottom of the meniscus exactly at the center of the calibration mark, if viewed edge on.

A volumetric flask is used to prepare a solution of a known volume. Most commonly, a reagent is weighed into a volumetric flask, dissolved, and diluted to the mark. This way the mass of reagent and the final volume of the solution are both known. The solid should first be dissolved in *less* liquid than the flask is calibrated to contain. Then more liquid is added, and the solution is again mixed. The final adjustment to the mark should be done with as much well-mixed solution as possible in the flask. This minimizes the change in volume upon mixing pure liquid with solution already in the flask. The final drops of liquid should be added with a pipet, *not a squirt bottle*, for best control. After adjusting the solution to its final volume the cap should be held firmly in place and the flask *inverted 20 or more times* to assure complete mixing.

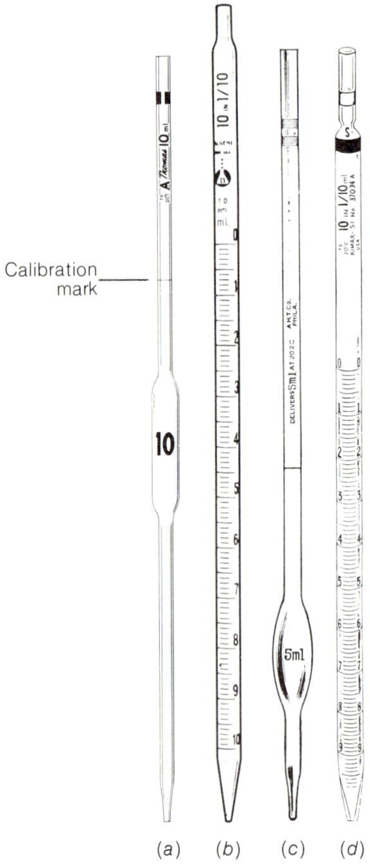

2-5 PIPETS AND SYRINGES

Pipets are used to deliver known volumes of liquid. Four common types are shown in Figure 2-9. The transfer pipet, which is the most accurate, is calibrated to deliver one fixed volume. The last drop of liquid does not drain out of the pipet; it is meant to be left in the pipet. *It should not be blown out*. The measuring pipet is calibrated to deliver a variable volume, as indicated by the difference between the volumes indicated before and after delivery. The measuring pipet in Figure 2-9 could, for example, be used to deliver

Figure 2-9
Some common pipets. (a) Transfer. (b) Measuring (Mohr). (c) Ostwald–Folin (blow out last drop). (d) Serological (blow out last drop). [Courtesy A. H. Thomas Co., Philadelphia, Pa.]

Do not blow out the last drop from a transfer pipet.

5.6 mL by starting delivery at the 1 mL mark and ending at the 6.6 mL mark. The Ostwald–Folin pipet is similar to the transfer pipet, except that the last drop *should* be blown out. When in doubt about whether the pipet is made for blowout, consult the manufacturer's catalog. Serological pipets are measuring pipets calibrated all the way to the tip. The serological pipet in Figure 2-9 will deliver 10 mL when the last drop is blown out.

In general, transfer pipets are more accurate than measuring pipets. The tolerances for class A transfer pipets are given in Table 2-3. To deliver the calibrated volume, bring the bottom of the meniscus to the center of the calibration mark, as shown in Figure 2-8.

Using a Transfer Pipet

Using a rubber bulb, *not your mouth*, suck liquid up past the mark. Quickly put your index finger over the end of the pipet in place of the bulb. The liquid should still be above the mark at this time. Pressing the pipet against the bottom of the vessel while removing the rubber bulb helps prevent liquid from draining while you get your finger in place.[†] Wipe the excess liquid off the outside of the pipet with a clean tissue. *Touch the tip of the pipet to the side of a beaker*, and drain the liquid until the bottom of the meniscus just reaches the center of the mark. The pipet must touch the beaker during the draining so that extra liquid is not hanging from the tip of the pipet when the meniscus reaches the mark. Any liquid outside of the pipet will be drawn onto the wall of the beaker. Now transfer the pipet to the desired receiving vessel and drain it *while holding the tip to the wall of the vessel*. After the pipet stops draining, hold it against the wall for a few seconds more to be sure that everything that can drain has drained. *Do not blow out the last drop*. The pipet should be nearly vertical to ensure that the proper amount of liquid has drained. When you finish with a pipet, it should be rinsed with distilled water or placed in a pipet container to soak until it is cleaned. Solutions should never be allowed to dry inside a pipet because removing deposits is very difficult.

Delivering Small Volumes

Plastic micropipets, such as that shown in Figure 2-10, are used to deliver microliter volumes (1 μL = 10^{-6} L). The liquid is contained entirely in a disposable plastic tip. Pipets are available covering the range 1–1 000 μL, in fixed or variable volumes. The accuracy is 1–2%, and the precision may reach 0.5%. Accuracy refers to how close the delivered volume is to the desired volume. Precision refers to the reproducibility of replicate deliveries.

For delivering very small variable volumes of liquid, a microliter **syringe**

[†] G. Deckey (*J. Chem. Ed.*, **57**, 526 (1980)) describes a way to use a rubber bulb and an Eppendorf pipet tip to apply suction to a pipet whose outer diameter is smaller than the hole in the rubber bulb.

2-5 PIPETS AND SYRINGES

Table 2-3
Tolerances of Class A transfer pipets

Volume (mL)	Tolerance (mL)
0.5	±0.006
1	±0.006
2	±0.006
3	±0.01
4	±0.01
5	±0.01
10	±0.02
15	±0.03
20	±0.03
25	±0.03
50	±0.05
100	±0.08

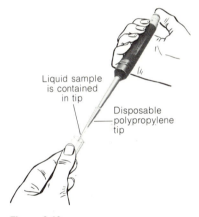

Figure 2-10
A fixed-volume Eppendorf micropipet. [Courtesy A. H. Thomas Co., Philadelphia, Pa.]

The composition of the container (glass, plastic, steel, etc.) is important when handling small volumes. It takes very little impurity from the container wall to significantly affect the composition of 1 μL of solution.

Figure 2-11
A Hamilton syringe with a total volume of 1 μL and divisions separated by 0.01 μL. The barrel is made of glass, and the needle is stainless steel. In this particular syringe, the entire sample is contained in the needle. In larger syringes, the liquid is held in the barrel. [Courtesy A. H. Thomas Co., Philadelphia, Pa.]

(Figure 2-11) is excellent. Syringes are available in a wide range of volumes, with accuracy and precision close to 1%.

2-6 FILTRATION

In **gravimetric analysis,** the mass of product from a reaction is measured to determine how much unknown was present. Precipitates for gravimetric analysis are collected by filtration, washed, and then dried. If the precipitate does not need to be ignited, it is most conveniently collected in a fritted-glass funnel, such as that shown in Figure 2-12. This funnel has a porous glass disk that allows liquid to pass through, but retains solid particles. In gravimetric analysis, the filter crucible is dried and weighed before the precipitate is collected. After the product is collected and dried in the crucible, the crucible and its contents are weighed together to determine the mass of precipitate. The filter crucible is usually used with suction provided by a water aspirator, as shown in Figure 2-13. The liquid from which a substance precipitates or crystallizes is called the **mother liquor.** The liquid that passes through a filter is called the **filtrate.**

Figure 2-12
A fritted-glass Gooch filter crucible. [Courtesy A. H. Thomas Co., Philadelphia, Pa.]

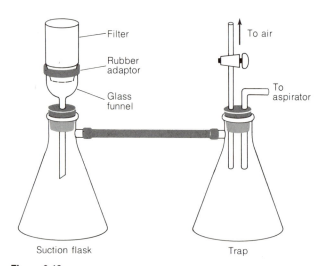

Figure 2-13
Setup for filtration with a glass Gooch filter crucible. The purpose of the trap is to prevent the possible backup of tap water from the aspirator into the suction flask.

Some gravimetric procedures require isolating a precipitate, then **igniting** (heating) it at high temperature to convert it to a well-defined product of known composition. For example, Fe^{3+} is precipitated as a poorly defined hydrated form of $Fe(OH)_3$ and ignited to Fe_2O_3 before weighing. When a gravimetric precipitate is to be ignited, it is collected in *ashless* filter paper that leaves little residue when oxidized at high temperature. The filter paper is folded in quarters, one corner torn off, and the paper placed in a conical glass funnel (Figure 2-14). The paper should fit snugly and be seated with a

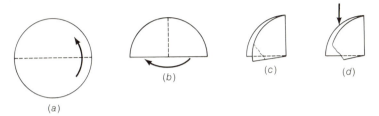

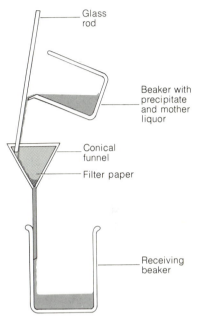

Figure 2-14
Folding filter paper for a conical funnel. (a) Fold the paper in half. (b) Then fold it in half again. (c) Tear off a corner. (d) Open the side that has not been torn when fitting the paper in the funnel.

little distilled water. When liquid is poured into the funnel, an unbroken stream of liquid should fill the stem of the funnel. The weight of the liquid in the stem helps to speed the filtration.

The correct procedure for filtration is shown in Figure 2-15. The liquid containing suspended precipitate is poured down a glass rod into the filter. The rod helps prevent splattering or dripping down the side of the beaker. Particles adhering to the beaker or rod can be dislodged with a **rubber policeman** (a glass rod with a flattened piece of rubber attached to the end) and transferred with a jet of liquid from a squirt bottle containing distilled water or an appropriate wash liquid. Particles that remain in the beaker may be wiped onto a small piece of moist filter paper and the paper placed in the filter, to be ignited with the bulk of the precipitate.

Figure 2-15
Procedure for filtering a precipitate.

2-7 DRYING

Reagents, precipitates, and glassware are conveniently dried in an electric oven, usually maintained at about 110°C. (Some reagents or precipitates require other drying temperatures.) When a filter crucible is to be brought to constant mass prior to a gravimetric analysis, the crucible should be dried for one hour or longer, then cooled in a desiccator. The crucible is weighed and then heated again for about 30 minutes. When successive weighings agree to within 0.3 mg, the filter is considered to have been dried "to constant mass." Drying solid reagents or filter crucibles in an oven should be done using a beaker and watchglass (Figure 2-16) to prevent dust from falling into the reagent.

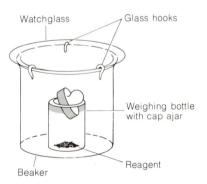

Figure 2-16
Use of a beaker and watchglass to keep dust out of a reagent while it is drying in the oven.

Dust is an important source of contamination in analytical chemistry. It is good practice to cover all vessels on the benchtop whenever possible.

After a reagent or crucible has been dried at high temperature, it is cooled to room temperature in a **desiccator,** a closed chamber that contains a drying agent **(desiccant).** The interface between the lid and the body of the desiccator is greased to make an airtight seal. Two common types of desiccators are shown in Figure 2-17. The desiccant is placed beneath the perforated disk near the bottom of the chamber. Data on the efficacy of various common drying agents are given in Table 2-4. In addition to those listed,

Table 2-4
Efficiencies of drying agents

Agent	Formula	Water left in atmosphere (μg H_2O/L)
Magnesium perchlorate, anhydrous	$Mg(ClO_4)_2$	0.2
"Anhydrone"	$Mg(ClO_4)_2 \cdot 1-1.5H_2O$	1.5
Barium oxide	BaO	2.8
Alumina	Al_2O_3	2.9
Phosphorous pentoxide	P_4O_{10}	3.6
Lithium perchlorate, anhydrous	$LiClO_4$	13
Calcium chloride (dried at 127°C)	$CaCl_2$	67
Calcium sulfate ("Drierite")	$CaSO_4$	67
Silica gel	SiO_2	70
Ascarite	NaOH on asbestos	93
Sodium hydroxide	NaOH	513
Barium perchlorate	$Ba(ClO_4)_2$	599
Calcium oxide	CaO	656
Magnesium oxide	MgO	753
Potassium hydroxide	KOH	939

Note: Moist nitrogen was passed over each desiccant, and the water remaining in the gas was condensed and weighed.
SOURCE: A. I. Vogel, *A Textbook of Quantitative Inorganic Analysis*, 3rd ed. (New York: Wiley, 1961), p. 178.

Figure 2-17
(a) Ordinary desiccator. (b) Vacuum desiccator. The vacuum desiccator can be evacuated through the sidearm at the top and then sealed by rotating the joint with the sidearm. Drying is more efficient at low pressure. [Courtesy A. H. Thomas Co., Philadelphia, Pa.]

98% sulfuric acid is also a common and efficient desiccant. After placing a hot object in a desiccator, leave the lid cracked open for a minute or two until the object has cooled slightly. This prevents the lid from popping open when the air inside is warmed by the hot object. An object being cooled prior to weighing should be given about 30 minutes to reach room temperature. The correct way to open a desiccator is to slide the lid sideways until it can be removed. The greased seal prevents you from opening the desiccator with a direct upward pull on the lid.

Summary

This chapter deals with the basic equipment and manipulations of "wet" chemical analysis. The analytical balance should be treated as a delicate piece of equipment, and buoyancy corrections should be employed in exact work. Burets should be read in a reproducible manner and drained slowly for best results. Always interpolate between the markings to obtain accuracy one decimal place beyond the graduations. Volumetric flasks are used to prepare solutions with a known volume. Transfer pipets are used to deliver a fixed volume of liquid, while measuring pipets, which are less accurate, deliver variable volumes. Proper techniques for filtering solutions or collecting precipitates are described, and the relative effectiveness of various desiccants is given. The importance of maintaining a complete, accurate, and intelligible notebook cannot be overstated.

Terms to Understand

ashless filter paper	filtrate	parallax
buoyancy	gravimetric analysis	pipet
buret	hygroscopic	rubber policeman
constant mass	ignition	syringe
desiccant	meniscus	tare
desiccator	mother liquor	volumetric flask

Problems

2-1. Prepare a graph showing the buoyancy correction (expressed as a percent of the sample weight) versus the sample density. Calculate correction factors for the following densities (g/mL).
(a) 0.5　(c) 2　(e) 4　(g) 8　(i) 12
(b) 1　(d) 3　(f) 6　(h) 10　(j) 14
For comparison, look up the densities of the following substances: pentane, acetic acid, CCl_4, sulfur, sodium acetate, $AgNO_3$, Hg, Pb, PbO_2. For which substance will the buoyancy correction be least?

2-2. What do the symbols "TD" and "TC" mean on volumetric glassware?

2-3. Describe the poper procedure for preparing 250.0 mL of 0.1500 M K_2SO_4 with a volumetric flask.

2-4. Describe the correct procedure for transferring 5.00 mL of a liquid reagent using a transfer pipet.

2-5. Which pipet is more accurate: a transfer pipet or a measuring pipet?

2-6. What would you do differently when delivering 1.00 mL of liquid from a 1 mL serological pipet, as compared to using a 1 mL measuring pipet?

2-7. Which drying agent is more efficient: "Drierite" or phosphorus pentoxide?

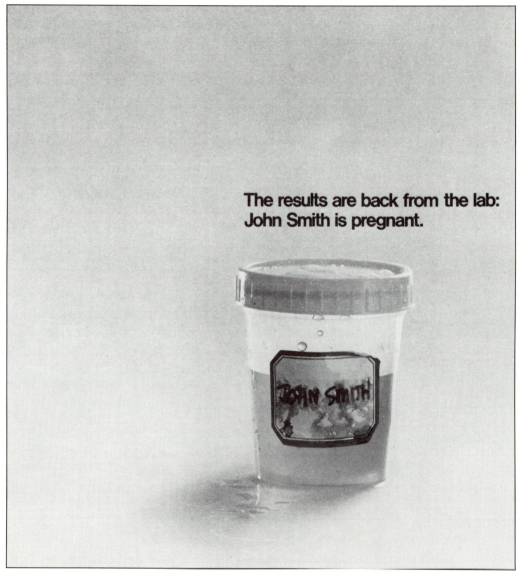

[Courtesy 3M Company, St. Paul, Minn.]

3 / Experimental Error

Not all laboratory errors are so monumental, but there is error associated with every kind of measurement. There is no way to measure the "true value" of anything. The best we can do in a chemical analysis is to apply carefully a technique that experience tells us is reliable. Also, we can measure a quantity in several ways using different methods to see if the measurements agree. You should always be cognizant of the error associated with a result and, therefore, of how reliable that result may be. In this chapter we deal with the relationship between errors in the individual measurements made during an experiment and the reliability of the final result.

3-1 SIGNIFICANT FIGURES

The number of **significant figures** is the minimum number of digits needed to write a given value in scientific notation without loss of accuracy. The number 142.7 has four significant figures, since it can be written as 1.427×10^2, and all four figures are needed to express fully the value. If you wrote 1.4270×10^2, you would be implying that you know the value of the digit after 7, which is not the case for the number 142.7. The number 1.4270×10^2 therefore has five significant figures.

Definition of significant figures.

The number 6.302×10^{-6} has four significant figures, since all four digits are necessary. You could write the same number as 0.000 006 302, which also has just *four* significant figures. The zeros to the left of the 6 are all merely holding decimal places. Since 0.000 006 302 may also be written as 6.302×10^{-6}, only four figures are necessary, and we say only four are significant. The number 92 500 is ambiguous with regard to significant figures. It could mean any of the following:

$$9.25 \times 10^4 \qquad \text{3 significant figures}$$

$$9.250 \times 10^4 \qquad \text{4 significant figures}$$

$$9.2500 \times 10^4 \qquad \text{5 significant figures}$$

3 / EXPERIMENTAL ERROR

Significant zeros below are **bold**:

106   0.010 **6**

0.106   0.106 **0**

It is preferable to write one of the three numbers above, instead of 92 500, to indicate how many figures are actually known.

Zeros are significant when they occur (1) in the middle of a number, or (2) at the end of a number on the right-hand side of a decimal point.

The last significant figure in a measured quantity always has some associated error. The minimum uncertainty would be ± 1 in the last digit. The scale of a Spectronic 20 spectrophotometer is drawn in Figure 3-1. The

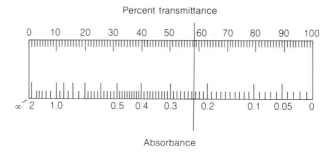

Figure 3-1
The scale of a Bausch and Lomb Spectronic 20 spectrophotometer. The percent transmittance is a linear scale, and the absorbance is a logarithmic scale.

needle shown in the figure appears to be at an absorbance value of 0.234. We say that there are three significant figures because the numbers 2 and 3 are completely certain and the number 5 is an estimate. The value could be read 0.233 or 0.235 by different people. The percent transmittance is near 58.3. Because the transmittance scale is smaller than the absorbance scale at this point, there is probably more uncertainty in the last digit of transmittance. A reasonable estimate of the uncertainty might be 58.3 ± 0.2. There are three significant figures in the number 58.3.

Always try to estimate the reading to the nearest tenth of the distance between scale divisions.

In general, when reading the scale of any apparatus, you should interpolate between the markings. It is usually possible to estimate to the nearest tenth of the distance between two marks. Thus on a 50 mL buret, which is graduated to 0.1 mL, you should read the level to the nearest 0.01 mL. When using a ruler calibrated in millimeters, estimate distances to the nearest tenth of a millimeter.

Arithmetic Operations

This section deals with determining the number of significant figures to keep in the answer after you have performed various arithmetic operations with your data.

Addition and subtraction

Usually the numbers involved have the same number of figures at the outset. In this case, the answer should go to the *same decimal place* as the individual numbers. For example:

$$1.362 \times 10^{-4}$$
$$+3.111 \times 10^{-4}$$
$$\overline{4.473 \times 10^{-4}}$$

The number of significant figures in the answer can sometimes either exceed or be less than that in the original data:

$$\begin{array}{cc} 5.345 & 7.26 \times 10^{14} \\ +6.728 & -6.69 \times 10^{14} \\ \hline 12.073 & 0.57 \times 10^{14} \end{array}$$

If the numbers being added do not have the same number of significant figures, we are usually limited by the least certain one. For example, in calculating the formula weight of KrF_2, the answer is known only to the second decimal place, as we are limited by our knowledge of the atomic weight of Kr.

$$\begin{array}{cl} 18.998\,403 & \text{(F)} \\ +18.998\,403 & \text{(F)} \\ +83.80 & \text{(Kr)} \\ \hline 121.796\,806 \end{array}$$

The number 121.796 806 should be rounded to 121.80 as the final answer.

Rules for rounding off numbers.

When rounding off, look at *all* the digits *beyond* the last place desired. In the example above, the digits 6 806 lie beyond the last significant decimal place. Since this number is more than halfway to the next higher digit, we round the 9 up to 10 (i.e., we round up to 121.80 instead of down to 121.79). If the insignificant figures were less than halfway, we would round down. For example, 121.794 8 is correctly rounded to 121.79.

In the special case where the number is exactly halfway, we round to the nearest *even* digit. Thus, 43.550 00 is rounded to 43.6, if we can only have three significant figures. If we are only retaining three figures, 1.425×10^{-9} becomes 1.42×10^{-9}. The number $1.425\,01 \times 10^{-9}$ would become 1.43×10^{-9}, since 501 is more than halfway to the next digit. The rationale for rounding to the nearest even digit is that it avoids systematically increasing or decreasing results through successive round-off errors. On the average, half of our round-offs will be up and half down.

When adding or subtracting numbers expressed in scientific notation, all numbers should first be expressed with the same exponent. For example, to add the numbers below we could write

For addition and subtraction, express all numbers using the same exponent, and align all numbers with respect to the decimal point. Terminate the answer according to the number of decimal places in the number with the fewest decimal places.

$$\begin{array}{ccl} 1.632 \times 10^5 & & 1.632 \quad\ \times 10^5 \\ +4.107 \times 10^3 & \Rightarrow & +0.041\,07 \times 10^5 \\ +0.984 \times 10^6 & & +9.84 \quad\ \times 10^5 \\ \hline & & 11.51 \quad\ \times 10^5 \end{array}$$

30

> *Challenge:* Show that the answer would still have four significant figures if all numbers were expressed as multiples of 10^4 instead of 10^5.

A more complete discussion of multiplication and division is reserved for the end of the chapter, after we have looked at relative errors.

A refresher on the algebra of logs and exponents can be found in Appendix A.

Number of figures in **mantissa** of $\log x$ = number of significant figures in x:

$$\log(\underset{\text{4 digits}}{\underline{5.403}} \times 10^{-8}) = -7.\underset{\text{4 digits}}{\underline{2674}}$$

The sum $11.51\,307 \times 10^5$ is rounded to 11.51×10^5 because the number 9.84×10^5 limits us to two decimal places when all numbers are expressed as multiples of 10^5.

Multiplication and division

In these operations we are normally limited to the number of digits contained in the number with the fewest significant figures. For example:

$$
\begin{array}{ccc}
3.26 \times 10^{-5} & 4.3179 \times 10^{12} & 34.60 \\
\underline{\times\, 1.78} & \underline{\times\, 3.6 \quad \times 10^{-19}} & \underline{\div\, 2.462\,87} \\
5.80 \times 10^{-5} & 1.6 \quad \times 10^{-6} & 14.05
\end{array}
$$

The power of ten has no influence on the number of figures that should be retained.

Logarithms and antilogarithms

The **logarithm** of a is the number, b, whose value is such that

$$a = 10^b \tag{3-1}$$

$$\log a = b \tag{3-2}$$

The number a is said to be the **antilogarithm** of b. A logarithm is composed of a **mantissa** and a **character**:

$$\log 339 = \underset{\substack{| \qquad | \\ \text{Character} \quad \text{Mantissa}}}{2.530}$$

The number 339 can be written 3.39×10^2. *The number of figures in the **mantissa** of log 339 should equal the number of significant figures in 339.* The logarithm of 339 is properly expressed as 2.530. The character, 2, corresponds to the exponent in 3.39×10^2.

To see that the third decimal place is the last significant place, consider the following results:

$$10^{2.531} = 340 \, (339.6)$$

$$10^{2.530} = 339 \, (338.8)$$

$$10^{2.529} = 338 \, (338.1)$$

The numbers in parentheses are the results prior to rounding to three figures. Changing the exponent by one digit in the third decimal place changes the answer by one digit in the last place of 339.

In converting a logarithm to its antilogarithm, *the number of significant figures in the antilogarithm should equal the number of figures in the **mantissa**.* Thus

$$\text{antilog}(-3.\underbrace{42}_{2 \text{ digits}}) = 10^{-3.\underbrace{42}_{2 \text{ digits}}} = \underbrace{3.8}_{2 \text{ digits}} \times 10^{-4}$$

Some examples showing the proper use of significant figures for logs and antilogs are given below.

$\log 0.001\,237 = -2.907\,6$ antilog $4.37 = 2.3 \times 10^4$

$\log 1\,237 = 3.092\,4$ $10^{4.37} = 2.3 \times 10^4$

$\log 3.2 = 0.51$ $10^{-2.600} = 2.51 \times 10^{-3}$

Number of figures in antilog $x\ (=10^x)=$ number of figures in **mantissa** of x:
$10^{6.142} = 1.39 \times 10^6$
3 digits   3 digits

3-2 SIGNIFICANT FIGURES AND GRAPHS

The rulings on a sheet of graph paper should be compatible with the number of significant figures of the coordinates. The graph in Figure 3-2a has reasonable rulings on which to graph the points (0.53, 0.65) and (1.08, 1.47).

The accuracy of a graph should be consistent with the accuracy of the data being plotted.

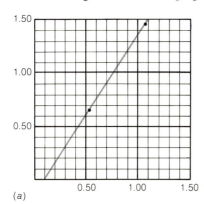
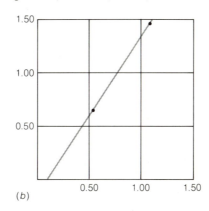

Figure 3-2
Graphs demonstrating choice of rulings in relation to significant figures in the data. The graph in part b does not have fine enough divisions to plot data accurate to the hundredths place.

There are rulings corresponding to every 0.1 unit, and it is easy to estimate the position of the 0.01 unit. The graph in Figure 3-2b is the same size but does not have fine enough rulings for estimating the position of the 0.01 unit.

In general, a graph must be at least as accurate as the data being plotted. Therefore, it is a good practice to use the most finely ruled graph paper available. Paper ruled with 10 lines per centimeter or 20 lines per inch is usually best. Plan the coordinates so that the data are spread over as much of the sheet of paper as possible.

3-3 TYPES OF ERRORS

Systematic Error

Systematic error is a consistent error that may be detected and corrected.

Experimental errors can be classified as either **systematic** or **random.** A systematic error, also called a **determinate** error, can in principle be discovered and corrected. One example would be using a pH meter that has been standardized incorrectly. Suppose you think that the pH of the buffer used to standardize the meter is 7.00, but it is really 7.08. If the meter is otherwise working properly, all of your pH readings will be 0.08 pH units too low. When you read a pH of 5.60, the actual pH of the sample is 5.68. This is a simple example of systematic error. It is always in the same direction and could be discovered by using another buffer of known pH to test the meter.

A slightly more complicated example of systematic error occurs with an uncalibrated buret. The manufacturer's tolerance for a Class A 50 mL buret is ± 0.05 mL. That is, when you believe that you have delivered 29.43 mL, the actual volume could be 29.40 mL and still be within the manufacturer's tolerance. One way to correct for this type of error is by constructing an experimental calibration curve, such as that shown in Figure 3-3. In this procedure distilled water is delivered from the buret

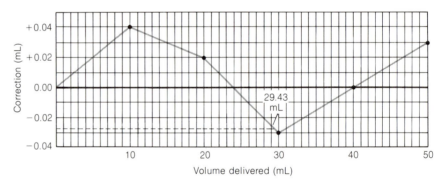

Figure 3-3
Calibration curve for a 50 mL buret.

into a flask and weighed. By knowing the density of water, we can determine the volume of water from its mass. The graph of correction factors assumes that the level of liquid is always close to the zero mark at the start of the titration. Using Figure 3-3, you would apply a correction factor of -0.03 mL to the measured value of 29.43 mL to reach the correct value of 29.40 mL.

The systematic error associated with using the buret whose calibration is shown in Figure 3-3 is a positive error in some regions and a negative error in others. The key feature of systematic error is that, with care and cleverness, you can detect it and correct it.

Random Error

Random error is also called **indeterminate** error. It arises from natural limitations on our ability to make physical measurements. As the name implies, random error is sometimes positive and sometimes negative. It is always present, cannot be corrected, and is the ultimate limitation on the determination of a quantity. One type of random error is that associated with reading a scale. Different people reading the absorbance or transmittance in Figure 3-1 would report a range of values reflecting their subjective interpolations between the markings. One person reading the same instrument several times probably would also report several different readings. Another type of indeterminate error may result from electrical noise in an instrument. In measuring a voltage, for example, the reading generally has a small fluctuation resulting from electrical instability of the meter itself. This sort of instability is normally random. Positive and negative fluctuations occur with approximately equal frequency and cannot be completely eliminated.

> Random error is due to the limitations of physical measurement and cannot be eliminated. A better experiment may reduce the magnitude of the random error, but cannot eliminate it entirely.

Precision and Accuracy

Precision is a measure of the reproducibility of a result. **Accuracy** refers to how close a measured value is to the "true" value.

The result of an experiment may be very reproducible, but wrong. For example, if you made an error in preparing a solution for a titration, the solution would not have the desired concentration. You might then do a series of highly reproducible titrations, but report an incorrect result because the concentration of solution was not as you intended. In such a case we would say that the precision of the result is high, but the accuracy is poor. Conversely, it is possible to make a series of poorly reproducible measurements clustered around the correct value. In this case, the precision is low but the accuracy is high. An ideal procedure will provide both precision and accuracy.

> Precision: reproducibility.
>
> Accuracy: nearness to the "truth."

Accuracy is defined as nearness to the "true" value. The word "true" is in quotes because somebody had to *measure* the "true" value, and there is error associated with *every* measurement. The "true" value is best obtained by an experienced worker using a well-tested procedure. It is desirable to test the result using different procedures, because, while each method might be precise, systematic error could lead to poor agreement between methods. Good agreement among several methods affords us some confidence, but never proof, that the results are correct.

Absolute and Relative Error

Absolute error is an expression of the margin of uncertainty associated with a measurement. If the estimated uncertainty in reading a perfectly calibrated buret is ± 0.02 mL, we call the quantity ± 0.02 mL the absolute error or absolute uncertainty associated with the reading.

> We use the terms *error* and *uncertainty* interchangeably.

Relative error is an expression comparing the size of the absolute error to the size of its associated measurement. The relative error of a buret

34

3 / EXPERIMENTAL ERROR

reading of 12.35 ± 0.02 mL is

$$\text{Relative error} = \frac{\text{Absolute error}}{\text{Magnitude of measurement}} \qquad (3\text{-}3)$$

$$= \frac{0.02 \text{ mL}}{12.35 \text{ mL}} = 0.002$$

The percent relative error is simply

$$\text{Percent relative error} = 100 \times \text{Relative error} \qquad (3\text{-}4)$$

In the above example the percent relative error is 0.2%.

A constant absolute error leads to a smaller relative error as the magnitude of the measurement increases. If the uncertainty in reading a buret is constant at ± 0.02 mL, the relative error is 0.2% for a volume of 10 mL and 0.1% for a volume of 20 mL.

3-4 PROPAGATION OF ERROR

It is usually possible to estimate or measure the random error associated with a particular measurement, such as the length of an object or the temperature of a solution. The error could be based on your estimate of how well you can read an instrument or on your experience with a particular method. When possible, error is usually expressed as the *standard deviation* of a series of replicate measurements. The discussion that follows applies only to random error; it is assumed that any systematic error has been detected and corrected.

Standard deviation will be defined and discussed in the next chapter.

In most experiments, it is necessary to perform arithmetic operations on several numbers, each of which has its associated random error. The most likely error in the result is not simply the sum of the individual errors, because some of these are likely to be positive and some negative. We expect a certain amount of cancellation of errors.

Addition and Subtraction

Suppose you wish to perform the following arithmetic, in which the experimental uncertainties are given in parentheses:

$$
\begin{array}{r}
1.76 \, (\pm 0.03) \leftarrow e_1 \\
+ \, 1.89 \, (\pm 0.02) \leftarrow e_2 \\
- \, 0.59 \, (\pm 0.02) \leftarrow e_3 \\
\hline
3.06 \, (\pm e_4)
\end{array}
\qquad (3\text{-}5)
$$

The arithmetic answer is 3.06, but what is the error associated with this result?

We start by calling the three errors e_1, e_2, and e_3. For addition and subtraction, the error in the answer is obtained by manipulating the *absolute errors* of the individual terms:

$$e_4 = \sqrt{e_1^2 + e_2^2 + e_3^2} \tag{3-6}$$

For the sum in Equation 3-5 we can write

$$e_4 = \sqrt{(0.03)^2 + (0.02)^2 + (0.02)^2} = 0.04_1 \tag{3-7}$$

The absolute error associated with the sum is ± 0.04, and we can write the answer as 3.06 ± 0.04. Although there is only one significant figure in the error, we wrote it initially as 0.04_1, with the first insignificant figure subscripted. The reason for retaining one or more insignificant figures is to avoid introducing round-off errors into later calculations through the number 0.04_1. The insignificant figure was subscripted to remind us where the last significant figure should be at the conclusion of the calculations.

If we wish to express the percent relative error in the sum of Equation 3-5, we may write

$$\text{Percent relative error} = \frac{0.04_1}{3.06} \times 100 = 1._3\% \tag{3-8}$$

The error, 0.04_1, is $1._3\%$ of the result, 3.06. The subscript, 3, in $1._3\%$ is not significant. It would now be sensible to drop the insignificant figures and express the final result as

$$3.06 \, (\pm 0.04) \qquad \text{(Absolute error)}$$

or

$$3.06 \, (\pm 1\%) \qquad \text{(Relative error)}$$

Multiplication and Division

For multiplication and division we first convert all errors to percent relative errors (or relative errors). Then we calculate the error of the product or quotient as follows:

$$\%e_4 = \sqrt{(\%e_1)^2 + (\%e_2)^2 + (\%e_3)^2} \tag{3-9}$$

For example, consider the operations below:

$$\frac{1.76 \, (\pm 0.03) \times 1.89 \, (\pm 0.02)}{0.59 \, (\pm 0.02)} = 5.6_4 \pm \, ? \tag{3-10}$$

For addition and subtraction, use absolute error.

For addition and subtraction, use absolute errors. The relative error can be found at the end of the calculation.

Some advice: Retain one or more extra insignificant figures until you have finished your entire calculation. Then round the final answer to the correct number of figures. If you are using a calculator, just keep all of the digits in the calculator until you need to express the final answer.

For multiplication and division, use relative error.

3 / EXPERIMENTAL ERROR

First convert all of the absolute errors to percent relative errors:

$$\frac{1.76\,(\pm 1._7\%) \times 1.89\,(\pm 1._1\%)}{0.59\,(\pm 3._4\%)} = 5.6_4 \pm \;? \tag{3-11}$$

Then find the relative error of the answer using Equation 3-9:

$$\%e_4 = \sqrt{(1._7)^2 + (1._1)^2 + (3._4)^2} = 4._0\% \tag{3-12}$$

The answer is $5.6_4\,(\pm 4._0\%)$.

To convert the relative error to absolute error, find $4._0\%$ of the answer:

$$4._0\% \times 5.6_4 = 0.04_0 \times 5.6_4 = 0.2_3 \tag{3-13}$$

The answer is $5.6_4\,(\pm 0.2_3)$. Finally, we drop all of the figures that are not significant. The result may be expressed as

For multiplication and division, use relative errors. The absolute error can be found at the end of the calculation.

$$5.6\,(\pm 0.2) \qquad \text{(Absolute error)}$$

$$5.6\,(\pm 4\%) \qquad \text{(Relative error)}$$

There are only two significant figures because we are limited by the denominator, 0.59, in the original problem.

Mixed Operations

A general treatment of propagation of error in complex calculations (including logs and exponents) can be found in Appendix C.

As a final example, consider the combined operations below:

$$\frac{[1.76\,(\pm 0.03) - 0.59\,(\pm 0.02)]}{1.89\,(\pm 0.02)} = 0.619_0 \pm \;? \tag{3-14}$$

First work out the difference in brackets, using absolute errors:

$$1.76\,(\pm 0.03) - 0.59\,(\pm 0.02) = 1.17 \pm 0.03_6 \tag{3-15}$$

since $\sqrt{(0.03)^2 + (0.02)^2} = 0.03_6$.

Then convert to relative errors:

$$\frac{1.17\,(\pm 0.03_6)}{1.89\,(\pm 0.02)} = \frac{1.17\,(\pm 3._1\%)}{1.89\,(\pm 1._1\%)} = 0.619_0\,(\pm 3._3\%) \tag{3-16}$$

since $\sqrt{(3._1)^2 + (1._1)^2} = 3._3$.

The relative error in the result is $3._3\%$. The absolute error is $0.03_3 \times 0.619_0 = 0.02_0$. The final answer could be written as

$$0.619\,(\pm 0.02_0) \qquad \text{(Absolute error)}$$

or

$$0.619 \ (3._3\%) \qquad \text{(Relative error)}$$

Since the error spans the last *two* places of the result, it would also be reasonable to write the result as

$$0.62 \ (\pm 0.02)$$

or

$$0.62 \ (\pm 3\%)$$

The result of a calculation ought to be written in a manner consistent with the uncertainty in the result.

Comment on Significant Figures

The number of figures used to express a calculated result should be consistent with the uncertainty in that result. For example, the ratio

$$\frac{0.002 \ 364 \ (\pm 0.000 \ 003)}{0.025 \ 00 \ (\pm 0.000 \ 05)} = 0.094 \ 6 \ (\pm 0.000 \ 2)$$

is properly expressed with *three* significant figures, even though the original data have four figures. *The first uncertain figure of the answer is the last significant figure.* The ratio

$$\frac{0.002 \ 664 \ (\pm 0.000 \ 003)}{0.025 \ 00 \ (+0.000 \ 05)} = 0.106 \ 6 \ (\pm 0.000 \ 2)$$

is expressed with *four* figures because the error occurs in the fourth place. The quotient

$$\frac{0.821 \ (\pm 0.002)}{0.803 \ (\pm 0.002)} = 1.022 \ (\pm 0.004)$$

is expressed with *four* figures even though the dividend and divisor each have *three* figures.

The first uncertain figure should be the last significant figure.

Summary

The number of significant figures in a value is the minimum number of digits needed to write the value in scientific notation. The first uncertain digit in a calculated result should be the last significant digit. In addition and subtraction, the last significant figure is determined by the decimal place of the least certain number. In multiplication and division, the number of figures is usually limited by the factor with the least number of digits. The number of figures in the mantissa of the logarithm of a quantity should equal the number of significant figures in the quantity. Random error mainly affects the precision (reproducibility) of a result, while systematic error mainly affects the accuracy (nearness to the "true" value). To analyze propagation of error in addition and subtraction, we use absolute errors: $e_3 = \sqrt{e_1^2 + e_2^2}$. In multiplication and division, propagation of error is analyzed using relative errors: $\%e_3 = \sqrt{\%e_1^2 + \%e_2^2}$. Always retain more digits than necessary during a calculation, and round off to the appropriate number of digits at the end.

Terms to Understand

absolute error	mantissa
accuracy	precision
antilogarithm	random error
character	relative error
determinate error	significant figure
indeterminate error	systematic error
logarithm	

Exercises

3-A. Write each answer with a reasonable number of figures. Find the absolute and percent relative error for each answer.

(a) $[12.41 \, (\pm 0.09) \div 4.16 \, (\pm 0.01)]$
   $\times \, 7.068 \, 2 \, (\pm 0.000 \, 4) = \, ?$

(b) $3.26 \, (\pm 0.10) \times 8.47 \, (\pm 0.05) - 0.18 \, (\pm 0.06) = \, ?$

(c) $6.843 \, (\pm 0.008) \times 10^4 \div [2.09 \, (\pm 0.04) -$
   $1.63 \, (\pm 0.01)] = \, ?$

3-B. Suppose that you have a bottle of aqueous solution labeled "53.4 $(\pm 0.4)\%$ (wt/wt) NaOH—density = 1.52 (± 0.01) g/mL."

(a) How many milliliters of 53.4 % NaOH are needed to prepare 2.000 L of 0.169 M NaOH?

(b) If the error in delivering the NaOH is ± 0.10 mL, calculate the absolute uncertainty in the molarity (0.169 M). You may assume negligible uncertainty in the molecular weight of NaOH and in the final volume, 2.000 L.

3-C. Consider a solution containing 37.0 $(\pm 0.5)\%$ (wt/wt) HCl in water. The density of the solution is 1.18 (± 0.01) g/mL. To deliver 0.050 0 $(\pm 2\%)$ mol of HCl requires 4.18 $(\pm x)$ mL of solution. Find x.

Caution: In this problem you have been given the error in the *answer* to a calculation. You need to find the error in one factor along the way in the calculation. Be sure to propagate errors in the right direction. For example, if $a = b \cdot c$, then $\%e_a^2 = \%e_b^2 + \%e_c^2$. If you are given $\%e_a$, the error $\%e_c$ must be *smaller* than $\%e_a$, and $\%e_c^2 = \%e_a^2 - \%e_b^2$.

Problems

3-1. How many significant figures are there in
(a) 1.903 0 (b) 0.039 10 (c) 1.40 × 10⁴

3-2. Round each number to the number of significant figures indicated.
(a) 1.236 7 to 4 figures
(b) 1.238 4 to 4 figures
(c) 0.135 2 to 3 figures
(d) 2.051 1 to 2 figures
(e) 2.005 0 to 3 figures

3-3. Round each number to three significant figures.
(a) 0.216 74 (b) 0.216 5 (c) 0.216 500 3

3-4. Write each answer with the correct number of significant figures.
(a) $1.0 + 2.1 + 3.4 + 5.8 = 12.300\ 0$
(b) $106.9 - 31.4 = 75.500\ 0$
(c) $107.868 - (2.113 \times 10^2) + (5.623 \times 10^3) = 5\ 519.568$
(d) $(26.14/37.62) \times 4.38 = 3.043\ 413$
(e) $(26.14/37.62 \times 10^8) \times (4.38 \times 10^{-2}) = 3.043\ 413 \times 10^{-10}$
(f) $(26.14/3.38) + 4.2 = 11.933\ 7$
(g) $\log(3.98 \times 10^4) = 4.599\ 9$
(h) $10^{-6.31} = 4.897\ 79 \times 10^{-7}$

3-5. Write the answers with the correct number of figures.
(a) $\log(4.218 \times 10^{12}) = ?$
(b) $\text{antilog}(-3.22) = ?$
(c) $10^{2.384} = ?$

3-6. Calculate the formula weight of (a) $BaCl_2$ and (b) $C_{31}H_{32}O_8N_2$ using the correct number of significant figures.

3-7. Rewrite the number 3.123 56 ($\pm 0.167\ 89\%$) in the forms (a) Number ($\pm$ Absolute error) and (b) Number ($\pm$ Percent relative error). Use a reasonable number of figures in each expression.

3-8. Find the absolute and percent relative error for each calculation. Express the answers with a reasonable number of significant figures.
(a) $9.23\ (\pm 0.03) + 4.21\ (\pm 0.02) - 3.26\ (\pm 0.06) = ?$
(b) $91.3\ (\pm 1.0) \times 40.3\ (\pm 0.2)/21.1\ (\pm 0.2) = ?$
(c) $[4.97\ (\pm 0.05) - 1.86\ (\pm 0.01)]/21.1\ (\pm 0.2) = ?$

3-9. Each target in the figure shows where arrows have struck. Match the letter of the target with the description below.

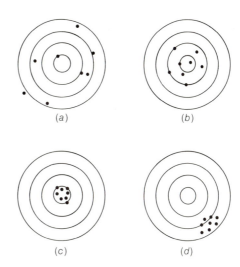

(a) Accurate and precise
(b) Accurate but not precise
(c) Precise but not accurate
(d) Neither precise nor accurate

3-10. The constant ℏ (read "h bar") is defined as $h/2\pi$, where h is Planck's constant ($6.626\ 176 \pm 0.000\ 036) \times 10^{-34}$ J s). Calculate the value and absolute uncertainty of ℏ. The number 2 is an integer (infinitely accurate) and π is also an exact number. The first ten digits of π are 3.141 592 653.

3-11. The value of Boltzmann's constant (k) listed on the cover of the book is calculated from the quotient R/N, where R is the gas constant and N is Avogadro's number. If the uncertainty in k is $0.000\ 044 \times 10^{-23}$ J K⁻¹ and the uncertainty in R is 0.000 26 J mol⁻¹ K⁻¹, find the uncertainty in N. Remember that the value of k is obtained by measuring R and N.

4 / Statistics

Since all real measurements contain experimental error, it is never possible to be completely certain of a result. Nevertheless, we still seek to answer such questions as "Is my red blood cell count today higher than its usual value?" If today's count is twice as high as its usual value, answering this question is trivial. But what if the counts are

"Normal" days	Today
5.1	5.6×10^6 cells/μL
5.3	
4.8 $\times 10^6$ cells/μL	
5.4	
5.2	

The number 5.6 is higher than the five normal values, but the random variation in normal values might lead us to expect that 5.6 will be observed on some "normal" days.

The study of statistics addresses the question of how to deal with experimental results. We will never be able to answer a question with complete certainty, but we are able to say that the value 5.6×10^6 cells/μL is expected to be observed one time in twenty on normal days. There is a 5% probability that 5.6×10^6 represents a normal count and a 95% probability that it represents an elevated count.

The best we can do is assess the probability that today's count is elevated. We cannot say that it *is* or *is not* elevated.

4-1 GAUSSIAN ERROR CURVE

When the variation in a set of experimental data is strictly random, the data approximate a bell-shaped curve, as illustrated in Figure 4-1. In this hypothetical case, a manufacturer has tested the lifetimes of 4 768 electric light bulbs. The bar graph in Figure 4-1 shows the number of bulbs having a lifetime in each 20 hour interval. The smooth curve is the **Gaussian** or

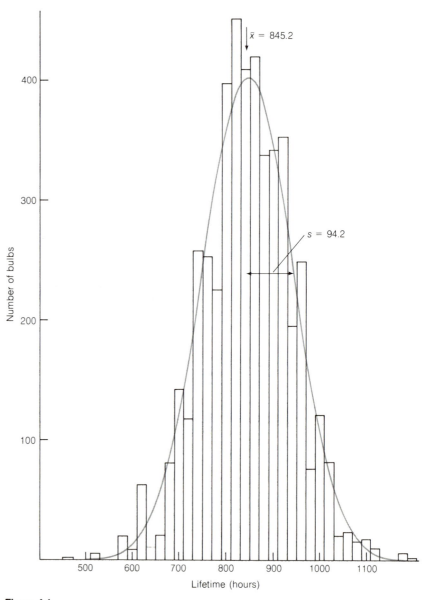

Figure 4-1
Bar graph and normal error curve describing the lifetime of a hypothetical set of electric bulbs. The smooth curve has the same mean, standard deviation, and area as does the bar graph.

normal error curve that best fits the data. In any finite set of data, there will be variation from the Gaussian curve. As the number of data points increases, the data should more closely approach the smooth curve.

Mean Value and Standard Deviation

The data for light bulb lifetimes, and the corresponding Gaussian curve, are characterized by two parameters. The arithmetic **mean** or **average** value, $\bar{x}$, is defined as

$$\text{Mean} = \text{Average} = \bar{x} = \frac{\sum_i x_i}{n} \qquad (4\text{-}1)$$

where each x_i is the lifetime of an individual bulb. The average is the sum of the measured values divided by n, the total number of values. In Figure 4-1, the mean value is indicated by the arrow at a value of 845.2 hours.

The **standard deviation,** s, measures how closely the data are clustered about the mean.

$$\text{Standard deviation} = s = \sqrt{\frac{\sum_i (x_i - \bar{x})^2}{n - 1}} \qquad (4\text{-}2)$$

For the data in Figure 4-1, $s = 94.2$ hr. The significance of s is that the smaller the standard deviation, the more closely the data are centered about the mean (Figure 4-2). A set of light bulbs having a small standard deviation in lifetime must be more uniformly manufactured than a set with a large standard deviation.

For an infinite set of data, the mean is called μ (the population mean) and the standard deviation is called σ (the population standard deviation). We can never measure μ and σ, but the values of $\bar{x}$ and s approach μ and σ as the number of measurements increases.

The quantity $n - 1$ in Equation 4-2 is called the **degrees of freedom** of the system. The square of the standard deviation is called the **variance**.

EXAMPLE

To illustrate the use of Equations 4-1 and 4-2, suppose that four measurements are made: 821, 783, 834, and 855 hours. The average is

$$\bar{x} = \frac{821 + 783 + 834 + 855}{4} = 823._2 \text{ hr}$$

To avoid round-off errors, we generally retain one more significant figure for the average and standard deviation than was present in the original data. The standard deviation is

The mean gives the center of the distribution. The standard deviation measures the width of the distribution.

An experimental technique that produces a small standard deviation is more reliable (precise) than one that produces a large standard deviation, provided that they are equally accurate.

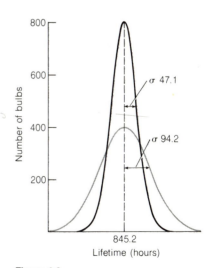

Figure 4-2
Gaussian curves for two sets of light bulbs, one having a standard deviation one-half that of the other. The number of bulbs described by each curve is the same.

$$s = \sqrt{\frac{(821-823.2)^2 + (783-823.2)^2 + (834-823.2)^2 + (855-823.2)^2}{(4-1)}} = 30._3 \text{ hr}$$

The average and standard deviation should both end at the *same decimal place*. For $\bar{x} = 823.2$, the value $s = 30.3$ is reasonable, but $s = 30.34$ is not.

Other terms

We now define some quantities that do not apply directly to the normal error curve, but whose definitions you ought to know. The **median** is the value above and below which there is an equal number of data points. For an odd number of points, the median is the middle one. Therefore 3 is the median of 1, 2, 3, 7, 8. For an even number of points, the median is halfway between the two center values. Thus the median of 1, 2, 3, and 6 is 2.5. The **range** (or **spread**) is the difference between the highest and lowest values. The range of 126.2, 127.5, 127.1, 125.9, and 126.4 is (127.5–125.9) = 1.6. The **geometric mean** of n numbers is

$$\text{Geometric mean} = \sqrt[n]{\prod_i x_i} \tag{4-3}$$

The symbol $\prod$ means the product of all the values. The geometric mean of 2, 4, 9, 13, and 29 is $(2 \cdot 4 \cdot 9 \cdot 13 \cdot 29)^{1/5} = 7.7$.

Standard Deviation and Probability

The formula for a Gaussian curve is

$$y = \frac{1}{\sigma\sqrt{2\pi}} e^{-(x-\mu)^2/2\sigma^2} \tag{4-4}$$

where e (2.718 28 ...) is the base of the natural logarithm. To describe a finite set of data, we approximate μ by $\bar{x}$ and σ by s. A graph of Equation 4-4 is shown in Figure 4-3, in which we have set $\sigma = 1$ and $\mu = 0$ for simplicity. In general, the most probable value of x is $x = \mu$ and the curve is symmetric about $x = \mu$. The relative probability of making a particular measurement is proportional to the ordinate (y value) for the value of x. Thus in Figure 4-3 the maximum probability for any measurement occurs at $x = \mu = 0$. The probability of measuring the value $x = 1$ is $0.242/0.399 = 0.607$ times the probability of measuring the value $x = 0$. Table 4-1 gives ordinate and area values for the Gaussian curve.

The probability of measuring x in a certain *range* is proportional to the *area* of that range. For example, the probability of observing x between -2 and -1 is 0.136. This corresponds to the shaded area in Figure 4-3. The area under each portion of the Gaussian curve is given in Table 4-1. Since

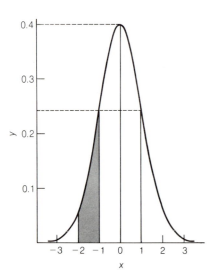

Figure 4-3
A Gaussian error curve in which $\mu = 0$ and $\sigma = 1$.

Table 4-1
Ordinate and area for the normal error curve

x	$y = \dfrac{1}{\sqrt{2\pi}}e^{-x^2/2}$		x	$y = \dfrac{1}{\sqrt{2\pi}}e^{-x^2/2}$		x	$y = \dfrac{1}{\sqrt{2\pi}}e^{-x^2/2}$	
	y	Area		y	Area		y	Area
0.0	0.398 9	0.000 0	1.4	0.149 7	0.419 2	2.8	0.007 9	0.497 4
0.1	0.397 0	0.039 8	1.5	0.129 5	0.433 2	2.9	0.006 0	0.498 1
0.2	0.391 0	0.079 3	1.6	0.110 9	0.445 2	3.0	0.004 4	0.498 7
0.3	0.381 4	0.117 9	1.7	0.094 1	0.455 4	3.1	0.003 3	0.499 0
0.4	0.368 3	0.155 4	1.8	0.079 0	0.464 1	3.2	0.002 4	0.499 3
0.5	0.352 1	0.191 5	1.9	0.065 6	0.471 3	3.3	0.001 7	0.499 5
0.6	0.333 2	0.225 8	2.0	0.054 0	0.477 3	3.4	0.001 2	0.499 7
0.7	0.312 3	0.258 0	2.1	0.044 0	0.482 1	3.5	0.000 9	0.499 8
0.8	0.289 7	0.288 1	2.2	0.035 5	0.486 1	3.6	0.000 6	0.499 8
0.9	0.266 1	0.315 9	2.3	0.028 3	0.489 3	3.7	0.000 4	0.499 9
1.0	0.242 0	0.341 3	2.4	0.022 4	0.491 8	3.8	0.000 3	0.499 9
1.1	0.217 9	0.364 3	2.5	0.017 5	0.493 8	3.9	0.000 2	0.500 0
1.2	0.194 2	0.384 9	2.6	0.013 6	0.495 3	4.0	0.000 1	0.500 0
1.3	0.171 4	0.403 2	2.7	0.010 4	0.496 5			

Note: The area refers to the area between $x = 0$ and $x =$ the value in the table. Thus the area from $x = 0$ to $x = 1.4$ is 0.419 2. The area from $x = -0.7$ to $x = 0$ is the same as from $x = 0$ to $x = 0.7$. The area from $x = -0.5$ to $x = +0.3$ is $(0.191\ 5 + 0.117\ 9) = 0.309\ 4$. The total area between $x = -\infty$ and $x = +\infty$ is unity. A more complete table can be found in any edition of the *Handbook of Chemistry and Physics* (Boca Raton, Fla.: CRC Press).

the probability of making all measurements must be unity, the area under the whole curve from $x = -\infty$ to $x = +\infty$ must be unity. The number $1/\sigma\sqrt{2\pi}$ in Equation 4-4 is called the *normalization factor*. It guarantees that the area under the entire curve is unity.

EXAMPLE

Suppose the manufacturer offers to replace free of charge any bulb that burns out in less than 600 hours. What fraction of the bulbs should be kept available as replacements?

To answer this question we express the desired interval in multiples of the standard deviation. Then we find the area of the interval using Table 4-1. The number 600 is $(845.2 - 600)/94.2 = 2.60$ times the standard deviation. The area under the curve between the mean value and -2.60 standard deviations is given as 0.495 3 in Table 4-1. Since the entire area from $-\infty$ to the mean value is 0.500 0, the area from $-\infty$ to -2.60 must be 0.004 7. That is, the area to the left of 600 hours in Figure 4-1 is only 0.47% of the entire area under the normal error curve. Only 0.47% of the bulbs are expected to fail in less than 600 hours.

EXAMPLE

What fraction of the bulbs is expected to have a lifetime between 900 and 1 000 hours?

To answer this question *we need to find the fraction of the area of the Gaussian curve between $x = 900$ and $x = 1\ 000$ hours*. Since $\bar{x} = 845.2$ and $s = 94.2$, the number 900 lies (900 − 845.2)/94.2 = 0.582 standard deviations to the right of the mean value. The number 1 000 lies 1.643 standard deviations to the right of the mean value. Figure 4-4 shows that we can find the area between 900 and 1 000 hours by first finding the area from $\bar{x}$ to 1 000 hours and then subtracting the area from $\bar{x}$ to 900 hours.

Table 4-1 can be used to find the area from $\bar{x}$ to 900 hours (= $\bar{x}$ + 0.582 s) as follows. The area up to 0.5 s is 0.191 5. The area up to 0.6 s is 0.225 8. We can estimate the area up to 0.582 by *linear interpolation*:

$$\text{Area between 0.500 s and 0.582 s} = \underbrace{\left(\frac{0.582 - 0.500}{0.600 - 0.500}\right)}_{\substack{\text{Fraction of the} \\ \text{way between} \\ \text{0.5 s and 0.6 s}}} \underbrace{(0.225\,8 - 0.191\,5)}_{\substack{\text{Area between} \\ \text{0.5 s and 0.6 s}}} = 0.028\,1$$

Area between 0 and 0.582 = Area between 0 and 0.5 + Area between 0.5 and 0.582

Area between 0 and 0.582 = 0.191 5 + 0.028 1 = 0.219 6

To find the area between $\bar{x}$ and 1 000 hrs (= $\bar{x}$ + 1.643 s) we interpolate between x = 1.6 and x = 1.7 to find

Area between 0 and 1.643 = 0.449 6

The area between 900 and 1 000 hours is 0.449 6 − 0.219 6 = 0.230 0. That is, 23% of the bulbs are expected to have a lifetime between 900 and 1 000 hours.

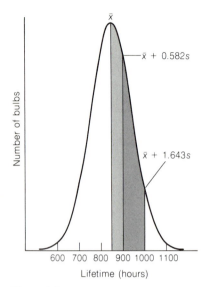

Figure 4-4
Use of the Gaussian curve to find the fraction of bulbs with a lifetime between 900 and 1 000 hours.

Range	Percent of measurements
$\mu \pm 1\sigma$	68.3%
$\mu \pm 2\sigma$	95.5%
$\mu \pm 3\sigma$	99.7%

The significance of the standard deviation is that it measures the width of the normal error curve. The larger the value of σ, the broader the curve. In Figure 4-3, 68.3% of the area falls under $\mu \pm 1\sigma$. That is, more than $\frac{2}{3}$ of the measurements are expected to lie within one standard deviation of the mean. In a normal error curve. 95.5% of the area lies within $\mu \pm 2\sigma$, and 99.7% of the area lies within $\mu \pm 3\sigma$ (Table 4-1).

Suppose that you use two different experimental methods to measure the percent of sulfur in coal; method A has a standard deviation of 0.4%, and method B has a standard deviation of 1.1%. You can expect that $\frac{2}{3}$ of measurements from method A will lie within 0.4% of the mean. For method B, $\frac{2}{3}$ of measurements will lie within 1.1% of the mean.

4-2 STUDENT'S *t*

Student's *t* is a valuable statistical tool used to measure probability. We use it most frequently to express confidence intervals and for comparing results obtained in different experiments.

Confidence Intervals

From a limited number of measurements it is impossible to find the true population mean, μ, or the true standard deviation, σ. What we can deter-

Table 4-2
Values of Student's t

Degrees of freedom	Confidence level (%)				
	50	80	90	95	99
1	1.000	3.078	6.314	12.706	63.657
2	0.816	1.886	2.920	4.303	9.925
3	0.765	1.638	2.353	3.182	5.841
4	0.741	1.533	2.132	2.776	4.604
5	0.727	1.476	2.015	2.571	4.032
6	0.718	1.440	1.943	2.447	3.707
7	0.711	1.415	1.895	2.365	3.500
8	0.706	1.397	1.860	2.306	3.355
9	0.703	1.383	1.833	2.262	3.250
10	0.700	1.372	1.812	2.228	3.169
15	0.691	1.341	1.753	2.131	2.947
20	0.687	1.325	1.725	2.086	2.845
∞	0.674	1.282	1.645	1.960	2.576

Note: When calculating confidence intervals, σ may be substituted for s in Equation 4-5 if you have a great deal of experience with a particular method and have therefore determined its "true" population standard deviation. If σ is used instead of s, the value of t to use in Equation 4-5 comes from the bottom row of Table 4-2.

mine are $\bar{x}$ and s. The **confidence interval** is an expression stating that the true mean, μ, is likely to lie within a certain distance of the measured mean, $\bar{x}$. The confidence interval is given by

$$\mu = \bar{x} \pm \frac{ts}{\sqrt{n}} \tag{4-5}$$

where s is the measured standard deviation, n is the number of observations, and t is a number called Student's t. Values of t for various confidence levels are given in Table 4-2.

"Student" was the pseudonym of W. S. Gossett, who published the classic paper on this subject (*Biometrika*, **6**, 1 [1908]). Gossett's employer, the Guinness Breweries of Ireland, restricted publications for proprietary reasons. Because of the importance of Gossett's work, he was allowed to publish it, but under an assumed name.

EXAMPLE

Suppose that the percent carbohydrate content of a glycoprotein (a protein with sugars attached to it) is determined to be 12.6, 11.9, 13.0, 12.7, and 12.5 in replicate analyses. Find the 50% and 95% confidence intervals for the carbohydrate content.

First we calculate $\bar{x} = 12.5_4\%$ and $s = 0.4_0\%$ for the five measurements. To calculate the 50% confidence interval, look up t in Table 4-2 under 50% and across from *four* degrees of freedom. (Recall that degrees of freedom $= n - 1$.) The value of t is 0.741, so the 50% confidence interval is

$$\mu = \bar{x} \pm \frac{ts}{\sqrt{n}} = 12.5_4 \pm \frac{(0.741)(0.4_0)}{\sqrt{5}} = 12.5_4 \pm 0.1_3$$

48

4 / STATISTICS

The 95% confidence interval is

$$\mu = \bar{x} \pm \frac{ts}{\sqrt{n}} = 12.5_4 \pm \frac{(2.776)(0.4_0)}{\sqrt{5}} = 12.5_4 \pm 0.5_0$$

These calculations mean that there is a 50% chance that the true mean, μ, lies within the range 12.4_1–12.6_7. There is a 95% chance that μ lies within the range 12.0_4–13.0_4.

EXAMPLE

Suppose that you wish to test the validity of a new analytical technique. You prepare a sample containing 0.031 9% Ni and analyze it four times with your new method. You observe values of 0.032 9, 0.032 2, 0.033 0, and 0.032 3. Does your method yield a value "significantly" higher than the known value?

To answer this question, we can calculate the value of μ at several confidence intervals. First we calculate $\bar{x} = 0.032\ 6_0$ and $s = 0.000\ 4_1$. Plugging into Equation 4-5 gives

$$90\% \text{ Confidence:} \quad \mu = 0.032\ 6_0 \pm \frac{(2.353)(0.000\ 4_1)}{\sqrt{4}} = 0.032\ 6_0 \pm 0.000\ 4_8$$

$$95\% \text{ Confidence:} \quad \mu = 0.032\ 6_0 \pm \frac{(3.182)(0.000\ 4_1)}{\sqrt{4}} = 0.032\ 6_0 \pm 0.000\ 6_5$$

$$99\% \text{ Confidence:} \quad \mu = 0.032\ 6_0 \pm \frac{(5.841)(0.000\ 4_1)}{\sqrt{4}} = 0.032\ 6_0 \pm 0.001\ 1_9$$

The known value, 0.031 9%, lies approximately at the limit of the 95% confidence interval (which covers the range $0.032\ 6_0 \pm 0.000\ 6_5 = 0.031\ 9_5$ to $0.033\ 2_5$). Therefore, you can be 95% confident that the new method produces a high value. That is, if $\mu = 0.031\ 9$, the value $\bar{x} = 0.032\ 6$ will be observed in only about 5% of experiments. The chances are 19 out of 20 that μ is really greater than 0.031 9 for your new method. Because 0.031 9 is within the 99% confidence interval ($0.031\ 4_1$ to $0.033\ 7_9$), you cannot be 99% confident that your new method produces a high result.

Statistical tests only give us probabilities. They do not relieve us of the responsibility of interpreting our results.

Statistical tests do not relieve you of the ultimate subjective decision to accept or reject a conclusion. The tests only provide guidance in the form of probabilities. In the example above, it would probably be reasonable to conclude that the new method is systematically high. However, with only four values, it would be wise to run the analysis several more times to confirm your conclusion.

Comparison of Means

We sometimes need to compare the results of two tests to see if they are "the same" or "different" from each other. For this purpose, we perform the t test using Student's t. The discussion that follows assumes that the population standard deviation (σ) for each method is essentially the same.

Comparing replicate measurements

4-2 STUDENT'S t

For the two sets of data we calculate a value of t using the formula

$$t = \frac{\bar{x}_1 - \bar{x}_2}{s} \sqrt{\frac{n_1 n_2}{n_1 + n_2}} \qquad (4\text{-}6)$$

where

$$s = \sqrt{\frac{\displaystyle\sum_{\text{set 1}} (x_i - \bar{x}_1)^2 + \sum_{\text{set 2}} (x_j - \bar{x}_2)^2}{n_1 + n_2 - 2}} \qquad (4\text{-}7)$$

The value of s is a *pooled standard deviation* making use of both sets of data. The value of t from Equation 4-6 is to be compared to the value of t in Table 4-2 for $n_1 + n_2 - 2$ degrees of freedom. *If the calculated t is greater than the tabulated t, the two results are significantly different at the confidence level in question.*

If $t_{\text{calculated}} > t_{\text{table}}$, the difference is significant.

EXAMPLE

$^{14}CO_2$ may be used as a radioactive tracer to study metabolism in plants. Suppose that a compound isolated from a plant exhibited 28, 32, 27, 39, and 40 radioactive decays per minute. A blank sample used to measure the background of the radiation counter (due to electrical noise and background radiation) gave 28, 21, 28, and 20 counts per minute. It appears that the isolated compound gives more counts than background. Can we be 95% confident that the compound is indeed radioactive?

The averages for the two sets of data are respectively $33._2$ and $24._2$ counts per minute. The pooled s is calculated from Equation 4-7 as follows:

$$s = \sqrt{\frac{\begin{array}{c}(28 - 33._2)^2 + (32 - 33._2)^2 + (27 - 33._2)^2 + (39 - 33._2)^2 + (40 - 33._2)^2 \\ + (28 - 24._2)^2 + (21 - 24._2)^2 + (28 - 24._2)^2 + (20 - 24._2)^2\end{array}}{5 + 4 - 2}}$$

$$s = 5._4$$

Equation 4-6 then gives

$$t = \frac{33._2 - 24._2}{5._4} \sqrt{\frac{5 \cdot 4}{5 + 4}} = 2._{48}$$

The calculated value of t is bigger than the value 2.365 listed in Table 4-2 for the seven degrees of freedom and 95% confidence. Therefore, we can be 95% confident that the observed counts are higher than background and that radioactive ^{14}C has been incorporated into the compound. Note that the calculated t (2.48) is less than the t for 99% confidence (3.500). Therefore, the probability that the two sets of counts are different lies between 95% and 99%.

Table 4-3
Comparison of two methods for measuring cholesterol

	Cholesterol Content (g/L)		
Plasma sample	Method A	Method B	Difference (d_i)
1	1.46	1.42	0.04
2	2.22	2.38	−0.16
3	2.84	2.67	0.17
4	1.97	1.80	0.17
5	1.13	1.09	0.04
6	2.35	2.25	0.10
			$\bar{d} = +0.06_0$

Comparing individual differences

Suppose that the cholesterol content of six sets of human blood plasma is measured by two different techniques. The results are listed in Table 4-3. Each of the six plasma samples is a different sample with a different cholesterol content. Method B gives a lower result than method A in five out of the six samples. Is method B systematically different from method A?

To answer this question, we perform a t test on the individual *differences*, d_i, as follows:

$$t = \frac{\bar{d}}{s_d} \sqrt{n} \tag{4-8}$$

where

$$s_d = \sqrt{\frac{\sum (d_i - \bar{d})^2}{n - 1}} \tag{4-9}$$

and n is the number of pairs of data (6 in this case). For the data in Table 4-3, the standard deviation, s_d, of the differences is calculated to be

$$s_d = \sqrt{\frac{(0.04 - 0.06_0)^2 + (-0.16 - 0.06_0)^2 + (0.17 - 0.06_0)^2 + (0.17 - 0.06_0)^2 + (0.04 - 0.06_0)^2 + (0.10 - 0.06_0)^2}{6 - 1}}$$

$$s_d = 0.12_2$$

Putting this value into Equation 4-8 gives

$$t = \frac{0.06_0}{0.12_2} \sqrt{6} = 1.20$$

The calculated value of t lies between the tabulated values of t at 50% and 80% confidence for five degrees of freedom in Table 4-2. That is, there is more than a 50% chance, but less than an 80% chance, that the two methods are systematically different. It would be reasonable to conclude that the two techniques are *not* significantly different.

When applying t tests, most people regard differences as significant when they occur in the 90–95% confidence range. A confidence level of 99% is considered highly significant.

$$t_{calculated} = 1.20$$

From Table 4-2, for $6 - 1 = 5$ degrees of freedom, we find:

$$t(50\% \text{ confidence}) = 0.727$$

$$t(80\% \text{ confidence}) = 1.476$$

4-3 DEALING WITH BAD DATA

Sometimes one datum appears to be inconsistent with the remaining data. When this happens, you are faced with the decision of whether to retain the questionable point or to throw it away as unreliable. The Q test is used to help make this decision.

Consider the five results 12.53, 12.56, 12.47, 12.67, and 12.48. Is 12.67 a "bad point"? To apply the Q test, we arrange the data in order of increasing value and calculate Q defined as

$$Q = \frac{\text{Gap}}{\text{Range}} \tag{4-10}$$

$$\text{Gap} = 0.11$$

12.47　　12.48　　12.53　　12.56　　　　　　(12.67)—Questionable value (too high?)

$$\text{Range} = 0.20$$

The **gap** is the difference between the questionable point and the nearest value.

If Q (observed) > Q (tabulated), the questionable point should be discarded. For the numbers above, $Q = 0.11/0.20 = 0.55$. Referring to Table 4-4, the critical value of Q at the 90% confidence limit is 0.64. *Since the observed Q is smaller than the tabulated Q, the questionable point should be retained.* That is, there is more than a 10% chance that the value 12.67 is a member of the same population as the other four numbers.

If Q (observed) > Q (tabulated), discard the questionable point.

Table 4-4
Values of Q for rejection of data

Q (90% confidence)	0.94	0.76	0.64	0.56	0.51	0.47	0.44	0.41
Number of observations	3	4	5	6	7	8	9	10

Note: Q = Gap/Range. If Q (observed) > Q (tabulated), the value in question may be rejected with 90% confidence.

SOURCE: R. B. Dean and W. J. Dixon, *Anal. Chem.*, **23**, 636 (1951). Reprinted in part with permission from *Analytical Chemistry*. Copyright 1951 American Chemical Society. This reference also provides useful recipes for the rapid estimation of statistical parameters for small sets of data ($n \leq 10$).

Table 4-5
Spectrophotometer readings for protein analysis by the Lowry method

Sample	Absorbance of three independent samples			Range	Average with all data
0 μg	0.099	0.099	0.100	0.001	0.099_3
5 μg	0.185	0.187	0.188	0.003	0.188_7
10 μg	0.282	0.272	0.272	0.010	0.275_3
15 μg	**0.392**	0.345	0.347	0.047	0.361_2
20 μg	0.425	0.425	0.430	0.005	0.426_7
25 μg	0.483	0.488	0.496	0.013	0.489_0

The Q test is fairly stringent and not particularly helpful for small sets of data ($n < 5$). If you strive to retain anything that has more than a 10% chance of being real, you may accumulate some "bad" data. *Good common sense and the fortitude to repeat a questionable experiment are usually more valuable than any statistical test.*

Some real data from a spectrophotometric analysis are given in Table 4-5. In this analysis a color is developed that is in proportion to the amount of protein present in the sample. Scanning across the three absorbance values for each size of sample, the number 0.392 seems clearly out of line and should be rejected. The absorbance values in Table 4-5 show that the range for the 15 μg sample is much bigger than the range for the other samples. The number 0.392 is inconsistent with the other values observed for the same sample. The nearly linear relation between the average values of absorbance up to the 20 μg sample indicates that the value 0.392 is in error (Figure 4-5).

It is reasonable to ask whether all three absorbances for the 25 μg sample are low for some unknown reason, since this point falls below the straight line in Figure 4-5. The answer to this question can be discovered only by repetition of the experiment. Many repetitions of this particular analysis show that the 25 μg point is consistently below the straight line and there is nothing "wrong" with these data in Table 4-5.

What is the difference between the Q test and a confidence interval? The confidence interval applies to the *mean*, whereas the Q test applies to individual data. For the five points 12.53, 12.56, 12.47, 12.67, and 12.48, we find $\bar{x} = 12.54_2$, $s = 0.08_0$, and

$$90\% \text{ Confidence:} \quad \mu = 12.54_2 \pm 0.07_6$$

Student's *t* tells us that there is *less* than a 10% chance that the mean value lies above $12.54_2 + 0.07_6 = 12.61_8$. The Q test tells us that there is *more* than a 10% chance that the individual value 12.67 is a normal part of the data set, differing from the other points through purely random error.

A common-sense approach to rejection of data.

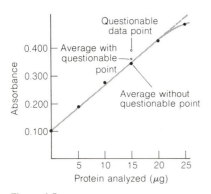

Figure 4-5
Graph of the average values of absorbance in Table 4-5.

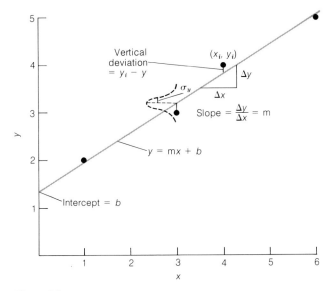

Figure 4-6
An example of least-squares curve fitting. The points (1, 2) and (6, 5) do not fall exactly on the solid line, but they are too close to the line to show their deviations. The Gaussian curve drawn over the point (3, 3) is a schematic indication of the fact that each value of y_i is normally distributed about the straight line. That is, the most probable value of y will fall on the line, but there is a finite probability of measuring y some distance from the line.

4-4 FINDING THE "BEST" STRAIGHT LINE

Often we seek to draw the "best" straight line through a set of experimental data, as in Figure 4-5. This section provides a very important recipe for finding that "best" line.

Method of Least Squares

The method of least squares assumes that the errors in the y values are substantially greater than the errors in the x values.[†] This condition is often true and applies to Figure 4-5, since the variation in absorbance is normally much greater than the uncertainty in concentration. A second assumption is that the uncertainties (standard deviations) in all of the y values are similar.

Suppose we seek to draw the best straight line through the points in Figure 4-6 by minimizing the vertical deviations between the points and the

---

[†] The case in which both the x and y coordinates have substantial uncertainty is treated in a very readable article by D. York, *Can. J. Phys.*, **44,** 1079 (1966). A general approach to least-squares curve fitting using a computer can be found in an article by W. E. Wentworth, *J. Chem. Ed.*, **42,** 96, 162 (1965). Articles dealing with the statistical significance of least-squares parameters can be found in M. D. Pattengill and D. E. Sands, *J. Chem. Ed.*, **56,** 244 (1979) and E. Heilbronner, *J. Chem. Ed.*, **56,** 240 (1979).

54

4 / STATISTICS

The equation of a straight line is discussed in Appendix B.

line. The reason for minimizing only the vertical deviations is the assumption that uncertainties in the y values are much greater than uncertainties in the x values.

Let the equation of the line be

$$y = mx + b \tag{4-11}$$

in which m is the slope and b is the intercept. The vertical deviation for the point (x_i, y_i) will be given by $y_i - y$, where y is the ordinate of the straight line when $x = x_i$.

$$\text{Vertical deviation} \equiv d_i = y_i - y = y_i - (mx_i + b) \tag{4-12}$$

Some of the deviations are positive and some are negative. Since we wish to minimize the magnitude of the deviations irrespective of their signs, we can square all of the deviations so that we are dealing only with positive numbers.

$$d_i^2 = (y_i - y)^2 = (y_i - mx_i - b)^2 \tag{4-13}$$

Because we seek to minimize the squares of the deviations, this is called the method of least squares. It can be shown that minimizing the squares of the deviations (rather than simply their magnitudes) corresponds to assuming that the set of y values is the most probable set.

Finding the values of m and b that minimize the sum of the squares of the vertical deviations involves some calculus, which we will omit. We will express the final solution for slope and intercept in terms of **determinants,** which are a neat way of expressing certain products. The determinant $\begin{vmatrix} e & f \\ g & h \end{vmatrix}$ has the value $eh - fg$. So, for example,

$$\begin{vmatrix} 6 & 5 \\ 4 & 3 \end{vmatrix} = 6 \cdot 3 - 5 \cdot 4 = -2 \tag{4-14}$$

The slope and intercept of the "best" straight line are found to be

Least-squares slope and intercept

$$m = \begin{vmatrix} \sum x_i y_i & \sum x_i \\ \sum y_i & n \end{vmatrix} \div D \tag{4-15}$$

$$b = \begin{vmatrix} \sum (x_i^2) & \sum x_i y_i \\ \sum x_i & \sum y_i \end{vmatrix} \div D \tag{4-16}$$

where the number D is given by

$$D = \begin{vmatrix} \sum (x_i^2) & \sum x_i \\ \sum x_i & n \end{vmatrix} \tag{4-17}$$

and n is the number of points.

Table 4-6
Calculations for least-squares analysis

x_i	y_i	$x_i y_i$	x_i^2	$d_i (= y_i - mx_i - b)$	d_i^2
1	2	2	1	0.038 47	0.001 479 9
3	3	9	9	−0.192 29	0.036 975
4	4	16	16	0.192 33	0.036 991
6	5	30	36	−0.038 43	0.001 476 9
$\sum x_i = 14$	$\sum y_i = 14$	$\sum x_i y_i = 57$	$\sum (x_i^2) = 62$		$\sum (d_i^2) = 0.076\ 923$

Now we apply Equations 4-15 and 4-16 to the points in Figure 4-6 to find the slope and intercept of the best straight line through the four points. The work is set out in Table 4-6. Noting that $n = 4$ and putting the various sums into the determinants in Equations 4-15, 4-16 and 4-17 gives

$$m = \begin{vmatrix} 57 & 14 \\ 14 & 4 \end{vmatrix} \div \begin{vmatrix} 62 & 14 \\ 14 & 4 \end{vmatrix} = \frac{32}{52} = 0.615\ 38 \qquad (4\text{-}18)$$

$$b = \begin{vmatrix} 62 & 57 \\ 14 & 14 \end{vmatrix} \div \begin{vmatrix} 62 & 14 \\ 14 & 4 \end{vmatrix} = \frac{70}{52} = 1.346\ 15 \qquad (4\text{-}19)$$

The equation of the best straight line through the points in Figure 4-6 is therefore

$$y = 0.615\ 38\ x + 1.346\ 15 \qquad (4\text{-}20)$$

We avoid the question of how many significant figures should be associated with m and b until we calculate the uncertainties in the slope and intercept in the next section.

How Reliable Are Least-Squares Parameters?

To estimate the errors (expressed as standard deviations) in the slope and intercept, an error analysis must be performed on Equations 4-15 and 4-16. Since the errors in m and b are related to the error in measuring each value of y, we need first to estimate the standard deviation that describes the population of y values. This standard deviation, σ_y, characterizes the little Gaussian curve inscribed in Figure 4-6.

We will estimate σ_y, the population standard deviation of all y values, by calculating s_y, the standard deviation, for the four measured values of y. The deviation of each value of y_i from the center of its Gaussian curve is just $d_i = y_i - y = y_i - (mx_i + b)$ (Equation 4-12). The standard deviation

56

4 / STATISTICS

of these vertical deviations will be given by an equation analogous to Equation 4-2:

$$\sigma_y \approx s_y = \sqrt{\frac{\sum(d_i - \bar{d})^2}{(\text{Degrees of freedom})}} \tag{4-21}$$

But the average deviation, $\bar{d}$, is zero for the best straight line. So the numerator of Equation 4-21 reduces to $\sum(d_i^2)$.

The degrees of freedom is the number of independent pieces of information available. In Equation 4-2 we set the degrees of freedom equal to $n - 1$. The rationale for so doing is that we began with n degrees of freedom, but one degree of freedom is "lost" in determining the average value ($\bar{x}$). That is, only $n - 1$ pieces of information are available in addition to the average value. If you know $n - 1$ values and you also know the average value, then you can calculate the nth value.

How does this apply to Equation 4-21? We began with n points. But two degrees of freedom were "used up" in determining the slope and intercept of the best line. Therefore, the degrees of freedom is $n - 2$. Equation 4-21 should be written

$$\sigma_y \approx s_y = \sqrt{\frac{\sum(d_i^2)}{n - 2}} \tag{4-22}$$

where d_i is given by Equation 4-12.

The error analysis for Equations 4-15 and 4-16 uses the methods of Appendix C and leads to the following results:

$$\text{Standard deviation of slope and intercept} \begin{cases} \sigma_m^2 = \dfrac{\sigma_y^2 n}{D} & (4\text{-}23) \\[2em] \sigma_b^2 = \dfrac{\sigma_y^2 \sum(x_i^2)}{D} & (4\text{-}24) \end{cases}$$

where σ_m is our estimate of the standard deviation of the slope, σ_b is our estimate of the standard deviation of the intercept, σ_y is given by Equation 4-22, and D is given by Equation 4-17.

Now we can finally address the question of significant figures for the slope and intercept of the line in Figure 4-6. In Table 4-6 it can be seen that $\sum(d_i^2) = 0.076\,923$. Putting this number into Equation 4-22 gives

$$\sigma_y^2 \approx s_y^2 = \frac{0.076\,923}{4 - 2} = 0.038\,462 \tag{4-25}$$

Now, we can plug numbers into Equations 4-23 and 4-24 to find

$$\sigma_m^2 = \frac{\sigma_y^2 n}{D} = \frac{(0.038\,462)(4)}{52} = 0.002\,958\,6 \tag{4-26}$$

$$\sigma_b^2 = \frac{\sigma_y^2 \sum (x_i^2)}{D} = \frac{(0.038\ 462)(62)}{52} = 0.045\ 859 \qquad (4\text{-}27)$$

or

$$\sigma_m = 0.054\ 39 \qquad \text{and} \qquad \sigma_b = 0.214\ 15$$

Combining the results for m, σ_m, b and σ_b we can write

Slope: $\quad \dfrac{0.615\ 38}{\pm 0.054\ 39} = 0.62 \pm 0.05$

The first digit of the uncertainty is the last significant figure.

Intercept: $\quad \dfrac{1.346\ 15}{\pm 0.214\ 15} = 1.3 \pm 0.2$

where the uncertainties represent one standard deviation. The choice of the last significant figure of slope and intercept is dictated by the decimal place of the first digit of the standard deviation. *The first uncertain figure is the last significant figure.*

A Practical Example

What good is all of this? One real application of a least-squares analysis involves the determination of protein concentration using the absorbance values in Table 4-5. To obtain the numbers in this table, analysis of a series of known protein standards was performed. Each standard leads to a certain absorbance measured with a spectrophotometer. Figure 4-5 shows that the standards containing from 0 to 20 μg of protein appear to fall on a straight line. These standards provide fourteen absorbance values in Table 4-6, with the value 0.392 omitted. Using all fourteen values, the least-squares parameters for Figure 4-5 are calculated to be

Least squares in action.

$$m = 0.016\ 3_0 \qquad \sigma_m = 0.000\ 2_2$$

$$b = 0.104_0 \qquad \sigma_b = 0.002_6$$

$$\sigma_y = 0.005_9$$

Now suppose that the absorbance of an unknown sample is found to be 0.246. How many micrograms of protein does it contain, and what error is associated with the answer? Note that in Figure 4-5 the y axis is absorbance, and the x axis is micrograms of protein. Solving for concentration gives

$$x = \frac{y - b}{m} \qquad (4\text{-}28)$$

Follow the rules regarding propagation of error for subtraction and division.

$$= \frac{0.246\ (\pm 0.005_9) - 0.104_0\ (\pm 0.002_6)}{0.016\ 3_0\ (\pm 0.000\ 2_2)}$$

$$= \frac{0.142_0\,(\pm 0.006_4)}{0.016\ 3_0\,(\pm 0.000\ 2_2)}$$

$$= \frac{0.142_0\,(\pm 4._5\%)}{0.016\ 3_0\,(\pm 1._3\%)}$$

$$= 8.7_1\,(\pm 4._7\%) = 8.7\,(\pm 0.4)\ \mu g \text{ of protein}$$

Operating on the standard deviations with the usual rules for propagation of error leads us to calculate that the unknown has 8.7 (± 0.4) μg of protein. The estimate of uncertainty is the standard deviation associated with the number 8.7.

A final caution is in order. Many people have calculators or computer programs that perform least-squares curve fitting automatically. If you do not first make a graph of your data, you will not have the opportunity to reject bad data. Mindless use of a computer program may not do justice to your hard-earned data. *The operation in which a human being evaluates his or her data should never be sacrificed.*

Summary

The results of many measurements of an experimental quantity follow a Gaussian distribution, provided the errors are purely random. The measured mean, $\bar{x}$, approaches the true mean, μ, as the number of measurements becomes very large. The broader the distribution, the greater is σ, the standard deviation. For a limited number of measurements, the standard deviation is given by the formula $s = \sqrt{[\sum(x_i - \bar{x})^2]/(n - 1)}$. About $\frac{2}{3}$ of all measurements lie within $\pm 1\sigma$ of the mean value and 95% lie within $\pm 2\sigma$. The probability of observing a value within a certain interval is proportional to the area of that interval, given in Table 4-1.

Student's t is used to find confidence intervals ($\mu = \bar{x} \pm ts/\sqrt{n}$) and to compare means. To compare two sets of replicate measurements, we use the formula $t = (1/s)(\bar{x}_1 - \bar{x}_2)\sqrt{n_1 n_2/(n_1 + n_2)}$. To compare individual differences in a series of measurements made by two methods, we use the formula $t = \bar{d}\sqrt{n}/s_d$. The Q test is used to reject bad data, but common sense and repetition of a questionable experiment are even better ideas. The method of least squares is used to find the slope and intercept of the best straight line through a series of points. The standard deviations of slope and intercept are used in analyzing the error associated with an experimental measurement.

Terms to Understand

average	normal error curve
confidence interval	Q test
determinant	range
Gaussian error curve	slope
geometric mean	spread
intercept	standard deviation
mean	t test
median	variance

Exercises

4-A. For the numbers 116.0, 97.9, 114.2, 106.8, and 108.3, find the mean, standard deviation, median, geometric mean, range, and 90% confidence interval for the mean. Using the Q test, should the number 97.9 be discarded?

4-B. Suppose that data were collected for 10 000 sets of automobile brakes. The mileage at which each set had been 80% worn through was recorded. The average was 62 700, and the standard deviation was 10 400 miles.
   (a) What fraction of brakes is expected to be 80% worn in less than 45 800 miles?
   (b) What fraction is expected to be 80% worn at a mileage between 60 000 and 70 000 miles?

4-C. It is found from a reliable assay that the ATP (adenosine triphosphate) content of a certain type of cell is 112 μmol/100 mL. You have developed a new assay, which gave the following values for replicate analyses: 117, 119, 111, 115, 120 μmol/100 mL. The average value is 116.4. Can you be 90% confident that your method produces a high value? Can you be 99% confident?

4-D. The Ca content of a powdered mineral sample was analyzed five times by each of two methods, with similar standard deviations:

Ca (percent composition)

Method 1:	0.027 1, 0.028 2, 0.027 9, 0.027 1, 0.027 5
Method 2:	0.027 1, 0.026 8, 0.026 3, 0.027 4, 0.026 9

Are the mean values significantly different at the 90% confidence level?

4-E. A common procedure for protein determination is the dye-binding assay of Bradford.[†] In this method, a dye binds to the protein, and the color of the dye changes from brown to blue. The amount of blue color is proportional to the amount of protein present. Some real data are given below:

Protein (μg):   0.00   9.36   18.72   28.08   37.44
Absorbance
  at 595 nm: 0.466 0.676   0.883   1.086   1.280

   (a) Using the method of least squares, determine the equation of the best straight line through these points. Use the standard deviation of the slope and intercept to express the equation in the form $y = [m(\pm\sigma_m)]x + [b(\pm\sigma_b)]$ with a reasonable number of significant figures.
   (b) Make a graph showing the experimental data and the straight line calculated in part a.
   (c) An unknown protein sample gave an absorbance of 0.973. Calculate the number of micrograms of protein in the unknown, and estimate its uncertainty (expressed as a standard deviation).

[†] M. Bradford, *Anal. Biochem.*, **72**, 248 (1976).

Problems

4-1. (a) Calculate the fraction of bulbs in Figure 4-1 expected to have a lifetime greater than 1 000 hours.
   (b) Calculate the fraction expected to have a lifetime between 800 and 900 hours.

4-2. Write the equation of the smooth Gaussian curve in Figure 4-1. What is the value of y when $x = 1\,000$ hours? See if your calculated value agrees with the value on the graph. Remember that the curve represents the results of 4 768 measurements, and each bar on the graph corresponds to a 20-hour interval.

4-3. The time needed for a certain process to occur was measured five times and found to be 14.48, 14.57, 14.59, 14.32, and 14.52 seconds.

   (a) Can the number 14.32 be rejected as bad data at the 90% confidence level?
   (b) Including the value 14.32 in the data set, calculate the fraction of measurements expected between 14.55 and 14.60 seconds if a large number of measurements are made.

4-4. The percent of an additive in gasoline was measured six times with the following results: 0.13, 0.12, 0.16, 0.17, 0.20, 0.11%. Find the 90% and 99% confidence intervals for the percent of the additive.

4-5. If you measure a quantity four times and the standard deviation is 1.0% of the average, can you be at least 90% confident that the true value is within 1.2% of the measured average?

4-6. The Ti content of five different ore samples (each with a different Ti content) was measured by each of two methods:

	Sample				
	A	B	C	D	E
Method 1:	0.013 4	0.014 4	0.012 6	0.012 5	0.013 7
Method 2:	0.013 5	0.015 6	0.013 7	0.013 7	0.013 6

Do the two analytical techniques give results that are significantly different at the 90% confidence level?

4-7. The calcium content of a person's urine was determined on two different days.

Day	Average Ca (mg/L)	Number of measurements
1	238	4
2	255	5

The analytical method applied to many samples yields a standard deviation known to be 14 mg/L. Are the two measurements above significantly different at the 90% confidence level?

4-8. Two methods were used to measure the specific activity (units of enzyme activity per milligram of protein) of an enzyme. One unit of enzyme activity is defined as the amount of enzyme that catalyzes the formation of one micromole of product per minute under specified conditions.

Enzyme activity (five replications)

| Method 1: | 139 | 147 | 160 | 158 | 135 |
| Method 2: | 148 | 159 | 156 | 164 | 159 |

Is the mean value of method 1 significantly different from the mean value of method 2 at the 90% confidence level? At the 80% confidence level?

4-9. Using the Q test, should the value 216 be rejected from the set of results 192, 216, 202, 195, 204?

4-10. Using the Q test, what is the largest number (n) that should be retained in the set 63, 65, 68, 72, n?

4-11. Use the method of least squares to calculate the equation of the best straight line going through the points (1, 3), (3, 2), and (5, 0). Express your answer in the form $y = [m(\pm\sigma_m)]x + [b(\pm\sigma_b)]$ with a reasonable number of significant figures.

4-12. Consider a suspension of cells containing 4.13 (± 0.09) $\times 10^{-13}$ moles of cells per liter. (This is $\sim 2.5 \times 10^8$ cells/mL.) Call this total concentration of cells C_t. Each cell has n equivalent binding sites per cell to which the hormone H can bind. The dissociation constant, K, is the equilibrium constant for the reaction below:

$$CH \overset{K}{\rightleftharpoons} C + H$$

where C is a cell, H is the hormone, and CH is hormone attached to the cell. To determine the number of binding sites, n, per cell, one usually varies the concentration of H while maintaining a fixed concentration of cells, C_t. The concentrations of free hormone (H_f) and bound hormone (H_b) are measured in each experiment. A graph of H_b/H_f against H_b is then drawn. This is called a *Scatchard plot*. It can be shown that

$$\frac{H_b}{H_f} = -\left(\frac{1}{K}\right)H_b + \frac{nC_t}{K}$$

That is, a graph of H_b/H_f against H_b should give a straight line.

The results of a hypothetical experiment are shown below. The slope and intercept were calculated by the method of least squares and have *not* been reduced to a reasonable number of significant figures. The uncertainties shown in parentheses are one standard deviation.

Intercept: $b = 3.138\ 6\ (\pm 0.289\ 9) \times 10^{-3}$

Slope: $m = -1.364\ 5\ (\pm 0.132\ 8) \times 10^6\ \text{M}^{-1}$

Find the value of n and the absolute uncertainty (standard deviation) of n. Express your answer with a reasonable number of significant figures. Do not try to assign significant figures until the *end* of the calculation.

5 / Solubility of Ionic Compounds

In this chapter we will study some of the simplest equilibria of ionic compounds. We begin with a review of important aspects of chemical equilibrium that will be useful throughout the remainder of the text.

5-1 REVIEW OF CHEMICAL EQUILIBRIUM

For the reaction

$$aA + bB \rightleftharpoons cC + dD \qquad (5\text{-}1)$$

it is customary to write the **equilibrium constant,** K, in the form

$$K = \frac{[C]^c[D]^d}{[A]^a[B]^b} \qquad (5\text{-}2)$$

The equilibrium constant is more correctly expressed as a ratio of *activities*, rather than concentrations. We reserve a discussion of activity for the next chapter.

where the small letters denote stoichiometric coefficients and each capital letter stands for a chemical species. The symbol $[A]$ stands for the concentration of A.

In the thermodynamic derivation of the equilibrium constant, each quantity in Equation 5-2 is expressed as the *ratio* of the concentration of each species to its concentration in its **standard state.** For solutes the standard state is 1 M. For gases the standard state is 1 atm, and for solids and liquids the standard states are the pure solid or liquid. It is understood (but rarely written) that the term $[A]$ in Equation 5-2 really means $[A]/(1 \text{ M})$ if A is a

62

5 / SOLUBILITY OF IONIC COMPOUNDS

Equilibrium constants are dimensionless.

solute. If D is a gas, $[D]$ really means (pressure of D in atmospheres)/(1 atm). To emphasize that $[D]$ means pressure of D, we will usually write P_D in place of $[D]$. The terms of Equation 5-2 are actually dimensionless, therefore all equilibrium constants are dimensionless.

For the ratios $[A]/(1 \text{ M})$ and $[D]/(1 \text{ atm})$ to be dimensionless, $[A]$ *must* be expressed in moles per liter (M), and $[D]$ *must* be expressed in atmospheres. If C were a pure liquid or solid, the ratio $[C]/$(concentration of C in its standard state) would be unity (1) because the standard state is the pure liquid or solid. If $[C]$ is solvent, the concentration is so close to that of pure liquid C that the value of $[C]$ is still essentially 1.

The take-home lesson is this. When evaluating an equilibrium constant:

1. The concentrations of solutes should be expressed as moles per liter.

2. The concentrations of gases should be expressed in atmospheres.

3. The concentrations of pure solids, pure liquids, and solvents are omitted because they are unity.

Equilibrium constants are dimensionless, but you must use units of M for solutes and atm for gases.

These conventions are arbitrary, but you must use them if you wish to use tabulated values of equilibrium constants, standard reduction potentials, and free energies.

Manipulation of Equilibrium Constants

Consider the reaction

Throughout this text it is understood that all species in chemical equations are in aqueous solution, unless otherwise specified.

$$HA \rightleftharpoons H^+ + A^- \qquad K_1 = \frac{[H^+][A^-]}{[HA]} \qquad (5\text{-}3)$$

If the direction of a reaction is reversed, the new value of K *is simply the reciprocal of the original value of* K.

$$H^+ + A^- \rightleftharpoons HA \qquad K_1' = \frac{[HA]}{[H^+][A^-]} = 1/K_1 \qquad (5\text{-}4)$$

If two reactions are added, the new value of K *is the product of the two individual values:*

$$
\begin{array}{ll}
HA \rightleftharpoons H^+ + A^- & K_1 \\
\underline{H^+ + C \rightleftharpoons CH^+} & K_2 \\
HA + C \rightleftharpoons A^- + CH^+ & K_3
\end{array}
$$

If a reaction is reversed, $K' = 1/K$. If two reactions are added, $K = K_1 K_2$.

$$K_3 = K_1 K_2 = \frac{[H^+][A^-]}{[HA]} \cdot \frac{[CH^+]}{[H^+][C]} = \frac{[A^-][CH^+]}{[HA][C]} \qquad (5\text{-}5)$$

If n reactions are added, the overall equilibrium constant is the product of all n individual equilibrium constants.

EXAMPLE

The equilibrium constant for the reaction

$$H_2O \rightleftharpoons H^+ + OH^-$$

is called $K_w (= [H^+][OH^-])$ and has the value 1.0×10^{-14} at 25°C. Given that

$$NH_3(aq) + H_2O \rightleftharpoons NH_4^+ + OH^- \qquad K_{NH_3} = 1.8 \times 10^{-5}$$

find the equilibrium constant for the reaction

$$NH_4^+ \rightleftharpoons NH_3(aq) + H^+$$

The third reaction can be obtained by reversing the second reaction and adding it to the first reaction:

$$
\begin{array}{lll}
\cancel{H_2O} \rightleftharpoons H^+ + \cancel{OH^-} & K_1 = K_w \\
NH_4 + \cancel{OH^-} \rightleftharpoons NH_3(aq) + \cancel{H_2O} & K_2 = 1/K_{NH_3} \\
\hline
NH_4^+ \rightleftharpoons H^+ + NH_3(aq) & K_3 = K_w \cdot \dfrac{1}{K_{NH_3}} = 5.6 \times 10^{-10}
\end{array}
$$

Equilibrium and Thermodynamics

Enthalpy

The change in **enthalpy** (ΔH) for a reaction is the heat absorbed when the reaction takes place under constant applied pressure. The *standard enthalpy change* ($\Delta H°$) refers to the heat absorbed when all reactants and products are in their standard states.[†] For Reaction 5-6, $\Delta H° = -75.15 \text{ kJ mol}^{-1}$ at 25°C.

$$HCl(g) \rightleftharpoons H^+(aq) + Cl^-(aq) \qquad (5\text{-}6)$$

The negative sign of $\Delta H°$ indicates that heat is given off by the products when Reaction 5-6 occurs. This means that the solution becomes warmer when the reaction occurs. For other reactions, ΔH is positive, which means that heat is absorbed by reactants as they react. That is, the solution gets colder during the reaction. A reaction for which ΔH is positive is said to be **endothermic.** If ΔH is negative, the reaction is **exothermic.**

$\Delta H = (+)$	$\Delta H = (-)$
Heat is absorbed.	Heat is liberated.
endothermic	exothermic

[†] The precise definition of the standard state contains subtleties beyond the scope of this text. For Reaction 5-6 the standard state of H^+ or Cl^- is the hypothetical state in which each ion is present at a concentration of 1 M but behaves as if it were in an infinitely dilute solution. That is, the standard concentration is 1 M, but the standard physical behavior is what would be observed in a very dilute solution in which each ion is unaffected by surrounding ions. For a discussion of this rather important point, see W. J. Moore, *Physical Chemistry*, 4th ed., (Englewood Cliffs, N.J.: Prentice-Hall, 1972), pp. 305–312.

64

5 / SOLUBILITY OF IONIC COMPOUNDS

Entropy

The **entropy** (S) of a substance is a measure of its "disorder," which we will not attempt to define in a quantitative way. The greater the disorder, the greater the entropy. In general, a gas is more disordered (has higher entropy) than a liquid, which is more disordered than a solid. Ions in aqueous solution are normally more disordered than in their solid salt. For example, for Reaction 5-7, we find $\Delta S° = +76 \ JK^{-1} \ mol^{-1}$ at 25°C.

$$KCl(s) \rightleftharpoons K^+(aq) + Cl^-(aq) \tag{5-7}$$

This means that a mole of $K^+(aq)$ plus a mole of $Cl^-(aq)$ is more disordered than a mole of $KCl(s)$. (The dimensions of entropy are energy/temperature per mole of reactant.) However, for Reaction 5-6, $\Delta S° = -131.5 \ JK^{-1} \ mol^{-1}$ at 25°C. The aqueous ions are less disordered than gaseous HCl.

Free energy

There is a tendency in nature for a system to seek its lowest energy and highest entropy. A chemical reaction is driven toward the formation of products by a *negative* value of ΔH (heat given off) and/or a *positive* value of ΔS (entropy increases). If ΔH is negative and ΔS is positive, the reaction is clearly favored. If ΔH is positive and ΔS is negative, the reaction is clearly disfavored.

If ΔH and ΔS are both positive or both negative, what decides whether a reaction will be favored? The change in **Gibbs free energy** (ΔG) is the arbiter between opposing tendencies of ΔH and ΔS. At constant temperature

$$\Delta G = \Delta H - T\Delta S \tag{5-8}$$

The effects of entropy and enthalpy are combined in Equation 5-8. *A reaction is favored if ΔG is negative.*

For Reaction 5-6, $\Delta H°$ favors the reaction and $\Delta S°$ disfavors it. To find the net result, we must evaluate $\Delta G°$:

Note that 25.00°C = 298.15 K.

$$\Delta G° = \Delta H° - T\Delta S° = (-75.15 \times 10^3 \ J) - (298.15 \ K)(-131.5 \ JK^{-1})$$

$$\Delta G° = -35.94 \ kJ \ mol^{-1} \tag{5-9}$$

$\Delta G°$ is negative and the reaction is favored under standard conditions. The favorable enthalpy change is greater than the unfavorable entropy change in this case.

The relation between $\Delta G°$ and the equilibrium constant for a reaction is

$$K = e^{-\Delta G°/RT} \tag{5-10}$$

where R is the gas constant ($8.314 \ 41 \ JK^{-1} \ mol^{-1}$) and T is in kelvins. For

Reaction 5-6, we find

$$K = e^{-(-35.94 \times 10^3 \text{ J mol}^{-1})/(8.314\,41 \text{ JK}^{-1} \text{ mol}^{-1})(298.15 \text{ K})}$$

$$K = 1.98 \times 10^6 \tag{5-11}$$

The equilibrium constant for Reaction 5-6 is large. $HCl(g)$ is therefore very soluble in water and is nearly completely ionized to H^+ and Cl^- when it dissolves.

To summarize, ΔG takes into account the tendency for heat to be liberated (ΔH negative) and disorder to increase (ΔS positive). A reaction is favored ($K > 1$) if ΔG° is negative. A reaction is disfavored ($K < 1$) if ΔG° is positive. A reaction for which ΔG is positive is said to be **endergonic.** A reaction for which ΔG is negative is called **exergonic.**

$\Delta G = (+)$

Reaction is disfavored.

endergonic

$\Delta G = (-)$

Reaction is favored.

exergonic

Le Châtelier's Principle

Suppose that a system at equilibrium is subjected to a change that disturbs the system. **Le Châtelier's principle** states that the direction in which the system proceeds back to equilibrium is such that the change is partially offset.

To see what this means, consider Reaction 5-12:

$$BrO_3^- + 2Cr^{3+} + 4H_2O \rightleftharpoons Br^- + Cr_2O_7^{2-} + 8H^+ \tag{5-12}$$

for which the equilibrium constant is given by

$$K = \frac{[Br^-][Cr_2O_7^{2-}][H^+]^8}{[BrO_3^-][Cr^{3+}]^2} = 1 \times 10^{11} \text{ at } 25^\circ C \tag{5-13}$$

In one particular equilibrium state of this system, the following concentrations exist:

$$[H^+] = 5.0 \text{ M} \qquad [Cr_2O_7^{2-}] = 0.10 \text{ M} \qquad [Cr^{3+}] = 0.003\,0 \text{ M}$$

$$[Br^-] = 1.0 \text{ M} \qquad [BrO_3^-] = 0.043 \text{ M}$$

Suppose that the equilibrium is disturbed by adding dichromate to the solution to increase the concentration of $[Cr_2O_7^{2-}]$ from 0.10 to 0.20 M. In what direction will the reaction proceed to reach equilibrium? According to the principle of Le Châtelier, the reaction should go back to the left to partially offset the increase in dichromate, which appears on the right side of Equation 5-12. We can verify this algebraically by setting up the **reaction quotient,** Q, which has the same form as the equilibrium constant. The only difference is that Q is evaluated with whatever concentrations happen to exist in the solution. When the system reaches equilibrium, $Q = K$. For

Reaction quotient.

5 / SOLUBILITY OF IONIC COMPOUNDS

Reaction 5-12,

$$Q = \frac{(1.0)(0.20)(5.0)^8}{(0.043)(0.003\ 0)^2} = 2 \times 10^{11} > K \qquad (5-14)$$

If $Q < K$, the reaction must proceed to the right to reach equilibrium. If $Q > K$, the reaction must proceed to the left to reach equilibrium.

Since $Q > K$, the reaction must go to the left to decrease the numerator and increase the denominator, until $Q = K$.

In general:

1. If a reaction is at equilibrium and products are added (or reactants are removed), the reaction goes to the left.
2. If a reaction is at equilibrium and reactants are added (or products are removed), the reaction goes to the right.

When the temperature of a system is changed, so is the equilibrium constant. Equations 5-8 and 5-10 can be combined to predict the effect of temperature on K:

$$K = e^{-\Delta G^\circ/RT} = e^{-(\Delta H^\circ - T\,\Delta S^\circ)/RT} = e^{(-\Delta H^\circ/RT + \Delta S^\circ/R)}$$

$$K = e^{-\Delta H^\circ/RT} \cdot e^{\Delta S^\circ/R} \qquad (5-15)$$

The term $e^{\Delta S^\circ/R}$ is independent of T. The term $e^{-\Delta H^\circ/RT}$ increases with increasing temperature if ΔH° is positive. The term $e^{-\Delta H^\circ/RT}$ decreases with increasing temperature if ΔH° is negative. Equation 5-15 therefore tells us:

1. The equilibrium constant of an endothermic reaction increases if the temperature is raised.
2. The equilibrium constant for an exothermic reaction decreases if the temperature is raised.

These statements can be understood in terms of Le Châtelier's principle as follows. Consider an endothermic reaction:

Heat can be treated as if it were a reactant in an endothermic reaction and a product in an exothermic reaction.

$$\text{Heat + Reactants} \rightleftharpoons \text{Products} \qquad (5-16)$$

If the temperature is raised, heat is added to the system. Reaction 5-16 proceeds to the right to partially offset this change.[†]

Let the buyer beware.

In dealing with equilibrium problems, we are making thermodynamic predictions, not kinetic predictions. We are calculating what must happen for a system to reach equilibrium, but not how long it will take. Some reactions are over in an instant; others will not reach equilibrium in a million years.

[†] The solubility of the vast majority of ionic compounds increases with temperature, despite the fact that the heat of solution (ΔH^0) is negative for about half of them. A discussion of this seeming contradiction can be found in G. M. Bodner, *J. Chem. Ed.*, **57**, 117 (1980).

5-2 SOLUBILITY PRODUCT

The **solubility product** is the equilibrium constant for the reaction in which a solid salt dissolves to give its constituent ions in solution.[†] The concentration of the solid is omitted from the equilibrium constant because the solid is in its standard state. A table of solubility products can be found in Appendix F.

As a typical example, consider the dissolution of mercurous chloride (Hg_2Cl_2) in water. The reaction is

$$Hg_2Cl_2(s) \rightleftharpoons Hg_2^{2+} + 2Cl^- \qquad (5\text{-}17)$$

for which

$$K_{sp} = [Hg_2^{2+}][Cl^-]^2 = 1.2 \times 10^{-18} \qquad (5\text{-}18)$$

A solution containing excess, undissolved solid is said to be **saturated** with that solid. The solution contains all of the solid capable of being dissolved under the prevailing conditions. What will be the concentration of Hg_2^{2+} in a solution saturated with Hg_2Cl_2?

In Equation 5-17 we see that two Cl^- ions are produced for each Hg_2^{2+} ion. If the concentration of dissolved Hg_2^{2+} is x M, the concentration of dissolved Cl^- must be $2x$ M.

	$Hg_2Cl_2(s) \rightleftharpoons$	Hg_2^{2+}	$2Cl^-$
Initial concentration:	solid	0	0
Final concentration:	solid	x	$2x$

Putting these values of concentration into the solubility product gives

$$[Hg_2^{2+}][Cl^-]^2 = (x)(2x)^2 = 1.2 \times 10^{-18} \qquad (5\text{-}19)$$

$$4x^3 = 1.2 \times 10^{-18}$$

$$x = 6.7 \times 10^{-7} \text{ M}$$

The concentration of Hg_2^{2+} ion is calculated to be 6.7×10^{-7} M, and the concentration of Cl^- is $(2)(6.7 \times 10^{-7}) = 13.4 \times 10^{-7}$ M.

The physical meaning of the solubility product is this: If an aqueous solution is left in contact with excess solid Hg_2Cl_2, the solid will dissolve until the condition $[Hg_2^{2+}][Cl^-]^2 = K_{sp}$ is satisfied. Thereafter, the amount of undissolved solid remains constant. Unless excess solid remains, there

> The equilibrium constant omits the pure solid because its concentration is unchanged.

> Box 5-1 contains some information about Hg_2^{2+}.

> To find the cube root of a number with a calculator, raise the number to the 0.333 333 33.... power.

[†] The solubility product does not tell the entire story on the solubility of slightly soluble salts. For many, the dissolution of *undissociated* species is as important as dissolution accompanied by dissociation. That is, the salt $MX(s)$ can give $MX(aq)$ as well as $M^+(aq)$ and $X^-(aq)$.

Box 5-1 THE MERCUROUS ION

Mercury has three common oxidation states:

1. Hg(0) is metallic mecury. Hg is one of only two elements (the other being Br) that is a liquid at 298 K. Ga metal melts just above room temperature, at 303 K.
2. Hg(I) is the mercurous ion, which is a dimer with a bond length close to 250 pm (2.50 Å).

$$[\text{Hg–Hg}]^{2+}$$

When mercurous salts dissolve, they do not give monatomic Hg^+ in solution. Hg_2^{2+} remains present as a diatomic ion.

3. Hg(II) is the mercuric ion, which exists as Hg^{2+}.

Hg(I) and Hg(II) exist in equilibrium in the presence of Hg(l):

$$Hg_2^{2+} \rightleftharpoons Hg^{2+} + Hg(l)$$

$$K = [Hg^{2+}]/[Hg_2^{2+}] = 0.011$$

In the presence of excess Hg(l), Hg^{2+} is reduced to Hg_2^{2+}. In the absence of Hg(l), Hg_2^{2+} **disproportionates** to give Hg^{2+} and Hg(l). A disproportionation is a reaction in which an element in a particular oxidation state reacts with itself to produce products in higher and lower oxidation states. In the reaction above Hg(I) gives Hg(II) and Hg(0).

The mercurous ion is stable in the presence of only a limited number of anions. Many anions stabilize Hg(II), causing Hg(I) to disproportionate. For example:

$$Hg_2^{2+} + 2OH^- \rightarrow HgO(s) + Hg(l) + H_2O$$

$$Hg_2^{2+} + S^{2-} \rightarrow HgS(s) + Hg(l)$$

$$Hg_2^{2+} + 2CN^- \rightarrow Hg(CN)_2(aq) + Hg(l)$$

is no guarantee that $[Hg_2^{2+}][Cl^-]^2 = K_{sp}$. If Hg_2^{2+} and Cl^- are mixed together (with appropriate counterions) such that the product $[Hg_2^{2+}][Cl^-]^2$ exceeds K_{sp}, then Hg_2Cl_2 will precipitate.

5-3 COMMON ION EFFECT

What will be the concentration of Hg_2^{2+} in a solution containing 0.030 M NaCl saturated with Hg_2Cl_2? The concentration table now looks like this:

	$Hg_2Cl_2(s) \rightleftharpoons$	$Hg_2^{2+} +$	$2Cl^-$
Initial concentration:	solid	0	0.030
Final concentration:	solid	x	$2x + 0.030$

The initial concentration of Cl^- is due to the dissolved NaCl (which is completely dissociated into Na^+ and Cl^-). The final concentration of Cl^- is due to contributions from NaCl and Hg_2Cl_2.

The proper solubility equation is

$$[Hg_2^{2+}][Cl^-]^2 = (x)(2x + 0.030)^2 = K_{sp} \qquad (5\text{-}20)$$

But think about the size of x. In the previous example, $x = 6.7 \times 10^{-7}$ M, which is pretty small compared to 0.030 M. In the present example, we anticipate that x will be even smaller than 6.7×10^{-7}, by Le Châtelier's principle. Addition of a product (Cl^- in this case) to Reaction 5-17 displaces the reaction toward the left. There will be less dissolved Hg_2^{2+} than in the absence of a second source of Cl^-. This application of Le Châtelier's principle is called the **common ion effect.** *A salt will be less soluble if one of its constituent ions is already present in the solution.*

Common ion effect: A salt is less soluble if one of its ions is already present in the solution.

Getting back to Equation 5-20, we expect that $2x \ll 0.030$. As a good approximation, we will ignore $2x$ in comparison to 0.030. The equation simplifies to

$$(x)(0.030)^2 = K_{sp} = 1.2 \times 10^{-18} \qquad (5\text{-}21)$$

$$x = 1.3 \times 10^{-15}$$

The answer shows that it was clearly justified to omit $2x$ ($= 2.6 \times 10^{-15}$) to solve the problem. The answer also illustrates the common ion effect. In the absence of Cl^-, the solubility of Hg_2^{2+} was 6.7×10^{-7} M. In the presence of 0.030 M Cl^-, the solubility of Hg_2^{2+} is reduced to 1.3×10^{-15} M.

It is important to confirm at the end of the calculation that the approximation $2x \ll 0.030$ is valid. Look at Box 5-2 for more on approximations.

What is the Cl^- concentration in a solution in which the concentration Hg_2^{2+} is somehow *fixed* at 1.0×10^{-9} M? Our concentration table looks like this now:

	$Hg_2Cl_2(s) \rightleftharpoons$	Hg_2^{2+}	$+ \ 2Cl^-$
Initial concentration:	0	1.0×10^{-9}	0
Final concentration:	solid	1.0×10^{-9}	x

$[Hg_2^{2+}]$ is not x in this example, so there is no reason to set $[Cl^-] = 2x$. The problem is solved by plugging each concentration into the solubility product:

$$[Hg_2^{2+}][Cl^-]^2 = K_{sp}.$$

$$(1.0 \times 10^{-9})(x)^2 = 1.2 \times 10^{-18} \qquad (5\text{-}22)$$

$$x = [Cl^-] = 3.5 \times 10^{-5} \text{ M}$$

Box 5-2 THE LOGIC OF APPROXIMATIONS

It is hopelessly complicated to solve most real problems in chemistry (and other sciences) without the aid of judicious numerical approximations. For example, in solving the equation

$$(x)(2x + 0.030)^2 = 1.2 \times 10^{-18}$$

we used the approximation that $2x \ll 0.030$, and therefore solved the much simpler equation

$$(x)(0.030)^2 = 1.2 \times 10^{-18}$$

Whenever you use an approximation you are *assuming* that the approximation is true. *If the assumption is true, it will not create a contradiction. If the assumption is false, it will lead to a contradiction.*

You may object to this reasoning, feeling "How can the truth of an assumption be tested by using the assumption?" Suppose you wish to test the statement, "Gail can swim 100 meters." To see if the statement is true you can *assume* that it is true. If Gail can swim 100 meters, then you could dump her in the middle of a lake with a radius of 100 meters and expect her to swim to shore. If she comes ashore alive, then your assumption was correct and no contradiction is created. If she does not make it to shore, there is a contradiction. Your assumption must have been wrong. There are only two possibilities: Either the assumption is correct and using it is correct, or the assumption is wrong and using it is wrong. You can test an assumption by using it and seeing if you are right or wrong when you are done.

Gail's lake

Here are some numerical examples illustrating the logic of assumptions used to solve numerical equations:

5-4 SEPARATION BY PRECIPITATION

It is sometimes useful to separate one substance from another by precipitating one from solution. As an example, consider a solution containing plumbous (Pb^{2+}) and mercurous (Hg_2^{2+}) ions, each at a concentration of 0.010 M. Each forms an insoluble iodide, but the mercurous iodide is considerably less soluble, as indicated by the smaller value of K_{sp}:

$$PbI_2(s) \rightleftharpoons Pb^{2+} + 2I^- \qquad K_{sp} = 7.9 \times 10^{-9} \qquad (5\text{-}23)$$

$$Hg_2I_2(s) \rightleftharpoons Hg_2^{2+} + 2I^- \qquad K_{sp} = 1.1 \times 10^{-28} \qquad (5\text{-}24)$$

Example 1. $(x)(3x + 0.01)^3 = 10^{-12}$
$\quad\quad\quad (x)(0.01)^3 = 10^{-12}$ (assuming $3x \ll 0.01$)
$\quad\quad\quad x = 10^{-12}/(0.01)^3 = 10^{-6}$
$\quad\quad\quad$ No contradiction: $3x = 3 \times 10^{-6} \ll 0.01$.
$\quad\quad\quad$ The assumption is true.

Example 2. $(x)(3x + 0.01)^3 = 10^{-8}$
$\quad\quad\quad (x)(0.01)^3 = 10^{-8}$ (assuming $3x \ll 0.01$)
$\quad\quad\quad x = 10^{-8}/(0.01)^3 = 0.01$.
$\quad\quad\quad$ A contradiction: $3x = 0.03 > 0.01$.
$\quad\quad\quad$ The assumption is false.

In Example 2 the assumption leads to a contradiction, so the assumption cannot be correct. When this happens you must solve the quartic equation $x(3x + 0.01)^3 = 10^{-8}$.

You can try to solve the quartic equation exactly, but approximate methods are usually easier. For the equation $x(3x + .01)^3 = 10^{-8}$, one quick procedure is trial-and-error guessing. The first guess, $x = 0.01$, comes from the (incorrect) assumption that $3x \ll 0.01$ made in Example 2.

guess	$x(3x + .01)^3$	
$x = 0.01$	6.4×10^{-7}	
$x = 0.005$	7.8×10^{-8}	
$x = 0.002$	8.2×10^{-9}	
$x = 0.002\,4$	1.22×10^{-8}	
$x = 0.002\,2$	1.006×10^{-8}	
$x = 0.002\,18$	9.86×10^{-9}	
$x = 0.002\,19$	9.96×10^{-9}	best value of x with 3 figures

Example 3. $(x)(3x + 0.01)^3 = 10^{-9}$
$\quad\quad\quad (x)(0.01)^3 = 10^{-9}$ (assuming $3x \ll 0.01$)
$\quad\quad\quad x = 10^{-9}/(0.01)^3 = 10^{-3}$

In this case $3x = 0.003$. This is less than 0.01, but not a great deal less. *Whether or not the assumption is adequate depends on your purposes.* If you need an answer accurate to within a factor of 2, the approximation is acceptable. If you need an answer accurate to 1%, the approximation is not adequate. The correct answer is 6.06×10^{-4}.

Is it possible to *completely* separate the Pb^{2+} and Hg_2^{2+} by selectively precipitating the latter with iodide?

Complete separation means whatever you want it to. *You* must define what you mean by complete.

To answer this, we must first define complete separation. "Complete" can mean anything we choose. Let us ask whether 99.99% of the Hg_2^{2+} can be precipitated without causing Pb^{2+} to precipitate. That is, we seek to reduce the Hg_2^{2+} concentration to 0.01% of its original value without precipitating Pb^{2+}. The concentration of Hg_2^{2+} will be 0.01% of 0.010 M = 1.0×10^{-6} M. The minimum concentration of I^- needed to effect this precipitation is calculated as follows:

72

5 / SOLUBILITY OF IONIC COMPOUNDS

$$Hg_2I_2(s) \rightleftharpoons Hg_2^{2+} + 2I^-$$

Initial concentration:	0	0.010	0
Final concentration:	solid	$\underbrace{1.0 \times 10^{-6}}$	x

0.01% of 0.010 M

x is the concentration of I^- needed to reduce $[Hg_2^{2+}]$ to 10^{-6} M.

$$[Hg_2^{2+}][I^-]^2 = K_{sp}$$

$$(1.0 \times 10^{-6})(x)^2 = 1.1 \times 10^{-28} \qquad (5\text{-}25)$$

$$x = [I^-] = 1.0 \times 10^{-11} \text{ M}$$

Will this concentration of I^- cause Pb^{2+} to precipitate? We answer this by seeing whether the solubility product of PbI_2 is exceeded.

$$Q = [Pb^{2+}][I^-]^2 = (0.010)(1.0 \times 10^{-11})^2$$

$$= 1.0 \times 10^{-24} < K_{sp} \text{ (for } PbI_2) \qquad (5\text{-}26)$$

The reaction quotient, Q, is 1.0×10^{-24}, which is less than K_{sp} ($= 7.9 \times 10^{-9}$). Therefore, the Pb^{2+} will not precipitate and the "complete" separation of Pb^{2+} and Hg_2^{2+} is feasible. We predict that adding I^- to a solution of Pb^{2+} and Hg_2^{2+} will precipitate virtually all of the mercurous ion before any plumbous ion precipitates.

Life should be so easy. What we have just done is to make a (valid) thermodynamic prediction. If the system comes to equilibrium, we can achieve the desired separation. However, occasionally one substance **coprecipitates** with the other. For example, some Pb^{2+} might become attached to the surface of the Hg_2I_2 crystal, or might even occupy sites within the crystal. The calculation above says that the separation is worth trying. *But only an experiment can show whether the separation will actually work.*

> *Question:* If you want to know whether a small amount of Pb^{2+} coprecipitates with Hg_2I_2, should you measure the Pb^{2+} concentration in the mother liquor (the solution) or the Pb^{2+} concentration in the precipitate? Which measurement is more sensitive? By "sensitive," we mean responsive to a small amount of coprecipitation.

5-5 COMPLEX FORMATION

If the anion, X, precipitates the metal, M, it is often observed that at a high concentration of X, MX redissolves. One example is Pb^{2+} in the presence of excess I^-. The solubility product for PbI_2 is

$$PbI_2(s) \xrightarrow{K_{sp}} Pb^{2+} + 2I^- \qquad K_{sp} = [Pb^{2+}][I^-]^2 = 7.9 \times 10^{-9} \quad (5\text{-}27)$$

Besides forming $PbI_2(s)$, Pb^{2+} and I^- can react to form the **complex ions** PbI_n^{2-n}.

The notation for these equilibrium constants is discussed in Box 5-3.

$$Pb^{2+} + I^- \xrightarrow{K_1} PbI^+ \qquad K_1 = [PbI^+]/[Pb^{2+}][I^-] = 1.0 \times 10^2 \qquad (5\text{-}28)$$

$$Pb^{2+} + 2I^- \xrightarrow{\beta_2} PbI_2(aq) \quad \beta_2 = [PbI_2(aq)]/[Pb^{2+}][I^-]^2 = 1.4 \times 10^3 \quad (5\text{-}29)$$

$$Pb^{2+} + 3I^- \xrightarrow{\beta_3} PbI_3^- \qquad \beta_3 = [PbI_3^-]/[Pb^{2+}][I^-]^3 = 8.3 \times 10^3 \qquad (5\text{-}30)$$

$$Pb^{2+} + 4I^- \xrightleftharpoons{\beta_4} PbI_4^{2-} \quad \beta_4 = [PbI_4^{2-}]/[Pb^{2+}][I^-]^4 = 3.0 \times 10^4 \quad (5\text{-}31)$$

The species $PbI_2(aq)$ in Reaction 5-29 is *dissolved* PbI_2, containing two iodine atoms bound to a lead atom. Reaction 5-29 is *not* the reverse of Reaction 5-23, since the latter deals with solid PbI_2.

In the species PbI^+, PbI_2, PbI_3^-, and PbI_4^{2-}, iodide is said to be the **ligand** of Pb. A ligand is any atom or group of atoms attached to the species of interest.

At low I^- concentrations, the solubility of lead is governed by the solubility of $PbI_2(s)$. However, at high I^- concentrations, Reactions 5-28 through 5-31 are driven to the right (Le Châtelier's principle), and the total concentration of dissolved lead is considerably greater than that of Pb^{2+} alone (Figure 5-1).

A most useful characteristic of chemical equilibrium is that *all equilibrium conditions are satisfied simultaneously*. If we somehow know the concentration of I^-, we can calculate the concentration of Pb^{2+} using Equation 5-27 regardless of whether there are zero, four, or 1 006 other reactions involving Pb^{2+}. *The concentration of Pb^{2+} that satisfies any one of the equilibria must satisfy all of the equilibria. There can be only one concentration of Pb^{2+} in the solution.*

It is instructive to consider the composition of the solution when the final concentration of I^- is at 0.001 0 M and again at 1.0 M. From Equation 5-27, we calculate

$$[Pb^{2+}] = K_{sp}/[I^-]^2 = 7.9 \times 10^{-3} \text{ M} \quad (5\text{-}32)$$

if $[I^-] = 10^{-3}$ M. From Equations 5-28 through 5-31, we can then calculate

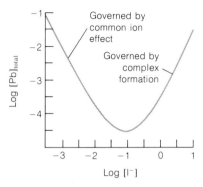

Figure 5-1
Total concentration of dissolved lead as a function of the concentration of free iodide. To the left of the minimum, $[Pb]_{total}$ is governed by the solubility product for $PbI_2(s)$. As $[I^-]$ is increased, $[Pb]_{total}$ decreases because of the common ion effect. At high values of $[I^-]$, $PbI_2(s)$ begins to redissolve because of its reaction with I^- to form soluble complex ions.

Box 5-3 NOTATION FOR FORMATION CONSTANTS

Formation constants are the equilibrium constants for complex ion formation. The *stepwise* formation constants, designated K_i, are defined as follows:

$$M + X \xrightleftharpoons{K_1} MX \quad K_1 = [MX]/[M][X]$$

$$MX + X \xrightleftharpoons{K_2} MX_2 \quad K_2 = [MX_2]/[MX][X]$$

$$MX_{n-1} + X \xrightleftharpoons{K_n} MX_n \quad K_n = [MX_n]/[MX_{n-1}][X]$$

The *overall* or *cumulative* formation constants are denoted β_i:

$$M + 2X \xrightleftharpoons{\beta_2} MX_2 \quad \beta_2 = [MX_2]/[M][X]^2$$

$$M + nX \xrightleftharpoons{\beta_n} MX_n \quad \beta_n = [MX_n]/[M][X]^n$$

A useful relationship is that $\beta_n = K_1 K_2 \cdots K_n$.

the concentrations of the other lead-containing species:

$$[PbI^+] = K_1[Pb^{2+}][I^-] = (1.0 \times 10^2)(7.9 \times 10^{-3})(1.0 \times 10^{-3}) \quad (5\text{-}33)$$

$$= 7.9 \times 10^{-4} \text{ M}$$

$$[PbI_2(aq)] = \beta_2[Pb^{2+}][I^-]^2 = 1.1 \times 10^{-5} \text{ M} \quad (5\text{-}34)$$

$$[PbI_3^-] = \beta_3[Pb^{2+}][I^-]^3 = 6.6 \times 10^{-8} \text{ M} \quad (5\text{-}35)$$

$$[PbI_4^{2-}] = \beta_4[Pb^{2+}][I^-]^4 = 2.4 \times 10^{-10} \text{ M} \quad (5\text{-}36)$$

If, instead, we take $[I^-] = 1.0$ M, then the concentrations are calculated to be

$$[Pb^{2+}] = 7.9 \times 10^{-9} \text{ M} \qquad [PbI_3^-] = 6.6 \times 10^{-5} \text{ M}$$

$$[PbI^+] = 7.9 \times 10^{-7} \text{ M} \qquad [PbI_4^{2-}] = 2.4 \times 10^{-4} \text{ M}$$

$$[PbI_2(aq)] = 1.1 \times 10^{-5} \text{ M}$$

The total concentration of dissolved lead is

$$[Pb]_{total} = [Pb^{2+}] + [PbI^+] + [PbI_2(aq)] + [PbI_3^-] + [PbI_4^{2-}] \quad (5\text{-}37)$$

When $[I^-] = 10^{-3}$ M, $[Pb]_{total} = 8.7 \times 10^{-3}$ M, of which 91% is Pb^{2+}. As $[I^-]$ increases, $[Pb]_{total}$ decreases by the common ion effect operating in Reaction 5-27. However, at sufficiently high $[I^-]$, complex formation takes over and $[Pb]_{total}$ increases (Figure 5-1). When $[I^-] = 1.0$ M, $[Pb]_{total} = 3.2 \times 10^{-4}$ M, of which 76 % is PbI_4^{2-}.

Comment on Equilibrium Constants

If you look up the equilibrium constant of a chemical reaction in two different books, there is an excellent chance that the values will be different (sometimes by a factor of ten or more). This happens because the constant may have been determined under different conditions by different investigators, perhaps using different techniques.

One of the most common sources of variation in the reported value of K is the ionic composition of the solution. It is important to note whether K is reported for a particular, stated ionic composition or whether the value has been extrapolated to zero ionic strength. If you need to use an equilibrium constant for your own work, you should choose a value of K measured under conditions as close as possible to those you will employ. If the value of K is critical to your experiment, it is best to measure K yourself under the precise conditions of your experiment.

Some of the most useful compilations of equilibrium constants are found in the following books: L. G. Sillén and A. E. Martell, *Stability*

Constants of Metal–Ion Complexes (Chemical Society, London, Special Publications No. 17 and 25, 1964 and 1971); A. E. Martell and R. M. Smith, *Critical Stability Constants* (New York: Plenum Press, 1974), a multivolume collection.

At equilibrium, the forward and reverse rates of a chemical reaction are equal. For the reaction $aA + bB \rightleftharpoons cC + dD$, the equilibrium constant is $K = [C]^c[D]^d/[A]^a[B]^b$. Solute concentrations should be expressed in moles per liter; gas concentrations should be in atmospheres; and the concentrations of pure solids, liquids, and solvents are omitted. If the direction of a reaction is changed, $K' = 1/K$. If two reactions are added, $K_3 = K_1K_2$. The value of the equilibrium constant can be calculated from the free energy change for a chemical reaction: $K = e^{-\Delta G^\circ/RT}$. Le Châtelier's principle predicts the effect on a chemical reaction when reactants or products are added, or temperature is changed. The reaction quotient, Q, is used to tell whether a system is at equilibrium, or how it must change to reach equilibrium.

The solubility product, which is the equilibrium constant for the dissolution of a solid salt, is used to calculate the solubility of a salt in aqueous solution. If one of the ions of that salt is already present in the solution, the solubility of the salt is decreased (the common ion effect). You should be able to evaluate the feasibility of selectively precipitating one ion from a solution containing other ions. At high concentration of ligand, a precipitated metal ion may redissolve because of the formation of soluble complex ions. This chapter demonstates the important skills of setting up and solving equilibrium problems, and of using and testing numerical approximations.

Terms to Understand

common ion effect	exothermic
complex ion	Gibbs free energy
coprecipitation	Le Châtelier's principle
disproportionation	ligand
endergonic	overall formation constant
endothermic	reaction quotient
enthalpy	saturated solution
entropy	solubility product
equilibrium constant	standard state
exergonic	stepwise formation constant

Exercises

5-A. Consider the following equilibria in which all ions are aqueous:

(1) $Ag^+ + Cl^- \rightleftharpoons AgCl(aq)$ $K = 2.0 \times 10^3$
(2) $AgCl(aq) + Cl^- \rightleftharpoons AgCl_2^-$ $K = 9.3 \times 10^1$
(3) $AgCl(s) \rightleftharpoons Ag^+ + Cl^-$ $K = 1.8 \times 10^{-10}$

(a) Calculate the numerical value of the equilibrium constant for the reaction $AgCl(s) \rightleftharpoons AgCl(aq)$.
(b) Calculate the concentration of $AgCl(aq)$ in equilibrium with excess undissolved solid AgCl.
(c) Find the numerical value of K for the reaction $AgCl_2^- \rightleftharpoons AgCl(s) + Cl^-$.

5-B. Reaction 5-12 is allowed to come to equilibrium in a solution initially containing 0.0100 M BrO_3^-, 0.0100 M Cr^{3+} and 1.00 M H^+.
  (a) Set up an equation analogous to Eq. 5-20 to find the concentrations at equilibrium.
  (b) Noting that $K = 1 \times 10^{11}$ for Reaction 5-12, it is reasonable to assume that the reaction will go nearly all the way "to completion." That is, we expect both the concentration of Br^- and of $Cr_2O_7^{2-}$ to be close to .00500 M at equilibrium. (Why?) Using these concentrations for Br^- and $Cr_2O_7^{2-}$, find the concentrations of BrO_3^- and Cr^{3+} at equilibrium. Note that Cr^{3+} is the *limiting reagent* in this problem. The reaction uses up the Cr^{3+} before consuming all of the BrO_3^-.

5-C. How many grams of lanthanum iodate, $La(IO_3)_3$ (which dissociates into La^{3+} and $3IO_3^-$), will dissolve in 250.0 mL of (a) water and (b) 0.050 M $LiIO_3$?

5-D. Which will be more soluble (moles of metal dissolved per liter of solution) in water?
  (a) $Ba(IO_3)_2$ ($K_{sp} = 1.5 \times 10^{-9}$) or $Ca(IO_3)_2$ ($K_{sp} = 7.1 \times 10^{-7}$)
  (b) $TlIO_3$ ($K_{sp} = 3.1 \times 10^{-6}$) or $Sr(IO_3)_2$ ($K_{sp} = 3.3 \times 10^{-7}$)

5-E. Fe(III) can be precipitated from acidic solution by addition of OH^- to form $Fe(OH)_3(s)$. At what concentration of OH^- will the concentration of Fe(III) be reduced to 1.0×10^{-10} M? If Fe(II) is used instead, what concentration of OH^- is necessary to reduce the Fe(II) concentration to 1.0×10^{-10} M?

5-F. It is desired to perform a 99% complete separation of 0.010 M Ca^{2+} and 0.010 M Ce^{3+} by precipitation with oxalate ($C_2O_4^{2-}$). Given the solubility products below, is this feasible?

$$CaC_2O_4 \cdot H_2O \quad K_{sp} = 1.3 \times 10^{-8}$$
$$Ce_2(C_2O_4)_3 \cdot 9H_2O \quad K_{sp} = 3 \times 10^{-29}$$

5-G. For a solution of Ni^{2+} and ethylenediamine, the following equilibrium constants apply at 20°C:

$$Ni^{2+} + H_2NCH_2CH_2NH_2 \rightleftharpoons Ni(en)^{2+} \quad \log K_1 = 7.52$$
ethylenediamine (abbreviated en)

$$Ni(en)^{2+} + en \rightleftharpoons Ni(en)_2^{2+} \quad \log K_2 = 6.32$$
$$Ni(en)_2^{2+} + en \rightleftharpoons Ni(en)_3^{2+} \quad \log K_3 = 4.49$$

Calculate the concentration of free Ni^{2+} in a solution prepared by mixing 0.100 mol of en plus 1.00 mL of 0.0100 M Ni^{2+} and diluting to 1.00 L. *Hint:* Start by assuming that all of the Ni is in the form $Ni(en)_3^{2+}$.

Problems

5-1. Write the expression for the equilibrium constant for each of the following reactions. Write the pressure of a gaseous molecule, X, as P_X.
  (a) $3Ag^+(aq) + PO_4^{3-}(aq) \rightleftharpoons Ag_3PO_4(s)$
  (b) $C_6H_6(l) + \frac{15}{2}O_2(g) \rightleftharpoons 3H_2O(l) + 6CO_2(g)$
  (c) $Cl_2(g) + 2OH^-(aq) \rightleftharpoons Cl^-(aq) + OCl^-(aq) + H_2O(l)$
  (d) $Hg(l) + I_2(g) \rightleftharpoons HgI_2(s)$

5-2. For the reaction $2A(g) + B(aq) + 3C(l) \rightleftharpoons D(s) + 3E(g)$, the concentrations at equilibrium are found to be

  A: 2.8×10^3 Pa   D: 16.5 M
  B: 1.2×10^{-2} M   E: 3.6×10^4 torr
  C: 12.8 M

Find the numerical value of the equilibrium constant that would appear in a conventional table of equilibrium constants.

5-3. The solubility product for $BaF_2(s)$ is given by $[Ba^{2+}][F^-]^2 = K_{sp}$. This equation is plotted on the graph below.

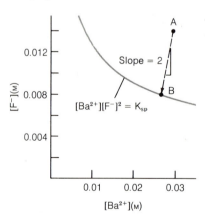

At the point A, $[Ba^{2+}][F^-]^2 > K_{sp}$ and the solution is said to be *supersaturated*. It contains more solute

than should be present at equilibrium. Suppose that a supersaturated solution with composition A on the graph is prepared by some means. Is the following statement true or false? "The composition of the solution at equilibrium will be given by the point B, which is connected to A by a line of slope = 2."

5-4. From the equilibrium constants below, calculate the equilibrium constant for the reaction $HO_2CCO_2H \rightleftharpoons 2H^+ + C_2O_4^{2-}$. All species in this problem are aqueous.

$$\underset{\text{oxalic acid}}{\overset{\text{OO}}{\overset{||\ ||}{HOCCOH}}} \rightleftharpoons H^+ + \overset{\text{OO}}{\overset{||\ ||}{HOCCO^-}}$$

$$K_1 = 5.6 \times 10^{-2}$$

$$\overset{\text{OO}}{\overset{||\ ||}{HOCCO^-}} \rightleftharpoons H^+ + \underset{\text{oxalate}}{\overset{\text{OO}}{\overset{||\ ||}{^-OCCO^-}}}$$

$$K_2 = 5.4 \times 10^{-5}$$

5-5. From the equations

$$HOCl \rightleftharpoons H^+ + OCl^-$$

$$K = 3.0 \times 10^{-8}$$

$$HOCl + OBr^- \rightleftharpoons HOBr + OCl^-$$

$$K = 15$$

find K for the reaction $HOBr \rightleftharpoons H^+ + OBr^-$. All species are aqueous.

5-6. The solubility product for CuCl is 3.2×10^{-7}. The equilibrium constant for the reaction

$$Cu(s) + Cu^{2+} \rightleftharpoons 2Cu^+$$

is 9.6×10^{-7}. Calculate the equilibrium constant for the reaction

$$Cu(s) + Cu^{2+} + 2Cl^- \rightleftharpoons 2CuCl(s)$$

5-7. For the sum of two reactions, we know that $K_3 = K_1 \cdot K_2$.

$$
\begin{array}{ll}
A + B \rightleftharpoons C + D & K_1 \\
\underline{D + E \rightleftharpoons B + F} & K_2 \\
A + E \rightleftharpoons C + F & K_3
\end{array}
$$

Show that this implies that $\Delta G_3^0 = \Delta G_1^0 + \Delta G_2^0$.

5-8. The equilibrium constant for the reaction $H_2O \rightleftharpoons H^+ + OH^-$ is 1.0×10^{-14} at 25°C. What is the value of K for the reaction $4H_2O \rightleftharpoons 4H^+ + 4OH^-$?

5-9. For the reaction $HCO_3^- \rightleftharpoons H^+ + CO_3^{2-}$, $\Delta G^0 = +59.0$ kJ/mol at 298.15 K. Find the value of K for the reaction.

5-10. For the reaction $Mg^{2+} + Cu(s) \rightleftharpoons Mg(s) + Cu^{2+}$, $K = 10^{-92}$ and $\Delta S^0 = +18$ JK^{-1} mol^{-1}.
  (a) Under standard conditions, is the reaction endergonic or exergonic? The term "standard conditions" means that reactants and products are in their standard states.
  (b) Under standard conditions, is the reaction endothermic or exothermic?

5-11. When $BaCl_2 \cdot H_2O(s)$ is dried in an oven, it loses gaseous water:

$$BaCl_2 \cdot H_2O(s) \rightleftharpoons BaCl_2(s) + H_2O(g)$$

$$\Delta H^0 = 63.11 \text{ kJ mol}^{-1} \text{ at } 25°C$$

$$\Delta S^0 = +148 \text{ JK}^{-1} \text{ mol}^{-1} \text{ at } 25°C$$

  (a) Calculate the vapor pressure of gaseous H_2O above $BaCl_2 \cdot H_2O$ at 298 K.
  (b) Assuming that ΔH^0 and ΔS^0 are not temperature-dependent (a poor assumption), estimate the temperature at which the vapor pressure of $H_2O(g)$ above $BaCl_2 \cdot H_2O(s)$ will be 1 atm.

5-12. The equilibrium constant for the reaction of ammonia with water has the following values in the range 5–10°C:

$$NH_3 + H_2O \rightleftharpoons NH_4^+ + OH^-$$

$$K = 1.479 \times 10^{-5} \text{ at } 5°C$$

$$K = 1.570 \times 10^{-5} \text{ at } 10°C$$

  (a) Assuming that ΔH^0 and ΔS^0 are constant in the interval 5–10°C, use Equation 5-15 to find the value of ΔH^0 for the reaction in this temperature range.
  (b) Describe how Equation 5-15 could be used to make a linear graph to determine ΔH^0, if ΔH^0 and ΔS^0 were constant over some range of temperature.

5-13. The formation of tetrafluoroethylene from its elements is highly exothermic.

$$\underset{\text{fluorine}}{2F_2(g)} + \underset{\text{graphite}}{2C(s)} \rightleftharpoons \underset{\text{tetrafluoroethylene}}{F_2C{=}CF_2(g)}$$

(a) If a mixture of F_2, graphite, and C_2F_4 is at equilibrium in a closed container, will the reaction go to the right or to the left if F_2 is added?

(b) Will it go to the right or left if rare bacteria from the planet Teflon are added? These bacteria eat C_2F_4 and make Teflon for their cell walls. Teflon has the structure·

$$\cdots\!-\!\underset{\underset{F}{|}}{\overset{\overset{F}{|}}{C}}\!-\!\underset{\underset{F}{|}}{\overset{\overset{F}{|}}{C}}\!-\!\underset{\underset{F}{|}}{\overset{\overset{F}{|}}{C}}\!-\!\underset{\underset{F}{|}}{\overset{\overset{F}{|}}{C}}\!-\!\underset{\underset{F}{|}}{\overset{\overset{F}{|}}{C}}\!-\!\underset{\underset{F}{|}}{\overset{\overset{F}{|}}{C}}\!-\!\cdots$$

(c) Will it go to the right or left if solid graphite is added? (Neglect any effect of increased pressure due to the decreased volume in the vessel when solid is added.)

(d) Will it go to the right or left if the container is crushed to one-eighth its original volume?

(e) Will it go to the right or left if the container is heated?

5-14. Suppose that the following reaction has come to equilibrium:

$$Br_2(l) + I_2(s) + 4Cl^-(aq) \rightleftharpoons 2Br^-(aq) + 2ICl_2^-(aq)$$

If more $I_2(s)$ is added, will the concentration of ICl_2^- in the aqueous phase increase, decrease, or remain unchanged?

5-15. For the reaction $H_2(g) + Br_2(g) \rightleftharpoons 2HBr(g)$, $K = 7.2 \times 10^{-4}$ at 1 362 K and ΔH^0 is positive. A vessel is charged with 48.0 Pa HBr, 1 370 Pa H_2, and 3 310 Pa Br_2 at 1 362 K.

(a) Will the reaction proceed to the left or right to reach equilibrium?

(b) Calculate the pressures (in pascals) of each species in the vessel at equilibrium.

(c) If the mixture at equilibrium is compressed to half its original volume, will the reaction proceed to the left or right to reestablish equilibrium?

(d) If the mixture at equilibrium is heated from 1 362 to 1 407 K, will HBr be formed or consumed in order to reestablish equilibrium?

5-16. How many grams of PbI_2 will dissolve in 0.500 L of (a) water and (b) 0.063 4 M NaI?

5-17. Express the solubility of $AgIO_3$ in 10.0 mM KIO_3 as a fraction of its solubility in pure water.

5-18. Ag^+ at a concentration of 10–100 ppb (ng/mL) is an effective disinfectant for swimming pool water. However, the concentration should not exceed a few hundred parts per billion for human health. One way to maintain an appropriate concentration of Ag^+ is to add a slightly soluble silver salt to the pool. For each salt below, calculate the concentration of Ag^+ (in parts per billion) that would exist at equilibrium.

$$AgCl: \quad K_{sp} = 1.8 \times 10^{-10}$$

$$AgBr: \quad K_{sp} = 5.0 \times 10^{-13}$$

$$AgI: \quad K_{sp} = 8.3 \times 10^{-17}$$

5-19. If a solution containing 0.10 M Cl^-, Br^-, I^-, and CrO_4^{2-} is treated with Ag^+, in what order will the anions precipitate?

5-20. A solution contains 0.050 0 M Ca^{2+} and 0.030 0 M Ag^+. Can 99% of either ion be precipitated by addition of sulfate, without precipitating the other metal ion? What will be the concentration of Ca^{2+} when Ag_2SO_4 begins to precipitate?

5-21. (a) Calculate the solubility (milligrams per liter) of $Zn_2Fe(CN)_6$ in distilled water. This salt dissociates as follows:

$$Zn_2Fe(CN)_6(s) \rightleftharpoons 2Zn^{2+} + Fe(CN)_6^{4-}$$

$$\text{ferrocyanide}$$

$$K_{sp} = 2.1 \times 10^{-16}$$

(b) Calculate the concentration of ferrocyanide in a solution of 0.040 M $ZnSO_4$ saturated with $Zn_2Fe(CN)_6$. $ZnSO_4$ will be completely dissociated into Zn^{2+} and SO_4^{2-}.

(c) What concentration of $K_4Fe(CN)_6$ should be added to a suspension of solid $Zn_2Fe(CN)_6$ in water to give $[Zn^{2+}] = 5.0 \times 10^{-7}$ M?

5-22. Given the equilibria below, calculate the concentrations of each zinc-containing species in a solution saturated with $Zn(OH)_2(s)$ and containing $[OH^-]$ at a fixed concentration of 3.2×10^{-7} M.

$$Zn(OH)_2(s): \quad K_{sp} = 3.0 \times 10^{-16}$$

$$Zn(OH)^+: \quad \beta_1 = 2.5 \times 10^4$$

$$Zn(OH)_3^-: \quad \beta_3 = 7.2 \times 10^{15}$$

$$Zn(OH)_4^{2-}: \quad \beta_4 = 2.8 \times 10^{15}$$

5-23. For the reaction below, $K = 6.9 \times 10^{-18}$.

$$Cu(OH)_{1.5}(SO_4)_{0.25}(s) \rightleftharpoons Cu^{2+} + \tfrac{3}{2}OH^- + \tfrac{1}{4}SO_4^{2-}$$

What will be the concentration of Cu^{2+} in a solution saturated with $Cu(OH)_{1.5}(SO_4)_{0.25}$?

5-24. Consider the following equilibria:

$$AgCl(s) \rightleftharpoons Ag^+ + Cl^- \quad K_{sp} = 1.8 \times 10^{-10}$$

$$AgCl(s) + Cl^- \rightleftharpoons AgCl_2^- \quad K_2 = 1.5 \times 10^{-2}$$

$$AgCl_2^- + Cl^- \rightleftharpoons AgCl_3^- \quad K_3 = 0.49$$

Find the total concentration of silver-containing species in a silver-saturated, aqueous solution containing the following concentrations of Cl^-:
(a) 0.010 M (b) 0.20 M (c) 2.0 M

5-25. The planet Aragonose (which is made mostly of the mineral aragonite, whose composition is $CaCO_3$) has an atmosphere containing methane and carbon dioxide, each at a pressure of 0.10 atm. The oceans are saturated with aragonite and have a concentration of H^+ equal to 1.8×10^{-7} M. Given the following equilibria, how many grams of calcium are contained in 2.00 L of Aragonose sea water?

$$CaCO_3(s, \text{aragonite}) \rightleftharpoons Ca^{2+}(aq) + CO_3^{2-}(aq)$$

$$K_{sp} = 6.0 \times 10^{-9}$$

$$CO_2(g) \rightleftharpoons CO_2(aq)$$

$$K_{CO_2} = 3.4 \times 10^{-2}$$

$$CO_2(aq) + H_2O(l) \rightleftharpoons HCO_3^-(aq) + H^+(aq)$$

$$K_1 = 4.4 \times 10^{-7}$$

$$HCO_3^-(aq) \rightleftharpoons H^+(aq) + CO_3^{2-}(aq)$$

$$K_2 = 4.7 \times 10^{-11}$$

Hint: Don't be too upset. Reverse the first reaction, add all the reactions together, and see what cancels.

5-26. (a) Calculate the ratio $[Pb^{2+}]/[Sr^{2+}]$ in a solution of distilled water saturated with PbF_2 *and* SrF_2.

(b) *Caution:* This question may damage your sanity. Calculate the concentrations of Pb^{2+}, Sr^{2+}, and F^- in the solution above. This problem is best solved by numerical trial and error. Begin by guessing a value of $[Pb^{2+}]$. Then find the values of $[F^-]$ and $[Sr^{2+}]$ from the relations

$$[F^-] = \sqrt{K_{sp} (\text{for } PbF_2)/[Pb^{2+}]}$$

$$[Sr^{2+}] = K_{sp} (\text{for } SrF_2)/[F^-]^2$$

When you have guessed the correct value of $[Pb^{2+}]$, the relation $[F^-] = 2[Pb^{2+}] + 2[Sr^{2+}]$ will be satisfied. Find the best answer with two significant figures.

6 / Activity

In Chapter 5 we wrote the equilibrium constant for the reaction $aA + bB \rightleftharpoons cC + dD$ in the form

$$K = \frac{[C]^c[D]^d}{[A]^a[B]^b} \qquad (6\text{-}1)$$

This statement is not strictly correct, because the quotient $[C]^c[D]^d/[A]^a[B]^b$ is not constant under all conditions. In this chapter we will see how the concentrations should be replaced by *activities*, and we will see how activities are used.

6-1 ROLE OF IONIC STRENGTH IN IONIC EQUILIBRIA

Equilibria involving ionic species are affected by the presence of all ions in a solution. The most useful measure of the total concentration of ions in a solution is the **ionic strength,** μ, defined as

$$\mu = \frac{1}{2} \sum_i c_i z_i^2 \qquad (6\text{-}2)$$

where c_i is the concentration of the ith species and z_i is its charge. The sum extends over *all* ions in solution.

EXAMPLE

Find the ionic strength of (a) 0.10 M $NaNO_3$, (b) 0.10 M Na_2SO_4, and (c) 0.020 M KBr plus 0.030 M $ZnSO_4$.

(a) $\mu = \frac{1}{2}\{[Na^+] \cdot (+1)^2 + [NO_3^-] \cdot (-1)^2\}$
$= \frac{1}{2}\{(0.10) \cdot 1 + (0.10) \cdot 1\} = 0.10$ M

(b) $\mu = \frac{1}{2}\{[Na^+] \cdot (+1)^2 + [SO_4^{2-}] \cdot (-2)^2\}$
$= \frac{1}{2}\{(0.20) \cdot 1 + (0.10) \cdot 4\} = 0.30$ M

Note that $[Na^+] = 0.20$ M because there are 2 moles of Na^+ per mole of Na_2SO_4.

(c) $\mu = \frac{1}{2}\{[K^+] \cdot (+1)^2 + [Br^-] \cdot (-1)^2 + [Zn^{2+}] \cdot (+2)^2 + [SO_4^{2-}] \cdot (-2)^2\}$
$= \frac{1}{2}\{(0.020) \cdot 1 + (0.020) \cdot 1 + (0.030) \cdot 4 + (0.030) \cdot 4\} = 0.14$ M

Electrolyte	Molarity	Ionic strength
1:1	M	M
2:1	M	3M
3:1	M	6M
2:2	M	4M

$NaNO_3$ is called a 1:1 electrolyte because the cation and anion both have a charge of 1. For 1:1 electrolytes the ionic strength equals the molarity. For any other stoichiometry (such as the 2:1 electrolyte, Na_2SO_4), the ionic strength is greater than the molarity.

Effect of Ionic Strength on Solubility of Salts

Consider a saturated solution of $Hg_2(IO_3)_2$ in distilled water. Based on the solubility product, we expect the concentration of mercurous ion to be 6.9×10^{-7} M:

$$Hg_2(IO_3)_2(s) \rightleftharpoons Hg_2^{2+} + 2IO_3^- \qquad K_{sp} = 1.3 \times 10^{-18} \qquad (6\text{-}3)$$

$$K_{sp} = [Hg_2^{2+}][IO_3^-]^2 = x(2x)^2$$

$$x = [Hg_2^{2+}] = 6.9 \times 10^{-7} \text{ M}$$

This concentration is indeed observed when $Hg_2(IO_3)_2$ is dissolved in distilled water.

However, a seemingly strange effect is observed when a salt such as KNO_3 is added to the solution. Neither K^+ nor NO_3^- reacts with either Hg_2^{2+} or IO_3^-. Yet, if 0.050 M KNO_3 is added to the saturated solution of $Hg_2(IO_3)_2$, more solid dissolves until the concentration of Hg_2^{2+} has increased by about 50% (to 1.0×10^{-6} M).

It turns out to be a general observation that adding any "inert" salt (such as KNO_3) to any sparingly soluble salt (such as $Hg_2(IO_3)_2$) increases the solubility of the latter. By "inert," we mean a salt whose ions do not react with the compound of interest.

Addition of an "inert" salt increases the solubility of an ionic compound.

Explanation of Increased Solubility

How can we explain the increased solubility induced by the addition of ions to the solution? Consider one particular Hg_2^{2+} ion and one particular IO_3^- ion in the solution. The IO_3^- ion is surrounded by cations (K^+, Hg_2^{2+}) and anions (NO_3^-, IO_3^-) in its region of solution. However, on the average, there will be more cations than anions around any one chosen anion. This is because cations are attracted to the anion, but anions are repelled. This gives rise to a region of net positive charge around any particular anion. We call this region the **ionic atmosphere**. Ions are continually diffusing into and out of the ionic atmosphere. The net charge in the atmosphere, averaged over time, is less than the charge of the anion at the center. A similar phenomenon accounts for a region of negative charge surrounding any cation in solution. This is shown schematically in Figure 6-1.

The ionic atmosphere serves to attenuate the attraction between ions in solution. The cation plus its negative atmosphere has less positive charge than the cation alone. The anion plus its atmosphere has less negative charge than the anion alone. The net attraction between the cation-plus-atmosphere and the anion-plus-atmosphere is smaller than it would be between pure cation and anion in the absence of ionic atmospheres. *The greater the ionic strength of a solution, the higher the charge in the ionic atmosphere. This means that each ion-plus-atmosphere contains less charge, and there is less attraction between any particular cation and anion.*

Increasing the ionic strength therefore reduces the attraction between any particular Hg_2^{2+} ion and any IO_3^- ion, as compared to their attraction for each other in distilled water. The effect is to reduce their tendency to come together, thereby increasing the solubility of $Hg_2(IO_3)_2$.

Increasing ionic strength promotes the dissociation of compounds into ions. Thus, each reaction below is driven to the right if the ionic strength is raised from, say, 0.01 to 0.1 M.

An anion is surrounded by more cations than anions. A cation is surrounded by more anions than cations.

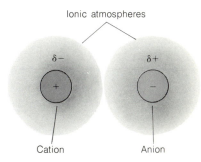

Figure 6-1
Schematic representation of the ionic atmospheres surrounding any ions in solution. The charges of the atmospheres ($\delta+$ and $\delta-$) are less than the charges of the central ions.

Ion dissociation is increased by increasing the ionic strength. Demonstration 6-1 shows this effect in the case of Reaction 6-5.

$$Hg_2(IO_3)_2(s) \rightleftharpoons Hg_2^{2+} + 2IO_3^- \qquad (6\text{-}3)$$

$$\text{phenol} \longrightarrow OH \rightleftharpoons \text{phenolate} \longrightarrow O^- + H^+ \qquad (6\text{-}4)$$

$$Fe(SCN)^{2+} \rightleftharpoons Fe^{3+} + \underset{\text{thiocyanate}}{SCN^-} \qquad (6\text{-}5)$$

6-2 ACTIVITY COEFFICIENTS

The equilibrium "constant" in Equation 6-1 does not predict that there should be any effect of ionic strength on ionic reactions. To account for the effect of ionic strength, the concentrations should be replaced by **activities**:

$$\underset{\substack{\uparrow \\ \text{Activity} \\ \text{of C}}}{\mathcal{A}_C} = \underset{\substack{\uparrow \\ \text{Concentration} \\ \text{of C}}}{[C]} \underset{\substack{\nwarrow \\ \text{Activity coefficient} \\ \text{of C}}}{\gamma_C} \qquad (6\text{-}6)$$

Do not confuse the terms *activity* and *activity coefficient*.

The activity of each species is its concentration multiplied by its **activity coefficient**.

Demonstration 6-1   EFFECT OF IONIC STRENGTH ON ION DISSOCIATION[†]

This experiment demonstrates the effect of ionic strength on the dissociation of the red ferric thiocyanate complex:

$$Fe(SCN)^{2+} \rightleftharpoons Fe^{3+} + SCN^- \tag{6-5}$$

$$\text{red} \qquad \text{pale yellow} \quad \text{colorless}$$

Prepare a solution of 1 mM $FeCl_3$ by dissolving 0.27 g of $FeCl_3 \cdot 6H_2O$ in 1 L of water containing 3 drops of 15 M (concentrated) HNO_3. The purpose of the acid is to slow the precipitation of Fe^{3+}, which occurs in a few days and necessitates the preparation of fresh solution for this demonstration.

To demonstrate the effect of ionic strength on Reaction 6-5, mix 300 mL of the 1 mM $FeCl_3$ solution with 300 mL of 1.5 mM NH_4SCN or KSCN. Divide the pale red solution into two equal portions and add 12 g of KNO_3 to one of them to increase the ionic strength to 0.4 M. As the KNO_3 dissolves, the red $Fe(SCN)^{2+}$ complex dissociates and the color fades noticeably (see Color Plates 1 and 2).

Adding a few crystals of NH_4SCN or KSCN to either solution drives the reaction toward formation of $Fe(SCN)^{2+}$, intensifying the red color. This demonstrates Le Châtelier's principle.

[†] D. R. Driscoll, *J. Chem. Ed.*, **56**, 603 (1979). An excellent experiment (or demonstration) dealing with the effect of ionic strength on the acid–base dissociation reaction of bromcresol green can be found in J. A. Bell (ed.), *Chemical Principles in Practice* (Reading, Mass.: Addision-Wesley, 1967), pp. 105–108, and in R. W. Ramette, *J. Chem. Ed.*, **40**, 252 (1963).

The correct form of the equilibrium constant is

The "real" equilibrium constant.

$$K = \frac{\mathscr{A}_C^c \mathscr{A}_D^d}{\mathscr{A}_A^a \mathscr{A}_B^b} = \frac{[C]^c \gamma_C^c [D]^d \gamma_D^d}{[A]^a \gamma_A^a [B]^b \gamma_B^b} \tag{6-7}$$

Equation 6-7 allows for the effect of ionic strength on a reaction such as 6-3 because the activity coefficients depend on ionic strength.

For Reaction 6-3 the equilibrium constant is

$$K = \mathscr{A}_{Hg_2^{2+}} \mathscr{A}_{IO_3^-}^2 = [Hg_2^{2+}] \gamma_{Hg_2^{2+}} [IO_3^-]^2 \gamma_{IO_3^-}^2 \tag{6-8}$$

If the concentrations of Hg_2^{2+} and IO_3^- are to increase with increasing ionic strength, the values of the activity coefficients must decrease with increasing ionic strength.

At low ionic strength, activity coefficients approach unity, and the thermodynamic equilibrium constant, Equation 6-7, approaches the "concentration" equilibrium constant, Equation 6-1. One way to measure thermodynamic equilibrium constants is to measure the concentration ratio, Equation 6-1, at successively lower ionic strengths and then extrapolate to zero ionic strength. Very commonly, tabulated equilibrium constants are not true thermodynamic constants but just the concentration ratio, Equation 6-1, measured under a particular set of conditions.

Activity Coefficients of Ions

6-2 ACTIVITY COEFFICIENTS

Detailed treatment of the ionic atmosphere model leads to the **extended Debye–Hückel equation,** relating activity coefficients to ionic strength:

$$\log \gamma = \frac{-0.51 z^2 \sqrt{\mu}}{1 + (\alpha \sqrt{\mu}/305)} \quad \text{(at 25°C)} \tag{6-9}$$

In Equation 6-9, γ is the activity coefficient of an ion of charge $\pm z$ and size α (pm) in an aqueous medium of ionic strength μ. The equation works fairly well for $\mu \leq 0.1$ M.

pm = picometers = 10^{-12} m

The value of α is the effective hydrated radius of the ion plus its tightly bound sheath of water molecules. Small, highly charged ions bind solvent molecules more tightly and have *larger* hydrated radii than do larger or less highly charged ions. F^-, for example, has a hydrated radius greater than that of I^- (Figure 6-2). Each anion attracts solvent molecules mainly by electrostatic interaction between the negative ion and the positive pole of the H_2O dipole:

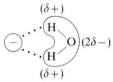

Table 6-1 lists the sizes and calculated activity coefficients of many common ions. The table is arranged according to the size (α) and charge of the

Table 6-1
Activity coefficients

	Ion size (α, pm)	\multicolumn{5}{c}{Ionic strength (μ, M)}			
Ion		0.001	0.005	0.01	0.05	0.1
\multicolumn{7}{c}{Charge = ±1}
H^+	900	0.967	0.933	0.914	0.86	0.83
	800	0.966	0.931	0.912	0.85	0.82
	700	0.965	0.930	0.909	0.845	0.81
Li^+	600	0.965	0.929	0.907	0.835	0.80
	500	0.964	0.928	0.904	0.83	0.79
$Na^+, CdCl^+, ClO_2^-, IO_3^-, HCO_3^-,$ $H_2PO_4^-, HSO_3^-, H_2AsO_4^-,$ $Co(NH_3)_4(NO_2)_2^+$	450	0.964	0.928	0.902	0.82	0.775
	400	0.964	0.927	0.901	0.815	0.77
$OH^-, F^-, SCN^-, OCN^-, HS^-, ClO_3^-,$ $ClO_4^-, BrO_3^-, IO_4^-, MnO_4^-$	350	0.964	0.926	0.900	0.81	0.76
$K^+, Cl^-, Br^-, I^-, CN^-, NO_2^-, NO_3^-$	300	0.964	0.925	0.899	0.805	0.755
$Rb^+, Cs^+, NH_4^+, Tl^+, Ag^+$	250	0.964	0.924	0.898	0.80	0.75

(continued)

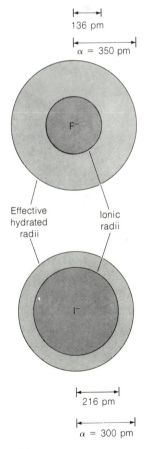

Figure 6-2
Ionic and hydrated radii of fluoride and iodide.

Table 6-1 (continued)

Ion	Ion size (α, pm)	Ionic strength (μ, M)				
		0.001	0.005	0.01	0.05	0.1
Charge = ±2						
Mg^{2+}, Be^{2+}	800	0.872	0.755	0.69	0.52	0.45
	700	0.872	0.755	0.685	0.50	0.425
Ca^{2+}, Cu^{2+}, Zn^{2+}, Sn^{2+}, Mn^{2+}, Fe^{2+}, Ni^{2+}, Co^{2+}	600	0.870	0.749	0.675	0.485	0.405
Sr^{2+}, Ba^{2+}, Cd^{2+}, Hg^{2+}, S^{2-}, $S_2O_4^{2-}$, WO_4^{2-}	500	0.868	0.744	0.67	0.465	0.38
Pb^{2+}, CO_3^{2-}, SO_3^{2-}, MoO_4^{2-}, $Co(NH_3)_5Cl^{2+}$, $Fe(CN)_5NO^{2-}$	450	0.867	0.742	0.665	0.455	0.37
Hg_2^{2+}, SO_4^{2-}, $S_2O_3^{2-}$, $S_2O_6^{2-}$, $S_2O_8^{2-}$, SeO_4^{2-}, CrO_4^{2-}, HPO_4^{2-}	400	0.867	0.740	0.660	0.445	0.355
Charge = ±3						
Al^{3+}, Fe^{3+}, Cr^{3+}, Sc^{3+}, Y^{3+}, In^{3+}, lanthanides[†]	900	0.738	0.54	0.445	0.245	0.18
	500	0.728	0.51	0.405	0.18	0.115
PO_4^{3-}, $Fe(CN)_6^{3-}$, $Cr(NH_3)_6^{3+}$, $Co(NH_3)_6^{3+}$, $Co(NH_3)_5H_2O^{3+}$	400	0.725	0.505	0.395	0.16	0.095
Charge = ±4						
Th^{4+}, Zr^{4+}, Ce^{4+}, Sn^{4+}	1100	0.588	0.35	0.255	0.10	0.065
$Fe(CN)_6^{4-}$	500	0.57	0.31	0.20	0.048	0.021

Sizes of some organic ions (α, pm): Charge = ±1

$HCOO^-$, $H_2citrate^-$, $CH_3NH_3^+$, $(CH_3)_2NH_2^+$	350
$NH_3^+CH_2COOH$, $(CH_3)_3NH^+$, $C_2H_5NH_3^+$	400
CH_3COO^-, CH_2ClCOO^-, $(CH_3)_4N^+$, $(C_2H_5)_2NH_2^+$, $NH_2CH_2COO^-$	450
$CHCl_2COO^-$, CCl_3COO^-, $(C_2H_5)_3NH^+$, $(C_3H_7)NH_3^+$	500
$C_6H_5COO^-$, $C_6H_4OHCOO^-$, $C_6H_4ClCOO^-$, $C_6H_5CH_2COO^-$, $CH_2=CHCH_2COO^-$, $(CH_3)_2CHCHCOO^-$, $(C_2H_5)_4N^+$, $(C_3H_7)_2NH_2^+$	600
$[OC_6H_2(NO_3)_3]^-$, $(C_3H_7)_3NH^+$, $CH_3OC_6H_4COO^-$	700
$(C_6H_5)_2CHCOO^-$, $(C_3H_7)_4N^+$	800

Charge = ±2

$(COO)_2^{2-}$, $Hcitrate^{2-}$	450
$H_2C(COO)_2^{2-}$, $(CH_2COO)_2^{2-}$, $(CHOHCOO)_2^{2-}$	500
$C_6H_4(COO)_2^{2-}$, $H_2C(CH_2COO)_2^{2-}$, $(CH_2CH_2COO)_2^{2-}$	600
$[OOC(CH_2)_5COO]^{2-}$, $[OOC(CH_2)_6COO]^{2-}$, Congo red anion^{2-}	700

Charge = ±3

$Citrate^{3-}$	500

[†] Lanthanides are elements 57–71 in the periodic table.
SOURCE: Data from J. Kielland, *J. Amer. Chem. Soc.*, **59**, 1675 (1937).

ion. All ions of the same size and charge appear on the same line and have the same activity coefficients. At the bottom of the table is a list giving the sizes of some organic ions. To find the activity coefficient of these ions, refer to the appropriate size and charge in the main body of the table.

Over the range of ionic strength from zero to 0.1 M, the effect of each variable on activity coefficients is as follows:

1. As the ionic strength increases, the activity coefficient decreases. For all ions, γ approaches unity as μ approaches zero.
2. As the charge of the ion increases, the departure of the activity coefficient from unity increases. Activity corrections are much more important for an ion with a charge of ± 3 than for one with a charge of ± 1. Note that the activity coefficients in Table 6-1 depend on the magnitude of the charge, but not on its sign.
3. The smaller the hydrated radius of the ion, the more important activity effects become.

Effects 1 and 2 for ions of hydrated radius 500 pm are shown in Figure 6-3.

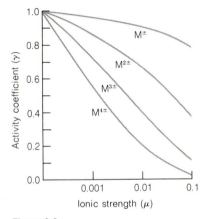

Figure 6-3
Dependence of activity coefficient on ionic strength for ions with a hydrated radius of 500 pm. Note that the abscissa is logarithmic.

EXAMPLE
Find the activity coefficient of Hg_2^{2+} in a solution of 0.033 M $Hg_2(NO_3)_2$.
The ionic strength is

$$\mu = \tfrac{1}{2}([Hg_2^{2+}] \cdot 2^2 + [NO_3^-] \cdot (-1)^2)$$
$$= \tfrac{1}{2}([0.033] \cdot 4 + [0.066] \cdot 1) = 0.10 \text{ M}$$

In Table 6-1, Hg_2^{2+} is listed under the charge ± 2 and has a size of 400 pm. $\gamma = 0.355$ when $\mu = 0.1$ M.

EXAMPLE
Calculate the activity coefficient of H^+ when $\mu = 0.025$ M.
We can do this by two methods. The one we will use for the remainder of this text is to interpolate between values listed in Table 6-1. H^+ is the first entry in Table 6-1.

$$H^+: \mu = 0.01 \quad 0.025 \quad 0.05$$
$$\gamma = 0.914 \quad ? \quad 0.86$$

By linear interpolation we find the value of γ when $\mu = 0.025$ M as follows:

$$\gamma = \underbrace{0.914}_{\substack{\text{Value of} \\ \gamma \text{ when} \\ \mu = 0.01}} - \underbrace{\left(\frac{0.025 - 0.01}{0.05 - 0.01}\right)}_{\substack{\text{Fraction of} \\ \text{the way} \\ \text{between 0.01} \\ \text{and 0.05}}} \underbrace{(0.914 - 0.86)}_{\substack{\text{Difference in} \\ \gamma \text{ between 0.01} \\ \text{and 0.05}}} = 0.89_4$$

A more correct and more tedious calculation uses Equation 6-9. Table 6-1 gives a value of 900 pm for the size of H^+:

$$\log \gamma_{H^+} = \frac{(-0.51)(1^2)\sqrt{0.025}}{1 + (900\sqrt{0.025}/305)} = -0.054\,98$$

$$\gamma_{H^+} = 0.88_1$$

The difference between this calculated value and the interpolated value is less than 2%.

Activity Coefficients of Nonionic Compounds

Neutral molecules, such as benzene or acetic acid, are not surrounded by an ionic atmosphere because they have no charge. To a good approximation, their activity coefficients are unity when the ionic strength is less than 0.1 M. For all problems in this text, we will set $\gamma = 1$ for neutral molecules. That is, *the activity of a neutral molecule will be assumed to be equal to its concentration.*

For neutral species, $\mathscr{A} \approx [C]$.

For gaseous reactants such as H_2, the activity is written

$$\mathscr{A}_{H_2} = P_{H_2}\gamma_{H_2} \qquad (6\text{-}10)$$

where P_{H_2} is pressure in atmospheres. For most gases at or below 1 atm, $\gamma \approx 1$. For all gases, *we will assume that $\mathscr{A} = P(atm)$.* The activity of a gas is called its **fugacity,** and the activity coefficient is called the *fugacity coefficient.* Deviation of gas behavior from the ideal gas law results in deviation of the fugacity coefficient from unity.

For gases, $\mathscr{A} \approx P\,(atm)$.

High Ionic Strengths

For the dissociation of acetic acid, we can write

$$\text{CH}_3\overset{\overset{\displaystyle O}{\|}}{\text{C}}\text{OH}(aq) \rightleftharpoons \text{CH}_3\overset{\overset{\displaystyle O}{\|}}{\text{C}}\text{O}^- + \text{H}^+ \qquad (6\text{-}11)$$

$$K = \frac{[\text{CH}_3\text{CO}_2^-]\gamma_{\text{CH}_3\text{CO}_2^-}[\text{H}^+]\gamma_{H^+}}{[\text{CH}_3\text{CO}_2\text{H}]} \qquad (6\text{-}12)$$

where $\gamma_{\text{CH}_3\text{CO}_2\text{H}}$ has been set equal to unity. Equation 6-12 can be rearranged to show the quotient of concentrations:

$$R = \frac{[\text{CH}_3\text{CO}_2^-][\text{H}^+]}{[\text{CH}_3\text{CO}_2\text{H}]} = \frac{K}{\gamma_{\text{CH}_3\text{CO}_2^-}\gamma_{H^+}} \qquad (6\text{-}13)$$

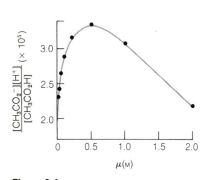

Figure 6-4
Value of R in Equation 6-13 as a function of increasing KCl concentration in an aqueous solution of acetic acid. [Data from H. S. Harned and F. C. Hickey, *J. Amer. Chem. Soc.*, **59**, 2303 (1937).]

The extended Debye–Hückel equation 6-9 predicts that each γ will decrease as μ increases. The quotient R should therefore increase as the ionic strength increases.

Experimental values of R are shown in Figure 6-4. As the ionic strength is increased by addition of KCl, R increases up to an ionic strength near

0.5 M. Above this ionic strength the quotient R decreases. The only way this can happen is if the activity coefficients in Equation 6-13 *increase* above an ionic strength of 0.5 M.

This effect is general. Above an ionic strength of approximately 1 M, most activity coefficients increase. The ionic atmosphere model cannot account for this. A model that is successful for ionic strengths as high as 3 M is based on the change in dielectric constant of the solution in the immediate vicinity of each ion in the solution.[†] We will not attempt any activity coefficient calculations for ionic strengths above 0.1 M.

At high ionic strength, γ increases with increasing μ.

Mean Activity Coefficient

The theories dealing with activity coefficients derive coefficients for individual ions. In experiments to date, ions are available only in pairs. It has not been possible to measure the activity coefficient of one ion at a time. Instead, the **mean activity coefficient,** $\gamma_\pm$, is derived from various experiments. For a salt with the stoichiometry (cation)$_m$(anion)$_n$, the mean activity coefficient is related to the individual activity coefficients by the equation

$$\gamma_\pm^{m+n} = \gamma_+^m \gamma_-^n \tag{6-14}$$

where γ_+ is the activity coefficient of the cation and γ_- is the activity coefficient of the anion.

Prediction of $\gamma_\pm$ for $La(NO_3)_3$ from the individual coefficients for La^{3+} and NO_3^- in Table 6-1 at $\mu = 0.1$ M:

$$\gamma_\pm = [(0.18)^1(0.755)^3]^{1/4} = 0.53$$

Observed value = 0.59.

6-3 USING ACTIVITY COEFFICIENTS

This section presents three examples illustrating the proper use of activity coefficients in equilibrium calculations. The general prescription is trivial: *Write each equilibrium constant with activities in place of concentrations. Use the ionic strength of the solution to find the correct values of activity coefficients.*

In the first example, we will calculate the concentration of Ca^{2+} in a 0.012 5 M solution of $MgSO_4$ saturated with CaF_2. The relevant equilibrium is

$$CaF_2(s) \rightleftharpoons Ca^{2+} + 2F^- \qquad K_{sp} = 3.9 \times 10^{-11} \tag{6-15}$$

The equilibrium expression is set up in the usual way:

$$CaF_2(s) \rightleftharpoons Ca^{2+} + 2F^-$$

Initial concentration:	solid	0	0
Final concentration:	solid	x	$2x$

[†] L. W. Bahe, *J. Phys. Chem.*, **76**, 1062 (1972); L. W. Bahe and D. Parker, *J. Amer. Chem. Soc.*, **97**, 5664 (1975).

$$K = \mathcal{A}_{Ca^{2+}}\mathcal{A}_{F^-}^2 = [Ca^{2+}]\gamma_{Ca^{2+}}[F^-]^2\gamma_F^2 \tag{6-16}$$

$$= [x]\gamma_{Ca^{2+}}[2x]^2\gamma_{F^-}^2$$

To find the values of γ to use in Equation 6-16, we need to calculate the ionic strength. The ionic strength is due to the dissolved $MgSO_4$ *and* the dissolved CaF_2. However, K_{sp} for CaF_2 is quite small, and we will start by guessing that the contribution of CaF_2 to the ionic strength will be negligible. For 0.012 5 M $MgSO_4$, we calculate

$$\mu = \tfrac{1}{2}([0.012\ 5]\cdot 2^2 + [0.012\ 5]\cdot(-2)^2) = 0.050\ 0\ \text{M} \tag{6-17}$$

Using $\mu = 0.050\ 0$ M, we look at Table 6-1 to find $\gamma_{Ca^{2+}} = 0.485$ and $\gamma_{F^-} = 0.81$. Substituting into Equation 6-16 gives

Note that $[F^-]$ and γ_{F^-} are both squared.

$$3.9 \times 10^{-11} = [x](0.485)[2x]^2(0.81)^2 \tag{6-18}$$

$$x = [Ca^{2+}] = 3.1 \times 10^{-4}\ \text{M}$$

> *Question:* What is the value of μ, including the contribution of CaF_2?

Our assumption was correct; the contribution of CaF_2 to the ionic strength is negligible.

As the next example, let's find the concentration of Ca^{2+} in a 0.050 M NaF solution saturated with CaF_2. The ionic strength is again 0.050 M, this time due to the presence of NaF.

	$CaF_2(s) \rightleftharpoons$	Ca^{2+}	$+$	$2F^-$
Initial concentration:	solid	0		0.050
Final concentration:	solid	x		$2x + 0.050$

Assuming that $2x \ll 0.050$, we solve the problem as follows:

$$K_{sp} = [Ca^{2+}]\gamma_{Ca^{2+}}[F^-]^2\gamma_{F^-}^2$$

> *Question:* Why is the solubility of CaF_2 lower in this solution than in $MgSO_4$ solution?

$$3.9 \times 10^{-11} = (x)(0.485)(0.050)^2(0.81)^2 \tag{6-19}$$

$$x = [Ca^{2+}] = 4.9 \times 10^{-8}$$

As a final example, we calculate the solubility of LiF in distilled water:

$$\text{LiF}(s) \rightleftharpoons \text{Li}^+ + \text{F}^- \qquad K_{sp} = 1.7 \times 10^{-3} \tag{6-20}$$

The ionic strength is determined by the concentration of LiF dissolved in the solution. We do not know this concentration yet. As a first approximation, we will calculate the concentrations of Li^+ and F^- by neglecting activity coefficients. Calling x_1 our first approximation for $[Li^+]$ (and $[F^-]$), we can write

$$K_{sp} \approx [\text{Li}^+][\text{F}^-] = x_1^2 \Rightarrow x_1 = [\text{Li}^+] = [\text{F}^-] = 0.041 \tag{6-21}$$

As a second approximation, we will assume that $\mu = 0.041$ M, which is the result of our first approximation. Interpolating in Table 6-1 gives $\gamma_{Li^+} = 0.851$ and $\gamma_{F^-} = 0.830$ for $\mu = 0.041$ M. Putting these values into the expression for K_{sp} gives

This is a method of successive approximations. Each cycle of calculations is called one iteration.

$$K_{sp} = [Li^+]\gamma_{Li^+}[F^-]\gamma_{F^-} = [x_2](0.851)[x_2](0.830) \Rightarrow x_2 = 0.049 \text{ M} \quad (6\text{-}22)$$

As a third approximation, we assume that $\mu = 0.049$ M. Using this to find new values of activity coefficients in Table 6-1 gives

$$K_{sp} = [x_3](0.837)[x_3](0.812) \Rightarrow x_3 = 0.050 \text{ M} \quad (6\text{-}23)$$

Using $\mu = 0.050$ M gives a fourth approximation:

$$K_{sp} = [x_4](0.835)[x_4](0.81) \Rightarrow x_4 = 0.050 \text{ M} \quad (6\text{-}24)$$

The fourth answer is the same as the third answer. We have reached a self-consistent result, which must therefore be correct.

As LiF dissolves, it increases the ionic strength and increases its own solubility.

Summary

The true thermodynamic equilibrium constant for the reaction $aA + bB \rightleftharpoons cC + dD$ is $K = \mathscr{A}_C^c \mathscr{A}_D^d / \mathscr{A}_A^a \mathscr{A}_B^b$, where $\mathscr{A}_i$ is the activity of the ith species. The activity is the product of the concentration and the activity coefficient: $\mathscr{A}_i = c_i \gamma_i$. For nonionic compounds and gases, we will assume that $\gamma_i = 1$. For ionic species the activity coefficient depends on the ionic strength, defined as $\mu = \frac{1}{2}\sum c_i z_i^2$. The activity coefficient decreases as the ionic strength increases, at least for low ionic strengths. The extent of dissociation of ionic compounds increases with ionic strength because the ionic atmosphere of each ion diminishes the attraction of ions for each other. You should be able to perform equilibrium calculations correctly using activities instead of concentrations. You should be able to estimate activity coefficients by interpolation in Table 6-1.

Terms to Understand

activity
activity coefficient
Debye–Hückel equation
fugacity

hydrated radius
ionic atmosphere
ionic strength
mean activity coefficient

Exercises

6-A. Calculate the ionic strength of
    (a) 0.02 M KBr
    (b) 0.02 M Cs_2CrO_4
    (c) 0.02 M $MgCl_2$ plus 0.03 M $AlCl_3$

6-B. Find the activity (not the activity coefficient) of the $(CH_3CH_2CH_2)_4N^+$ ion in a solution containing 0.005 0 M $(CH_3CH_2CH_2)_4N^+Br^-$ plus 0.005 0 M $(CH_3)_4N^+Cl^-$.

6-C. Find the solubility of AgSCN (moles of Ag^+/L) in
(a) 0.060 M KNO_3 (b) 0.060 M KSCN

6-D. Find the concentration of OH^- in a solution of 0.025 M $CaCl_2$ saturated with $Mn(OH)_2$.

6-E. Calculate the value of $\gamma_\pm$ for $MgCl_2$ at a concentration of 0.020 M from the data in Table 6-1.

Problems

6-1. Which statements are true: In the ionic strength range 0–0.1 M, activity coefficients *decrease* with
(a) increasing ionic strength
(b) increasing ionic charge
(c) decreasing hydrated radius

6-2. Suggest a reason why the hydrated radii decrease in the order $Sn^{4+} > In^{3+} > Cd^{2+} > Rb^+$.

6-3. Calculate the ionic strength of
(a) 0.000 2 M $La(IO_3)_3$
(b) 0.02 M $CuSO_4$
(c) 0.01 M $CuCl_2$ + 0.05 M $(NH_4)_2SO_4$ + 0.02 M $Gd(NO_3)_3$

6-4. Find the activity coefficient of each ion at the indicated ionic strength.
(a) SO_4^{2-} ($\mu = 0.01$ M) (c) Eu^{3+} ($\mu = 0.1$ M)
(b) Sc^{3+} ($\mu = 0.005$ M) (d) $(CH_3CH_2)_3NH^+$ ($\mu = 0.05$ M)

6-5. Calculate the activity coefficient of Zn^{2+} when $\mu = 0.083$ M by
(a) using Equation 6-9
(b) using linear interpolation with Table 6-1

6-6. The observed mean activity coefficient ($\gamma_\pm$) for HCl at a concentration of 0.005 M is 0.93. What is the value of $\gamma_\pm$ calculated from Table 6-1?

6-7. The equilibrium constant for dissolution in water of a nonionic compound, such as diethyl ether $(CH_3CH_2OCH_2CH_3)$, can be written

$$\text{ether } (l) \rightleftharpoons \text{ether } (aq) \qquad K = [\text{ether}]\gamma_{\text{ether}}$$

At low ionic strength, $\gamma \approx 1$ for all nonionic compounds. At high ionic strength, ether and most other neutral molecules can be *salted out* of aqueous solution. That is, when a high concentration (typically > 1 M) of an ionic compound (such as NaCl) is added to an aqueous solution, neutral molecules usually become *less* soluble. Does the activity coefficient, γ_{ether}, increase or decrease at high ionic strength?

6-8. Calculate the solubility of Hg_2Br_2 (expressed as moles of Hg_2^{2+} per liter in
(a) 0.000 33 M $Mg(NO_3)_2$
(b) 0.003 3 M $Mg(NO_3)_2$
(c) 0.033 M $Mg(NO_3)_2$

6-9. Find the concentration of Ba^{2+} in a 0.033 3 M $Mg(IO_3)_2$ solution saturated with $Ba(IO_3)_2$.

7 / Systematic Treatment of Equilibrium

Many chemical equilibrium problems are exceedingly complex because of the large number of chemical reactions and species involved. This text focuses on some of the most common (and simplest) types of equilibria. Even so, there are times when we must resort to a powerful procedure that allows us to deal with a range of equilibrium problems from the simplest to the very difficult. In favorable cases, this *systematic treatment of equilibrium* can be applied with little arithmetic difficulty. In complicated cases, computers must be employed to solve the equations.

The basic procedure in the systematic treatment is to write as many algebraic equations as there are unknowns (species) in the problem. Having n equations and n unknowns, the problem can, in principle, be solved. Sometimes the solution is simple, but more often it is not. The n equations are generated by writing all of the chemical equilibrium conditions plus two more: the balances of charge and of mass. The following two sections describe these latter relationships.

7-1 CHARGE BALANCE

The **charge balance** is an algebraic statement of electroneutrality of the solution. That is, *the sum of the positive charges in solution equals the sum of the negative charges in solution.*

Consider a sulfate ion with concentration 0.016 7 M. Since the charge on SO_4^{2-} is -2, the charge contributed by sulfate is $(-2)(0.016\ 7) = -0.033\ 4$ M. In general, an ion with a charge of $\pm n$ and a concentration $[A]$ will contribute $\pm n[A]$ to the charge of the solution. The charge balance equates the magnitude of the total positive charge to the magnitude of the total negative charge.

Solutions must have zero total charge.

94

7 / SYSTEMATIC TREATMENT OF EQUILIBRIUM

The coefficient in each term in the charge balance equals the magnitude of the charge on each ion.

Suppose that a solution contains the following ionic species: H^+, OH^-, K^+, $H_2PO_4^-$, HPO_4^{2-}, and PO_4^{3-}. The charge balance is written

$$[H^+]+[K^+]=[OH^-]+[H_2PO_4^-]+2[HPO_4^{2-}]+3[PO_4^{3-}] \quad (7\text{-}1)$$

This statement says that the total charge contributed by H^+ and K^+ equals the magnitude of the charge contributed by all of the anions on the right side of the equation. *The coefficient in front of each species always equals the magnitude of the charge on the ion.* This is because a mole of, say, PO_4^{3-} contributes three moles of negative charge. If $[PO_4^{3-}] = 0.01$ M, the negative charge is $3[PO_4^{3-}] = 3(0.01) = 0.03$ M.

The charge balance equation (7-1) appears unbalanced to many people. "The right side of the equation has much more charge than the left side!" you may think. But that is wrong.

For example, consider a solution prepared by weighing out 0.025 0 mol of KH_2PO_4 plus 0.030 0 mol of KOH and diluting to 1.00 L. The concentrations of the species at equilibrium are calculated to be

$$[H^+] = 3.9 \times 10^{-12} \text{ M} \qquad [H_2PO_4^-] = 1.4 \times 10^{-6} \text{ M}$$

$$[K^+] = 0.055\,0 \text{ M} \qquad [HPO_4^{2-}] = 0.022\,6 \text{ M}$$

$$[OH^-] = 0.002\,6 \text{ M} \qquad [PO_4^{3-}] = 0.002\,4 \text{ M}$$

This calculation, which you should be able to do when you have finished studying acids and bases, takes into account the reaction of OH^- with $H_2PO_4^-$ to produce HPO_4^{2-} and PO_4^{3-}.

Are the charges balanced? Yes indeed. Plugging into Equation 7-1, we find

$$[H^+]+[K^+]=[OH^-]+[H_2PO_4^-]+2[HPO_4^{2-}]+3[PO_4^{3-}]$$

$$3.9 \times 10^{-12}+0.055\,0=0.002\,6+1.4 \times 10^{-6}+2(0.022\,6)+3(0.002\,4) \quad (7\text{-}2)$$

$$0.055\,0=0.055\,0$$

For the force between beakers of "charged solutions," see Problem 7-13.

The total positive charge (to three figures) is 0.055 0 M, and the total negative charge is also 0.055 0 M. The charges must be balanced in every solution. Otherwise your beaker with excess positive charge would glide across the lab bench and smash into another beaker with excess negative charge.

The general form of the charge balance for any solution is

$\sum$ [positive charges] $= \sum$ [negative charges]. *Activity coefficients do not appear in the charge balance.* The charge contributed by 0.1 M H^+ is *exactly* 0.1 M. Think about this.

$$n_1[C_1] + n_2[C_2] + \cdots = m_1[A_1] + m_2[A_2] + \cdots \quad (7\text{-}3)$$

where $[C_i]$ = concentration of the ith cation

$\qquad n_i$ = charge of the ith cation

$\qquad [A_i]$ = concentration of the ith anion

$\qquad m_i$ = magnitude of the charge of the ith anion

EXAMPLE

Write the charge balance for a solution containing H_2O, H^+, OH^-, ClO_4^-, $Fe(CN)_6^{3-}$, CN^-, Fe^{3+}, Mg^{2+}, CH_3OH, HCN, NH_3, and NH_4^+.

The correct equation is

$$[H^+] + 3[Fe^{3+}] + 2[Mg^{2+}] + [NH_4^+] = [OH^-] + [ClO_4^-] + 3[Fe(CN)_6^{3-}] + [CN^-]$$

Neutral species (H_2O, HCN, CH_3OH, and NH_3) do not appear in the charge balance.

7-2 MASS BALANCE

The **mass balance,** also called the material balance, is a statement of the conservation of matter. The mass balance states that *the sum of the amounts of all species containing a particular atom (or group of atoms) must equal the amount of that atom (or group) delivered to the solution.* It is easier to see this through particular examples than by a general statement.

Suppose that a solution is prepared by dissolving 0.050 mol of acetic acid in water to give a total volume of 1.00 L. The acetic acid will partially dissociate into acetate:

$$\underset{\text{acetic acid}}{CH_3CO_2H} \rightleftharpoons \underset{\text{acetate}}{CH_3CO_2^-} + H^+ \qquad (7\text{-}4)$$

The mass balance is a statement of the conservation of matter. It really refers to conservation of atoms, not to mass.

The mass balance is simply a statement that the sum of the amount of dissociated and undissociated acid must equal the amount of acid put into the solution.

$$\text{Mass balance:} \quad 0.050 \text{ M} = [CH_3CO_2H] + [CH_3CO_2^-] \qquad (7\text{-}5)$$

Phosphoric acid (H_3PO_4) can dissociate into $H_2PO_4^-$, HPO_4^{2-}, and PO_4^{3-}. The mass balance for a solution prepared by dissolving 0.025 0 moles of H_3PO_4 in 1.00 L is

$$0.025\,0 \text{ M} = [H_3PO_4] + [H_2PO_4^-] + [HPO_4^{2-}] + [PO_4^{3-}] \qquad (7\text{-}6)$$

Activity coefficients do not appear in the mass balance. The concentration of each species gives an exact count of the number of atoms of that species.

EXAMPLE

Write the mass balances for K^+ and for phosphate in a solution prepared by mixing 0.025 0 mol KH_2PO_4 plus 0.030 0 mol KOH and diluting to 1.00 L.

The total concentration of K^+ is 0.025 0 M + 0.030 0 M, so one trivial mass balance is

$$[K^+] = 0.055\,0 \text{ M}$$

The total concentration of *all forms* of phosphate is 0.025 0 M. The mass balance for phosphate is

$$[H_3PO_4] + [H_2PO_4^-] + [HPO_4^{2-}] + [PO_4^{3-}] = 0.025\,0 \text{ M}$$

7 / SYSTEMATIC TREATMENT OF EQUILIBRIUM

Now consider a solution prepared by dissolving $La(IO_3)_3$ in water:

$$La(IO_3)_3(s) \xrightleftharpoons{K_{sp}} La^{3+} + 3IO_3^- \tag{7-7}$$

iodate

We do not know how much La^{3+} or IO_3^- is dissolved, but we do know that there must be three iodate ions for each lanthanum ion dissolved. The mass balance is

$$[IO_3^-] = 3[La^{3+}] \tag{7-8}$$

We have already used this sort of relation in solubility problems. Perhaps the following table will jog your memory:

	$La(IO_3)_3(s) \rightleftharpoons$	La^{3+}	$+ \ 3IO_3^-$
Initial concentration:	solid	0	0
Final concentration:	solid	x	$3x$

When we set $[IO_3^-] = 3x$, we are saying that $[IO_3^-] = 3[La^{3+}]$.

EXAMPLE

Write the mass balance for a saturated solution of the slightly soluble salt Ag_3PO_4, which produces PO_4^{3-} plus $3Ag^+$ when it dissolves.

If the phosphate in solution remained as PO_4^{3-}, we could write

$$[Ag^+] = 3[PO_4^{3-}]$$

because three silver ions are produced for each phosphate ion. However, since phosphate reacts with water to give HPO_4^{2-}, $H_2PO_4^-$, and H_3PO_4, the correct mass balance is

Atoms of Ag = 3 (atoms of P).

$$[Ag^+] = 3\{[PO_4^{3-}] + [HPO_4^{2-}] + [H_2PO_4^-] + [H_3PO_4]\}$$

That is, the number of atoms of Ag^+ must equal three times the total number of atoms of phosphorus, regardless of how many species contain phosphorus atoms.

7-3 SYSTEMATIC TREATMENT OF EQUILIBRIUM

General Prescription

Step 1. Write all the pertinent chemical reactions.

Step 2. Write the charge balance.

Step 3. Write the mass balance.

Step 4. Write the equilibrium constant for each chemical reaction. This is the only step in which activity coefficients figure.

Step 5. Count the equations and unknowns. At this point you should have as many equations as unknowns (chemical species). If not, you must either write more equilibria or fix some concentrations at known values.

Step 6. By hook or by crook, solve for all of the unknowns.

Steps 1 and 6 are usually the heart of the problem. Knowing (or guessing) what chemical equilibria exist in a given solution requires a fair degree of chemical intuition. In this text you will usually be given some help with Step 1. Unless we know all of the relevant equilibria, it is not possible to correctly calculate the composition of a solution. Because of not knowing all the chemical reactions, we undoubtedly oversimplify many equilibrium problems.

Step 6 is a mathematical problem, not a chemical problem. In some cases it is easy to solve for all of the unknowns, but for most problems a computer must be employed and/or approximations must be made. How the systematic treatment of equilibrium works is best understood by studying some examples.

Ionization of Water

The dissociation of water into H^+ and OH^- occurs in every aqueous solution, and the equilibrium constant is called K_w.

$$H_2O \xrightleftharpoons{K_w} H^+ + OH^- \qquad K_w = 1.0 \times 10^{-14} \text{ at } 25°C \qquad (7\text{-}9)$$

Let us apply the systematic treatment of equilibrium to find the concentrations of H^+ and OH^- in pure water.

Step 1. Pertinent reactions. The only reaction is 7-9.

Step 2. Charge balance. The only ions are H^+ and OH^-, so the charge balance is

$$[H^+] = [OH^-] \qquad (7\text{-}10)$$

Step 3. Mass balance. In Reaction 7-9 we see that one H^+ ion is generated each time one OH^- ion is made. The mass balance is simply

$$[H^+] = [OH^-]$$

which is the same as the charge balance for this particularly trivial system.

Step 4. Write the equilibrium constants. The only one is

$$K_w = [H^+]\gamma_{H^+}[OH^-]\gamma_{OH^-} = 1.0 \times 10^{-14} \qquad (7\text{-}11)$$

This is the only step in which activity coefficients are introduced into the problem.

98

7 / SYSTEMATIC TREATMENT OF EQUILIBRIUM

It requires n equations to solve for n unknowns.

Step 5. Count the equations and unknowns. We have two equations, 7-10 and 7-11, and two unknowns, $[H^+]$ and $[OH^-]$.

Step 6. Solve.

Now we must stop and decide what to do about the activity coefficients. Later, it will be our custom to ignore them unless the calculation requires an accurate result. In the present problem, we anticipate that the ionic strength of pure water will be very low (since the only ions are small amounts of H^+ and OH^-). Therefore, it is reasonable to suppose that γ_{H^+} and γ_{OH^-} are both unity, since $\mu \approx 0$.

Recall that γ approaches 1 as μ approaches 0.

Putting the equality $[H^+] = [OH^-]$ into Equation 7-11 gives

$$[H^+]\gamma_{H^+}[OH^-]\gamma_{OH^-} = 1.0 \times 10^{-14}$$

$$[H^+] \cdot 1 \cdot [H^+] \cdot 1 = 1.0 \times 10^{-14} \tag{7-12}$$

$$[H^+] = 1.0 \times 10^{-7} \text{ M}$$

Since $[H^+] = [OH^-]$, $[OH^-] = 1.0 \times 10^{-7}$ M also. We have solved the problem.

The ionic strength is 10^{-7} M. The assumption that $\gamma_{H^+} = \gamma_{OH^-} = 1$ is good.

At this point we should define the term **pH** which you have undoubtedly used before. The correct definition of pH is

Correct definition of pH.

$$pH \equiv -\log \mathscr{A}_{H^+} = -\log([H^+]\gamma_{H^+}) \tag{7-13}$$

Note that pH is the negative logarithm of the *activity* of H^+, not the negative logarithm of the concentration of H^+. A pH meter responds to $\mathscr{A}_{H^+}$, not $[H^+]$.

In the problem we just solved,

$$pH = -\log \mathscr{A}_{H^+} = -\log[H^+]\gamma_{H^+}$$
$$= -\log(1.0 \times 10^{-7})(1) \tag{7-14}$$
$$= 7.00$$

The pH of pure water is 7.00 at 25°C.

We will have more to say about pH in Chapter 10. For now it suffices to remind you that solutions with pH < 7 are **acidic** and solutions with pH > 7 are **basic**. A low pH means a high concentration of H^+. A high pH means a low concentration of H^+.

A note about activity coefficients

While it is proper to write all equilibrium constants in terms of activities, the algebraic complexity of manipulating the activity coefficients often obscures the chemistry of a problem. For the remainder of this text we will omit activity coefficients unless there is a particular point to be made with them. There will be occasional problems in which activity is used. This provides you with an occasional reminder of the activity concept. Alternatively, there is no loss in continuity, even if the activity problems are skipped.

By neglecting activity coefficients, we are in effect assigning the value unity to all activity coefficients. This makes the working definition of pH

$$pH = -\log[H^+] \tag{7-15}$$

Our usual working definition of pH.

According to Equation 7-15, the pH of $10^{-2.00}$ M H^+ is 2.00.

Solubility of Hg_2Cl_2

Let's study one application of the systematic treatment of equilibrium by calculating the concentration of Hg_2^{2+} in a saturated solution of Hg_2Cl_2.

Step 1. Pertinent reactions. The two reactions that come to mind are

$$Hg_2Cl_2(s) \xrightarrow{K_{sp}} Hg_2^{2+} + 2Cl^- \tag{7-16}$$

$$H_2O \xrightarrow{K_w} H^+ + OH^- \tag{7-17}$$

The equilibrium, 7-17, exists in every aqueous solution.

Step 2. Charge balance. Equating positive and negative charges gives

$$[H^+] + 2[Hg_2^{2+}] = [Cl^-] + [OH^-] \tag{7-18}$$

Multiply $[Hg_2^{2+}]$ by two because one mole of this ion has two moles of charge.

Step 3. Mass balance. There are two mass balances in this system. One is the trivial statement that $[H^+] = [OH^-]$, since both arise only from the ionization of water. If there were any other reactions involving H^+ or OH^-, we could not automatically say $[H^+] = [OH^-]$. A slightly more interesting mass balance is

$$[Cl^-] = 2[Hg_2^{2+}] \tag{7-19}$$

Two Cl^- ions are produced for each Hg_2^{2+} ion.

Step 4. The equilibrium constants are

$$[Hg_2^{2+}][Cl^-]^2 = 1.2 \times 10^{-18} \tag{7-20}$$

$$[H^+][OH^-] = 1.0 \times 10^{-14} \tag{7-21}$$

We have neglected the activity coefficients in these equations.

Step 5. There are four equations (7-18 to 7-21) and four unknowns: $[H^+]$, $[OH^-]$, $[Hg_2^{2+}]$, and $[Cl^-]$.

Step 6. Since we have not written any chemical reactions between the ions produced by H_2O and the ions produced by Hg_2Cl_2, there are really two separate problems. One is the trivial problem of water ionization, which we already solved:

$$[H^+] = [OH^-] = 1.0 \times 10^{-7} \text{ M} \tag{7-22}$$

A slightly more interesting problem is the Hg_2Cl_2 equilibrium. Noting that $[Cl^-] = 2[Hg_2^{2+}]$, we can write

100

7 / SYSTEMATIC TREATMENT OF EQUILIBRIUM

$$[Hg_2^{2+}][Cl^-]^2 = [Hg_2^{2+}](2[Hg_2^{2+}])^2 = K_{sp} \qquad (7\text{-}23)$$

$$[Hg_2^{2+}] = (K_{sp}/4)^{1/3} = 6.7 \times 10^{-7} \text{ M}$$

Satisfy yourself that the calculated concentrations fulfil the charge balance condition.

This result is exactly what you would have found by writing a table of concentrations and solving by the methods of Chapter 5.

7-4 DEPENDENCE OF SOLUBILITY ON pH

We are now in a position to study some cases that are not trivial. In this section we present two examples in which there are *coupled equilibria*. That is, the product of one reaction is a reactant in the next reaction.

Solubility of CaF_2

Let's set up the equations needed to find the solubility of CaF_2 in water. There are three pertinent reactions in this system. First $CaF_2(s)$ dissolves:

$$CaF_2(s) \xrightleftharpoons{K_{sp}} Ca^{2+} + 2F^- \qquad (7\text{-}24)$$

The fluoride ion can then react with water to give $HF(aq)$:

$$F^- + H_2O \xrightleftharpoons{K_B} HF(aq) + OH^- \qquad (7\text{-}25)$$

The equilibrium constant is called K_B because F^- is behaving as a **B**ase when it removes H^+ from H_2O.

Finally, for every aqueous solution we can write

$$H_2O \xrightleftharpoons{K_w} H^+ + OH^- \qquad (7\text{-}26)$$

If Reaction 7-25 occurs, the solubility of CaF_2 is greater than what is predicted by the solubility product. The reason is that a product of Reaction 7-24 is consumed in Reaction 7-25. According to Le Châtelier's principle, Reaction 7-24 will be driven to the right. The systematic treatment of equilibrium allows us to find the net effect of all three reactions.

Step 1. The pertinent reactions are 7-24 through 7-26.

Step 2. Charge balance.

$$[H^+] + 2[Ca^{2+}] = [OH^-] + [F^-] \qquad (7\text{-}27)$$

Step 3. Mass balance. If all fluoride remained in the form F^-, we could write $[F^-] = 2[Ca^{2+}]$ from the stoichiometry of Reaction 7-24. But some F^- reacts to give HF. The total moles of fluorine atoms is equal to the sum of F^- plus HF. The mass balance is

$$\underbrace{[F^-] + [HF]}_{\substack{\text{Total concentration} \\ \text{of fluorine}}} = 2[Ca^{2+}] \qquad (7\text{-}28)$$

Step 4. Equilibria: $K_{sp} = [Ca^{2+}][F^-]^2 = 3.9 \times 10^{-11}$ (7-29)

$$K_B = \frac{[HF][OH^-]}{[F^-]} = 1.5 \times 10^{-11}$$ (7-30)

$$K_w = [H^+][OH^-] = 1.0 \times 10^{-14}$$ (7-31)

Step 5. We have five equations (7-27 through 7-31) and five unknowns:

$[H^+], [OH^-], [Ca^{2+}], [F^-]$, and $[HF]$.

The final step is to solve the problem, which is no simple matter for these five equations. Instead, let us ask a simpler question: "What will be the concentrations of $[Ca^{2+}]$, $[F^-]$, and $[HF]$ if the pH were somehow *fixed* at the value 3.00?" To fix the pH at 3.00 means that $[H^+] = 1.0 \times 10^{-3}$ M.

Once we know the value of $[H^+]$, a simple procedure for solving all of the equations is the following: Setting $[H^+] = 1.0 \times 10^{-3}$ M in Equation 7-31 gives

$$[OH^-] = \frac{K_w}{[H^+]} = 1.0 \times 10^{-11}$$ (7-32)

Putting this value of $[OH^-]$ into Equation 7-30 gives

$$\frac{[HF]}{[F^-]} = \frac{K_B}{[OH^-]} = 1.5$$

$$[HF] = 1.5\,[F^-]$$ (7-33)

Substituting this expression for $[HF]$ into the mass balance (7-28) gives

$$[F^-] + [HF] = 2[Ca^{2+}]$$

$$[F^-] + 1.5[F^-] = 2[Ca^{2+}]$$

$$[F^-] = 0.80[Ca^{2+}]$$ (7-34)

Finally, we use this value of $[F^-]$ in the solubility product (7-29):

$$[Ca^{2+}][F^-]^2 = K_{sp}$$

$$[Ca^{2+}](0.80[Ca^{2+}])^2 = K_{sp}$$

$$[Ca^{2+}] = 3.9 \times 10^{-4}\,\text{M}$$ (7-35)

A graph of the pH-dependence of the concentrations of Ca^{2+}, F^-, and HF is shown in Figure 7-1. At high pH there is very little HF, so $[F^-] \approx 2[Ca^{2+}]$. At low pH there is very little F^-, so $[HF] \approx 2[Ca^{2+}]$. The concentration of Ca^{2+} increases at low pH, because Reaction 7-24 is drawn to

If you insist on solving the problem, see Box 7-1.

The pH could be fixed at a desired value by adding a buffer, which is discussed in Chapter 11.

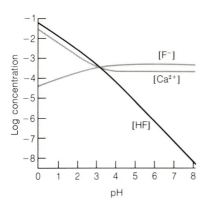

Figure 7-1
pH-dependence shown by the concentrations of Ca^{2+}, F^-, and HF in a saturated solution of CaF_2. Note the logarithmic scale.

Challenge: Use the concentration of Ca^{2+} that we just calculated to show that the concentrations of $[F^-]$ and $[HF^-]$ are 3.1×10^{-4} M and 4.7×10^{-4} M, respectively.

High pH means there is a low concentration of H^+. Low pH means that there is a high concentration of H^+.

Box 7-1 ALL RIGHT, DAN, HOW WOULD YOU REALLY SOLVE THE CaF₂ PROBLEM?

Suppose that we did not simplify the CaF_2 problem by specifying a fixed pH. How can we find the composition of the system if it simply contains CaF_2 dissolved in water? A systematic guessing procedure is a pretty good way to do it.

First let's use a little intuition. F^- is a pretty weak base ($K_B = 1.5 \times 10^{-11}$, Equation 7-25) and there will not be very much F^- dissolved (since $K_{sp} = 3.9 \times 10^{-11}$, Equation 7-24). Therefore, a first guess is that the pH is in the neighborhood of 7.

Guessing that the pH is 7.00, we can calculate the concentration of each species and see if the charge balance equation (7-27) is satisfied. It is valid to use the charge balance now because we have not fixed the pH by addition of some other reagent. Assuming pH = 7.00, the concentrations of Ca^{2+} and F^- are calculated to be 2.14×10^{-4} and 4.27×10^{-4} M, respectively.

$$[H^+] + 2[Ca^{2+}] = [OH^-] + [F^-] \qquad (7\text{-}27)$$

$$10^{-7} + 2(2.14 \times 10^{-4}) > 10^{-7} + 4.27 \times 10^{-4}$$

A pH of 7.00 almost satisfies the charge balance. Raising the pH will increase the right side of the equation above and bring it into balance. A little trial-and-error guessing shows that a pH of 7.11 comes closest to satisfying the charge balance. When the charge balance is satisfied we *must* have found the correct pH and composition of the solution. Although this procedure is cumbersome to do by hand, it is well-suited to a computer.

---

the right by the reaction of F^- with H^+ to make HF. It is a general phenomenon that salts of basic ions will become more soluble at low pH, because the basic ions react with H^+. Some examples of basic ions are F^-, OH^-, S^{2-}, CO_3^{2-}, $C_2O_4^{2-}$, and PO_4^{3-}. For an example using the interaction of pH, solubility, and tooth decay, see Box 7-2.

Finally, you should realize that the charge balance equation (7-27) is no longer valid if the pH is fixed by external means. To adjust the pH, an ionic compound must necessarily have been added to the solution. Equation 7-27 is therefore incomplete, since it omits those ions. We did not use Equation 7-27 to solve the problem.

Fixing the pH invalidates the original charge balance. There exists a new charge balance, but we do not know enough to write an equation expressing this fact.

Solubility of HgS

The mineral cinnabar consists of red HgS. When it dissolves in water, some reactions that may occur are

$$HgS(s) \rightleftharpoons Hg^{2+} + S^{2-} \qquad K_{sp} = 5 \times 10^{-54} \qquad (7\text{-}36)$$

$$S^{2-} + H_2O \rightleftharpoons HS^- + OH^- \qquad K_{B1} = 0.80 \qquad (7\text{-}37)$$

$$HS^- + H_2O \rightleftharpoons H_2S(aq) + OH^- \qquad K_{B2} = 1.1 \times 10^{-7} \qquad (7\text{-}38)$$

$$H_2O \rightleftharpoons H^+ + OH^- \qquad K_w = 1.0 \times 10^{-14} \qquad (7\text{-}39)$$

We are ignoring the equilibrium $H_2S(aq) \rightleftharpoons H_2S(g)$.

Box 7-2 pH AND TOOTH DECAY

The enamel covering of teeth contains the mineral *hydroxyapatite*, a calcium hydroxyphosphate. This slightly soluble mineral will dissolve in acid, because both the PO_4^{3-} and OH^- react with H^+:

$$Ca_{10}(PO_4)_6(OH)_2 + 14H^+ \rightleftharpoons 10Ca^{2+} + 6H_2PO_4^- + 2H_2O$$

$\qquad\qquad$ hydroxyapatite

The decay-causing bacteria that adhere to teeth produce lactic acid from the metabolism of sugar.

$$\underset{\text{}}{CH_3\overset{OH}{\underset{|}{C}}HCO_2H} \quad \text{lactic acid}$$

The lactic acid lowers the pH at the surface of the tooth to less than 5. At any pH below about 5.5, hydroxyapatite dissolves and tooth decay occurs, as shown in the electron micrographs below:

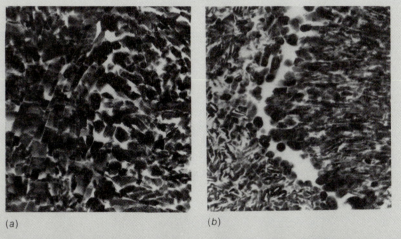

(a) (b)

(a) Transmission electron micrograph of normal human tooth enamel, showing crystals of hydroxyapatite. (b) Transmission electron micrograph of decayed enamel, showing regions where the mineral has been dissolved by acid. [Courtesy D. B. Scott, J. W. Simmelink, and V. K. Nygaard, Case Western Reserve University, School of Dentistry. Originally published in *J. Dent. Res.*, **53,** 165(1974).]

Fluoride inhibits tooth decay because it forms *fluorapatite*, $Ca_{10}(PO_4)_6F_2$, which is less soluble and more acid-resistant than hydroxyapatite.

104

7 / SYSTEMATIC TREATMENT OF EQUILIBRIUM

Because S^{2-} is a strong base, it reacts with H_2O to give HS^-, thereby drawing Reaction 7-36 to the right and increasing the solubility of HgS. Now let's set up the equations to find the composition of a saturated solution of HgS.

Step 1. The pertinent reactions are 7-36 through 7-39. It is worth restating that if there are any other significant reactions, the calculated composition will be wrong. We are necessarily limited by our knowledge of the chemistry of the system.

Step 2. Charge balance. If the pH is not adjusted by external means, the charge balance is

$$2[Hg^{2+}] + [H^+] = 2[S^{2-}] + [HS^-] + [OH^-] \qquad (7\text{-}40)$$

Step 3. Mass balance. For each atom of Hg in solution, there is one atom of S. If all of the S remained as S^{2-}, the mass balance would be $[Hg^{2+}] = [S^{2-}]$. But S^{2-} reacts with water to give HS^- and H_2S. The mass balance is

$$[Hg^{2+}] = [S^{2-}] + [HS^-] + [H_2S] \qquad (7\text{-}41)$$

Step 4. Equilibria.

$$K_{sp} = [Hg^{2+}][S^{2-}] = 5 \times 10^{-54} \qquad (7\text{-}42)$$

$$K_{B_1} = \frac{[HS^-][OH^-]}{[S^{2-}]} = 0.80 \qquad (7\text{-}43)$$

$$K_{B_2} = \frac{[H_2S][OH^-]}{[HS^-]} = 1.1 \times 10^{-7} \qquad (7\text{-}44)$$

$$K_w = [H^+][OH^-] = 1.0 \times 10^{-14} \qquad (7\text{-}45)$$

Step 5. There are six equations (7-40 through 7-45) and six unknowns: $[Hg^{2+}]$, $[S^{2-}]$, $[HS^-]$, $[H_2S]$, $[H^+]$, and $[OH^-]$.

Step 6. Solve.

Once again the problem is very difficult to solve. We will simplify matters by assuming that the pH is *fixed* at 8.00 by some external means. This means that the charge balance (7-40) is invalid, but we can use the remaining equations to find the composition of the system.

If the pH is 8.00, Equation 7-45 tells us that $[OH^-] = 1.0 \times 10^{-6}$ M. Putting this value into Equation 7-44 gives

The charge balance is invalid because we are adding new ions to adjust the pH to 8.00.

$$[H_2S] = \frac{K_{B2}[HS^-]}{[OH^-]} = 0.11[HS^-] \qquad (7\text{-}46)$$

Substituting the value of $[OH^-]$ into Equation 7-43 gives

$$[HS^-] = \frac{K_{B1}[S^{2-}]}{[OH^-]} = 8.0 \times 10^5[S^{2-}] \qquad (7\text{-}47)$$

Now we can use these values of $[H_2S]$ and $[HS^-]$ in the mass balance (7-41) to write

$$[Hg^{2+}] = [S^{2-}] + [HS^-] + [H_2S]$$

$$[Hg^{2+}] = [S^{2-}] + 8.0 \times 10^5[S^{2-}] + 0.11[HS^-]$$

$$[Hg^{2+}] = [S^{2-}] + 8.0 \times 10^5[S^{2-}] + (0.11)(8.0 \times 10^5)[S^{2-}]$$

$$[Hg^{2+}] = [S^{2-}](8.88 \times 10^5) \qquad (7\text{-}48)$$

Putting this relation between $[Hg^{2+}]$ and $[S^{2-}]$ into the solubility product solves the problem:

$$K_{sp} = [Hg^{2+}][S^{2-}]$$

$$K_{sp} = [Hg^{2+}]\left(\frac{[Hg^{2+}]}{8.88 \times 10^5}\right) \Rightarrow [Hg^{2+}] = 2.1 \times 10^{-24} \text{ M} \qquad (7\text{-}49)$$

A graph showing the effect of pH on the composition of a saturated solution of HgS is shown in Figure 7-2. Below pH ≈ 6, H_2S is the predominant form of sulfur in solution, so $[Hg^{2+}] \approx [H_2S]$. Above pH ≈ 8, HS^- is the predominant form, and $[Hg^{2+}] \approx [HS^-]$. The solubility of HgS increases as the pH is lowered since S^{2-} reacts with H^+, drawing Reaction 7-36 to the right.

We can use the procedure above to find the composition of the solution at any given pH. If we happen to guess the true pH of the system, then the charge balance (7-40) will be satisfied. Since HgS has such a small solubility product, so little will be dissolved that it will have very little effect on the pH of the water. A reasonable first guess for the pH is 7.00. Successive guesses with the aid of a computer show that the pH of a saturated solution of HgS is, in fact, 7.00.

Challenge: Show that the other concentrations are $[S^{2-}] = 2.4 \times 10^{-30}$ M, $[HS^-] = 1.9 \times 10^{-24}$ M, and $[H_2S] = 2.1 \times 10^{-25}$ M.

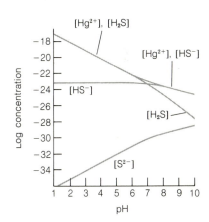

Figure 7-2
pH-dependence shown by the concentrations of species in a saturated solution of HgS. Note the logarithmic scale.

Recapitulation

In the previous section we dealt with the solubility of salts of the type MA, where M is a metal and A is a basic anion. A can react with water to give HA, H_2A, ..., etc. A general procedure for finding the composition of the system at a given pH is the following:

Step 1. Set up all of the equations (except the charge balance) using the systematic treatment of equilibrium.

Step 2. Use the concentration of OH^- to write expressions for each protonated form of the anion (H_nA) in terms of A.

Step 3. Substitute the values of H_nA from Step 2 into the mass balance to derive a relation between M and A.

Step 4. Put the relation between M and A into the solubility product to solve for the concentrations of M and A and of all other species.

106

7 / SYSTEMATIC TREATMENT OF EQUILIBRIUM

Comments

Although the systematic treatment is always a valid approach to equilibrium problems, it is not always the handiest approach. In this text we will use whatever method proves to be most convenient. It will be extremely helpful to you to learn the shortcuts as they are introduced. When all else fails, the systematic treatment cannot lead you astray, even though it may not be the easiest method.

In all approaches to all equilibrium problems, we are ultimately limited by how much of the system's chemistry is understood. Unless we know all of the relevant equilibria, it is not possible to correctly calculate the composition of a solution. From ignorance of some of the chemical reactions, we undoubtedly oversimplify many equilibrium problems.

Summary

In the systematic treatment of equilibrium, we write down all of the pertinent equilibrium expressions, as well as the charge and mass balances. The charge balance states that the sum of all positive charge in solution equals the sum of all negative charge. The mass balance states that the sum of the moles of all forms of an element in solution must equal the moles of that element delivered to the solution. We make certain that we have as many equations as unknowns, then set about to solve for the concentration of each species, using algebra, insight, approximations, computers, magic, or anything else. We applied this procedure to salts of basic anions and discovered that their solubilities increase at low pH because the anion is protonated in acidic solution.

Terms to Understand

acidic solution

basic solution

charge balance

mass balance

pH

Exercises

7-A. Write the charge balance for a solution prepared by dissolving CaF_2 in H_2O. Consider that CaF_2 may give Ca^{2+}, F^-, and CaF^+.

7-B. (a) Write the mass balance for a solution of $CaCl_2$ in water, where the aqueous species are Ca^{2+} and Cl^-.

(b) Write a mass balance where the aqueous species are Ca^{2+}, Cl^-, and $CaCl^+$.

7-C. (a) Write a mass balance for a saturated solution of CaF_2 in water in which the reactions are

$$CaF_2(s) \rightleftharpoons Ca^{2+} + 2F^-$$

$$F^- + H^+ \rightleftharpoons HF(aq)$$

(b) Write a mass balance for CaF_2 in water where, in addition to the previous equations, the following reaction occurs:

$$HF(aq) + F^- \rightleftharpoons HF_2^-$$

7-D. Write a mass balance for an aqueous solution of $Ca_3(PO_4)_2$, where the aqueous species are Ca^{2+}, PO_4^{3-}, HPO_4^{2-}, $H_2PO_4^-$, and H_3PO_4.

7-E. (a) Find the concentrations of Ag^+, CN^-, and HCN in a saturated solution of AgCN whose pH is somehow fixed at 9.00. Consider the following equilibria:

$$AgCN(s) \rightleftharpoons Ag^+ + CN^-$$

$$K_{sp} = 2.2 \times 10^{-16}$$

$$CN^- + H_2O \rightleftharpoons HCN(aq) + OH^-$$

$$K_B = 1.6 \times 10^{-5}$$

(b) *Activity problem:* Answer the above question, but using activity coefficients to solve it correctly. Assume that the ionic strength is fixed at 0.10 M by addition of an inert salt. When using activities, the statement that the pH is 9.00 means that $-\log[H^+]\gamma_{H^+} = 9.00$.

7-F. Calculate the solubility of ZnC_2O_4 (g/L) in a solution held at pH 3.00. Consider the equilibria

$$ZnC_2O_4(s) \rightleftharpoons Zn^{2+} + C_2O_4^{2-}$$

oxalate

$$K_{sp} = 7.5 \times 10^{-9}$$

$$C_2O_4^{2-} + H_2O \rightleftharpoons HC_2O_4^- + OH^-$$

$$K_{B1} = 1.8 \times 10^{-10}$$

$$HC_2O_4^- + H_2O \rightleftharpoons H_2C_2O_4 + OH^-$$

$$K_{B2} = 1.8 \times 10^{-13}$$

Problems

7-1. State in words the meaning of the charge balance equation.

7-2. State in words the meaning of a mass balance equation.

7-3. Why are activity coefficients excluded from the charge and mass balances?

7-4. Write a charge balance for a solution containing H^+, OH^-, Ca^{2+}, HCO_3^-, CO_3^{2-}, $Ca(HCO_3)^+$, $Ca(OH)^+$, K^+, and ClO_4^-.

7-5. Write a charge balance for a solution of H_2SO_4 in water if the H_2SO_4 ionizes to HSO_4^- and SO_4^{2-}.

7-6. For a 0.1 F solution of sodium acetate (Na^+ $CH_3CO_2^-$) in water, one mass balance is simply $[Na^+] = 0.1$ M. Write a mass balance involving acetate.

7-7. Calculate the concentration of each ion in a solution of 4.0×10^{-8} M $Mg(OH)_2$, which is completely dissociated to Mg^{2+} and OH^-.

7-8. Calculate the concentrations of Ca^{2+}, F^-, and HF in a saturated aqueous solution of CaF_2 held at pH 2.00.

7-9. A certain metal salt of acrylic acid has the formula $M(H_2C\!\!=\!\!CHCO_2)_2$. Find the concentration of M^{2+} in a saturated aqueous solution of this salt in which $[OH^-]$ is maintained at the value 1.8×10^{-10} M. The equilibria are

$$M(CH_2\!\!=\!\!CHCO_2)_2(s) \rightleftharpoons M^{2+} + 2H_2C\!\!=\!\!CHCO_2^-$$

$$K_{sp} = 6.3 \times 10^{-14}$$

$$H_2C\!\!=\!\!CHCO_2^- + H_2O \rightleftharpoons H_2C\!\!=\!\!CHCO_2H + OH^-$$

acrylic acid

$$K_B = 1.8 \times 10^{-10}$$

7-10. Consider a saturated aqueous solution of the slightly soluble salt $R_3NH^+Br^-$, where R is an organic group.

$$R_3NH^+Br^-(s) \rightleftharpoons R_3NH^+ + Br^- \quad K_{sp} = 4.0 \times 10^{-8}$$

$$R_3NH^+ \rightleftharpoons R_3N + H^+ \quad K_A = 2.3 \times 10^{-9}$$

Calculate the solubility (moles per liter) of $R_3NH^+Br^-$ in a solution maintained at pH 9.50.

7-11. (a) Use the systematic treatment of equilibrium to find how many moles of PbO will dissolve in a 1.00 L solution in which the pH is fixed at 10.50. Consider the equilibrium involving Pb^{2+} to be

$$PbO(s) + H_2O \rightleftharpoons Pb^{2+} + 2OH^-$$

$$K = 5.0 \times 10^{-16}$$

(b) Answer the same question as in part a, but also consider the reaction

$$Pb^{2+} + H_2O \rightleftharpoons PbOH^+ + H^+$$

$$K_A = 1.3 \times 10^{-8}$$

(c) *Activity problem:* Answer the same question as in part a, now using activity coefficients. Assume μ is fixed at 0.050 M.

7-12. Calculate the molarity of Ag^+ in a saturated aqueous solution of Ag_3PO_4 at pH 6.00 if the equilibria are

$$Ag_3PO_4(s) \rightleftharpoons 3Ag^+ + PO_4^{3-}$$

$$K_{sp} = 2.8 \times 10^{-18}$$

$$PO_4^{3-} + H_2O \rightleftharpoons HPO_4^{2-} + OH^-$$

$$K_{B1} = 2.3 \times 10^{-2}$$

$$HPO_4^{2-} + H_2O \rightleftharpoons H_2PO_4^- + OH^-$$

$$K_{B2} = 1.6 \times 10^{-7}$$

$$H_2PO_4^- + H_2O \rightleftharpoons H_3PO_4 + OH^-$$

$$K_{B3} = 1.4 \times 10^{-12}$$

7-13. This problem demonstrates what would happen if charge balance did not exist in a solution. The force (newtons) between two charges of magnitude q_1 and q_2 (coulombs) is given by

$$Force = -(8.988 \times 10^9) \frac{q_1 q_2}{r^2}$$

where r is the distance (meters) between the two charges. What is the force (in newtons and pounds) between two beakers separated by 1.5 m if one beaker contains 250 mL of solution with 1.0×10^{-6} M excess negative charge and the other has 250 mL of 1.0×10^{-6} M excess positive charge? Note that there are 9.648×10^4 coulombs per mole of charge and 0.224 8 pounds per newton.

8 / Gravimetric Analysis

Gravimetric analysis encompasses a variety of techniques in which the mass of a product is used to determine the quantity of the original analyte. Because mass can be measured very accurately, gravimetric methods are among the most accurate in analytical chemistry.

8-1 EXAMPLES OF GRAVIMETRIC ANALYSIS

A familiar example is the determination of Cl^- by precipitation with Ag^+.

$$Ag^+ + Cl^- \rightarrow AgCl(s) \tag{8-1}$$

The weight of AgCl produced tells us how much Cl^- was originally present.

EXAMPLE

A 10.00 mL solution containing Cl^- was treated with excess $AgNO_3$ to precipitate 0.436 8 g of AgCl. What was the molarity of Cl^- in the unknown?

The formula weight of AgCl is 143.321. A precipitate weighing 0.463 8 g contains

$$\frac{0.436\ 8 \text{ g of AgCl}}{143.321 \text{ g of AgCl/mol of AgCl}} = 3.048 \times 10^{-3} \text{ mol of AgCl}$$

Since one mole of AgCl contains one mole of Cl^-, there must have been 3.048×10^{-3} mol of Cl^- in the unknown.

$$[Cl^-] = \frac{3.048 \times 10^{-3} \text{ mol}}{0.010\ 00 \text{ L}} = 0.304\ 8 \text{ M}$$

---

8 / GRAVIMETRIC ANALYSIS

Exceedingly careful gravimetric analysis employing AgCl was used by T. W. Richards and his colleagues to determine the atomic weights of Ag, Cl, and N to six-figure accuracy.[†] This Nobel Prize-winning research formed the basis for accurate determination of the atomic weights of many other elements.

Combustion Analysis

Combustion methods are useful for the analysis of carbon, hydrogen, nitrogen, sulfur, and halogens.

The most widely used form of gravimetry is **combustion analysis,** used to determine the carbon and hydrogen content of organic compounds.[‡] Because the technique is accurate and applicable to a very wide range of substances, most chemists consider this form of elemental analysis to be a necessary step in the characterization and identification of a new compound. The procedure requires specialized equipment and is commonly carried out in commercial laboratories. A simplified diagram of the apparatus is shown in Figure 8-1.

The sample is heated in an oxygen atmosphere, and the partially combusted product is passed through catalysts at elevated temperature to complete the oxidation of the material to CO_2 and H_2O. Useful catalysts include Pt gauze, CuO, PbO_2, and MnO_2. After the catalyst, sometimes additional reagents are added to remove halogen or sulfur compounds. The combustion products are flushed through a chamber containing P_4O_{10} ("phosphorus pentoxide"), which absorbs the water, and then through a chamber of Ascarite (NaOH on asbestos), which absorbs the CO_2. The increase in weight of each chamber tells how much carbon and hydrogen, respectively, was produced. A guard tube downstream of the two chambers prevents atmospheric H_2O or CO_2 from entering the chambers to be weighed.

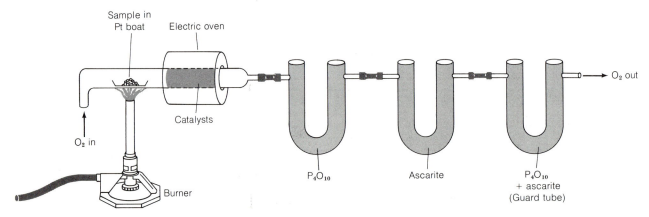

Figure 8-1
Combustion analysis of carbon and hydrogen.

[†] T. W. Richards, *Chem. Rev.*, **1**, 1 (1925).
[‡] A comprehensive discussion of this technique can be found in N. H. Furman, ed., *Standard Methods of Chemical Analysis*, 6th ed., Vol. 1 (Princeton, N.J.: van Nostrand, 1962), pp. 279–287.

EXAMPLE

An organic compound weighing 5.714 mg produced 14.414 mg of CO_2 and 2.529 mg of H_2O upon combustion. Find the weight percent of C and H in the sample.

One mole of CO_2 contains one mole of carbon. Therefore,

$$\text{Moles of C in sample} = \text{Moles of } CO_2 \text{ produced}$$

$$= \frac{14.414 \times 10^{-3} \text{ g } CO_2}{44.010 \text{ g/mol } CO_2} = 3.275 \times 10^{-4} \text{ mol}$$

$$\text{Mass of C in sample} = (3.275 \times 10^{-4} \text{ mol C})(12.011 \text{ g/mol C}) = 3.934 \text{ mg}$$

$$\text{Weight percent of C} = \frac{3.934 \text{ mg C}}{5.714 \text{ mg sample}} \times 100 = 68.84\%$$

One mole of H_2O contains *two* moles of H. Therefore,

$$\text{Moles of H in sample} = 2(\text{moles of } H_2O \text{ produced})$$

$$= 2\left(\frac{2.529 \times 10^{-3} \text{ g } H_2O}{18.015\ 2 \text{ g/mol } H_2O}\right) = 2.808 \times 10^{-4} \text{ mol}$$

$$\text{Mass of H in sample} = (2.808 \times 10^{-4} \text{ mol H})(1.007\ 9 \text{ g/mol H}) = 2.830 \times 10^{-4} \text{ g}$$

$$\text{Weight percent of H} = \frac{0.283\ 0 \text{ mg H}}{5.714 \text{ mg sample}} \times 100 = 4.952\%$$

The last calculation uses the conversion $(2.830 \times 10^{-4} \text{ g})(1\ 000 \text{ mg/g}) = 0.283\ 0 \text{ mg}$.

The nitrogen content of the compound may also be determined by a modification known as the Dumas method. In this procedure the sample is mixed with powdered CuO and ignited in a stream of CO_2, decomposing the compound to H_2O, CO_2, N_2, and some nitrogen oxides. The latter are reduced to N_2 by a hot Cu catalyst downstream. The gases then enter a gas buret filled with concentrated KOH solution to absorb all CO_2. The volume of gas finally collected is due to the evolved N_2.

8-2 PRECIPITATION PROCESS

The ideal product of a gravimetric analysis should be very insoluble, easily filterable, very pure, and possess a known and constant composition. While few substances meet all these requirements, appropriate techniques can help optimize the properties of gravimetric precipitates.

Gravimetric analyses will give rise to systematic error if any of these criteria are not met.

Solubility

The solubility of a precipitate may be decreased by cooling the solution, since the solubility of most compounds decreases as the temperature is lowered. Alternatively, the solvent may be changed. The solubility of many organic salts can be decreased by decreasing the polarity of the solvent to less than that of pure water.

Demonstration 8-1 COLLOIDS AND DIALYSIS

Colloids are particles with diameters in the range 1–100 nm. They are larger than what we usually think of as molecules, but too small to precipitate. Colloids remain in solution indefinitely, suspended by the Brownian motion of the solvent.

Heat one beaker containing 200 mL of distilled water to 70–90°C, and leave an identical beaker of water at room temperature. Add 1 mL of 1 M $FeCl_3$ to each beaker and stir. The warm solution turns brown-red in a few seconds, while the cold solution remains yellow (Color Plate 3). The yellow color is characteristic of low molecular weight Fe(III) compounds. The red color results from colloidal aggregates of Fe(III) ions held together by hydroxide, oxide, and some chloride ions. These particles have a molecular weight of $\sim 10^5$, a diameter of ~ 10 nm and contain $\sim 10^3$ atoms of Fe.[†]

To demonstrate the size of colloidal particles, we can perform a **dialysis** experiment. In dialysis, two solutions are separated by a *semipermeable membrane*. A semipermeable membrane is one with holes through which some molecules, but not others, can diffuse. The common dialysis tubing available from most scientific supply houses is made of cellulose and has pore diameters of 1–5 nm.[‡] Small molecules can diffuse through these pores, but large molecules (such as proteins or colloids) cannot.

Pour some of the brown-red colloidal Fe solution into a dialysis tube knotted at one end, then tie off the other end. Drop this into a flask of distilled water to show that the color remains entirely within the bag even after several days (Color Plate 4). For comparison, an identical bag containing a dark blue solution of 1 M $CuSO_4 \cdot 5H_2O$ can be left in another flask of water. The blue color of the Cu^{2+} will diffuse out of the bag, and the entire solution in the flask will be a uniform light blue color in 24 hours.

[†] R. N. Silva, *Rev. Pure and Appl. Chem.*, **22,** 115 (1972); K. M. Towe and W. F. Bradley, *J. Colloid Interface Sci.*, **24,** 384 (1967); T. G. Spiro, S. E. Allerton, J. Renner, A. Terzis, R. Bils, and P. Saltman, *J. Amer. Chem. Soc.*, **88,** 2721 (1966).

[‡] Tubing such as catalog number 3787–D4, sold by A. H. Thomas Co., P.O. Box 779, Philadelphia, PA 19105, is adequate for this demonstration.

Filterability

See Demonstration 8-1 for an experiment with colloids and dialysis.

The particles of the product need to be large enough so that the precipitate does not clog the filter or, worse, pass through the filter. The most desirable particles are, of course, crystals. At the other extreme is a **colloidal suspension** of particles so small that they remain indefinitely suspended and pass right through most filters. Colloids are particles with diameters in the range 1–100 nm. The conditions of the precipitation have much to do with the resulting particle size.

Crystal growth

Crystallization is generally considered as occurring in two phases: nucleation and particle growth. **Nucleation** is the process whereby molecules in solution randomly come together and form small aggregates. Particle growth involves the addition of more molecules to the nucleus to form a crystal. When a

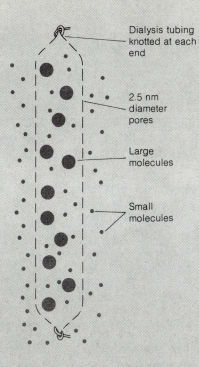

Dialysis tubing knotted at each end

2.5 nm diameter pores

Large molecules

Small molecules

Large molecules remain trapped inside a dialysis bag, while small molecules can diffuse through the membrane in both directions.

Dialysis is used to treat patients suffering from kidney failure. Blood is run over a dialysis membrane having a very large surface area. The small molecules of metabolic waste products that build up in blood diffuse across the membrane and are diluted into a large volume of liquid going out to waste. Protein molecules, which are a necessary part of the blood plasma, are too large to cross the membrane and are retained in the blood.

solution contains more solute than should be present at equilibrium, the solution is said to be **supersaturated. Relative supersaturation** is expressed as $(Q - S)/S$, where Q is the concentration of solute actually present and S is the concentration at equilibrium. The more substance that is dissolved, the greater the supersaturation.

The rate of nucleation has been found to depend more on relative supersaturation than does the rate of particle growth. That is, in a highly supersaturated solution, nucleation proceeds faster than particle growth. The result is a suspension of very tiny particles or, worse, a colloid. In a less supersaturated solution, nucleation is not so rapid, and the nuclei have a chance to grow into larger, more tractable particles.

To decrease supersaturation, and thereby promote particle growth, three techniques may be employed:

1. The temperature is raised to increase S and thereby decrease the relative supersaturation. (Most substances are more soluble in warm solution than in cold solution.)

Low relative supersaturation promotes increased particle size.

114

8 / GRAVIMETRIC ANALYSIS

Precipitant refers to the reagent which is added to cause the precipitation to occur.

2. The precipitant is added slowly with vigorous mixing, to avoid a local, highly supersaturated condition where the stream of precipitant first enters the analyte.

3. The volume of solution is kept large so that the concentrations of analyte and precipitant are low.

Controlled precipitation

It is sometimes possible to control the solubility of a precipitate through chemical means, such as pH or complexing ion control. For example, CaC_2O_4 is commonly precipitated from hot acidic solution, in which the salt is more soluble by virtue of the reaction of $C_2O_4^{2-}$ with H^+:

$$CaC_2O_4(s) \xrightleftharpoons{K_{sp}} Ca^{2+} + C_2O_4^{2-} \tag{8-2}$$

calcium oxalate

$$C_2O_4^{2-} + H^+ \rightleftharpoons HC_2O_4^- \tag{8-3}$$

Gradually raising the pH drives Reaction 8-3 to the left. This, in turn, drives Reaction 8-2 to the left.

Homogeneous precipitation

This is one of the nicest tricks for controlling supersaturation. In this technique, the precipitating agent is generated slowly by means of a chemical reaction. The most commonly employed reagent is urea, which slowly decomposes in boiling water to produce OH^-:

$$\underset{\substack{\text{urea}}}{H_2N-\overset{\overset{\textstyle O}{\|}}{C}-NH_2} + 3H_2O \xrightarrow{\text{heat}} CO_2 + 2NH_4^+ + 2OH^- \tag{8-4}$$

By this means the OH^- concentration of a solution can be raised very gradually. For example, one precipitation where slow hydroxide formation greatly enhances the particle size is that of ferric formate:

$$(H_2N)_2CO + 3H_2O \xrightarrow{\text{heat}} CO_2 + 2NH_4^+ + 2OH^- \tag{8-5}$$

$$OH^- + H\overset{\overset{\textstyle O}{\|}}{C}OH \rightarrow HCO_2^- + H_2O \tag{8-6}$$

formic acid   formate

$$3HCO_2^- + Fe^{3+} \rightarrow Fe(HCO_2)_3 \cdot nH_2O \tag{8-7}$$

ferric formate

Urea hydrolysis can also be used to drive Reactions 8-3 and 8-2 very slowly to the left, producing easily filterable crystals of CaC_2O_4. Table 8-1 lists some reagents employed in homogeneous precipitation.

Table 8-1
Some common reagents used for homogeneous precipitation

Precipitant	Reagent	Reaction	Some elements precipitated
OH^-	Urea	$(H_2N)_2CO + 3H_2O \rightarrow CO_2 + 2NH_4^+ + 2OH^-$	Al, Ga, Th, Bi, Fe, Sn
OH^-	Potassium cyanate	$HOCN + 2H_2O \rightarrow NH_4^+ + CO_2 + OH^-$ hydrogen cyanate	Cr, Fe
S^{2-}	Thioacetamide[†]	$CH_3\overset{\overset{S}{\|\|}}{C}NH_2 + H_2O \rightarrow CH_3\overset{\overset{O}{\|\|}}{C}NH_2 + H_2S$	Sb, Mo, Cu, Cd
SO_4^{2-}	Sulfamic acid	$H_3\overset{+}{N}SO_3^- + H_2O \rightarrow NH_4^+ + SO_4^{2-} + H^+$	Ba, Ca, Sr, Pb
SO_4^{2-}	Dimethyl sulfate[‡]	$(CH_3O)_2\overset{\overset{O}{\|\|}}{S}{=}O + 2H_2O \rightarrow 2CH_3OH + SO_4^{2-} + 2H^+$	Ba, Ca, Sr, Pb
$C_2O_4^{2-}$	Dimethyl oxalate	$CH_3O\overset{\overset{OO}{\|\|}}{CC}OCH_3 + 2H_2O \rightarrow 2CH_3OH + C_2O_4^{2-} + 2H^+$	Ca, Mg, Zn
PO_4^{3-}	Trimethyl phosphate	$(CH_3O)_3P{=}O + 3H_2O \rightarrow 3CH_3OH + PO_4^{3-} + 3H^+$	Zr, HF
CrO_4^{2-}	Chromic ion plus Bromate	$2Cr^{3+} + BrO_3^- + 5H_2O \rightarrow 2CrO_4^{2-} + Br^- + 10H^+$	Pb
8-Hydroxyquinoline	8-Acetoxyquinoline		Al, U, Mg, Zn

[†] The H_2S liberated is volatile and very toxic; it should only be handled in a well-vented hood.
[‡] Dimethyl sulfate is extremely toxic and should be handled only in a well-vented hood.

Precipitation in the presence of an electrolyte

It is desirable to precipitate most ionic compounds in the presence of added electrolyte. To understand the reason for this, we must discuss how tiny colloidal crystallites *coagulate* (come together) into larger particles (crystals). To illustrate this, we will discuss the formation of AgCl, which is commonly formed in the presence of ~ 0.1 M HNO_3.

Figure 8-2 is a schematic drawing of a colloidal particle of AgCl growing in a solution containing excess Ag^+, H^+, and NO_3^-. The surface of the particle has an excess positive charge due to the **adsorption** of extra silver ions on exposed chloride ion. (To be *adsorbed* means to be attached to the surface. In contrast, **absorption** involves penetration beyond the surface, to the inside.) The positively charged surface attracts anions and repels cations from the *ionic atmosphere* surrounding the particle. The positively charged particle and the negatively charged ionic atmosphere, together, are called the **electric double layer.**

Recall that an *electrolyte* is a compound which dissociates into ions when it dissolves.

While it is common to find the excess common ion adsorbed on the crystal surface, it is also possible to find other ions selectively adsorbed. In the presence of citrate and sulfate there is more citrate than sulfate adsorbed on a particle of $BaSO_4$.

8 / GRAVIMETRIC ANALYSIS

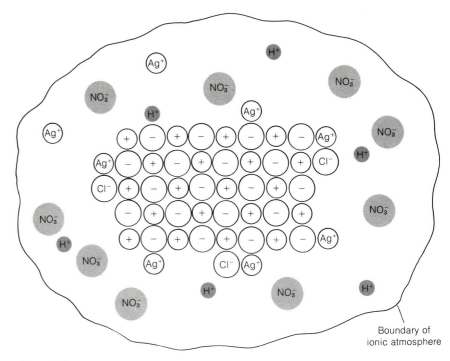

Figure 8-2
Schematic diagram showing a colloidal particle of AgCl growing in a solution containing excess Ag^+, H^+, and NO_3^-. The particle has a net positive charge because of adsorbed Ag^+ ions. The region of solution surrounding the particle is called the ionic atmosphere. It has a net negative charge, since the particle attracts anions and repels cations.

In order for two colloidal particles of AgCl to coalesce, they must collide with each other. However, the negatively charged ionic atmospheres of the particles repel each other. The particles, therefore, must have enough kinetic energy to overcome this electrostatic repulsion before they can coalesce.

Two steps can be taken to promote coalescence of the particles. One is to heat the solution, thereby increasing the particles' kinetic energy. The other is to increase the concentration of electrolyte (HNO_3 in this case). The greater the concentration of electrolyte, the less volume the ionic atmospheres occupy, and the closer together the two particles can come before their electrostatic repulsion becomes significant. This is why most gravimetric precipitations are done in the presence of an electrolyte.

Digestion

Following precipitation, most gravimetric procedures call for a period of standing in the presence of the mother liquor, usually with heating. This treatment, called **digestion,** promotes slow recrystallization of the precipitate. During this process the particle size usually increases, and impurities tend to be expelled from the crystal.

Purity

117

8-2 PRECIPITATION PROCESS

Types of impurities

Impurities bound to the surface of a crystal are said to be *adsorbed*. Impurities held within the crystal (*absorbed* impurities) are classified as either **inclusions** or **occlusions.** Inclusions are impurity ions that randomly occupy sites in the crystal lattice normally occupied by ions that belong in the crystal. Inclusions are more likely when the impurity ion has a similar size and charge to one of the ions that belongs to the product. Occlusions are pockets of impurity that are literally trapped inside the growing crystal.

Adsorbed, occluded, and included impurities are said to be **coprecipitated.** That is, the impurity is precipitated along with the desired product, even though the solubility of the impurity has not been exceeded. Coprecipitation tends to be worst in colloidal precipitates (which have a large surface area), such as $BaSO_4$, $Al(OH)_3$, and $Fe(OH)_3$. Many procedures call for washing away the mother liquor, redissolving the precipitate, and **reprecipitating** the product. During the second precipitation the concentration of impurities in the solution is lower than during the first precipitation, and the degree of coprecipitation therefore tends to be lower. Occasionally a trace component is intentionally isolated by being coprecipitated with a major component of the solution. The precipitate used to collect the trace component is said to be a **gathering agent,** and the process is called **gathering.**

Sometimes a precipitate forms in a pure state, but, while it is standing in the mother liquor, impurities form on the product. This is called **postprecipitation** and usually involves a supersaturated impurity that does not readily crystallize. After a period of time the impurity crystallizes and contaminates the desired product. An example is MgC_2O_4 postprecipitating in the presence of CaC_2O_4.

Washing is an important step in a gravimetric analysis. When the precipitate is collected on a filter, there are always droplets of liquid containing excess solute adhering to the solid. If the solution containing $AgNO_3$ were not washed away from the solid AgCl, the $AgNO_3$ would still be present after drying and the precipitate would weigh too much.

Some precipitates can simply be washed with water to remove the solute. However, many precipitates require electrolyte to maintain their coherence. For these, the ionic atmospheres are still needed to neutralize the surface charges of the tiny particles. If the electrolyte is washed away with water, the charged solid particles repel each other and the product breaks up. This breaking up is called **peptization** and can actually result in loss of the product through the filter. Silver chloride is an example of a precipitate that will peptize if washed with water. Therefore, AgCl is washed with dilute HNO_3 to remove excess $AgNO_3$ and to prevent peptization. The electrolyte used for washing must be volatile, so that it will be lost during drying. Some common volatile electrolytes are HNO_3, HCl, NH_4NO_3, NH_4Cl, and $(NH_4)_2CO_3$.

Reprecipitation improves the purity of some precipitates.

Ammonium chloride, for example, decomposes as follows when it is heated:

$$NH_4Cl(s) \rightarrow NH_3(g) + HCl(g)$$

Composition of Product

The final product must have a known, stable composition. If it is **hygroscopic** (picks up water from the air), it will be difficult to weigh accurately. Many precipitates contain a variable quantity of water and must be dried to achieve a constant composition. Some precipitates are **ignited** (heated strongly) to change the chemical form. Conditions of heating or ignition vary with each gravimetric analysis.

An example of how the composition depends on heating temperature is shown in Figure 8-3. The curve in this figure was obtained by a *thermobalance*, and the technique is called **thermogravimetric analysis.** In this procedure a sample is heated, and its change in mass is measured as a function of temperature. We see that calcium salicylate decomposes in the following stages:

$$\text{(salicylate)} \cdot H_2O \longrightarrow \text{(salicylate)} \longrightarrow \text{Ca complex} \longrightarrow CaCO_3 \longrightarrow CaO$$

(8-8)

Clearly the composition of the product to be weighed depends on the temperature and, usually, the duration of heating.

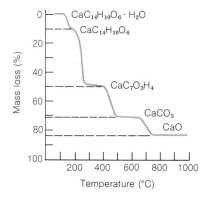

Figure 8-3
Thermogravimetric curve for calcium salicylate. [G. Liptay, ed., *Atlas of Thermoanalytical Curves* (London: Heyden and Son, 1976).]

Many gravimetric precipitates are made to produce a stable product of definite composition by being ignited at high temperature in an oven or over a flame. Two examples are $Fe(HCO_2)_3 \cdot nH_2O$, which is ignited at 850°C for one hour to give Fe_2O_3; and $Mg(NH_4)PO_4 \cdot 6H_2O$, which is ignited at 1 100°C to give $Mg_2P_2O_7$.

8-3 SCOPE OF GRAVIMETRIC ANALYSIS

The only major equipment needed for gravimetric analysis is an accurate balance. It is one of the oldest analytical techniques, and gravimetric analyses were developed for many elements and compounds long before alternative methods were available. Gravimetric procedures are usually very accurate, but more tedious than other methods.

To indicate the scope of gravimetric analysis, some analytically useful precipitations of common cations and anions are listed in Table 8-2. A few common organic precipitating agents are listed in Table 8-3. Because most precipitants (agents that cause precipitation) are not specific, conditions usually need to be carefully controlled in order to selectively precipitate one species in the presence of interfering substances. Often the potentially interfering substance must be separated from the analyte prior to analysis. It is important to be cognizant of potential interference in all analytical techniques, not just gravimetric analysis.

Separation of interfering species from the analyte is a necessary step in many analytical procedures

8-4 CALCULATIONS OF GRAVIMETRIC ANALYSIS

We will now examine some examples that illustrate how to relate the weight of a gravimetric precipitate to the quantity of the original analyte. The general approach is to relate the moles of product to the moles of reactant.

EXAMPLE

The piperazine content of an impure commercial grade of piperazine can be determined by precipitating and weighing the diacetate:[†]

$$:N \hspace{1em} N: + 2CH_3CO_2H \longrightarrow \left(H\overset{+}{N} \hspace{1em} \overset{+}{N}H \right)(CH_3CO_2^-)_2 \ (s) \qquad (8\text{-}9)$$

| piperazine F.W. 84.121 | acetic acid F.W. 60.053 | piperazine diacetate F.W. 204.227 |

In one experiment 0.3126 g of the sample was dissolved in 25 mL of acetone, and 1 mL of acetic acid was added. After five minutes, the precipitate was filtered, washed with acetone, dried at 110°C, and found to weigh 0.712 1 g. What is the weight percent of piperazine in the commercial material?

For each mole of piperazine in the reaction, one mole of product will be formed. The

---

[†] G. W. Latimer, Jr., *J. Chem. Ed.*, **43,** 148 (1966); G. R. Bond, *Anal. Chem.*, **32,** 1332 (1962).

Table 8-2
Representative gravimetric analyses

Species analyzed	Precipitated form	Form weighed	Some interfering species
K^+	$KB(C_6H_5)_4$	$KB(C_6H_5)_4$	NH_4^+, Ag^+, Hg^{2+}, Tl^+, Rb^+, Cs^+
Mg^{2+}	$Mg(NH_4)PO_4 \cdot 6H_2O$	$Mg_2P_2O_7$	Many metals except Na^+ and K^+
Ca^{2+}	$CaC_2O_4 \cdot H_2O$	$CaCO_3$ or CaO	Many metals except Mg^{2+}, Na^+, K^+
Ba^{2+}	$BaSO_4$	$BaSO_4$	Na^+, K^+, Li^+, Ca^{2+}, Al^{3+}, Cr^{3+}, Fe^{3+}, Sr^{2+}, Pb^{2+}, NO_3^-
Ti^{4+}	$TiO(5,7\text{-dibromo-8-hydroxyquinoline})_2$	Same	Fe^{3+}, Zr^{4+}, Cu^{2+}, $C_2O_4^{2-}$, citrate, HF
VO_4^{3-}	Hg_3VO_4	V_2O_5	Cl^-, Br^-, I^-, SO_4^{2-}, CrO_4^{2-}, AsO_4^{3-}, PO_4^{3-}
Cr^{3+}	$PbCrO_4$	$PbCrO_4$	Ag^+, NH_4^+
Mn^{2+}	$Mn(NH_4)PO_4 \cdot H_2O$	$Mn_2P_2O_7$	Many metals
Fe^{3+}	$Fe(HCO_2)_3$	Fe_2O_3	Many metals
Co^{2+}	$Co(1\text{-nitroso-2-naphtholate})_3$	$CoSO_4$ (by reaction with H_2SO_4)	Fe^{3+}, Pd^{2+}, Zr^{4+}
Ni^{2+}	$Ni(\text{dimethylglyoximate})_2$	Same	Pd^{2+}, Pt^{2+}, Bi^{3+}, Au^{3+}
Cu^{2+}	$Cu_2(SCN)_2$	$Cu_2(SCN)_2$	NH_4^+, Pb^{2+}, Hg^{2+}, Ag^+
Zn^{2+}	$Zn(NH_4)PO_4 \cdot H_2O$	$Zn_2P_2O_7$	Many metals
Ce^{4+}	$Ce(IO_3)_4$	CeO_2	Th^{4+}, Ti^{4+}, Zr^{4+}
Al^{3+}	$Al(8\text{-hydroxyquinolate})_3$	Same	Many metals
Sn^{4+}	$Sn(\text{cupferron})_4$	SnO_2	Cu^{2+}, Pb^{2+}, As(III)
Pb^{2+}	$PbSO_4$	$PbSO_4$	Ca^{2+}, Sr^{2+}, Ba^{2+}, Hg^{2+}, Ag^+, HCl, HNO_3
NH_4^+	$NH_4B(C_6H_5)_4$	$NH_4B(C_6H_5)_4$	K^+, Rb^+, Cs^+
Cl^-	AgCl	AgCl	Br^-, I^-, SCN^-, S^{2-}, $S_2O_3^{2-}$, CN^-
Br^-	AgBr	AgBr	Cl^-, I^-, SCN^-, S^{2-}, $S_2O_3^{2-}$, CN^-
I^-	AgI	AgI	Cl^-, Br^-, SCN^-, S^{2-}, $S_2O_3^{2-}$, CN^-
SCN^-	$Cu_2(SCN)_2$	$Cu_2(SCN)_2$	NH_4^+, Pb^{2+}, Hg^{2+}, Ag^+
CN^-	AgCN	AgCN	Cl^-, Br^-, I^-, SCN^-, S^{2-}, $S_2O_3^{2-}$
F^-	$(C_6H_5)_3SnF$	$(C_6H_5)_3SnF$	Many metals (except alkali metals), SiO_4^{4-}, CO_3^{2-}
ClO_4^-	$KClO_4$	$KClO_4$	
SO_4^{2-}	$BaSO_4$	$BaSO_4$	Na^+, K^+, Li^+, Ca^{2+}, Al^{3+}, Cr^{3+}, Fe^{3+}, Sr^{2+}, Pb^{2+}, NO_3^-
PO_4^{3-}	$Mg(NH_4)PO_4 \cdot 6H_2O$	$Mg_2P_2O_7$	Many metals except Na^+, K^+
NO_3^-	Nitron nitrate	Nitron nitrate	ClO_4^-, I^-, SCN^-, CrO_4^{2-}, ClO_3^-, NO_2^-, Br^-, $C_2O_4^{2-}$
CO_3^{2-}	CO_2 (by acidification)	CO_2	(The liberated CO_2 is trapped with Ascarite and weighed.)

Table 8-3
Common organic precipitating agents

Name	Structure	Some ions precipitated
Dimethylglyoxime		Ni^{2+}, Pd^{2+}, Pt^{2+}
Cupferron		Fe^{3+}, VO_2^+, Ti^{4+}, Zr^{4+}, Ce^{4+}, Ga^{3+}, Sn^{4+}
8-Hydroxyquinoline (oxine)		Mg^{2+}, Zn^{2+}, Cu^{2+}, Cd^{2+}, Pb^{2+}, Al^{3+}, Fe^{3+}, Bi^{3+}, Ga^{3+}, Th^{4+}, Zr^{4+}, UO_2^{2+}, TiO^{2+}
Salicylaldoxime		Cu^{2+}, Pb^{2+}, Bi^{3+}, Zn^{2+}, Ni^{2+}, Pd^{2+}
1-Nitroso-2-naphthol		Co^{2+}, Fe^{3+}, Pd^{2+}, Zr^{4+}
Nitron		NO_3^-, ClO_4^-, BF_4^-, WO_4^{2-}
Sodium tetraphenylborate	$Na^+B(C_6H_5)_4^-$	K^+, Rb^+, Cs^+, NH_4^+, Ag^+, organic ammonium ions
Tetraphenylarsonium chloride	$(C_6H_5)_4As^+Cl^-$	$Cr_2O_7^{2-}$, MnO_4^-, ReO_4^-, MoO_4^{2-}, WO_4^{2-}, ClO_4^-, I_3^-

122

8 / GRAVIMETRIC ANALYSIS

moles of product will therefore be

$$\text{Moles of product} = \frac{0.712\,1\,\text{g}}{204.227\,\text{g/mol}} = 3.487 \times 10^{-3}\,\text{mol}$$

This many moles of piperazine corresponds to

$$\text{Grams of piperazine} = (3.487 \times 10^{-3}\,\text{mol})\left(84.121\,\frac{\text{g}}{\text{mol}}\right) = 0.293\,3\,\text{g}$$

which gives

$$\text{Percent of piperazine in analyte} = \frac{0.293\,3\,\text{g}}{0.312\,6\,\text{g}} \times 100 = 93.83\%$$

An alternative (but equivalent) way to work this problem is to realize that 204.227 g (1 mol) of product will be formed for every 84.121 g (1 mol) of piperazine analyzed. Since 0.712 1 g of product was formed, the amount of reactant is given by

$$\frac{x\,\text{g piperazine}}{0.712\,1\,\text{g product}} = \frac{84.121\,\text{g piperazine}}{204.227\,\text{g product}}$$

or

$$x\,\text{g piperazine} = \left(\frac{84.121}{204.227}\right)0.712\,1\,\text{g} = 0.293\,3\,\text{g}$$

The quantity 84.121/204.227 is often referred to as a *gravimetric factor* relating the amount of starting material to the amount of product.

As a practical note, if you were performing this analysis, it would be important to know or to determine that the impurities in the piperazine are not precipitated under the conditions of this experiment.

EXAMPLE

The nickel content in steel can be determined gravimetrically. First dissolve the alloy in 12 M HCl and neutralize in the presence of citrate ion, which maintains the iron in solution near neutral pH. Warming the slightly basic solution in the presence of dimethylglyoxime (DMG) causes the red dimethylglyoxime complex of nickel to precipitate quantitatively. The product is filtered, washed with cold water, and dried at 110°C.

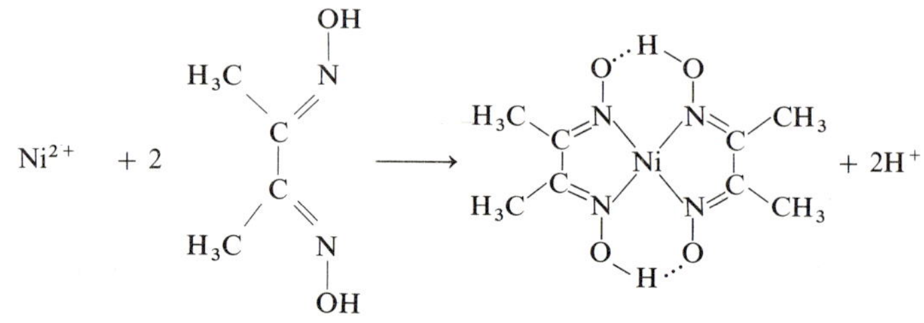

$$\text{Ni}^{2+} + 2\ \text{(dimethylglyoxime)} \longrightarrow \text{bis(dimethylglyoximate)nickel(II)} + 2\text{H}^+ \quad (8\text{-}10)$$

F.W. 58.71   dimethylglyoxime   bis(dimethylglyoximate)nickel(II)
              F.W. 116.12   F.W. 288.93

Suppose that the nickel content of the steel is known to be near 3% and you wish to analyze 1.0 g of the steel. What volume of 1.0% alcoholic dimethylglyoxime solution should be used to give a 50% excess of DMG for the analysis? Assume that the density of the alcohol solution is 0.79 g/mL.

Since the Ni content is around 3%, 1.0 g of steel will contain about 0.03 g of Ni, which corresponds to

$$\frac{0.03 \text{ g}}{58.71 \text{ g/mol}} = 5.11 \times 10^{-4} \text{ mol of Ni}$$

This amount of metal requires

$$2(5.11 \times 10^{-4} \text{ mol})(116.12 \text{ g of DMG/mol}) = 0.119 \text{ g of DMG}$$

since one mole of Ni^{2+} requires 2 moles of DMG. A 50% excess of DMG would be $(1.5)(0.119 \text{ g}) = 0.178 \text{ g}$. This much DMG is contained in

$$\left(\frac{0.178 \text{ g DMG}}{0.010 \text{ g DMG/g of solution}}\right) = 17.8 \text{ g of solution}$$

which occupies a volume of

$$\frac{17.8 \text{ g}}{0.79 \text{ g/mL}} = 23 \text{ mL}$$

In the calculation above we made use of the fact that 1.0% DMG means that 1.0 g (not 1.0 mL) of solution contains 0.010 g of DMG.

If the analysis of 1.163 4 g of steel resulted in formation of 0.179 5 g of precipitate, what is the percentage of Ni in the steel?

For each mole of Ni in the steel, one mole of precipitate will be formed. Therefore, 0.179 5 g of precipitate corresponds to

$$\frac{0.179 \text{ 5 g}}{288.93 \text{ g/mol}} = 6.213 \times 10^{-4} \text{ mol of Ni(DMG)}_2$$

The Ni in the alloy must therefore be

$$(6.213 \times 10^{-4} \text{ mol of Ni})(58.71 \text{ g/mol of Ni}) = 0.036 \text{ 47 g}$$

The weight percent of Ni in steel is

$$\frac{0.036 \text{ 47 g of Ni}}{1.163 \text{ 4 g of steel}} \times 100 = 3.135\%$$

A slightly simpler way to approach this problem comes from realizing that 58.71 g of Ni (1 mol) would give 288.93 g (1 mol) of product. Calling the weight of Ni in the sample x, we can write

$$\frac{\text{Ni analyzed}}{\text{Product formed}} = \frac{x}{0.179 \text{ 5}} = \frac{58.71}{288.93} \Rightarrow \text{Ni} = 0.036 \text{ 47 g}$$

8 / GRAVIMETRIC ANALYSIS

EXAMPLE

A mixture made up of the 8-hydroxyquinoline complexes of aluminum and magnesium weighed 1.084 3 g. We do not know how much of each complex is in the mixture. When ignited in a Bunsen burner flame, the mixture decomposed, leaving a residue of Al_2O_3 and MgO weighing 0.134 4 g.

F.W. 459.441   F.W. 312.611   F.W. 101.961   F.W. 40.304

Find the weight percent of $Al(C_9H_6NO)_3$ in the original mixture.

We will abbreviate the 8-hydroxyquinoline anion as Q. Letting the weight of AlQ_3 be x and the weight of MgQ_2 be y, we can write

$$\underset{\substack{\text{Weight of}\\ AlQ_3}}{x} + \underset{\substack{\text{Weight of}\\ MgQ_2}}{y} = 1.0843 \text{ g}$$

The moles of Al will be $x/459.441$, and the moles of Mg will be $y/312.611$. The moles of Al_2O_3 will be one-half of the total moles of Al, since it takes two moles of Al to make one mole of Al_2O_3.

$$\text{Moles of } Al_2O_3 = \left(\frac{1}{2}\right)\frac{x}{459.441}$$

The moles of MgO will equal the moles of $Mg = y/312.611$. Now we can write

$$\overbrace{\underbrace{\left(\frac{1}{2}\right)\frac{x}{459.441}}_{\text{moles}}\underbrace{(101.961)}_{\text{g/mole}}}^{\text{Weight of } Al_2O_3} + \overbrace{\underbrace{\frac{y}{312.611}}_{\text{moles}}\underbrace{(40.304)}_{\text{g/mole}}}^{\text{Weight of MgO}} = 0.134\ 4 \text{ g}$$

Substituting $y = 1.084\ 3 - x$ into the equation above gives

$$\left(\frac{1}{2}\right)\left(\frac{x}{459.441}\right)(101.961) + \left(\frac{1.084\ 3 - x}{312.611}\right)(40.304) = 0.134\ 4 \text{ g}$$

from which we find $x = 0.300\ 3$ g, which is 27.70% of the original mixture.

EXAMPLE

A mixture of benzene and acetonitrile weighing 6.423 mg is analyzed by combustion analysis. If the CO_2 formed weighs 16.396 mg, find the weight percent of acetonitrile in the mixture.

$$\text{benzene} + CH_3CN \xrightarrow{\;O_2\;} CO_2 + H_2O$$

benzene (C_6H_6)　　acetonitrile　　　carbon dioxide
F.W. 78.114　　　　F.W. 41.053　　　　F.W. 44.010

Call the weight of acetonitrile x and the weight of benzene y. We can say that

$$x + y = 6.423 \times 10^{-3} \text{ g} \tag{8-11}$$

We know that one mole of CO_2 is produced for each mole of carbon in the unknown. The moles of CO_2 produced are

$$\frac{16.396 \times 10^{-3} \text{ g}}{44.010 \text{ g/mol}} = 0.372\ 55 \text{ mmol of } CO_2$$

Each mole of C_6H_6 contains six moles of carbon, and each mole of CH_3CN contains two moles of carbon. Equating moles of carbon in the mixture to moles of carbon in the product, we can write

$$\text{Moles of } CO_2 \text{ produced} = \underbrace{6\left(\underbrace{\frac{y}{78.114}}_{\substack{\text{Moles of} \\ C_6H_6}}\right)}_{\substack{\text{Moles of C in} \\ C_6H_6}} + \underbrace{2\left(\underbrace{\frac{x}{41.053}}_{\substack{\text{Moles of} \\ CH_3CN}}\right)}_{\substack{\text{Moles of C in} \\ CH_3CN}}$$

Now we use the value 0.372 55 mmol of CO_2 produced, and make the substitution $y = 6.423 \times 10^{-3} - x$ from Equation 8-11:

$$0.372\ 55 \times 10^{-3} = 6\left(\frac{6.423 \times 10^{-3} - x}{78.114}\right) + 2\left(\frac{x}{41.053}\right)$$

$$x = 4.300 \text{ mg}$$

$$= 66.95\% \text{ of original mixture}$$

Summary

Gravimetric analysis is based on the formation of a weighable product whose mass can be related to the mass of analyte. Most commonly, one ion is precipitated by the addition of a suitable counterion. The ideal product should be insoluble, easily filterable, pure, and possess a known composition. Measures taken to reduce supersaturation and promote the formation of large, easily filtered particles include (1) raising the temperature during precipitation, (2) slow addition and vigorous mixing of reagents, (3) maintaining a large sample volume, and (4) use of homogeneous precipitation. Colloid formation is the extreme of undesirability in gravimetric analysis. After a precipitate forms, it is usually digested in the mother liquor at an elevated temperature to promote particle growth and recrystallization. All precipitates are then filtered and washed; some must be washed with a volatile electrolyte to prevent peptization. In the final step, the precipitate is heated to dryness or ignited to achieve a reproducible, stable composition. All gravimetric calculations seek to relate moles of product to moles of analyte.

Terms to Understand

absorption	gathering	peptization
adsorption	gravimetric analysis	postprecipitation
colloid	homogeneous precipitation	precipitant
combustion analysis	hygroscopic	relative supersaturation
coprecipitation	ignition	reprecipitation
dialysis	inclusion	thermogravimetric analysis
digestion	nucleation	
electric double layer	occlusion	

Exercises

8-A. An organic compound with a molecular weight of 417 was analyzed for ethoxyl (CH_3CH_2O—) groups using the reactions

$$ROCH_2CH_3 + HI \rightarrow ROH + CH_3CH_2I$$

$$CH_3CH_2I + Ag^+ + H_2O \rightarrow AgI(s) + CH_3CH_2OH$$

A 25.42 mg sample of compound produced 29.03 mg of AgI. How many ethoxyl groups are there in each molecule?

8-B. A 0.649 g sample containing only K_2SO_4 (F.W. 174.27) and $(NH_4)_2SO_4$ (F.W. 132.14) was dissolved in water and treated with $Ba(NO_3)_2$ to precipitate all of the SO_4^{2-} as $BaSO_4$ (F.W. 233.40). Find the weight percent of K_2SO_4 in the sample if 0.977 g of precipitate was formed.

8-C. Consider a mixture of the two solids $BaCl_2 \cdot 2H_2O$ and KCl, in an unknown ratio. (The notation $BaCl_2 \cdot 2H_2O$ means that a crystal is formed with two water molecules for each $BaCl_2$.) When the unknown is heated to 160°C for one hour, the water of crystallization is driven off:

$$BaCl_2 \cdot 2H_2O(s) \xrightarrow{160°C} BaCl_2(s) + 2H_2O(g)$$

A sample originally weighing 1.783 9 g weighed 1.562 3 g after heating. Calculate the weight percent of Ba, K, and Cl in the original sample.

8-D. A mixture containing only aluminum tetrafluoroborate, $Al(BF_4)_3$ (F.W. 287.39), and magnesium nitrate, $Mg(NO_3)_2$ (F.W. 148.31), weighed 0.282 8 g. It was dissolved in 1% aqueous HF solution and treated with nitron solution to precipitate a mixture of nitron tetrafluoroborate and nitron nitrate weighing 1.322 g. Calculate the weight percent of Mg in the original solid mixture.

nitron
$C_{20}H_{16}N_4$
F.W. 312.37

nitron tetrafluoroborate
$C_{20}H_{17}N_4BF_4$
F.W. 400.18

nitron nitrate
$C_{20}H_{17}N_5O_3$
F.W. 375.39

Problems

8-1. (a) What is the difference between absorption and adsorption?

(b) How is an inclusion different from an occlusion?

8-2. State four desirable properties of a gravimetric precipitate.

8-3. Why is a high relative supersaturation undesirable in a gravimetric precipitation?

8-4. What measures can be taken to decrease the relative supersaturation during a precipitation?

8-5. Why are many ionic precipitates washed with electrolyte solution instead of pure water?

8-6. Why is it less desirable to wash $AgCl$ precipitate with aqueous $NaNO_3$ than with HNO_3 solution?

8-7. Write a balanced equation for the combustion of benzoic acid $(C_6H_5CO_2H)$ to give CO_2 and H_2O. How many milligrams of CO_2 and of H_2O will be produced by the combustion of 4.635 mg of benzoic acid?

8-8. A mixture weighing 7.290 mg contained only cyclohexane, C_6H_{12} (F.W. 84.161), and oxirane, C_2H_4O (F.W. 44.053). When analyzed by combustion analysis, 21.999 mg of CO_2 (F.W. 44.010) was produced. Find the weight percent of oxirane in the mixture.

8-9. How many milliliters of 2.15% alcoholic dimethylglyoxime should be used to provide a 50.0% excess for Reaction 8-10 with 0.998 4 g of steel containing 2.07% Ni? Assume that the density of the dimethylglyoxime solution is 0.790 g/mL.

8-10. A 0.050 02 g sample of impure piperazine contained 71.29% piperazine. How many g of product will be formed when this sample is analyzed by Reaction 8-9?

8-11. Referring to Figure 8-3, what is the product when calcium salicylate monohydrate is heated to 550°C or 1 000°C? Using the formula weights of these products, calculate what mass is expected to remain when 0.635 6 g of calcium salicylate monohydrate is heated to 550°C or 1 000°C.

8-12. Twenty dietary iron tablets with a total mass of 22.131 g were ground and mixed thoroughly. Then 2.998 g of the powder was dissolved in HNO_3 and heated to convert all of the iron to Fe(III). Addition of NH_3 caused quantitative precipitation of $Fe_2O_3 \cdot xH_2O$, which was ignited to give 0.264 g of Fe_2O_3 (F.W. 159.69). What is the average mass of $FeSO_4 \cdot 7H_2O$ (F.W. 278.01) in each tablet?

8-13. A 1.475 g sample containing NH_4Cl, K_2CO_3, and inert ingredients was dissolved to give 0.100 L of solution. A 25.0 mL aliquot was acidified and treated with excess sodium tetraphenylborate, $Na^+B(C_6H_5)_4^-$, to precipitate K^+ and NH_4^+ ions completely:

$$(C_6H_5)_4B^- + K^+ \rightarrow (C_6H_5)_4BK(s)$$
$$(C_6H_5)_4B^- + NH_4^+ \rightarrow (C_6H_5)_4BNH_4(s)$$

The resulting precipitate amounted to 0.617 g. A fresh 50.0 mL aliquot of the original solution was made alkaline and heated to drive off all of the NH_3:

$$NH_4^+ + OH^- \rightarrow NH_3(g) + H_2O$$

It was then acidified and treated with sodium tetraphenylborate to give 0.554 g of precipitate. Find the weight percent of NH_4Cl and of K_2CO_3 in the original solid.

8-14. A mixture containing only Al_2O_3 and Fe_2O_3 weighs 2.019 g. When heated under a stream of H_2, the Al_2O_3 is unchanged, but the Fe_2O_3 is converted to metallic Fe plus $H_2O(g)$. If the residue weighs 1.774 g, what is the weight percent of Fe_2O_3 in the original mixture?

8-15. A solid mixture weighing 0.548 5 g contained only ferrous ammonium sulfate and ferrous chloride. The sample was dissolved in 1 M H_2SO_4, oxidized to Fe(III) with H_2O_2, and precipitated with cupferron. The ferric cupferron complex was ignited to produce 0.167 8 g of ferric oxide (Fe_2O_3, F.W. 159.69). Calculate the weight percent of Cl in the original sample.

$FeSO_4 \cdot (NH_4)_2SO_4 \cdot 6H_2O$
ferrous ammonium sulfate
F.W. 392.13

$FeCl_2 \cdot 6H_2O$
ferrous chloride
F.W. 234.84

cupferron
F.W. 155.16

8-16. A 2.000 g sample of a solid mixture containing only $PbCl_2$ (F.W. 278.1), $CuCl_2$ (F.W. 134.45), and KCl (F.W. 74.55) was dissolved in water to give 100.0 mL of solution. First 50.00 mL of the unknown solution was treated with sodium piperidine dithiocarbamate to precipitate 0.726 8 g of lead piperidine dithio-

128 carbamate:

$$Pb^{2+} + 2 \ \langle \ N{-}CS_2^- \ \rightarrow$$

lead piperidine dithiocarbamate
F.W. 527.74

Then 25.00 mL of the unknown solution was treated with iodic acid to precipitate 0.838 8 g of $Pb(IO_3)_2$

and $Cu(IO_3)_2$.

$$Cu^{2+} + 2IO_3^- \rightarrow Cu(IO_3)_2$$
F.W. 413.35

$$Pb^{2+} + 2IO_3^- \rightarrow Pb(IO_3)_2$$
F.W. 557.0

Calculate the weight percent of Cu in the unknown mixture.

9 / Precipitation Titrations

The use of precipitation reactions for gravimetric analysis is not very popular because of the time, labor, and skill required. However, many precipitation reactions can be adapted for accurate, easily performed titrations in which we measure the volume of precipitant needed for complete reaction. This tells us how much analyte was present. We begin this chapter with a general discussion of volumetric analysis and then proceed to the details of precipitation titrations.

9-1 PRINCIPLES OF VOLUMETRIC ANALYSIS

In **volumetric analysis** the volume of reagent **(titrant)** needed to react with **analyte** (that being analyzed) is measured. In a **titration,** increments of reagent are added to the analyte until their reaction is complete. The most common procedure is to deliver titrant from a *buret*, as shown in Figure 9-1.

Titrations can be based on any kind of chemical reaction. The principal requirements for the reaction are that it be *complete* (have a large equilibrium constant) and *rapid*. The most common titrations are based on acid–base, oxidation–reduction, complex formation, or precipitation reactions.

> Titration reactions should ideally be complete and rapid.

Methods of determining when the reaction is complete are discussed in several places in this text. The usual methods involve observing an indicator color change, monitoring a spectrophotometric absorbance change, or detecting changes in current or potential between a pair of electrodes immersed in the analyte solution. An **indicator** is a compound that has a physical property (usually color) that changes abruptly near the equivalence point. The change is caused by the sudden disappearance of analyte or appearance of titrant at the equivalence point.

> Indicators are discussed in the following sections:
>
> Precipitation indicators—9-6
> Acid–base indicators—12-5
> Metal ion indicators—14-4
> Redox indicators—17-3

The **equivalence point** is that point in the titration at which the quantity of titrant added is the exact amount necessary for stoichiometric reaction

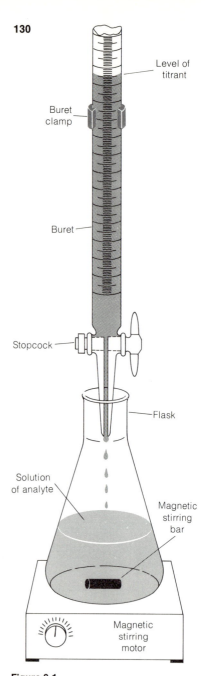

Figure 9-1
Typical setup for a titration. The analyte is contained in the flask, and the titrant in the buret. The magnetic stirring bar is a magnet coated with Teflon, which is inert to almost all solutions. The bar is spun by a rotating magnet inside the stirring motor.

with the analyte. For example, oxalic acid reacts with permanganate in hot acidic solution as follows:

$$5\text{HOCCOH} + 2\text{MnO}_4^- + 6\text{H}^+ \longrightarrow 10\text{CO}_2 + 2\text{Mn}^{2+} + 8\text{H}_2\text{O} \quad (9\text{-}1)$$

oxalic acid (colorless)   permanganate (purple)   (colorless)   (colorless)

Suppose the analyte contains 5.000 mmol of oxalate. The equivalence point is reached when 2.000 mmol of MnO_4^- has been added, since *two moles of MnO_4^- are required for five moles of oxalic acid.*

The equivalence point is the ideal result we seek in a titration. What we actually measure is the **endpoint.** The endpoint is marked by a sudden change in a physical or chemical property of the solution. In this particular case, the most convenient endpoint is the appearance of the purple color of permanganate in the flask. Up to the equivalence point, all of the added permanganate is consumed by the oxalate, and the analyte solution remains colorless. After the equivalence point, the purple MnO_4^- ion builds up and, as soon as it can be seen, marks the endpoint. The better your eyes are, the closer will be the endpoint to the equivalence point. Here the endpoint cannot exactly equal the equivalence point because extra MnO_4^-, beyond that needed to react with oxalic acid, is required to show the purple color.

The difference between the endpoint and the equivalence point is an inescapable **titration error.** By choosing an appropriate physical property, in which a change is easily observed (such as indicator color, optical absorbance of a reactant or product, pH, or conductivity), it is possible to observe and record the endpoint very close to the true equivalence point. It is also usually possible to estimate the titration error using a **blank titration.** In the example above, a solution containing no oxalic acid could be titrated with MnO_4^- under conditions otherwise identical to those of the original experiment. The amount of MnO_4^- needed to show a purple color in the blank is subtracted from the amount observed in the oxalic acid titration. Subtracting the blank reading from the observed endpoint reading accurately locates the equivalence point.

The validity of an analytical procedure depends on knowing the amount of one of the reactants used. In our discussion so far, we have assumed that we know the concentration of the titrant in the buret. We might know this because a weighed amount of pure reagent was dissolved in a known volume of solution. In such a case, we call the reagent a **primary standard,** since it is pure enough to be weighed and used directly. To be useful, a primary standard should be 99.9% pure, or better. It should not decompose under ordinary storage, and it should be stable when dried (by heating or vacuum). The minimum purity of many primary standards is certified by the National Bureau of Standards.

In most cases, the titrant is not available as a primary standard. Instead, a solution having the approximate desired concentration is used to titrate a weighed, primary standard. By this procedure, we **standardize** the solution to be used in the analysis. In all cases, the validity of an analytical result ultimately depends on knowing the composition of some primary standard.

In a **direct titration,** titrant is added to the analyte until the reaction is complete. Occasionally it is more convenient to perform a **back titration.** This is done by *adding* a known *excess* of reagent to the analyte. Then a second reagent is used to titrate the *excess* of the first reagent. Back titrations are most useful when the endpoint of the back titration is clearer than the endpoint of the direct titration or when an excess of the first reagent is necessary for complete reaction with the analyte.

A reagent is *standardized* by titration against a primary standard.

9-2 CALCULATIONS OF VOLUMETRIC ANALYSIS

In volumetric analysis our calculations serve to *relate the moles of titrant to the moles of analyte.* If the reaction can be written

$$tT + aA \rightarrow \text{Products} \qquad (9\text{-}2)$$

then t moles of titrant (T) are required to react with a moles of analyte (A). The moles of titrant delivered are

$$\text{Moles of T} = V_T M_T \qquad (9\text{-}3)$$

where V_T is the volume of T and M_T is the molarity of T in the buret. The moles of A that must have been present to react with this many moles of T are

$$\text{Moles of A} = \frac{a}{t}(\text{Moles of T}) \qquad (9\text{-}4)$$

Calculations:

$$tT + aA \rightarrow \text{Products}$$
$$\text{Moles of T} = V_T M_T$$
$$\text{Moles of A} = \frac{a}{t}(\text{Moles of T})$$

since a moles of A react with t moles of T. For example, if the reaction were $2T + 3A \rightarrow \text{Products}$, then at the equivalence point

$$\text{Moles of A} = \frac{3}{2}(\text{Moles of T}) \qquad (9\text{-}5)$$

EXAMPLE

The chloride content of blood serum, cerebrospinal fluid, or urine can be measured by titration of the chloride with mercuric ion:

$$Hg^{2+} + 2Cl^- \rightarrow HgCl_2(aq) \qquad (9\text{-}6)$$

When the reaction is complete, excess Hg^{2+} appears in the solution. This is detected with the indicator diphenylcarbazone, which forms a violet-blue complex with Hg^{2+}.

diphenylcarbazone

132

9 / PRECIPITATION TITRATIONS

Mercuric nitrate was standardized by titrating a solution containing 147.6 mg of NaCl in about 25 mL of water. It required 28.06 mL of mercuric nitrate solution to reach the violet-blue indicator endpoint. When this same Hg^{2+} solution was used to titrate 2.000 mL of urine, 22.83 mL was required. (a) Calculate the molarity of Hg^{2+} in the titrant and (b) find the concentration of Cl^- (mg/mL) in the urine.

(a) The moles of Cl^- in 147.6 mg of NaCl are

$$\frac{147.6 \times 10^{-3} \text{ g of NaCl}}{58.443 \text{ g of NaCl/mole}} = 2.526 \times 10^{-3} \text{ mol}$$

According to Reaction 9-6, two moles of Cl^- are required for each mole of Hg^{2+}. The moles of Hg^{2+} used in the reaction must have been

$$\text{Moles of } Hg^{2+} = \tfrac{1}{2}(\text{Moles of } Cl^-) = 1.263 \times 10^{-3} \text{ mol}$$

This many moles are contained in 28.06 mL, so the molarity of Hg^{2+} is

$$\text{Molarity of } Hg^{2+} = \frac{1.263 \times 10^{-3} \text{ mol}}{28.06 \times 10^{-3} \text{ L}} = 0.045\,01 \text{ M}$$

As a point of useful information, you should be able to use the units mg, mL, and mmol together in the same way you use g, L, and mol together:

$$\frac{147.6 \text{ mg}}{58.443 \text{ mg/mmol}} = 2.526 \text{ mmol}$$

$$\frac{(\tfrac{1}{2})(2.526 \text{ mmol})}{28.06 \text{ mL}} = 0.045\,01 \frac{\text{mmol}}{\text{mL}} = 0.045\,01 \frac{\text{mol}}{\text{L}}$$

(b) In the titration of 2.000 mL of urine, 22.83 mL of Hg^{2+} was required. The moles of Hg^{2+} used were

$$\text{Moles of } Hg^{2+} = (22.83 \times 10^{-3} \text{ L})(0.045\,01 \text{ M}) = 1.028 \times 10^{-3} \text{ mol}$$

Since one mole of Hg^{2+} reacts with two moles of Cl^-, the moles of Cl^- in the 2.000 mL sample must have been

$$\text{Moles of } Cl^- = 2(\text{Moles of } Hg^{2+}) = 2.056 \times 10^{-3} \text{ mol}$$

The quantity of Cl^- in one milliliter is half of this value, or 1.028×10^{-3} mol. This much Cl^- weighs

$$(1.028 \times 10^{-3} \text{ mol})(35.453 \text{ g of Cl/mol}) = 36.45 \text{ mg}$$

EXAMPLE

The calcium content of urine can be determined by the following procedure:

1. Ca^{2+} is precipitated as calcium oxalate in basic solution:

$$Ca^{2+} + C_2O_4^{2-} \rightarrow Ca(C_2O_4) \cdot H_2O(s)$$

$$\text{oxalate} \qquad \text{calcium oxalate}$$

2. After washing the precipitate to remove any free oxalate, the solid is dissolved in acid.
3. The dissolved oxalic acid is heated to 60°C and titrated with standardized potassium permanganate until the purple endpoint is observed:

$$5H_2C_2O_4 + 2MnO_4^- + 6H^+ \rightarrow 10CO_2 + 2Mn^{2+} + 8H_2O$$

(colorless)   (purple)   (colorless)   (colorless)

Suppose that 0.356 2 g of $Na_2C_2O_4$ is dissolved in a total volume of 250.0 mL. If 10.00 mL of this solution requires 48.36 mL of $KMnO_4$ solution for titration, find the molarity of the permanganate solution.

The concentration of the oxalate solution is

$$\frac{0.356\ 2 \text{ g of } Na_2C_2O_4/(134.00 \text{ g } Na_2C_2O_4/\text{mol})}{0.250\ 0 \text{ L}} = 0.010\ 63 \text{ M}$$

The moles of $C_2O_4^{2-}$ in 10.00 mL are

$$\left(0.010\ 63\ \frac{\text{mol}}{\text{L}}\right)(0.010\ 0 \text{ L}) = 1.063 \times 10^{-4} \text{ mol} = 0.106\ 3 \text{ mmol}$$

Since two moles of MnO_4^- are required for five moles of oxalate, the moles of MnO_4^- delivered must have been

$$\text{Moles of } MnO_4^- = \frac{2}{5}(\text{moles of } C_2O_4^{2-}) = 0.042\ 53 \text{ mmol}$$

The concentration of MnO_4^- is

$$\text{Molarity of } MnO_4^- = \frac{0.042\ 53 \text{ mmol}}{48.36 \text{ mL}} = 8.795 \times 10^{-4} \text{ M}$$

The key step in this calculation is to note that five moles of oxalate require two moles of permanganate. You should look at the answer to be sure that the moles of MnO_4^- are *less* than the moles of oxalate. If not, you probably used the factor $\frac{5}{2}$ instead of $\frac{2}{5}$ in the calculation.

Suppose that the calcium in a 5.00 mL urine sample is precipitated with excess oxalate, and the excess oxalate is washed away from the solid. After dissolving the precipitate in acid, it required 16.17 mL of the above standard MnO_4^- solution to titrate the oxalate. Find the molar concentration of Ca^{2+} in the urine.

In 16.17 mL of MnO_4^- there are $(0.016\ 17 \text{ L})(8.795 \times 10^{-4} \text{ mol/L}) = 1.422 \times 10^{-5} \text{ mol}$ of MnO_4^-. This will react with

$$\text{Moles of oxalate} = \tfrac{5}{2}(\text{Moles of } MnO_4^-) = 3.555 \times 10^{-5} \text{ mol}$$

Since there is one oxalate ion for each calcium ion in $CaC_2O_4 \cdot H_2O$, there must have been 3.555×10^{-5} mol of Ca^{2+} in 5.00 mL of urine. The concentration of Ca^{2+} is

$$\frac{3.555 \times 10^{-5} \text{ mol}}{5.00 \times 10^{-3} \text{ L}} = 0.007\ 11 \text{ M}$$

EXAMPLE

The most accurate and reliable analysis of protein uses the Kjeldahl (pronounced kel′-dall) nitrogen determination, which is applicable to a wide variety of organic compounds. The sample is first chemically digested with boiling H_2SO_4 containing a catalyst that converts all of the organic nitrogen to NH_4^+. The solution is made basic, converting the ammonium ion to NH_3, which is distilled out into a *known excess* of aqueous HCl. The HCl in excess of the amount needed to react with NH_3 is then titrated with NaOH to determine how much NH_3 was collected.

$$\text{Digestion:} \qquad \text{N (in protein)} \to NH_4^+ \qquad (9\text{-}7)$$

$$\text{Distillation of } NH_3\text{:} \qquad NH_4^+ + OH^- \to NH_3(g) + H_2O \qquad (9\text{-}8)$$

$$\text{Collection of } NH_3 \text{ in HCl:} \qquad NH_3 + H^+ \to NH_4^+ \qquad (9\text{-}9)$$

$$\text{Titration of unreacted HCl:} \qquad H^+ + OH^- \to H_2O \qquad (9\text{-}10)$$

Suppose that 0.500 mL of a protein solution is analyzed by the Kjeldahl procedure. The protein is known to contain 16.2% (wt/wt) nitrogen (a typical figure for proteins). The liberated ammonia was collected in 10.00 mL of 0.021 40 M HCl, and the unreacted acid required 3.26 mL of 0.019 8 M NaOH for complete titration. Find the concentration of protein (milligrams of protein per milliliter) in the original solution.

The total moles of HCl in the flask receiving the NH_3 is

$$(10.00 \text{ mL})(0.021 \ 40 \text{ mmol/mL}) = 0.214 \ 0 \text{ mmol of HCl}$$

The NaOH required for titration of the unreacted HCl is

$$(3.26 \text{ mL})(0.019 \ 8 \text{ mmol/mL}) = 0.064 \ 6 \text{ mmol}$$

The difference between the moles of HCl and the moles of NaOH must equal the moles of NH_3 participating in Reaction 9-9.

$$\text{mmol } NH_3 = 0.214 \ 0 - 0.064 \ 6 = 0.149 \ 4 \text{ mmol}$$

Since one mole of nitrogen in the protein gives rise to one mole of NH_3, there must have been 0.149 4 mmol of nitrogen in the protein sample. This much nitrogen weighs

$$(0.149 \ 4 \text{ mmol})\left(14.006 \ 7 \ \frac{\text{mg of N}}{\text{mmol}}\right) = 2.093 \text{ mg of nitrogen}$$

If the protein contains 16.2% (wt/wt) nitrogen, there must be

$$\text{mg of protein} = \frac{2.093 \text{ mg of N}}{0.162 \text{ mg of N/mg of protein}} = 12.9 \text{ mg}$$

This corresponds to

$$\frac{12.9 \text{ mg of protein}}{0.500 \text{ mL}} = 25.8 \text{ mg of protein/mL}$$

EXAMPLE

A sample weighing 0.362 9 g contained only Fe_2O_3 (F.W. 159.69) and MgO (F.W. 40.304). It was dissolved in 250.0 mL of acid to give Fe^{3+} and Mg^{2+} in solution. A 25.00 mL *aliquot* (a fancy word for *portion*) was treated with 25.00 mL of 0.053 79 M EDTA. (EDTA is a merciful abbreviation for ethylenediaminetetraacetic acid, which forms a strong 1:1 complex with most metal ions having a charge other than +1.)

$$^-O_2CCH_2 \diagdown \atop ^-O_2CCH_2 \diagup NCH_2CH_2N \diagup^{CH_2CO_2^-} \atop \diagdown_{CH_2CO_2^-} \quad + \ M^{n+} \to M(EDTA)^{n-4}$$

$$\text{EDTA} \qquad\qquad \text{metal} \qquad \text{complex}$$

One mole of EDTA combines with *one mole* of metal ion, regardless of the charge of the metal ion. The pH was then raised to near 10 with ammonia buffer, and the excess unreacted EDTA was *back-titrated* with 16.56 mL of 0.036 87 M Zn^{2+} solution. The presence of excess zinc ion (beyond that needed for reaction with EDTA) is marked by a color change caused by an indicator that forms a colored metal complex. Find the weight percent of Fe_2O_3 in the solid mixture.

Since one mole of EDTA is consumed by each mole of each metal, *the total moles of metal must equal the total moles of EDTA*. The EDTA used is

$$\left(0.053\ 79\ \frac{\text{mmol}}{\text{mL}}\right)(25.00\ \text{mL}) = 1.344\ 8\ \text{mmol of EDTA}$$

The moles of Zn^{2+} needed to titrate the excess, unreacted EDTA are

$$\left(0.036\ 87\ \frac{\text{mmol}}{\text{mL}}\right)(16.56\ \text{mL}) = 0.610\ 6\ \text{mmol of } Zn^{2+}$$

Because mmol of EDTA = mmol of $(Fe^{3+} + Mg^{2+} + Zn^{2+})$,

$$\text{mmol of } (Fe^{3+} + Mg^{2+}) = \text{mmol of EDTA} - \text{mmol of } Zn^{2+}$$

$$= 0.734\ 2\ \text{mmol}$$

Since the 25.00 mL aliquot of $(Fe^{3+} + Mg^{2+})$ represents one-tenth of the original solid sample, we can say that the total $Fe^{3+} + Mg^{2+}$ in the solid sample is 7.342 mmol.

Calling the moles of iron x and the moles of magnesium y, we can write

$$x + y = 7.342\ \text{mmol}$$

Since each mole of Fe_2O_3 contains two moles of Fe, the moles of Fe_2O_3 must equal half of the moles of Fe. Because the mass of $(Fe_2O_3 + MgO)$ must equal 0.362 9 g, we can write

$$\underbrace{\left(\frac{x}{2}\ \text{mol}\right)\left(159.69\ \frac{\text{g}}{\text{mol}}\right)}_{\text{Mass of } Fe_2O_3} + \underbrace{(y\ \text{mol})\left(40.304\ \frac{\text{g}}{\text{mol}}\right)}_{\text{Mass of MgO}} = 0.362\ 9\ \text{g}$$

Substituting the value $y = (7.342 \times 10^{-3}\ \text{mol} - x)$ into the above equation allows us to solve for x.

$$x = 1.694\ \text{mmol of iron}$$

9 / PRECIPITATION TITRATIONS

The weight of Fe_2O_3 in the solid is

$$\left(\frac{x}{2}\text{ mol}\right)\left(159.69 \frac{\text{g}}{\text{mol}}\right) = 0.135\ 3 \text{ g of } Fe_2O_3$$

The weight percent of Fe_2O_3 in the solid is

$$100 \times \frac{0.135\ 3 \text{ g of } Fe_2O_3}{0.362\ 9 \text{ g of sample}} = 37.28\%$$

9-3 EXAMPLE OF A PRECIPITATION TITRATION

A sensitive (but not very precise) determination of sulfate is by precipitation titration with Ba^{2+}:

$$SO_4^{2-} + Ba^{2+} \rightarrow BaSO_4(s) \qquad (9\text{-}11)$$

The best means of detecting the endpoint is to measure the light scattered by particles of precipitate. A liquid containing very fine particles that scatter light is said to be **turbid** (Box 9-1).

In the $BaSO_4$ precipitation, the turbidity (light scattering) increases until the equivalence point is reached. The abrupt leveling-off of the scattering marks the endpoint. Since scattering is very dependent on particle size, the reaction must be carried out in a very reproducible manner to obtain precise results. A glycerol–alcohol mixture is used to help stabilize the particles and prevent the rapid settling of a mass of solid.

Alternatively, the endpoint can be detected using an indicator that forms a colored precipitate with Ba^{2+}. In this procedure, excess standard $BaCl_2$ is added to precipitate all of the sulfate. The indicator disodium rhodizonate is added to give a red precipitate with excess Ba^{2+}.

rhodizonate

(colorless in acidic solution)   (red precipitate)

The back titration is necessary because the free indicator is not stable under the conditions of the titration.

Back titration with standard SO_4^{2-} converts the red precipitate to colorless $BaSO_4$. The disappearance of the last trace of color marks the endpoint.

9-4 THE SHAPE OF A PRECIPITATION TITRATION CURVE

Deriving a theoretical titration curve will help you to appreciate what is happening in precipitation titrations. The titration curve is a graph showing how the concentration of one of the reactants varies as titrant is added.

Box 9-1 TURBIDIMETRY AND NEPHELOMETRY

A suspension of particles scatters light and is said to be *turbid*. In **turbidimetry** the suspension is placed in a spectrophotometer cuvette and the "absorbance" is measured with an ordinary spectrophotometer.[†] Light is not really absorbed by the solution, but is scattered in all directions and does not reach the detector.

The apparent absorption usually obeys an equation analogous to Beer's law over some limited range of concentration:

$$\text{``A''} = \log_{10} \frac{P_0}{P} = kbc$$

where A is the apparent absorbance, P_0 is the radiant power of the incident light, P is the radiant power of emergent light, k is a constant, b is the path length, and c is the "concentration" of precipitate. The value of k is determined empirically with a series of standards. The equation above can be rearranged to the form

$$\text{``T''} = \frac{P}{P_0} = e^{-\tau b}$$

where T is the "transmittance" of the turbid solution and τ is called the **turbidity coefficient.**

In **nephelometry** the light scattered at 90° to the incident beam by the turbid solution is measured. The power of the scattered beam is usually proportional to the concentration of particles over some limited range of concentrations. An empirical curve is used to relate the intensity of the scattered beam to the concentration of particles.

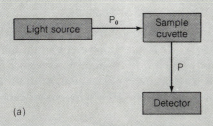

(a) Schematic diagram of nephelometer.   (b) Monitek digital nephelometer. [Courtesy A. H. Thomas Co., Philadelphia, Pa.]

In turbidimetry we measure the *fraction* of light scattered (the "transmittance"). This is independent of the intensity of the light source. In nephelometry we measure the *absolute intensity* of the scattered light. Even though the fraction of scattered light is constant, the observed intensity of scattered light increases if the source intensity is increased. Therefore, the sensitivity of nephelometry can be raised simply by increasing the power of the light source or by using a more sensitive detector. Being more sensitive than turbidimetry, nephelometry is more useful at very low concentrations of analyte.

Turbidimetric and nephelometric titrations are less precise than others because endpoint detection depends on particle size, which is not very reproducible. However, the sensitivity is excellent; sulfate can be analyzed at the parts-per-million level.

[†] The basic concepts of absorption spectroscopy, including Beer's law, will be used from time to time in this text before its formal introduction in Chapter 20. We assume that most readers are familiar with this material. If you are not, Sections 20-1 and 20-2 will give you enough background to understand all that we do with spectroscopy prior to Chapter 20.

138

9 / PRECIPITATION TITRATIONS

The correct definition of the p function.

Our usual working definition of the p function.

V_e = Volume of titrant at equivalence point.

Rather than plotting concentration directly, it is more useful to plot the p function, defined as

$$pX \equiv -\log_{10}\mathscr{A}_X = -\log[X]\gamma_X \qquad (9\text{-}13)$$

where $\mathscr{A}_X$ is the activity of X, $[X]$ is the concentration of X, and γ_X is the activity coefficient of X. If we neglect activity coefficients, as will usually be our custom, we can consider the p function to be given by

$$pX \approx -\log[X] \qquad (9\text{-}14)$$

Consider the titration of 25.00 mL of 0.100 0 M I^- with 0.050 00 M Ag^+

$$I^- + Ag^+ \rightarrow AgI(s) \qquad (9\text{-}15)$$

and suppose that we are monitoring the Ag^+ concentration with an electrode. Reaction 9-15 is the reverse of the dissolution of $AgI(s)$, whose solubility product is rather small:

$$AgI(s) \rightleftharpoons Ag^+ + I^- \qquad K_{sp} = [Ag^+][I^-] = 8.3 \times 10^{-17} \qquad (9\text{-}16)$$

Since the equilibrium constant for the titration reaction (9-15) is large (K for Reaction 9-15 = $1/K_{sp} = 1.2 \times 10^{16}$), the equilibrium lies far to the right. It is reasonable to say that each aliquot of Ag^+ reacts completely with I^-, leaving a very tiny amount of Ag^+ in solution. At the equivalence point, there will be a sudden increase in the Ag^+ concentration because all of the I^- has been consumed and we are now adding Ag^+ directly to the solution.

What volume of Ag^+ solution is needed to reach the equivalence point? To calculate this volume, which we designate V_e (volume of titrant at the equivalence point), first note that one mole of Ag^+ is required for each mole of I^-.

$$\text{Moles of } I^- = \text{Moles of } Ag^+$$

$$(0.025\,00\text{L})(0.100\,0 \text{ mol } I^-/\text{L}) = (V_e)(0.050\,00 \text{ mol } Ag^+/\text{L}) \qquad (9\text{-}17)$$

$$V_e = 0.050\,00 \text{ L} = 50.00 \text{ mL}$$

An easy way to find V_e in this example is to note that the concentration of Ag^+ is half of the concentration of I^-. Therefore, the volume of Ag^+ required will be twice as great as the volume of I^-, and 25.00 mL of I^- will require 50.00 mL of Ag^+.

The titration curve has three distinct regions. The calculations are different depending on whether we are before, at, or after the equivalence point. We will consider each region separately.

Before the Equivalence Point

Consider the point at which the volume of Ag^+ added is 10.00 mL. Since there are more moles of I^- than Ag^+ at this point, virtually all of the Ag^+ is "used up" to make $AgI(s)$. We want to find the very small concentration of Ag^+ remaining in solution after reaction with I^-. One way to do this is to imagine that Reaction 9-15 has gone to completion and that some AgI redissolves (Reaction 9-16). The solubility of Ag^+ will be determined by the concentration of free I^- remaining in the solution:

$$[Ag^+] = \frac{K_{sp}}{[I^-]} \tag{9-18}$$

When $V < V_e$, the concentration of unreacted I^- regulates the solubility of AgI.

The free I^- is overwhelmingly due to the I^- that has not been precipitated by 10.00 mL of Ag^+. By comparison, the I^- due to dissolution of $AgI(s)$ is negligible.

So let's find the concentration of unprecipitated I^-. The moles of I^- remaining in solution will be

$$\text{Moles of } I^- = \text{Original moles of } I^- - \text{Moles of } Ag^+ \text{ added} \tag{9-19}$$

$$= (0.025\,00\ L)(0.100\ mol/L) - (0.010\,00\ L)(0.050\,00\ mol/L)$$

$$= 0.002\,000\ mol\ I^-$$

Since the volume is 0.035 00 L (25.00 mL + 10.00 mL), the concentration is

$$[I^-] = \frac{0.002\,000\ mol\ I^-}{0.035\,00\ L} = 0.057\,14\ \text{M} \tag{9-20}$$

The concentration of Ag^+ in equilibrium with this much I^- is given by

$$[Ag^+] = \frac{K_{sp}}{[I^-]} = \frac{8.3 \times 10^{-17}}{0.057\,14} = 1.4_5 \times 10^{-15}\ \text{M} \tag{9-21}$$

The p function we seek is given by

$$pAg^+ = -\log[Ag^+] = 14.84 \tag{9-22}$$

Two details are worth noting. First, we have neglected activity coefficients in this calculation. A more rigorous calculation would include activity coefficients in Equations 9-21 and 9-22. Second, there are two significant figures in the concentration of Ag^+ because there are two significant figures in K_{sp}. The two figures in $[Ag^+]$ translate into two figures in the *mantissa* of the p function. The p function is therefore correctly written as 14.84.

The step-by-step calculation outlined above is a safe, but tedious, way to find the concentration of I^- at this point in the titration. We will now

$$\log(1.4_5 \times 10^{-15}) = 14.84$$

Two significant figures

Two digits in mantissa

If you need a review of significant figures in logarithms, see Section 3-1.

140

9 / PRECIPITATION TITRATIONS

examine a streamlined procedure that is well worth learning. The streamlined calculation requires us to bear in mind that $V_e = 50.00$ mL. When 10.00 mL of Ag^+ has been added, the reaction is one-fifth complete because 10.00 mL out of the 50.00 mL of Ag^+ needed for complete reaction has been added. Therefore, four-fifths of the I^- remains unreacted. If there were no dilution, the concentration of I^- would be four-fifths of its original value. However, the original volume of 25.00 mL has been increased to 35.00 mL. If no I^- had been consumed, the concentration would be the original value of $[I^-]$ times $(25.00)/(35.00)$. Accounting for both the reaction and the dilution, we can write

Streamlined calculation, *well worth using.*

$$[I^-] = \underbrace{\left(\frac{4.000}{5.000}\right)}_{\substack{\text{Fraction} \\ \text{remaining}}} \underbrace{(0.100\ 0\ \text{M})}_{\substack{\text{Original} \\ \text{concentration}}} \underbrace{\left(\frac{25.00}{35.00}\right)}_{\substack{\text{Dilution} \\ \text{factor}}} = 0.057\ 14\ \text{M} \qquad (9\text{-}23)$$

(Original volume of I^-; Total volume of solution)

This is the same result found from Equation 9-20.

Just to reinforce the streamlined procedure, let's calculate pAg^+ when $V_{Ag^+} = 49.00$ mL. (V_{Ag^+} is the volume of Ag^+ added from the buret.) Since $V_e = 50.00$ mL, the fraction of I^- reacted is 49.00/50.00, and the fraction remaining is 1.00/50.00. The total volume is $25.00 + 49.00 = 74.00$ mL.

$$[I^-] = \left(\frac{1.00}{50.00}\right)(0.100\ 0\ \text{M})\left(\frac{25.00}{74.00}\right) = 6.76 \times 10^{-4}\ \text{M} \qquad (9\text{-}24)$$

$$[Ag^+] = K_{sp}/[I^-] = 1.2_3 \times 10^{-13}\ \text{M} \qquad (9\text{-}25)$$

$$pAg^+ = -\log[Ag^+] = 12.91 \qquad (9\text{-}26)$$

The value $pAg^+ = 12.91$ shows that the concentration of Ag^+ is negligible compared to the concentration of unreacted I^-. This is true even though the titration is 98% complete.

At the Equivalence Point

Now we have added exactly enough Ag^+ to react with all of the I^-. You may imagine that all of the AgI precipitates and some redissolves to give equal concentrations of Ag^+ and I^-. Ag^+ is found very simply:

When $V = V_e$, the concentration of Ag^+ is determined by the solubility of pure AgI.

$$[Ag^+][I^-] = K_{sp}$$

$$(x)(x) = 8.3 \times 10^{-17} \Rightarrow x = 9.1 \times 10^{-9} \qquad (9\text{-}27)$$

$$pAg^+ = -\log x = 8.04 \qquad (9\text{-}28)$$

This value of pAg is independent of the original concentration or volumes.

After the Equivalence Point

Now the concentration of Ag^+ is determined almost completely by the amount of Ag^+ added after the equivalence point. Virtually all of the Ag^+ added before the equivalence point has precipitated as AgI. Suppose that $V_{Ag^+} = 52.00$ mL. The amount added past the equivalence point is 2.00 mL. The calculation proceeds as follows:

$$\text{Moles of } Ag^+ = (0.002\ 00\ \text{L})(0.050\ 00\ \text{mol } Ag^+/\text{L}) = 0.000\ 100\ \text{mol} \quad (9\text{-}29)$$

$$[Ag^+] = (0.000\ 100\ \text{mol})/(0.077\ 00\ \text{L}) = 1.30 \times 10^{-3}\ \text{M} \quad (9\text{-}30)$$

Total volume $= 77.00$ mL

$$pAg^+ = 2.89 \quad (9\text{-}31)$$

> When $V > V_e$, the concentration of Ag^+ is determined by the excess Ag^+ added from the buret.

It would be justified to use three significant figures for the mantissa of pAg^+, since there are now three significant figures in $[Ag^+]$. To be consistent with our earlier results, we will retain only two figures. *For consistency, we will generally express all p functions in this text with two decimal places.*

A somewhat streamlined calculation can save time. The concentration of Ag^+ in the buret is 0.050 00 M, and 2.00 mL of this solution is being diluted to $(25.00 + 52.00) = 77.00$ mL. Hence $[Ag^+]$ is given by

$$[Ag^+] = \underbrace{(0.050\ 00\ \text{M})}_{\substack{\text{Original} \\ \text{concentration} \\ \text{of } Ag^+}} \underbrace{\left(\frac{2.00}{77.00}\right)}_{\substack{\text{Dilution} \\ \text{factor}}} = 1.30 \times 10^{-3}\ \text{M} \quad (9\text{-}32)$$

Volume of excess Ag^+

Total volume of solution

> Streamlined calculation.

The Shape of the Titration Curve

Figure 9-2 shows the complete titration curve and also illustrates the effect of reactant concentrations on the titration. The equivalence point is the steepest point of the curve. It is the point of maximum slope (a negative slope in this case) and is therefore an inflection point:

$$\text{Steepest slope:} \quad \frac{dy}{dx} \text{ reaches its greatest value}$$

$$\text{Inflection point:} \quad \frac{d^2y}{dx^2} = 0$$

In titrations involving 1:1 stoichiometry of reactants, the equivalence point is the steepest point of the titration curve. This is true of acid–base, complexometric, and redox titrations as well. For stoichiometries other than 1:1, such as the reaction $2Ag^+ + CrO_4^- \rightarrow Ag_2CrO_4(s)$, the curve is not symmetric near the equivalence point. The equivalence point is not at the

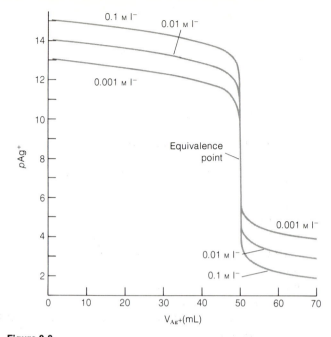

Figure 9-2
Titration curves showing the effect of concentration.
Outer curve: 25.000 mL of 0.100 0 M I⁻ titrated with 0.050 00 M Ag⁺.
Middle curve: 25.00 mL of 0.010 00 M I⁻ titrated with 0.005 000 M Ag⁺.
Inner curve: 25.00 mL of 0.001 000 M I⁻ titrated with 0.000 500 0 M Ag⁺.

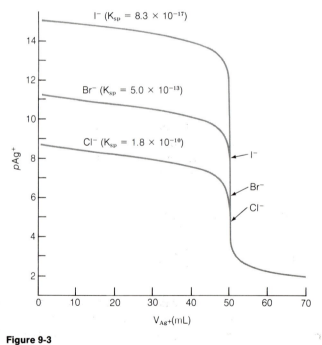

Figure 9-3
Titration curves showing the effect of K_{sp}. Each curve is calculated for 25.00 mL of 0.100 0 M halide titrated with 0.050 00 M Ag⁺. Equivalence points are marked by arrows.

center of the steepest section of the curve, and it is not an inflection point. In practice, titration curves are steep enough so that the steepest point is almost always taken as a good estimate of the equivalence point, regardless of the stoichiometry.

Figure 9-3 illustrates the role of the magnitude of K_{sp} in the titration of halide ions. It is clear that the least soluble product (AgI) gives the sharpest change at the equivalence point. However, even for AgCl, the curve is steep enough to locate the equivalence point with very little error. The larger the equilibrium constant for any titration reaction, the more pronounced will be the change in concentration near the equivalence point.

At the equivalence point, the titration curve is steepest for the least soluble precipitates.

EXAMPLE

A solution containing 25.00 mL of 0.041 32 M $Hg_2(NO_3)_2$ was titrated with 0.057 89 M KIO_3.

$$Hg_2^{2+} + 2IO_3^- \rightarrow Hg_2(IO_3)_2(s)$$

iodate

The solubility product for $Hg_2(IO_3)_2$ is 1.3×10^{-18}. Calculate the concentration of Hg_2^{2+} ion in the solution (a) after addition of 34.00 mL of KIO_3, (b) after addition of 36.00 mL of KIO_3, and (c) at the equivalence point.

The reaction requires two moles of IO_3^- per mole of Hg_2^{2+}. The volume of iodate needed to reach the equivalence point is found as follows:

$$\text{Moles of } IO_3^- = 2 \, (\text{Moles of } Hg_2^{2+}) \quad \begin{cases} \text{Notice which side of} \\ \text{the equation has the 2!} \end{cases}$$

$$(V_e)(0.057\,89 \text{ M}) = 2 \, (25.00 \text{ mL})(0.041\,32 \text{ M}) \Rightarrow V_e = 35.69 \text{ mL}$$

(a) When V = 34.00 mL, the precipitation of Hg_2^{2+} is not yet complete.

$$[Hg_2^{2+}] = \underbrace{\left(\frac{35.69 - 34.00}{35.69}\right)}_{\substack{\text{Fraction} \\ \text{remaining}}} \underbrace{(0.041\,32)}_{\substack{\text{Original} \\ \text{concentration} \\ \text{of } Hg_2^{2+}}} \overbrace{\underbrace{\left(\frac{25.00}{25.00 + 34.00}\right)}_{\substack{\text{Dilution} \\ \text{factor}}}}^{\text{Original volume of } Hg_2^{2+}} = 8.29 \times 10^{-4} \text{ M}$$

Total volume of solution

(b) When V = 36.00 mL, the precipitation is complete. We have gone $(36.00 - 35.69) = 0.31$ mL *past* the equivalence point. The concentration of excess IO_3^- in the solution is

$$[IO_3^-] = \underbrace{(0.057\,89)}_{\substack{\text{Original} \\ \text{concentration} \\ \text{of } IO_3^-}} \overbrace{\underbrace{\left(\frac{0.31}{25.00 + 36.00}\right)}_{\substack{\text{Dilution} \\ \text{factor}}}}^{\text{Volume of excess } IO_3^-} = 2.9 \times 10^{-4} \text{ M}$$

Total volume of solution

The concentration of Hg_2^{2+} in equilibrium with solid $Hg_2(IO_3)_2$ plus this much IO_3^- is

$$[Hg_2^{2+}] = \frac{K_{sp}}{[IO_3^-]^2} = \frac{1.3 \times 10^{-18}}{(2.9 \times 10^{-4})^2} = 1.5 \times 10^{-11} \text{ M}$$

(c) At the equivalence point, we have added only enough IO_3^- to react with all of the Hg_2^{2+}. We can write

$$Hg_2(IO_3)_2(s) \rightleftharpoons Hg_2^{2+} + 2IO_3^-$$
$$ x 2x$$

$$(x)(2x)^2 = K_{sp} \Rightarrow x = [Hg_2^{2+}] = 6.9 \times 10^{-7} \text{ M}$$

9-5 TITRATION OF A MIXTURE

When a mixture is titrated, the product with the *smaller* K_{sp} precipitates first.

If a mixture of two precipitable ions is titrated, the less soluble precipitate will be formed first. If the two solubility products are sufficiently different, the first precipitation will be nearly complete before the second commences.

Consider the titration with $AgNO_3$ of a solution containing KI and KCl. Since $K_{sp}(AgI) \ll K_{sp}(AgCl)$, the first Ag^+ added will precipitate AgI. Further addition of Ag^+ continues to precipitate I^- with no effect on Cl^-. When precipitation of I^- is almost complete, the concentration of Ag^+ abruptly increases. Then, when the concentration of Ag^+ is high enough, AgCl begins to precipitate and $[Ag^+]$ levels off again. Finally, when the Cl^- is consumed, another abrupt change in $[Ag^+]$ occurs. Qualitatively, we expect to see two breaks in the titration curve. The first corresponds to the AgI equivalence point, and the second to the AgCl equivalence point.

The precipitation of I^- and Cl^- with Ag^+ produces two distinct breaks in the titration curve. The first corresponds to the reaction of I^- and the second to the reaction of Cl^-.

Figure 9-4 shows an experimental curve for this titration. The apparatus

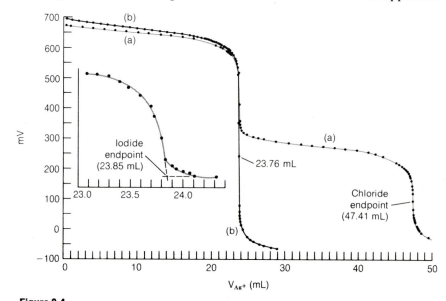

Figure 9-4

(a) Experimental titration curve for 40.0 mL of 0.050 2 M KI plus 0.050 0 M KCl titrated with 0.084 5 M $AgNO_3$. The inset is an expanded view of the region near the first equivalence point. (b) Also shown is the titration curve for 20.00 mL of 0.100 4 M I^- titrated with 0.084 5 M Ag^+.

9-5 TITRATION OF A MIXTURE

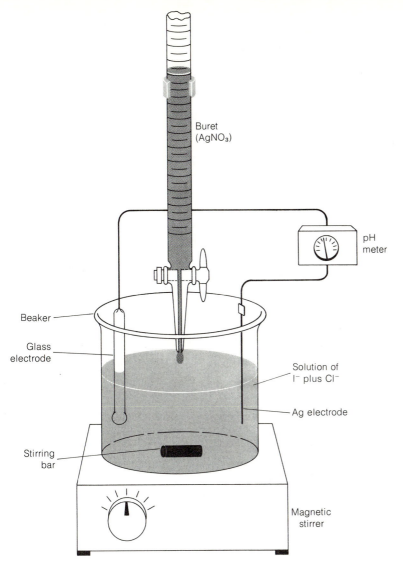

Figure 9-5
Apparatus for measuring the titration curves in Figure 9-4. The silver electrode responds to changes in the Ag^+ concentration, and the glass electrode provides a constant reference potential in this experiment. The measured potential changes by approximately 59 mV for each factor-of-ten change in $[Ag^+]$. All solutions, including $AgNO_3$, were maintained at pH 2.0 using 0.010 M sulfate buffer prepared from H_2SO_4 and KOH.

See Section 24-10 for details of this procedure.

used to measure the curve is shown in Figure 9-5, and the theory of how this system measures Ag^+ concentration is discussed in Chapter 16.

The I^- endpoint is taken as the intersection of the steep and nearly horizontal curves shown in the inset of Figure 9-4. The reason for using the intersection is that the precipitation of I^- is not quite complete as Cl^- begins to precipitate. Therefore, the end of the steep portion (the intersection) is a better approximation of the equivalence point than is the middle of the steep section. The Cl^- endpoint is taken as the midpoint of the second steep section,

Before Cl^- precipitates, the calculations for AgI precipitation are just as they were in Section 9-4.

at 47.41 mL. The moles of Cl^- in the unknown correspond to the moles of Ag^+ delivered between the first and second endpoints. That is, it requires 23.76 mL of Ag^+ to precipitate I^-, and $(47.41 - 23.76)$ mL of Ag^+ to precipitate Cl^-.

Comparing the I^-/Cl^- and pure I^- titration curves in Figure 9-4 shows that the iodide endpoint is 0.38% too high in the I^-/Cl^- titration. We expect the first endpoint at 23.76 mL, but it is observed at 23.85 mL. There are two factors contributing to this high value. One is experimental error, which is always present. This is as likely to be positive as negative. However, the endpoint in some titrations, especially Br^-/Cl^- titrations, is found to be systematically 0–3% high, depending on conditions. This has been attributed to a small amount of **coprecipitation** of AgCl with AgBr. Even though the solubility product of AgCl has not been exceeded, a little Cl^- precipitates with the Br^- and carries down a corresponding amount of Ag^+. A high concentration of an anion such as nitrate has been found to minimize the coprecipitation.

The second endpoint in Figure 9-4 corresponds to the total precipitation of both halides. It is observed at the expected value of V_{Ag^+}. The concentration of Cl^-, found from the *difference* between the two endpoints, will be slightly low in Figure 9-4, because the first endpoint is slightly high.

9-6 ENDPOINT DETECTION

Three techniques are commonly employed to detect the endpoint in precipitation titrations:

1. Potentiometric methods—using electrodes, as in Figure 9-5. These techniques are discussed in Chapter 16.
2. Indicator methods—discussed below.
3. Light scattering methods—turbidimetry and nephelometry, described in Box 9-1.

In the remainder of this section we will discuss three types of indicator methods applied to the titration of Cl^- with Ag^+. Titrations with Ag^+ are called **argentometric** titrations. The three indicator methods are:

1. **Mohr titration**—formation of a colored precipitate at the endpoint.
2. **Volhard titration**—formation of a soluble, colored complex at the endpoint.
3. **Fajans titration**—adsorption of a colored indicator on the precipitate at the endpoint.

Mohr Titration

In this procedure, Cl^- is titrated with Ag^+ in the presence of CrO_4^{2-} (chromate, dissolved as Na_2CrO_4).

Titration reaction: $$Ag^+ + Cl^- \rightarrow AgCl(s) \quad \text{(white)} \qquad (9\text{-}33)$$

Endpoint reaction: $$2Ag^+ + CrO_4^{2-} \rightarrow Ag_2CrO_4(s) \quad \text{(red)} \quad \text{(9-34)}$$

The AgCl precipitates before Ag_2CrO_4. The color of AgCl is white; dissolved CrO_4^{2-} is yellow; and Ag_2CrO_4 is red. The AgCl endpoint is indicated by the first appearance of red Ag_2CrO_4. Reasonable control of CrO_4^{2-} concentration and pH are required for Ag_2CrO_4 precipitation to occur at the desired point in the titration.

Since a certain amount of Ag_2CrO_4 is necessary for visual detection, the color is not seen until after the true equivalence point. We can correct for this titration error in two ways. One is by means of a blank titration with no chloride present. The volume of Ag^+ needed to form detectable red color is then subtracted from V_{Ag^+} in the Cl^- titration. Alternatively, we can standardize the $AgNO_3$ by the Mohr method, using a standard NaCl solution and conditions similar to those for the titration of unknown. The Mohr method is useful for Cl^-, Br^-, and CN^-, but not for I^- or SCN^- (thiocyanate).

> Titration error arises because some Ag^+ is needed to form a detectable amount of red precipitate.

Volhard Titration

This is actually a procedure for the titration of Ag^+. To determine Cl^-, a back titration is necessary. First, the Cl^- is precipitated by a known, excess quantity of standard $AgNO_3$.

$$Ag^+ + Cl^- \rightarrow AgCl(s) \quad \text{(9-35)}$$

The AgCl is isolated, and the excess Ag^+ is titrated with standard KSCN in the presence of Fe^{3+}.

> Since the Volhard method is a titration of Ag^+, it can be adapted for the determination of any anion that forms an insoluble silver salt.

$$Ag^+ + SCN^- \rightarrow AgSCN(s) \quad \text{(9-36)}$$

When all of the Ag^+ has been consumed, the SCN^- reacts with Fe^{3+} to form a red complex.

$$Fe^{3+} + SCN^- \rightarrow FeSCN^{2+} \quad \text{(9-37)}$$
$$\text{(red)}$$

The appearance of the red color signals the endpoint. Knowing how much SCN^- was required for the back titration tells us how much Ag^+ was left over from the reaction with Cl^-. Since the total amount of Ag^+ is known, the amount consumed by Cl^- can then be calculated.

In the analysis of Cl^- by the Volhard method, the endpoint slowly fades because AgCl is more soluble than AgSCN. The AgCl slowly dissolves and is replaced by AgSCN. To prevent this from happening, two techniques are commonly used. One is to filter off the AgCl and titrate only the Ag^+ in the filtrate. An easier procedure is to shake a few milliliters of nitrobenzene ($C_6H_5NO_2$) with the precipitated AgCl prior to the back titration. Nitrobenzene coats the AgCl and effectively isolates it from attack by SCN^-. Br^- and I^-, whose silver salts are *less* soluble than AgSCN, may be titrated by the Volhard method without isolating the silver halide precipitate.

9 / PRECIPITATION TITRATIONS

See Figure 8-2 and its associated discussion for a description of the *electric double layer* surrounding a particle of precipitate.

Fajans Titration

This procedure uses an **adsorption indicator.** To see how this works, we must recall the electrical phenomena associated with precipitate formation. When Ag^+ is added to Cl^-, there will be excess Cl^- ions in solution prior to the equivalence point. Some Cl^- is selectively adsorbed on the AgCl surface, imparting a negative charge to the crystal surface. After the equivalence point, there is excess Ag^+ in solution. Adsorption of the Ag^+ cations on the crystal surface creates a positive charge on the particles of precipitate. The abrupt change from negative charge to positive charge occurs at the equivalence point.

The common adsorption indicators are anionic dyes, which are attracted to the positively charged particles of precipitate produced immediately after the equivalence point. The adsorption of the negatively charged dye on the positively charged surface changes the color of the dye by interactions that are not well understood. The color change signals the endpoint in the titration. Because the indicator reacts with the precipitate surface, it is desirable to have as much surface area as possible. This means performing the titration under conditions that tend to keep the particles as small as possible. (Small particles have more surface area than an equal volume of large particles.) This condition is the opposite of that required for gravimetric analysis, in which large, easily filterable particles are desired. Low electrolyte concentration helps to prevent coagulation of the precipitate and maintain small particle size.

The indicator most commonly used for AgCl is dichlorofluorescein.

dichlorofluorescein

tetrabromofluorescein
(eosin)

This dye has a greenish-yellow color in solution, but turns pink when it is adsorbed on AgCl (see Demonstration 9-1). Because the indicator is a weak acid, and must be present in its anionic form, the pH of the reaction must be controlled. The dye eosin is useful in the titration of Br^-, I^-, and SCN^-. It gives a sharper endpoint than dichlorofluorescein and is more sensitive (i.e., less halide is required for titration). It cannot be used for AgCl because the eosin anion is more strongly bound to AgCl than is Cl^- ion. Eosin will bind to the AgCl crystallites even before the particles become positively charged.

In all argentometric titrations, but especially with adsorption indicators, strong light (such as daylight through a window) should be avoided. Light causes decomposition of the silver salts, and adsorbed indicators are especially light-sensitive.

Demonstration 9-1   FAJANS TITRATION

The Fajans titration of Cl^- with Ag^+ convincingly demonstrates the utility of indicator endpoints in precipitation titrations. Dissolve 0.5 g of NaCl plus 0.15 g of dextrin in 400 mL of water. The purpose of the dextrin is to retard coagulation of the AgCl precipitate. Add 1 mL of dichlorofluorescein indicator. The indicator solution contains 1 mg/mL of dichlorofluorescein in 95% aqueous ethanol or 1 mg/mL of the sodium salt in water. Titrate the NaCl solution with a solution containing 2 g of $AgNO_3$ in 30 mL. About 20 mL are required to reach the endpoint.

Color Plate 5 shows the yellow color of the indicator in the NaCl solution prior to the titration. Color Plate 6 shows the milky white appearance of the AgCl suspension during titration, before the endpoint is reached. The pink suspension in Color Plate 7 appears at the endpoint, when the anionic indicator becomes adsorbed to the cationic particles of precipitate.

Table 9-1
Some applications of precipitation titrations

Species analyzed	Notes
	MOHR METHOD
Cl^-, Br^-	Ag_2CrO_4 endpoint is used.
	VOLHARD METHOD
Br^-, I^-, SCN^-, CNO^-, AsO_4^{3-}	Precipitate removal is unnecessary.
Cl^-, PO_4^{3-}, CN^-, $C_2O_4^{2-}$, CO_3^{2-}, S^{2-}, CrO_4^{2-}	Precipitate removal required.
BH_4^-	Back titration of Ag^+ left after reaction with BH_4^-: $$BH_4^- + 8Ag^+ + 8OH^- = 8Ag(s) + H_2BO_3^- + 5H_2O$$
K^+	K^+ is first precipitated with a known excess of $(C_6H_5)_4B^-$. Remaining $(C_6H_5)_4B^-$ is precipitated with a known excess of Ag^+. Unreacted Ag^+ is then titrated with SCN^-.
	FAJANS METHOD
Cl^-, Br^-, I^-, SCN^-, $Fe(CN)_6^{4-}$	Titration with Ag^+. Detection with such dyes as fluorescein, dichlorofluorescein, eosin, bromophenol blue.
F^-	Titration with $Th(NO_3)_4$ to produce ThF_4. Endpoint detected with alizarin red S.
Zn^{2+}	Titration with $K_4Fe(CN)_6$ to produce $K_2Zn_3[Fe(CN)_6]_2$. Endpoint detection with diphenylamine.
SO_4^{2-}	Titration with $Ba(OH)_2$ in 50% (vol/vol) aqueous methanol using alizarin red S as indicator.
Hg_2^{2+}	Titration with NaCl to produce Hg_2Cl_2. Endpoint detected with bromophenol blue.
PO_4^{3-}, $C_2O_4^{2-}$	Titration with $Pb(CH_3CO_2)_2$ to give $Pb_3(PO_4)_2$ or PbC_2O_4. Endpoint detected with dibromofluorescein (PO_4^{3-}) or fluorescein ($C_2O_4^{2-}$).

Scope of Indicator Methods

To indicate the scope of indicator methods in precipitation titrations, a few applications are listed in Table 9-1. While the Mohr and Volhard methods are specifically applicable to argentometric titrations, the Fajans method is used for a wider variety of titrations. Because the Volhard titration is carried out in acidic solution (typically 0.2 M HNO_3), it avoids certain interferences that would affect other titrations. The silver salts of such anions as CO_3^{2-}, $C_2O_4^{2-}$, and AsO_4^{3-} are soluble in acidic solution and thus do not interfere with the analysis.

Summary

In a titration the volume of titrant required for complete reaction is used to calculate the quantity of analyte present. If the stoichiometry of the reaction is $tT + aA \rightarrow$ products, the moles of titrant delivered are $V_T M_T$, and the moles of analyte are (a/t)(Moles of titrant). We approximate the equivalence point by measuring the endpoint, at which a sudden change in a physical property (such as indicator color, pH, electrode potential, conductivity, or absorbance) occurs. The difference between the measured endpoint and the true equivalence point, called the titration error, can be estimated by performing a blank titration. The utility of any titration depends on knowledge of the titrant concentration. This is determined by dissolving a known mass of primary standard in a known volume of solution. Alternatively, titrant can be standardized by titrating a known mass of primary standard. In a direct titration, titrant is added to analyte and the volume of titrant is measured. In a back titration, a known excess of reagent is added to analyte, and the excess is titrated with a second standard reagent.

In a precipitation titration, analyte is precipitated by the titrant. The endpoint can be detected by potentiometric, light-scattering, or indicator methods. The most general indicator method (the Fajans titration) is based on the adsorption of a charged indicator by the charged surface of the precipitate after the equivalence point. The Volhard titration, used to measure Ag^+, is based on the reaction of Fe^{3+} with SCN^- after the precipitation of AgSCN is complete. The Mohr titration utilizes the precipitation of colored Ag_2CrO_4 after the argentometric titration of Cl^- or Br^-.

The concentrations of reactants and products during a precipitation titration are calculated in three ways. Before the endpoint, there is excess analyte. The concentration of titrant can be found from the solubility product of the precipitate and the known concentration of excess analyte. At the equivalence point, the concentrations of both reactants are governed by the dissociation equilibrium of product. After the equivalence point, the concentration of analyte can be determined from the solubility product of precipitate and the known concentration of excess titrant.

Terms to Understand

adsorption indicator

analyte

argentometric titration

back titration

blank titration

coprecipitation

direct titration

endpoint

equivalence point

Fajans titration

indicator

Mohr titration

nephelometry

primary standard

standardization

titrant

titration error

turbidimetry

turbidity

turbidity coefficient

Volhard titration

volumetric analysis

Exercises

9-A. Ascorbic acid (vitamin C) reacts with I_3^- according to the following reaction:

HOCH$_2$CCH ... C=O + I_3^- ⟶

ascorbic acid

HOCH$_2$CCH ... C=O + 3I$^-$ + 2H$^+$

dehydroascorbic acid

Starch is used as an indicator in the reaction. The endpoint is marked by the appearance of a deep blue starch–iodine complex when the first drop of un-reacted I_3^- remains in the solution.

(a) If 29.41 mL of I_3^- solution is required to react with 0.197 0 g of pure ascorbic acid, what is the molarity of the I_3^- solution?

(b) A vitamic C tablet containing ascorbic acid plus an inert binder was ground to a powder, and 0.424 2 g was titrated by 31.63 mL of I_3^-. Find the weight percent of ascorbic acid in the tablet.

9-B. A solid sample weighing 0.237 6 g contained only malonic acid (F. W. 104.06) and anilinium chloride (F. W. 129.59). If it required 34.02 mL of 0.087 71 M NaOH to neutralize the sample, find the weight percent of each component in the solid mixture. The reactions are

$$CH_2(CO_2H)_2 + 2OH^- \rightarrow CH_2(CO_2^-)_2 + H_2O$$

malonic acid   malonate

—NH$_3^+$Cl$^-$ + OH$^-$ ⟶

anilinium chloride

—NH$_2$ + H$_2$O + Cl$^-$

aniline

9-C. Suppose that 50.0 mL of 0.080 0 M KSCN is titrated with 0.040 0 M Cu$^+$. The solubility product of CuSCN is 4.8×10^{-15}. At each of the following volumes of titrant, calculate pCu$^+$, and construct a graph of pCu$^+$ versus milliliters of Cu$^+$ added: 0.10, 10.0, 25.0, 50.0, 75.0, 95.0, 99.0, 99.9, 100.0, 100.1, 101.0, 110.0 mL.

9-D. A 40.0 mL solution of 0.040 0 M Hg$_2$(NO$_3$)$_2$ was titrated with 60.0 mL of 0.100 M KI to precipitate Hg$_2$I$_2(s)$ ($K_{sp} = 4.5 \times 10^{-29}$).

(a) What volume of KI is needed to reach the equivalence point?

(b) Calculate the ionic strength of the solution when 60.0 mL of KI has been added.

(c) *Without ignoring activity*, calculate pHg$_2^{2+}$ ($\equiv -\log \mathscr{A}_{Hg_2^{2+}}$) when 60.0 mL of KI has been added.

9-E. Construct a graph of pAg$^+$ versus milliliters of Ag$^+$ for the titration of 40.00 mL of solution containing 0.050 00 M Br$^-$ and 0.050 00 M Cl$^-$. The titrant is 0.084 54 M AgNO$_3$. Calculate pAg$^+$ at the following volumes: 2.00, 10.00, 22.00, 23.00, 24.00, 30.00, 40.00, second equivalence point, 50.00 mL.

9-F. Consider the titration of 50.00 (± 0.05) mL of a mixture of I$^-$ and SCN$^-$ with 0.068 3 (± 0.000 1) M Ag$^+$. The first equivalence point is observed at 12.6 (± 0.4) mL, and the second occurs at 27.7 (± 0.3) mL.

(a) Find the molarity and the uncertainty in molarity of thiocyanate in the original mixture.

(b) Suppose that the uncertainties listed above are all the same, except that the uncertainty of the first equivalence point (12.6 $\pm$? mL) is variable. What is the maximum uncertainty (milliliters) of the first equivalence point if the uncertainty in SCN$^-$ molarity is to be $\leq 4.0\%$?

9-G. Sketch graphs showing how (a) the turbidimetric "transmittance" and (b) the nephelometric scattered-light intensity would vary during the titration of SO$_4^{2-}$ by Ba^{2+}. Indicate the endpoint in each case.

Problems

9-1. Explain the statement: "The validity of an analytical result ultimately depends on knowing the composition of some primary standard."

9-2. Distinguish between the terms *endpoint* and *equivalence point*.

9-3. For Reaction 9-1, how many milliliters of 0.165 0 M $KMnO_4$ are needed to react with 108.0 mL of 0.165 0 M oxalic acid? How many milliliters of 0.165 0 M oxalic acid are required to react with 108.0 mL of 0.165 0 M $KMnO_4$?

9-4. How many milliliters of 0.100 M KI are needed to react with 40.0 mL of 0.040 0 M $Hg_2(NO_3)_2$ if the reaction is $Hg_2^{2+} + 2I^- \rightarrow Hg_2I_2(s)$?

9-5. How many milligrams of oxalic acid dihydrate $(H_2C_2O_4 \cdot 2H_2O)$ will react with 1.00 mL of 0.027 3 M ceric sulfate $(Ce(SO_4)_2)$ if the reaction is $H_2C_2O_4 + 2Ce^{4+} \rightarrow 2CO_2 + 2Ce^{3+} + 2H^+$?

9-6. Sulfur can be determined by combustion analysis, which produces a mixture of SO_2 and SO_3. The gas stream is passed through H_2O_2 to convert these products to H_2SO_4, which is titrated with standard base. When 6.123 mg of a substance was burned, the resulting H_2SO_4 required 3.01 mL of 0.015 76 M NaOH for titration. What is the weight percent of sulfur in the sample?

9-7. The Kjeldahl procedure (Reactions 9-7 through 9-10) was used to analyze 256 μL of a solution containing 37.9 mg protein/mL. The liberated NH_3 was collected in 5.00 mL of 0.033 6 M HCl, and the remaining acid required 6.34 mL of 0.010 M NaOH for complete titration. What is the weight percent of nitrogen in the protein?

9-8. Ammonia reacts with hypobromite (OBr^-) according to the following reaction:

$$2NH_3 + 3OBr^- \rightarrow N_2 + 3Br^- + 3H_2O$$

What is the molarity of a hypobromite solution if 1.00 mL of the OBr^- solution reacts with 1.69 mg of NH_3?

9-9. Arsenious oxide (As_2O_3) is available in pure form and is a useful (and poisonous) primary standard for standardizing many oxidizing agents, such as MnO_4^-. The As_2O_3 is first dissolved in base and then titrated with MnO_4^- in acidic solution. A small amount of iodide (I^-) or iodate (IO_3^-) is used to catalyze the reaction between H_3AsO_3 and MnO_4^-.

The reactions are

$$As_2O_3 + 4OH^- \rightleftharpoons 2HAsO_3^{2-} + H_2O$$

$$HAsO_3^{2-} + 2H^+ \rightleftharpoons H_3AsO_3$$

$$5H_3AsO_3 + 2MnO_4^- + 6H^+ \rightleftharpoons 5H_3AsO_4 + 2Mn^{2+}$$
$$+ 3H_2O$$

(a) A 3.214 g aliquot of $KMnO_4$ (F.W. 158.034) was dissolved in 1.000 L of water, heated to cause any reactions with impurities to occur, cooled, and filtered. What is the theoretical molarity of this solution if no MnO_4^- was consumed by impurities?

(b) What mass of As_2O_3 (F.W. 197.84) would be just sufficient to react with 25.00 mL of the $KMnO_4$ solution in part a?

(c) It was found that 0.146 8 g of As_2O_3 required 29.98 mL of $KMnO_4$ solution for the faint color of unreacted MnO_4^- to appear. In a blank titration, 0.03 mL of MnO_4^- was required to produce enough color to be seen. Calculate the molarity of the permanganate solution.

9-10. A 20.00 mL solution containing Co^{2+} was treated with 25.00 mL of 0.050 23 M EDTA to consume all of the Co^{2+} and leave excess EDTA. The excess EDTA was titrated by 4.32 mL of 0.049 37 M Zn^{2+}. Calculate the molarity of Co^{2+} in the original solution.

9-11. A cyanide solution with a volume of 12.73 mL was treated with 25.00 mL of Ni^{2+} solution (containing excess Ni^{2+}) to convert the cyanide to tetracyanonickelate:

$$4CN^- + Ni^{2+} \rightarrow Ni(CN)_4^{2-}$$

The excess Ni^{2+} was then titrated with 10.15 mL of 0.013 07 M EDTA.

$$Ni^{2+} + EDTA^{4-} \rightarrow Ni(EDTA)^{2-}$$

$Ni(CN)_4^{2-}$ does not react with EDTA. If 39.35 mL of EDTA was required to react with 30.10 mL of the original Ni^{2+} solution, calculate the molarity of CN^- in the 12.73 mL cyanide sample.

9-12. A 0.238 6 g sample contained only NaCl and KBr. It was dissolved in water and required 48.40 mL of 0.048 37 M $AgNO_3$ for complete titration of both halides. Calculate the weight percent of Br in the solid sample.

9-13. Consider the titration of 25.00 mL of 0.082 30 M KI with 0.051 10 M $AgNO_3$. Calculate pAg^+ at the following volumes of added $AgNO_3$:
(a) 39.00 mL (b) V_e (c) 44.30 mL

9-14. A 30.00 mL solution containing an unknown amount of I^- was treated with 50.00 mL of 0.365 0 M $AgNO_3$. The precipitated AgI was filtered off, and the filtrate (plus Fe^{3+}) was titrated with 0.287 0 M KSCN. When 37.60 mL had been added, the solution turned red. How many milligrams of I^- were present in the original solution?

9-15. Co^{2+} can be analyzed by treatment with a known excess of thiocyanate in the presence of pyridine:

$$Co^{2+} + 4 \ \langle \text{⬡} \rangle N: + \ 2SCN^- \longrightarrow$$

pyridine   thiocyanate

$$Co(C_5H_5N)_4(SCN)_2(s)$$

The precipitate is filtered off, and the SCN^- content of the filtrate is determined by Volhard titration. A 25.00 mL unknown solution was treated with 3 mL of pyridine and 25.00 mL of 0.102 8 M KSCN in a 250 mL volumetric flask. The solution was diluted to the mark, mixed, and filtered. After discarding the first few milliliters of filtrate, 50.0 mL of filtrate was acidified with HNO_3 and treated with 5.00 mL of 0.105 5 M $AgNO_3$. After addition of Fe^{3+} indicator, the excess Ag^+ required 3.76 mL of 0.102 8 M KSCN to reach the Volhard endpoint. Calculate the Co^{2+} concentration in the unknown.

9-16. Why does the surface charge of a precipitate particle change sign at the equivalence point?

9-17. Examine the procedure in Table 9-1 for the Fajans titration of Zn^{2+}. Do you expect the charge on the precipitate to be positive or negative after the equivalence point?

9-18. Derive an algebraic relation between the turbidity coefficient and the constant k in Box 9-1. What values of τ give apparent transmittances of 90.0% and 10.0% if $b = 1.00$ cm? What value of τ gives an absorbance of 1.00? (You may wish to use the relation $\ln x = (\log x)(\ln 10)$ derived in Appendix A.)

9-19. A solution of volume 25.00 mL containing 0.031 1 M $Na_2C_2O_4$ was titrated with 0.025 7 M $La(ClO_4)_3$ to precipitate lanthanum oxalate:

$$2La^{3+} + 3C_2O_4^{2-} \rightarrow La_2(C_2O_4)_3(s)$$

lanthanum   oxalate

(a) What volume of $La(ClO_4)_3$ is required to reach the equivalence point?
(b) Calculate the value of pLa^{3+} when 10.00 mL of $La(ClO_4)_3$ has been added to the sodium oxalate.

9-20. A mixture having a volume of 10.00 mL and containing 0.100 0 M Ag^+ and 0.100 0 M Hg_2^{2+} was titrated with 0.100 0 M KCN to precipitate $Hg_2(CN)_2$ and AgCN.
(a) Calculate pCN^- at each of the following volumes of added KCN: 5.00, 10.00, 15.00, 19.90, 20.10, 25.00, 30.00, 35.00 mL.
(b) Should any AgCN be precipitated at 19.90 mL?

9-21. A solution containing 10.00 mL of 0.100 M LiF was titrated with 0.010 0 M $Th(NO_3)_4$ to precipitate ThF_4.
(a) What volume of $Th(NO_3)_4$ is needed to reach the equivalence point?
(b) What is the ionic strength of the solution when 1.00 mL of $Th(NO_3)_4$ has been added?
(c) Neglecting activities and using the value of K_{sp} for ThF_4 in Appendix F, calculate pTh^{4+} when 1.00 mL of $Th(NO_3)_4$ has been added to the LiF.

9-22. Suppose that 20.00 mL of 0.100 0 M Br^- is titrated with 0.080 0 M $AgNO_3$ and the endpoint is detected by the Mohr method. The equivalence point occurs at 25.00 mL. If the titration error is to be $\leq 0.1\%$, then the Ag_2CrO_4 should precipitate between 24.975 and 25.025 mL. Calculate the upper and lower limits for the CrO_4^{2-} concentration in the titration solution at the equivalence point if the precipitation of Ag_2CrO_4 is to commence between 24.975 and 25.025 mL.

10 / Introduction to Acids and Bases

The next four chapters present a fairly detailed description of acid–base chemistry. Understanding the behavior of acids and bases is essential in virtually every field of science involving chemistry. It would be difficult to have a meaningful discussion of almost anything from protein biosynthesis to the weathering of rocks without understanding the chemistry of acids and bases. If your major field of interest is *not* chemistry, Chapters 10–13 are probably the most important parts of this book for you.

10-1 WHAT ARE ACIDS AND BASES?

For aqueous chemistry, an **acid** is most usefully defined as a substance that increases the concentration of H_3O^+ when added to water. Conversely, a **base** decreases the concentration of H_3O^+ in aqueous solution. The formula H_3O^+, called the **hydronium ion,** is a more accurate description of the species we have been calling H^+ so far in this text. The hydronium ion is a combination of H^+ (a proton; that is, a hydrogen atom that has lost an electron) with H_2O. Since a decrease in H_3O^+ concentration necessarily requires an increase in OH^- concentration, a base is a substance that increases the concentration of OH^- in aqueous solution.

> *Question:* Why can we say that a decrease in H_3O^+ concentration necessarily requires an increase in OH^- concentration?

Two Classifications of Acid–Base Behavior

In 1923 Brønsted and Lowry classified acids as proton donors and bases as proton acceptors. These definitions include the one given above. For example, HCl is an acid (a proton donor) and it increases the concentration of H_3O^+ in water:

Brønsted–Lowry definitions:

acid—proton donor
base—proton acceptor

$$HCl + H_2O \rightleftharpoons H_3O^+ + Cl^- \qquad (10\text{-}1)$$

10 / INTRODUCTION TO ACIDS AND BASES

The Brønsted and Lowry definition does not require that H_3O^+ be formed. This definition can therefore be extended to nonaqueous solvents and even to the gas phase:

$$HCl(g) \; + \; NH_3(g) \rightleftharpoons NH_4^+Cl^-(s) \qquad (10\text{-}2)$$

hydrochloric   ammonia   ammonium
acid   chloride

(acid)   (base)   (salt)

Any ionic solid, such as ammonium chloride, is called a **salt.** In a formal sense, a salt can be thought of as the product of an acid–base reaction. When an acid and a base react, they are said to **neutralize** each other. Most salts are **strong electrolytes.** This means that they dissociate completely into their component ions when dissolved in water. Thus, ammonium chloride gives NH_4^+ and Cl^- in aqueous solution:

$$NH_4^+Cl^-(s) \rightarrow NH_4^+(aq) + Cl^-(aq) \qquad (10\text{-}3)$$

Lewis definitions:

    acid—electron pair acceptor
    base—electron pair donor

Another definition of acid–base behavior was put forth by G. N. Lewis in the 1920s. He defined an acid as an electron-pair acceptor and a base as an electron-pair donor. The Lewis definitions are more general and include those given above, extending the concept of acid and base to species that need not have a reactive H^+. The simplest application of this classification is to the reaction between H_3O^+ and OH^- in aqueous solution:

Room to   A pair of   A chemical
accept   electrons to   bond
electrons   be donated

The pair of electrons donated by OH^- is accepted by H^+ (which dissociates from H_3O^+), forming a bond.

The Lewis definitions can apply to reactions that have little resemblance to aqueous acid–base reactions. For example, the reaction of antimony pentachloride with chloride in certain nonaqueous solvents can be thought of as an acid–base reaction:

antimony   hexachloroantimonate
pentachloride   ion

The antimony pentachloride is acting as a **Lewis acid,** accepting Cl^-. The chloride ion donates a pair of electrons and is therefore a **Lewis base.**

For the remainder of this text, when we speak of acids and bases, we are speaking of **Brønsted and Lowry acids and bases.** *The acids we speak of are proton donors, and the bases are proton acceptors.* Reactions such as 10-5 are important, but we usually think of them as complex ion equilibria, rather than as acid–base reactions.

Conjugate Acids and Bases

The products of any reaction between an acid and base may also be classified as acids and bases:

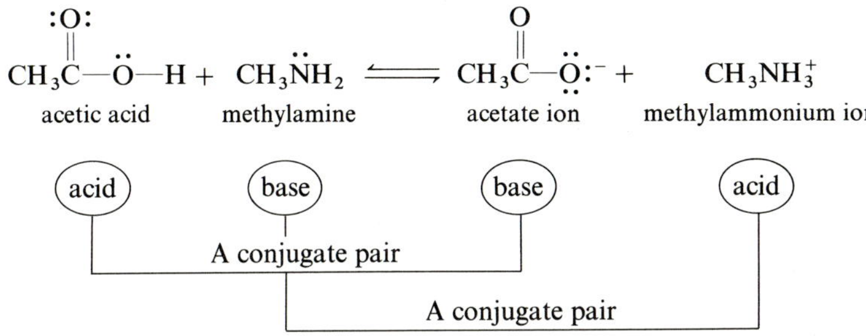

$$CH_3C\overset{:O:}{\|}\ddot{O}-H + CH_3\ddot{N}H_2 \rightleftharpoons CH_3C\overset{O}{\|}\ddot{O}:^- + CH_3NH_3^+ \qquad (10\text{-}6)$$

acetic acid   methylamine   acetate ion   methylammonium ion

(acid)   (base)   (base)   (acid)

A conjugate pair

A conjugate pair

> Conjugate acids and bases are related by the gain or loss of one proton.

Acetate is a base because it can accept a proton to make acetic acid. The methylammonium ion is an acid because it can donate a proton and become methylamine. Acetic acid and the acetate ion are said to be a **conjugate acid–base pair.** Methylamine and the methylammonium ion are likewise conjugate. *Conjugate acids and bases are related to each other by the gain or loss of one H^+.*

Nature of H^+ and OH^-

For simplicity, we will generally write H^+, not H_3O^+, as the formula for an ionized hydrogen atom (a proton) in aqueous solution. It is certain that the proton does not exist by itself in water. The simplest formula, which is found in some crystalline salts, is H_3O^+. For example, crystals of the compound perchloric acid monohydrate consist of tetrahedral perchlorate ions and pyramidal hydronium (also called hydroxonium) ions.

> We will write H^+ when we really mean H_3O^+.

$$HClO_4 \cdot H_2O \text{ is really } \left[H \cdots \overset{\cdot\cdot}{\underset{H}{\overset{|}{O}}} \overset{H}{} \right]^+ \left[O \cdots \overset{O}{\underset{O}{\overset{|}{Cl}}} O \right]^-$$

hydronium   perchlorate

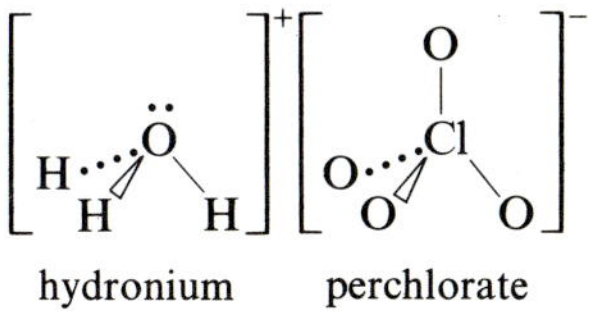

10 / INTRODUCTION TO ACIDS AND BASES

Bond enthalpy refers to the heat needed to break a chemical bond in the gas phase.

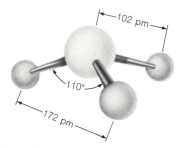

Figure 10-1
Typical structure of the H_3O^+ ion, as found in several crystals.

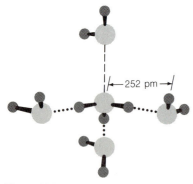

Figure 10-2
Typical coordination environment of H_3O^+ in dilute aqueous solution. Three molecules of water are bound by very strong hydrogen bonds, and one molecule (at the top) is held by much weaker ion–dipole interactions.

The formula "$HClO_4 \cdot H_2O$" is a way of specifying the composition of the substance when we are ignorant of its structure. A more accurate formula would be $H_3O^+ClO_4^-$.

The average dimensions of the H_3O^+ cation that occurs in many crystals are shown in Figure 10-1. The bond enthalpy of the OH bond of H_3O^+ is 544 kJ/mol, some 84 kJ/mol greater than the OH bond enthalpy in H_2O. In aqueous solution the H_3O^+ cation is tightly associated with three molecules of water through exceptionally strong hydrogen bonds (Figure 10-2). The O–H$\cdots$O hydrogen-bonded distance of 252 pm may be compared to an O–H$\cdots$O distance of 283 pm between hydrogen-bonded water molecules. In liquid SO_2, where the H_3O^+ cation is not surrounded by H_2O molecules, H_3O^+ appears to be planar, or nearly so.[†]

Less is known of the structure of OH^- in solution, but the ion $H_3O_2^-$ ("$OH^- \cdot H_2O$") has been observed by x-ray crystallography.[‡] The central O–H–O linkage contains the shortest hydrogen bond involving H_2O that has ever been observed.

$$\begin{array}{c} H \\ O\text{—}H\text{—}O \\ H \longleftarrow 229\ pm \longrightarrow \end{array}$$

We will ordinarily write H^+ in most chemical equations, though we really mean H_3O^+. To emphasize certain chemistry of water, we may write H_3O^+. For example, water can be either an acid or a base. Water is an acid with respect to methoxide,

$$H_2\ddot{O} + CH_3\ddot{O}{:}^- \rightleftharpoons H{-}\ddot{O}{:}^- + CH_3{-}\ddot{O}{-}H \qquad (10\text{-}7)$$
$$\text{water} \quad \text{methoxide} \quad\quad \text{hydroxide} \quad\quad \text{methanol}$$

but a base with respect to hydrogen bromide:

$$H_2O + HBr \rightleftharpoons H_3O^+ + Br^- \qquad (10\text{-}8)$$
$$\text{water} \quad \text{hydrogen} \quad \text{hydronium} \quad \text{bromide}$$
$$ \text{bromide} \quad\quad \text{ion}$$

Water also undergoes **self-ionization** (also called **autoprotolysis**), in which it acts as both an acid and a base:

$$H_2O + H_2O \rightleftharpoons H_3O^+ + OH^- \qquad (10\text{-}9)$$

or

$$H_2O \rightleftharpoons H^+ + OH^- \qquad (10\text{-}10)$$

Reaction 10-10 should be taken to mean the same as Reaction 10-9.

[†] G. D. Mateescu and G. M. Benedikt, *J. Amer. Chem. Soc.*, **101**, 3959 (1979).
[‡] K. Abu-Dari, K. N. Raymond, and D. P. Freyberg, *J. Amer. Chem. Soc.*, **101**, 3688 (1979).

Other **protic solvents** (solvents with a reactive H^+) also undergo autoprotolysis. An example is acetic acid:

$$2CH_3\overset{\displaystyle O}{\overset{\|}{C}}OH \rightleftharpoons CH_3\overset{OH}{\underset{OH}{C^+}} + CH_3\overset{\displaystyle O}{\overset{\|}{C}}\!-\!O^- \text{ (in acetic acid)} \quad (10\text{-}11)$$

The extent of these reactions of water or of acetic acid is very small. The *autoprotolysis constants* (equilibrium constants) for Reactions 10-10 and 10-11 are 1.0×10^{-14} and 3.5×10^{-15}, respectively, at 25°C.

10-2 pH

The autoprotolysis constant for H_2O is given the special symbol K_w:

$$H_2O \xrightleftharpoons{K_w} H^+ + OH^- \quad (10\text{-}12)$$

The rigorous meaning of K_w is

$$K_w = \mathscr{A}_{H^+} \cdot \mathscr{A}_{OH^-} = [H^+]\gamma_{H^+}[OH^-]\gamma_{OH^-} \quad (10\text{-}13)$$

As is true of all equilibrium constants, K_w varies with temperature (Table 10-1). The value of K_w at 25.00°C (298.15 K) is 1.008×10^{-14}.

> Examples of *protic* solvents:
>
> H_2O and CH_3CH_2OH.
>
> Examples of *aprotic* solvents (solvents without acidic protons):
>
> $CH_3CH_2OCH_2CH_3$ and CH_3CN.

> Recall that H_2O (the solvent) is omitted from the equilibrium constant. The value $K_w = 1.0 \times 10^{-14}$ at 25°C will be accurate enough for our purposes in this text.

EXAMPLE

Calculate the activity and concentration of H^+ and OH^- in pure water at 25°C.

The stoichiometry of Reaction 10-12 tells us that H^+ and OH^- are produced in a 1:1 mole ratio. Their concentrations must be equal. Calling each concentration x, we can write

$$K_w = 1.0 \times 10^{-14} = (x)\gamma_{H^+}(x)\gamma_{OH^-}$$

But the ionic strength of pure water is so small that it is reasonable to expect that $\gamma_{H^+} = \gamma_{OH^-} = 1$. Using these values in the above equation gives

$$1.0 \times 10^{-14} = x^2 \Rightarrow x = 1.0 \times 10^{-7} \text{ M}$$

The concentrations of H^+ and OH^- are both 1.0×10^{-7} M. Their activities are also both 1.0×10^{-7} because each activity coefficient is very close to 1.00.

EXAMPLE

Calculate the activity and concentration of H^+ and OH^- in 0.10 M KCl at 25°C.

Reaction 10-12 tells us that $[H^+] = [OH^-]$. However, the values of γ in Equation 10-13 are not equal. The ionic strength of 0.10 M KCl is 0.10 M. According to Table 6-1, the activity coefficients of H^+ and OH^- are 0.83 and 0.76, respectively, when the ionic

Table 10-1
Temperature-dependence of K_w

Temperature (°C)	$K_w^\dagger$	$pK_w^\ddagger$
0	1.139×10^{-15}	14.943 5
5	1.846×10^{-15}	14.733 8
10	2.920×10^{-15}	14.534 6
15	4.505×10^{-15}	14.346 3
20	6.809×10^{-15}	14.166 9
24	1.000×10^{-14}	14.000 0
25	1.008×10^{-14}	13.996 5
30	1.469×10^{-14}	13.833 0
35	2.089×10^{-14}	13.680 1
40	2.912×10^{-14}	13.534 8
45	4.018×10^{-14}	13.396 0
50	5.474×10^{-14}	13.261 7
55	7.296×10^{-14}	13.136 9
60	9.614×10^{-14}	13.017 1

$\dagger$ $K_w = \mathscr{A}_{H^+} \cdot \mathscr{A}_{OH^-}$. The last digit is not significant.
$\ddagger$ $pK_w = -\log K_w$. The last digit is not significant.

160

10 / INTRODUCTION TO ACIDS AND BASES

strength is 0.10 M. Putting these values into Equation 10-13 gives

$$K_w = [H^+]\gamma_{H^+}[OH^-]\gamma_{OH^-}$$

$$1.0 \times 10^{-14} = (x)(0.83)(x)(0.76)$$

$$x = 1.26 \times 10^{-7} \text{ M}$$

The concentrations of H^+ and OH^- are equal and are both greater than 1.0×10^{-7} M. The activities of H^+ and OH^- are not equal in this solution:

$$\mathscr{A}_{H^+} = [H^+]\gamma_{H^+} = (1.26 \times 10^{-7})(0.83) = 1.05 \times 10^{-7} \quad (10\text{-}14)$$

$$\mathscr{A}_{OH^-} = [OH^-]\gamma_{OH^-} = (1.26 \times 10^{-7})(0.76) = 0.96 \times 10^{-7} \quad (10\text{-}15)$$

The pH Scale

The definition of **pH** is

The correct definition of pH.

$$pH = -\log \mathscr{A}_{H^+} = -\log[H^+]\gamma_{H^+} \quad (10\text{-}16)$$

When we measure pH with a pH meter, we are measuring the negative log of the hydrogen ion *activity*, not the concentration. To simplify life, we will usually use the incomplete form, Equation 10-17, when we speak of pH.

As a general rule, neglect activity coefficients unless told otherwise when working problems in this text.

$$pH \approx -\log[H^+] \quad (10\text{-}17)$$

EXAMPLE

Using activity coefficients, find the pH of pure water at 25°C.

In the previous section we calculated $[H^+] = [OH^-] = 1.0 \times 10^{-7}$ M in pure water. We also reasoned that $\gamma_{H^+} = \gamma_{OH^-} = 1.00$, since the ionic strength is quite low. Therefore,

$$pH = -\log[H^+]\gamma_{H^+} = -\log(1.0 \times 10^{-7})(1.00) = 7.00$$

EXAMPLE

Using activity coefficients, find the pH of 0.10 M KCl at 25°C.

We found in Equation 10-14 that $\mathscr{A}_{H^+} = 1.05 \times 10^{-7}$ for this solution. Therefore,

$$pH = -\log \mathscr{A}_{H^+} = -\log(1.05 \times 10^{-7}) = 6.98$$

The pH of pure water changed from 7.00 to 6.98 because we added 0.10 M KCl to it. KCl is not an acid or a base. The small change in pH was an effect of activity, not of acidity. The pH change of just 0.02 units lies at the current limit of accuracy of pH measurements and is hardly significant. However, the H^+ *concentration* in 0.10 M KCl is 26% greater than the H^+ concentration in pure water.

At 25°C, any solution containing a concentration of H^+ higher than 10^{-7} M is said to be **acidic**. Any solution with a concentration of H^+ lower than 10^{-7} M is said to be **basic**. An acidic solution has a pH below 7 and a basic solution has a pH above 7, at 25°C.

pH   −1   0   1   2   3   4   5   6   7   8   9   10   11   12   13   14   15

⟵———— Acidic ————⟶ ↑ ⟵———— Basic ————⟶

Neutral

A solution is acidic if $[H_3O^+] > [OH^-]$. A solution is basic if $[H_3O^+] < [OH^-]$.

Although the pH of most solutions is in the range 0–14, these are not the limits of pH. A pH of -1.00, for example, means $-\log \mathscr{A}_{H^+} = -1.00$, or $\mathscr{A}_{H^+} = 10$. This activity is easily attained in a concentrated solution of a strong acid. Because activity coefficients are greater than 1 at high ionic strength, the pH of 12.0 M HCl is approximately -2.8.

pH is usually measured with a glass electrode, whose operation is described in Chapter 16. Some methods of extending pH measurements to very low or very high values (where the glass electrode is useless) are discussed in Box 12-2.

Is there such a thing as pure water? In most labs, the answer is "No." Distilled water from the tap in most labs is acidic because it contains CO_2 from the atmosphere. The CO_2 is an acid by virtue of the reaction

$$CO_2 + H_2O \rightleftharpoons \underset{\text{bicarbonate}}{HCO_3^-} + H^+ \tag{10-18}$$

Carbon dioxide can be largely removed by first boiling the water and then protecting it from the atmosphere. Alternatively, the water may be distilled under an inert atmosphere, such as pure N_2.

About a hundred years ago, careful measurements of the conductivity of water were made by Friedrich Kohlrausch and his students. In removing ionic impurities from their water, they found it necessary to distill the water *42 consecutive times* under vacuum to reduce conductivity to a limiting value.

Table 10-2 should be memorized.

10-3   STRENGTHS OF ACIDS AND BASES

Acids and bases are commonly classified as strong or weak, depending on whether they react "completely" or only "partly" to produce H^+ or OH^-. Since there is a continuous range of possibilities for "partial" reaction, there is no sharp distinction between weak and strong. However, some compounds react so completely that they are easily considered strong acids or bases, and, by convention, anything else is termed weak.

Strong Acids and Bases

The most common strong acids and bases are listed in Table 10-2. By definition, a strong acid or base is completely dissociated in aqueous solution. That is, the equilibrium constants for Reactions 10-19 and 10-20 are very large.

Table 10-2
Common strong acids and bases

	Acids
HCl	Hydrochloric acid (hydrogen chloride)
HBr	Hydrogen bromide
HI	Hydrogen iodide
H_2SO_4[†]	Sulfuric acid
HNO_3	Nitric acid
$HClO_4$	Perchloric acid

	Bases
Formula	Name
LiOH	Lithium hydroxide
NaOH	Sodium hydroxide
KOH	Potassium hydroxide
RbOH	Rubidium hydroxide
CsOH	Cesium hydroxide
R_4NOH[‡]	Quarternary ammonium hydroxide

[†] For H_2SO_4, only the first proton ionization is complete. Dissociation of the second proton has an equilibrium constant of 1.0×10^{-2}.

[‡] This is a general formula for any hydroxide salt of an ammonium cation containing four organic groups. An example is tetrabutylammonium hydroxide: $(CH_3CH_2CH_2CH_2)_4N^+OH^-$.

$$HCl(aq) \rightleftharpoons H^+ + Cl^- \qquad (10\text{-}19)$$

$$KOH(aq) \rightleftharpoons K^+ + OH^- \qquad (10\text{-}20)$$

Virtually no undissociated HCl or KOH exists in aqueous solution. Demonstration 10-1 shows one consequence of the strong acid behavior of HCl.

Notice that, although the hydrogen halides HCl, HBr, and HI are strong acids, HF is *not* a strong acid. Box 10-1 gives an explanation of this unexpected observation. For most practical purposes, the hydroxides of the alkaline metals (Mg^{2+}, Ca^{2+}, Sr^{2+}, and Ba^{2+}) may be considered strong bases, although they are far less soluble than alkali metal hydroxides and also have some tendency to form MOH^+ complexes (Table 10-3).

The leveling effect

The nature of a particular solvent determines how strong the strongest acids or bases may be in that solvent. For an aqueous solution, the strongest

Table 10-3
Equilibria of alkaline metal hydroxides

$$M(OH)_2(s) \rightleftharpoons M^{2+} + 2OH^-$$
$$K_{sp} = [M^{2+}][OH^-]^2$$
$$M^{2+} + OH^- \rightleftharpoons MOH^+$$
$$K_1 = [MOH^+]/[M^{2+}][OH^-]$$

Metal	$\log K_{sp}$	$\log K_1$
Mg^{2+}	-11.15	2.58
Ca^{2+}	-5.19	1.30
Sr^{2+}	—	0.82
Ba^{2+}	—	0.64

Note: 25°C and ionic strength $= 0$.

Demonstration 10-1 THE HCl FOUNTAIN

The complete dissociation of HCl into H^+ and Cl^- makes $HCl(g)$ extremely soluble in water.

$$HCl(g) \rightleftharpoons HCl(aq) \qquad (10\text{-}21)$$

$$HCl(aq) \rightleftharpoons H^+(aq) + Cl^-(aq) \qquad (10\text{-}22)$$

Net reaction: $\quad HCl(g) \rightleftharpoons H^+(aq) + Cl^-(aq) \qquad (10\text{-}23)$

Since the equilibrium of Reaction 10-22 lies far to the right, it pulls Reaction 10-21 to the right as well.

> *Challenge:* The standard free energy change ($\Delta G°$) for Reaction 10-23 is -37.0 kJ mol^{-1}. Show that the equilibrium constant is 3.1×10^6.

The extreme solubility of $HCl(g)$ in water is the basis for the HCl fountain, set up as shown on the facing page. In Figure a an inverted 250 mL round-bottom flask containing air is set up with its inlet tube leading to a source of $HCl(g)$ and its outlet tube directed into an inverted bottle of water. As HCl is admitted to the flask, air is displaced. When the bottle is filled with air the flask is filled mostly with $HCl(g)$.

The hoses are disconnected and replaced with a beaker of indicator and a rubber bulb (Figure b). For an indicator we use slightly alkaline xylenol orange, which is red or purple above pH 7.5 and yellow below pH 6.0. When ~ 1 mL of water is squirted from the rubber bulb into the flask, a vacuum is created and indicator solution is drawn up into the flask, making a fascinating fountain (Color Plates 8–10).

possible acid is H_3O^+ (H^+), and the strongest possible base is OH^-. All acids (proton donors) stronger than H_3O^+ will protonate water to make H_3O^+. All bases stronger than OH^- remove H^+ from H_2O to make OH^-. The reduction of all acid strengths greater than H_3O^+ to that of H_3O^+, as well as the reduction of all base strengths greater than that of OH^- to that of OH^-, is called the **leveling effect**. To see why leveling occurs, consider the ethoxide ion, which is a stronger base (proton acceptor) than hydroxide. When dissolved in water, ethoxide reacts with water to give hydroxide:

$$\underset{\text{ethoxide}}{CH_3CH_2O^-} + H_2O \rightleftharpoons \underset{\text{ethanol}}{CH_3CH_2OH} + OH^- \qquad (10\text{-}24)$$

Leveling effect: The strongest acid that can exist in water is H_3O^+; the strongest base that can exist in water is OH^-.

If hydroxide were a stronger base than ethoxide, the equilibrium in Reaction 10-24 would lie to the left instead of to the right.

Although we will deal primarily with aqueous solutions, it is instructive to realize that acid and base strengths that are not distinguished in aqueous solution may be distinguishable in nonaqueous solvents. For example,

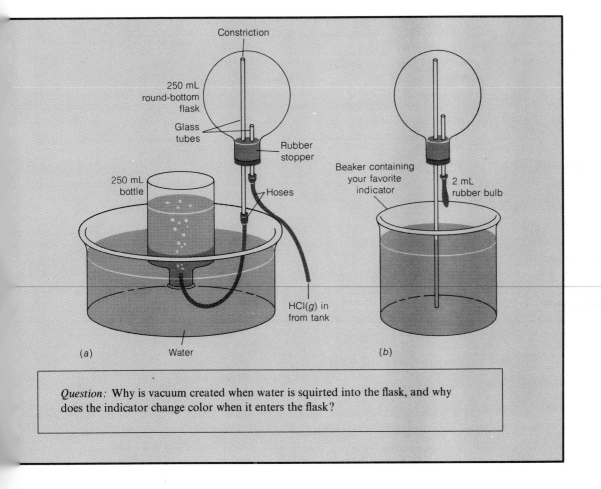

Question: Why is vacuum created when water is squirted into the flask, and why does the indicator change color when it enters the flask?

Box 10-1   THE STRANGE BEHAVIOR OF HYDROFLUORIC ACID[†]

The hydrogen halides HCl, HBr, and HI are all strong acids, which means that the reactions

$$HX(aq) + H_2O \rightarrow H_3O^+ + X^-$$

(X = Cl, Br, I) all go to completion. Why, then, does HF behave as a weak acid?
  The answer is curious. First, HF *does* completely give up its proton to H_2O:

$$HF(aq) \rightarrow \underset{\substack{\text{hydronium} \\ \text{ion}}}{H_3O^+} + \underset{\substack{\text{fluoride} \\ \text{ion}}}{F^-}$$

But fluorine forms the strongest hydrogen bonds of any element. The hydronium ion remains tightly associated with the fluoride ion through a hydrogen bond. We call such an association an **ion pair.**

$$\underset{\text{An ion pair}}{H_3O^+ + F^- \rightleftharpoons F^- \cdots H_3O^+}$$

Ion pairs are common in nonaqueous solvents, which cannot promote ion dissociation as well as water. But the hydronium–fluoride ion pair is unusual for aqueous solutions.
  Thus HF does not behave as a strong acid because the F^- and H_3O^+ ions remain associated with each other. Dissolving one mole of the strong acid HCl in water creates one mole of free H_3O^+. Dissolving one mole of the "weak" acid HF in water creates very little free H_3O^+.

[†] P. A. Giguère, *J. Chem. Ed.*, **56**, 571 (1979).

In acetic acid solution, $HClO_4$ is a stronger acid than HCl, but in aqueous solution they are leveled to the strength of H_3O^+.

perchloric acid is a stronger acid than hydrochloric acid *in acetic acid solvent,* and neither acid is completely dissociated:

$$\underset{\substack{\text{perchloric} \\ \text{acid}}}{HClO_4} + \underset{\substack{\text{acetic acid} \\ \text{(solvent)}}}{CH_3CO_2H} \rightleftharpoons CH_3CO_2H_2^+ + ClO_4^- \qquad K = 1.3 \times 10^{-5}$$

$$\tag{10-25}$$

$$\underset{\substack{\text{hydrochloric} \\ \text{acid}}}{HCl} + CH_3CO_2H \rightleftharpoons CH_3CO_2H_2^+ + Cl^- \qquad K = 5.8 \times 10^{-8}$$

$$\tag{10-26}$$

Weak Acids and Bases

All weak acids, HA, react with water according to Equation 10-27:

$$HA + H_2O \xrightarrow{\;K_A\;} H_3O^+ + A^- \tag{10-27}$$

which means exactly the same as

$$HA \xrightleftharpoons{K_A} H^+ + A^- \qquad (10\text{-}28)$$

The equilibrium constant for both of these equations is called K_A, the **acid dissociation constant.**

$$K_A = \frac{[H^+][A^-]}{[HA]} \qquad (10\text{-}29)$$

By definition, a weak acid is one that is only partially dissociated in water. This means that K_A is "small" for a weak acid.

Weak bases, B, react with water according to Equation 10-30:

$$B + H_2O \xrightleftharpoons{K_B} BH^+ + OH^- \qquad (10\text{-}30)$$

The equilibrium constant, K_B, is usually called the base hydrolysis or **base "dissociation" constant.**

$$K_B = \frac{[BH^+][OH^-]}{[B]} \qquad (10\text{-}31)$$

By definition, a weak base is one for which K_B is "small."

Common classes of weak acids and bases

Acetic acid and methylamine are typical of common weak acids and bases.

$$CH_3\overset{\overset{\displaystyle O}{\|}}{C}OH \rightleftharpoons CH_3\overset{\overset{\displaystyle O}{\|}}{C}O^- + H^+ \qquad K_A = 1.75 \times 10^{-5} \qquad (10\text{-}32)$$

acetic acid (HA)   acetate (A$^-$)

$$CH_3\overset{\cdot\cdot}{N}H_2 + H_2O \rightleftharpoons CH_3\overset{+}{N}H_3 + OH^- \qquad K_B = 4.4 \times 10^{-4} \qquad (10\text{-}33)$$

methylamine (B)   methylammonium ion (BH$^+$)

Acetic acid is representative of the carboxylic acids, which have the general structure

$$R\!-\!\overset{\overset{\displaystyle O}{\|}}{C}\!-\!OH$$

a carboxylic acid

where R is an organic substituent. *All* **carboxylic acids** *are weak acids, ar all* **carboxylate anions** *are weak bases.*

$$R\!-\!\overset{\overset{\displaystyle O}{\|}}{C}\!-\!O^-$$

a carboxylate anion

165

10-3 STRENGTHS OF ACIDS AND BASES

Of course, if we were not neglecting activity coefficients we should write

$$K_A = \frac{[H^+]\gamma_{H^+}[A^-]\gamma_{A^-}}{[HA]\gamma_{HA}}$$

166

10 / INTRODUCTION TO ACIDS AND BASES

Methylamine is a representative amine, a nitrogen-containing compound.

$$R\overset{..}{N}H_2 \qquad \text{a primary amine} \qquad RNH_3^+ \;\Big\rangle$$

$$R_2\overset{..}{N}H \qquad \text{a secondary amine} \qquad R_2NH_2^+ \;\Big\rbrace \; \begin{array}{c}\text{ammonium}\\\text{ions}\end{array}$$

$$R_3\overset{..}{N} \qquad \text{a tertiary amine} \qquad R_3NH^+ \;\Big/$$

Amines *are weak bases, and* **ammonium ions** *are weak acids.* The "parent" of all amines is ammonia, NH_3. When a base such as methylamine reacts with water, the product formed is the conjugate acid. That is, the methylammonium ion produced in Reaction 10-33 is a weak acid:

$$\underset{BH^+}{CH_3\overset{+}{N}H_3} \overset{K_A}{\rightleftharpoons} \underset{B}{CH_3\overset{..}{N}H_2} + H^+ \qquad K_A = 2.3 \times 10^{-11} \quad (10\text{-}34)$$

Although we will usually write a **base** as **B** and an **acid** as **HA,** it is important to realize that **BH⁺** is also an **acid** and **A⁻** is also a **base.**

Methylammonium chloride is a weak acid because
1. It dissociates into $CH_3NH_3^+$ and Cl^-.
2. $CH_3NH_3^+$ is a weak acid, being conjugate to CH_3NH_2, a weak base.
3. Cl^- has no basic properties. It is conjugate to HCl, a strong acid. That is, HCl dissociates completely.

The methylammonium ion is the conjugate acid of methylamine.

You should learn to recognize whether a compound will have acidic or basic properties. Note carefully that the salt methylammonium chloride, for example, will dissociate completely in aqueous solution to give the methylammonium cation and the chloride anion.

$$\underset{\substack{\text{methylammonium}\\\text{chloride}}}{CH_3\overset{+}{N}H_3Cl^-(s)} \to CH_3\overset{+}{N}H_3(aq) + Cl^-(aq) \qquad (10\text{-}35)$$

The methylammonium ion, being the conjugate acid of methylamine, is a weak acid (Reaction 10-34). The chloride ion is neither an acid nor a base. It is the conjugate base of HCl, a strong acid. This means that *Cl^- has virtually no tendency to associate with H^+*, or else HCl would not be a strong acid. We can predict that a solution of methylammonium chloride will be acidic, since the methylammonium ion is an acid and Cl^- is not a base.

Challenge: Phenol is a weak acid. Show that a solution of the ionic compound, potassium phenolate $(C_6H_5O^-K^+)$, will be basic.

Polyprotic acids and bases

These are compounds that can donate or accept more than one proton. For example, oxalic acid is diprotic and phosphate is tribasic:

$$\underset{\substack{\text{oxalic}\\\text{acid}}}{\overset{\overset{\text{O O}}{\|\;\|}}{HOCCOH}} \rightleftharpoons H^+ + \underset{\substack{\text{monohydrogen}\\\text{oxalate}}}{\overset{\overset{\text{O O}}{\|\;\|}}{{}^-OCCOH}} \qquad K_{A1} = 5.60 \times 10^{-2} \quad (10\text{-}36)$$

$$\overset{\overset{\text{O O}}{\|\;\|}}{{}^-OCCOH} \rightleftharpoons H^+ + \underset{\text{oxalate}}{\overset{\overset{\text{O O}}{\|\;\|}}{{}^-OCCO^-}} \qquad K_{A2} = 5.42 \times 10^{-5} \quad (10\text{-}37)$$

$$\text{phosphate} + H_2O \rightleftharpoons \text{monohydrogen phosphate} + OH^- \qquad K_{B1} = 2.3 \times 10^{-2} \qquad (10\text{-}38)$$

$$\text{(dihydrogen phosphate)} + H_2O \rightleftharpoons \text{dihydrogen phosphate} + OH^- \qquad K_{B2} = 1.59 \times 10^{-7} \qquad (10\text{-}39)$$

$$\text{(phosphoric acid)} + H_2O \rightleftharpoons \text{phosphoric acid} + OH^- \qquad K_{B3} = 1.42 \times 10^{-12} \qquad (10\text{-}40)$$

167

10-3 STRENGTHS OF ACIDS AND BASES

Notation for acid and base equilibrium constants: K_{AI} refers to the acidic species with the most protons and K_{BI} refers to the basic species with the least protons. The subscript A in acid dissociation constants will usually be omitted.

The standard notation for successive acid dissociation constants of a polyprotic acid is K_1, K_2, K_3, and so on. That is, the subscript A is usually omitted. We will retain or omit the subscript as dictated by clarity. For successive base hydrolysis constants, we will retain the subscript B. The examples above illustrate that K_{A1} *(or K_1) refers to the acidic species with the most protons, and K_{B1} refers to the basic species with the least number of protons.* Carbonic acid, the diprotic carboxylic acid derived from CO_2, is described in Box 10-2.

pK

The term **pK** refers to the negative logarithm of the equilibrium constant K. Let's write a few p functions:

$$pK_w = -\log K_w = -\log[H^+][OH^-] \qquad (10\text{-}41)$$

$$pK_A = -\log K_A = -\log \frac{[A^-][H^+]}{[HA]} \qquad (10\text{-}42)$$

$$pK_B = -\log K_B = -\log \frac{[BH^+][OH^-]}{[B]} \qquad (10\text{-}43)$$

As a number gets bigger, its p function decreases, and vice versa. Comparing formic and benzoic acids, the former is a stronger acid, with a bigger dissociation constant and a smaller pK_A than benzoic acid.

$$\text{HCOH} \rightleftharpoons H^+ + HCO_2^- \qquad K_A = 1.80 \times 10^{-4} \qquad (10\text{-}44)$$
$$\text{formic acid} \qquad \text{formate} \qquad \mathbf{pK_A = 3.745}$$

As K_A increases, pK_A decreases. The smaller is pK_A, the stronger is the acid.

Box 10-2 CARBONIC ACID[†]

Carbonic acid is the acidic form of dissolved carbon dioxide.

$$CO_2(g) \rightleftharpoons CO_2(aq) \qquad K = \frac{[CO_2(aq)]}{P_{CO_2}} = 0.034\ 4$$

$$CO_2(aq) + H_2O \rightleftharpoons \underset{\substack{\text{carbonic} \\ \text{acid}}}{\overset{\displaystyle \text{HO}\overset{\displaystyle O}{\underset{\displaystyle \|}{C}}\text{OH}}{}} \qquad K = \frac{[H_2CO_3]}{[CO_2(aq)]} \approx 1.3 \times 10^{-3}$$

$$H_2CO_3 \rightleftharpoons \underset{\text{bicarbonate}}{HCO_3^-} + H^+ \qquad K_1$$

$$HCO_3^- \rightleftharpoons \underset{\text{carbonate}}{CO_3^{2-}} + H^+ \qquad K_2$$

Its behavior as a diprotic acid appears anomalous at first, because the value of K_{A1} is about 10^2 to 10^4 times smaller than K_A for other carboxylic acids.

$$H_2CO_3 \qquad K_{A1} = 4.45 \times 10^{-7}$$

$$CH_3CO_2H \qquad K_A = 1.75 \times 10^{-5} \qquad HCO_2H \qquad K_A = 1.80 \times 10^{-4}$$

$$\text{acetic acid} \qquad\qquad\qquad\qquad\qquad \text{formic acid}$$

$$N\equiv CCH_2CO_2H \qquad K_A = 3.37 \times 10^{-3} \qquad HOCH_2CO_2H \qquad K_A = 1.48 \times 10^{-4}$$

$$\text{cyanoacetic acid} \qquad\qquad\qquad\qquad\qquad \text{glycolic acid}$$

The reason for this seeming anomaly is not that H_2CO_3 is unusual but, rather, the value commonly given for K_{A1} applies to the equation

$$\text{All dissolved } CO_2 \rightleftharpoons HCO_3^- + H^+$$
$$(= CO_2(aq) + H_2CO_3)$$

$$K_{A1} = \frac{[HCO_3^-][H^+]}{[CO_2(aq) + H_2CO_3]} = 4.45 \times 10^{-7}$$

Only about 0.2% of dissolved CO_2 is in the form H_2CO_3. When the true value of $[H_2CO_3]$ is used, instead of the value $[H_2CO_3 + CO_2(aq)]$, the value of the equilibrium constant becomes

$$K_{A1} = \frac{[HCO_3^-][H^+]}{[H_2CO_3]} = 2 \times 10^{-4}$$

The hydration of CO_2 and dehydration of H_2CO_3 are surprisingly slow reactions, which can be demonstrated readily in a classroom. Living cells utilize the enzyme *carbonic anhydrase* to speed the rate at which H_2CO_3 and CO_2 equilibrate, in order to process this key metabolite.

[†] M. Kern, *J. Chem. Ed.*, **37**, 14 (1960); H. S. Harned and R. Davis, Jr., *J. Amer. Chem. Soc.*, **65**, 2030 (1943).

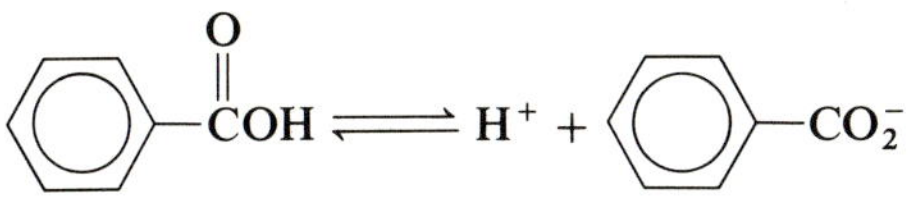

$$\langle\!\!\!\!\bigcirc\!\!\!\!\rangle\!-\!\overset{\overset{\displaystyle O}{\|}}{C}OH \rightleftharpoons H^+ + \langle\!\!\!\!\bigcirc\!\!\!\!\rangle\!-\!CO_2^- \qquad K_A = 6.25 \times 10^{-5} \quad (10\text{-}45)$$
$$pK_A = 4.202$$

benzoic acid

Relation Between K_A and K_B

A most important relationship exists between K_A and K_B of a conjugate acid–base pair in aqueous solution. We will derive this result using the acid, HA, and its conjugate base, A^-.

$$HA \rightleftharpoons H^+ + A^- \qquad K_A = \frac{[H^+][A^-]}{[HA]} \qquad (10\text{-}46)$$

$$A^- + H_2O \rightleftharpoons HA + OH^- \qquad K_B = \frac{[HA][OH^-]}{[A^-]} \qquad (10\text{-}47)$$

$$\overline{\qquad\qquad\qquad H_2O \rightleftharpoons H^+ + OH^- \qquad K_w = K_A \cdot K_B \qquad\qquad\qquad}$$

$$= \frac{[H^+][A^-]}{[HA]} \frac{[HA][OH^-]}{[A^-]} \quad (10\text{-}48)$$

When Reactions 10-46 and 10-47 are added, their equilibrium constants must be multiplied, giving a most useful result:

$$\boxed{K_A \cdot K_B = K_w} \qquad (10\text{-}49)$$

$K_A \cdot K_B = K_w$ for a conjugate acid–base pair in aqueous solution.

Equation 10-49 applies to any acid and its conjugate base in aqueous solution.

EXAMPLE

K_A for acetic acid is 1.75×10^{-5} (Reaction 10-32). Find K_B for the acetate ion.
    The solution is trivial:

$$K_B = K_w/K_A = 1.0 \times 10^{-14}/1.75 \times 10^{-5} = 5.7 \times 10^{-10}$$

EXAMPLE

K_B for methylamine is 4.4×10^{-4} (Reaction 10-33). Find pK_A for the methylammonium ion.
    Once again:

$$K_A = K_w/K_B = 2.3 \times 10^{-11}$$

$$pK_A = -\log K_A = 10.64$$

170

10 / INTRODUCTION TO ACIDS AND BASES

For a diprotic acid, we can derive results relating each of two acids and their conjugate bases:

$$H_2A \rightleftharpoons H^+ + HA^- \quad K_{A1} \qquad\qquad HA^- \rightleftharpoons H^+ + A^{2-} \quad K_{A2}$$

$$\underline{HA^- + H_2O \rightleftharpoons H_2A + OH^- \quad K_{B2}} \qquad \underline{A^{2-} + H_2O \rightleftharpoons HA^- + OH^- \quad K_{B1}}$$

$$H_2O \rightleftharpoons H^+ + OH^- \quad K_w \qquad\qquad H_2O \rightleftharpoons H^+ + OH^- \quad K_w$$

The final results are

$$K_{A1} \cdot K_{B2} = K_w \qquad\qquad (10\text{-}50)$$

$$K_{A2} \cdot K_{B1} = K_w \qquad\qquad (10\text{-}51)$$

Challenge: Derive the following results for a triprotic acid:

$$K_{A1} \cdot K_{B3} = K_w \qquad\qquad (10\text{-}52)$$

$$K_{A2} \cdot K_{B2} = K_w \qquad\qquad (10\text{-}53)$$

$$K_{A3} \cdot K_{B1} = K_w \qquad\qquad (10\text{-}54)$$

Weak Is Conjugate to Weak

The conjugate base of a weak acid is a weak base. The conjugate acid of a weak base is a weak acid. Let's examine these statements in some detail. Consider a weak acid, HA, with $K_A = 10^{-4}$. The conjugate base, A^-, has $K_B = K_w/K_A = 10^{-10}$. That is, if HA is a weak acid, A^- is a weak base. If K_A were 10^{-5}, then K_B would be 10^{-9}. We see that, as HA becomes a weaker acid, A^- becomes a stronger base. Conversely, the greater the acid strength of HA, the less the base strength of A^-. However, if either A^- or HA is weak, so is its conjugate. If HA is strong (such as HCl), its conjugate base (Cl^-) is *so* weak that it is not a base at all.

The conjugate base of a weak acid is a weak base. The conjugate acid of a weak base is a weak acid. *Weak is conjugate to weak.*

Using Appendix G

A table of acid dissociation constants appears in Appendix G. Each compound is shown in its *fully protonated form*. Methylamine, for example, is shown as $CH_3NH_3^+$, which is really the methylammonium ion. The value of K_A (2.3×10^{-11}) given for methylamine is actually K_A for the methylammonium ion. To find K_B for methylamine, we write $K_B = K_w/K_A = 1.0 \times 10^{-14}/2.3 \times 10^{-11} = 4.3 \times 10^{-4}$.

For polyprotic acids and bases, several K_A values are given. Pyridoxal phosphate is given in its fully protonated form as follows:

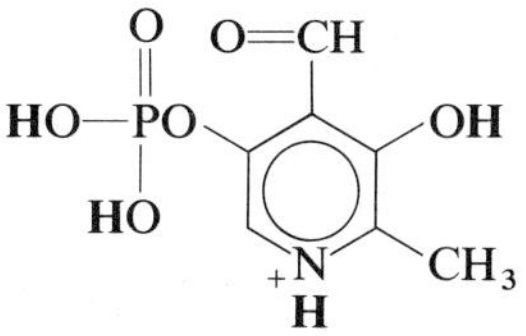

171

10-3 STRENGTHS OF ACIDS AND BASES

	pK$_A$	K$_A$
	1.4 (POH)	0.04
	3.44 (POH)	3.6×10^{-4}
	6.01 (NH)	9.8×10^{-7}
	8.45 (OH)	3.5×10^{-9}

This means that pK_1 (1.4) is for dissociation of one of the phosphate protons, and pK_2 (3.44) is for the other phosphate proton. The third most acidic group is the NH group, for which $pK_3 = 6.01$. Finally, the phenolic (OH) group is the least acidic ($pK_4 = 8.45$).

Summary

Acids are proton donors, and bases are proton acceptors. An acid increases the concentration of H_3O^+ in aqueous solution, and a base increases the concentration of OH^-. An acid–base pair related through the gain or loss of a single proton is described as conjugate. When a proton is transferred from one molecule to another molecule of a protic solvent, the reaction is called autoprotolysis. The leveling effect describes the fact that the strongest acid or base that can exist in a protic solvent is the protonated or deprotonated form of that solvent.

The definition of pH is: $pH = -\log \mathscr{A}_{H^+}$, which will commonly be used in the form $pH \approx -\log[H^+]$. The definition of pK is: $pK = -\log K$, where K is the equilibrium constant for any reaction. K_A refers to the equilibrium constant for the dissociation of an acid: $HA + H_2O \rightleftharpoons H_3O^+ + A^-$. K_B is the base hydrolysis constant for the reaction $B + H_2O \rightleftharpoons BH^+ + OH^-$. When K_A or K_B is large, the acid or base is said to be strong; otherwise the acid or base is weak. The common strong acids and bases are listed in Table 10-2. The most common weak acids are carboxylic acids (RCO_2H), and the most common weak bases are amines (R_3N). The conjugate base of a weak acid is a weak base, and conversely, the conjugate acid of a weak base is a weak acid. Carboxylate anions (RCO_2^-) are therefore weak bases, and ammonium ions (R_3NH^+) are weak acids. For a conjugate acid–base pair, $K_A \cdot K_B = K_w$. For polyprotic acids we denote the successive acid dissociation constants as $K_{A1}, K_{A2}, K_{A3}, \ldots$, or just $K_1, K_2, K_3, \ldots$. For polybasic species we denote successive hydrolysis constants $K_{B1}, K_{B2}, K_{B3}, \ldots$. For a polyprotic system with n protons, the relation between the ith K_A and the $(n - i)$th K_B is $K_{Ai} \cdot K_{B(n-i)} = K_w$.

Terms to Understand

acid dissociation constant (K_A)
amine
ammonium ion
autoprotolysis
base "dissociation" constant (K_B)
Brønsted acid
Brønsted base
carboxylate anion
carboxylic acid
conjugate acid–base pair
hydronium ion

ion pair
leveling effect
Lewis acid
Lewis base
neutralization
pH
pK
polyprotic acids and bases
protic solvent
salt

Exercises

10-A. If each of the following is dissolved in water, will the solution be acidic, basic, or neutral?
  (a) Na^+Br^-
  (b) $Na^+CH_3CO_2^-$
  (c) $NH_4^+Cl^-$
  (d) K_3PO_4
  (e) $(CH_3)_4N^+Cl^-$
  (f) $(CH_3)_4N^+$ ⬡$-CO_2^-$

10-B. Succinic acid dissociates as follows:

$$HOCCH_2CH_2COH \xrightleftharpoons{K_1}$$

$$HOCCH_2CH_2CO^- + H^+ \qquad pK_1 = 4.21$$

$$HOCCH_2CH_2CO^- \xrightleftharpoons{K_2}$$

$$^-OCCH_2CH_2CO^- + H^+ \qquad pK_2 = 5.64$$

Calculate K_{B1} and K_{B2} (not pK) for the following reactions:

$$^-OCCH_2CH_2CO^- + H_2O \xrightleftharpoons{K_{B1}}$$

$$HOCCH_2CH_2CO^- + OH^-$$

$$HOCCH_2CH_2CO^- + H_2O \xrightleftharpoons{K_{B2}}$$

$$HOCCH_2CH_2COH + OH^-$$

10-C. Histidine is a triprotic amino acid:

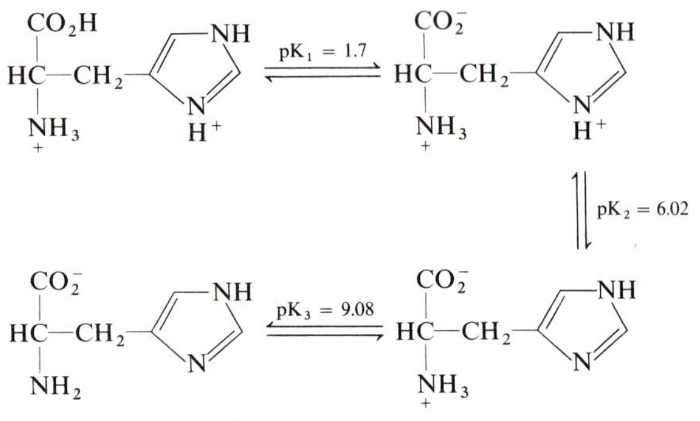

What is the value of the equilibrium constant for the reaction

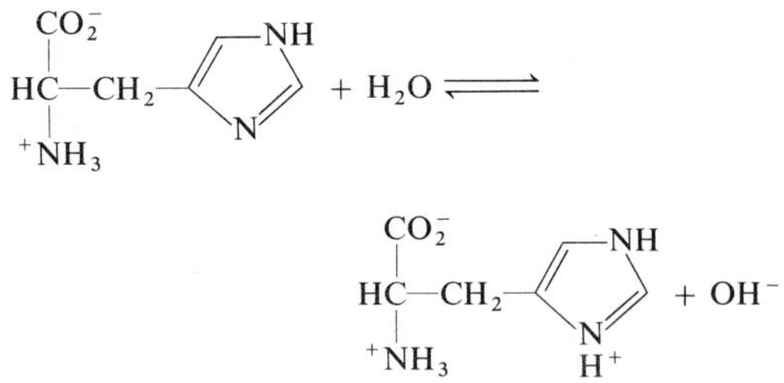

10-D. Using the values of K_w in Table 10-1, calculate the pH of distilled water at 0°C, 20°C, and 40°C. Would the inclusion of activity coefficients affect your answers?

Problems

10-1. Make a list of the common strong acids and strong bases. Memorize this list.

10-2. Use electron dot structures to show why tetramethylammonium hydroxide is an ionic compound. That is, show why the hydroxide is not covalently bound to the rest of the molecule.

10-3. Why does "pure" distilled water (at 25°C) from the tap in most labs have a pH below 7?

10-4. Explain the statement that the strongest acid that can exist in water is H_3O^+. What is the term used to describe this phenomenon?

10-5. Write the definitions of acids and bases advanced (a) by Brønsted and Lowry (b) by Lewis

10-6. Identify the Lewis acids in the following reactions:
  (a) $BF_3 + NH_3 \rightleftharpoons F_3\bar{B}-\overset{+}{N}H_3$
  (b) $F^- + AsF_5 \rightleftharpoons AsF_6^-$

10-7. Identify the Brønsted acids among the reactants in the following reactions:

(a) $NaHSO_3 + NaOH \rightleftharpoons Na_2SO_3 + H_2O$

(b) $KCN + HI \rightleftharpoons HCN + KI$

(c) $PO_4^{3-} + H_2O \rightleftharpoons HPO_4^{2-} + OH^-$

10-8. Write the autoprotolysis reaction of H_2SO_4.

10-9. Using activities correctly, calculate the pH and concentration of H^+ in pure water containing 0.050 M LiBr at 25°C.

10-10. (a) Using only K_{sp} in Table 10-3, how many moles of $Ca(OH)_2$ will dissolve in 1.00 L of water?

(b) How will the solubility calculated in part a be affected by the K_1 reaction in Table 10-3?

10-11. Although KOH, RbOH, and CsOH show no evidence of association between metal and hydroxide in aqueous solution, Li^+ and Na^+ do form complexes with OH^-:

$$Li^+ + OH^- \rightleftharpoons LiOH(aq) \quad K_1 = \frac{[LiOH(aq)]}{[Li^+][OH^-]}$$

$$= 0.83$$

$$Na^+ + OH^- \rightleftharpoons NaOH(aq) \quad K_1 = 0.20$$

Calculate the fraction of sodium in the form $NaOH(aq)$ in 1 F NaOH.

10-12. Write the K_A reaction for trichloroacetic acid, Cl_3CCO_2H, and for the anilinium ion,

10-13. Write the K_B reactions for pyridine and for sodium 2-mercaptoethanol.

pyridine   sodium 2-mercaptoethanol

10-14. Write the K_A *and* K_B reactions of $NaHCO_3$.

10-15. Identify the conjugate acid–base pairs in the following reactions:

(a) $H_2NCH_2CH_2NH_2 + H_2O \rightleftharpoons$

ethylenediamine

$$H_3\overset{+}{N}CH_2CH_2NH_2 + OH^-$$

(b) $H_3\overset{+}{N}CH_2CH_2\overset{+}{N}H_3 + H_2O \rightleftharpoons$

$$H_3\overset{+}{N}CH_2CH_2NH_2 + H_3O^+$$

(c)

benzoic acid   pyridine

benzoate   pyridinium

10-16. Write the equations for the stepwise acid–base reactions of the following ions in water. Write the correct symbol (e.g., K_{B1}) for the equilibrium constant for each reaction.

(a) $H_3\overset{+}{N}CH_2CH_2\overset{+}{N}H_3$

ethylenediammonium ion

(b) $^-OCCH_2CO^-$ (with two $C=O$ groups)

malonate ion

10-17. Which is a stronger acid, a or b?

(a) Cl_2HCCOH (with $C=O$) $pK_A = 1.30$

dichloroacetic acid

(b) ClH_2CCOH (with $C=O$) $pK_A = 2.865$

chloroacetic acid

Which is a stronger base, c or d?

(c) H_2NNH_2 $pK_B = 5.52$

hydrazine

(d) $H_2N\overset{O}{C}NH_2$ $pK_B = 13.82$

urea

10-18. Write the K_B reaction of CN^-. Given that K_A for HCN is 6.2×10^{-10}, calculate pK_B for CN^-.

10-19. Write the K_{A2} reaction of phosphoric acid (H_3PO_4) and the K_{B2} reaction of disodium oxalate ($Na_2C_2O_4$).

10-20. From the K_B values for phosphate in Equations 10-38 through 10-40, calculate the three pK_A values of phosphoric acid.

11 / Acid–Base Equilibria

In this chapter we will examine the equilibria necessary to understand the behavior of acids and bases. A firm grasp of this material is fundamental to understanding both the control of pH and the information available from acid–base titrations, as well as the more advanced topics presented in the next two chapters.

11-1 STRONG ACIDS AND BASES

What could be easier than calculating the pH of 0.10 M HBr? Since HBr is a **strong acid,** it is completely dissociated. So $[H^+] = 0.10$ M and

$$pH = -\log(0.10) = 1.00 \qquad (11\text{-}1)$$

Table 10-2 gives a list of strong acids and bases.

EXAMPLE

Equation 11-1 uses our customary tactic of ignoring activity coefficients. Now calculate the pH of 0.10 M HBr properly, using activity coefficients.

The ionic strength of 0.10 M HBr is 0.10 M, at which the activity coefficient of H^+ is 0.83 (Table 6-1). The pH is given by

$$pH = -\log[H^+]\gamma_{H^+} = -\log(0.10)(0.83) = 1.08$$

The activity correction is not very large, and we will generally continue to ignore activity coefficients in our calculations.

176

11 / ACID–BASE EQUILIBRIA

If you know $[OH^-]$, you can always find $[H^+]$, since $[H^+] = K_w/[OH^-]$.

The Dilemma

How do we calculate the pH of 0.10 M KOH? Since KOH is a strong base, it is completely dissociated, and $[OH^-] = 0.10$ M. Using $K_w = [H^+][OH^-]$, we write

$$[H^+] = \frac{K_w}{[OH^-]} = \frac{1.0 \times 10^{-14}}{0.10} = 1.0 \times 10^{-13} \qquad (11\text{-}2)$$

$$pH = -\log[H^+] = 13.00$$

To find the pH of other concentrations of KOH is pretty trivial:

$[OH^-]$ (M)	$[H^+]$ (M)	pH
$10^{-3.00}$	$10^{-11.00}$	11.00
$10^{-4.00}$	$10^{-10.00}$	10.00
$10^{-5.00}$	$10^{-9.00}$	9.00

A generally useful relation is that

$pH + pOH = 14.00$

$$pH + pOH = pK_w = 14.00 \text{ at } 25°C \qquad (11\text{-}3)$$

Now for the dilemma. What is the pH of 1.0×10^{-8} M KOH? Applying our usual reasoning we calculate

$$[H^+] = K_w/(1.0 \times 10^{-8}) = 1.0 \times 10^{-6} \Rightarrow pH = 6.00$$

But how can the base KOH produce an acidic solution (pH < 7) when dissolved in pure water? It's impossible.

The Cure

Clearly, there is something wrong with our calculation. In particular, we have not considered the contribution of OH^- from the ionization of water. In pure water, $[OH^-] = 1.0 \times 10^{-7}$ M, which is greater than the amount of KOH added to the solution.

To handle this problem, we resort to the systematic treatment of equilibrium discussed in Chapter 7. The procedure is to write the charge and mass balances, as well as all relevant equilibria. The species in the solution are K^+, OH^-, and H^+. All of the H^+ and some of the OH^- come from dissociation of water. The remainder of the OH^- comes from KOH. The charge balance is

$$[K^+] + [H^+] = [OH^-] \qquad (11\text{-}4)$$

The mass balance is rather trivial in this case:

$$[K^+] = 1.0 \times 10^{-8} \text{ M} \tag{11-5}$$

The only equilibrium equation to consider is

$$[H^+][OH^-] = K_w \tag{11-6}$$

There are three equations and three unknowns ($[H^+]$, $[OH^-]$, $[K^+]$), so we have enough information to solve the problem.

Since we are seeking the pH, let's set $[H^+] = x$. Using the values $[K^+] = 1.0 \times 10^{-8}$ M and $[H^+] = x$, and substituting into Equation 11-4, we find

$$[OH^-] = [K^+] + [H^+] = 1.0 \times 10^{-8} + x \tag{11-7}$$

Using this value of $[OH^-]$ in the K_w equilibrium allows us to solve the problem:

$$[H^+][OH^-] = K_w$$
$$(x)(1.0 \times 10^{-8} + x) = 1.0 \times 10^{-14} \tag{11-8}$$
$$x^2 + (1.0 \times 10^{-8})x - (1.0 \times 10^{-14}) = 0$$
$$x = \frac{-1.0 \times 10^{-8} \pm \sqrt{(10^{-8})^2 - 4(1)(-1.0 \times 10^{-14})}}{2(1)}$$
$$x = 9.6 \times 10^{-8} \quad \text{or} \quad -1.1 \times 10^{-7} \text{ M}$$

Solution of a quadratic equation:

$$ax^2 + bx + c = 0$$
$$x = \frac{-b \pm \sqrt{b^2 - 4ac}}{2a}$$

Rejecting the negative solution, we conclude that

$$[H^+] = 9.6 \times 10^{-8} \text{ M}$$
$$\text{pH} = -\log[H^+] = 7.02 \tag{11-9}$$

This pH is eminently reasonable, since a solution of 1.0×10^{-8} M KOH should be very slightly basic.

Figure 11-1 shows the pH calculated for different concentrations of a strong base or a strong acid dissolved in water. We can think of these curves in terms of three regions:

1. If the concentration is "high" ($\gtrsim 10^{-6}$ M), the pH has the value we would calculate by just considering the concentration of added H^+ or OH^-. That is, the pH of $10^{-5.00}$ M KOH *is* 9.00.
2. If the concentration is "low" ($\lesssim 10^{-8}$ M) the pH is 7.00. We have not added enough acid or base to significantly affect the pH of the water itself.
3. At intermediate concentrations ($\sim 10^{-6}$–10^{-8} M), the effects of water ionization and the added acid or base are comparable. Only in this region is it necessary to do a systematic equilibrium calculation.

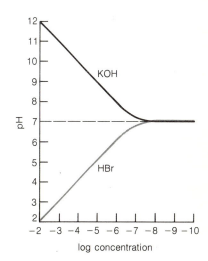

Figure 11-1
Graph showing the calculated pH as a function of the concentration of a strong acid or strong base dissolved in water.

178

11 / ACID–BASE EQUILIBRIA

Case 1 is the only practical case. Unless you protected, say, a 10^{-7} M KOH solution from the air, the pH would be overwhelmingly governed by the dissolved CO_2, not the 10^{-7} M KOH. To make a pH near 7, you should use a buffer, not a strong acid or base. You do not need to use the systematic treatment of equilibrium to find the pH of any practical concentration of a strong acid or base.

Water Almost Never Produces 10^{-7} M H^+ and 10^{-7} M OH^-

Any acid or base suppresses water ionization. This is an application of Le Châtelier's principle.

The common misconception that dissociation of water always produces 10^{-7} M H^+ and 10^{-7} M OH^- is true *only* in pure water with no added acids or bases. In a $10^{-4.00}$ M solution of HBr, for example, the pH is 4.00. The concentration of OH^- is

$$[OH^-] = K_w/[H^+] = 10^{-10.00} \text{ M} \qquad (11\text{-}10)$$

> *Question:* What concentrations of H^+ and OH^- are produced by H_2O dissociation in 10^{-2} M NaOH?

But the only source of $[OH^-]$ is the dissociation of water. If water produces only $10^{-10.0}$ M OH^-, it must also be producing only $10^{-10.00}$ M H^+, since it makes one H^+ for every OH^-. In a $10^{-4.00}$ M HBr solution, water dissociation produces only $10^{-10.00}$ M OH^- and $10^{-10.00}$ M H^+.

11-2 WEAK ACIDS

In this section we will calculate the pH and composition of a solution containing a weak acid—one that is not completely dissociated. We should begin by reviewing the definitions of K_A and pK_A:

$$HA \rightleftharpoons H^+ + A^- \qquad (11\text{-}11)$$

Of course you know that K_A is really $\mathscr{A}_{H^+}\mathscr{A}_{A^-}/\mathscr{A}_{HA}$. Is the definition of pK_A in Equation 11-13 correct? (Yes.)

$$K_A = \frac{[H^+][A^-]}{[HA]} \qquad (11\text{-}12)$$

$$pK_A \equiv -\log K_A \qquad (11\text{-}13)$$

A Typical Problem

The acetyl ($CH_3\overset{\text{O}}{\overset{\|}{C}}$—) derivative of *o*-hydroxybenzoic acid is the active ingredient in aspirin.

acetylsalicylic acid

Let's compare the ionization of *o*-hydroxybenzoic acid and *p*-hydroxybenzoic acid.

o-hydroxybenzoic acid
(salicylic acid)
$pK_A = 2.97$

p-hydroxybenzoic acid
$pK_A = 4.58$

It is thought that the reason that the *o*-hydroxy acid is more than an order of magnitude stronger than the *p*-hydroxy acid is that the conjugate base of the *o*-hydroxy acid is stabilized by strong intramolecular hydrogen bonding.

$$\text{(11-14)}$$

Such intramolecular hydrogen bonding between the hydroxyl and carboxyl groups is not possible in the *para* isomer because the two functional groups are too far apart. Since the *ortho* isomer is a stronger acid than the *para* isomer, we expect that a solution of the former should have a lower pH than an equimolar solution of the latter.

Our problem is to find the pH of a solution of the weak acid, HA, given the formal concentration of HA and the value of K_A. Let us call the formal concentration F. One way to attack this problem is by the systematic treatment of equilibrium discussed in Chapter 7.

> Formal concentration is the total number of moles of a compound dissolved in a liter. The formal concentration of a weak acid refers to the total amount of HA placed in the solution, regardless of the fact that some has changed into A^-.

Charge balance: $[H^+] = [A^-] + [OH^-]$ $\quad\quad$ (11-15)

Mass balance: $F = [A^-] + [HA]$ $\quad\quad$ (11-16)

Equilibria: $HA \rightleftharpoons H^+ + A^-$ $\quad K_A = \dfrac{[H^+][A^-]}{[HA]}$ (11-17)

$H_2O \rightleftharpoons H^+ + OH^-$ $\quad K_w = [H^+][OH^-]$ (11-18)

There are four independent equations and four unknowns ($[A^-]$, $[HA]$, $[H^+]$, $[OH^-]$), so the problem is solved if we can just do the necessary algebra.

The problem is that it's not so easy to solve these four simultaneous equations. If you combine them, you will discover that a cubic equation results. At this point the chemist steps in and cries, "Wait! There is no reason to solve a cubic equation. We can make an excellent, simplifying approximation. (Besides, I have trouble solving cubic equations.)"

In a solution of any respectable weak acid, the concentration of H^+ due to acid dissociation will be much greater than the concentration due to water dissociation. When HA dissociates, it produces A^-. When H_2O dissociates, it produces OH^-. If the acid dissociation is much greater than the water dissociation, we can say $[A^-] \gg [OH^-]$, and Equation 11-15 reduces to

$$[H^+] \approx [A^-] \quad\quad (11\text{-}19)$$

Using Equations 11-16, 11-17, and 11-19, we can set up a solution. First set $[H^+] = x$. Equation 11-19 says that $[A^-] = x$. Equation 11-16 says that $[HA] = F - [A^-] = F - x$. Putting these values into Equation 11-17 gives

180

11 / ACID–BASE EQUILIBRIA

$$K_A = \frac{[H^+][A^-]}{[HA]} = \frac{(x)(x)}{F - x} \tag{11-20}$$

It will now be useful to put real numbers into the problem. Let $F = 0.050\,0$ M and $K_A = 1.07 \times 10^{-3}$ for o-hydroxybenzoic acid. Equation 11-20 can be readily solved, since it is just a quadratic equation.

$$\frac{x^2}{F - x} = 1.07 \times 10^{-3} \tag{11-21}$$

$$x^2 + (1.07 \times 10^{-3})x - 5.35 \times 10^{-5} = 0$$

$$x = 6.80 \times 10^{-3} \text{ (negative root rejected)}$$

$$[H^+] = [A^-] = x = 6.80 \times 10^{-3} \text{ M} \tag{11-22}$$

$$[HA] = F - x = 0.043\,2 \text{ M} \tag{11-23}$$

$$pH = -\log x = 2.17 \tag{11-24}$$

Was the approximation $[H^+] \approx [A^-]$ justified? The calculated pH is 2.17, which means that $[OH^-] = K_w/[H^+] = 1.47 \times 10^{-12}$.

> In a solution of a weak acid, H^+ is derived almost entirely from the weak acid, not from H_2O dissociation.

$[A^-]$ (from HA dissociation) $= 6.80 \times 10^{-3}$ M
(implies that $[H^+]$ from HA dissociation $= 6.80 \times 10^{-3}$ M)

$[OH^-]$ (from H_2O dissociation) $= 1.47 \times 10^{-12}$ M
(implies that $[H^+]$ from H_2O dissociation $= 1.47 \times 10^{-12}$ M)

The assumption that H^+ is derived mainly from HA is excellent.

> We will express pH values to the 0.01 decimal place in this text.

For uniformity, we will calculate pH values to the 0.01 decimal place, regardless of what is justified by significant figures. It is important to retain all the digits in your calculator during the solution of a quadratic equation. In the quadratic formula, the term b^2 is often nearly equal to $4ac$, and, if you do not keep all the digits, the subtraction $b^2 - 4ac$ may generate garbage instead of a real answer.

A Better Way

> The method of successive approximations.

Solving Equation 11-21 with the quadratic formula is reasonable and not very difficult. However, an even simpler approach to solving the equation is by *successive approximations*. This method is generally quicker than using the quadratic formula. It also leads to fewer mistakes, and it can be extended to higher-order (such as cubic) equations.

Our first step is to write Equation 11-21 in the form

$$x = \sqrt{(1.07 \times 10^{-3})(0.050\,0 - x)} \tag{11-25}$$

As a first approximation, we will neglect x on the right-hand side. That is, we are supposing that $x \ll 0.0500$, which might be a good or a bad approximation. Neglecting x on the right gives

$$x_1 = \sqrt{(1.07 \times 10^{-3})(0.0500)} = 7.31 \times 10^{-3} \qquad (11\text{-}26)$$

x_1 is our first approximation. We then plug x_1 into the right side of Equation 11-25 to get a second approximation, x_2.

$$x_2 = \sqrt{(1.07 \times 10^{-3})(0.0500 - 7.31 \times 10^{-3})} = 6.76 \times 10^{-3} \qquad (11\text{-}27)$$

Continuing,

$$x_3 = \sqrt{(1.07 \times 10^{-3})(0.0500 - 6.76 \times 10^{-3})} = 6.80 \times 10^{-3} \qquad (11\text{-}28)$$

$$x_4 = \sqrt{(1.07 \times 10^{-3})(0.0500 - 6.80 \times 10^{-3})} = 6.80 \times 10^{-3} \qquad (11\text{-}29)$$

In four iterations we have come to an answer that is constant to three figures. The third iteration agrees with the exact solution to three places. The first guess (ignoring x on the right-hand side) gave an error of 7.6% in the value of x.

> When successive approximations agree with each other, you are done.

Good advice

When faced with an equation of the type of 11-21, try using successive approximations. Usually the method converges to a constant solution in a few cycles. If successive answers oscillate between a high value and a low value, guess a number between them and try it out. When you have the right value of x, it will regenerate itself when you plug it back into the equation

$$x_n = \sqrt{K_A(F - x_{n-1})} \qquad (11\text{-}30)$$

On those few occasions when the approximations fail to converge, you will have to solve the quadratic equation.

> *Challenge:* Solve the equations $x^2/(0.0100 - x) = 4.00 \times 10^{-6}$ and $x^2/(0.0100 - x) = 4.00 \times 10^{-3}$ by using the quadratic equation and by the method of successive approximations. Show that both methods give the same answer.

Return to Chemistry

The **fraction of dissociation,** α, of a weak acid is defined as the fraction that is in the form A^-. It is given by

$$\alpha = \frac{[A^-]}{[A^-] + [HA]} = \frac{x}{(x) + (F - x)} = \frac{x}{F} \qquad (11\text{-}31)$$

For 0.0500 M o-hydroxybenzoic acid, we find

$$\alpha = \frac{6.80 \times 10^{-3} \text{ M}}{0.0500 \text{ M}} = 0.136 \qquad (11\text{-}32)$$

182

11 / ACID–BASE EQUILIBRIA

That is, the acid is 13.6% dissociated at a formal concentration of 0.050 0 M.

The variation of α with formal concentration is shown in Figure 11-2. All **weak electrolytes** (compounds that are only partially dissociated) dissociate more as they are diluted. (Demonstration 11-1 illustrates some properties of weak electrolytes.) It can be seen in Figure 11-2 that *o*-hydroxybenzoic acid is more dissociated than *p*-hydroxybenzoic acid at the same formal concentration. This is reasonable, since the *ortho* isomer is a stronger acid than the *para* isomer.

Essence of Weak-Acid Problems

The way to do it.

When faced with a weak-acid problem, you should immediately realize that $[H^+] = [A^-] = x$ and proceed to set up and solve the equation

$$\frac{[H^+][A^-]}{[HA]} = \frac{x^2}{F - x} = K_A \tag{11-33}$$

where F is the formal concentration of HA. The approximation $[H^+] = [A^-]$ would be poor only if the acid is outrageously dilute ($\leq 10^{-6}$ M) or ridiculously weak. Neither of these conditions constitutes a practical problem.

EXAMPLE
Find the pH of 0.100 M trimethylammonium chloride.

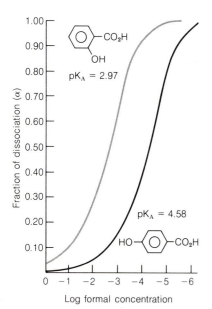

Figure 11-2
The fraction of dissociation of a weak electrolyte increases as the electrolyte is diluted.

trimethylammonium chloride

We must first realize that salts of this type are *completely dissociated* to give $(CH_3)_3NH^+$ and Cl^-. We then recognize that the trimethylammonium ion is a weak acid, being the conjugate acid of trimethylamine, $(CH_3)_3N$, a typical weak organic base. Cl^- has no basic or acidic properties and should be ignored. Looking in Appendix G, we find the trimethylammonium ion listed under the name trimethylamine, but drawn as the trimethylammonium ion. The value of pK_A is 9.800, so

$$K_A = 10^{-pK_A} = 1.58 \times 10^{-10}$$

From here everything is downhill.

$$(CH_3)_3NH^+ \rightleftharpoons (CH_3)_3N + H^+$$
$$ F - x \qquad\quad x \qquad\quad x$$

$$\frac{x^2}{0.100 - x} = 1.58 \times 10^{-10}$$

$$x = 3.97 \times 10^{-6} \text{ M} \Rightarrow pH = 5.40$$

Demonstration 11-1 CONDUCTIVITY OF WEAK ELECTROLYTES

The relative conductivity of strong and weak acids is directly related to their different degrees of dissociation in aqueous solution. To demonstrate conductivity we use the equipment shown below, but any kind of buzzer or light bulb could easily be substituted for the electric horn.[†] The voltage required will depend on the buzzer or light chosen.

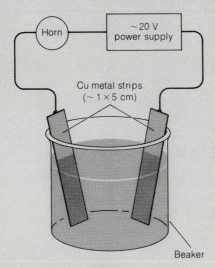

Apparatus for demonstrating conductivity of electrolyte solutions.

When a conducting solution is placed in the beaker, the horn can be heard. We can first show that distilled water or sucrose solutions are nonconductive. Solutions of the strong electrolytes NaCl or HCl are conductive. We compare strong and weak electrolytes, by demonstrating that 1 mM HCl gives a loud sound, while 1 mM acetic acid gives little or no sound. With 10 mM acetic acid the strength of the sound varies noticeably as the electrodes are moved away from each other in the beaker.

[†] The horn used in this demonstration is a Mallory Sonalert® Audible Signal, Model SC628, available from many electronic suppliers or from P. R. Mallory & Co., Box 1284, Indianapolis, IN 46206.

11-3 WEAK BASES

The treatment of weak bases is almost the same as that of weak acids. The usual notation for the base reaction is

$$B + H_2O = BH^+ + OH^- \qquad (11\text{-}34)$$

$$K_B = \frac{[BH^+][OH^-]}{[B]} \qquad (11\text{-}35)$$

184

11 / ACID–BASE EQUILIBRIA

If we suppose that the dissociation of H_2O is negligible compared to Reaction 11-34, then we can say that just about all of the OH^- in the solution comes from Reaction 11-34. Setting $[OH^-] = x$, we must also set $[BH^+] = x$, since one BH^+ is produced for each OH^-. If the formal concentration of base $(= [B] + [BH^+])$ is called F, we can set

$$[B] = \text{F} - [BH^+] = \text{F} - x \tag{11-36}$$

Plugging these values into Equation 11-35, we get

A weak-base problem has the same algebra as a weak-acid problem, except $K = K_B$ and $x = [OH^-]$.

$$K_B = \frac{(x)(x)}{\text{F} - x} \tag{11-37}$$

which looks a lot like a weak-acid problem, except that now $x = [OH^-]$.

Standard Weak-Base Problem

Let's work one standard weak-base problem, using a commonly encountered base, cocaine, as an example.

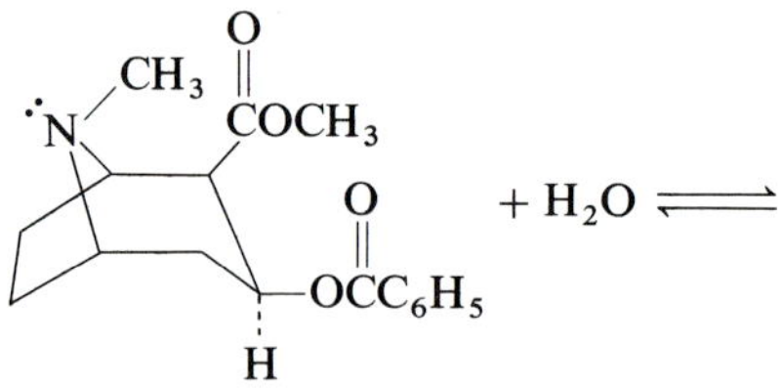

cocaine

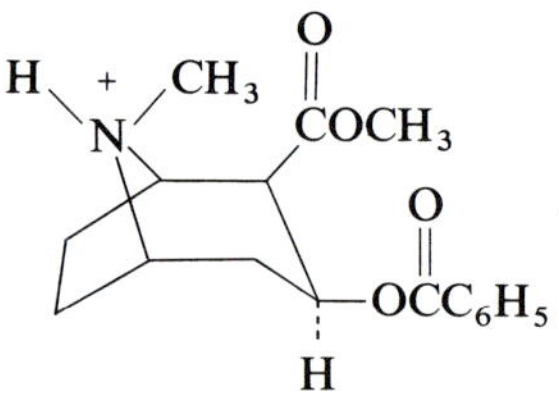

$$\text{B} + \text{H}_2\text{O} \rightleftharpoons \text{BH}^+ + \text{OH}^- \qquad K_B = 2.6 \times 10^{-6} \tag{11-38}$$

If the formal concentration of cocaine is 0.037 2 M, the problem can be formulated and solved as follows:

$$\text{B} \quad + \quad \text{H}_2\text{O} \rightleftharpoons \text{BH}^+ + \text{OH}^- \tag{11-39}$$
$$0.037\ 2 - x \qquad\qquad x \qquad x$$

$$\frac{x^2}{0.037\ 2 - x} = 2.6 \times 10^{-6} \Rightarrow x = 3.1_0 \times 10^{-4} \tag{11-40}$$

Since $x = [OH^-]$, we can write

$$[H^+] = K_w/[OH^-] = 1.0 \times 10^{-14}/3.1_0 \times 10^{-4} = 3.2_2 \times 10^{-11} \quad (11\text{-}41)$$

$$pH = -\log[H^+] = 10.49 \quad (11\text{-}42)$$

This is a reasonable pH for a weak base.

What fraction (α) of cocaine has reacted with water in this solution? We can formulate α as

$$\alpha = \frac{[BH^+]}{[BH^+] + [B]} = \frac{x}{F} = 0.008\ 3 \quad (11\text{-}43)$$

> *Question:* Based on the final answer, was it justified to neglect water dissociation as a source of OH^-? What concentration of OH^- is produced by H_2O dissociation in this solution?

Only 0.83% of the base has reacted.

Conjugate Acids and Bases, Revisited

In Chapter 10 it was stressed that **the conjugate base of a weak acid is a weak base,** and **the conjugate acid of a weak base is a weak acid.** We also derived an exceedingly important relation between K_A and K_B for a conjugate pair:

HA and A^- are a conjugate acid–base pair. So are BH^+ and B.

$$K_A \cdot K_B = K_w \quad (11\text{-}44)$$

In Section 11-2 we considered o- and p-hydroxybenzoic acids, designated HA. Now consider their conjugate bases. For example, the salt sodium o-hydroxybenzoate will dissolve to give the Na^+ cation (which has no acid–base chemistry) and the o-hydroxybenzoate anion, which is a weak base.

The acid–base chemistry is the reaction of o-hydroxybenzoate with water:

$$\underset{\substack{A^- \\ F-x}}{\text{⬡}-CO_2^-} + H_2O \rightleftharpoons \underset{\substack{HA \\ x}}{\text{⬡}-CO_2H} + \underset{x}{OH^-} \quad (11\text{-}45)$$

$$\frac{x^2}{F-x} = K_B \quad (11\text{-}46)$$

gives

o-hydroxybenzoate

in aqueous solution

Using the values of K_A for the *ortho* and *para* isomers, we can calculate the K_B values of the conjugate bases.

Isomer of hydroxybenzoic acid	K_A	K_B
ortho	1.07×10^{-3}	9.35×10^{-12}
para	2.63×10^{-5}	3.80×10^{-10}

Putting each value of K_B into Equation 11-46, and letting $F = 0.050\,0$ M, we find

$$\text{pH of } 0.050\,0 \text{ M } o\text{-hydroxybenzoate} = 7.83$$

$$\text{pH of } 0.050\,0 \text{ M } p\text{-hydroxybenzoate} = 8.64$$

These are reasonable pH values for solutions of weak bases. Furthermore, as expected, the conjugate base of the stronger acid is the weaker base.

EXAMPLE
Find the pH of 0.10 M ammonia.
  When ammonia is dissolved in water, its reaction is

$$\underset{\substack{\text{ammonia}\\F-x}}{NH_3} + H_2O \rightleftharpoons \underset{\substack{\text{ammonium ion}\\x}}{NH_4^+} + \underset{x}{OH^-} \quad K_B$$

In Appendix G we find the ammonium ion, NH_4^+, listed next to ammonia. pK_A for the ammonium ion is given as 9.244. Therefore, K_B for NH_3 is

$$K_B = K_w/K_A = 10^{-14.00}/10^{-9.244} = 1.75 \times 10^{-5}$$

To find the pH of 0.10 M NH_3 we set up and solve the equation

$$\frac{[NH_4^+][OH^-]}{[NH_3]} = \frac{x^2}{0.10 - x} = K_B = 1.75 \times 10^{-5}$$

$$x = [OH^-] = 1.3_1 \times 10^{-3} \text{ M}$$

$$[H^+] = K_w/[OH^-] = 7.6_1 \times 10^{-12} \text{ M}$$

$$pH = -\log[H^+] = 11.12$$

11-4 BUFFERS

A buffered solution is one that resists changes in pH when acids or bases are added. The **buffer** consists of a mixture of an acid and its conjugate base. The importance of buffers in all areas of science is overwhelming. Biochemists and other life scientists are particularly concerned with buffers because the proper functioning of any biological system is critically dependent upon pH. For example, Figure 11-3 shows how the rate of the enzyme-catalyzed Reaction 11-47 varies with pH.

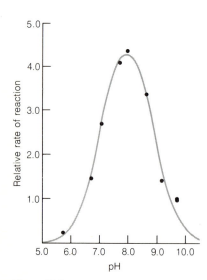

Figure 11-3
Graph showing the pH-dependence of the rate of cleavage of the C–N bond in Reaction 11-47. [M. L. Bender, G. E. Clement, F. J. Kézdy, and H. A. Heck, *J. Amer. Chem. Soc.*, **86**, 3680 (1964).]

$$\begin{array}{c} \text{N-acetyl-L-tryptophan} \\ \text{amide} \end{array} + H_2O \xrightarrow{\text{α-chymotrypsin}} \begin{array}{c} \text{n-acetyl-L-tryptophan} \end{array} + NH_3 \quad (11\text{-}47)$$

N-acetyl-L-tryptophan amide

n-acetyl-L-tryptophan   ammonia

In the absence of the enzyme α-chymotrypsin, the rate of this reaction is negligible over the same pH range, 5–10. The efficiency of the enzyme is highly dependent upon pH. For any organism to survive, it must control the pH of each subcellular compartment so that each of its enzyme-catalyzed reactions may proceed at the proper rate.

Mixing a Weak Acid and Its Conjugate Base

The purpose of this section is to show that *if you mix A moles of a weak acid with B moles of its conjugate base, the moles of acid remain close to A and the moles of base remain close to B.* Very little reaction occurs to change either concentration.

When you mix a weak acid with a weak base, you get what you mix!

To understand why this should be so, look at the K_A and K_B reactions in terms of Le Châtelier's principle. Consider an acid with $pK_A = 4.00$ and its conjugate base with $pK_B = 10.00$. We will calculate the fraction of acid that dissociates in a 0.100 M solution of HA.

$$\begin{array}{cccc} HA & \rightleftharpoons H^+ + A^- & pK_A = 4.00 & (11\text{-}48) \\ 0.100 - x & x \quad\quad x & & \end{array}$$

$$\frac{x^2}{F - x} = K_A \Rightarrow x = 3.11 \times 10^{-3} \quad\quad (11\text{-}49)$$

$$\text{Fraction of dissociation} = \alpha = \frac{x}{F} = 0.031\,1 \quad\quad (11\text{-}50)$$

The acid is only 3.11% dissociated under these conditions. In a solution containing 0.100 mol of A^- dissolved in 1.00 L, the extent of reaction of A^- with water is even smaller.

$$\begin{array}{cccc} A^- & + \quad H_2O \rightleftharpoons HA + OH^- & pK_B = 10.00 & (11\text{-}51) \\ 0.100 - x & \quad\quad x \quad\quad x & & \end{array}$$

$$\frac{x^2}{F - x} = K_B \Rightarrow x = 3.16 \times 10^{-6} \qquad (11\text{-}52)$$

$$\text{Fraction of association} = \alpha = \frac{x}{F} = 3.16 \times 10^{-5} \qquad (11\text{-}53)$$

This approximation breaks down for dilute solutions or at extremes of pH. We will test the validity of the approximation later in this section.

Pure HA dissociates very little, and adding extra A^- to the solution will make the HA dissociate even less. Similarly, A^- does not react very much with water, and adding extra HA makes A^- react even less. If 0.050 moles of A^- plus 0.036 moles of HA are added to water, there will be close to 0.050 moles of A^- and close to 0.036 moles of HA in the solution at equilibrium.

Henderson–Hasselbalch Equation

The central equation dealing with buffers is the **Henderson–Hasselbalch equation,** which is merely a rearranged form of the K_A equilibrium expression.

$$HA \rightleftharpoons H^+ + A^-$$

$$K_A = \frac{[H^+][A^-]}{[HA]}$$

$$\log K_A = \log \frac{[H^+][A^-]}{[HA]} = \log[H^+] + \log \frac{[A^-]}{[HA]}$$

$$\underbrace{-\log[H^+]} = \underbrace{-\log K_A} + \log \frac{[A^-]}{[HA]}$$

There is nothing in the Henderson–Hasselbalch equation that was not in the K_A equilibrium expression. Each is a rearranged form of the other.

$$\boxed{\quad pH \quad = \quad pK_A \quad + \log \frac{[A^-]}{[HA]} \quad} \qquad (11\text{-}54)$$

The Henderson–Hasselbalch equation tells us the pH of a solution, provided we know the ratio of the concentrations of conjugate acid and base, as well as pK_A for the acid. If a solution is prepared from the weak base, B, and its conjugate acid, the analogous equation is

$$pH = pK_A + \log \frac{[B]}{[BH^+]} \quad \left(\begin{array}{c}pK_A \text{ applies to}\\ \text{this acid!}\end{array}\right) \qquad (11\text{-}55)$$

where pK_A is the acid dissociation constant of the weak acid BH^+. The important features of Equations 11-54 and 11-55 are that the base (A^- or B) appears in the numerator of both equations, and the equilibrium constant of the reaction is K_A of the acid in the denominator.

Challenge: Show that if activities are not neglected, the correct form of the Henderson–Hasselbalch equation is

$$pH = pK_A + \log \frac{[A^-]\gamma_{A^-}}{[HA]\gamma_{HA}} \qquad (11\text{-}56)$$

Properties of the Henderson–Hasselbalch equation

In Equation 11-54 you can see that if $[A^-] = [HA]$, $pH = pK_A$.

$$pH = pK_A + \log \frac{[A^-]}{[HA]} = pK_A + \log 1 = pK_A \qquad (11\text{-}57)$$

When $[A^-] = [HA]$, $pH = pK_A$.

Regardless of how complex a solution may be, whenever $pH = pK_A$, $[A^-]$ must equal $[HA]$. This is true because *all equilibria must be satisfied simultaneously in any solution at equilibrium.* If there are ten different acids and bases in the solution, the ten forms of Equation 11-54 must all give the same pH, because **there can be only one concentration of H^+ in a solution.**

Another feature of the Henderson–Hasselbalch equation is that for every power of ten change in the ratio $[A^-]/[HA]$, the pH changes by one unit. As the base (A^-) increases, the pH goes up. As the acid (HA) increases, the pH goes down. This is shown in Table 11-1. For any conjugate acid–base pair, you can say, for example, that if $pH = pK_A - 1$, ten-elevenths is in the form HA and one-eleventh is in the form A^-.

Table 11-1
Change of pH with change of $[A^-]/[HA]$

$[A^-]/[HA]$	pH
100:1	$pK_A + 2$
10:1	$pK_A + 1$
1:1	pK_A
1:10	$pK_A - 1$
1:100	$pK_A - 2$

EXAMPLE

Sodium hypochlorite (NaOCl, the active ingredient of almost all bleaches) was dissolved in a solution buffered to pH 6.20. Find the ratio $[OCl^-]/[HOCl]$ in this solution.

In Appendix G we find that $pK_A = 7.53$ for hypochlorous acid, HOCl. Since the pH is known, the ratio $[OCl^-]/[HOCl]$ can be calculated from the Henderson–Hasselbalch equation

$$HOCl \rightleftharpoons H^+ + OCl^-$$

$$pH = pK_A + \log \frac{[OCl^-]}{[HOCl]}$$

$$6.20 = 7.53 + \log \frac{[OCl^-]}{[HOCl]}$$

$$-1.33 = \log \frac{[OCl^-]}{[HOCl]}$$

$$10^{-1.33} = 10^{\log([OCl^-]/[HOCl])} = \frac{[OCl^-]}{[HOCl]}$$

$$0.047 = \frac{[OCl^-]}{[HOCl]}$$

Note that finding the ratio $[OCl^-]/[HOCl]$ only requires knowing the pH. We do not need to know what else is in the solution, how much NaOCl was added, or the volume of the solution.

190

11 / ACID–BASE EQUILIBRIA

A Buffer in Action

For illustration, we will work with a very widely used buffer called "tris," which is short for tris(hydroxymethyl)aminomethane.

$$
\underset{\substack{\text{BH}^+ \\ \text{p}K_A = 8.075}}{\underset{\text{HOCH}_2\quad\text{CH}_2\text{OH}}{\overset{\overset{+}{\text{NH}_3}}{\underset{\displaystyle\text{C}}{\text{HOCH}_2\text{''''}}}}}
\;\rightleftharpoons\;
\underset{\substack{\text{B} \\ \text{(This form is "tris.")}}}{\underset{\text{HOCH}_2\quad\text{CH}_2\text{OH}}{\overset{\text{NH}_2}{\underset{\displaystyle\text{C}}{\text{HOCH}_2\text{''''}}}}}
\;+\;\text{H}^+ \qquad (11\text{-}58)
$$

In Appendix G we find $\mathrm{p}K_A$ for the conjugate acid of tris to be 8.075. An example of a salt containing the BH^+ cation is tris hydrochloride, which is really $\mathrm{BH}^+\mathrm{Cl}^-$. The formula weight of tris is 121.136, and the formula weight of tris hydrochloride is 157.597. When $\mathrm{BH}^+\mathrm{Cl}^-$ is dissolved in water, it dissociates completely to BH^+ plus Cl^-.

EXAMPLE

Find the pH of a solution prepared by dissolving 12.43 g of tris plus 4.67 g of tris hydrochloride in 1.00 L of water.

The concentrations of B and BH^+ added to the solution are

$$[\text{B}] = \frac{12.43 \text{ g/L}}{121.136 \text{ g/mol}} = 0.102\,6 \text{ M}$$

$$[\text{BH}^+] = \frac{4.67 \text{ g/L}}{157.597 \text{ g/mol}} = 0.029\,6 \text{ M}$$

Assuming that what we mixed stays in the same form, we can simply plug these concentrations into the Henderson–Hasselbalch equation to find the pH.

$$\text{pH} = \text{p}K_A + \log\frac{[\text{B}]}{[\text{BH}^+]} = 8.075 + \log\frac{0.102\,6}{0.029\,6} = 8.61$$

The pH of a buffer is nearly independent of volume.

Notice that *the volume of solution is irrelevant to finding the pH*, since the volume cancels in the numerator and denominator of the log term:

$$\text{pH} = \text{p}K_A + \log\frac{\text{Moles of B}/\cancel{\text{volume of solution}}}{\text{Moles of BH}^+/\cancel{\text{volume of solution}}}$$

$$\text{pH} = \text{p}K_A + \log\frac{\text{Moles of B}}{\text{Moles of BH}^+} \qquad (11\text{-}59)$$

The pH in the example above would be 8.61 whether the volume was 1.00, 0.63, or 2.41 L.

Box 11-1 STRONG PLUS WEAK REACT COMPLETELY

A strong acid reacts with a weak base essentially "completely" because the equilibrium constant is large.

$$B + H^+ \rightleftharpoons BH^+ \qquad K = 1/K_A \text{ (for } BH^+\text{)}$$

weak   strong
base   acid

If B is tris(hydroxymethyl)aminomethane, the equilibrium constant for reaction with HCl is

$$K = 1/K_A = 1/10^{-8.075} = 1.2 \times 10^8$$

A strong base reacts "completely" with a weak acid because the equilibrium constant is, again, very large.

$$OH^- + HA \rightleftharpoons A^- + H_2O \qquad K = 1/K_B \text{ (for } A^-\text{)}$$

strong   weak
base   acid

If HA is acetic acid, the equilibrium constant for reaction with NaOH is

$$K = 1/K_B = (K_A \text{ for HA})/K_w = 1.7 \times 10^9$$

The reaction of a strong acid with a strong base is even more complete, because the equilibrium constant is $1/K_w$.

$$H^+ + OH^- \rightleftharpoons H_2O \qquad K = 1/K_w$$

strong   strong
acid   base

If you mix a strong acid, a strong base, a weak acid and a weak base, the strong acid and base will react with each other until one is used up. The remainder of the strong acid or base will then react with the weak base or weak acid.

EXAMPLE

If we add 12.0 mL of 1.00 M HCl to the solution used in the previous example, what will be the new pH?

The key to this problem is to realize that *when a strong acid is added to a weak base, they react completely to give* BH^+. (You can read more about this important statement in Box 11-1.) In the present example we are adding 12.0 mL of 1.00 M HCl, which contains $(0.0120 \text{ L})(1.00 \text{ mol/L}) = 0.0120$ mol of H^+. This much H^+ will consume 0.0120 mol of B to create 0.0120 mol of BH^+. This is shown conveniently in a little table:

192

11 / ACID–BASE EQUILIBRIA

$$
\begin{array}{ccccc}
 & B & + & H^+ & \rightarrow & BH^+ \\
 & \text{(tris)} & & \text{(from HCl)} & &
\end{array}
$$

Initial moles:	0.102 6	0.012 0	0.029 6
Final moles:	0.090 6	—	0.041 6
	$\overbrace{(0.102\ 6 - 0.012\ 0)}$		$\overbrace{(0.029\ 6 + 0.012\ 0)}$

The table contains enough information for us to calculate the pH.

$$pH = pK_A + \log \frac{\text{Moles of B}}{\text{Moles of BH}^+}$$

$$= 8.075 + \log \frac{0.090\ 6}{0.041\ 6} = 8.41$$

The volume of the solution is irrelevant, as usual.

Question: Does the pH change in the right direction when HCl is added?

A buffer resists changes in pH . . .

The example above illustrates that *the pH of a buffer does not change very much when a strong acid or base is added.* Addition of 12.0 mL of 1.00 M HCl changed the pH from 8.61 to 8.41. Addition of 12.0 mL of 1.00 M HCl to 1.00 L of unbuffered solution would have lowered the pH to 1.93.

. . . because the buffer consumes the added acid or base.

Why does a buffer resist changes in pH? It does so because the strong acid or base is consumed by B or BH^+. If you add HCl to tris, B is converted to BH^+. If you add NaOH, BH^+ is converted to B. As long as we don't use up the B or BH^+ (by adding too much HCl or NaOH) the log term of the Henderson–Hasselbalch equation does not change very much and the pH does not change very much. Demonstration 11-2 provides a nice illustration of how buffers work. The buffer has its maximum capacity to resist changes of pH when $pH = pK_A$. We will return to this point later.

EXAMPLE

How many mL of 0.500 M NaOH should be added to 10.0 g of tris hydrochloride to give a pH of 7.60 in a final volume of 250 mL?

The number of moles of tris hydrochloride in 10.0 g is (10.0 g)/(157.597 g/mol) = 0.063 5. We can make a table to help solve the problem.

Reaction with OH^-:	BH^+	+	OH^-	$\rightarrow$	B
Initial moles:	0.063 5		x		—
Final moles:	0.063 5 − x		—		x

The Henderson–Hasselbalch equation allows us to find x, since we know pH and pK_A.

$$pH = pK_A + \log \frac{\text{Moles of B}}{\text{Moles of BH}^+}$$

$$7.60 = 8.075 + \log \frac{x}{0.063\,5 - x}$$

$$-0.475 = \log \frac{x}{0.063\,5 - x}$$

$$10^{-0.475} = \frac{x}{0.063\,5 - x} \Rightarrow x = 0.015\,9 \text{ mol}$$

This many moles of NaOH is contained in

$$\frac{0.015\,9 \text{ mol}}{0.500 \text{ mol/L}} = 0.031\,8 \text{ L} = 31.8 \text{ mL}$$

Notice that the volume of buffer solution (250 mL) was not used anywhere in answering the question.

Preparing a Buffer in Real Life!

If you really wanted to prepare a tris buffer of pH 7.60, you would *not* do it by calculating what to mix. Suppose that you wish to prepare 1.00 L of buffer containing 0.100 M tris at a pH of 7.60. You have available solid tris hydrochloride and approximately 1 M NaOH. Here's how to do it:

1. Weigh out 0.100 mol of tris hydrochloride and dissolve it in a beaker containing about 800 mL of water.
2. Place a pH electrode in the solution and monitor the pH.
3. Add NaOH until the pH is exactly 7.60.
4. Transfer the solution to a volumetric flask and wash the beaker a few times. Add the washings to the volumetric flask.
5. Dilute to the mark and mix.

You do not mix calculated quantities, though a quick calculation is helpful so that you have some idea of how much will be needed.

Reasons why a calculation would be wrong:
1. You might have ignored activity coefficients.
2. The temperature might not be just right.
3. The approximations that $[HA] = F_{HA}$ and $[A^-] = F_{A^-}$ could be in error.
4. The pK_A reported for tris in your favorite table is probably not what you would measure in your lab.
5. You will probably make an arithmetic error anyway.

Buffer Capacity

The **buffer capacity,** β (also called **buffer intensity**), is defined as

$$\beta = \frac{dC_B}{dpH} = -\frac{dC_A}{dpH} \tag{11-60}$$

where C_A and C_B are the number of moles of strong acid or base per liter

Demonstration 11-2   HOW BUFFERS WORK

A buffer resists changes in pH because the added acid or base is consumed by the buffer. As the buffer is used up it becomes less resistant to changes in pH.

In this demonstration,[†] a mixture containing approximately a 10:1 mole ratio of $HSO_3^-:SO_3^{2-}$ is prepared. Since pK_A for HSO_3^- is 7.2, the pH should be approximately

$$pH = pK_A + \log \frac{[SO_3^{2-}]}{[HSO_3^-]} = 7.2 + \log \frac{1}{10} = 6.2$$

When formaldehyde is added, the net reaction is the consumption of HSO_3^-, but not of SO_3^{2-}.

$$H_2C{=}O \;+\; HSO_3^- \;\rightarrow\; H_2C\!\!\begin{array}{c} O^- \\ SO_3H \end{array} \;\rightarrow\; H_2C\!\!\begin{array}{c} OH \\ SO_3^- \end{array} \qquad (A)$$

formaldehyde   bisulfite

$$H_2C{=}O + SO_3^{2-} \;\rightarrow\; H_2C\!\!\begin{array}{c} O^- \\ SO_3^- \end{array}$$

sulfite

$$\qquad (B)$$

$$H_2C\!\!\begin{array}{c} O^- \\ SO_3^- \end{array} \;+\; HSO_3^- \;\rightarrow\; H_2C\!\!\begin{array}{c} OH \\ SO_3^- \end{array} \;+\; SO_3^{2-}$$

(In sequence A bisulfite is consumed directly. In sequence B the net reaction is destruction of HSO_3^-, with no change in the SO_3^{2-} concentration.)

We can prepare a table showing how the pH should change as the HSO_3^- reacts.

Percent of reaction completed	$[SO_3^{2-}]:[HSO_3^-]$	Calculated pH
0	1:10	6.2
90	1:1	7.2
99	1:0.1	8.2
99.9	1:0.01	9.2
99.99	1:0.001	10.2

[†] F. B. Dutton and G. Gordon in H. N. Alyea and F. B. Dutton, eds., *Tested Demonstrations in Chemistry* 6th ed. (Easton, Pa.: Journal of Chemical Education, 1965), p. 147; R. L. Barrett, *J. Chem. Ed.*, **32**, 78 (1955).

needed to produce a unit change in pH. Buffer capacity is a positive number. The larger the value of β, the more resistant the solution is to pH change.

At the top of Figure 11-4 is a graph showing C_B versus pH for a solution containing 0.100 F HA with $pK_A = 5.00$. The ordinate (C_B) is the formal concentration of strong base needed to be mixed with 0.100 F HA to give the indicated pH. For example, a solution containing 0.050 F OH^- plus 0.100 F HA would have a pH of 5.00 (neglecting activities).

The lower graph in Figure 11-4 shows the buffer capacity as a function

You can see that through 90% completion the pH should rise by just 1 pH unit. In the next 9% of the reaction the pH will rise by another unit. At the end of the reaction the change in pH should be very abrupt.

In the formaldehyde clock reaction, formaldehyde is added to a solution containing HSO_3^-, SO_3^{2-}, and phenolphthalein indicator. Phenolphthalein is colorless below a pH of ~8.5 and red above this pH. What is observed is that the solution remains colorless for more than a minute. Suddenly the pH shoots up and the liquid turns pink. Monitoring the pH with a glass electrode gave the results below.

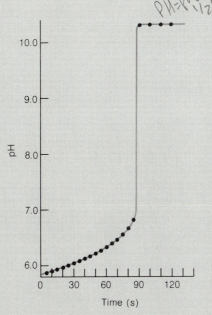

Graph of pH versus time in the formaldehyde clock reaction.

Procedure: All solutions should be fresh. Prepare a solution of formaldehyde by diluting 9 mL of 37% (wt/wt) formaldehyde to 100 mL. Dissolve 1.5 g of $NaHSO_3$ and 0.18 g of Na_2SO_3 in 400 mL of water, and add ~1 mL of phenolphthalein indicator solution (Table 12-3). Add 23 mL of the formaldehyde solution to the well-stirred buffer solution to initiate the clock reaction. The time of reaction can be adjusted by changing the temperature, concentrations, or volume.

of pH for the same system of HA plus strong base. The lower curve in Figure 11-4 is the derivative of the upper curve. The most notable feature of buffer capacity is that it reaches a maximum when $pH = pK_A$. That is, a buffer is most effective in resisting changes in pH when $pH = pK_A$ (that is, when $[HA] = [A^-]$).

In choosing a buffer for an experiment, you should seek one whose pK_A is as close as possible to the desired pH. *The useful pH range of a buffer is usually considered to be $pK_A \pm 1$ pH unit.* Outside this range there is not

Choose a buffer whose pK_A is close to the desired pH.

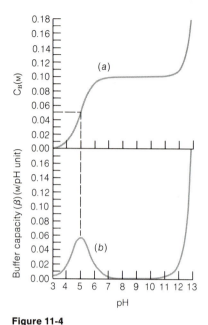

Figure 11-4
(a) C_B versus pH for a solution containing 0.100 F HA with $pK_A = 5.00$. (b) Buffer capacity versus pH for the same system. This curve is the derivative of the curve in part a.

Changing ionic strength changes pH.

Changing temperature changes pH.

What you mix is not what you get in dilute solutions or at extremes of pH.

enough of either the weak acid or the weak base to react with added base or acid. Clearly, the buffer capacity can be increased by increasing the concentration of the buffer.

The buffer capacity curve in Figure 11-4 continues upward at high pH (and at low pH, which is not shown in the figure) simply because there is a high concentration of OH^-. Addition of a small amount of acid or base to a large amount of OH^- (or H^+) will not have a very great effect on pH. A solution of high pH is well buffered by the H_2O/OH^- conjugate acid–conjugate base pair. A solution of low pH is well buffered by the H_3O^+/H_2O conjugate acid–conjugate base pair.

Table 11-2 lists the pK_A values of several common buffers. These are widely used in biochemistry. The measurement of pH with glass electrodes, and the buffers used by the U.S. National Bureau of Standards to define the pH scale, are described in Chapter 16.

Limitations of Buffers

Activity coefficients

The correct Henderson–Hasselbalch Equation 11-56 includes activity coefficients. Failure to include activity coefficients is the principal reason why our calculated pH values will not be in perfect agreement with measured pH values. The activity coefficients also predict that the pH of a buffer will vary with ionic strength. Adding an inert salt (such as NaCl) will change the pH of a buffer. This effect can be significant for highly charged species, such as citrate or phosphate. When a 0.5 M stock solution of phosphate buffer at pH 6.6 is diluted to 0.05 M, the pH rises to 6.9. This is a rather significant effect of changing ionic strength.

Temperature

Most buffers exhibit a noticeable dependence of pK_A on temperature. Tris has an exceptionally large dependence, approximately -0.031 pK_A units per degree, near room temperature. A solution of tris made up to pH 8.08 at 25°C will have pH $\approx$ 8.7 at 4° and pH $\approx$ 7.7 at 37°C.

When what you mix is not what you get

In a dilute solution, or at extremes of pH, the molar concentrations of HA and A^- in solution are not equal to their formal concentrations. This can be seen as follows. Suppose we mix F_{HA} moles of HA and F_{A^-} moles of A^-. The equilibria are

$$HA \rightleftharpoons H^+ + A^- \qquad K_A \qquad (11\text{-}61)$$

$$A^- + H_2O \rightleftharpoons HA + OH^- \qquad K_B \qquad (11\text{-}62)$$

Table 11-2
Structures and pK$_A$ values for some commonly used buffers

Name	Structure	pK$_A$ ($\sim25°C$)
Phosphoric acid	H_3PO_4	2.15 (pK$_1$)
Citric acid	$HO_2CCH_2\overset{\overset{\displaystyle OH}{\vert}}{\underset{\underset{\displaystyle CO_2H}{\vert}}{C}}CH_2CO_2H$	3.13 (pK$_1$)
Formic acid	HCO_2H	3.74
Succinic acid	$HO_2CCH_2CH_2CO_2H$	4.21 (pK$_1$)
Citric acid	$H_2(citrate)^-$	4.76 (pK$_2$)
Acetic acid	CH_3CO_2H	4.76
Succinic acid	$H(succinate)^-$	5.64 (pK$_2$)
2-(N-morpholino)ethane-sulfonic acid (MES)	$O\!\!-\!\!\!\overset{+}{N}HCH_2CH_2SO_3^-$	6.15
Cacodylic acid	$(CH_3)_3AsO_2H$	6.19
Citric acid	$H(citrate)^{2-}$	6.40 (pK$_3$)
N-2-acetamidoiminodiacetic acid (ADA)	$H_2N\overset{\overset{\displaystyle O}{\parallel}}{C}CH_2\overset{+}{N}H\!\!\overset{\nearrow CH_2CO_2^-}{\underset{\searrow CH_2CO_2H}{}}$	6.60
1,3-bis[tris(hydroxymethyl)-methylamino]propane (BIS-TRIS propane)	$(HOCH_2)_3C\overset{+}{N}H_2(CH_2)_3NHC(CH_2OH)_3$	6.80
Piperazine-N,N'-bis(2-ethane-sulfonic acid) (PIPES)	$^-O_3SCH_2CH_2\overset{+}{N}H\!\!-\!\!H\overset{+}{N}CH_2CH_2SO_3^-$	6.80
N-2-acetamido-2-aminoethane-sulfonic acid (ACES)	$H_2N\overset{\overset{\displaystyle O}{\parallel}}{C}CH_2\overset{+}{N}H_2CH_2CH_2SO_3^-$	6.90
Imidazole hydrochloride	$\overset{+}{HN}\!\!-\!\!\!\!\underset{\underset{\displaystyle H}{N}}{}\!\!\!\!\quad Cl^-$	6.99
3-(N-morpholine)propane-sulfonic acid (MOPS)	$O\!\!-\!\!\!\overset{+}{N}HCH_2CH_2CH_2SO_3^-$	7.20
Phosphoric acid	$H_2PO_4^-$	7.20 (pK$_2$)
N-tris(hydroxymethyl)methyl-2-aminoethanesulfonic acid (TES)	$(HOCH_2)_3C\overset{+}{N}H_2CH_2CH_2SO_3^-$	7.50
N-2-hydroxyethylpiperazine-N'-2-ethanesulfonic acid (HEPES)	$HOCH_2CH_2N\!\!-\!\!\!\overset{+}{N}HCH_2CH_2SO_3^-$	7.55

Note: The protonated form of each molecule is shown. Acidic hydrogen atoms are shown in **bold** type.

(continued)

Table 11-2 (*continued*)

Name	Structure	pK$_A$ (~25°C)
N-2-hydroxyethylpiperazine-N′-3-propanesulfonic acid (HEPPS)	HOCH$_2$CH$_2$N◯$\overset{+}{N}$HCH$_2$CH$_2$CH$_2$SO$_3^-$	8.00
N-tris(hydroxymethyl)methyl-glycine (TRICINE)	(HOCH$_2$)$_3$C$\overset{+}{N}$H$_2$CH$_2$CO$_2^-$	8.15
Glycine amide, hydrochloride	H$_3$$\overset{+}{N}CH_2\overset{\overset{\displaystyle O}{|}}{C}NH_2$ Cl$^-$	8.20
Tris(hydroxymethyl)amino-methane hydrochloride (TRIS hydrochloride)	(HOCH$_2$)$_3$C$\overset{+}{N}$H$_3$ Cl$^-$	8.08
N,N-bis(2-hydroxyethyl)-glycine (BICINE)	(HOCH$_2$CH$_2$)$_2\overset{+}{N}$HCH$_2$CO$_2^-$	8.35
Glycylglycine	H$_3\overset{+}{N}$CH$_2\overset{\overset{\displaystyle O}{\|}}{C}$NHCH$_2CO_2^-$	8.40
Boric acid	B(OH)$_3$	9.24 (pK$_1$)
Cyclohexylaminoethane-sulfonic acid (CHES)	◯—$\overset{+}{N}$H$_2$CH$_2$CH$_2$SO$_3^-$	9.50
3-(cyclohexylamino)propane-sulfonic acid (CAPS)	◯—$\overset{+}{N}$H$_2$CH$_2$CH$_2$CH$_2$SO$_3^-$	10.40
Phosphoric acid	HPO$_4^-$	12.35 (pK$_3$)
Boric acid	OB(OH)$_2^-$	12.74 (pK$_2$)

Reaction 11-61 decreases the concentration of HA, and Reaction 11-62 increases the concentration of HA. For each mole of H$^+$ made by Reaction 11-61, HA decreases by one mole. For each mole of OH$^-$ made in Reaction 11-62, HA increases by one mole. The total concentration of HA in the solution is therefore

$$[HA] = F_{HA} - [H^+] + [OH^-] \qquad (11\text{-}63)$$

(Equation 11-63 neglects any contribution of H$_2$O dissociation to the concentrations of H$^+$ and OH$^-$.) By similar reasoning we can write

$$[A^-] = F_{A^-} + [H^+] - [OH^-] \qquad (11\text{-}64)$$

In our work so far, we have assumed that $[HA] \approx F_{HA}$ and $[A^-] \approx F_{A^-}$, and we used these values in the Henderson–Hasselbalch equation. A more rigorous procedure is to use the values given by Equations 11-63 and 11-64. We see that if F_{HA} or F_{A^-} is small, or if $[H^+]$ or $[OH^-]$ is large, the approximations $[HA] \approx F_{HA}$ and $[A^-] \approx F_{A^-}$ are not good. In acidic solutions $[H^+] \gg [OH^-]$, so $[OH^-]$ can be ignored in Equations 11-63 and 11-64. In basic solutions $[H^+]$ can be neglected.

EXAMPLE

What will be the pH if 0.010 0 mol of HA (with $pK_A = 2.00$) and 0.010 0 mol of A^- are dissolved in 1.00 L?

Since the solution will be acidic (pH $\approx pK_A = 2.00$), we can neglect the $[OH^-]$ terms in Equations 11-63 and 11-64. Setting $[H^+] = x$, we can use the K_A equation to calculate the value of $[H^+]$.

$$HA \rightleftharpoons H^+ + A^-$$

$$0.010\,0 - x \qquad x \qquad 0.010\,0 + x$$

$$\frac{[H^+][A^-]}{[HA]} = \frac{(x)(0.010\,0 + x)}{(0.010\,0 - x)} = 10^{-2.00} \Rightarrow x = 0.004\,14 \text{ M}$$

$$pH = \log[H^+] = 2.38$$

The concentrations of HA and A^- are not what we mixed:

$$[HA] = F_{HA} - [H^+] = 0.005\,86 \text{ M}$$

$$[A^-] = F_{A^-} + [H^+] = 0.014\,1 \text{ M}$$

In this example, HA is too strong and the concentrations too low for HA and A^- to be equal to their formal concentrations.

The Henderson–Hasselbalch equation is a true statement. It is just a rearrangement of the K_A equilibrium expression, which is always true. What is an approximation is the pair of statements, $[HA] \approx F_{HA}$ and $[A^-] \approx F_{A^-}$.

> The Henderson–Hasselbalch equation is *always* true.

Summary

A buffer consists of a mixture of a weak acid and its conjugate base. The buffer is most useful when pH $\approx pK_A$. Over a reasonable range of concentration, the pH of a buffer is nearly independent of concentration. A buffer resists changes in pH because it reacts with added acids or bases. If too much acid or base is added, the buffer will be consumed and no longer resist changes in pH.

11-5 DIPROTIC ACIDS AND BASES

Polyprotic acids and bases are those that can donate or accept more than one proton. The twenty common amino acids, which are the building blocks of proteins, are all polyprotic. Most are **diprotic,** which means that their acid–base chemistry involves two protons.

11 / ACID–BASE EQUILIBRIA

200

The general structure of the natural **amino acids** is

ammonium group $\longrightarrow$ $\overset{+}{H_3N}$

carboxyl group $\longrightarrow$ ^-O-C — CH—R

$$\overset{+}{H_3N}\diagdown CH-R \qquad ^-O-\overset{\|}{\underset{O}{C}}$$

A zwitterion is a molecule with positive and negative groups.

where R is a different group for each compound. The carboxyl group, drawn above in its ionized (basic) form, is a stronger acid than the ammonium group. Therefore, the nonionic form rearranges spontaneously to the **zwitterion**

$$H_2N\diagdown CH-R \longrightarrow \overset{+}{H_3N}\diagdown CH-R$$
$$HO_2C\diagup \qquad \qquad ^-O_2C\diagup$$

zwitterion

At low pH, both the ammonium group and the carboxyl group are protonated. At high pH, neither is protonated. The acid dissociation constants of the amino acids are listed in Table 11-3 where each compound is drawn in its fully protonated form.

In our discussion, we will take as a specific example the amino acid leucine, designated HL.

$$H_3\overset{+}{N}CHCO_2H \xrightleftharpoons{pK_{A1}=2.329} H_3\overset{+}{N}CHCO_2^- \xrightleftharpoons{pK_{A2}=9.747} H_2NCHCO_2^- \qquad (11\text{-}65)$$

$$H_2L^+ \qquad \qquad \underset{HL}{\text{leucine}} \qquad \qquad L^-$$

The equilibrium constants refer to the following reactions

We customarily omit the subscript $_A$ in K_{A1} and K_{A2}. We will always write the subscript $_B$ in K_{B1} and K_{B2}.

$$H_2L^+ \rightleftharpoons HL + H^+ \qquad K_{A1} \equiv K_1 \qquad (11\text{-}66)$$

$$HL \rightleftharpoons L^- + H^+ \qquad K_{A2} \equiv K_2 \qquad (11\text{-}67)$$

$$L^- + H_2O \rightleftharpoons HL + OH^- \qquad K_{B1} \qquad (11\text{-}68)$$

$$HL + H_2O \rightleftharpoons H_2L^+ + OH^- \qquad K_{B2} \qquad (11\text{-}69)$$

You recall that the relations between the acid and base equilibrium constants are

These are Equations 10-50 and 10-51.

$$K_{A1} \cdot K_{B2} = K_w \qquad (11\text{-}70)$$

$$K_{A2} \cdot K_{B1} = K_w \qquad (11\text{-}71)$$

We now set out to calculate the pH and composition of individual solutions of $0.050\ 0$ M H_2L^+, $0.050\ 0$ M HL, and $0.050\ 0$ M L^-. Our methods are

Table 11-3
Acid dissociation constants of amino acids

Amino acid	Structure[†]	Carboxylic acid[‡]	Ammonium group[‡]	Substituent[‡]
Alanine	NH_3^+ — CH—CH_3 — CO_2H	$pK_A = 2.348$	$pK_A = 9.867$	
Arginine	NH_3^+ — CH—$CH_2CH_2CH_2NHC$($\overset{+}{N}H_2$)(NH_2) — CO_2H	$pK_A = 1.823$	$pK_A = 8.991$	$(pK_A = 12.48)$
Asparagine	NH_3^+ — CH—$CH_2\overset{O}{\overset{\|}{C}}NH_2$ — CO_2H	$pK_A = 2.14^§$	$pK_A = 8.72^§$	
Aspartic acid	NH_3^+ — CH—CH_2CO_2H — CO_2H	$pK_A = 1.990$	$pK_A = 10.002$	$pK_A = 3.900$
Cysteine	NH_3^+ — CH—CH_2SH — CO_2H	$(pK_A = 1.71)$	$pK_A = 10.77$	$pK_A = 8.36$
Glutamic acid	NH_3^+ — CH—$CH_2CH_2CO_2H$ — CO_2H	$pK_A = 2.23$	$pK_A = 9.95$	$pK_A = 4.42$
Glutamine	NH_3^+ — CH—$CH_2CH_2\overset{O}{\overset{\|}{C}}NH_2$ — CO_2H	$pK_A = 2.17^§$	$pK_A = 9.01^§$	
Glycine	NH_3^+ — CH—H — CO_2H	$pK_A = 2.350$	$pK_A = 9.778$	
Histidine	NH_3^+ — CH—CH_2—(imidazole ring, NH, $\overset{+}{N}H$) — CO_2H	$pK_A = 1.7^§$	$pK_A = 9.08^§$	$pK_A = 6.02^§$
Isoleucine	NH_3^+ — CH—CH(CH_3)(CH_2CH_3) — CO_2H	$pK_A = 2.319$	$pK_A = 9.754$	

[†] The acidic protons are shown in **bold** type. Each amino acid is written in its fully protonated form.
[‡] pK_A values refer to 25°C and zero ionic strength unless marked by [§]. Values considered to be uncertain are enclosed in parentheses.
[§] For these entries the ionic strength is 0.1 M, and the constant refers to a product of concentrations instead of activities.
SOURCE: Data from A. E. Martell and R. M. Smith, *Critical Stability Constants*, Vol. 1 (New York: Plenum Press 1974).

(continued)

Table 11-3 (*continued*)

Amino acid	Structure[†]	Carboxylic acid[‡]	Ammonium group[‡]	Substituent[‡]
Leucine	$\overset{\overset{+}{N}H_3}{C}H-CH_2CH(CH_3)_2$ (CO_2H)	$pK_A = 2.329$	$pK_A = 9.747$	
Lysine	$\overset{\overset{+}{N}H_3}{C}H-CH_2CH_2CH_2CH_2\overset{+}{N}H_3$ (CO_2H)	$pK_A = 2.04^{\S}$	$pK_A = 9.08^{\S}$	$pK_A = 10.69^{\S}$
Methionine	$\overset{\overset{+}{N}H_3}{C}H-CH_2CH_2SCH_3$ (CO_2H)	$pK_A = 2.20^{\S}$	$pK_A = 9.05^{\S}$	
Phenylalanine	$\overset{\overset{+}{N}H_3}{C}H-CH_2-C_6H_5$ (CO_2H)	$pK_A = 2.20$	$pK_A = 9.31$	
Proline	($H_2\overset{+}{N}$ pyrrolidine ring, HO_2C)	$pK_A = 1.952$	$pK_A = 10.640$	
Serine	$\overset{\overset{+}{N}H_3}{C}H-CH_2OH$ (CO_2H)	$pK_A = 2.187$	$pK_A = 9.209$	
Threonine	$\overset{\overset{+}{N}H_3}{C}H-CH(CH_3)(OH)$ (CO_2H)	$pK_A = 2.088$	$pK_A = 9.100$	
Tryptophan	$\overset{\overset{+}{N}H_3}{C}H-CH_2-\text{(indole)}$ (CO_2H)	$pK_A = 2.35^{\S}$	$pK_A = 9.33^{\S}$	
Tyrosine	$\overset{\overset{+}{N}H_3}{C}H-CH_2-C_6H_4-OH$ (CO_2H)	$pK_A = 2.17^{\S}$	$pK_A = 9.19$	$pK_A = 10.47$
Valine	$\overset{\overset{+}{N}H_3}{C}H-CH(CH_3)_2$ (CO_2H)	$pK_A = 2.286$	$pK_A = 9.718$	

general. They do not depend on the charge type of the acids and bases. That is, we would use the same procedure to find the pH of the diprotic acids H_2A, where A is anything, or H_2L^+, where L is leucine.

Acidic Form, H_2L^+

Easy stuff.

A salt such as leucine hydrochloride contains the protonated species, H_2L^+, which can dissociate twice, as indicated in Reaction 11-65. Since $K_1 = 4.69 \times 10^{-3}$, H_2L^+ is a weak acid. HL is an even weaker acid, since $K_2 = 1.79 \times 10^{-10}$. It appears that the H_2L^+ will dissociate only partly, and the resulting HA^- will hardly dissociate at all. For this reason, we make the (superb) approximation that a solution of H_2L^+ behaves as a monoprotic acid, with $K_A = K_1$.

With this approximation, the calculation of the pH of $0.050\,0$ M H_2L^+ is a trivial matter.

$$\underset{0.050\,0-x}{H_3NCHCO_2H} \rightleftharpoons \underset{x}{H_3NCHCO^{2-}} + \underset{x}{H^+}$$

$$\underset{0.050\,0-x}{H_2L^+} \rightleftharpoons \underset{x}{HL} + \underset{x}{H^+} \qquad K_A = K_1 = 4.69 \times 10^{-3} \qquad (11\text{-}72)$$

H_2L^+ can be treated as monoprotic, with $K_A = K_{A1}$.

$$\frac{x^2}{F - x} = K_A \Rightarrow x = 1.31 \times 10^{-2} \text{ M}$$

$$[HL] = x = 1.31 \times 10^{-2} \text{ M} \qquad (11\text{-}73)$$

$$[H^+] = x = 1.31 \times 10^{-2} \text{ M} \Rightarrow pH = 1.88 \qquad (11\text{-}74)$$

$$[H_2L^+] = F - x = 3.69 \times 10^{-2} \text{ M} \qquad (11\text{-}75)$$

What is the concentration of L^- in the solution? We have already assumed that it is very small, but it cannot be zero. We can calculate $[L^-]$ using the K_{A2} equation (11-67), with the concentrations of HL and H^+ from Equations 11-73 and 11-74.

$$K_{A2} = \frac{[H^+][L^-]}{[HL]}$$

$$[L^-] = \frac{K_{A2}[HL]}{[H^+]} \qquad (11\text{-}76)$$

$$[L^-] = \frac{(1.79 \times 10^{-10})(1.31 \times 10^{-2})}{(1.31 \times 10^{-2})} = 1.79 \times 10^{-10} \text{ M} \,(= K_{A2})$$

204

11 / ACID–BASE EQUILIBRIA

Since we have made the approximation that $[H^+] = [HL]$ in Equation 11-72, it follows from Equation 11-76 that $[L^-] = K_{A2} = 1.79 \times 10^{-10}$ M.

Our approximation is confirmed by this last result. The concentration of L^- is about eight orders of magnitude smaller than that of HL. As a source of protons, the dissociation of HL is indeed negligible compared to the dissociation of H_2L^+. For most diprotic acids, K_1 is sufficiently larger than K_2 for this approximation to be valid. In the example just treated, even if K_2 were ten times less than K_1, the value of $[H^+]$ calculated by ignoring the second ionization would be in error by only 4%. The error in pH would be only 0.01 pH unit. *In summary, a solution of a diprotic acid behaves as a solution of a monoprotic acid, with $K_A = K_{A1}$.*

More easy stuff.

Basic Form, L^-

The fully basic species, L^-, would be found in a salt such as sodium leucinate, which could be prepared by treating leucine with an equimolar quantity of NaOH. Dissolving sodium leucinate in water gives a solution of L^-, the fully basic species. The two K_B values for this dibasic anion are

$$K_{B1} = K_w/K_{A2} = 5.59 \times 10^{-5} \qquad (11\text{-}77)$$

$$K_{B2} = K_w/K_{A1} = 2.13 \times 10^{-12} \qquad (11\text{-}78)$$

Hydrolysis refers to the reaction of anything with water. Specifically, the reaction $L^- + H_2O \rightleftharpoons HL + OH^-$ is called hydrolysis.

These numbers tell us that L^- will not hydrolyze very much to give HL. Further, the resulting HL is such a weak base that hardly any further reaction to make H_2L^+ will occur.

As before, then, it is reasonable to treat L^- as a monobasic species, with $K_B = K_{B1}$. The results of this (fantastic) approximation are outlined below.

L^- can be treated as monobasic, with $K_B = K_{B1}$.

$$H_2NCHCO_2^- + H_2O \rightleftharpoons H_3\overset{+}{N}CHCO_2^- + OH^-$$

$$L^- \quad + H_2O \rightleftharpoons \quad HL \quad + OH^-$$

$$0.0500 - x \qquad\qquad x \qquad\quad x$$

$$K_B = K_{B1} = \frac{K_w}{K_{A2}} = 5.59 \times 10^{-5} \quad (11\text{-}79)$$

$$\underset{(.05)}{\frac{x^2}{F - x}} = 5.59 \times 10^{-5} \Rightarrow x = 1.64 \times 10^{-3} \text{ M}$$

$$[HL] = x = 1.64 \times 10^{-3} \text{ M} \qquad (11\text{-}80)$$

$$[H^+] = K_w/x = 6.08 \times 10^{-12} \text{ M} \Rightarrow pH = 11.22 \qquad (11\text{-}81)$$

$$[\text{L}^-] = \text{F} - x = 4.84 \times 10^{-2} \text{ M} \qquad (11\text{-}82)$$

The concentration of H_2L^+ can be found from the K_{B2} (or K_{A1}) equilibrium.

$$K_{B2} = \frac{[H_2L^+][OH^-]}{[HL]} = \frac{[H_2L^+]x}{x} = [H_2L^+] \qquad (11\text{-}83)$$

We find that $[H_2L^+] = K_{B2} = 2.13 \times 10^{-12}$ M, and the approximation that $[H_2L^+]$ is insignificant compared to $[HL]$ is well justified. In summary, if there is any reasonable separation between K_{A1} and K_{A2} (and therefore between K_{B1} and K_{B2}), *a solution of the fully basic form of a diprotic acid can be treated as monobasic, with* $K_B = K_{B1}$.

Intermediate Form, HL

A tougher problem.

A solution prepared from leucine (HL) is more complicated than either H_2L^+ or L^-, because HL is both an acid and a base.

$$HL \rightleftharpoons H^+ + L^- \qquad K_A = K_{A2} = 1.79 \times 10^{-10} \qquad (11\text{-}84)$$

$$HL + H_2O \rightleftharpoons H_2L^+ + OH^- \qquad K_B = K_{B2} = 2.13 \times 10^{-12} \qquad (11\text{-}85)$$

HL is both an acid and a base.

A molecule that can both donate and accept a proton is said to be **amphiprotic.** The acid dissociation reaction (11-84) has a larger equilibrium constant than the base association reaction (11-85), so we expect that a solution of leucine will be acidic.

However, we cannot simply ignore Reaction 11-85. It turns out that both reactions proceed to a nearly equal extent for the following reason: Reaction 11-84 produces one mole of H^+ for each mole of L^-. The mole of H^+ reacts with a mole of OH^- from Reaction 11-85, thereby driving reaction 11-85 to the right. The number of molecules of HL reacting by each path is nearly equal, with Reaction 11-84 dominating just slightly.

To treat this case correctly, we must resort to the systematic method of Chapter 7. The procedure is applied to leucine, whose intermediate form (HL) has no net charge. However, the results we obtain apply to the intermediate form of *any* diprotic acid, regardless of its charge.

Our problem deals with 0.050 0 M leucine, in which both Reactions 11-84 and 11-85 can happen. The charge balance is

$$[H^+] + [H_2L^+] = [L^-] + [OH^-] \qquad (11\text{-}86)$$

which can be rearranged to

$$[H_2L^+] - [L^-] + [H^+] - [OH^-] = 0 \qquad (11\text{-}87)$$

We see from Equation 11-66 that we can replace $[H_2L^+]$ with $[HL][H^+]/K_1$,

206

11 / ACID–BASE EQUILIBRIA

and from Equation 11-67 that we can replace $[L^-]$ with $[HL]\, K_2/[H^+]$. The K_w equation tells us that we can always write $[OH^-] = K_w/[H^+]$. Putting all of these values into Equation 11-87 gives

$$\frac{[HL][H^+]}{K_1} - \frac{[HL]\, K_2}{[H^+]} + [H^+] - \frac{K_w}{[H^+]} = 0 \qquad (11\text{-}88)$$

Equation 11-88 can now be solved for $[H^+]$. First multiply all terms by $[H^+]$:

$$\frac{[HL][H^+]^2}{K_1} - [HL]\, K_2 + [H^+]^2 - K_w = 0$$

Then factor out $[H^+]^2$ and rearrange:

$$[H^+]^2 \left(\frac{[HL]}{K_1} + 1\right) = K_2\, [HL] + K_w$$

$$[H^+]^2 = \frac{K_2\, [HL] + K_w}{\dfrac{[HL]}{K_1} + 1} \qquad (11\text{-}89)$$

Multiplying the numerator and denominator of Equation 11-89 by K_1, and taking the square root of both sides gives

$$[H^+] = \sqrt{\frac{K_1 K_2\, [HL] + K_1 K_w}{K_1 + [HL]}} \qquad (11\text{-}90)$$

Up to this point we have made no approximations, except to neglect activity coefficients in the equilibria. We have solved for $[H^+]$ in terms of known constants plus the single unknown, $[HL]$. Where do we proceed from here?

The missing insight!

A chemist promptly gallops down from the mountains on her white stallion to provide the missing insight: "The major species will be HL, because it is both a weak acid and a weak base. Neither Reaction 11-84 nor Reaction 11-85 will go very far. For the concentration of HL in Equation 11-90, you can simply substitute the value 0.050 0 M."

Taking the chemist's advice, we write Equation 11-90 in its most useful form:

K_1 and K_2 in Equation 11-91 are both *acid* dissociation constants (K_{A1} and K_{A2}).

$$\boxed{\;[H^+] \approx \sqrt{\frac{K_1 K_2 F + K_1 K_w}{K_1 + F}}\;} \qquad (11\text{-}91)$$

where F is the formal concentration of HL ($= 0.050\,0$ M in the present case).

At long last we can calculate the pH of $0.050\,0$ M leucine with Equation 11-91.

$$[H^+] = \sqrt{\frac{(4.69 \times 10^{-3})(1.79 \times 10^{-10})(0.050\,0) + (4.69 \times 10^{-3})(1.0 \times 10^{-14})}{4.69 \times 10^{-3} + 0.050\,0}}$$

$$[H^+] = 8.76 \times 10^{-7}\,\text{M} \Rightarrow pH = 6.06 \tag{11-92}$$

The concentrations of H_2L^+ and L^- can be found from Equations 11-66 and 11-67, using $[H^+] = 8.76 \times 10^{-7}$ M and $[HL] = 0.050\,0$ M.

Equation 11-66

$$[H_2L^+] = \frac{[H^+][HL]}{K_1} = \frac{(8.76 \times 10^{-7})(0.050\,0)}{4.69 \times 10^{-3}} = 9.34 \times 10^{-6}\,\text{M} \tag{11-93}$$

Equation 11-67

$$[L^-] = \frac{K_2[HL]}{[H^+]} = \frac{(1.79 \times 10^{-10})(0.050\,0)}{8.76 \times 10^{-7}} = 1.02 \times 10^{-5}\,\text{M} \tag{11-94}$$

Was the approximation $[HL] \approx 0.050\,0$ M a good one? It certainly was, because $[H_2L^+]$ (9.34×10^{-6} M) and $[L^-]$ (1.02×10^{-5} M) are quite small in comparison to 0.050 0 M. Nearly all of the leucine remained in the form HL. Note also that $[H_2L^+]$ is nearly equal to $[L^-]$; this confirms that Reactions 11-84 and 11-85 proceed almost equally, even though K_A is 84 times bigger than K_B for leucine.

If $[H_2L^+] + [L^-]$ is not much less than $[HL]$, and if you wish to refine your values of $[H_2L^+]$ and $[L^-]$, the method in Box 11-2 can be used.

We will usually find that Equation 11-91 is a fair-to-excellent approximation. It applies to the intermediate form of any diprotic acid, regardless of its charge type.

An even simpler form of Equation 11-91 results from two conditions that usually obtain. First, if $K_2F \gg K_w$, the second term in the numerator of Equation 11-91 can be dropped.

$$[H^+] \approx \sqrt{\frac{K_1K_2F + K_1K_w}{K_1 + F}}$$

Then, if $K_1 \ll F$, the first term in the denominator can also be neglected.

$$[H^+] \approx \sqrt{\frac{K_1K_2F}{K_1 + F}}$$

Canceling F in the numerator and denominator gives

$$[H^+] \approx \sqrt{K_1K_2} \tag{11-95}$$

or

$$\log[H^+] \approx \tfrac{1}{2}(\log K_1 + \log K_2)$$

$$-\log[H^+] \approx -\tfrac{1}{2}(\log K_1 + \log K_2)$$

Box 11-2 MORE ON SUCCESSIVE APPROXIMATIONS

To further extol the virtues and utility of the method of successive approximations, we now consider a case in which the concentration of the intermediate (amphiprotic) species is not very close to F, the formal concentration of the solution. This happens when K_1 and K_2 are not very far apart, and F is small. Consider a solution of 1.00×10^{-3} F HM^-, where HM^- is the intermediate form of malic acid.

$$\underset{\substack{\text{malic acid} \\ H_2M}}{\text{HO}\diagdown\overset{CO_2H}{\underset{CO_2H}{}}} \quad \underset{pK_1 = 3.40}{\overset{K_1 = 4.0 \times 10^{-4}}{\rightleftharpoons}} \quad \underset{HM^-}{\text{HO}\diagdown\overset{CO_2^-}{\underset{CO_2H}{}}} \quad \underset{pK_2 = 5.05}{\overset{K_2 = 8.9 \times 10^{-6}}{\rightleftharpoons}} \quad \underset{M^{2-}}{\text{HO}\diagdown\overset{CO_2^-}{\underset{CO_2^-}{}}}$$

As a first approximation, we assume that $[HM^-] \approx 1.00 \times 10^{-3}$ M. Plugging this value into Equation 11-91, we calculate first approximations for $[H^+]$, $[H_2M]$, and $[M^{2-}]$.

$$[H^+]_1 = \sqrt{\frac{K_1 K_2 (0.001\ 00) + K_1 K_w}{K_1 + (0.001\ 00)}} = 5.04 \times 10^{-5} \text{ M}$$

$$\Rightarrow [H_2M]_1 = 1.26 \times 10^{-4} \text{ M and } [M^{2-}]_1 = 1.77 \times 10^{-4} \text{ M}$$

Clearly, $[H_2M]$ and $[M^{2-}]$ are not negligible in comparison to F = 1.00×10^{-3} M, so we need to revise our estimate of $[HM^-]$. As a second approximation, use

$$[HM^-]_2 = F - [H_2M]_1 - [M^{2-}]_1$$

$$= 0.001\ 00 - 0.000\ 126 - 0.000\ 177$$

$$= 0.000\ 697 \text{ M}$$

The pH of the intermediate form of a diprotic acid is roughly midway between the two pK_A values and is almost independent of concentration.

$$\boxed{pH \approx \frac{pK_1 + pK_2}{2}} \qquad (11\text{-}96)$$

Equation 11-96 is a good one to keep in your head. It is not as exact as Equation 11-91, but it is usually pretty close. Equation 11-96 gives a pH of 6.04 for leucine, compared to pH = 6.06 from Equation 11-91. Equation 11-96 says that **the pH of the intermediate form of a diprotic acid is roughly midway between pK_1 and pK_2, regardless of the formal concentration.**

Using the value $[HM^-]_2 = 0.000\,697$ in Equation 11-90 gives

$$[H^+]_2 = \sqrt{\frac{K_1 K_2 (0.000\,697) + K_1 K_w}{K_1 + 0.000\,697}} = 4.76 \times 10^{-5}\ \text{M}$$

$$\Rightarrow [H_2M]_2 = 8.29 \times 10^{-5}\ \text{M and}\ [M^{2-}] = 1.30 \times 10^{-4}\ \text{M}$$

The values of $[H_2M]_2$ and $[M^{2-}]_2$ can be used to calculate a third approximation for $[HM^-]$:

$$[HM^-]_3 = F - [H_2M]_2 - [M^{2-}]_2 = 0.000\,787\ \text{M}$$

Plugging $[HM^-]_3$ into Equation 11-90 gives

$$[H^+]_4 = 4.85 \times 10^{-5}$$

and the procedure can be repeated to get

$$[H^+]_5 = 4.83 \times 10^{-5}$$

We are homing in on an estimate of $[H^+]$ in which the uncertainty is already less than 1%. This is more accuracy than is justified by the accuracy of the equilibrium constants, K_1 and K_2. The fifth approximation for $[H^+]$ gives pH = 4.32, compared to pH = 4.30 for the first approximation and pH = 4.23 using the formula pH $\approx (pK_1 + pK_2)/2$. Considering the uncertainty in pH measurements, all of this calculation was hardly worth the effort. However, the concentration of $[HM^-]$ is 0.000 768 M, which is 23% less than the original estimate ($[HM^-] \approx F = 0.001\,00$ M).

EXAMPLE

Potassium acid phthalate (KHP) is a salt of the intermediate form of phthalic acid. Calculate the pH of 0.10 M and of 0.010 M KHP.

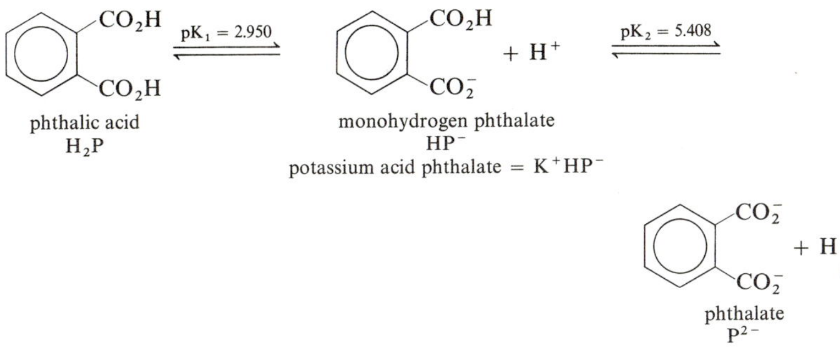

phthalic acid
H_2P

monohydrogen phthalate
HP^-
potassium acid phthalate $= K^+HP^-$

phthalate
P^{2-}

(11-97)

The pH of potassium acid phthalate is calculated to be $(pK_1 + pK_2)/2 = 4.18$, regardless of concentration, using Equation 11-96. Using Equation 11-91, we calculate pH = 4.18 for 0.10 M K^+HP^- and pH = 4.20 for 0.010 M K^+HP^-.

210

11 / ACID–BASE EQUILIBRIA

A good way to do it.

Good advice

When faced with the intermediate form of a diprotic acid, use Equation 11-91 to calculate the pH. As a check on your calculation, the answer should be very close to $(pK_1 + pK_2)/2$.

Summary

The way to calculate the pH and composition of solutions prepared from different forms of a diprotic acid (H_2A, HA^-, or A^{2-}) is outlined below. A summary of calculations for leucine is given in Table 11-4.

Solution of H_2A

(a) Treat H_2A as a monoprotic acid with $K_A = K_1$. This gives $[H^+]$, $[HA^-]$ and $[H_2A]$.

$$H_2A \xrightleftharpoons{K_1} H^+ + HA^-$$

$$F - x \qquad x \qquad x$$

$$\frac{x^2}{F - x} = K_1$$

(b) Use the K_2 equilibrium to solve for $[A^{2-}]$ using values of $[H^+]$ and $[HA^-]$ from part a.

$$[A^{2-}] = \frac{K_2[HA^-]}{[H^+]} = K_2$$

Solution of HA^-

(a) Use the approximation $[HA^-] \approx F$ and find the pH with Equation 11-91.

$$[H^+] = \sqrt{\frac{K_1 K_2 F + K_1 K_w}{K_1 + F}}$$

The pH should be close to $(pK_1 + pK_2)/2$.

Table 11-4
Summary of calculations for leucine

Solution	pH	$[H^+]$(M)	$[H_2L^+]$(M)	$[HL]$(M)	$[L^-]$(M)
0.0500 M H_2A	1.88	1.31×10^{-2}	3.69×10^{-2}	1.31×10^{-2}	1.79×10^{-10}
0.0500 M HA^-	6.06	8.76×10^{-7}	9.34×10^{-6}	5.00×10^{-2}	1.02×10^{-5}
0.0500 M A^{2-}	11.22	6.08×10^{-12}	2.13×10^{-12}	1.64×10^{-3}	4.84×10^{-2}

(b) Using $[H^+]$ from part a and $[HA^-] \approx F$, solve for H_2A and A^{2-} using the K_1 and K_2 equilibria.

$$[H_2A] = \frac{[HA^-][H^+]}{K_1}$$

$$[A^{2-}] = \frac{K_2[HA^-]}{[H^+]}$$

Solution of A^{2-}

(a) Treat A^{2-} as monobasic, with $K_B = K_{B1} = K_w/K_{A2}$. This gives $[A^{2-}]$, $[HA^-]$, and $[H^+]$.

$$A^{2-} + H_2O \xrightleftharpoons{K_{B1}} HA^- + OH^-$$

$$\begin{array}{ccc} F - x & x & x \end{array}$$

$$\frac{x^2}{F - x} = K_{B1} = K_w/K_{A2}$$

$$[H^+] = K_w/[OH^-] = K_w/x$$

(b) Use the K_1 equilibrium to solve for $[H_2A]$, using the values of $[HA^-]$ and $[H^+]$ from part a.

$$[H_2A] = \frac{[HA^-][H^+]}{K_{A1}} = \frac{[HA^-]K_w}{K_{A1}[OH^-]} = K_{B2}$$

Diprotic Buffers

A buffer made from a diprotic (or polyprotic) acid is treated just as a buffer made from a monoprotic acid. For the acid H_2A we can write the two Henderson–Hasselbalch equations below, both of which are *always* necessarily true. We can use either one whenever we please.

$$pH = pK_1 + \log \frac{[HA^-]}{[H_2A]} \tag{11-98}$$

$$pH = pK_2 + \log \frac{[A^{2-}]}{[HA^-]} \tag{11-99}$$

All Henderson–Hasselbalch equations are always true for a solution at equilibrium.

EXAMPLE

Find the pH of a solution prepared by dissolving 1.00 g of potassium acid phthalate and 1.20 g of disodium phthalate in 50.0 mL of water.

The structures of monohydrogen phthalate and phthalate were shown in Reaction 11-97. The formula weights are $KHP = C_8H_5O_4K = 204.223$ and $Na_2P = C_8H_4O_4Na_2 = 210.097$. The pH is given by

$$pH = pK_2 + \log \frac{[P^{2-}]}{[HP^-]} = 5.408 + \log \frac{1.20/210.097}{1.00/204.223} = 5.47$$

We used pK_2 because K_2 is the acid dissociation constant of HP^-, which appears in the denominator of the Henderson–Hasselbalch equation. Notice that the volume of solution was not used in answering the question.

EXAMPLE

How many mL of 0.800 M KOH should be added to 3.38 g of oxalic acid to give a pH of 4.40 when diluted to 500 mL?

$$\begin{array}{cc}
\overset{\displaystyle O \quad O}{\underset{\displaystyle \text{HOCCOH}}{\| \; \|}} & \text{Formula weight} = 90.036 \\
\text{oxalic acid} & pK_1 = 1.252 \\
(H_2Ox) & pK_2 = 4.266
\end{array}$$

We know that a 1:1 mole ratio of $HOx^- : Ox^{2-}$ would have $pH = pK_2 = 4.266$. If the pH is to be 4.40 there must be more Ox^{2-} than HOx^- present. We must add enough base to convert all of the H_2Ox to HOx^- plus enough additional base to convert the right amount of HOx^- into Ox^{2-}.

$$H_2Ox + OH^- \rightarrow HOx^- + H_2O$$
$$pH \approx (pK_1 + pK_2)/2 = 2.76$$
$$HOx^- + OH^- \rightarrow Ox^{2-} + H_2O$$
a 1:1 mixture would have
$$pH = pK_2 = 4.27$$

In 3.38 g of H_2Ox there are $0.037\,5_4$ mol. The volume of 0.800 M KOH needed to react with this much H_2Ox to make HOx^- is

$$(\text{Volume of KOH})(0.800 \text{ M}) = 3.75_4 \times 10^{-2} \text{ mol}$$

$$\text{Volume} = 46.9_2 \text{ mL}$$

To produce a pH of 4.40 requires

$$HOx^- \quad + \quad OH^- \rightarrow Ox^{2-}$$

Initial moles:	$0.037\,5_4$	x	—
Final moles:	$0.037\,5_4 - x$	—	x

$$pH = pK_2 + \log \frac{[Ox^{2-}]}{[HOx^-]}$$

$$4.40 = 4.266 + \log \frac{x}{0.037\,5_4 - x} \Rightarrow x = 0.021\,6_4 \text{ mol}$$

The volume of KOH needed to deliver $0.021\,6_4$ moles is $0.021\,6_4$ mol$/0.800$ M $= 27.0_5$ mL. The total volume of KOH needed to bring the pH up to 4.40 is $46.9_2 + 27.0_5 = 73.9_7$ mL.

11-6 POLYPROTIC ACIDS AND BASES

The treatment of diprotic acids and bases can be extended to polyprotic systems. By way of review we write the pertinent equilibria for a triprotic system below.

$$H_3A \rightleftharpoons H_2A^- + H^+ \qquad K_{A1} \equiv K_1 \qquad (11\text{-}100)$$

$$H_2A^- \rightleftharpoons HA^{2-} + H^+ \qquad K_{A2} \equiv K_2 \qquad (11\text{-}101)$$

$$HA^{2-} \rightleftharpoons A^{3-} + H^+ \qquad K_{A3} \equiv K_3 \qquad (11\text{-}102)$$

$$A^{3-} + H_2O \rightleftharpoons HA^{2-} + OH^- \qquad K_{B1} = K_w/K_{A3} \qquad (11\text{-}103)$$

$$HA^{2-} + H_2O \rightleftharpoons H_2A^- + OH^- \qquad K_{B2} = K_w/K_{A2} \qquad (11\text{-}104)$$

$$H_2A^- + H_2O \rightleftharpoons H_3A + OH^- \qquad K_{B3} = K_w/K_{A1} \qquad (11\text{-}105)$$

> The acid–base equilibria for a triprotic system:
> $$K_{Bi} = K_w/K_{A(3-i)}$$

Triprotic systems are treated as follows:

1. H_3A is treated as a monoprotic weak acid, with $K_A = K_1$.

2. H_2A^- is treated as the intermediate form of a diprotic acid.

$$[H^+] \approx \sqrt{\frac{K_1K_2F + K_1K_w}{K_1 + F}} \qquad (11\text{-}106)$$

> The K values in Equations 11-106 and 11-107 are K_A values for the triprotic acid.

3. HA^{2-} is also treated as the intermediate form of a diprotic acid. However, HA^{2-} is "surrounded" by H_2A^- and A^{3-}, so the equilibrium constants to use in Equation 11-91 are K_2 and K_3, instead of K_1 and K_2.

$$[H^+] \approx \sqrt{\frac{K_2K_3F + K_2K_w}{K_2 + F}} \qquad (11\text{-}107)$$

4. A^{3-} is treated as monobasic, with $K_B = K_{B1} = K_w/K_{A3}$.

EXAMPLE

Find the pH of 0.10 M H_3His^{2+}, 0.10 M H_2His^+, 0.10 M HHis, and 0.10 M His^-, where His stands for the amino acid histidine.

11 / ACID–BASE EQUILIBRIA

214

His⁻ ⇌ histidine (HHis) with $pK_3 = 9.08$

0.10 M H_3His^{2+}: Treating this as a monoprotic acid we write

$$H_3His^{2+} \rightleftharpoons H_2His^+ + H^+$$

$$F - x \qquad x \qquad x$$

$$\frac{x^2}{F - x} = K_1 = 2 \times 10^{-2} \Rightarrow x = 3._{58} \times 10^{-2} \text{ M} \Rightarrow pH = 1.45$$

0.10 M H_2His^+: Using Equation 11-106, we write

$$[H^+] = \sqrt{\frac{(2 \times 10^{-2})(9.5 \times 10^{-7})(.10) + (2 \times 10^{-2})(1.0 \times 10^{-14})}{2 \times 10^{-2} + 0.10}}$$

$$= 1.2_6 \times 10^{-4} \text{ M} \Rightarrow pH = 3.90$$

Note that $(pK_1 + pK_2)/2 = 3.86$.

0.10 M HHis: Using Equation 11-107, we write

$$[H^+] = \sqrt{\frac{(9.5 \times 10^{-7})(8.3 \times 10^{-10})(0.10) + (9.5 \times 10^{-7})(1.0 \times 10^{-14})}{9.5 \times 10^{-7} + 0.10}}$$

$$= 2.8_1 \times 10^{-8} \Rightarrow pH = 7.55$$

Note that $(pK_1 + pK_2)/2 = 7.55$.

0.10 M His⁻: Treating this as monobasic, we can write

$$His^- + H_2O \rightleftharpoons HHis + OH^-$$

$$F - x \qquad\qquad x \qquad x$$

$$\frac{x^2}{F - x} = K_{B1} = \frac{K_w}{K_{A3}} = 1.2 \times 10^{-5} \Rightarrow x = 1.0_9 \times 10^{-3} \text{ M}$$

$$pH = -\log(K_w/x) = 11.04$$

There are only three forms of acids and bases: acidic, basic, and intermediate (amphiprotic) species.

In this chapter we have reduced most acid-base problems to a few common types. When you encounter an acid or base you should first write down (or think about) the acid–base chemistry of that species. Decide whether you are dealing with an acidic, basic, or intermediate form. Then do the appropriate arithmetic to answer the question at hand.

Summary

Acid–base systems fall into several categories:

Strong acids or bases. For practical concentrations ($\gtrsim 10^{-6}$ M), pH or pOH can be found by inspection. When the concentration is near 10^{-7} M, we use the systematic treatment of equilibrium to calculate pH. At still lower concentrations, the pH is 7.00, set by autoprotolysis of the solvent.

Weak acids. For the reaction $HA \rightleftharpoons H^+ + A^-$, we set up and solve the equation $K_A = x^2/(F - x)$, where $[H^+] = [A^-] = x$, and $[HA] = F - x$. The fraction of dissociation is given by $\alpha = [A^-]/([HA] + [A^-]) = x/F$.

Weak bases. For the reaction $B + H_2O \rightleftharpoons BH^+ + OH^-$, we set up and solve the equation $K_B = x^2/(F - x)$, where $[OH^-] = [BH^+] = x$, and $[B] = F - x$. Of course, the conjugate acid of a weak base is a weak acid, and the conjugate base of a weak acid is a weak base. For a conjugate acid–base pair, $K_A \cdot K_B = K_w$.

Diprotic systems. These problems are divided into three categories:

1. The fully acidic form, H_2A, behaves as a monoprotic acid, $H_2A \rightleftharpoons H^+ + HA^-$, for which we solve the equation $K_{A1} = x^2/(F - x)$, where $[H^+] = [HA^-] = x$, and $[H_2A] = F - x$. After finding $[HA^-]$ and $[H^+]$, the value of $[A^{2-}]$ can be found from the K_{A2} equilibrium condition.

2. The fully basic form, A^{2-}, behaves as a base, $A^{2-} + H_2O \rightleftharpoons HA^- + OH^-$, for which we solve the equation $K_{B1} = x^2/(F - x)$, where $[OH^-] = [HA^-] = x$, and $[A^{2-}] = F - x$. After finding these concentrations, $[H_2A]$ can be found from the K_{A1} or K_{B2} equilibrium conditions.

3. The intermediate (amphiprotic) form, HA^-, is both

an acid and a base. Its pH is given by

$$[H^+] = \sqrt{\frac{K_1 K_2 F + K_1 K_w}{K_1 + F}}$$

where K_1 and K_2 are acid dissociation constants for H_2A, and F is the formal concentration of the intermediate. In most cases this equation reduces to the form $pH \approx (pK_1 + pK_2)/2$, in which pH is independent of concentration.

In triprotic systems there are two intermediate forms. The pH of each is found with an equation analogous to that for the intermediate form of a diprotic system. Triprotic systems also have one fully acidic and one fully basic form; these can be treated as monoprotic for the purpose of calculating pH.

Buffers. A buffer is a mixture of a weak acid and its conjugate base. It resists changes in pH because it reacts with added acid or base. The pH is given by the Henderson–Hasselbalch equation

$$pH = pK_A + \log \frac{[A^-]}{[HA]}$$

where pK_A applies to the species in the denominator. The concentrations of HA and A^- are essentially unchanged from those used to prepare the solution. The pH of a buffer is nearly independent of dilution, but the buffer capacity increases as the concentration of buffer increases. The maximum buffer capacity is found at $pH = pK_A$, and the useful range of a buffer is approximately $pH = pK_A \pm 1$. The Henderson–Hasselbalch equation can be used for polyprotic systems, as long as a conjugate acid–base pair is used in the log term.

Terms to Understand

amino acid
amphiprotic
buffer
buffer capacity
conjugate acid–base pair
diprotic acid
fraction of association, α (of a base)

fraction of dissociation, α (of an acid)
Henderson–Hasselbalch equation
hydrolysis
polyprotic acid
strong electrolyte
weak electrolyte
zwitterion

Exercises

11-A. Using activity coefficients correctly, find the pH of 1.0×10^{-2} M NaOH.

11-B. Calculate the pH of
(a) 1.0×10^{-8} M HBr
(b) 1.0×10^{-8} M H_2SO_4 (The H_2SO_4 dissociates completely to $2H^+$ plus SO_4^{2-} at this low concentration.)

11-C. What is the pH of a solution prepared by dissolving 1.23 g of 2-nitrophenol in 0.250 L?

11-D. The pH of 0.010 M o-cresol is 6.05. Find pK_A for this weak acid.

o-cresol

11-E. Calculate the limiting value of the fraction of dissociation (α) of a weak acid ($pK_A = 5.00$) as the concentration of HA approaches zero. Repeat the same calculation for $pK_A = 9.00$.

11-F. Find the pH of 0.050 M sodium butanoate (the sodium salt of butanoic acid, also called butyric acid).

11-G. The pH of 0.10 M ethylamine is 11.80.
(a) Without using the appendix, find K_B for ethylamine.
(b) Using the results of part a, calculate the pH of 0.10 M ethylammonium chloride.

11-H. Which of the following bases would be most suitable for preparing a buffer of pH 9.00?
(a) NH_3 (ammonia, $K_B = 1.75 \times 10^{-5}$)
(b) $C_6H_5NH_2$ (aniline, $K_B = 3.99 \times 10^{-10}$)
(c) H_2NNH_2 (hydrazine, $K_B = 3.0 \times 10^{-6}$)
(d) C_5H_5N (pyridine, $K_B = 1.69 \times 10^{-9}$)

11-I. A solution contains 63 different conjugate acid–base pairs. Among them is acrylic acid and acrylate ion, with the ratio [acrylate]/[acrylic acid] = 0.75. What is the pH of the solution?

$$H_2C{=}CHCO_2H \qquad pK_A = 4.25$$
acrylic acid

11-J. (a) How many grams of $NaHCO_3$ (F.W. 84.007) must be added to 4.00 g of K_2CO_3 (F.W. 138.206) to give a pH of 10.80 in 500 mL of water?
(b) What will be the pH if 100 mL of 0.100 M HCl are added to the solution in part a?

(c) How many milliliters of 0.320 M HNO_3 should be added to 4.00 g of K_2CO_3 to give a pH of 10.00 in 250 mL?

11-K. How many milliliters of 0.800 M KOH should be added to 3.38 g of oxalic acid to give a pH of 2.40 when diluted to 500 mL?

11-L. Find the pH and the concentrations of H_2SO_3, HSO_3^-, and SO_3^{2-} in each of the following solutions:
(a) 0.050 M H_2SO_3
(b) 0.050 M $NaHSO_3$
(c) 0.050 M Na_2SO_3

11-M. Calculate the pH of a 0.010 M solution of each amino acid in the form drawn below:

(a) glutamine
(b) cysteine$^-$

(c) arginine

11-N. A solution with an ionic strength of 0.10 M containing 0.010 0 M phenylhydrazine has a pH of 8.13. Using activity coefficients correctly, find pK_A for the phenylhydrazinium ion found in phenylhydrazine hydrochloride. Assume that $\gamma_{BH^+} = 0.80$.

phenylhydrazine
$\equiv$ B

phenylhydrazine hydrochloride
$\equiv$ BH^+Cl^-

Problems

11-1. Explain the following statement: The Henderson–Hasselbalch equation is *always* true. What may not be correct are the values of $[A^-]$ and $[HA]$ that we choose to use in the equation.

11-2. Calculate the pH of the following solutions.
(a) 1.0×10^{-3} M HBr (b) 1.0×10^{-2} M KOH

11-3. In dilute solution $Ca(OH)_2$ dissociates completely to $Ca^{2+} + 2OH^-$. Find the pH of 1.0×10^{-7} M $Ca(OH)_2$
(a) at 25°C (b) at 35°C

11-4. Write the chemical reaction whose equilibrium constant is
(a) K_A for benzoic acid
(b) K_B for benzoate ion
(c) K_B for aniline
(d) K_A for anilinium ion

benzoic acid — potassium benzoate

aniline — anilinium bromide

11-5. Find the pH and concentrations of $(CH_3)_3N$ and $(CH_3)_3NH^+$ in a 0.060 M solution of trimethylamine.

11-6. Find the pH and concentrations of $(CH_3)_3N$ and $(CH_3)_3NH^+$ in a 0.060 M solution of trimethylammonium chloride.

11-7. Find the pH of 0.050 M NaCN.

11-8. Barbituric acid dissociates as follows:

barbituric acid HA $\qquad$ A$^-$ $\qquad$ $K_A = 9.8 \times 10^{-5}$

(a) Calculate the pH and fraction of dissociation of $10^{-2.00}$ M barbituric acid.
(b) Calculate the pH and fraction of dissociation of $10^{-10.00}$ M barbituric acid.

(c) Calculate the pH and fraction of association of $10^{-2.00}$ M potassium barbiturate.

11-9. Calculate the pH of 0.085 0 M pyridinium bromide, $C_5H_5NH^+Br^-$.

11-10. A 0.045 0 M solution of benzoic acid has a pH of 2.78. Calculate pK_A for this acid.

11-11. If a 0.10 M solution of a base has pH = 9.28, find K_B for the base.

11-12. Compound **A** below reacts with H_2O as follows:

A

The equilibrium constant (in aqueous methanol solution) is $10^{-5.4}$. Suppose that this same equilibrium constant applies in pure water. Find the pH of a 0.020 M solution of **A**.

11-13. Calculate the fraction of association (α) for 1.00×10^{-1}, 1.00×10^{-2}, and 1.00×10^{-12} M sodium acetate. Does α increase or decrease with dilution?

11-14. The temperature-dependence of pK_A for acetic acid is given in the table below. Is the dissociation of this acid endothermic or exothermic?
(a) at 5°C (b) at 45°C

Temperature (°C)	pK_A	Temperature (°C)	pK_A
0	4.780 7	30	4.757 0
5	4.769 6	35	4.762 5
10	4.762 2	40	4.768 8
15	4.758 2	45	4.777 3
20	4.756 2	50	4.787 0
25	4.756 0		

11-15. Write the Henderson–Hasselbalch equation for a solution of formic acid. Calculate the ratio $[HCO_2^-]/[HCO_2H]$ at the following values of pH: (a) 3.000 (b) 3.745 (c) 4.000

11-16. Which of the following acids would be most suitable for preparing a buffer of pH 3.10?
(a) hydrogen peroxide
(b) propanoic acid
(c) cyanoacetic acid
(d) 4-aminobenzenesulfonic acid

11-17. Phosphate, present to an extent of 0.01 M, is one of the main buffers in blood plasma, whose pH is 7.45. Would phosphate be as useful if the plasma pH were 8.5?

11-18. (a) Describe the correct procedure for preparing 0.250 L of 0.050 0 M HEPES (Table 11-2), pH 7.45.
(b) Would you need NaOH or HCl to bring the pH to 7.45?

11-19. Calculate how many milliliters of 0.626 M KOH should be added to 5.00 g of HEPES (Table 11-2) to give a pH of 7.40.

11-20. (a) Calculate the pH of a solution prepared by dissolving 10.0 g of tris (Equation 11-58) plus 10.0 g of tris hydrochloride in 0.250 L of water.
(b) What will be the pH if 10.5 mL of 0.500 M NaOH is added?

11-21. Use Equations 11-63 and 11-64 to find the concentrations of HA and A^- in a solution prepared by mixing 0.002 000 mol of acetic acid plus 0.004 000 mol of sodium acetate in 1.000 L of water.

11-22. Given that pK_B for nitrite ion (NO_2^-) is 10.85, find the ratio $[HNO_2]/[NO_2^-]$ in a solution of sodium nitrite at:
(a) pH 2.00
(b) pH 10.00

11-23. How many milliliters of 0.246 M HNO_3 should be added to 213 mL of 0.006 66 M ethylamine to give a pH of 10.52?

11-24. (a) Write the chemical reactions whose equilibrium constants are K_B and K_A for imidazole and imidazole hydrochloride, respectively.
(b) Calculate the pH of a solution prepared by mixing 1.00 g of imidazole with 1.00 g of imidazole hydrochloride and diluting to 100.0 mL.
(c) Calculate the pH of the solution if 2.30 mL of 1.07 M $HClO_4$ is added to the solution.

(d) How many milliliters of 1.07 M $HClO_4$ should be added to 1.00 g of imidazole to give a pH of 6.993?

11-25. Calculate the pH of a solution prepared by mixing 0.080 0 mol of chloracetic acid plus 0.040 0 mol of sodium chloroacetate in 1.00 L of water.
(a) First do the calculation assuming that the concentrations of HA and A^- equal their formal concentrations.
(b) Then do the calculation using the real values of $[HA]$ and $[A^-]$ in the solution.
(c) Using first your head, and then the Henderson–Hasselbalch equation, find the pH of a solution prepared by dissolving all of the following in one beaker containing a total volume of 1.00 L: 0.180 mol of $ClCH_2CO_2H$, 0.020 mol of $ClCH_2CO_2Na$, 0.080 mol HNO_3, and 0.080 mol $Ca(OH)_2$. Assume that $Ca(OH)_2$ dissociates completely.

11-26. Calculate the pH of a solution prepared by mixing 0.010 0 mol of the base B $(K_B = 10^{-2.00})$ with 0.020 0 mol of BH^+Br^- and diluting to 1.00 L. First calculate the pH assuming $[B] = 0.010 0$ and $[BH^+] = 0.020 0$ M. Compare this answer to the pH calculated without making such an assumption. Which calculation is more correct?

11-27. Write the chemical reactions whose equilibrium constants are K_{B1} and K_{B2} for the amino acid, proline. Find the values of K_{B1} and K_{B2}.

11-28. H_2O is the *intermediate form* of a diprotic acid. Write the formula for that diprotic acid and give the values of pK_1 and pK_2.

11-29. Calculate the pH of 0.300 M piperazine. Calculate the concentration of each form of piperazine in this solution.

11-30. We will abbreviate malonic acid, $CH_2(CO_2H)_2$, as H_2M. Find the pH and concentrations of H_2M, HM^-, and M^{2-} in each of the following solutions:
(a) 0.100 M H_2M
(b) 0.100 M NaHM
(c) 0.100 M Na_2M

11-31. Starting with the fully protonated species, write the stepwise acid dissociation reactions of the amino acids glutamic acid and tyrosine. Be sure to remove the protons in the correct order.

11-32. Find the pH and the concentration of each species of lysine in a solution of 0.010 0 M lysine·HCl (lysine monohydrochloride).

11-33. (a) Calculate the ratio $[H_3PO_4]/[H_2PO_4^-]$ in $0.050\,0$ M KH_2PO_4.

(b) Find the same ratio for $0.050\,0$ M K_2HPO_4.

11-34. Using activity coefficients correctly, calculate the pH of

(a) 0.050 M HBr   (b) 0.050 M NaOH

11-35. Using activity coefficients correctly, calculate the fraction of dissociation, α, of 50.0 mM hydroxybenzene (phenol) in 0.050 M LiBr. Assume that the size of $C_6H_5O^-$ is 600 pm.

11-36. (a) Using activity coefficients correctly, calculate the pH of a solution containing a $2.00{:}1.00$ mole ratio of $HC^{2-}{:}C^{3-}$, where H_3C is citric acid. Assume that the ionic strength is 0.010 M.

(b) What will be the pH if the ionic strength is raised to 0.10 M and the mole ratio $HC^{2-}{:}C^{3-}$ is kept constant?

11-37. Use the method of Box 11-2 to calculate the concentrations of H^+, H_2Ox, HOx^-, and Ox^{2-} in $0.001\,00$ M monosodium oxalate (NaHOx).

11-38. In this problem we will calculate the pH of the intermediate form of a diprotic acid correctly, taking activities into account.

(a) Derive Equation 11-90 for a solution of potassium acid phthalate (KHP in Reaction 11-97). Do not neglect activity coefficients in this derivation.

(b) Calculate the pH of 0.050 M KHP using the results of part a. Assume that the sizes of both HP^- and P^{2-} are 600 pm.

11-39. Write down, but do not attempt to solve, the exact equations needed to calculate the composition of one liter of solution containing F_1 mol of HCl, F_2 mol of disodium ascorbate (Na_2A, the salt of a weak acid whose two K_A values may be called K_1 and K_2), and F_3 mol of trimethylamine (a weak base, B, whose equilibrium constant should be called K_B). Include activity coefficients wherever appropriate.

12 / Acid–Base Titrations

Acid–base titrations are used routinely in virtually every field of chemistry. Figure 12-1 shows experimental data for the titration of the enzyme ribonuclease, with either strong acid or strong base. Ribonuclease is a protein with 124 amino acids. Its function is to cleave molecules of ribonucleic acid (RNA). Near pH 9.6, the enzyme has no net charge. Sixteen amino acids of the neutral enzyme can be protonated by titration with acid, and twenty can lose protons through titration with base.

From the shape of the titration curve in Figure 12-1 it is possible to deduce the approximate value of pK_A for each titratable group. This, in turn, provides insight into the immediate environment of that amino acid in its place in the protein. In the case of ribonuclease it was found that three tyrosine residues exhibit "normal" values of pK_A (~ 9.95), and three others have $pK_A > 12$. The interpretation is that three tyrosine groups are accessible to solvent and OH^-, while three are buried inside the protein where they cannot be easily titrated. The solid lines in Figure 12-1 are calculated from the values of pK_A deduced for all of the titratable groups.

In this chapter we will discuss the shapes of titration curves. We will not deal with molecules as complicated as ribonuclease, but the principles we discuss can be applied to any complex molecule.

12-1 TITRATION OF STRONG ACID WITH STRONG BASE

For each type of titration studied in this chapter *our goal is to construct a graph showing how the pH changes as titrant is added*. If you can do this, then you understand what is happening during the titration, and you will be able to interpret an experimental titration curve.

The first step in each case is to write the chemical reaction between titrant and analyte. We will then use that reaction to calculate the composi-

First write the reaction between *titrant* and *analyte*.

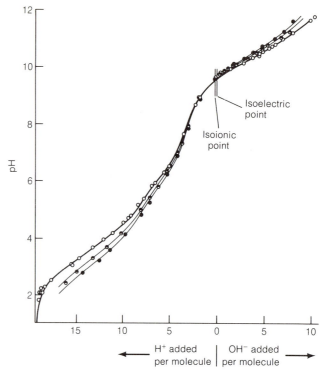

Figure 12-1
Acid–base titration of the enzyme ribonuclease. Circles are experimental points for the following ionic strengths: ● = 0.01 M, ◓ = 0.03 M, ○ = 0.15 M. The abscissa represents moles of acid or base added per mole of enzyme. The isoelectric and isoionic points are discussed in Chapter 13. [C. T. Tanford and J. D. Hauenstein, *J. Amer. Chem. Soc.*, **78**, 5287 (1956).]

tion and pH after each addition of titrant. As a simple example, let's focus our attention on the titration of 50.00 mL of 0.020 00 M KOH with 0.100 0 M HBr. The chemical reaction between titrant and analyte is merely

The titration reaction.

$$H^+ + OH^- \rightarrow H_2O \qquad (12\text{-}1)$$

Since the equilibrium constant for this reaction is $1/K_w = 10^{14}$, it is fair to say that it "goes to completion." *Any amount of H^+ added will consume a stoichiometric amount of OH^-.*

One useful beginning is to calculate the volume of HBr needed to reach the equivalence point (V_e).

$$\underbrace{(V_e(\text{mL}))(0.100\ 0\ \text{M})}_{\substack{\text{mmol of HBr} \\ \text{at equivalence} \\ \text{point}}} = \underbrace{(50.00\ \text{mL})(0.020\ 00\ \text{M})}_{\substack{\text{mmol of } OH^- \\ \text{being titrated}}} \Rightarrow V_e = 10.00\ \text{mL} \qquad (12\text{-}2)$$

It is helpful to bear in mind that when 10.00 mL of HBr has been added, the titration is complete. Prior to this point there will be excess, unreacted OH^- present. After V_e there will be excess H^+ in the solution.

In the titration of any strong base with any strong acid, there are three regions of the titration curve that represent different kinds of calculations:

1. Before reaching the equivalence point the pH is determined by the excess OH^- in the solution.
2. At the equivalence point the H^+ is just sufficient to react with all of the OH^- to make H_2O. The pH is determined by the dissociation of water.
3. After the equivalence point, pH is determined by the excess H^+ in the solution.

We will do one sample calculation for each region. The results of the calculations are shown in Table 12-1 and plotted in Figure 12-2.

Table 12-1
Calculation of the titration curve for 50.00 mL of 0.020 00 M KOH treated with 0.100 0 M HBr

	mL HBr added (V_a)	Concentration of unreacted OH^- (M)	Concentration of excess H^+ (M)	pH
Region 1	0.00	0.020 0		12.30
	1.00	0.017 6		12.24
	2.00	0.015 4		12.18
	3.00	0.013 2		12.12
	4.00	0.011 1		12.04
	5.00	0.009 09		11.95
	6.00	0.007 14		11.85
	7.00	0.005 26		11.72
	8.00	0.003 45		11.53
	9.00	0.001 69		11.22
	9.50	0.000 840		10.92
	9.90	0.000 167		10.22
	9.99	0.000 016 6		9.22
Region 2	10.00	—	—	7.00
Region 3	10.01		0.000 016 7	4.78
	10.10		0.000 166	3.78
	10.50		0.000 826	3.08
	11.00		0.001 64	2.79
	12.00		0.003 23	2.49
	13.00		0.004 76	2.32
	14.00		0.006 25	2.20
	15.00		0.007 69	2.11
	16.00		0.009 09	2.04

12-1 TITRATION OF STRONG ACID WITH STRONG BASE

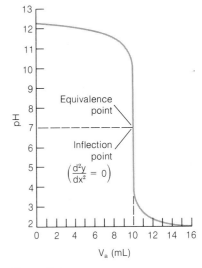

Figure 12-2
Calculated titration curve for the reaction of 50.00 mL of 0.020 00 M KOH with 0.100 0 M HBr. The equivalence point is also an inflection point.

Region 1: Before the Equivalence Point

Before the equivalence point, there is excess OH^-.

When 3.00 mL of HBr have been added the reaction is three-tenths complete, since 10.00 mL of HBr are required to reach the equivalence point. The fraction of OH^- left unreacted is seven-tenths. The concentration of OH^- remaining in the flask is

$$[OH^-] = \underbrace{\left(\frac{10.00 - 3.00}{10.00}\right)}_{\substack{\text{Fraction} \\ \text{of } OH^- \\ \text{remaining}}}\underbrace{(0.020\,00)}_{\substack{\text{Initial} \\ \text{concentration} \\ \text{of } OH^-}}\underbrace{\left(\frac{50.00}{50.00 + 3.00}\right)}_{\substack{\text{Dilution} \\ \text{factor}}} = 0.013\,2 \text{ M} \quad (12\text{-}3)$$

where the numerator 50.00 is the Initial volume of OH^- and the denominator $50.00 + 3.00$ is the Total volume of solution.

$$[H^+] = K_w/[OH^-] = 1.0 \times 10^{-14}/0.013\,2$$

$$= 7.5_8 \times 10^{-13} \text{ M} \Rightarrow pH = 12.12 \quad (12\text{-}4)$$

Equation 12-3 is an example of a type of calculation that was introduced in Section 9-4 in connection with precipitation titrations. This equation tells us that the concentration of OH^- is equal to a certain fraction of the initial concentration, with a correction for dilution. The dilution factor equals the initial volume of OH^- divided by the total volume of solution.

In Table 12-1 and Figure 12-2 the volume of acid added is designated V_a. The values of pH in Table 12-1 are all expressed to the 0.01 decimal place, regardless of what is justified by significant figures. We will do this throughout the text for the sake of consistency and also because that is the usual limit of accuracy in pH measurements.

> *Challenge:* Using a setup similar to Equation 12-3, calculate $[OH^-]$ when 6.00 mL of HBr have been added. Check your pH against the value in Table 12-1.

Region 2: At the Equivalence Point

This is the equivalence point, where enough H^+ has been added to react with all of the OH^-. We could prepare the same solution by dissolving KBr in water. The pH is determined by the dissociation of water:

$$H_2O \rightleftharpoons \underset{x}{H^+} + \underset{x}{OH^-} \quad (12\text{-}5)$$

At the equivalence point, pH = 7.00, but *only* in a strong acid–strong base reaction.

$$K_w = x^2 \Rightarrow x = 1.00 \times 10^{-7} \text{ M} \Rightarrow pH = 7.00 \quad (12\text{-}6)$$

The pH at the equivalence point in the titration of any strong base (or acid) with acid (or base) will be 7.00.

As we will discover in Section 12-2, *the pH is **not** 7.00 at the equivalence point in the titration of weak acids or bases.* The pH is 7.00 only if the titration involves a strong acid and a strong base.

Region 3: After the Equivalence Point

Beyond the equivalence point we are adding excess HBr to the solution. The concentration of excess H^+ at, say, 10.50 mL is given by

$$[H^+] = \underbrace{(0.100\,0)}_{\substack{\text{Initial} \\ \text{concentration} \\ \text{of } H^+}}\underbrace{\left(\frac{\overbrace{0.50}^{\substack{\text{Volume of} \\ \text{excess } H^+}}}{\underbrace{50.00 + 10.50}_{\substack{\text{Total volume} \\ \text{of solution}}}}\right)}_{\substack{\text{Dilution} \\ \text{factor}}} = 8.26 \times 10^{-4}\ \text{M} \qquad (12\text{-}7)$$

After the equivalence point, there is excess H^+.

$$pH = -\log[H^+] = 3.08 \qquad (12\text{-}8)$$

At $V_a = 10.50$ mL, there is an excess of just $V_a - V_e = 10.50 - 10.00 = 0.50$ mL of HBr. That is the reason why 0.50 appears in the dilution factor.

Titration Curve

The complete titration curve is shown in Figure 12-2. Characteristic of all analytically useful titrations is a sudden change in pH near the equivalence point. The equivalence point is where the slope (dpH$/d$V$_a$) is greatest. It is therefore an inflection point. To repeat an important statement, the pH at the equivalence point is 7.00 *only* in a strong acid–strong base titration. If one or both of the reactants is weak, the equivalence point pH is *not* 7.00, and might be quite far from 7.00.

12-2   TITRATION OF WEAK ACID WITH STRONG BASE

This kind of titration allows us to put all of our knowledge of acid–base chemistry to work. The example we will examine is the titration of 50.00 mL of 0.020 00 M MES with 0.100 0 M NaOH. MES is an abbreviation for 2-(N-morpholino)ethanesulfonic acid, which is a weak acid with $pK_A = 6.15$.
The *titration reaction* is

$$O\underset{\text{HA}}{\overset{+}{N}HCH_2CH_2SO_3^-} + OH^- \longrightarrow O\underset{A^-}{NCH_2CH_2SO_3^-} + H_2O \qquad (12\text{-}9)$$

MES, $pK_A = 6.15$

Always start by writing the titration reaction.

A look at Reaction 12-9 will show you that it is the reverse reaction of K_B for the base A^-. Therefore the equilibrium constant for Reaction 12-9 is $K = 1/K_B = 1/(K_w/K_A \text{ (for HA)}) = 7.1 \times 10^7$. The equilibrium constant

226

12 / ACID–BASE TITRATIONS

Strong + Weak → Complete reaction.

is so large that we may say that the reaction goes to completion after each addition of OH^-. As we saw in Box 11-1, *strong plus weak react completely.*

It is very helpful first to calculate the volume of base (V_b) needed to reach the equivalence point.

$$\underbrace{(V_b(mL))(0.100\,0\ \text{M})}_{\text{mmol of base}} = \underbrace{(50.00\ \text{mL})(0.020\,00\ \text{M})}_{\text{mmol of HA}} \Rightarrow V_b = 10.00\ \text{mL} \quad (12\text{-}10)$$

The titration calculations for this problem are of four types:

1. Before any base is added, the solution contains just HA in water. This is a weak-acid problem in which the pH is determined by the equilibrium

$$HA \xrightarrow{K_A} H^+ + A^- \quad\quad (12\text{-}11)$$

2. From the first addition of NaOH until immediately before the equivalence point, there is a mixture of unreacted HA plus the A^- produced by Reaction 12-9. *A buffer!* We can use the Henderson–Hasselbalch equation to find the pH.

3. At the equivalence point "all" of the HA has been converted to A^-. The problem is the same as if the solution had been made by merely dissolving A^- in water. We have a weak-base problem in which the pH is determined by Reaction 12-12.

$$A^- + H_2O \xrightarrow{K_B} HA + OH^- \quad\quad (12\text{-}12)$$

4. Beyond the equivalence point, excess NaOH is being added to a solution of A^-. To a good approximation, the pH is determined by the strong base. We will calculate the pH as if we had simply added the excess NaOH to water. We will neglect the very small effect from having A^- present as well.

Region 1: Before Base Is Added

The initial solution contains just the *weak acid* HA.

Before adding any base, we have a solution of 0.020 00 M HA with $pK_A = 6.15$. This is simply a weak-acid problem.

$$HA \rightleftharpoons H^+ + A^- \quad\quad K_A = 10^{-6.15}$$
$$F - x \quad\quad x \quad\quad x$$

$$\frac{x^2}{0.020\,0 - x} = K_A \Rightarrow x = 1.19 \times 10^{-4} \Rightarrow pH = 3.93 \quad (12\text{-}13)$$

Region 2: Before the Equivalence Point

Once we begin to add OH^-, a mixture of HA and A^- is created by the titration reaction (12-9). This mixture is a buffer whose pH can be calculated with the Henderson–Hasselbalch equation (11-54) once we know the ratio $[A^-]/[HA]$.

Suppose we wish to calculate the ratio $[A^-]/[HA]$ when 3.00 mL of OH^- have been added. Since $V_e = 10.00$ mL, we have added enough base to react with three-tenths of the HA. We can make a table showing the relative concentrations before and after the reaction:

Titration reaction: $\qquad\qquad\qquad$ $HA + OH^- \rightarrow A^- + H_2O$

Relative initial quantities (HA $\equiv$ 1): $\quad$ 1 $\qquad \frac{3}{10}$ $\qquad$ —

Relative final quantities: $\qquad\qquad\qquad \frac{7}{10} \qquad$ — $\qquad \frac{3}{10}$

Once we know the *ratio* $[A^-]/[HA]$ in any solution, we know its pH:

$$pH = pK_A + \log \frac{[A^-]}{[HA]} = 6.15 + \log \frac{3/10}{7/10} = 5.78 \qquad (12\text{-}14)$$

The point at which the volume of titrant is $\frac{1}{2}V_e$ is a special one in any titration.

$$HA + OH^- \rightarrow A^- + H_2O$$

Relative initial quantities: $\quad$ 1 $\qquad \frac{1}{2}$ $\qquad$ —

Relative final quantities: $\quad \frac{1}{2}$ $\qquad$ — $\qquad \frac{1}{2}$

$$pH = pK_A + \log \frac{1/2}{1/2} = pK_A$$

When the volume of titrant is $\frac{1}{2}V_e$, $pH = pK_A$ for the acid HA (neglecting activity coefficients). If you have an experimental titration curve you can find the approximate value of pK_A by reading the pH when $V_b = \frac{1}{2}V_e$, where V_b is the volume of added base. (To find the true value of pK_A requires a knowledge of the ionic strength and activity coefficients.)

In considering the amount of OH^- needed to react with HA, you need not worry about the small amount of HA that dissociates to A^- in the absence of OH^-. The reason for this is explained in Box 12-1.

Advice. As soon as you recognize a mixture of HA and A^- in any solution, *you have a buffer!* You can calculate the pH if you can find the ratio $[A^-]/[HA]$.

$$pH = pK_A + \log \frac{[A^-]}{[HA]}$$

Learn to recognize buffers! They lurk in every corner of acid–base chemistry.

12-2 TITRATION OF WEAK ACID WITH STRONG BASE

Before the equivalence point, there is a mixture of HA and A^-, which is a *buffer*.

When $V_b = \frac{1}{2}V_e^\bullet$, $pH = pK_A$. This is a landmark point in any titration.

Box 12-1 THE ANSWER TO A NAGGING QUESTION

Consider the titration of 100 mL of a 1.00 M solution of the weak acid HA with 1.00 M NaOH. The equivalence point occurs at $V_b = 100$ mL. At any point between $V_b = 0$ and $V_b = 100$ mL part of the HA will be converted to A^- and part will be left as HA. If 10 mL of NaOH had been added, we would calculate the pH as follows:

$$\text{Titration reaction:} \quad HA + OH^- \rightarrow A^- + H_2O$$

Initial mmol:	100	10	—	—
Final mmol:	90	—	10	—

$$pH = pK_A + \log\frac{[A^-]}{[HA]} = pK_A + \log\frac{10}{90}$$

Upon some reflection you might think "Wait a minute! The initial solution contains some A^- in equilibrium with the HA. There must be less than 100 mmol of HA and more than 0 mmol of A^- at the start. Doesn't this make our answer wrong?"

It does not, and the reason is easy to see with some numbers. Suppose that the HA dissociates to give 1 mmol of A^- before any NaOH is added.

$$HA \xrightleftharpoons{K_A} H^+ + A^- \tag{1}$$
$$\text{Initial solution:} \quad 99 \text{ mmol} \qquad 1 \text{ mmol} \quad 1 \text{ mmol}$$

The solution contains 1 mmol of H^+, 1 mmol of A^-, and 99 mmol of HA.

When 10 mmol of OH^- is added, 1 mmol reacts with the strong acid, H^+, and 9 mmol is left to react with the weak acid, HA.

$$H^+ + OH^- \rightarrow H_2O$$
$$1 \text{ mmol} \quad 1 \text{ mmol} \quad 1 \text{ mmol}$$

$$HA + OH^- \rightarrow A^- \tag{3}$$
$$9 \text{ mmol} \quad 9 \text{ mmol} \quad 9 \text{ mmol}$$

The resulting solution contains 1 mmol of A^- from Reaction 1 and 9 mmol of A^- from Reaction 3. The total A^- is 10 mmol. The total HA is 90 mmol. These are exactly the same numbers that we calculated by ignoring the initial dissociation of HA.

Moral: The moral of this box is that you can treat a mixture of HA plus OH^- as if no dissociation of HA occurred before adding the NaOH.

Region 3: At the Equivalence Point

At the equivalence point the quantity of NaOH is exactly enough to consume the HA.

At the equivalence point, HA has been converted to A^-, a *weak base.*

Titration reaction:		$HA + OH^- \rightarrow A^- + H_2O$	
Initial relative quantities:	1	1	—
Final relative quantities:	—	—	1

The resulting solution contains "just" A^-. We could have prepared the same solution by dissolving the salt Na^+A^- in distilled water. *A solution of Na^+A^- is merely a solution of a weak base.*

To compute the pH of a weak base we write the reaction of the weak base with water:

$$A^- + H_2O \rightleftharpoons HA + OH^- \qquad K_B = K_w/K_A \qquad (12\text{-}15)$$
$$F - x \qquad\qquad x \qquad x$$

The only tricky point is that the formal concentration of A^- is no longer 0.020 00 M, which was the initial concentration of HA. The A^- has been diluted by NaOH from the buret:

$$F' = (0.020\,00)\left(\underbrace{\frac{50.00}{50.00 + 10.00}}\right) = 0.016\,7 \text{ M} \qquad (12\text{-}16)$$

Initial volume of HA

Total volume of solution

Initial concentration of HA

Dilution factor

With this value of F' we can solve the problem:

$$\frac{x^2}{F' - x} = K_B = \frac{K_w}{K_A} = 1.43 \times 10^{-8} \Rightarrow x = 1.54 \times 10^{-5} \text{ M} \quad (12\text{-}17)$$

$$pH = -\log[H^+] = -\log(K_w/x) = 9.18 \qquad (12\text{-}18)$$

> The pH is higher than 7 at the equivalence point in the titration of a weak acid with a strong base.

The pH at the equivalence point in this titration is 9.18. **It is not 7.00.** The equivalence point pH will *always* be above 7 for the titration of a weak acid, since the acid is converted to its conjugate base at the equivalence point.

Region 4: After the Equivalence Point

Now we are adding NaOH to a solution of A^-. The NaOH is such a stronger base than A^- that it is a fair approximation to say that the pH is determined by the concentration of excess OH^- in the solution.

Let's calculate the pH when $V_b = 10.10$ mL. This is just 0.10 mL past V_e. The concentration of excess OH^- is

> Here we assume the pH is governed by the excess OH^-.

$$[OH^-] = (0.100\,0)\left(\underbrace{\frac{0.10}{50.00 + 10.10}}\right) = 1.66 \times 10^{-4} \qquad (12\text{-}19)$$

Volume of excess OH^-

Total volume of solution

Initial concentration of OH^-

Dilution factor

> *Challenge:* Compare the concentration of OH^- due to excess titrant at $V_b = 10.10$ mL to the concentration of OH^- due to hydrolysis of A^-. Satisfy yourself that it is fair to neglect the contribution of A^- to the pH after the equivalence point.

$$pH = -\log(K_w/[OH^-]) = 10.22 \qquad (12\text{-}20)$$

Titration Curve

A summary of the calculations for the titration of MES with NaOH is shown in Table 12-2. The calculated titration curve appears in Figure 12-3. The curve has two easily identified points. One is the equivalence point, which is the steepest part of the curve. The other landmark is the point where $V_b = \frac{1}{2}V_e$ and $pH = pK_A$. This latter point is also an inflection point, having the minimum slope.

If you look back at Figure 11-4, you will note that the maximum *buffer capacity* occurs when $pH = pK_A$. That is, the solution is most resistant to pH changes when $pH = pK_A$ (and $V_b = \frac{1}{2}V_e$); the slope (dpH/dV_b) is therefore at its minimum.

Figure 12-4 shows how the titration curve depends upon the acid dissociation constant of HA. As K_A decreases, the abruptness of the inflection near the equivalence point decreases, until the equivalence point becomes too shallow to detect. A similar phenomenon occurs as the concentration of analyte and/or titrant decreases. *It is not practical to titrate an acid or base when it is too weak or too dilute.*

Landmark points in a titration:

At $V_b = V_e$, curve is steepest.
At $V_b = \frac{1}{2}V_e$, $pH = pK_A$
and the slope is minimal.

The *buffer capacity* measures the ability to resist changes in pH.

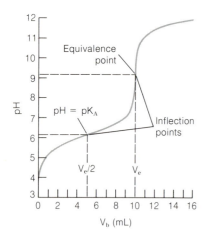

Figure 12-3
Calculated titration curve for the reaction of 50.00 mL of 0.020 00 M MES with 0.100 0 M NaOH.

Table 12-2
Calculation of the titration curve for 50.00 mL of 0.020 00 M MES treated with 0.100 0 M NaOH

	mL base added (V_b)	pH
Region 1	0.00	3.93
Region 2	0.50	4.87
	1.00	5.20
	2.00	5.55
	3.00	5.78
	4.00	5.97
	5.00	6.15
	6.00	6.33
	7.00	6.52
	8.00	6.75
	9.00	7.10
	9.50	7.43
	9.90	8.15
Region 3	10.00	9.18
Region 4	10.10	10.22
	10.50	10.91
	11.00	11.21
	12.00	11.50
	13.00	11.67
	14.00	11.79
	15.00	11.88
	16.00	11.95

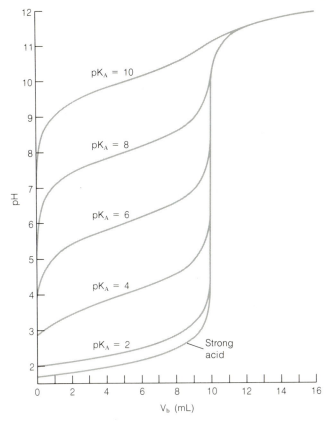

Figure 12-4
Calculated curves showing the titration of 50.0 mL of 0.020 0 M HA with 0.100 M NaOH. As the acid becomes weaker, the change in slope at the equivalence point becomes less abrupt.

12-3 TITRATION OF WEAK BASE WITH STRONG ACID

This problem is just the reverse of the titration of a weak acid with a strong base. The *titration reaction* is

$$B + H^+ \rightarrow BH^+ \tag{12-21}$$

Since the reactants are a weak base and a strong acid, the reaction goes essentially to completion after each addition of acid. There are four distinct regions in the titration:

1. Before acid is added, the solution contains just the weak base, B, in water. The pH is determined by the K_B reaction:

$$\underset{F-x}{B} + H_2O \underset{}{\overset{K_B}{\rightleftharpoons}} \underset{x}{BH^+} + \underset{x}{OH^-} \tag{12-22}$$

When $V_a = 0$, we have a *weak-base* problem.

232

12 / ACID–BASE TITRATIONS

When $0 < V_a < V_e$, we have a *buffer*.

When $V_a = V_e$, the solution contains the *weak acid*, BH^+.

For $V_a > V_e$, there is excess *strong acid*.

2. Between the initial point and the equivalence point, there is a mixture of B and BH^+—*a buffer!* The pH is computed using

$$pH = pK_A \text{ (for } BH^+) + \log \frac{[B]}{[BH^+]} \qquad (12\text{-}23)$$

In adding acid (increasing V_a), we reach the special point where $V_a = \frac{1}{2}V_e$ and $pH = pK_A$ (for BH^+). As before, pK_A (and therefore pK_B) can be determined easily from the titration curve.

3. At the equivalence point, B has been converted into BH^+, a weak acid. The pH is calculated by considering the acid dissociation reaction of BH^+.

$$BH^+ \rightleftharpoons B + H^+ \qquad K_A = K_w/K_B \qquad (12\text{-}24)$$
$$ F' - x \quad\;\; x \quad\;\; x$$

The formal concentration of BH^+, F', is not the same as the original formal concentration of B, since some dilution has occurred. Since the solution contains BH^+ at the equivalence point, it is acidic. *The pH at the equivalence point must be below 7.*

4. After the equivalence point, there is excess H^+ in the solution. We treat this problem by considering only the concentration of excess H^+ and neglecting the contribution of a weak acid, BH^+.

EXAMPLE

Consider the titration of 25.00 mL of 0.083 64 M pyridine with 0.106 7 M HCl.

$$\text{pyridine} \qquad K_B = 1.69 \times 10^{-9}$$

The titration reaction is

and the equivalence point occurs at 19.60 mL:

$$\underbrace{(V_e(\text{mL}))(0.106\ 7\ \text{M})}_{\text{mmol of HCl}} = \underbrace{(25.00\ \text{mL})(0.083\ 64\ \text{M})}_{\text{mmol of pyridine}} \Rightarrow V_e = 19.60\ \text{mL}$$

Find the pH when $V_a = 4.63$ mL.

Part of the pyridine has been neutralized, so there is a mixture of pyridine and pyridinium ion—*a buffer*. The fraction of pyridine that has been titrated is $4.63/19.60 = 0.236$, since it takes 19.60 mL to titrate the whole sample. The fraction of pyridine remaining is $(19.60 - 4.63)/19.60 = 0.764$. The pH is

$$pH = pK_A \left(= -\log \frac{K_w}{K_B} \right) + \log \frac{[B]}{[BH^+]}$$

$$= 5.23 + \log \frac{0.764}{0.236} = 5.74$$

12-4 TITRATIONS IN DIPROTIC SYSTEMS

The principles developed for titrations of monoprotic acids and bases are readily extended to titrations of polyprotic acids and bases. These principles are demonstrated in this section by a few sample calculations.

A Typical Case

The upper curve in Figure 12-5 is calculated for the titration of 10.0 mL of 0.100 M base (B) with 0.100 M HCl. The base is dibasic, with $pK_{B1} = 4.00$ and $pK_{B2} = 9.00$. The titration curve has reasonably sharp breaks at both equivalence points, corresponding to the reactions

$$B + H^+ \rightarrow BH^+ \quad (12\text{-}25)$$

$$BH^+ + H^+ \rightarrow BH_2^{2+} \quad (12\text{-}26)$$

The volume at the first equivalence point is 10.0 mL because

$$\underbrace{V_e(\text{mL})(0.100 \text{ M})}_{\text{mmol of HCl}} = \underbrace{(10.0 \text{ mL})(0.100\ 0 \text{ M})}_{\text{mmol of B}} \quad (12\text{-}27)$$

$$V_e = 10.0 \text{ mL}$$

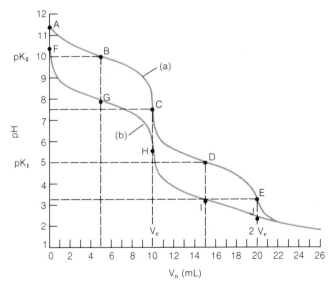

Figure 12-5
(a) Titration of 10.0 mL of 0.100 M base ($pK_{B1} = 4.00$, $pK_{B2} = 9.00$) with 0.100 M HCl. The two equivalence points are labeled C and E. Points B and D are the half-neutralization points, whose pH values equal pK_{A2} and pK_{A1}, respectively. (b) Titration of 10.0 mL of 0.100 M nicotine ($pK_{B1} = 6.15$, $pK_{B2} = 10.85$) with 0.100 M HCl.

234

12 / ACID–BASE TITRATIONS

$V_{e_2} = 2V_{e_1}$, always.

The volume at the second equivalence point must be $2V_e$, since Reaction 12-26 requires exactly the same number of moles of HCl as Reaction 12-25.

The calculation of pH at each point along the curve is very similar to the corresponding point in the titration of a monobasic compound. We will examine each point, A through E in Figure 12-5.

Point A

Before any acid is added, the solution contains just B, a weak base, whose pH is governed by Reaction 12-28.

$$\text{B} \quad + \quad \text{H}_2\text{O} \xrightleftharpoons{K_{B1}} \text{BH}^+ + \text{OH}^- \qquad (12\text{-}28)$$
$$0.100 - x \qquad\qquad\qquad x \qquad x$$

Recall that the fully basic form of a dibasic compound can be treated as if it were monobasic. (The K_{B2} reaction can be neglected.)

$$\frac{x^2}{0.100 - x} = 1.00 \times 10^{-4} \Rightarrow x = 3.11 \times 10^{-3}$$

$$[\text{H}^+] = K_w/x \Rightarrow \text{pH} = 11.49$$

Point B

At any point between A (the initial point) and C (the first equivalence point), we have a buffer containing B and BH^+. Point B is halfway to the equivalence point, so $[\text{B}] = [\text{BH}^+]$. The pH is calculated with the Henderson–Hasselbalch equation *for the weak acid, BH^+*, whose acid dissociation constant is K_{A2} for BH_2^{2+}. The value of K_{A2} is

Of course, you remember that

$$K_{B1} = K_w/K_{A2}$$
$$K_{B2} = K_w/K_{A1}$$

$$K_{A2} = K_w/K_{B1} = 10^{-10.00} \qquad (12\text{-}29)$$

To calculate the pH at point B, we write

$$\text{pH} = \text{p}K_{A2} + \log \frac{[\text{B}]}{[\text{BH}^+]} = 10.00 + \log 1 = 10.00 \qquad (12\text{-}30)$$

So the pH at point B is just $\text{p}K_{A2}$.

To calculate the ratio $[\text{B}]/[\text{BH}^+]$ at any point in the buffer region, just find what fraction of the way from point A to point C the titration has progressed. For example, if $V_a = 1.5$ mL, then

$$\frac{[\text{B}]}{[\text{BH}^+]} = \frac{8.5}{1.5} \qquad (12\text{-}31)$$

because 10.0 mL are required to reach the equivalence point and we have added just 1.5 mL. The pH at $V_a = 1.5$ mL is given by

$$\text{pH} = 10.00 + \log \frac{8.5}{1.5} = 10.75 \qquad (12\text{-}32)$$

Point C

At the first equivalence point, B *has been converted to* BH^+, *the intermediate form of the diprotic acid*, BH_2^{2+}. BH^+ *is both an acid and a base.* As described in Section 11-5, the pH is given by

$$[H^+] \approx \sqrt{\frac{K_1 K_2 F + K_1 K_w}{K_1 + F}} \tag{12-33}$$

where K_1 and K_2 are the acid dissociation constants of BH_2^{2+}.

The formal concentration (F) of BH^+ is calculated by considering the dilution of the original solution of B.

$$F = \underbrace{(0.100 \text{ M})}_{\substack{\text{Original} \\ \text{concentration} \\ \text{of B}}} \underbrace{\left(\frac{10.0}{20.0}\right)}_{\substack{\text{Dilution} \\ \text{factor}}} = 0.050\ 0 \text{ M} \tag{12-34}$$

Initial volume of B / Total volume of solution

Plugging all of the numbers into Equation 12-33 gives

$$[H^+] = \sqrt{\frac{(10^{-5})(10^{-10})(0.050\ 0) + (10^{-5})(10^{-14})}{10^{-5} + 0.050\ 0}} = 3.16 \times 10^{-8} \tag{12-35}$$

$$pH = 7.50$$

Note that in this example $pH = (pK_{A1} + pK_{A2})/2$.

Point D

At any point between C and E, we can consider the solution to be a buffer containing BH^+ (the base) and BH_2^{2+} (the acid). When $V_a = 15.0$ mL, $[BH^+] = [BH_2^{2+}]$ and

$$pH = pK_{A1} + \log \frac{[BH^+]}{[BH_2^{2+}]} = 5.00 + \log 1 = 5.00 \tag{12-36}$$

Point E

This is the second equivalence point, at which the solution is formally the same as one prepared by dissolving BH_2Cl_2 in water. The formal concentration of BH_2^{2+} is

$$F = (0.100 \text{ M})\left(\frac{10.0}{30.0}\right) = 0.033\ 3 \text{ M} \tag{12-37}$$

Original volume of B / Total volume of solution

12-4 TITRATIONS IN DIPROTIC SYSTEMS

BH^+ is the *intermediate form* of a diprotic acid.

$$pH \approx \tfrac{1}{2}(pK_1 + pK_2)$$

Challenge: Show that if V_a were 17.2 mL, the ratio in the log term would be

$$\frac{[BH^+]}{[BH_2^{2+}]} = \frac{20.0 - 17.2}{17.2 - 10.0} = \frac{2.8}{7.2}$$

236

12 / ACID–BASE TITRATIONS

At the second equivalence point, we have made BH_2^{2+}, which can be treated as a monoprotic weak acid.

The pH is determined by the acid dissociation reaction of BH_2^{2+}.

$$BH_2^{2+} \rightleftharpoons BH^+ + H^+ \qquad K_{A1} = \frac{K_w}{K_{B2}} \qquad (12\text{-}38)$$

$$\phantom{BH_2^{2+}} F - x \qquad x \qquad x$$

$$\frac{x^2}{0.033\,3 - x} = 1.0 \times 10^{-5} \Rightarrow x = 5.72 \times 10^{-4} \Rightarrow pH = 3.24$$

Beyond the second equivalence point ($V_a > 20.0$ mL), the pH of the solution can be calculated from the volume of strong acid added to the solution. For example, at $V_a = 25.00$ mL, there is an excess of 5.00 mL of 0.100 M HCl in a total volume of $10.00 + 25.00 = 35.00$ mL. The pH is found by writing

$$[H^+] = (0.100 \text{ M})\left(\frac{5.00}{35.00}\right) = 1.43 \times 10^{-2} \text{ M} \Rightarrow pH = 1.85 \quad (12\text{-}39)$$

Blurred Endpoints

Under some conditions the endpoint is obscured.

The titrations of many diprotic acids or bases do show two clear endpoints, as in the upper curve of Figure 12-5. However, other titrations may not show both endpoints, as in the lower curve of Figure 12-5. This latter curve is calculated for the titration of 10.0 mL of 0.100 M nicotine ($pK_{B1} = 6.15$, $pK_{B2} = 10.85$) with 0.100 M HCl. The two reactions are

$$(12\text{-}40)$$

nicotine (B) $\qquad\qquad$ BH^+

$$(12\text{-}41)$$

BH^+ $\qquad\qquad$ BH_2^{2+}

Both reactions occur, but there is almost no perceptible break at the second equivalence point, simply because BH_2^{2+} is too strong an acid. (Or, equivalently, BH^+ is too weak a base.)

The pH values at points F, G, H, I, and J in Figure 12-5 are calculated in just the same manner used for points A, B, C, D, and E. However, as the acidic end of the titration is approached (say, pH $\lesssim$ 3), the approximation that all of the HCl reacts with BH^+ to give BH_2^{2+} is not true. The acidity of BH_2^{2+} is great enough for it to exist partially dissociated in equilibrium with substantial free H^+. To calculate the pH correctly, we use the systematic treatment of equilibrium. It was in this manner, using a computer, that the lower portion of the titration curve was calculated.

The pH at point J is correctly calculated using the same procedure as for point E. That is, at point J the solution contains the salt $[BH_2^{2+}][Cl^-]_2$, regardless of whether it was created by a titration or by dissolving BH_2Cl_2 in water. The pH calculated for point J is 2.35.

As an example of what should alert you to a poor approximation, suppose we try calculating the pH at $V_a = 21.0$ mL. We will use the approximation that the pH is determined by the excess HCl.

$$[H^+] \approx \underbrace{(0.100 \text{ M})}_{\substack{\text{Initial} \\ \text{concentration} \\ \text{of HCl}}} \underbrace{\left(\frac{\overset{\text{Volume of excess HCl}}{1.0}}{\underset{\text{Total volume of solution}}{31.0}}\right)}_{\substack{\text{Dilution} \\ \text{factor}}} = 3.2 \times 10^{-3} \text{ M} \Rightarrow pH = 2.49 \quad (12\text{-}42)$$

An approximation that did not work.

We calculate that pH = 2.49 for this value of V_a, beyond the second equivalence point. But the pH at the equivalence point was correctly calculated to be 2.35, *which is lower than 2.49*. The pH cannot turn around and start climbing. If the pH at point J is correct, the pH at $V_a = 21.0$ mL must be wrong. The problem is that we have neglected the contribution to $[H^+]$ made by dissociation of BH_2^{2+}, which is a fairly strong acid ($pK_{A1} = 3.15$). The systematic treatment of equilibrium gives a value of pH = 2.19 at $V_a = 21.0$ mL.

One moral of this section is that whenever an acid is too strong (low pK_A) or a base too weak (high pK_B), the titration curve may show no evident break. This is true for monoprotic or polyprotic systems. A second lesson is that the approximations used in our calculations tend to be misleading at very high or very low pH. The approximations also become poorer as the formal concentration of acid or base being titrated becomes smaller.

In the titration curve for ribonuclease shown in Figure 12-1, there is a continual change in pH, with no clear breaks. The reason is that there are 29 groups being titrated in the pH interval shown. The 29 endpoints are so close together that a nearly uniform rise results. The curve can be analyzed to find the many pK_A values, but this requires a computer, and the individual pK values will not be determined very precisely.

12-5 FINDING THE ENDPOINT

Titrations are most commonly performed either to find out how much analyte is present or to measure equilibrium constants of the analyte. To find out how much analyte is present requires a knowledge of V_e, the volume of titrant at the equivalence point. The most popular ways of determining the equivalence point involve indicators or pH measurement with the glass electrode.

238

12 / ACID–BASE TITRATIONS

An indicator is an acid or a base whose different protonated forms have different colors.

Indicators

What is an indicator?

An acid–base **indicator** is itself an acid or base whose different protonated species have different colors. An example is thymol blue, which has two useful color changes.

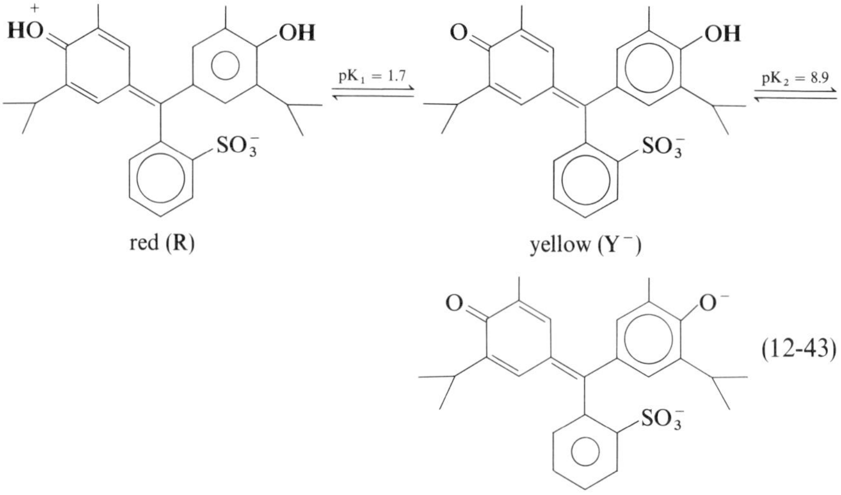

red (R)

yellow (Y⁻)

(12-43)

blue (B²⁻)

Below pH 1.7, the predominant species is red; between pH 1.7 and pH 8.9, the predominant species is yellow; and above pH 8.9, the predominant species is blue. For simplicity, we designate the three species R, Y^-, and B^{2-}, respectively. The sequence of color changes for thymol blue is shown in Color Plate 11.

The equilibrium between R and Y^- can be written

$$R \rightleftharpoons Y^- + H^+$$

$$K_1 = \frac{[Y^-][H^+]}{[R]}$$

$$pH = pK_1 + \log \frac{[Y^-]}{[R]} \qquad (12\text{-}44)$$

At $pH = 1.7 \ (= pK_1)$, there will be a 1:1 mixture of the yellow and red species, which will appear orange. As a very crude rule of thumb we may say that the solution will appear red when $[Y^-]/[R] \lesssim 1/10$ and yellow when $[Y^-]/[R] \gtrsim 10/1$. From the Henderson–Hasselbalch equation (12-44) we can see that the solution will be red when $pH \approx pK_1 - 1$ and yellow when $pH \approx pK_1 + 1$. In tables of indicator colors, thymol blue is listed as red below pH 1.2 and yellow above pH 2.8. By comparison, the pH values

pH	$[Y^-]/[R]$	Color
0.7	1:10	red
1.7	1:1	orange
2.7	10:1	yellow

239

> ### Demonstration 12-1 INDICATORS AND THE ACIDITY OF CO$_2$
>
> This demonstration is just plain fun to watch.[†] Fill two one-liter graduated cylinders with 900 mL of water and place a magnetic stirring bar in each. Add 10 mL of 1 M NH$_3$ to each. Then put 2 mL of phenolphthalein indicator solution in one and 2 mL of bromothymol blue indicator solution in the other. Both indicators will have the color of their basic species.
>
> Drop a few chunks of dry ice (solid CO$_2$) in each cylinder. As the CO$_2$ bubbles through each cylinder the solutions become more acidic. First the pink phenolphthalein color disappears. After some time the pH drops just low enough for bromothymol blue to change from blue to its green intermediate color. The pH does not go low enough to turn the indicator to its yellow color.
>
> Add about 20 mL of 6 M HCl to *the bottom* of each cylinder, using a length of tygon tubing attached to a funnel. Then stir each solution for a few seconds on a magnetic stirrer. Explain what happens. The sequence of events in this demonstration is shown in Color Plates 12–16.
>
> ---
>
> [†] A set of fascinating indicator demonstrations using universal indicator (a mixed indicator with many color changes) is described in J. T. Riley, *J. Chem. Ed.*, **54**, 29 (1977).

predicted by our rule of thumb are 0.7 and 2.7. Between pH 1.2 and pH 2.8, the indicator exhibits varying shades of orange. While most indicators have a single color change, thymol blue undergoes another transition, from yellow to blue, between pH 8.0 and pH 9.6. In this range, various shades of green would be seen. The interpretation of color varies among individuals, and indicator colors are no exception. Acid–base indicator color changes form the basis of Demonstration 12-1.

Choosing an indicator

A titration curve for which pH = 5.54 at the equivalence point is shown in Figure 12-6. An indicator with a color change near this pH would be useful in determining the endpoint of the titration. You can see on the graph in Figure 12-6 that the pH drops steeply (from 7 to 4) over a small volume interval. Therefore, any indicator with a color change in this pH interval would provide a fair approximation to the equivalence point. The closer the point of color change is to pH 5.54, the more accurate will be the endpoint. The difference between the observed endpoint (color change) and the true equivalence point is called the **indicator error.**

A list of some common indicators is given in Table 12-3. A rather large number of indicators in the table would provide a useful endpoint for the titration in Figure 12-6. For example, if bromocresol purple were used, we would use the purple-to-yellow color change as the endpoint. The last trace of purple should disappear near pH 5.2, which is quite close to the true equivalence point in Figure 12-6. If bromocresol green were used as the indicator, a color change from blue to green (= yellow + blue) would mark the endpoint.

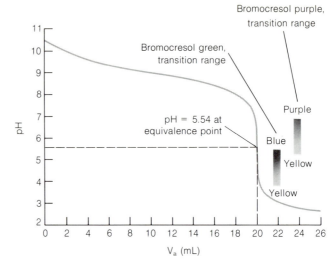

Figure 12-6
Calculated titration curve for the reaction of 100 mL of 0.010 0 M base ($pK_B = 5.00$) with 0.050 0 M HCl.

Choose an indicator whose color change comes as close as possible to the theoretical pH of the equivalence point.

In general, *we seek an indicator whose transition range overlaps the steepest part of the titration curve as closely as possible.* The steepness of the titration curve near the equivalence point in Figure 12-6 ensures that the indicator error caused by the noncoincidence of the endpoint and equivalence point will not be large. For example, if the indicator endpoint were at pH 6.4 (instead of 5.54), the error in V_e would be only 0.25% in this particular case.

Other uses of indicators

Indicators can be used to measure pH.

Equation 12-44 implies that if we know the pK of an indicator and the concentrations of both forms, we can measure pH with the indicator. This approach allows us to extend the idea of pH beyond the range over which glass electrodes function, and also to nonaqueous solvents. Box 12-2 tells how this is done.

Potentiometric Endpoint Detection

An alternative way to measure the equivalence point of a titration is to continuously monitor the pH with a glass electrode and pH meter. Since the meter measures the electric potential across the glass membrane of the electrode, we refer to this as a *potentiometric measurement*. The theory of such measurements is described in Chapter 16.

Figure 12-7 shows experimental results for the titration of a hexaprotic weak acid (H_6A) with NaOH. Because the compound is very difficult to purify, only a tiny amount was available for titration. Just 1.430 mg was dissolved in 1.00 mL of water and titrated with microliter quantities of 0.065 92 M NaOH, delivered with a Hamilton syringe.

Table 12-3
Some common indicators

Indicator	Transition range (pH)	Acid color	Base color	Preparation
Methyl violet	0.0–1.6	Yellow	Blue	0.05% in H_2O
Cresol red	0.2–1.8	Red	Yellow	0.1 g in 26.2 mL 0.01 M NaOH. Then add ~225 mL H_2O.
Thymol blue	1.2–2.8	Red	Yellow	0.1 g in 21.5 mL 0.01 M NaOH. Then add ~225 mL H_2O.
Cresol purple	1.2–2.8	Red	Yellow	0.1 g in 26.2 mL 0.01 M NaOH. Then add ~225 mL H_2O.
Erythrosine, disodium	2.2–3.6	Orange	Red	0.1% in H_2O
Methyl orange	3.1–4.4	Red	Orange	0.01% in H_2O
Congo red	3.0–5.0	Violet	Red	0.1% in H_2O
Ethyl orange	3.4–4.8	Red	Yellow	0.1% in H_2O
Bromocresol green	3.8–5.4	Yellow	Blue	0.1 g in 14.3 mL 0.01 M NaOH. Then add ~225 mL H_2O.
Methyl red	4.8–6.0	Red	Yellow	0.02 g in 60 mL ethanol. Then add 40 mL H_2O.
Chlorophenol red	4.8–6.4	Yellow	Red	0.1 g in 23.6 mL 0.01 M NaOH. Then add ~225 mL H_2O.
Bromocresol purple	5.2–6.8	Yellow	Purple	0.1 g in 18.5 mL 0.01 M NaOH. Then add ~225 mL H_2O.
p-Nitrophenol	5.6–7.6	Colorless	Yellow	0.1% in H_2O
Litmus	5.0–8.0	Red	Blue	0.1% in H_2O
Bromothymol blue	6.0–7.6	Yellow	Blue	0.1 g in 16 mL 0.01 M NaOH. Then add ~225 mL H_2O.
Phenol red	6.4–8.0	Yellow	Red	0.1 g in 28.2 mL 0.01 M NaOH. Then add ~225 mL H_2O.
Neutral red	6.8–8.0	Red	Orange	0.01 g in 50 mL ethanol. Then add 50 mL H_2O.
Cresol red	7.2–8.8	Yellow	Red	See above.
α-Naphtholphthalein	7.3–8.7	Yellow	Blue	0.1 g in 50 mL ethanol. Then add 50 mL H_2O.
Cresol purple	7.6–9.2	Yellow	Purple	See above.
Thymol blue	8.0–9.6	Yellow	Blue	See above.
Phenolphthalein	8.0–9.6	Colorless	Red	0.05 g in 50 mL ethanol. Then add 50 mL H_2O.
Thymolphthalein	8.3–10.5	Colorless	Blue	0.04 g in 50 mL ethanol. Then add 50 mL H_2O.
Alizarin yellow	10.1–12.0	Yellow	Orange-red	0.01% in H_2O
Nitramine	10.8–13.0	Colorless	Orange-brown	0.1 g in 70 mL ethanol. Then add 30 mL H_2O.
Tropaeolin O	11.1–12.7	Yellow	Orange	0.1% in H_2O

Box 12-2   WHAT DOES A NEGATIVE pH MEAN?

In the 1930s, Louis Hammett and his students devised a means to measure the basicity of very weak bases and the acidity of very strong acids. They began with a weak reference base, B, whose base strength could be measured in aqueous solution. An example is *p*-nitroaniline, whose protonated form has a pK_A of 0.99.

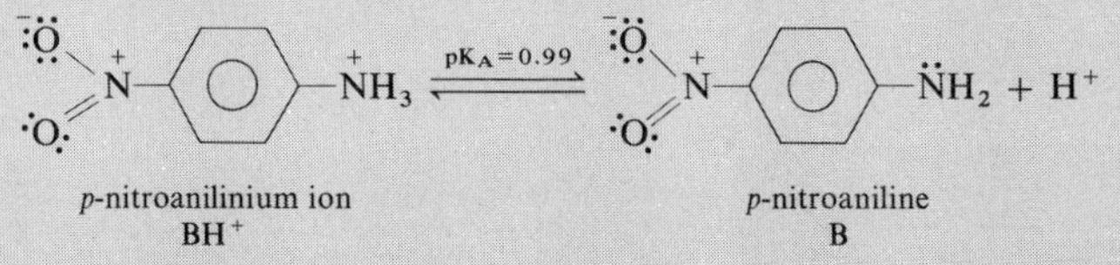

p-nitroanilinium ion
BH^+

p-nitroaniline
B

Suppose that some *p*-nitroaniline and a second base, C, are dissolved in a strong acid, such as 2 M HCl. The pK_A of CH^+ can be measured relative to that of BH^+ by first writing a Henderson–Hasselbalch equation for each acid:

$$pH = pK_A \text{ (for } BH^+) + \log \frac{[B]\gamma_B}{[BH^+]\gamma_{BH^+}}$$

$$pH = pK_A \text{ (for } CH^+) + \log \frac{[C]\gamma_C}{[CH^+]\gamma_{CH^+}}$$

Setting the two equations equal (since there is only one pH) gives

$$\underbrace{pK_A \text{ (for } CH^+) - pK_A \text{ (for } BH^+)}_{\Delta pK_A} = \log \frac{[B][CH^+]}{[C][BH^+]} + \log \frac{\gamma_B \gamma_{CH^+}}{\gamma_C \gamma_{BH^+}}$$

In solvents of high dielectric constant, the second term on the right, above, is close to zero because the ratio of activity coefficients is close to unity. Neglecting this last term gives the operationally useful result:

$$\Delta pK_A = \log \frac{[B][CH^+]}{[C][BH^+]}$$

That is, if you have a way to find the concentrations of B, BH^+, C, and CH^+, and you know the pK_A for BH^+, then you can find the pK_A for CH^+.

In practice, the concentrations can be measured spectrophotometrically, so pK_A for CH^+ can be determined. Then, using CH^+ as the reference, the pK_A for another compound, DH^+, can be measured. This procedure can be extended to measure the strengths of successively weaker bases, far too weak to be protonated in water.

The acidity of the solvent used to protonate the weak base, B, can be defined as H_0, which is analogous to the pH of an aqueous solution:

$$H_0 \equiv pK_A \text{ (for } BH^+) + \log \frac{[B]}{[BH^+]}$$

For dilute aqueous solutions, H_0 approaches pH. For concentrated solutions of strong acids, H_0 is considered to be a measure of the acid strength. H_0 is called the **Hammett acidity function.** A graph of H_0 versus acid concentration for several strong acids is shown below.

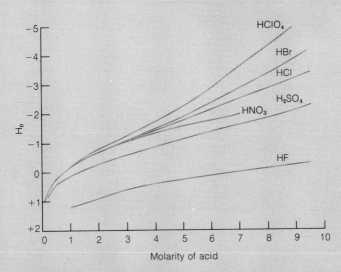

Hammett acidity function, H_0, for aqueous solutions of strong acids. [Data from M. A. Paul and F. A. Long, *Chem. Rev.*, **57**, 1(1957), which provides an informative review of acidity functions.]

When we refer to negative pH values, we are usually referring to H_0 values. For example, as measured by its ability to protonate very weak bases, 8 M $HClO_4$ has a "pH" close to -4. The figure shows why $HClO_4$ is considered to be a stronger acid than the other common strong acids, though they are all leveled to equal strength in water. The values of H_0 for several powerfully acidic solvents are given below.

Acid	Name	H_0
H_2SO_4 (100%)	Sulfuric acid	-11.93
$H_2SO_4 \cdot SO_3$	Fuming sulfuric acid (oleum)	-14.14
HSO_3F	Fluorosulfuric acid	-15.07
$HSO_3F + 10\% SbF_5$	"Super acid"	-18.94
$HSO_3F + 7\% SbF_5 \cdot 3SO_3$	—	-19.35

12 / ACID-BASE TITRATIONS

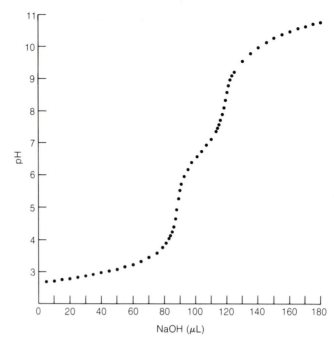

Figure 12-7
Experimental points in the titration of 1.430 mg of xylenol orange, a hexaprotic acid, dissolved in 1.00 mL of aqueous 0.10 M NaNO$_3$. The titrant was 0.065 92 M NaOH.

The curve in Figure 12-7 shows two clear breaks, near 90 and 120 µL, which correspond to titration of the *third* and *fourth* protons of H$_6$A.

$$H_4A^{2-} + OH^- \rightarrow H_3A^{3-} + H_2O \quad (\sim 90 \ \mu L \ \text{equivalence point})$$

$$H_3A^{3-} + OH^- \rightarrow H_2A^{4-} + H_2O \quad (\sim 120 \ \mu L \ \text{equivalence point})$$

The first two and last two equivalence points give unrecognizable endpoints, because they occur at too low or too high at pH.

The endpoint of an acid–base titration is taken as the point where the slope (dpH/dV) of the titration curve is greatest. How can we locate the steepest sections in Figure 12-7? A surprisingly accurate (and simple) way is to take a pencil and run it along the curve, feeling how the slope changes as you go. Most people can feel the inflection point at which the slope stops increasing and begins to decrease. A better, but much more laborious method, is to prepare a **Gran plot.**

The endpoint is the point of maximum slope.

Gran plot[†]

Consider the titration of a weak acid, HA, whose acid dissociation constant can be written as follows:

[†] More extensive discussions of the use and applications of the Gran plot can be found in G. Gran, *Analyst*, **77**, 661 (1952); and F. J. C. Rossotti and H. Rossotti, *J. Chem. Ed.*, **42**, 375 (1965).

$$HA \rightleftharpoons H^+ + A^- \qquad K_A = \frac{[H^+]\gamma_{H^+}[A^-]\gamma_{A^-}}{[HA]\gamma_{HA}} \qquad (12\text{-}45)$$

It will be necessary to include activity coefficients in this discussion because a pH electrode responds to hydrogen ion *activity*, not concentration.

At any point between the initial point and the endpoint of the titration it is usually a good approximation to say that each mole of NaOH converts one mole of HA into one mole of A^-. If we have titrated V_a mL of HA (whose formal concentration is F_a) with V_b mL of NaOH (whose formal concentration is F_b), we can write

Strong plus weak react completely.

$$[A^-] = \frac{\text{Moles of OH}^- \text{ delivered}}{\text{Total volume}} = \frac{V_b F_b}{V_b + V_a} \qquad (12\text{-}46)$$

$$[HA] = \frac{\text{Original moles of HA} - \text{Moles of OH}^-}{\text{Total volume}} = \frac{V_a F_a - V_b F_b}{V_a + V_b} \qquad (12\text{-}47)$$

Substituting the values of $[A^-]$ and $[HA]$ from Equations 12-46 and 12-47 into Equation 12-45 gives

$$K_A = \frac{[H^+]\gamma_{H^+} V_b F_b \gamma_{A^-}}{(V_a F_a - V_b F_b)\gamma_{HA}} \qquad (12\text{-}48)$$

which can be rearranged to

$$\underbrace{V_b[H^+]\gamma_{H^+}}_{10^{-pH}} = \frac{\gamma_{HA}}{\gamma_{A^-}} K_A \left(\frac{V_a F_a - V_b F_b}{F_b} \right) \qquad (12\text{-}49) \qquad \mathscr{A}_{H^+} = [H^+]\gamma_{H^+} = 10^{-pH}$$

The term on the left can be written $V_b 10^{-pH}$, since $[H^+]\gamma_{H^+} = 10^{-pH}$. The term in parentheses on the right is

$$\frac{V_a F_a}{F_b} - V_b = V_e - V_b \qquad (12\text{-}50) \qquad V_a F_a = V_e F_b \Rightarrow V_e = V_a F_a / F_b$$

Equation 12-49 can therefore be written in the form

$$V_b \cdot 10^{-pH} = \frac{\gamma_{HA}}{\gamma_{A^-}} K_A (V_e - V_b) \qquad (12\text{-}51)$$

A graph of $V_b \cdot 10^{-pH}$ versus V_b is called a **Gran plot.** If γ_{HA}/γ_{A^-} is constant, the graph should be a straight line with a slope of $-K_A\gamma_{HA}/\gamma_{A^-}$. The intercept on the V_b axis will be V_e. A Gran plot for the titration in Figure 12-7 is shown in Figure 12-8. Any units can be used for V_b, but the same units should be used on both axes. In Figure 12-8, V_b was expressed in microliters on both axes.

Plot $V_b 10^{-pH}$ versus V_b.
x intercept $= V_e$
Slope $= -K_A\gamma_{HA}/\gamma_{A^-}$

The beauty of a Gran plot is that it enables us to use data taken before the endpoint to find the endpoint. Other graphic methods (which we will

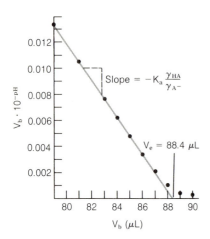

Figure 12-8
Gran plot for the data near the first equivalence point of Figure 12-7.

not discuss) require carefully collected data close to V_e. This is the hardest region of the titration curve in which to get accurate data, so these other graphic methods are often no better than estimating the steepest slope by eye (or "feel"). The slope of the Gran plot enables us to find K_A for the acid, HA.

The Gran function, $V_b \cdot 10^{-pH}$, does not actually go to zero, because 10^{-pH} is never zero. So the curve must be extrapolated to find V_e. The reason the function does not reach zero is that we have used the approximation that every mole of OH^- generates one mole of A^-. This approximation breaks down as V_b approaches V_e. The approximation that HA is undissociated can also be poor for acids of moderate strength in the early part of the titration. As a practical matter, only the linear portion of the Gran plot is used.

Another source of curvature in the Gran plot is changing ionic strength, which causes γ_{HA}/γ_{A^-} to vary. In Figure 12-7 this variation was avoided by having enough $NaNO_3$ present to maintain an essentially constant ionic strength throughout the titration. Even without added salt, the last 10–20% of data before V_e gives a fairly straight line because the ratio γ_{HA}/γ_{A^-} does not change very much.

Although we derived the Gran function for a monoprotic acid, the same plot ($V_b \cdot 10^{-pH}$ versus V_b) applies to polyprotic acids (such as H_6A in Figure 12-7). An analogous function can also be derived for the titration of a weak base with a strong acid.

Challenge: When a weak base, B, is titrated with a strong acid, the appropriate Gran function is

$$V_H \cdot 10^{+pH} = \left(\frac{1}{K_A} \cdot \frac{\gamma_{A^-}}{\gamma_{HA}}\right)(V_e - V_H) \tag{12-52}$$

where V_H is the volume of strong acid added, and K_A is the acid dissociation constant of BH^+. A graph of $V_H \cdot 10^{+pH}$ versus V_H should be a straight line with a slope of $-\gamma_{A^-}/\gamma_{HA} K_A$ and an intercept of V_e on the V_H axis. Use a procedure similar to the one above to derive Equation 12-52.

12-6 PRACTICAL NOTES

Several acids and bases can be obtained pure enough to be used as **primary standards**.[†] Some of them are listed in Table 12-4. Note that NaOH and KOH are not primary standards, for the reagent-grade materials contain carbonate (from reaction with atmospheric CO_2) and adsorbed water. Solutions of

---

[†] Instructions for purifying and using primary standards may be found in the following books: L. Meites, *Handbook of Analytical Chemistry* (New York: McGraw-Hill, 1963), pp. 3-32–3-35; I. M. Kolthoff and V. A. Stenger, *Volumetric Analysis*, Vol. 2 (New York: Wiley-Interscience, 1947); I. M. Kolthoff, E. B. Sandell, E. J. Meehan, and S. Bruckenstein, *Quantitative Chemical Analysis* (London: MacMillan, 1969), pp. 777–784.

Table 12-4   **247**
Some primary standards

Compound	Formula weight	Notes
potassium acid phthalate	204.223	The pure commercial material is dried at 105°C and used to standardize base. A phenolphthalein endpoint is satisfactory.
HCl hydrochloric acid	36.461	HCl and water distill as an *azeotrope* (a mixture) whose composition (~ 6 M) depends upon pressure. The composition is tabulated as a function of the pressure during distillation. See Problem 12-32 for more information.
$KH(IO_3)_2$ potassium hydrogen iodate	389.912	This is a strong acid, so any indicator with an endpoint between ~ 5 and ~ 9 is adequate.

Acids:

$$\text{(phthalate)} + OH^- \longrightarrow \text{(phthalate dianion)} + H_2O$$

Bases:

Compound	Formula weight	Notes
$H_2NC(CH_2OH)_3$ tris(hydroxymethylaminomethane) (also called "TRIS" or "THAM")	121.136	The pure commercial material is dried at 100–103°C and titrated with strong acid. The endpoint is in the range pH 4.5–5.
HgO mercuric oxide	216.59	Pure HgO is dissolved in a large excess of I^- or Br^-, whereupon $2OH^-$ are liberated:
Na_2CO_3 sodium carbonate	105.989	Primary standard grade Na_2CO_3 is commercially available. Alternatively, recrystallized $NaHCO_3$ can be heated for 1 hour at 260–270°C to produce pure Na_2CO_3. Sodium carbonate is titrated with acid to an end point of pH 4–5. Just before the endpoint, the solution is boiled to expel CO_2.
$Na_2B_4O_7 \cdot 10H_2O$ borax	381.367	The recrystallized material is dried in a chamber containing an aqueous solution saturated with NaCl and sucrose. This gives the decahydrate in pure form. The standard is titrated with acid to a methyl red endpoint.

$$H_2NC(CH_2OH)_3 + H^+ \rightarrow H_3\overset{+}{N}C(CH_2OH)_3$$

$$HgO + 4I^- + H_2O \rightarrow HgI_4^{2-} + 2OH^-$$

The base is titrated using an indicator endpoint.

$$\text{"}B_4O_7 \cdot 10H_2O^{2-}\text{"} + 2H^+ \rightarrow 4B(OH)_3 + 5H_2O$$

NaOH and KOH must be standardized against a primary standard. Potassium acid phthalate is among the most convenient compounds for this purpose.

Alkaline solutions (e.g., 0.1 M NaOH) must be protected from the atmosphere, or else they absorb CO_2:

> NaOH and KOH must be standardized with primary standards.

$$OH^- + CO_2 \rightarrow HCO_3^- \qquad (12\text{-}53)$$

CO_2 absorption changes the concentration of strong base over a period of time and decreases the extent of reaction near the endpoint in the titration

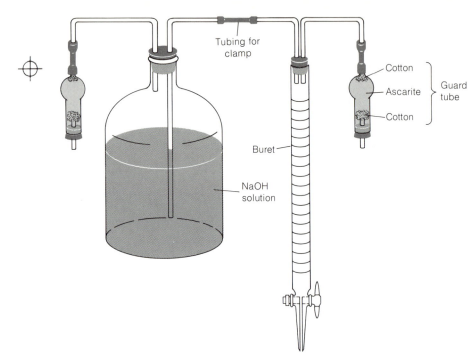

Figure 12-9
Setup for protecting alkaline solutions from atmospheric CO_2. The buret is filled by siphoning solution from the bottle without ever exposing the solution to the atmosphere.

of weak acids. If the solutions are kept in tightly capped polyethylene bottles, they may be used for about a week with little change. For longer periods, or for frequent use, the setup shown in Figure 12-9 is convenient. The ascarite (NaOH coated on asbestos) is a powerful CO_2 absorbent that protects the solution from air. A Gran plot can be used to estimate the carbonate content of an alkaline titrant.

Strongly basic solutions attack glass and are best stored in plastic containers. Such solutions should not be kept in a buret longer than necessary.

Strong base slowly dissolves glass.

12-7 TITRATIONS IN NONAQUEOUS SOLVENTS

After reading this far, you may think that water is the only solvent in the world. In fact, there are a great many chemical reactions that cannot be studied with aqueous solutions. In acid–base chemistry there are three common reasons why we might choose a nonaqueous solvent:

1. The reactants or products might be insoluble in water.
2. The reactants or products might react with water.
3. The analyte is too weak an acid or base to be titrated in water.

Reasons 1 and 2 are self-explanatory and will not be further discussed. Figure 12-4 gives an example illustrating reason 3. An acid with $pK_A \gtrsim 8$

does not give a distinct potentiometric endpoint and cannot be titrated in water. However, a very sharp endpoint might be observed if the same acid were titrated in a nonaqueous solvent.

You may recall from Chapter 10 that the strongest acid that can exist in water is H_3O^+, and the strongest base is OH^-. If an acid stronger than H_3O^+ is dissolved in water, it protonates H_2O to make H_3O^+. Because of this leveling effect, HCl and $HClO_4$ behave as if they had the same acid strength; both are *leveled* to H_3O^+.

$$HCl + H_2O \rightarrow H_3O^+ + Cl^- \qquad (12\text{-}54)$$

$$HClO_4 + H_2O \rightarrow H_3O^+ + Cl^- \qquad (12\text{-}55)$$

The leveling effect.

In a solvent less basic than H_2O, HCl and $HClO_4$ may not be leveled to the same strength.

$$HCl + \underset{\text{solvent}}{S} \rightleftharpoons HS^+ + Cl^- \qquad (12\text{-}56)$$

$$HClO_4 + S \rightleftharpoons HS^+ + ClO_4^- \qquad (12\text{-}57)$$

In acetic acid, for example, the equilibrium constants for Reactions 12-56 and 12-57 are 2.8×10^{-9} and 1.3×10^{-5}, respectively. That is, $HClO_4$ is a stronger acid than HCl in acetic acid solvent.

Figure 12-10 shows a titration curve for a mixture of five acids titrated with 0.2 M tetrabutylammonium hydroxide in methyl isobutyl ketone solvent.

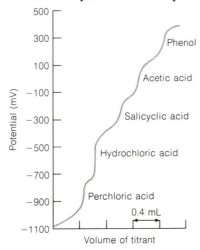

Figure 12-10
Titration of a mixture of acids with tetrabutylammonium hydroxide in methyl isobutyl ketone solvent. The measurements were made with a glass electrode and a platinum reference electrode. [D. B. Bruss and G. E. A. Wyld, *Anal. Chem.*, **29**, 232 (1957).]

Question: Where do you think the endpoint for the acid $H_3O^+ClO_4^-$ would come in Figure 12-10?

250

12 / ACID–BASE TITRATIONS

This solvent is not protonated to a great extent by any of the acids. It can be seen that perchloric acid is a stronger acid than HCl in this solvent as well.

Consider a base too weak to give a distinct endpoint when titrated with a strong acid in water.

$$B + H_3O^+ \rightleftharpoons BH^+ + H_2O \tag{12-58}$$

A base too weak to be titrated by H_3O^+ in water might be titrated by $HClO_4$ in acetic acid solvent.

The reason no endpoint is recognized is that the equilibrium constant for the titration reaction (12-58) is not large enough. If a stronger acid than H_3O^+ were available, the titration reaction might have a large enough equilibrium constant to give a distinct endpoint. If the same base were dissolved in acetic acid and titrated with $HClO_4$ in acetic acid, a clear endpoint might be observed. The reaction

$$B + HClO_4 \xrightleftharpoons{\text{in acetic acid}} \underbrace{BH^+ClO_4^-}_{\text{An ion pair}} \tag{12-59}$$

might have a large equilibrium constant, because $HClO_4$ is a much stronger acid than H_3O^+. (The product in Reaction 12-59 is written as an ion pair because acetic acid has too low a dielectric constant to allow ions to separate extensively.)

A wide variety of very weak acids and bases, or compounds insoluble or unstable in water, can be titrated in nonaqueous solution. An extensive literature of nonaqueous titrations exists.[†]

[†] J. S. Fritz, *Acid–Base Titrations in Nonaqueous Solvents* (Boston: Allyn and Bacon, 1973); J. Kucharsky and L. Safarik, *Titrations in Non-Aqueous Solvents* (New York: Elsevier, 1963); W. Huber, *Titrations in Nonaqueous Solvents* (New York: Academic Press, 1967); I. Gyenes, *Titration in Non-Aqueous Media* (Princeton, N.J.: van Nostrand, 1967).

Summary

In the titration of a strong acid with a strong base (or vice versa), the titration reaction is $H^+ + OH^- \rightarrow H_2O$, and the pH is determined by the concentration of excess unreacted analyte or titrant. The pH at the equivalence point is 7.00.

The titration curve of a weak acid with a strong base $(HA + OH^- \rightarrow A^- + H_2O)$ can be divided into four regions:

1. Before any base is added, the pH is determined by the acid dissociation reaction of the weak acid $(HA \rightleftharpoons H^+ + A^-)$.

2. Between the initial point and the equivalence point, there is a buffer made from A^- (which is equivalent to the moles of added base) and the excess unreacted HA. $pH = pK_A + \log([A^-]/[HA])$. At the special

point when $V_b = \frac{1}{2}V_e$, $pH = pK_A$ (neglecting activity coefficients).

3. At the equivalence point, the weak acid has been converted to its conjugate base, A^-, whose pH is governed by hydrolysis $(A^- + H_2O \rightleftharpoons HA + OH^-)$. The pH is necessarily above 7.00.

4. After the equivalence point, the pH is determined by the concentration of excess strong base.

The titration of a weak base with a strong acid has four regions analogous to those above. Before beginning, the pH is governed by hydrolysis of base, B. Between the initial and equivalence points, there is a buffer consisting of B plus BH^+. When $V_a = \frac{1}{2}V_e$, $pH = pK_A$ (for BH^+). At the equivalence point, B has been converted to its conjugate acid, BH^+, whose pH must be below 7.00. Beyond the equivalence point, the pH is governed by the concentration of excess strong acid titrant.

A titration of a diprotic acid (H_2A) with a strong base has six distinct regions:

1. The initial pH is determined by the dissociation of H_2A, which behaves as a monoprotic weak acid ($H_2A \rightleftharpoons H^+ + HA^-$).

2. Between the initial point and the first equivalence point, the solution is buffered by H_2A plus HA^-, whose pH is given by $pH = pK_{A1} + \log([HA^-]/[H_2A])$. When $V_b = \frac{1}{2}V_e$, $pH = pK_{A1}$.

3. At the first equivalence point, H_2A has been converted to HA^-, whose pH is given by

$$[H^+] = \sqrt{\frac{K_1 K_2 F' + K_1 K_w}{K_1 + F'}} \qquad pH \approx \frac{1}{2}(pK_{A1} + pK_{A2})$$

where F′ contains a correction for dilution of starting material.

4. Between the first and second equivalence points, there is a buffer consisting of HA^- and A^{2-}, whose pH is given by $pH = pK_{A2} + \log([A^{2-}]/[HA^-])$. When $V_b = \frac{3}{2}V_e$, $pH = pK_{A2}$.

5. At the second equivalence point, HA^- has been converted to its conjugate base, A^{2-}, whose pH is governed by hydrolysis ($A^{2-} + H_2O \rightleftharpoons HA^- + OH^-$).

6. Beyond the second equivalence point, the pH is governed by the concentration of excess strong base titrant.

In any titration, if the reactants are too dilute, or if the equilibrium constant for the titration reaction is not large enough, no sharp endpoint will be observed. When choosing an indicator, select one whose color transition range matches the pH of the equivalence point of the titration as closely as possible. A Gran plot, such as $V_b \cdot 10^{-pH}$ versus V_b, is a useful way to find the endpoint and the acid or base equilibrium constant. Acids or bases too weak to be titrated in H_2O may be titrated in a nonaqueous solvent in which the titrant is not leveled to the strength of H_3O^+ or OH^-.

Terms to Understand

Gran plot	indicator error
Hammett acidity function	leveling effect
indicator	primary standard

Exercises

12-A. Calculate the pH at each of the following points in the titration of 50.00 mL of 0.010 0 M NaOH with 0.100 M HCl. Volume of acid added: 0.00, 1.00, 2.00, 3.00, 4.00, 4.50, 4.90, 4.99, 5.00, 5.01, 5.10, 5.50, 6.00, 8.00, and 10.00 mL. Make a graph of pH versus volume of HCl added.

12-B. Calculate the pH at each point listed for the titration of 50.0 mL of 0.050 0 M formic acid with 0.050 0 M KOH. The points to calculate are $V_b = $ 0.0, 10.0, 20.0, 25.0, 30.0, 40.0, 45.0, 48.0, 49.0, 49.5, 50.0, 50.5, 51.0, 52.0, 55.0, and 60.0 mL. Draw a graph of pH versus V_b.

12-C. Calculate the pH at each point listed for the titration of 100.0 mL of 0.100 M cocaine (Equation 11-38) with 0.200 M HNO_3. The points to calculate are $V_a = $ 0.0, 10.0, 20.0, 25.0, 30.0, 40.0, 49.0, 49.9, 50.0, 50.1, 51.0, and 60.0 mL. Draw a graph of pH versus V_a.

12-D. Consider the titration of 50.0 mL of 0.050 0 M malonic acid with 0.100 M NaOH. Calculate the pH at each point listed and sketch the titration curve: $V_b = $ 0.0, 8.0, 12.5, 19.3, 25.0, 37.5, 50.0, and 56.3 mL.

12-E. Write the chemical reactions (including structures of reactants and products) that occur when histidine is titrated with perchloric acid. (Histidine is a molecule with no net charge.) A solution containing 25.0 mL of 0.050 0 M histidine was titrated with 0.050 0 M $HClO_4$. Calculate the pH at the following values of V_a: 0, 4.0, 12.5, 25.0, 26.0, 50.0 mL.

12-F. Select indicators from Table 12-3 that would be useful for the titrations in Figures 12-2, 12-3, and the second highest curve in Figure 12-4. Select a different indicator for each titration and state what color change you would use as the endpoint.

12-G. Acid–base indicators are themselves acids or bases. Consider an indicator, HIn, which dissociates according to the equation

$$HIn \xrightleftharpoons{K_A} H^+ + In^-$$

Suppose that the molar absorptivity, ε, is 2 080 for HIn and is 14 200 for In^-, at a wavelength of 440 nm.[†]

(a) Write the expression giving the absorbance at 440 nm of a solution containing HIn at a concentration $[HIn]$ and In^- at a concentration $[In^-]$. Assume the cell path-length is 1.00 cm. Note that absorbance is additive. The total absorbance is the sum of absorbances of all components.

(b) A solution containing the indicator at a formal concentration of 1.84×10^{-4} M is adjusted to pH 6.23 and found to exhibit an absorbance of 0.868 at 440 nm. Calculate pK_A for this indicator.

12-H. When 100.0 mL of a weak acid was titrated with 0.093 81 M NaOH, 27.63 mL was required to reach the equivalence point. The pH at the equivalence point was 10.99. What was the pH when only 19.47 mL of NaOH had been added?

12-I. A 0.100 M solution of the weak acid HA was titrated with 0.100 M NaOH. The pH measured when $V_b = \frac{1}{2}V_e$ was 4.62. Using activity coefficients correctly, calculate pK_A. The size of the A^- anion is 450 pm.

[†] This problem is an application of Beer's law, which you can read about in Sections 20-1 and 20-2.

Problems

12-1. Distinguish the terms *endpoint* and *equivalence point*.

12-2. At what point in the titration of a weak base with a strong acid is the maximum buffer capacity reached?

12-3. What is the equilibrium constant for the reaction between benzylamine and HCl?

12-4. A solution containing 50.0 mL of 0.031 9 M benzylamine was titrated with 0.050 0 M HCl. Calculate the pH at the following volumes of added acid: $V_a = 0$, 12.0, $\frac{1}{2}V_e$, 30.0, V_e, and 35.0 mL.

12-5. Consider the titration of the weak acid HA, with NaOH. At what fraction of V_e does $pH = pK_A - 1$? At what fraction of V_e does $pH = pK_A + 1$? Use these two points, plus $V_b = 0$, $\frac{1}{2}V_e$, V_e, and 1.2 V_e to sketch the titration curve for the reaction of 100 mL of 0.100 M anilinium bromide ("aminobenzene · HBr") with 0.100 M NaOH.

12-6. A solution of 100.0 mL of 0.040 0 M sodium propanoate (the sodium salt of propanoic acid) was titrated with 0.083 7 M HCl. Calculate the pH at the points $V_a = 0$, $\frac{1}{4}V_e$, $\frac{1}{2}V_e$, $\frac{3}{4}V_e$, V_e, and 1.1 V_e.

12-7. What is the pH at the equivalence point when 0.100 M hydroxyacetic acid is titrated with 0.050 0 M KOH?

12-8. Calculate the pH when 25.0 mL of 0.020 0 M 2-aminophenol has been titrated with 10.9 mL of 0.015 0 M $HClO_4$.

12-9. Calculate the pH at 10.0 mL intervals (from 0 to 100 mL) in the titration of 40.0 mL of 0.100 M piperazine with 0.100 M HCl. Make a graph of pH versus V_a.

12-10. Consider the titration of 50.0 mL of 0.100 M sodium glycinate with 0.100 M HCl.

(a) Calculate the pH at the second equivalence point.

(b) Show that our approximate method of calculations gives incorrect (physically unreasonable) values of pH at $V_a = 90.0$ and $V_a = 101.0$ mL.

12-11. A solution containing 0.100 M glutamic acid (the molecule with no net charge) was titrated with 0.025 0 M RbOH.

(a) Calculate the pH at the first equivalence point.

(b) Draw the structures of reactants and products.

12-12. (a) Find the pH of the solution when 0.010 0 M tyrosine is titrated to the first equivalence point with 0.004 00 M KOH.

(b) Draw the structures of reactants and products.

12-13. Find the pH of the solution when 0.010 0 M tyrosine is titrated to the equivalence point with 0.004 00 M $HClO_4$.

12-14. This problem deals with the amino acid cysteine, which we will abbreviate H_2C.

(a) A 0.030 0 M solution was prepared by dissolving dipotassium cysteine (K_2C) in water. Then

40.0 mL of this solution was titrated with 0.060 0 M HClO$_4$. Calculate the pH at the first equivalence point.
(b) Calculate the ratio $[C^{2-}]/[HC^-]$ in a solution of 0.050 0 M cysteinium bromide (the salt H$_3$C$^+$Br$^-$).

12-15. How many milliliters of 0.043 1 M NaOH should be added to 59.6 mL of 0.122 M leucine to obtain a pH of 8.00?

12-16. Would the indicator bromocresol green, with a transition range 3.8–5.4, ever be useful in the titration of a weak acid with a strong base?

12-17. Why would an indicator endpoint not be very useful in the titration curve for pK$_A$ = 10.00 in Figure 12-4?

12-18. What color do you expect to observe for cresol purple indicator (Table 12-3) at:
(a) pH 1.0 (b) pH 2.0 (c) pH 3.0

12-19. Cresol red has *two* transition ranges listed in Table 12-3. What color would you expect it to be at:
(a) pH 0 (b) pH 1 (c) pH 6 (d) pH 9

12-20. (a) What is the pH at the equivalence point when 0.030 0 M NaF is titrated with 0.060 0 M HClO$_4$?
(b) Why would an indicator endpoint probably not be useful in this titration?

12-21. A titration curve for Na$_2$CO$_3$ titrated with HCl is shown below. Suppose that *both* phenolphthalein and bromocresol green are present in the titration solution. State what colors you expect to observe at the following volumes of added HCl:
(a) 2 mL (b) 10 mL (c) 19 mL

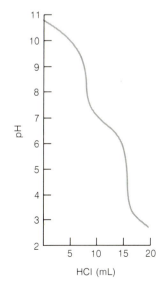

12-22. In the Kjeldahl nitrogen determination (Reactions 9-7 through 9-10), the final product is a solution of NH$_4^+$ ion in HCl solution. It is necessary to titrate the HCl without titrating the NH$_4^+$ ion.
(a) Calculate the pH of pure 0.010 M NH$_4$Cl.
(b) Select an indicator from Table 12-3 that would allow you to titrate HCl but not NH$_4^+$.

12-23. A solution was prepared by dissolving 0.194 7 g of HgO (Table 12-4) in 20 mL of water containing 4 g of KBr. Titration with HCl required 17.98 mL to reach a phenolphthalein endpoint. Calculate the molarity of the HCl.

12-24. How many grams of potassium acid phthalate should be weighed into a flask to standardize ~0.05 M NaOH if you wish to use ~30 mL of base for the titration?

12-25. Find the equilibrium constant for the reaction of MES (Table 11-2) with NaOH.

12-26. Calculate the pH of a solution made by mixing 50.00 mL of 0.100 M NaCN with
(a) 4.20 mL of 0.438 M HClO$_4$
(b) 11.82 mL of 0.438 M HClO$_4$
(c) What is the pH at the equivalence point with 0.438 M HClO$_4$?

12-27. When 22.63 mL of aqueous NaOH was added to 1.214 g of cyclohexylaminoethanesulfonic acid (F. W. 207.29, structure in Table 11-2) dissolved in 41.37 mL of water, the pH was 9.24. Calculate the molarity of the NaOH.

12-28. How many grams of dipotassium oxalate (F. W. 166.22) should be added to 20.0 mL of 0.800 M HClO$_4$ to give a pH of 4.40 when the solution is diluted to 500 mL?

12-29. Some properties of a particular indicator are given below:

$$\text{HIn} \quad \text{p}K_A = 7.95$$

$$\lambda_{max} = 395 \text{ nm}$$
$$\varepsilon_{395} = 1.80 \times 10^4 \text{ M}^{-1} \text{ cm}^{-1}$$
$$\varepsilon_{604} = 0$$

$$\text{In}^- + \text{H}^+$$
$$\lambda_{max} = 604 \text{ nm}$$
$$\varepsilon_{604} = 4.97 \times 10^4 \text{ M}^{-1} \text{ cm}^{-1}$$

A solution with a volume of 20.0 mL containing 1.40 × 10^{-5} M indicator plus 0.050 0 M benzene-1,2,3-tricarboxylic acid was treated with 20.0 mL of aqueous KOH. The resulting solution had an absorbance at 604 nm of 0.118 in a 1.00 cm cell. Calculate the molarity of the KOH solution.

12-30. Data for the titration of 100.0 mL of a weak acid by NaOH are given below. Find the equivalence point by preparing a Gran plot using data for the last 10% of the volume prior to V_e.

mL NaOH	pH	mL NaOH	pH
0.00	4.14	21.61	6.27
1.31	4.30	21.77	6.32
2.34	4.44	21.93	6.37
3.91	4.61	22.10	6.42
5.93	4.79	22.27	6.48
7.90	4.95	22.37	6.53
11.35	5.19	22.48	6.58
13.46	5.35	22.57	6.63
15.50	5.50	22.70	6.70
16.92	5.63	22.76	6.74
18.00	5.71	22.80	6.78
18.35	5.77	22.85	6.82
18.95	5.82	22.91	6.86
19.43	5.89	22.97	6.92
19.93	5.95	23.01	6.98
20.48	6.04	23.11	7.11
20.75	6.09	23.17	7.20
21.01	6.14	23.21	7.30
21.10	6.15	23.30	7.49
21.13	6.16	23.32	7.74
21.20	6.17	23.40	8.30
21.30	6.19	23.46	9.21
21.41	6.22	23.55	9.86
21.51	6.25		

12-31. Borax (Table 12-4) was used to standardize a solution of HNO_3. Titration of 0.261 9 g of borax required 21.61 mL. What is the molarity of the HNO_3?

12-32. Constant boiling aqueous HCl can be used as a primary standard for acid–base titrations. When $\sim 20\%$ (wt/wt) HCl is distilled, the composition of the distillate varies in a regular manner with the barometric pressure:

P (torr)	HCl[†] (g/100g of solution)
770	20.196
760	20.220
750	20.244
740	20.268
730	20.292

[†] The composition of distillate is from the paper of C. W. Foulk and M. Hollingsworth, *J. Am. Chem. Soc.*, **45**, 1223 (1923), with numbers corrected for the current values of atomic weights.

Suppose that constant boiling HCl was collected at a pressure of 746 torr.

(a) Make a graph of the data in the table above to find the weight percent of HCl collected at 746 torr.

(b) What weight of distillate (weighed in air, with stainless steel weights) should be dissolved in 1.000 0 L to give 0.100 00 M HCl? The density of distillate over the whole range in the table above is close to 1.096 g/mL. You will need this density to change the mass measured in vacuum to mass measured in air. See Section 2-2 for buoyancy corrections.

12-33. Use activity coefficients correctly to calculate the pH after 10.0 mL of 0.100 M trimethylammonium bromide was titrated with 4.0 mL of 0.100 M NaOH.

12-34. The base, B, is too weak to titrate in aqueous solution.

(a) Which solvent, pyridine or acetic acid, would be more suitable for titration of B with $HClO_4$?

(b) Which solvent would be more suitable for the titration of a very weak acid with tetrabutylammonium hydroxide?

13 / Advanced Topics in Acid–Base Chemistry

This chapter contains a collection of topics that build upon the three previous chapters. We begin with a discussion of how to think about the composition of an acid–base mixture without doing any calculations at all.

13-1 WHICH IS THE PRINCIPAL SPECIES?

We are often faced with the problem of identifying which species of acid, base, or intermediate is predominant under given conditions. A simple example is this: What is the principal form of benzoic acid in an aqueous solution at pH 8?

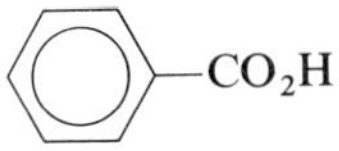

benzoic acid

$pK_A = 4.20$

The pK_A for benzoic acid is 4.20. This means that at pH 4.20 there would be a 1:1 mixture of benzoic acid (HA) and benzoate ion (A^-). At pH = pK_A + 1 (= 5.20) the ratio $[A^-]/[HA]$ is 10:1. At pH = pK_A + 2 (= 6.20), the ratio $[A^-]/[HA]$ is 100:1. As the pH increases the ratio $[A^-]/[HA]$ increases still further.

It is easy to see that for a monoprotic system the basic species (A^-) is the predominant form when pH > pK_A. The acidic species (HA) is the predominant form when pH < pK_A. The predominant form of benzoic acid at pH 8 is the benzoate anion ($C_6H_5CO_2^-$).

At pH = pK_A, $[A^-]$ = $[HA]$. This follows from the Henderson–Hasselbalch equation:

$$pH = pK_A + \log \frac{[A^-]}{[HA]}$$

pH	Major species
< pK_A	HA
> pK_A	A^-

255

13 / ADVANCED TOPICS IN ACID–BASE CHEMISTRY

EXAMPLE

What is the predominant form of ammonia in a solution at pH 7.0? Approximately what fraction is in this form?

In Appendix G we find that $pK_A = 9.24$ for the ammonium ion (NH_4^+, the conjugate acid of ammonia, NH_3). At pH = 9.24, $[NH_4^+] = [NH_3]$. Below pH 9.24, NH_4^+ will be the predominant form. Since pH = 7.0 is about 2 pH units below pK_A, the ratio $[NH_4^+]/[NH_3]$ will be around 100:1. More than 99% is in the form NH_4^+.

For polyprotic systems the reasoning is the same, but there are several values of pK_A. Consider oxalic acid (H_2Ox) with $pK_1 = 1.25$ and $pK_2 = 4.27$. At pH = pK_1, $[H_2Ox] = [HOx^-]$. At pH = pK_2, $[HOx^-] = [Ox^{2-}]$. We can make a little chart showing the major species in each pH region:

pH	Major species
$pH < pK_1$	H_2A
$pK_1 < pH < pK_2$	HA^-
$pH > pK_2$	A^{2-}

pH range	Predominant species
$pH < 1.25$	H_2Ox
$1.25 < pH < 4.27$	HOx^-
$pH > 4.27$	Ox^{2-}

EXAMPLE

The amino acid arginine has the following forms:

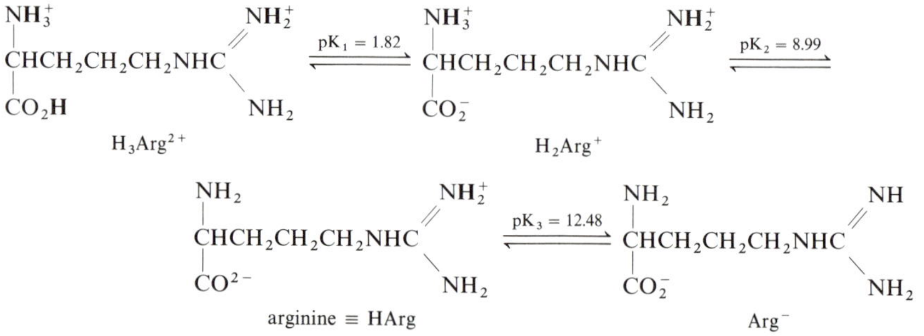

Note that the α-ammonium group (the one next to the carboxyl group) is more acidic than the substituent. This information comes from Appendix G. What is the principal form of arginine at pH 10.0? Approximately what fraction is in this form? Which is the second most abundant form at this pH?

We know that at pH = $pK_2 = 8.99$, $[H_2Arg^+] = [HArg]$. At pH = $pK_3 = 12.48$, $[HArg] = [Arg^-]$. At pH = 10.0, the major species is HArg. Since pH 10.0 is about one pH unit higher than pK_2, we can say that $[HArg]/[H_2Arg^+] \approx 10:1$. About 90% of arginine is in the form HArg. The second most important species is H_2Arg^+, which makes up about 10% of the arginine.

EXAMPLE

In the pH range 1.82 to 8.99, H_2Arg^+ is the principal form of arginine. Which is the second most prominent species at pH 6.0? At pH 5.0?

We know that the pH of the pure intermediate (amphiprotic) species, H_2Arg^+, is given by Equation 11-96:

$$\text{pH of } H_2Arg^+ \approx \tfrac{1}{2}(pK_1 + pK_2) = 5.40$$

Above pH 5.40 (and below pH = pK_2), we expect that HArg (the conjugate base of H_2Arg^+) will be the second most important species. Below pH 5.40 (and above pH = pK_1), we anticipate that H_3Arg^{2+} will be the second most important species.

The features of a triprotic system are sketched qualitatively in Figure 13-1. We determine the principal species by comparing the pH of the solution to the various pK_A values.

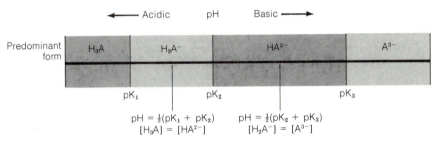

Figure 13-1
Schematic diagram showing the predominent form of a triprotic system (H_3A) in each pH interval.

13-2 FRACTIONAL COMPOSITION EQUATIONS

We will now derive equations that give the fraction of acid or base present in each possible form at a given pH. These equations are useful for the detailed solution of equilibrium problems involving acids and bases. They will also be used later in sections on EDTA titrations and electrochemistry.

Monoprotic Systems

First consider a solution of a monoprotic acid with a formal concentration F.

$$HA \xrightleftharpoons{K_A} H^+ + A^- \tag{13-1}$$

$$K_A = \frac{[H^+][A^-]}{[HA]} \tag{13-2}$$

$$\text{Mass balance:} \quad F = [HA] + [A^-] \tag{13-3}$$

Equation 13-3 can be rearranged to give

$$[A^-] = F - [HA] \tag{13-4}$$

which can be plugged into Equation 13-2 to give

$$K_A = \frac{[H^+](F - [HA])}{[HA]} \quad (13\text{-}5)$$

or

$$[HA] = \frac{[H^+]F}{[H^+] + K_A} \quad (13\text{-}6)$$

In a similar manner, we can derive the relationship

$$[A^-] = \frac{K_A F}{[H^+] + K_A} \quad (13\text{-}7)$$

α_0 = fraction of species in the form HA.

α_1 = fraction of species in the form A^-.

The *fraction* of molecules in the form HA is called α_0.

$$\alpha_0 = \frac{[HA]}{[HA] + [A^-]} = \frac{HA}{F} \quad (13\text{-}8)$$

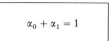

Dividing Equation 13-6 by F gives

$$\alpha_0 = \frac{[HA]}{F} = \frac{[H^+]}{[H^+] + K_A} \quad (13\text{-}9)$$

α_1 is the same thing we called the *fraction of dissociation* (α) on page 181.

In a similar manner, the fraction in the form A^- (designated α_1) can be obtained:

$$\alpha_1 = \frac{K_A}{[H^+] + K_A} \quad (13\text{-}10)$$

Figure 13-2 is a graph of Equations 13-9 and 13-10 for a system with $pK_A = 5.00$. At low pH almost all of the acid is in the form HA. At high pH almost everything is in the form A^-.

Diprotic Systems

Now we will derive the fractional composition equations for a diprotic system. The derivation follows the same pattern used for the monoprotic system.

$$H_2A \xrightleftharpoons{K_1} H^+ + HA^- \quad (13\text{-}11)$$

$$HA^- \xrightleftharpoons{K_2} H^+ + A^{2-} \quad (13\text{-}12)$$

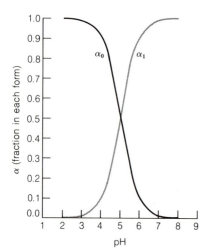

Figure 13-2
Fractional composition diagram of a monoprotic system with $pK_A = 5.00$.

$$K_1 = \frac{[H^+][HA^-]}{[H_2A]} \Rightarrow [HA^-] = [H_2A]\frac{K_1}{[H^+]} \quad (13\text{-}13)$$

$$K_2 = \frac{[H^+][A^{2-}]}{[HA^-]} \Rightarrow [A^{2-}] = [HA^-]\frac{K_2}{[H^+]} \quad (13\text{-}14)$$

$$= [H_2A]\frac{K_1 K_2}{[H^+]^2} \quad (13\text{-}15)$$

13-2 FRACTIONAL COMPOSITION EQUATIONS

The last equality comes from using Equation 13-13 to express $[HA^-]$ in terms of $[H_2A]$.

Mass balance: $F = [H_2A] + [HA^-] + [A^{2-}] \quad (13\text{-}16)$

$$F = [H_2A] + \frac{K_1}{[H^+]}[H_2A] + \frac{K_1 K_2}{[H^+]^2}[H_2A] \quad (13\text{-}17)$$

$$F = [H_2A]\left(1 + \frac{K_1}{[H^+]} + \frac{K_1 K_2}{[H^+]^2}\right) \quad (13\text{-}18)$$

For a diprotic system we designate the fraction in the form H_2A as α_0, the fraction in the form HA^- as α_1, and the fraction in the form A^{2-} as α_2. From the definition of α_0 we can write

α_0 = fraction of species in the form H_2A.
α_1 = fraction of species in the form HA^-.
α_2 = fraction of species in the form A^{2-}

$$\alpha_0 \equiv \frac{[H_2A]}{F} = \frac{[H^+]^2}{[H^+]^2 + [H^+]K_1 + K_1 K_2} \quad (13\text{-}19)$$

In a similar manner, we can derive Equations 13-20 and 13-21.

$$\boxed{\alpha_0 + \alpha_1 + \alpha_2 = 1}$$

$$\alpha_1 \equiv \frac{[HA^-]}{F} = \frac{K_1[H^+]}{[H^+]^2 + [H^+]K_1 + K_1 K_2} \quad (13\text{-}20)$$

$$\alpha_2 \equiv \frac{[A^{2-}]}{F} = \frac{K_1 K_2}{[H^+]^2 + [H^+]K_1 + K_1 K_2} \quad (13\text{-}21)$$

The general form of α for the polyprotic acid H_nA is

$\alpha_0 = [H^+]^n/D$
$\alpha_1 = K_1[H^+]^{n-1}/D$
$\alpha_j = K_1 K_2 \cdots K_j[H^+]^{n-j}/D$

where $D = [H^+]^n + K_1[H^+]^{n-1} + K_1 K_2[H^+]^{n-2} + \cdots + K_1 K_2 K_3 \cdots K_n$

Figure 13-3 shows α_0, α_1, and α_2 for fumaric acid, whose two pK_A values are only 1.5 units apart. The value of α_1 rises only to 0.72 because the two pK values are so close together. There is a substantial amount of both H_2A and A^{2-} in the region $pK_1 < pH < pK_2$.

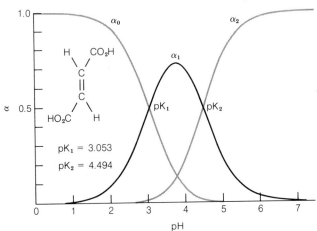

Figure 13-3
Fractional composition diagram for fumaric acid (*trans*-butenedioic acid).

260

**13 / ADVANCED TOPICS
IN ACID–BASE CHEMISTRY**

13-3 ISOELECTRIC AND ISOIONIC pH

Biochemists often refer to the isoelectric or isoionic pH of a polyprotic species, such as a protein. These terms can be readily understood in terms of a simple diprotic system, such as the amino acid alanine.

$$\underset{\substack{\text{alanine cation}\\ H_2A^+}}{\overset{CH_3}{\underset{|}{H_3\overset{+}{N}CHCO_2H}}} \rightleftharpoons \underset{\substack{\text{neutral zwitterion}\\ HA}}{\overset{CH_3}{\underset{|}{H_3\overset{+}{N}CHCO_2^-}}} + H^+ \qquad pK_1 = 2.35 \qquad (13\text{-}22)$$

$$\overset{CH_3}{\underset{|}{H_3\overset{+}{N}CHCO_2^-}} \rightleftharpoons \underset{\substack{\text{alanine anion}\\ A^-}}{\overset{CH_3}{\underset{|}{H_2NCHCO_2^-}}} + H^+ \qquad pK_2 = 9.87 \qquad (13\text{-}23)$$

Isoionic pH—the pH of the pure, neutral, polyprotic acid.

The **isoionic point** (or isoionic pH) is the pH obtained when the pure, neutral polyprotic acid (HA, the neutral zwitterion) is dissolved in water. The only ions in such a solution are H_2A^+, A^-, H^+, and OH^-. Most of the alanine is in the form HA.

Isoelectric pH—the pH at which the average charge of the polyprotic acid is zero.

The **isoelectric point** (or isoelectric pH) is the pH at which the *average* charge of the polyprotic acid is zero. At this pH most of the molecules are in the uncharged form HA. The concentrations of H_2A^+ and A^- are equal; therefore, the *average* charge of alanine will be zero. It is important to realize that the molecule is never present in only its neutral form. There will always be some H_2A^+ and some A^- in equilibrium with HA.

Alanine is the intermediate form of a polyprotic acid, so we use Equation 13-24 (which is the same as Equation 11-91 to find the pH).

The isoionic pH for alanine occurs when pure alanine is dissolved in water. Since alanine (HA) is the intermediate form of a diprotic acid (H_2A^+), the pH is given by

$$[H^+] = \sqrt{\frac{K_1K_2F + K_1K_w}{K_1 + F}} \qquad (13\text{-}24)$$

where F is the formal concentration of alanine. For 0.10 M alanine, the isoionic pH is

$$[H^+] = \sqrt{\frac{K_1K_2(0.10) + K_1K_w}{K_1 + (0.10)}} = 7.6 \times 10^{-7} \text{ M} \Rightarrow pH = 6.12 \qquad (13\text{-}25)$$

The isoionic pH shows a slight concentration dependence. For 0.010 M alanine, the isoionic pH is 6.19.

The isoelectric point is the pH at which the concentrations of H_2A^+ and A^- are equal, and the average charge of alanine is therefore zero. We can calculate the isoelectric pH by first writing expressions for the concentrations of the cation and anion.

$$[H_2A^+] = \frac{[HA][H^+]}{K_1} \qquad (13\text{-}26)$$

$$[A^-] = \frac{K_2[HA]}{[H^+]} \tag{13-27}$$

Setting $[H_2A^+] = [A^-]$, we find

$$\frac{[HA][H^+]}{K_1} = \frac{K_2[HA]}{[H^+]} \tag{13-28}$$

or

$$[H^+] = \sqrt{K_1 K_2} \text{ (at the isoelectric pH)} \tag{13-29}$$

which gives

$$\text{Isoelectric pH} = \frac{pK_1 + pK_2}{2} \tag{13-30}$$

The isoelectric point is midway between the two pK_A values "surrounding" the neutral, intermediate species.

For a diprotic amino acid, the isoelectric pH is halfway between the two pK_A values. The isoelectric pH of alanine is $\frac{1}{2}(2.35 + 9.87) = 6.11$. The isoelectric pH is not concentration-dependent.

You can see that the isoelectric and isoionic points for a polyprotic acid are very nearly the same. At the isoelectric pH, the average charge of the molecule is zero, thus $[H_2A^+] = [A^-]$ and $pH = \frac{1}{2}(pK_1 + pK_2)$. At the isoionic point, the pH is given by Equation 13-24, and $[H_2A^+]$ is not exactly equal to $[A^-]$. However, the pH is quite close to the isoelectric pH.

The relevance of isoelectric and isoionic points with respect to proteins is this: The *isoionic* pH of a protein is the pH of a solution containing the pure protein with no counterions except H^+ or OH^-. Proteins are usually isolated in a charged form together with various counterions (such as Na^+, NH_4^+, or Cl^-). If such a protein is **dialyzed** extensively against pure water, the pH in the protein compartment will approach the isoionic point provided the counterions are free to pass through the dialysis membrane. The *isoelectric* point is the pH at which the protein has no net charge. This property forms the basis for a very sensitive and important technique used in the preparative and analytical separation of proteins, as described in Box 13-1.

Dialysis was discussed in Demonstration 8-1.

13-4 REACTIONS OF WEAK ACIDS WITH WEAK BASES

In Chapters 11 and 12 we systematically dealt with the reactions between acids and bases in the combinations of strong-plus-strong and strong-plus-weak. These reactions were easy to treat because their equilibrium constants are large. Such reactions "go to completion."

The final coup would be the ability to deal with reactions between weak acids and weak bases. Calling the acid HA and the base B, we can write the reaction

$$HA + B \rightleftharpoons A^- + BH^+ \tag{13-31}$$

whose equilibrium constant can be calculated as follows:

$$HA \rightleftharpoons H^+ + A^- \qquad K = K_A$$

$$B + H^+ \rightleftharpoons BH^+ \qquad K = \frac{1}{K_A \text{ (for } BH^+)} = \frac{K_B}{K_w}$$

$$\overline{HA + B \rightleftharpoons A^- + BH^+} \qquad K = K_A K_B / K_w \qquad (13\text{-}32)$$

Box 13-1 ISOELECTRIC FOCUSSING

At its *isoelectric* point, a protein has no net charge. It will therefore not migrate in an electric field at its isoelectric pH. This is the basis of a very sensitive technique of protein separation called **isoelectric focussing.** A mixture of proteins is subjected to a strong electric field in a medium having a pH gradient. Positively charged molecules move toward the negative pole and negatively charged molecules move toward the positive pole. The proteins migrate in one direction or the other until they reach the region of their isoelectric pH. In this region they have no net charge and no longer move. Thus each protein in the mixture is focussed in one small region at its isoelectric pH.

An example of isoelectric focussing is shown on the following page. In this experiment a mixture of seven proteins (and, apparently, a host of impurities) was applied to a polyacrylamide gel containing a mixture of polyprotic compounds called *ampholytes*. Each end of the gel is placed in contact with a conducting solution and a potential of several hundred volts is applied across the length of the gel. The ampholytes migrate until they form a stable pH gradient (ranging from about pH 3 at one end of the gel to pH 10 at the other). The proteins migrate until they reach their isoelectric pH, at which point they have no net charge and cease migrating. If a molecule diffuses out of its isoelectric region, it becomes charged and immediately migrates back to its isoelectric zone in the gel. When the proteins finish migrating, the field is removed. The proteins are precipitated in place on the gel and stained with a dye to make their positions visible. The stained gel is shown at the bottom of the figure. A spectrophotometer scan of the dye peaks is shown on the graph, and a profile of measured pH is also plotted.

For weak acids and bases, K for Reaction 13-32 might be large or small.

It proves useful to divide the problems of weak bases and weak acids into three cases:

Case 1. K is large (K ≫ 1).
Case 2. K is not large.
Case 3. Equimolar mixture of HA and B, regardless of the size of K.

We will examine these cases one at a time.

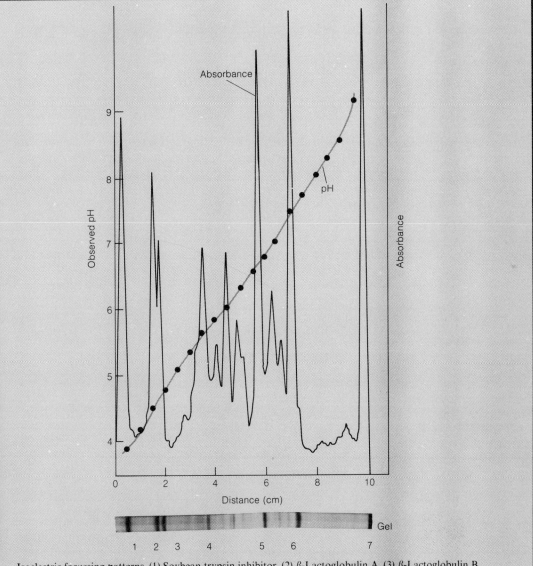

Isoelectric focussing patterns. (1) Soybean trypsin inhibitor. (2) β-Lactoglobulin A. (3) β-Lactoglobulin B. (4) Ovotransferrin. (5) Horse myoglobin. (6) Whale myoglobin. (7) Cytochrome C. [Courtesy Bio-Rad Laboratories, Richmond. Calif.]

Case 1: K Is Large ($K \gg 1$)

For a specific example, let's look at the pH of a mixture of chloroacetic acid and methylamine.

$$ClCH_2CO_2H \text{ (HA)} \qquad CH_3NH_2 \text{ (B)}$$

chloroacetic acid   methylamine

$$K_A = 1.36 \times 10^{-3} \qquad K_B = 4.4 \times 10^{-4}$$

$$pK_A = 2.86 \qquad pK_B = 3.36$$

Using Equation 13-32, we find the equilibrium constant for the acid–base reaction:

$$ClCH_2CO_2H + CH_3NH_2 \rightleftharpoons ClCH_2CO_2^- + CH_3NH_3^+ \quad (13\text{-}33)$$

$$K = K_A K_B / K_w = 6.0 \times 10^7$$

Since K is so large, it is fair to say that the reaction will go to completion. *When the reactants are mixed, they will proceed to make products until one of the reactants has been consumed.* That is, if K is large, the reaction between a weak acid and weak base can be treated just like the case in which one reactant is a strong acid or base.

> Case 1 is just like the weak-plus-strong combination.

EXAMPLE

What would be the pH if 100.0 mL of 0.050 0 M chloroacetic acid were mixed with 60.0 mL of 0.060 0 M methylamine?

We proceed as follows:

$$HA + B \rightleftharpoons A^- + BH^+$$

	HA	B	A⁻	BH⁺
Initial mmol:	5.00	3.60	—	—
Final mmol:	1.40	—	3.60	3.60

The little table above shows that we have created a known mixture of HA and A^-.

Aha! A buffer!

The pH is easily calculated from the appropriate Henderson–Hasselbalch equation:

$$pH = pK_A + \log \frac{[A^-]}{[HA]} = 2.86 + \log \frac{3.60}{1.40} = 3.27$$

Our analysis of Case 1 indicates that the reaction of a sufficiently strong "weak" base with a sufficiently strong "weak" acid can go "to completion." The lower curve in Figure 13-4 is calculated for the titration of chloroacetic acid by methylamine. Up to the equivalence point, the curve is identical to that of the titration of a weak acid with a strong base, shown in the upper curve.

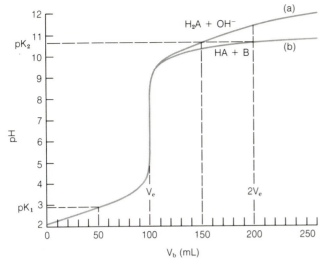

Figure 13-4
(a) Titration of 100 mL of 0.50 M H_2A ($pK_1 = 2.86$, $pK_2 = 10.64$) with 0.050 M NaOH. (b) Titration of 100 mL of the weak acid HA (0.50 M, $pK_A = 2.86$) with the weak base B (0.050 M, $pK_B = 3.36$).

13-4 REACTIONS OF WEAK ACIDS WITH WEAK BASES

Challenge: Figure 13-4 compares the titration of chloroacetic acid with methylamine to the titration of a diprotic acid (H_2A) with NaOH. Why does pK_2 intersect the upper curve at $\frac{3}{2}V_e$ and the lower curve at $2V_e$? On the lower curve, "pK_2" is pK_A for the acid, BH^+. If you can answer this question, you've come a long way.

Case 2: K Is Not Large

Suppose that 10.0 mL of 0.020 0 M ammonium chloride and 10.0 mL of 0.032 0 M trimethylamine were mixed together. What would be the pH?

NH_4^+	$(CH_3)_3N$
ammonium ion	trimethylamine
$pK_A = 9.244$	$pK_B = 4.200$
$K_A = 5.70 \times 10^{-10}$	$K_B = 6.31 \times 10^{-5}$

The acid–base reaction is

$$NH_4^+ + (CH_3)_3N \rightleftharpoons NH_3 + (CH_3)_3NH^+ \qquad (13\text{-}34)$$

$$K = K_A K_B / K_w = 3.60$$

and the equilibrium constant is only 3.60. This is not a large number. We cannot say that the reaction between this acid and base goes "to completion." A substantial quantity of unreacted starting material will exist in equilibrium with products.

To find the composition at equilibrium we can set up a little table of concentrations. When 10.0 mL of acid and 10.0 mL of base are mixed, they dilute each other by a factor of two. Therefore, the initial molarities are 0.010 0 M and 0.016 0 M. Since the products are formed in a 1:1 mole ratio, we can write

In Case 2 the limiting reactant is not used up.

266

**13 / ADVANCED TOPICS
IN ACID–BASE CHEMISTRY**

$$NH_4^+ + (CH_3)_3N \rightleftharpoons NH_3 + (CH_3)_3NH^+$$

Initial concentration (M): 0.010 0   0.016 0   —   —

Final concentration (M): 0.010 0 − x 0.016 0 − x x x

Putting these values into the equilibrium expression for Reaction 13-34 gives

$$\frac{x^2}{(0.010\,0 - x)(0.016\,0 - x)} = 3.60 \Rightarrow x = 0.007\,88 \qquad (13\text{-}35)$$

This value of x tells us that the concentration of each species is

$$[NH_4^+] = 0.002\,12 \text{ M} \qquad [(CH_3)_3N] = 0.008\,12$$

$$[NH_3] = 0.007\,88 \text{ M} \qquad [(CH_3)_3NH^+] = 0.007\,88$$

Aha! Two buffers!

The pH can be calculated by using either of the buffer systems present:

There are two buffer systems:
1. NH_4^+/NH_3
2. $(CH_3)_3NH^+/(CH_3)_3N$

$$pH = pK_A \text{ (for } NH_4^+) + \log\frac{[NH_3]}{[NH_4^+]} = 9.244 + \log\frac{0.007\,88}{0.002\,12} = 9.81$$

$$(13\text{-}36)$$

$$pH = pK_A \text{ (for } (CH_3)_3NH^+) + \log\frac{[(CH_3)_3N]}{[(CH_3)_3NH^+]}$$

$$= 9.800 + \log\frac{0.008\,12}{0.007\,88} = 9.81 \qquad (13\text{-}37)$$

If the value of x was calculated correctly in Equation 13-35, then Equations 13-36 and 13-37 must produce the same pH.

Case 3: Equimolar Mixture of HA and B

Before we treat this case, let's take another look at a diprotic acid, which has three forms: H_2A contains two protons, HA^- has one proton and A^{2-} has no protons. The intermediate form HA^-, can donate *or* accept one proton. This behavior led to the equation

$$[H^+] \approx \sqrt{\frac{K_1K_2F + K_1K_w}{K_1 + F}} \qquad (11\text{-}91)$$

to describe a solution of pure HA^-.

But an equimolar mixture of HA and B is very much like the intermediate form of a diprotic system. Let's draw the diprotic acid as HA_x⤳A_yH, where A_x and A_y are attached to each other (i.e., part of the same molecule). The three states of each system can be drawn as follows:

Diprotic acid	Proton-state diagram	Equimolar HA + B
$HA_x \text{\sim\sim} A_yH$	$\boxed{H \mid H}$ Two protons	$HA + BH^+$
$HA_x \text{\sim\sim} A_y^- \rightleftharpoons {}^- A_x \text{\sim\sim} A_yH$	$\boxed{H \mid } \rightleftharpoons \boxed{ \mid H}$ One proton	$HA + B \rightleftharpoons A^- + BH^+$
${}^- A_x \text{\sim\sim} A_y^-$	$\boxed{ \mid }$ No protons	$A^- + B$

The analogy between the intermediate species of a diprotic acid and an equimolar mixture of HA + B suggests that the pH of a mixture of HA + B can be calculated from the equation

$$[H^+] \approx \sqrt{\frac{K_1 K_2 F + K_1 K_w}{K_1 + F}} \tag{13-38}$$

> An equimolar mixture of HA and B behaves as the intermediate form of the imaginary diprotic acid $HA\text{\sim\sim}BH^+$.

where $K_1 = K_A$ for the stronger of the two acids HA and BH^+

$K_2 = K_A$ for the weaker of the two acids HA and BH^+

F = Formal concentration of either HA or B (they are equimolar)

With this in mind, consider a solution prepared by mixing 100 mL of 0.050 0 M chloroacetic acid with 100 mL of 0.050 0 M methylamine (Equation 13-33). This is an equimolar mixture of HA and B and should behave *just as the intermediate form of a diprotic acid.* The values of K_A for the two acids are

$$K_A \text{ for } ClCH_2CO_2H = 1.36 \times 10^{-3}$$

$$K_A \text{ for } CH_3NH_3^+ = 2.3 \times 10^{-11}$$

Calling the larger value K_1 and the smaller value K_2, we write

$$[H^+] = \sqrt{\frac{(1.36 \times 10^{-3})(2.3 \times 10^{-11})(0.025\,0) + (1.36 \times 10^{-3})(1.0 \times 10^{-14})}{1.36 \times 10^{-3} + 0.025\,0}}$$

$$[H^+] = 1.74 \times 10^{-7} \text{ M} \Rightarrow pH = 6.76 \tag{13-39}$$

We used $F = 0.025\,0$ M because each solution was diluted by the other when they were mixed.

To summarize, an equimolar mixture of a weak acid and a weak base behaves as the intermediate form of a polyprotic acid. *The pH of the equimolar mixture should be very nearly midway between pK_A for HA and pK_A for BH^+.*

> As a check, $\frac{1}{2}(pK_1 + pK_2) = \frac{1}{2}(2.865 + 10.64) = 6.75$.

EXAMPLE

Find the pH of 0.10 M ammonium bicarbonate.

Ammonium bicarbonate can be thought of as the first intermediate species of the *triprotic* system consisting of NH_4^+ and H_2CO_3:

$$NH_4^+: \quad pK_A = 9.244$$

$$H_2CO_3: \quad pK_1 = 6.352$$

$$pK_2 = 10.329$$

We can think of this triprotic system as follows:

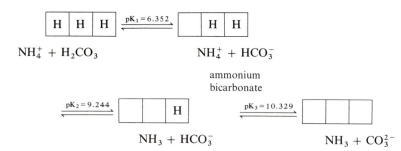

Since ammonium bicarbonate is "surrounded" by pK_1 and pK_2,

$$[H^+] = \sqrt{\frac{(10^{-6.352})(10^{-9.244})(0.10) + (10^{-6.352})(K_w)}{10^{-6.352} + 0.10}}$$

$$[H^+] = 1.59 \times 10^{-8} \text{ M} \Rightarrow pH = 7.80$$

The pH is midway between pK_1 (6.352) and pK_2 (9.244) for the imaginary triprotic system.

Summary

The principal species of a monoprotic or polyprotic system is found by comparing the pH with the various pK_A values. For $pH < pK_1$, the fully protonated species is the predominant form. For $pK_1 < pH < pK_2$ the form $H_{n-1}A^-$ is favored and at each successive pK value the next deprotonated species becomes the principal species. Finally, at pH values higher than the highest pK, the fully basic form (A^{n-}) is dominant. The fractional composition of a solution is expressed by the α equations, 13-9 and 13-10 for a monoprotic system, 13-19 through 13-21 for a diprotic system.

The isoelectric pH of a polyprotic species is that pH at which its net charge is zero. For a diprotic amino acid whose amphiprotic form is neutral, the isoelectric pH is given by $pH = \frac{1}{2}(pK_1 + pK_2)$. The isoionic pH of a polyprotic species is the pH that would exist in a solution containing only the ions derived from the neutral polyprotic species and from H_2O.

The reaction of a weak acid with a weak base can be written $HA + B \rightleftharpoons A^- + BH^+$, and the equilibrium constant expressed as $K_A K_B / K_w$. Such reactions fall into three classes:

1. If the equilibrium constant is large, the reaction proceeds to completion and can be treated as a strong-plus-weak titration.

2. If the equilibrium constant is not large, we must set up and solve an equilibrium problem to find the composition of the solution. Knowing the composition, we can use either of two Henderson–Hasselbalch equations to find the pH. Both equations are satisfied simultaneously.

3. An equimolar mixture of a weak acid and a weak base can be treated as the intermediate form of a polyprotic system, regardless of the magnitude of the equilibrium constant.

Terms to Understand

dialysis
isoelectric focussing

isoelectric point
isoionic point

Exercises

13-A. (a) Draw the structure of the predominant form (principal species) of 1,3-dihydroxybenzene at pH 9.00 and at pH 11.00.
  (b) What is the second most prominent species at each pH?
  (c) Calculate the percent in the major form at each pH.

13-B. Draw the structures of the predominant forms of glutamic acid and tyrosine at pH 9.0 and pH 10.0. What is the second most abundant species at each pH?

13-C. Calculate the isoionic pH of 0.010 M lysine.

13-D. Neutral lysine can be written HL. The other forms of lysine are H_3L^{2+}, H_2L^+, and L^-. The isoelectric point is the pH at which the *average* charge of lysine is zero. Therefore, at the isoelectric point $2[H_3L^{2+}] + [H_2L^+] = [L^-]$. Use this condition to calculate the isoelectric pH of lysine.

13-E. Solution A contains 0.050 0 M HF. Solution B contains 0.050 0 M ammonia.
  (a) Find the pH of a solution containing 20.0 mL of A plus 14.0 mL of B.
  (b) Find the pH of a solution containing 20.0 mL of B plus 14.0 mL of A.

13-F. Calculate the pH of 0.100 M
  (a) tetraethylammonium formate
  (b) triethylammonium formate

13-G. What is the pH when 25.0 mL of 0.010 M tyrosine is treated with 25.0 mL of 0.009 0 M ammonia?

13-H. Which solution will have the highest pH?
  (a) 0.010 M potassium acid phthalate
  (b) 0.025 M monosodium malonate
  (c) 0.030 M glycine
  (d) 0.016 M $(NH_4)(HCO_3)$
  (e) 0.013 M K_2HPO_4

13-I. A certain acid–base indicator exists in three colored forms:

$$H_2In \underset{}{\overset{pK_1 = 1.00}{\rightleftharpoons}} HIn^- \underset{}{\overset{pK_2 = 7.95}{\rightleftharpoons}} In^{2-}$$

$\lambda_{max} = 520$ nm $\lambda_{max} = 435$ nm $\lambda_{max} = 572$ nm

$\varepsilon_{520} = 5.00 \times 10^4$ $\varepsilon_{435} = 1.80 \times 10^4$ $\varepsilon_{572} = 4.97 \times 10^4$

  red   yellow   red

$\varepsilon_{435} = 1.67 \times 10^4$ $\varepsilon_{520} = 2.13 \times 10^3$ $\varepsilon_{520} = 2.50 \times 10^4$

$\varepsilon_{572} = 2.03 \times 10^4$ $\varepsilon_{572} = 2.00 \times 10^2$ $\varepsilon_{435} = 1.15 \times 10^4$

A solution containing 10.0 mL of 5.00×10^{-4} M indicator was mixed with 90.0 mL of 0.1 M phosphate buffer (pH 7.50). Calculate the absorbance of this solution at 435 nm in a 1.00 cm cell.[†]

[†] This problem is an application of Beer's law, which you can read about in Sections 20-1 and 20-2.

Problems

13-1. What is wrong with the following statement: At its isoelectric point, the charge on all molecules of a particular protein is zero.

13-2. (a) Derive equations for α_0, α_1, α_2, and α_3 for a triprotic system.
(b) Calculate the values of these fractions for phosphoric acid at pH 7.00.

13-3. Calculate the isoelectric and isoionic pH of 0.010 M threonine.

13-4. Draw the structure of the predominant form of pyridoxal-5-phosphate at pH 7.00.

13-5. What fraction of ethane-1,2-dithiol is in each form (H_2A, HA^-, A^{2-}) at pH 8.00? At pH 10.00?

13-6. Calculate α_0, α_1, and α_2 for *cis*-butenedioic acid at pH 1.00, 1.91, 6.00, 6.33, and 10.00.

13-7. (a) Write the acid–base reaction that occurs when pyridinium bromide is mixed with potassium 4-methylphenolate.
(b) Calculate the equilibrium constant for this reaction.

13-8. Calculate the pH that results from mixing 20.0 mL of 0.010 0 M pyridinium bromide with 23.0 mL of 0.010 0 M potassium 4-methylphenolate. (See Problem 13-7 for structures.)

13-9. Find the pH of a solution prepared by mixing 212 mL of 0.200 M acetic acid with 325 mL of 0.050 0 M sodium benzoate.

13-10. Calculate the pH of a solution prepared by mixing 50.0 mL of 0.100 M acetic acid with 50.0 mL of 0.100 M sodium benzoate.

13-11. Butanoic acid (0.100 M) was titrated with 0.100 M ethylamine. Calculate the pH when $V_b = \frac{1}{2}V_e$.

13-12. Butanoic acid (0.100 M) was titrated with 0.100 M aminobenzene. Calculate the pH when $V_b = \frac{1}{2}V_e$.

13-13. What is the pH of 0.100 M ethylammonium butanoate, $(CH_3CH_2NH_3^+)(CH_3CH_2CH_2CO_2^-)$?

13-14. What is the pH of 0.050 M pyridinium bisulfite $(C_5H_5NH^+HSO_3^-)$?

13-15. A solution was prepared using 10.0 mL of 0.100 M cacodylic acid plus 10.0 mL of 0.080 0 M NaOH. To this 1.00 mL of 1.27×10^{-6} M morphine was added. Calling morphine B, calculate the fraction of morphine present in the form BH^+.

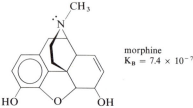

13-16. Calculate the pH of 0.100 M
(a) sodium bicarbonate   (b) ammonium acetate

13-17. What is the pH of a solution prepared by mixing 0.050 0 mol aspartic acid, 0.030 0 mol LiOH, and 0.030 0 mol H_2SO_4 in a total volume of 1.00 L?

14 / EDTA Titrations

Any chemical reaction that proceeds rapidly, has a well-defined stiochiometry, and a large equilibrium constant is of potential use for a titration. We have seen how precipitation and acid–base reactions meet these requirements. In this chapter we will examine how the formation of metal ion complexes can be used for analytical purposes.

14-1   METAL CHELATE COMPLEXES

Metal ions are **Lewis acids,** in that they can share electron pairs donated by ligands, which are therefore **Lewis bases.** Cyanide is termed a **monodentate** ligand because it binds to a metal ion through only one atom (the carbon atom). Most transition metal ions have room to bind six ligand atoms. A ligand that can bind to a metal ion through more than one ligand atom is said to be **multidentate.** It is also called a **chelating ligand** (pronounced keel′-ate-ing).

$$Ag^{\oplus} \quad + \quad 2:\overset{\ominus}{C}\equiv N: \quad \rightleftharpoons$$

Lewis acid (electron-pair acceptor) Lewis base (electron-pair donor)

$$[N\equiv C-Ag-C\equiv N]^{\ominus}$$

complex ion

One example of an important *tetradentate* ligand is adenosine triphosphate (ATP), which binds to divalent metal ions (such as Mg^{2+}, Mn^{2+}, Co^{2+}, and Ni^{2+}) through four of their six coordination positions (Figure 14-1). The fifth and sixth positions are occupied by water molecules. The biologically active form of ATP is generally the Mg^{2+} complex.

Although ATP forms some strong metal complexes, it is not analytically useful because it is very expensive and unstable. The synthetic aminocarboxylic acids shown in Figure 14-2 are some of the more common chelating agents for complexometric titrations. The nitrogen atoms and carboxylate oxygen atoms are the potential ligand atoms in these molecules. When these atoms bind to a metal ion, the ligand atoms lose their protons. Ethylenediaminetetraacetic acid (EDTA) is by far the most widely used chelating agent for titrations.

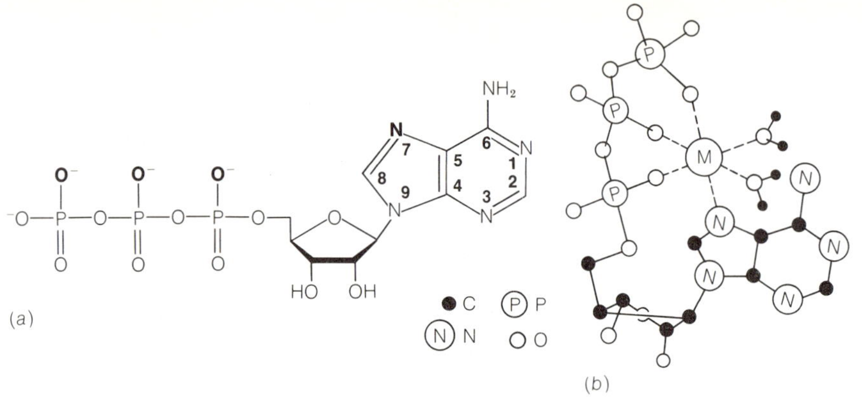

14 / EDTA TITRATIONS

Figure 14-1

(a) Structure of adenosine triphosphate (ATP), with ligand atoms shown in **bold** type. (b) Possible structure of a metal–ATP complex. There is controversy as to whether N_7 is bound directly to the metal or whether a molecule of water forms a hydrogen-bonded bridge between N_7 and the metal ion.

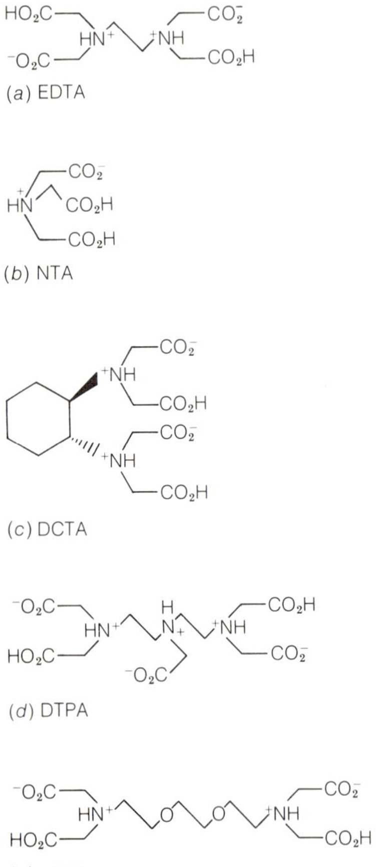

A titration based on formation of a complex ion is called a **complexometric titration.** The ligands in Figure 14-2 are especially useful because they form strong 1:1 complexes with many metal ions. The equilibrium constant for the reaction of a metal with a ligand is called the **formation constant, K_f,** also termed the **stability constant.** The reaction between *trans*-diaminohexaneteraacetic acid (DCTA) and Ni^{2+}, for example, has a formation constant of $10^{19.4}$.

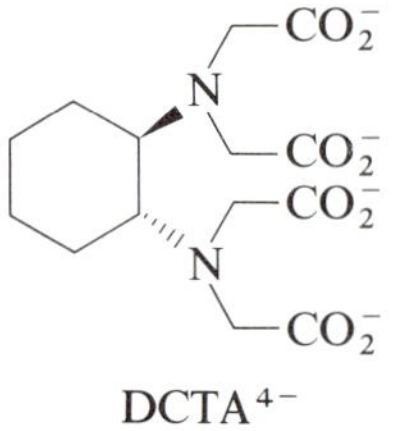

$$+ Ni^{2+} \xrightarrow{K_f = 10^{19.4}} Ni(DCTA)^{2-} \qquad (14-1)$$

The stoichiometry of the reaction between DCTA and a metal ion is 1:1, regardless of the charge of the metal. The only common ions that do not form strong complexes with the ligands in Figure 14-2 are univalent metal ions (Li^+, Na^+, K^+, etc.).

Chelate Effect

Multidentate ligands usually form stronger metal complexes than do similar monodentate ligands. For example, the reaction of Cd^{2+} with two molecules of ethylenediamine has a much larger equilibrium constant than its reaction with four molecules of methylamine.

$$Cd^{2+} + 2H_2\ddot{N}CH_2CH_2\ddot{N}H_2 \rightleftharpoons \left[\begin{array}{c} \end{array} \right]^{2+} \qquad K = 2 \times 10^{10} \qquad (14-2)$$

(a) EDTA

(b) NTA

(c) DCTA

(d) DTPA

(e) EGTA

Figure 14-2

Structures of some analytically useful chelating agents. (a) Ethylenediaminetetraacetic acid (also called ethylenedinitrilotetraacetic acid). (b) Nitrilotriacetic acid. (c) *trans*-Diaminohexanetetraacetic acid. (d) Diethylenetriaminepentaacetic acid. (e) *bis*(Aminoethyl)glycolether-N,N,N'.N'-tetraacetic acid.

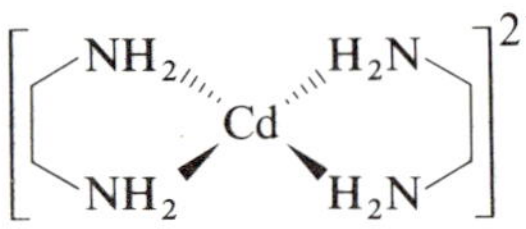

$$Cd^{2+} + 4CH_3\overset{..}{N}H_2 \rightleftharpoons \begin{bmatrix} CH_3NH_2 & & H_2NCH_3 \\ & Cd & \\ CH_3NH_2 & & H_2NCH_3 \end{bmatrix}^{2+} \qquad K = 3 \times 10^6$$

methylamine

$$(14\text{-}3)$$

This can be understood on thermodynamic grounds. The two tendencies that drive a chemical reaction are decreasing enthalpy (negative ΔH, liberation of heat) and increasing entropy (positive ΔS, more disorder). In Reactions 14-2 and 14-3, four Cd–N bonds are formed, and ΔH is about the same for both reactions.

However, Reaction 14-2 represents the coming together of *three* molecules (Cd^{2+} + 2 ethylenediamine), whereas in Reaction 14-3 *five* molecules (Cd^{2+} + 4 methylamine) are involved. There is more disorder associated with five molecules than with three molecules. If the enthalpy change (ΔH) of each reaction is about the same, the entropy change (ΔS) will favor Reaction 14-2 over 14-3.

The **chelate effect,** then, is the observation that multidentate ligands form more stable metal complexes than do similar monodentate ligands. The stability of the multidentate complex is mainly an entropy effect. The chelate effect is most pronounced for ligands such as EDTA or DCTA, which can occupy all six coordination sites about a metal ion. An important medical use of the chelate effect is discussed in Box 14-1.

> A reaction is favorable if $\Delta G > 0$.
>
> $$\Delta G = \Delta H - T\Delta S$$
>
> The reaction is favored by negative ΔH and positive ΔS.

> *Chelate effect:* A multidentate ligand forms stronger complexes than a similar monodentate ligand.

14-2 EDTA

Ethylenediaminetetraacetic acid (EDTA) is by far the most widely used chelator in analytical chemistry. It forms strong 1:1 complexes with most metal ions. By direct titration or through an indirect sequence of reactions, virtually every element of the periodic table can be analyzed with EDTA.

> *One* mole of EDTA reacts with *one* mole of metal ion.

Acid–Base Properties

EDTA is a hexaprotic system, which we will designate H_6Y^{2+}. Its acid–base properties are summarized by the following pK_A values:[†]

$$pK_1 = 0.0$$
$$pK_2 = 1.5$$
$$pK_3 = 2.0$$
$$pK_4 = 2.66$$
$$pK_5 = 6.16$$
$$pK_6 = 10.24$$

[†] pK_1 applies at 25°C, $\mu = 1.0$ M. The other values shown apply at 20°C, $\mu = 0.1$ M. A. E. Martell and R. M. Smith, *Critical Stability Constants*, Vol. 1 (New York: Plenum Press, 1974), p. 204.

Box 14-1 CHELATION THERAPY AND THALASSEMIA

Oxygen is carried in the human circulatory system by the iron-containing protein hemoglobin, which consists of two pairs of subunits, designated α and β. β-Thalassemia major is a genetic disease in which the β subunits of hemoglobin are not synthesized in adequate quantities. Children afflicted with this disease can survive only with frequent transfusions of normal red blood cells.

The problem with this treatment is that the patient accumulates 4–8 g of iron per year from the hemoglobin in the transfused cells. The body has no mechanism for excreting such large quantities of iron, so iron builds up in all tissues. Most thalassemia victims die by age 20 from the toxic effects of iron overload.

To help the body excrete excess iron, intensive chelation therapy is used. The most successful drug so far is the chelator *desferrioxamine B*, isolated from bacteria. The structure of the iron complex (ferrioxamine B) is shown below.

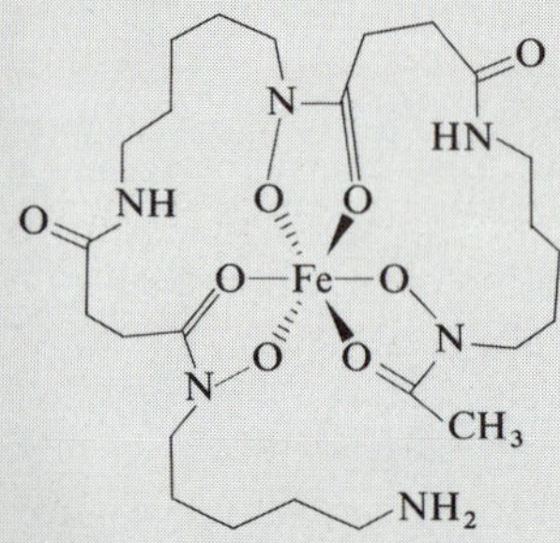

The ligand contains three hydroxamate groups

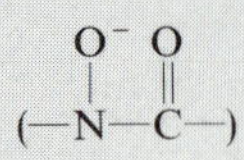

which occupy all six positions about the ferric ion. The formation constant for ferrioxamine B is $10^{30.6}$.

Used in conjunction with ascorbic acid—vitamin C, a reducing agent that reduces Fe(III) to the more soluble Fe(II)—desferrioxamine is able to remove several grams of iron per year from an overloaded patient. The ferrioxamine complex is excreted in the urine.

It is not yet known whether intensive chelation therapy can prolong the lives of thalassemia victims. The drug is expensive and must be taken by injection, because its molecules are too large to be absorbed through the intestine. Efforts are in progress to develop new iron chelators that can be taken orally.

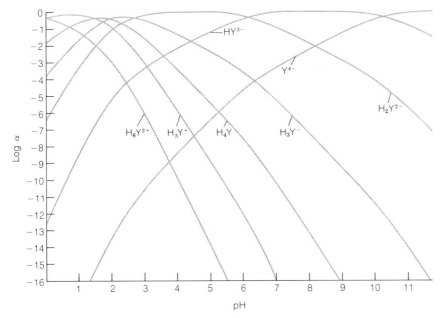

Figure 14-3
Fractional composition diagram for EDTA. Note that the ordinate is logarithmic.

The first four pK values apply to carboxyl protons, and the last two are for the ammonium protons. The neutral acid is tetraprotic, with the formula H_4Y. A commonly used reagent is the disodium salt, $Na_2H_2Y \cdot 2H_2O$.†

The logarithm of the fraction of EDTA in each of its protonated forms is shown in Figure 14-3. As in Section 13-2, we may define α for each species as the fraction of EDTA in that form. For example, $\alpha_{Y^{4-}}$ is defined as

$$\alpha_{Y^{4-}} = \frac{[Y^{4-}]}{[H_6Y^{2+}] + [H_5Y^+] + [H_4Y] + [H_3Y^-] + [H_2Y^{2-}] + [HY^{3-}] + [Y^{4-}]}$$

$$\alpha_{Y^{4-}} = \frac{[Y^{4-}]}{[\text{EDTA}]} \qquad (14\text{-}4) \qquad\qquad [Y^{4-}] = \alpha_{Y^{4-}}[\text{EDTA}]$$

where [EDTA] is the total concentration of all *free* EDTA species in the solution. By "free" we mean EDTA not complexed to metal ions. Following the derivation in Section 13-2, it can be shown that $\alpha_{Y^{4-}}$ is given by

---

† H_4Y can be dried at 140°C for two hours and used as a primary standard. It can be dissolved by adding NaOH solution from a plastic container. NaOH solution from a glass bottle should not be used because it contains alkaline earth metals leached from the glass. Reagent grade $Na_2H_2Y \cdot 2H_2O$ contains ~0.3% excess water. It may be used in this form with suitable correction for the mass of excess water, or dried to the composition $Na_2H_2Y \cdot 2H_2O$ at 80°C.

14 / EDTA TITRATIONS

$$\alpha_{Y^{4-}} = \frac{K_1 K_2 K_3 K_4 K_5 K_6}{[H^+]^6 + [H^+]^5 K_1 + [H^+]^4 K_1 K_2 + [H^+]^3 K_1 K_2 K_3 \\ + [H^+]^2 K_1 K_2 K_3 K_4 + [H^+] K_1 K_2 K_3 K_4 K_5 + K_1 K_2 K_3 K_4 K_5 K_6}$$

$$(14\text{-}5)$$

Table 14-1 gives values for $\alpha_{Y^{4-}}$ as a function of pH.

EDTA Complexes

The **formation constant, K_f,** of a metal–EDTA complex is the equilibrium constant for the reaction

> Equation 14-6 does not imply that Y^{4-} is the only species that reacts with M^{n+}. It simply says that the equilibrium constant is expressed in terms of the concentration of Y^{4-}.

$$M^{n+} + Y^{4-} \rightleftharpoons MY^{n-4} \qquad K_f = \frac{[MY^{n-4}]}{[M^{n+}][Y^{4-}]} \qquad (14\text{-}6)$$

Note that K_f is defined for reaction of the species Y^{4-} with the metal ion. This is only one of the seven different forms of free EDTA present in the solution. Table 14-2 shows that the formation constants for most EDTA

Table 14-1

Values of $\alpha_{Y^{4-}}$ for EDTA at 20°C and $\mu = 0.10$ M

pH	$\alpha_{Y^{4-}}$
0	1.3×10^{-23}
1	1.9×10^{-18}
2	3.3×10^{-14}
3	2.6×10^{-11}
4	3.8×10^{-9}
5	3.7×10^{-7}
6	2.3×10^{-5}
7	5.0×10^{-4}
8	5.6×10^{-3}
9	5.4×10^{-2}
10	0.36
11	0.85
12	0.98
13	1.00
14	1.00

Table 14-2

Formation constants for metal–EDTA complexes

Ion	$\log K_f$	Ion	$\log K_f$	Ion	$\log K_f$
Li^+	2.79	Mn^{3+}	25.3 (25°C)	Ce^{3+}	15.98
Na^+	1.66	Fe^{3+}	25.1	Pr^{3+}	16.40
K^+	0.8	Co^{3+}	41.4 (25°C)	Nd^{3+}	16.61
Be^{2+}	9.2	Zr^{4+}	29.5	Pm^{3+}	17.0
Mg^{2+}	8.79	Hf^{4+}	29.5 ($\mu = 0.2$)	Sm^{3+}	17.14
Ca^{2+}	10.69	VO^{2+}	18.8	Eu^{3+}	17.35
Sr^{2+}	8.73	VO_2^+	15.55	Gd^{3+}	17.37
Ba^{2+}	7.86	Ag^+	7.32	Tb^{3+}	17.93
Ra^{2+}	7.1	Tl^+	6.54	Dy^{3+}	18.30
Sc^{3+}	23.1	Pd^{2+}	18.5 (25°C, $\mu = 0.2$)	Ho^{3+}	18.62
Y^{3+}	18.09			Er^{3+}	18.85
La^{3+}	15.50			Tm^{3+}	19.32
V^{2+}	12.7	Zn^{2+}	16.50	Yb^{3+}	19.51
Cr^{2+}	13.6	Cd^{2+}	16.46	Lu^{3+}	19.83
Mn^{2+}	13.87	Hg^{2+}	21.7	Am^{3+}	17.8 (25°C)
Fe^{2+}	14.32	Sn^{2+}	18.3 ($\mu = 0$)	Cm^{3+}	18.1 (25°C)
Co^{2+}	16.31	Pb^{2+}	18.04	Bk^{3+}	18.5 (25°C)
Ni^{2+}	18.62	Al^{3+}	16.3	Cf^{3+}	18.7 (25°C)
Cu^{2+}	18.80	Ga^{3+}	20.3	Th^{4+}	23.2
Ti^{3+}	21.3 (25°C)	In^{3+}	25.0	U^{4+}	25.8
V^{3+}	26.0	Tl^{3+}	37.8 ($\mu = 1.0$)	Np^{4+}	24.6 (25°C, $\mu = 1.0$)
Cr^{3+}	23.4	Bi^{3+}	27.8		

Note: The stability constant is the equilibrium constant for the reaction $M^{n+} + Y^{4-} \rightleftharpoons MY^{n-4}$. Values in this table apply at 20°C, and ionic strength 0.1 M, unless otherwise noted.

SOURCE: Data from A. E. Martell and R. M. Smith, *Critical Stability Constants*, Vol. 1 (New York: Plenum Press, 1974), pp. 204–211.

complexes are quite large and tend to be larger for more positively charged metal ions.

In many complexes the EDTA ligand completely engulfs the metal ion, forming the six-coordinate species shown in Figure 14-4. In this structure, two nitrogen atoms occupy adjacent positions of the octahedrally coordinated metal ion. The other four positions are occupied by carboxyl oxygen atoms.

If you try to build a space-filling model of a six-coordinate metal–EDTA complex, you will find that there is considerable strain in the chelate rings. This strain is relieved if the oxygen ligands are drawn back toward the nitrogen atoms. Such distortion opens up a seventh coordination position, which becomes occupied by a water molecule, as shown in Figure 14-5. In some complexes, such as $Ca(EDTA)^{2-}$, the distortion is so great that two coordination positions are available and the metal ion becomes eight-coordinate.[†]

The formation constant can still be formulated as in Equation 14-6, even if there are water molecules attached to the product. This is true because the solvent (H_2O) is omitted from the reaction quotient.

○ Mn   ◐ O   ● C   ◉ N

Figure 14-4
Six-coordinate geometry of a metal–EDTA complex found in the compound $KMnEDTA \cdot 2H_2O$. [J. Stein, J. P. Fackler, Jr., G. J. McClune, J. A. Fee, and L. T. Chan, *Inorg. Chem.*, **18**, 3511 (1979).]

Conditional Formation Constant

The dissociation of a metal–EDTA complex into free metal ion and EDTA is very much analogous to the dissociation of a weak acid, HA, into H^+ and A^-.

$$MY^{n-4} \rightleftharpoons M^{n+} + Y^{4-} \qquad K = 1/K_f \qquad (14\text{-}7)$$

M^{n+} is analogous to H^+, and Y^{4-} is analogous to A^-. The equilibrium constant for the dissociation reaction is $1/K_f$.

An important difference between the dissociations of MY^{n-4} and HA is that only some of the EDTA remains in the form Y^{4-} after dissociation. At any given pH, the fraction of free EDTA in the form Y^{4-} is given by $\alpha_{Y^{4-}}$; that is,

$$[Y^{4-}] = \alpha_{Y^{4-}}[EDTA] \qquad (14\text{-}8)$$

where [EDTA] is the total concentration of *free* EDTA in all of its forms ($H_6Y^{2+}, H_5Y^+ \ldots Y^{4-}$).

The equilibrium constant for the dissociation reaction (14-7) can be written as follows:

$$K = \frac{1}{K_f} = \frac{[M^{n+}][Y^{4-}]}{[MY^{n-4}]} = \frac{[M^{n+}]\alpha_{Y^{4-}}[EDTA]}{[MY^{n-4}]} \qquad (14\text{-}9)$$

If the pH is fixed by a buffer, then $\alpha_{Y^{4-}}$ is a constant that can be combined

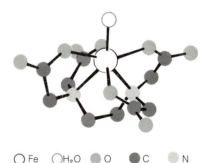

○ Fe   ○ H_2O ◐ O   ● C   ◉ N

Figure 14-5
Seven-coordinate geometry of $Fe(EDTA)(H_2O)^-$. Other metal ions that form seven-coordinate EDTA complexes include Ru^{3+}, Cr^{3+}, Co^{3+}, Mg^{2+}, and Mn^{2+}. [M. D. Lind, J. L. Hoard, M. J. Hamor, and T. A. Hamor, *Inorg. Chem.*, **3**, 34 (1964).]

Only some of the free EDTA is in the form Y^{4-}.

[†] R. L. Barnett and V. A. Uchtman, *Inorg. Chem.*, **18**, 2674 (1979).

14 / EDTA TITRATIONS

with K_f:

$$K' = \frac{1}{\alpha_{Y^{4-}} K_f} = \frac{[M^{n+}][EDTA]}{[MY^{n-4}]} \tag{14-10}$$

The number $\alpha_{Y^{4-}} K_f$ is called the **conditional formation constant** or *effective formation constant*. It describes the formation of MY^{n-4} at any particular pH.

The conditional formation constant is useful because it allows us to look at EDTA complex dissociation as if the dissociated EDTA were all in one form:

With the conditional formation constant, we can treat EDTA complex dissociation as if all the dissociated EDTA were in one form.

$$MY^{n-4} \rightleftharpoons M^{n+} + EDTA \qquad K' = \frac{1}{\alpha_{Y^{4-}} K_f} \tag{14-11}$$

At any given pH, we can find $\alpha_{Y^{4-}}$ and evaluate K'.

EXAMPLE

The formation constant in Table 14-2 for $Fe(EDTA)^-$ is $10^{25.1} = 1.3 \times 10^{25}$. Calculate the concentrations of free Fe^{3+} in solutions of 0.10 M $Fe(EDTA)^-$ at pH 8.00 and pH 2.00.

The dissociation reaction is

$$Fe(EDTA)^- \rightleftharpoons Fe^{3+} + EDTA \qquad K' = \frac{1}{\alpha_{Y^{4-}} K_f}$$

where EDTA on the right side of the equation refers to all forms of unbound EDTA. Using values of $\alpha_{Y^{4-}}$ from Table 14-1, we find

$$\text{At pH 8.00:} \quad K' = \frac{1}{(5.6 \times 10^{-3})(1.3 \times 10^{25})} = 1.4 \times 10^{-23}$$

$$\text{At pH 2.00:} \quad K' = \frac{1}{(3.3 \times 10^{-14})(1.3 \times 10^{25})} = 2.3 \times 10^{-12}$$

Since dissociation of $Fe(EDTA)^-$ must produce *equal quantities* of Fe^{3+} and EDTA, we can write

$$Fe(EDTA)^- \rightleftharpoons Fe^{3+} + EDTA$$

Initial concentration (M):	0.10	0	0
Final concentration (M):	0.10 − x	x	x

$$\frac{x^2}{0.10 - x} = K' = 1.4 \times 10^{-23} \text{ at pH 8.00}$$

$$= 2.3 \times 10^{-12} \text{ at pH 2.00}$$

Solving for $x (= [Fe^{3+}])$, we find $[Fe^{3+}] = 1.2 \times 10^{-12}$ M at pH 8.00 and 4.8×10^{-7} M at pH 2.00. *Using the conditional formation constant, we treat the dissociated EDTA as if it were a single species.*

You can see from the example that a metal–EDTA complex becomes less stable at lower pH. For a titration reaction to be effective, it must "go to completion." That is, its equilibrium constant must be large. Figure 14-6 shows how pH affects the titration of Ca^{2+} with EDTA. Below pH ≈ 8, the break at the endpoint is not sharp enough to allow accurate determination. This is because the conditional formation constant for CaY^{2-} is too small below pH 8.

Figure 14-7 gives the minimum pH needed for titration of many metal ions. The figure provides a strategy for the selective titration of one ion in the presence of another. For example, a solution containing both Fe^{3+} and Ca^{2+} could be titrated with EDTA at pH 4. At this pH, Fe^{3+} is titrated without interference from the Ca^{2+} ion.

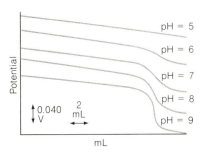

Figure 14-6
Titration of Ca^{2+} with EDTA as a function of pH. As the pH is lowered, the endpoint becomes less distinct. The potential was measured with mercury and calomel electrodes as described in Problem 16-B. [C. N. Reilley and R. W. Schmid, *Anal. Chem.*, **30**, 947 (1958).]

Control of pH can be used to select which metals will be titrated by EDTA and which will not.

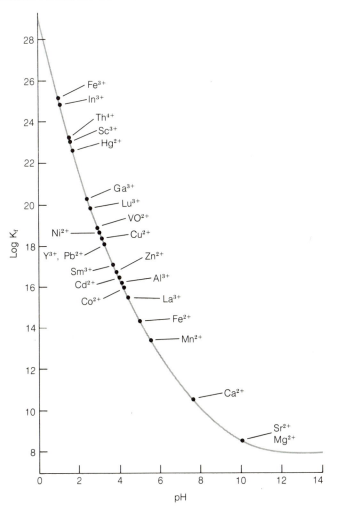

Figure 14-7
Minimum pH for effective titration of various metal ions by EDTA. The minimum pH was specified arbitrarily as the pH at which the conditional formation constant for each metal–EDTA complex is 10^6. [C. N. Reilley and R. W. Schmid, *Anal. Chem.*, **30**, 947 (1958).]

14-3 EDTA TITRATION CURVES

In this section we will calculate the concentration of free metal ion during the course of the titration of metal with EDTA. This is analogous to the titration of a strong acid by a weak base. The metal ion plays the role of H^+, and EDTA is the base. The titration reaction is

$$M^{n+} + EDTA \rightarrow MY^{n-4} \qquad K = \alpha_{Y^{4-}} K_f \qquad (14\text{-}12)$$

If $\alpha_{Y^{4-}} K_f$ is large, we can consider the reaction to be complete at each point in the titration.

The titration curve has three natural regions:

Region 1: Before the equivalence point

In this region there is excess M^{n+} left in solution after the EDTA has been consumed. The concentration of free metal ion is equal to the concentration of excess, unreacted M^{n+}. The dissociation of MY^{n-4} is negligible.

Region 2: At the equivalence point

There is exactly as much EDTA as metal in the solution. We can treat the solution as if it had been made by dissolving pure MY^{n-4}. Some free M^{n+} is generated by the slight dissociation of MY^{n-4}:

$$MY^{n-4} \rightleftharpoons M^{n+} + EDTA \qquad K' = \frac{1}{\alpha_{Y^{4-}} K_f} \qquad (14\text{-}13)$$

In Reaction 14-13, EDTA refers to the total concentration of free EDTA in all of its forms. At the equivalence point, $[M^{n+}] = [EDTA]$. The formal concentration of MY^{n-4} is equal to the initial concentration of metal ion, with a correction for dilution by titrant. In order to calculate K', we need to know $\alpha_{Y^{4-}}$, which depends on pH. It is therefore necessary to know the pH in order to calculate the shape of the EDTA titration curve.

Region 3: After the equivalence point

Now there is excess EDTA, and virtually all of the metal ion is in the form MY^{n-4}. Reaction 14-13 still governs the concentration of M^{n+}. However, the concentration of free EDTA can be equated to the concentration of excess EDTA added after the equivalence point.

Titration Calculations

Let's calculate the shape of the titration curve for the reaction of 50.0 mL of 0.050 0 M Mg^{2+} with 0.050 0 M EDTA. Suppose that the Mg^{2+} solution is buffered to pH 10.00. The titration reaction is

$$Mg^{2+} + EDTA \rightarrow MgY^{2-} \qquad K = \alpha_{Y^{4-}} K_f = (0.36)(6.2 \times 10^8) = 2.2 \times 10^8$$
$$(14\text{-}14)$$

14-3 EDTA TITRATION CURVES

The equivalence point will be 50.0 mL. Since K is large, it is reasonable to say that the reaction goes to completion with each addition of titrant. What we seek is to make a graph of $pMg^{2+} (= -\log[Mg^{2+}])$ versus mL of EDTA added.

Region 1: Before the equivalence point

Consider the case in which 5.0 mL of EDTA has been added. Since the equivalence point is 50.0 mL, one-tenth of the Mg^{2+} will be consumed and nine-tenths should remain.

> Before the equivalence point, there is excess unreacted M^{n+}.

$$[Mg^{2+}] = \underbrace{\left(\frac{50.0 - 5.0}{50.0}\right)}_{\substack{\text{Fraction} \\ \text{remaining} \\ (=9/10)}} \underbrace{(0.050\,0)}_{\substack{\text{Original} \\ \text{concentration} \\ \text{of } Mg^{2+}}} \underbrace{\left(\frac{50.0}{55.0}\right)}_{\substack{\text{Dilution} \\ \text{factor}}} \quad (14\text{-}15)$$

where 50.0 is the Initial volume of Mg^{2+} and 55.0 is the Total volume of solution.

$$[Mg^{2+}] = 0.040\,9\ \text{M} \Rightarrow pMg^{2+} = -\log[Mg^{2+}] = 1.39$$

In a similar manner, we could calculate pMg^{2+} for any volume of EDTA less than 50.0 mL.

Region 2: At the equivalence point

Virtually all of the metal is in the form MgY^{2-}. Assuming negligible dissociation, the concentration of MgY^{2-} is equal to the original concentration of Mg^{2+}, with a correction for dilution.

> At the equivalence point, the major species is MY^{n-4}. There are small and equal amounts of M^{n+} and EDTA in equilibrium with the complex.

$$[MgY^{2-}] = \underbrace{(0.050\,0\ \text{M})}_{\substack{\text{Original} \\ \text{concentration} \\ \text{of } Mg^{2+}}} \underbrace{\left(\frac{50.0}{100.0}\right)}_{\substack{\text{Dilution} \\ \text{factor}}} = 0.025\,0\ \text{M} \quad (14\text{-}16)$$

where 50.0 is the Initial volume of Mg^{2+} and 100.0 is the Total volume of solution.

The concentration of free Mg^{2+} is small and unknown. We can write

$$MgY^{4-} \rightleftharpoons Mg^{2+} + EDTA$$

Initial concentration (M):	0.025 0	—	—
Final concentration (M):	0.025 0 − x	x	x

$$\frac{[Mg^{2+}][EDTA]}{[MgY^{2-}]} = \frac{1}{\alpha_{Y^{4-}} K_f} = 4.5 \times 10^{-9} \quad (14\text{-}17)$$

> In Equation 14-17, [EDTA] refers to the total concentration of all forms of EDTA not bound to metal.

$$\frac{x^2}{0.025\,0 - x} = 4.5 \times 10^{-9} \Rightarrow x = 1.06 \times 10^{-5} \text{ M} \qquad (14\text{-}18)$$

$$pMg^{2+} = -\log x = 4.97$$

Region 3: After the equivalence point

After the equivalence point, virtually all of the metal is present as MY^{n-4}. There is a known excess of EDTA present. A small amount of free M^{n+} exists in equilibrium with the MY^{n-4} and EDTA.

In this region virtually all of the metal is in the form MgY^{2-}, and there is excess, unreacted EDTA. The concentrations of MgY^{2-} and excess EDTA are easily calculated. For example, at 51.0 mL there is 1.0 mL of excess EDTA:

$$[\text{EDTA}] = \underbrace{(0.050\,0)}_{\substack{\text{Original}\\\text{concentration}\\\text{of EDTA}}} \underbrace{\left(\frac{1.0}{101.0}\right)}_{\substack{\text{Dilution}\\\text{factor}}} = 4.95 \times 10^{-4} \text{ M} \qquad (14\text{-}19)$$

(Volume of excess EDTA; Total volume of solution)

$$[MgY^{2-}] = \underbrace{(0.050\,0)}_{\substack{\text{Original}\\\text{concentration}\\\text{of }Mg^{2+}}} \underbrace{\left(\frac{50.0}{101.0}\right)}_{\substack{\text{Dilution}\\\text{factor}}} = 2.48 \times 10^{-2} \text{ M} \qquad (14\text{-}20)$$

(Original volume of Mg^{2+}; Total volume of solution)

The concentration of Mg^{2+} is still governed by Equation 14-17:

$$\frac{[Mg^{2+}][\text{EDTA}]}{[MgY^{2-}]} = \frac{1}{\alpha_{Y^{4-}} K_f} \qquad (14\text{-}17)$$

$$\frac{[Mg^{2+}](4.95 \times 10^{-4})}{(2.48 \times 10^{-2})} = 4.5 \times 10^{-9} \qquad (14\text{-}21)$$

$$[Mg^{2+}] = 2.2 \times 10^{-7} \text{ M} \Rightarrow pMg^{2+} = 6.65$$

The same sort of calculation can be used for any volume past the equivalence point.

Titration curve

The calculated titration curve is shown in Figure 14-8. As in previous titrations, a distinct break occurs at the equivalence point, where the slope is greatest. For comparison, the titration curve for 0.050 0 M Zn^{2+} is also shown in Figure 14-8. The change in pZn^{2+} at the equivalence point is even greater than the change in pMg^{2+}, because the formation constant for ZnY^{2-} is greater than that of MgY^{2-}.

The completeness of reaction (and hence the sharpness of the equivalence point) is determined by the conditional formation constant, $\alpha_{Y^{4-}} K_f$, which is pH-dependent. Since $\alpha_{Y^{4-}}$ decreases drastically as the pH is lowered, the

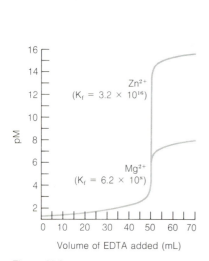

Figure 14-8
Theoretical titration curves for the reaction of 50.0 mL of 0.050 0 M metal ion with 0.050 0 M EDTA at pH 10.00.

pH is an important variable determining whether a titration is feasible. The endpoint is more distinct at high pH. However, the pH of a titration must not be so high that the metal hydroxide precipitates. The effect of pH on the titration of Ca^{2+} was shown in Figure 14-6.

The lower the pH, the less distinct is the endpoint.

Effect of Auxiliary Complexing Agents

The titration curve for Zn^{2+} in Figure 14-8 is not realistic because the pH is high enough to precipitate $Zn(OH)_2$ ($K_{sp} = 3.0 \times 10^{-16}$) before EDTA is added. The titration of Zn^{2+} is usually carried out in an ammonia buffer, which serves to complex the metal ion as well as to fix the pH.

$$Zn^{2+} + 4NH_3 \rightleftharpoons Zn(NH_3)_4^{2+} \quad (14\text{-}22)$$

$$K = \beta_4 = \frac{[Zn(NH_3)_4^{2+}]}{[Zn^{2+}][NH_3]^4} = 5.0 \times 10^8$$

The notation for the equilibrium constant (β) is described in Box 5-3.

The ammonia acts as an **auxiliary complexing agent**. It is a good enough ligand to prevent $Zn(OH)_2$ precipitation, but it is not so strong as to prevent ZnY^{2-} complex formation.

The titration of Zn^{2+} by EDTA with an ammonia buffer is best represented by Reaction 14-23, whose equilibrium constant is $K_f/\beta_4 = 6.4 \times 10^7$.

$$Zn(NH_3)_4^{2+} \rightleftharpoons Zn^{2+} + 4NH_3 \qquad K = 1/\beta_4$$

$$Zn^{2+} + Y^{4-} \rightleftharpoons ZnY^{2-} \qquad K = K_f$$

$$\overline{Zn(NH_3)_4^{2+} + Y^{4-} \rightleftharpoons ZnY^{2-} + 4NH_3 \qquad K = K_f/\beta_4} \quad (14\text{-}23)$$

Since the equilibrium constant of Reaction 14-23 is large, it is possible to titrate the ammonia complex of zinc with EDTA.

Figure 14-9 shows two curves for the titration of zinc with EDTA at pH 10.00. The buffer solutions have concentrations of NH_3 equal to 0.10 and 0.02 M, respectively. The magnitude of the change in pZn^{2+} near the equivalence point is not as great in the presence of NH_3 as in its absence (compare Figures 14-8 and 14-9). The greater the concentration of NH_3, the smaller the change of pZn^{2+} near the equivalence point. The reason for this behavior is that NH_3 is competing with EDTA for the metal ion. When using an auxiliary ligand, its amount must be kept below the level that would obliterate the endpoint of the titration.

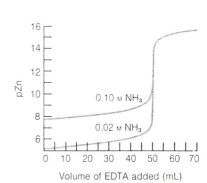

Figure 14-9
Titration curves for the reaction of 50.0 mL of 1.00×10^{-3} M Zn^{2+} with 1.00×10^{-3} M EDTA at pH 10.00 in the presence of two different concentrations of NH_3.

14-4 METAL ION INDICATORS

Several methods can be used to detect the endpoint in EDTA titrations. The most common technique is to use a metal ion indicator. Alternatively, a mercury electrode (described in Problem 16-B) can be used to produce titration curves such as those in Figure 14-6. A glass pH electrode can also

Endpoint detection methods:
1. Metal ion indicators
2. Mercury electrode
3. Glass (pH) electrode
4. Ion-selective electrode

14 / EDTA TITRATIONS

be used to follow the course of the titration, since H_2Y^{2-} releases $2H^+$ when it forms HY^{2-}. An ion-selective electrode (Section 16-5) for the metal being titrated is also a good endpoint detector.

A **metal ion indicator** is a compound whose color changes when it binds to a metal ion. Several common indicators are shown in Table 14-3. *For an indicator to be useful, it must bind metal less strongly than EDTA does.*

A typical analysis is illustrated by the titration of Mg^{2+} with EDTA using eriochrome black T as the indicator. We can write the reaction as follows:

The indicator must release its metal to EDTA.

$$MgIn + EDTA \rightarrow MgEDTA + In \qquad (14\text{-}24)$$

$$\text{(red)} \quad \text{(colorless)} \quad \text{(colorless)} \quad \text{(blue)}$$

A small amount of indicator (In) is added to the Mg^{2+} to form a red complex. As EDTA is added, it reacts first with free, colorless Mg^{2+} and then with the small amount of red MgIn complex. (The EDTA must therefore bind to Mg^{2+} better than the indicator binds to Mg^{2+}.) The change from the red of MgIn to the blue of unbound In signals the endpoint of the titration. Demonstration 14-1 illustrates this titration.

Demonstration 14-1 METAL ION INDICATOR COLOR CHANGES

We will demonstrate the color change associated with Reaction 14-24 and also show how a second dye can be added to a solution to produce a more easily detected color change.
Prepare the following solutions:

Eriochrome Black T: Dissolve 0.1 g of the solid in 7.5 mL of triethanolamine plus 2.5 mL of absolute ethanol.

Methyl Red: Dissolve 0.02 g in 60 mL of ethanol. Then add 40 mL of water.

Buffer: Add 142 mL of concentrated (14.5 M) aqueous ammonia to 17.5 g of ammonium chloride and dilute to 250 mL with water.

$MgCl_2$: 0.05 M

EDTA: 0.05 M $Na_2EDTA \cdot 2H_2O$

Prepare a solution containing 25 mL of $MgCl_2$, 5 mL of buffer and 300 mL of water. Add 6 drops of eriochrome black T indicator and titrate with EDTA. Note the color change from wine-red to pale blue at the endpoint (Color Plate 17).

For some observers the change of indicator color is not as sharp as desired. The colors can be affected by adding an "inert" dye whose color alters the appearance of the solution before and after the titration. Adding 3 mL of methyl red (or many other yellow dyes) produces an orange color prior to the endpoint and a green color after it. This sequence of colors is shown in Color Plate 18.

Table 14-3
Some common metal ion indicators

Name	Structure	pK_A	Color of Free Indicator	Color of Metal Ion Complex
Eriochrome black T	(H_2In^-)	$pK_2 = 6.3$ $pK_3 = 11.6$	H_2In^-—red HIn^{2-}—blue In^{3-}—orange	Wine-red
Calmagite	(H_2In^-)	$pK_2 = 8.1$ $pK_3 = 12.4$	H_2In^-—red HIn^{2-}—blue In^{3-}—orange	Wine-red
Murexide	(H_4In^-)	$pK_2 = 9.2$ $pK_3 = 10.9$	H_4In^-—red-violet H_3In^{2-}—violet H_2In^{3-}—blue	Yellow (with Co^{2+}, Ni^{2+}, Cu^{2+}) Red with Ca^{2+}
Xylenol orange	(H_3In^{3-})	$pK_2 = 2.32$ $pK_3 = 2.85$ $pK_4 = 6.70$ $pK_5 = 10.47$ $pK_6 = 12.23$	H_5In^-—yellow H_4In^{2-}—yellow H_3In^{3-}—yellow H_2In^{4-}—violet HIn^{5-}—violet In^{6-}—violet	Red
Pyridylazonaphthol (PAN)	(HIn)	$pK_A = 12.3$	HIn—orange-red In^-—pink	Red
Pyrocatechol violet	(H_3In^-)	$pK_1 = 0.2$ $pK_2 = 7.8$ $pK_3 = 9.8$ $pK_4 = 11.7$	H_4In—red H_3In^-—yellow H_2In^{2-}—violet HIn^{3-}—red-purple	Blue

286

14 / EDTA TITRATIONS

Most metal ion indicators are also acid–base indicators. Some pertinent pK_A values are listed in Table 14-3. Because the color of free indicator is pH-dependent, most indicators can be used only in certain pH ranges. For example, xylenol orange (pronounced zy′-leen-ol) changes from yellow to red when it binds to a metal ion at pH 5.5. This is an easy color change to observe. At pH 7.5 the change is from violet to red and rather difficult to see. A spectrophotometer can be used to measure an indicator color change, but it is more convenient if we can see it.

Some metal ion indicators are unstable. Solutions of azo indicators (compounds with —N≡N—bonds) deteriorate rapidly and should probably be prepared each week. Murexide solution should be prepared fresh each day.

For an indicator to be useful in the titration of a metal with EDTA, the indicator must give up its metal ion to the EDTA. If a metal does not freely dissociate from an indicator, the metal is said to *block* the indicator. Eriochrome black T is blocked by Cu^{2+}, Ni^{2+}, Co^{2+}, Cr^{3+}, Fe^{3+}, and Al^{3+}. It cannot be used as an indicator for the direct titration of any of these metals. Eriochrome black T can be used for a back titration, however. For example, excess standard EDTA can be added to Cu^{2+}. Then indicator is added and the excess EDTA is back-titrated with Mg^{2+}.

> *Question:* What will be the color change when the back titration is performed?

14-5 EDTA TITRATION TECHNIQUES

Because so many elements can be analyzed by titration with EDTA, there is an extensive literature dealing with many variations of the basic procedure.[†] In this section we discuss several important techniques.

Direct Titration

In a **direct titration,** analyte is titrated with standard EDTA. The analyte is buffered to an appropriate pH at which the conditional formation constant for the metal–EDTA complex is large enough to produce a sharp endpoint. Since most metal ion indicators are also acid–base indicators, they have different colors at different values of pH. An appropriate pH must be one at which the free indicator has a distinctly different color from the metal–indicator complex.

In many titrations an *auxiliary complexing agent,* such as ammonia, tartarate, citrate, or triethanolamine, is employed to prevent the metal ion from precipitating in the absence of EDTA. For example, the direct titration of Pb^{2+} is carried out in ammonia buffer at pH 10 in the presence of tartrate,

> The larger the effective formation constant, the more abrupt is the change in metal ion concentration at the endpoint.

[†] Some good sources for reading about EDTA titration techniques are G. Schwarzenbach and H. Flaschka, *Complexometric Titrations* (H. M. N. H. Irving, trans.) (London: Methuen, 1969); H. A. Flaschka, *EDTA Titrations* (New York: Pergamon Press, 1959); and C. N. Reilley, A. J. Barnard, Jr., and R. Puschel in L. Meites, ed., *Handbook of Analytical Chemistry* (New York: McGraw-Hill, 1963), pp. 3-76 to 3-234.

which complexes the metal ion and does not allow $Pb(OH)_2$ to precipitate. The lead–tartrate complex must be less stable than the lead–EDTA complex, or the titration would not be feasible.

287

14-5 EDTA TITRATION TECHNIQUES

Back Titration

In a **back titration** a known excess of EDTA is added to the analyte. The excess EDTA is then titrated with a standard solution of a second metal ion. A back titration is necessary if the analyte precipitates in the absence of EDTA, if it reacts too slowly with EDTA under titration conditions, or if it blocks the indicator. The metal ion used in the back titration should not displace the analyte metal ion from its EDTA complex.

EXAMPLE

Ni^{2+} can be analyzed by a back titration using standard Zn^{2+} at pH 5.5 with xylenol orange indicator. A solution containing 25.00 mL of Ni^{2+} in dilute HCl was treated with 25.00 mL of 0.052 83 M Na_2EDTA. The solution was neutralized with NaOH, and the pH was adjusted to 5.5 with acetate buffer. The solution turned yellow when a few drops of indicator were added. It was then titrated with 17.61 mL of 0.022 99 M Zn^{2+} to reach the red endpoint. What was the molarity of Ni^{2+} in the unknown?

The unknown was treated with 25.00 mL of 0.052 83 M EDTA, which contains

$$(25.00 \text{ mL})(0.052 83 \text{ M}) = 1.320 8 \text{ mmol of EDTA}$$

Back titration required 17.61 mL of 0.022 99 M Zn^{2+}:

$$(17.61 \text{ mL})(0.022 99 \text{ M}) = 0.404 9 \text{ mmol of } Zn^{2+}$$

Since one mole of EDTA reacts with one mole of any metal ion, there must have been $(1.320 8 \text{ mmol EDTA} - 0.404 9 \text{ mmol } Zn^{2+}) = 0.915 9 \text{ mmol } Ni^{2+}$. The concentration of Ni^{2+} was 0.915 9 mmol/25.00 mL = 0.036 64 M.

Preventing precipitation

Al^{3+} precipitates as $Al(OH)_3$ at pH 7 in the absence of EDTA. An acidic solution of Al^{3+} can be treated with excess EDTA, adjusted to pH 7–8 with sodium acetate, and boiled to ensure complete complexation of the ion. The Al^{3+}–EDTA complex is stable in solution at this pH. The solution is then cooled; eriochrome black T indicator is added; and back titration with standard Zn^{2+} is performed.

Preventing blocking of the indicator

An indicator is said to be *blocked* when it forms a metal ion complex whose stability constant is greater than that of the metal–EDTA complex. In this case no color change can be observed at the end of the titration. Blocking also

A metal ion cannot be titrated with EDTA if the metal ion binds to the indicator too strongly.

14 / EDTA TITRATIONS

occurs when the metal–indicator complex dissociates so slowly that the titration with EDTA cannot be carried out in a reasonable time. For example, Ni^{2+} forms a slowly dissociating complex with the indicator PAN (Table 14-3). It is therefore not feasible to add indicator to the free metal ion. Excess EDTA can be added to the Ni^{2+} and a back titration with Cu^{2+} performed.

Displacement Titrations

For metal ions that do not have a satisfactory indicator, a **displacement titration** may be feasible. In this procedure the analyte usually is treated with excess $Mg(EDTA)^{2-}$ to displace Mg^{2+}, which is later titrated with standard EDTA.

$$M^{n+} + MgY^{2-} \rightarrow MY^{n-4} + Mg^{2+} \qquad (14\text{-}25)$$

> *Challenge:* Calculate the equilibrium constant for Reaction 14-25 if M = Hg. Why is $Mg(EDTA)^{2-}$ used for a displacement titration?

Hg^{2+} is determined in this manner. The formation constant of $Hg(EDTA)^{2-}$ must be greater than the formation constant for $Mg(EDTA)^{2-}$, or else Reaction 14-25 will not work.

There is no suitable indicator for Ag^+. However, Ag^+ will displace Ni^{2+} from the tetracyanonickelate ion:

$$2Ag^+ + Ni(CN)_4^{2-} \rightarrow 2Ag(CN)_2^- + Ni^{2+} \qquad (14\text{-}26)$$

The liberated Ni^{2+} can then be titrated with EDTA to find out how much Ag^+ was added.

Indirect Titrations

Anions that form precipitates with certain metal ions may be analyzed with EDTA by **indirect titration.** For example, sulfate can be analyzed by precipitation with excess Ba^{2+} at pH 1. The $BaSO_4$ precipitate is filtered and washed. Boiling the precipitate with excess EDTA at pH 10 brings the Ba^{2+} back into solution as $Ba(EDTA)^{2-}$. The excess EDTA is back-titrated with Mg^{2+}.

Alternatively, an anion may be precipitated with excess metal ion. The precipitate is filtered and washed, and the excess metal ion in the filtrate is titrated with EDTA. Anions such as CO_3^{2-}, CrO_4^{2-}, S^{2-}, and SO_4^{2-} can be determined by indirect titration with EDTA.

Masking

> Masking is used to prevent one element from interfering in the analysis of another element. Masking is not restricted to EDTA titrations.

A **masking agent** is a reagent that protects some component of the analyte from reaction with EDTA. For example, Al^{3+} reacts with F^- to form the very stable complex AlF_6^{3-}. The Mg^{2+} in a mixture of Mg^{2+} and Al^{3+} can be titrated by first masking the Al^{3+} with F^-, leaving only the Mg^{2+} to react with EDTA.

Cyanide is a common masking agent that forms complexes with Cd^{2+}, Zn^{2+}, Hg^{2+}, Co^{2+}, Cu^+, Ag^+, Ni^{2+}, Pd^{2+}, Pt^{2+}, Fe^{2+}, and Fe^{3+}, but not with Mg^{2+}, Ca^{2+}, Mn^{2+}, or Pb^{2+}. If cyanide is first added to a solution containing Cd^{2+} and Pb^{2+}, only the Pb^{2+} is then able to react with EDTA. Fluoride can mask Al^{3+}, Fe^{3+}, Ti^{4+}, and Be^{2+}. Triethanolamine masks Al^{3+}, Fe^{3+}, and Mn^{2+}; and 2,3-dimercaptopropanol masks Bi^{3+}, Cd^{2+}, Cu^{2+}, Hg^{2+}, and Pb^{2+}.

In the routine determination of water "hardness" ($= Ca^{2+} + Mg^{2+}$ concentration), the sample is first treated with ascorbic acid and cyanide to reduce Fe^{3+} to Fe^{2+} and to mask Fe^{2+}, Cu^{+}, and several other minor metal ions in the water. Titration with EDTA at pH 10 in NH_3 buffer then gives the total $Ca^{2+} + Mg^{2+}$ in the water. The Ca^{2+} may be determined separately if the titration is carried out at pH 13 without NH_3. At this pH, $Mg(OH)_2$ precipitates and is inaccessible to the EDTA. Hydroxide thus serves to mask Mg^{2+} in this titration.

Demasking refers to the release of a metal ion from a masking agent. Cyanide complexes can be demasked by treatment with formaldehyde:

$$M(CN)_m^{n-m} + mH_2CO + mH^+ \longrightarrow mH_2C{\overset{OH}{\underset{CN}{\big\langle}}} + M^{n+} \qquad (14\text{-}27)$$

Thiourea reduces Cu^{2+} to Cu^{+} and masks Cu^{+}. Cu^{2+} can be liberated from the Cu(I)–thiourea complex by demasking with H_2O_2. The selectivity afforded by masking, demasking, and pH control allows individual components of complex mixtures of metal ions to be analyzed by EDTA titration.

N(CH₂CH₂OH)₃
triethanolamine

HOCH₂CHCH₂SH (with SH above)
2,3-dimercaptopropanol

thiourea
H_2N — C(=S) — NH_2

Summary

In a complexometric titration the reaction between analyte and titrant is one of complex ion formation, the equilibrium constant for which is called the formation constant (K_f). Chelating ligands (termed multidentate because they bind to a metal through more than one ligand atom) form more stable complexes than monodentate ligands. The reason for this chelate effect is that the entropy of complex formation favors the binding of one large ligand, rather than many small ligands. Synthetic multidentate aminocarboxylic acids, such as EDTA, have large metal-binding constants and are widely used in analytical chemistry.

Although EDTA is a hexaprotic system, the formation constant for complex formation is defined in terms of the form Y^{4-}. Since the fraction ($\alpha_{Y^{4-}}$) of free EDTA in the form Y^{4-} depends on pH, we define an effective (or conditional) formation constant as $\alpha_{Y^{4-}} \cdot K_f = [MY^{n-4}]/[M^{n+}][EDTA]$. This constant describes the hypothetical reaction $M^{n+} + EDTA \rightleftharpoons MY^{n-4}$, where EDTA refers to all forms of EDTA not bound to metal ion. Titration calculations fall into three categories. When excess unreacted M^{n+} is present, pM is calculated directly from

$pM = -\log[M^{n+}]$. When excess EDTA is present, we know both $[MY^{4-n}]$ and $[EDTA]$, so $[M^{n+}]$ can be calculated from the conditional formation constant. At the equivalence point, the condition $[M^{n+}] = [EDTA]$ allows us to solve for $[M^{n+}]$. EDTA titration curves become sharper as the formation constant increases and as the pH is raised. Auxiliary complexing agents, which compete with EDTA for the metal ion and thereby limit the sharpness of the titration curve, are sometimes necessary to keep the metal in solution.

For endpoint detection we commonly use metal ion indicators, a glass electrode, an ion-selective electrode, or a mercury electrode. When a direct titration is not suitable, because the analyte is unstable, reacts slowly with EDTA, or has no suitable indicator, a back titration of excess EDTA or a displacement titration of $MgEDTA^{2-}$ may be feasible. Masking is commonly used to prevent interference by certain species in complex solutions. Indirect EDTA titration procedures are available for the analysis of many anions or other species that do not react directly with the reagent.

Terms to Understand

auxiliary complexing agent	formation constant
back titration	indirect titration
blocking	Lewis acid
chelate effect	Lewis base
chelating ligand	masking agent
complexometric titration	metal ion indicator
conditional formation constant	monodentate
demasking	multidentate
direct titration	stability constant
displacement titration	

Exercises

14-A. The potassium ion in a $250.0(\pm0.1)$ mL water sample was precipitated with sodium tetraphenylborate:

$$K^+ + (C_6H_5)_4B^- \rightarrow KB(C_6H_5)_4(s)$$

The precipitate was filtered, washed, and dissolved in an organic solvent. Treatment of the organic solution with an excess of Hg(II)–EDTA then gave the following reaction:

$$4HgY^{2-} + (C_6H_5)_4B^- + 4H_2O \rightarrow$$
$$H_3BO_3 + 4C_6H_5Hg^+ + 4HY^{3-} + OH^-$$

The liberated EDTA was titrated with 28.73 (±0.03) mL of 0.043 7 $(\pm0.000\ 1)$ M Zn^{2+}. Find the concentration (and absolute uncertainty) of K^+ in the original sample.

14-B. A 25.00 mL sample of unknown containing Fe^{3+} and Cu^{2+} required 16.06 mL of 0.050 83 M EDTA for complete titration. A 50.00 mL sample of the unknown was treated with NH_4F to protect the Fe^{3+}. Then the Cu^{2+} was reduced and masked by addition of thiourea. Upon addition of 25.00 mL of 0.050 83 M EDTA, the Fe^{3+} was liberated from its fluoride complex and formed an EDTA complex. The excess EDTA required 19.77 mL of 0.018 83 M Pb^{2+} to reach an endpoint using xylenol orange. Find the concentration of Cu^{2+} in the unknown.

14-C. Calculate pGa^{3+} (to the 0.01 decimal place) at each of the following points in the titration of 50.0 mL of 0.040 0 M EDTA with 0.080 0 M $Ga(NO_3)_3$ at pH 4.00:

(a) 0.1	(d) 15.0	(f) 24.0	(h) 26.0
(b) 5.0	(e) 20.0	(g) 25.0	(i) 30.0 mL
(c) 10.0			

Make a graph of pGa^{3+} versus volume of titrant.

14-D. Calculate the concentration of H_2Y^{2-} at the equivalence point in Problem 14-C.

14-E. Suppose that 0.010 0 M Fe^{3+} is titrated with 0.005 00 M EDTA at pH 2.00.
(a) What is the concentration of free Fe^{3+} at the equivalence point?
(b) What is the ratio $[H_3Y^-]/[H_2Y^{2-}]$ in the solution when the titration is just 63.7% of the way to the equivalence point?

14-F. Consider the following equilibria:

$$Cu(NH_3)_4^{2+} \rightleftharpoons Cu^{2+} + 4NH_3$$
$$K = \frac{1}{\beta_4} = 4.8 \times 10^{-14}$$

$$Cu^{2+} + Y^{4-} \rightleftharpoons CuY^{2-}$$
$$K = K_f = 6.3 \times 10^{18}$$

$$\overline{Cu(NH_3)_4^{2+} + Y^{4-} \rightleftharpoons CuY^{2-} + 4NH_3}$$
$$K = 3.0 \times 10^5$$

Calculate pCu at each of the following points in the titration of 50.00 mL of 0.001 00 M Cu^{2+} with 0.001 00 M EDTA at pH 11.00 in a solution whose NH_3 concentration is somehow *fixed* at 0.100 M:
(a) 0 (b) 1.00 (c) 45.00 (d) 50.00 (e) 55.00 mL

14-G. Iminodiacetic acid (abbreviated H_2X in this problem) forms 2:1 complexes with many metal ions:

$$H_2\overset{+}{N}\underset{CH_2CO_2H}{\overset{CH_2CO_2H}{\diagdown}} \qquad H_3X^+$$

$$\alpha_{X^{2-}} \equiv \frac{[X^{2-}]}{[H_3X^+] + [H_2X] + [HX^-] + [X^{2-}]}$$

$$Cu^{2+} + 2X^{2-} \rightleftharpoons CuX_2^{2-} \qquad K = 3.5 \times 10^{16}$$

A solution of volume 25.0 mL containing 0.120 M iminodiacetic acid buffered to pH 7.00 was titrated with 25.0 mL of 0.050 0 M Cu^{2+}. Given that $\alpha_{X^{2-}} = 4.6 \times 10^{-3}$ at pH 7.00, calculate the concentration of Cu^{2+} in the resulting solution.

Problems

14-1. Explain why the change from red to blue in Reaction 14-24 occurs suddenly at the equivalence point instead of gradually throughout the entire titration.

14-2. List four methods for detecting the endpoint of an EDTA titration.

14-3. Give three circumstances in which an EDTA back titration might be necessary.

14-4. State (in words) what $\alpha_{Y^{4-}}$ means. Calculate $\alpha_{Y^{4-}}$ for EDTA at
(a) pH 3.50 (b) pH 10.50

14-5. A 50.0 mL aliquot of solution containing 0.450 g of $MgSO_4$ in 0.500 L required 37.6 mL of EDTA solution for titration. How many milligrams of $CaCO_3$ will react with 1.00 mL of this EDTA solution?

14-6. A 1.000 mL sample of unknown containing Co^{2+} and Ni^{2+} was treated with 25.00 mL of 0.038 72 M EDTA. Back titration with 0.021 27 M Zn^{2+} at pH 5 required 23.54 mL to reach the xylenol orange endpoint. A 2.000 mL sample of unknown was passed through an ion-exchange column that retards Co^{2+} more than Ni^{2+}. The Ni^{2+} that passed through the column was treated with 25.00 mL of 0.038 72 M EDTA and required 25.63 mL of 0.021 27 M Zn^{2+} for back titration. The Co^{2+} emerged from the column later. It, too, was treated with 25.00 mL of 0.038 72 M EDTA. How many milliliters of 0.021 27 M Zn^{2+} will be required for back titration?

14-7. Consider the titration of 25.0 mL of 0.020 0 M $MnSO_4$ with 0.010 0 M EDTA in a solution buffered to pH 8.00. Calculate pMn^{2+} at the following volumes of added EDTA:

(a) 0   (d) 49.0 (f) 50.0 (h) 55.0
(b) 20.0 (e) 49.9 (g) 50.1 (i) 60.0 mL
(c) 40.0

14-8. Using the same volumes as in Problem 14-7, calculate pCa for the titration of 25.00 mL of 0.020 00 M EDTA with 0.010 00 M $CaSO_4$ at pH 10.00.

14-9. Calculate pCo^{2+} at each of the following points in the titration of 25.00 mL of 0.020 26 M Co^{2+} by 0.038 55 M EDTA at pH 6.00:
(a) 12.00 mL (b) V_e (c) 14.00 mL

14-10. Calculate the molarity of HY^{3-} in a solution prepared by mixing 10.00 mL of 0.010 0 M $VOSO_4$, 9.90 mL of 0.010 0 M EDTA, and 10.0 mL of buffer with a pH of 4.00.

14-11. A 50.0 mL solution containing Ni^{2+} and Zn^{2+} was treated with 25.0 mL of 0.045 2 M EDTA to bind all of the metal. The excess unreacted EDTA required 12.4 mL of 0.012 3 M Mg^{2+} for complete reaction. An excess of the reagent 2,3-dimercapto-1-propanol was then added to displace the EDTA from zinc. Another 29.2 mL of Mg^{2+} was required for reaction with the liberated EDTA. Calculate the molarity of Ni^{2+} and Zn^{2+} in the original solution.

14-12. Cyanide can be determined indirectly by EDTA titration. A known excess of Ni^{2+} is added to the cyanide to form tetracyanonickelate:

$$4CN^- + Ni^{2+} \rightarrow Ni(CN)_4^{2-}$$

When the excess Ni^{2+} is titrated with standard EDTA, $Ni(CN)_4^{2-}$ does not react. In a cyanide analysis 12.7 mL of cyanide solution was treated with 25.0 mL of standard solution containing excess Ni^{2+} to form tetracyanonickelate. The excess Ni^{2+} required 10.1 mL of 0.013 0 M EDTA for complete reaction. In a separate experiment, 39.3 mL of 0.013 0 M EDTA was required to react with 30.0 mL of the standard Ni^{2+} solution. Calculate the molarity of CN^- in the 12.7 mL sample of unknown.

14-13. Sulfide ion was determined by indirect titration with EDTA. To a solution containing 25.00 mL of 0.043 32 M $Cu(ClO_4)_2$ plus 15 mL of 1 M acetate buffer (pH 4.5) was added 25.00 mL of unknown sulfide solution with vigorous stirring. The CuS precipitate was filtered and washed with hot water. Then ammonia was added to the filtrate (which contains excess Cu^{2+}) until the blue color of $Cu(NH_3)_4^{2+}$ was observed. Titration with 0.039 27 M EDTA required 12.11 mL to reach the murexide endpoint. Calculate the molarity of sulfide in the unknown.

14-14. A mixture of Mn^{2+}, Mg^{2+}, and Zn^{2+} was analyzed as follows: The 25.00 mL sample was treated with 0.25 g of $NH_3OH^+Cl^-$ (hydroxylammonium chloride, a reducing agent that maintains manga-nese in the $+2$ state), 10 mL of ammonia buffer (pH 10), a few drops of eriochrome black T indicator, and diluted to 100 mL. It was warmed to 40°C and titrated with 39.98 mL of 0.045 00 M EDTA to the blue point. Then 2.5 g of NaF was added to displace Mg^{2+} from its EDTA complex. The liberated EDTA required 10.26 mL of standard 0.020 65 M Mn^{2+} for complete titration. After this second endpoint was reached, 5 mL of 15% (wt/wt) aqueous KCN was added to displace Zn^{2+} from its EDTA complex. This time the liberated EDTA required 15.47 mL of standard 0.020 65 M Mn^{2+}. Calculate the number of milligrams of each metal (Mn^{2+}, Zn^{2+}, and Mg^{2+}) in the 25.00 mL sample of unknown.

14-15. *A brainbuster!* What is the ratio $[MgY^{2-}]/[NaY^{3-}]$ in a solution prepared by mixing 0.100 M Na_2EDTA with an equal volume of 0.100 M $Mg(NO_3)_2$? Assume that the pH is high enough that there is a negligible amount of unbound EDTA. You can approach this problem by realizing that nearly all Mg^{2+} will be bound to EDTA and nearly all Na^+ will be free.

15 / Fundamentals of Electrochemistry

A major branch of analytical chemistry makes use of electrical measurements for analytical purposes. In this chapter we will review fundamental concepts of electricity and electrochemical cells. The principles we develop will lay a foundation for the discussion of potentiometric measurements, electrogravimetric and coulometric analysis, polarography, and amperometric methods in the next four chapters.

15-1 BASIC CONCEPTS

A **redox reaction** involves transfer of electrons from one species to another. A species is said to be **oxidized** when it *loses electrons*. It is **reduced** when it *gains electrons*. An **oxidizing agent** (also called an **oxidant**) takes electrons from another substance and becomes reduced. A **reducing agent** (also called a **reductant**) gives electrons to another substance and is oxidized in the process. In the reaction

Oxidation—loss of electrons.
Reduction—gain of electrons.
Oxidizing agent—takes electrons.
Reducing agent—gives electrons.

$$\underset{\substack{\text{Oxidizing} \\ \text{agent}}}{Fe^{3+}} + \underset{\substack{\text{Reducing} \\ \text{agent}}}{Cu^+} \rightarrow Fe^{2+} + Cu^{2+} \tag{15-1}$$

Fe^{3+} is the oxidizing agent, since it takes an electron from Cu^+. Cu^+ is the reducing agent, since it gives an electron to Fe^{3+}. Fe^{3+} is reduced, and Cu^+ is oxidized as the reaction proceeds from left to right. A review of oxidation numbers and redox equation balancing can be found in Appendix D.

Relation Between Chemistry and Electricity

If the electrons involved in a redox reaction can be made to flow through an electric circuit, we can learn something about the reaction by studying the circuit. In Reaction 15-1, one electron must be transferred to oxidize one

294

15 / FUNDAMENTALS OF ELECTROCHEMISTRY

The *quantity* of electrons that flow from a reaction is proportional to the quantity of analyte that reacts.

The electric force (voltage) is related to the identity and concentrations of reactants and products.

atom of Cu^+ and to reduce one atom of Fe^{3+}. If we know how many moles of electrons are transferred from Cu^+ to Fe^{3+}, then we know how many moles of product have been formed.

The voltage produced by a cell in which Reaction 15-1 is occurring is related to the free energy change when electrons flow from Cu^+ to Fe^{3+}. The free energy change would be different if electrons were flowing, for example, from Sn^{2+} to Fe^{3+}. In techniques such as polarography, the voltage can be used to identify the reactants. Voltage is also related to the quantity of reactants and products present, as we shall see when we study the Nernst equation (Section 15-4).

Electrical Measurements

Charge

Faraday constant (F): $96\ 484.56$ C mol^{-1}

Electric charge is measured in **coulombs,** abbreviated C. The charge of a single electron is $1.602\ 189\ 2 \times 10^{-19}$C. One mole of electrons has a charge of $9.648\ 456 \times 10^4$ C; this number is called the **Faraday constant** and given the symbol F. The relation between coulombs and moles is therefore

$$
\begin{array}{ccccc}
q & = & n & \cdot & F \\[4pt]
\text{Coulombs} & & \text{Moles} & & \dfrac{\text{Coulombs}}{\text{Mole}}
\end{array}
\qquad (15\text{-}2)
$$

EXAMPLE

If 5.585 g of Fe^{3+} was reduced in Reaction 15-1, how many coulombs of charge must have been transferred from Cu^+ to Fe^{3+}?

First, we find that 5.585 g of Fe^{3+} equals 0.100 0 mol of Fe^{3+}. Since each Fe^{3+} ion requires one electron in Reaction 15-1, 0.100 0 mol of electrons must have been transferred. Using the Faraday constant, we find that 0.100 0 mol of electrons corresponds to

$$
(0.100\ 0 \text{ mol of } e^-)\left(9.648\ 456 \times 10^4 \ \frac{C}{\text{mol of } e^-}\right) = 9.648 \times 10^3 C
$$

Current

1 A $= 1$ C/s

The quantity of charge flowing each second through a circuit is called the **current.** The unit of current is the **ampere,** abbreviated A. A current of one ampere represents a charge flowing past a point in a circuit at a rate of one coulomb per second.

EXAMPLE

Suppose that electrons are forced into a platinum wire immersed in a solution containing Sn^{4+} (Figure 15-1), which is reduced to Sn^{2+} at a constant rate of 4.24 mmol/hr. How much current flows into the solution?

Two electrons are required to reduce *one* Sn^{4+} ion:

$$Sn^{4+} + 2e^- \rightarrow Sn^{2+}$$

If Sn^{4+} is reacting at a rate of 4.24 mmol/hr, electrons flow at a rate of 2(4.24) = 8.48 mmol/hr, which corresponds to

$$\frac{8.48 \text{ mmol/hr}}{3\,600 \text{ s/hr}} = 2.356 \times 10^{-3} \text{ mmol/s} = 2.356 \times 10^{-6} \text{ mol/s}$$

To find the current, we convert moles of electrons per second to coulombs per second:

$$\text{Current} = \frac{\text{Coulombs}}{\text{Second}} = \frac{q}{s} = \frac{nF}{s} = \frac{n}{s} F = \frac{\text{Moles}}{\text{Second}} \cdot F$$

$$= \left(2.356 \times 10^{-6} \frac{\text{mol}}{\text{s}}\right)\left(9.648\,456 \times 10^4 \frac{\text{C}}{\text{mol}}\right) = 0.227 \frac{\text{C}}{\text{s}} = 0.227 \text{ A}$$

A current of 0.227 A can also be expressed as 227 mA.

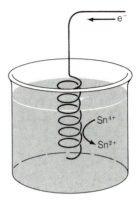

Figure 15-1
Electrons flowing into a coil of Pt wire at which Sn^{4+} ions in solution are reduced to Sn^{2+} ions. This process could not happen by itself because there is not a complete circuit. If Sn^{4+} is to be reduced at this Pt electrode, some other species must be oxidized at some other place.

In the example above, we encountered a Pt electrode. An **electrode** is any device that conducts electrons into or out of the chemical species involved in a redox reaction. Platinum is very commonly used as an *inert* conductor; it does not participate in the redox chemistry except as a conductor of electrons.

Voltage, work, and free energy

Electric potential (E) is measured in volts (V). The potential difference between two points is a measure of how much work is needed (or can be done) when one coulomb passes between those two points. Work has the dimensions of energy, whose units are joules (J).

The relation between work, potential (E), and charge (q) is

$$\boxed{\begin{array}{l}\text{Work} = E \cdot q \\ \text{Joules} = \text{Volts} \cdot \text{Coulombs}\end{array}} \qquad (15\text{-}3)$$

It costs energy to move like charges toward each other. Energy is released when opposite charges move toward each other.

One **joule** of energy is gained or lost when one *coulomb* of charge is moved through a potential difference of one volt. Equation 15-3 tells us that the dimensions of volts are joules/coulomb.

1 volt = 1 J/C

15 / FUNDAMENTALS OF ELECTROCHEMISTRY

EXAMPLE

How much work is required to move 2.36 mmol of electrons through a potential difference of 1.05 V?

To use Equation 15-3, we must convert moles of electrons to coulombs of charge. The relation is simply

$$q = n \cdot F = (2.36 \times 10^{-3} \text{ mol})(9.648 \times 10^4 \text{ C mol}^{-1}) = 2.277 \times 10^2 \text{ C}$$

The work required is

$$\text{Work} = E \cdot q = (1.05 \text{ V})(2.277 \times 10^2 \text{ C}) = 239 \text{ J}$$

The greater the potential difference between two points, the stronger will be the "push" on a charged particle traveling between those points. A 12 V battery will "push" electrons through a circuit eight times harder than a 1.5 V dry cell.

You might review Chapter 5 for a brief discussion of ΔG.

The free energy change (ΔG) for a chemical reaction conducted reversibly at constant temperature and pressure equals the maximum possible electrical work that can be done by the reaction on its surroundings.

$$\text{Work} = -\Delta G \tag{15-4}$$

The negative sign in Equation 15-4 indicates that the free energy of a system is considered as decreasing, when the work done on the surroundings is positive.

Combining Equations 15-2, 15-3, and 15-4 produces a relationship of utmost importance to chemistry:

$$\Delta G = -\text{Work} = -E \cdot q$$

Relation between free energy and electric potential:
$$\Delta G = -n\text{FE}$$

$$\boxed{\Delta G = -n\text{FE}} \tag{15-5}$$

Equation 15-5 relates the free energy change of a chemical reaction to the electrical potential that can be generated by the reaction.

Ohm's law

This fundamental law of physics states that the current flowing through a a circuit is directly proportional to the voltage and inversely proportional to the **resistance** (R) of the circuit.

Ohm's law: $I = E/R$. The greater the potential, the more current will flow. The greater the resistance, the less current will flow.

$$\boxed{I = E/R} \tag{15-6}$$

The units of resistance are ohms, given the Greek symbol Ω. A current of one ampere will flow through a circuit with a potential difference of one volt if the resistance of the circuit is one ohm.

Power

Power (P) is the work done per unit time. The SI unit of power is joules/s, better known as **watts** (W).

$$P = \frac{\text{Work}}{s} = \frac{E \cdot q}{s} = E \cdot \frac{q}{s} \qquad (15\text{-}7)$$

Since q/s is the current, I, we can write

$$P = E \cdot I \qquad (15\text{-}8)$$

A cell capable of delivering one ampere at a potential of one volt has a power output of one watt.

Power (watts) = Work per second

$$P = E \cdot I = (IR) \cdot I = I^2 R$$

$$P = E \cdot I = E \cdot \frac{E}{R} = E^2/R$$

EXAMPLE
A schematic diagram of a very simple circuit is shown in Figure 15-2. The battery has a potential of 3.0 V, and the resistor has a resistance of 100 Ω. We assume that the resistance of the wire connecting the battery and the resistor is negligible. How much current and how much power are delivered by the battery in this circuit?

The current flowing through this circuit is

$$I = E/R = 3.0 \text{ V}/100 \text{ Ω} = 0.030 \text{ A} = 30 \text{ mA}$$

The power produced by the battery must be

$$P = E \cdot I = (3.0 \text{ V})(0.030 \text{ A}) = 90 \text{ mW}$$

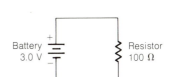

Figure 15-2
A simple electrical circuit with a battery and a resistor.

What happens to the energy needed to push electrons through the circuit in Figure 15-2? Ideally, the only place at which energy is lost is in the resistor. *The energy appears as heat in the resistor.* The power (90 mW) equals the rate at which heat is produced in the resistor.

15-2 GALVANIC CELLS

A **galvanic cell** is one that uses a *spontaneous* chemical reaction to generate electricity. In order to accomplish this, one reagent must be oxidized and another must be reduced. The two cannot be in contact, or electrons would simply flow directly from the reducing agent to the oxidizing agent. Instead, the oxidizing and reducing agents are physically separated, and electrons are forced to flow through an external circuit in order to go from one reactant to the other.

A galvanic cell uses a spontaneous chemical reaction to generate electricity.

A Cell in Action

Figure 15-3 shows a galvanic cell containing two electrodes suspended in an aqueous solution of 0.016 7 M CdCl$_2$. One electrode is a strip of cadmium metal; the other is a piece of metallic silver coated with solid AgCl. The chemical reactions occurring in this cell are

Oxidation: $$Cd(s) \rightleftharpoons Cd^{2+}(aq) + 2e^- \quad (15\text{-}9)$$

Reduction: $$2AgCl(s) + 2e^- \rightleftharpoons 2Ag(s) + 2Cl^-(aq) \quad (15\text{-}10)$$

Net reaction: $$Cd(s) + 2AgCl(s) \rightleftharpoons Cd^{2+}(aq) + 2Ag(s) + 2Cl^-(aq) \quad (15\text{-}11)$$

Reactions 15-9 and 15-10 are called **half-reactions** because they each contain free electrons and cannot be observed separately. The two half-reactions are written with equal numbers of electrons so that, when added together, their sum (the net reaction) contains no free electrons.

Oxidation of Cd metal—to produce Cd^{2+}(aq)—provides electrons that flow through the circuit to the Ag electrode in Figure 15-3. At the surface of the Ag electrode, Ag$^+$ (from AgCl) is reduced to Ag(s). The chloride from AgCl is left in solution. The free energy change for Reaction 15-11 is -150 kJ per mole of Cd. It is the energy liberated by this spontaneous reaction that provides the driving force pushing electrons through the circuit.

Recall that ΔG is *negative* for a spontaneous reaction.

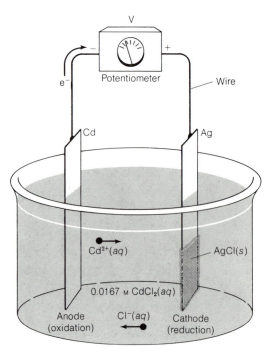

Figure 15-3
A simple galvanic cell. The potentiometer is a device for measuring voltage.

EXAMPLE

Calculate the voltage that would be measured by the potentiometer in Figure 15-3.

Since $\Delta G = -150$ kJ/mol of Cd, we can use Equation 15-5 (where n is the number of moles of electrons transferred in the balanced net reaction) to write

$$E = -\frac{\Delta G}{n\text{F}} = -\frac{-150 \times 10^3 \text{ J}}{(2 \text{ mol})\left(9.648 \times 10^4 \frac{\text{C}}{\text{mol}}\right)}$$

$$E = +0.777 \text{ J/C} = +0.777 \text{ V}$$

A spontaneous chemical reaction (negative ΔG) produces a *positive potential.*

Cell Conventions

Chemists define the electrode at which *oxidation* occurs as the **anode.** The **cathode** is the electrode at which *reduction* occurs. In Figure 15-3, Cd is the anode and Ag is the cathode.

When electrons flow into a potentiometer (voltmeter) through its negative terminal, the meter indicates a positive potential. In Figure 15-3, electrons flow from the Cd anode into the negative terminal of the meter. Therefore, the voltage is positive ($+0.777$ V). If the leads to the meter were reversed (with the Cd anode connected to the positive terminal of the meter), the potential would be -0.777 V.

We adopt the convention that the left-hand electrode of each cell is connected to the negative input terminal of the meter. This will show a positive voltage whenever oxidation occurs at the left-hand electrode. If reduction occurs at the left-hand electrode, it will register a negative voltage.

To calculate the cell potential, we *assume* that oxidation occurs at the left-hand electrode and reduction at the right-hand electrode. Thus, we assume that the left-hand electrode is the anode and the right-hand electrode is the cathode. This scheme generates a positive value of E if oxidation actually occurs at the left-hand electrode and a negative potential if reduction actually occurs there.

When you calculate potentials later in this chapter, do not be despondent if you find the potential of a cell to be negative. *Do not change the sign of the potential.* Leave it negative. A negative sign simply means that electrons are flowing from right to left through the potentiometer, instead of from left to right. The negative sign also means that all of the reactions have been written backwards as compared with the direction in which they actually proceed. *Do not change the direction of the reactions if the potential is negative.* Just realize that a negative potential implies that the reactions are spontaneous in the reverse direction.

Anode ↔ Oxidation

Cathode ↔ Reduction

The sign of the potential depends on how the input leads are connected to the potentiometer. We will *always* connect the negative input terminal of the meter to the left-hand electrode of the cell. This is equivalent to assuming that the left-hand electrode is the anode when we write the cell reaction.

To calculate the potential of a cell, *assume* that the left-hand electrode is the anode, at which oxidation occurs.

Salt Bridge

Consider the cell in Figure 15-4, in which the reactions are intended to be

$$\text{Anode:} \qquad \qquad \qquad Cd(s) \rightleftharpoons Cd^{2+}(aq) + 2e^- \qquad (15\text{-}12)$$

$$\underline{\text{Cathode:} \qquad 2Ag^+(aq) + 2e^- \rightleftharpoons 2Ag(s) \qquad \qquad (15\text{-}13)}$$

$$\text{Net reaction:} \quad Cd(s) + 2Ag^+(aq) \rightleftharpoons Cd^{2+}(aq) + 2Ag(s) \qquad (15\text{-}14)$$

The cell in Figure 15-4 is short-circuited.

Reaction 15-14 is spontaneous, but no current will flow through the circuit because Ag^+ ions are not forced to be reduced at the Ag electrode. The Ag^+ ions in solution can react directly at the Cd(s) surface, giving Reaction 15-14 with no flow of electrons through the circuit.

Okay. Let's try separating the two reactants, as in Figure 15-5. This time electrons will flow through the circuit for an instant as Reaction 15-14 begins. But after an instant, there will be a negative charge in the right half-cell (since electrons flowed into it) and a positive charge in the left half-cell (since electrons flowed out of it). The excess negative charge on the right-hand side will repel electrons trying to gain entry through the circuit. In an instant, the charge repulsion exactly counterbalances the driving force of the chemical reaction and no current can flow.

To fix the cell in Figure 15-5, we can insert a **salt bridge,** as shown in Figure 15-6. The salt bridge consists of a U-shaped tube filled with a gel

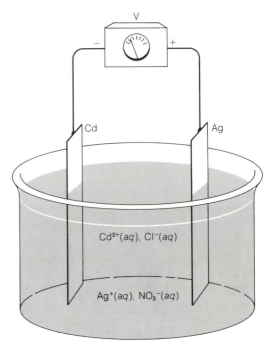

Figure 15-4
A cell that will not work. The solution contains $CdCl_2$ and $AgNO_3$.

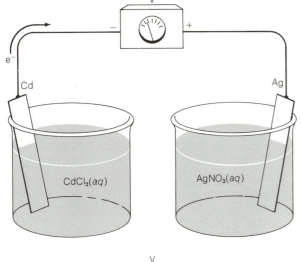

Figure 15-5
Another cell that will not work.

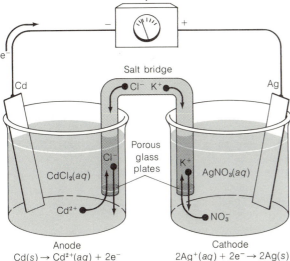

Figure 15-6
At last! A cell that works!

containing KCl (or any other electrolyte not involved in the cell reaction).† The ends of the bridge are covered with porous glass disks that allow ions to diffuse, but that minimize mixing of the solutions inside and outside the bridge. When the galvanic cell is operating, K^+ from the bridge migrates into the cathode compartment and NO_3^- migrates from the cathode into the bridge. The ion migration exactly offsets the charge buildup that would otherwise occur as electrons flow into the silver electrode. In the other half-cell, Cd^{2+} migrates into the bridge while Cl^- migrates into the anode compartment to compensate for the buildup of positive charge that would otherwise occur.

The net effect of a salt bridge is to maintain electroneutrality (no charge buildup) throughout the cell.

---

† A typical salt bridge is prepared by heating 3 g of agar with 30 g of KCl in 100 ml of water until a clear solution is obtained. The solution is poured into the U-tube and allowed to solidify to a gel. The bridge is stored in a solution of saturated aqueous KCl.

Demonstration 15-1 THE HUMAN SALT BRIDGE

A salt bridge consists of any ionic medium with a semipermeable barrier on each end. You can demonstrate a "proper" salt bridge by filling a U-tube with agar and KCl as described in the footnote on page 301. A suitable demonstration cell is shown below.

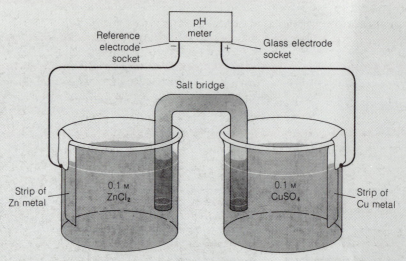

Setup for the galvanic cell.

The pH meter is a potentiometer whose negative terminal is the reference electrode socket.

You should be able to write the two half-reactions for this cell and use the Nernst equation in Section 15-4 to calculate the theoretical potential. Measure the potential with a conventional salt bridge. Then replace the salt bridge with one made of filter paper freshly soaked in NaCl solution and measure the potential again. Finally, replace the filter-paper salt bridge with two fingers of the same hand and measure the potential again. The human body is really just a bag of salt housed in a semipermeable membrane. The small differences in potential observed when the salt bridge is replaced can be attributed to the junction potential discussed in Section 16-3.

Line Notation

A shorthand notation is commonly used to describe electrochemical cells. Only two symbols are needed:

$$| \text{ phase boundary} \quad \| \text{ salt bridge}$$

The cell in Figure 15-3 is represented by

Boundary between separate phases: |

$$\text{Cd}(s) | \text{CdCl}_2(aq, 0.016\ 7\ \text{M}) | \text{AgCl}(s) | \text{Ag}(s)$$

Each phase boundary is indicated by a vertical line. The metallic electrodes are shown at the extreme left- and right-hand sides of the diagram.

The cell in Figure 15-6 can be written as follows:

$$Cd(s)\,|\,CdCl_2(aq)\,\|\,AgNO_3(aq)\,|\,Ag(s)$$

Salt bridge: ‖

The contents of the salt bridge need not be specified. When we evaluate the potential of a cell specified by a line diagram, we *assume* that the left-hand electrode is the anode. As before, this assumption could lead to either a positive or a negative cell potential.

15-3 STANDARD POTENTIALS

With each half-reaction, such as 15-12 or 15-13, we associate a **standard reduction potential** (E^0). The standard reduction potential for Reaction 15-13 is what would be observed for the hypothetical galvanic cell in Figure 15-7. The half-reaction of interest occurs in the right-hand cell, which is connected to the positive terminal of the potentiometer.

The left-hand cell is called the **standard hydrogen electrode** (S.H.E.). It consists of a catalytic Pt surface in contact with an acidic solution in which the activity of H^+ is unity while $H_2(g)$ (activity = 1, essentially 1 atm) is bubbled over the electrode. The half-reaction is

$$\tfrac{1}{2}H_2(g, \mathscr{A} = 1) \rightleftharpoons H^+(aq, \mathscr{A} = 1) + e^- \qquad (15\text{-}15)$$

For the standard hydrogen electrode, $E^0 = 0$.

The Pt takes a catalytic role and does not appear in the half-reaction.

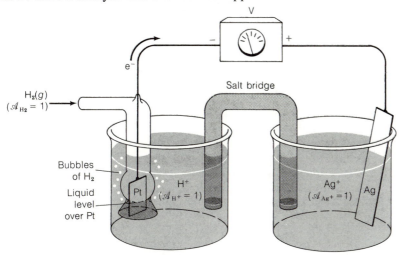

$Pt(s)\,|\,H_2(g, \mathscr{A} = 1)\,|\,H^+(aq, \mathscr{A} = 1)\,\|\,Ag^+(aq, \mathscr{A} = 1)\,|\,Ag(s)$

Figure 15-7
Cell used to measure the standard potential of the reaction $Ag^+ + e^- \rightleftharpoons Ag(s)$. This cell is hypothetical because it is usually not possible to adjust the activity of a species to 1.

15 / FUNDAMENTALS OF ELECTROCHEMISTRY

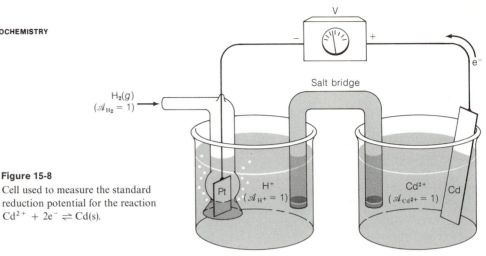

Figure 15-8
Cell used to measure the standard reduction potential for the reaction $Cd^{2+} + 2e^- \rightleftharpoons Cd(s)$.

$Pt(s)|H_2(g, \mathscr{A} = 1)|H^+(aq, \mathscr{A} = 1)\|Cd^{2+}(aq, \mathscr{A} = 1)|Cd(s)$

We *arbitrarily* assign a potential of zero to Reaction 15-15.[†] The potential measured by the meter in Figure 15-7 can therefore be *assigned* to the cathode half-reaction, which is

$$Ag^+ + e^- \rightleftharpoons Ag(s) \qquad (15\text{-}16)$$

> The standard potential is what would be measured if the activities of all reactants and products were unity.

The potential that would be measured in Figure 15-7 is +0.799 V. The positive sign tells us that electrons flow from left to right through the meter and that Reactions 15-15 and 15-16 are written in the correct directions.

The value +0.799 V is called the standard reduction potential (E^0) for Reaction 15-16 for two reasons. It is a *reduction* potential because it applies to the reduction of Ag^+. It is a *standard* potential because the activities of all reactants and products in the cell are unity. For solutes (Ag^+ and H^+), an activity of 1 corresponds very crudely to a concentration of 1 M. For a gas (H_2), $\mathscr{A} = 1$ corresponds very nearly to 1 atm. A solid (Ag) has an activity of 1, by definition. The superscript zero of E^0 means that all reagents are in their standard states.

> *Question:* What is the pH of the S.H.E.?

Of course, nothing is so simple. It is usually not possible to construct a cell in which all the activities are unity. For example, there may not be a salt soluble enough to give a 1 M solution. Even if there were, we do not have an accurate way to calculate the activity coefficient at such a high ionic strength. In real life, activities less than unity are used in both half-cells. The Nernst equation (in Section 15-4) can be used to extract the value of E^0 from the potential measured under nonstandard conditions.

> Nobody ever builds a *standard* hydrogen electrode. Instead, known hydrogen ion activities are obtained using National Bureau of Standards buffers.

In Figure 15-8, we see the hypothetical cell that would be used to measure

---

[†] A scale of absolute potentials has been proposed in G. P. Haight, Jr., *J. Chem. Ed.*, **53**, 693 (1976). The free energy change for Reaction 15-15 is estimated to be ~ 258 kJ/mol. Using the relation $\Delta G^0 = -nFE^0$, the potential of Reaction 15-15 is estimated at +2.67 V. This has the effect of adding +2.67 V to the standard reduction potentials in Appendix H to convert them to absolute potentials.

the standard reduction potential for the reaction

$$Cd^{2+} + 2e^- \rightleftharpoons Cd(s) \qquad (15\text{-}17)$$

In this case the measured potential is -0.402 V. The negative sign indicates that $Cd(s)$ is actually the anode in this cell and that electrons flow from right to left through the meter.

Table of Standard Potentials

The two galvanic cells we have just constructed allow us to establish a rudimentary table of standard reduction potentials.

	$E^0(V)$
$Ag^+ + e^- \rightleftharpoons Ag(s)$	$+0.799$
$2H^+ + 2e^- \rightleftharpoons H_2(g)$	0 (by convention)
$Cd^{2+} + 2e^- \rightleftharpoons Cd(s)$	-0.402

The potentials in this table tell us that electrons will flow from the standard hydrogen electrode (S.H.E.) to the $Ag^+ | Ag(s)$ cell under standard conditions. The table also tells us that electrons flow from the standard $Cd^{2+} | Cd(s)$ cell to the S.H.E. It follows that electrons would also flow from the $Cd^{2+} | Cd(s)$ cell to the $Ag^+ | Ag(s)$ cell if the two were connected (as in Figure 15-6).

Appendix H contains standard reduction potentials for many different reactions, arranged alphabetically by element. If the table were arranged according to descending value of E^0 (as in Table 15-1), we would find the strongest oxidizing agents at the upper left and the strongest reducing agents at the lower right.

> Remember that "standard conditions" means that all activities are unity.

Using E^0 Values

The information in Appendix H can be used to find the standard potential for any reaction composed of listed half-reactions. The potential of a balanced net reaction (in which no electrons appear) is given by

$$E^0(\text{net reaction}) = E^0(\text{cathode reaction}) - E^0(\text{anode reaction}) \quad (15\text{-}18)$$

EXAMPLE

Find the value of E^0 for the reaction

$$Cu(s) + PbF_2(s) \rightleftharpoons Cu^{2+} + 2F^- + Pb(s) \qquad (15\text{-}19)$$

We can break the net reaction into two half-reactions that appear in Appendix H.

$$Cu(s) \rightleftharpoons Cu^{2+} + 2e^- \qquad (15\text{-}20)$$

15 / FUNDAMENTALS OF ELECTROCHEMISTRY

Table 15-1
Ordered redox potentials

Oxidizing agent	Reducing agent	E^0 (V)
$F_2(g) + 2e^- \rightleftharpoons 2F^-$		2.87
$O_3(g) + 2H^+ + 2e^- \rightleftharpoons O_2(g) + H_2O$		2.07
$MnO_4^- + 8H^+ + 5e^- \rightleftharpoons Mn^{2+} + 4H_2O$		1.51
$Ag^2 + e^- \rightleftharpoons Ag(s)$		0.799
$Cu^{2+} + 2e^- \rightleftharpoons Cu(s)$		0.337
$2H^+ + 2e^- \rightleftharpoons H_2(g)$		0.000
$Cd^{2+} + 2e^- \rightleftharpoons Cd(s)$		-0.402
$K^+ + e^- \rightleftharpoons K(s)$		-2.925
$Li^+ + e^- \rightleftharpoons Li(s)$		-3.045

Oxidizing power increases (upward, left); *Reducing power increases* (downward, right).

> *Question:* The potential for the reaction $K^+ + e^- \rightleftharpoons K(s)$ is -2.925 V. This means that K^+ is a very poor oxidizing agent. (It does not readily accept electrons.) Does this potential imply that K^+ is therefore a good reducing agent?
>
> *Answer:* No!

$$PbF_2(s) + 2e^- \rightleftharpoons Pb(s) + 2F^- \tag{15-21}$$

The potential for Reaction 15-21 is given as -0.350 V in Appendix H. The potential for the reaction $Cu^{2+} + 2e^- \rightleftharpoons Cu(s)$ is $+0.337$ V. So we can write

$$
\begin{array}{ll}
Cu(s) \rightleftharpoons Cu^{2+} + 2e^- & E^0 = +0.337 \\
PbF_2(s) + 2e^- \rightleftharpoons Pb(s) + 2F^- & E^0 = -0.350 \\
\hline
Cu(s) + PbF_2(s) \rightleftharpoons Cu^{2+} + 2F^- + Pb(s) & E^0 = -0.350 - (0.337) = -0.687 \text{ V}
\end{array}
$$

The negative sign of E^0 simply means that the reaction would be spontaneous in the reverse direction under standard conditions.

EXAMPLE

Find the standard potential for the reaction

$$Cd(s) + 2Ag^+ \rightleftharpoons Cd^{2+} + 2Ag(s) \tag{15-22}$$

In Appendix H we find the following data:

$$Ag^+ + e^- \rightleftharpoons Ag(s) \qquad E^0 = +0.799 \tag{15-23}$$

$$Cd^{2+} + 2e^- \rightleftharpoons Cd(s) \qquad E^0 = -0.402 \tag{15-24}$$

After reversing Reaction 15-24 and multiplying Reaction 15-23 by 2, we can add them together:

$$2Ag^+ + 2e^- \rightleftharpoons 2Ag(s) \qquad E^0 = +0.799$$

$$\underline{Cd(s) \rightleftharpoons Cd^{2+} + 2e^- \qquad E^0 = -0.402}$$

$$Cd(s) + 2Ag^+ \rightleftharpoons Cd^{2+} + 2Ag(s) \qquad E^0 = 0.799 - (-0.402) = +1.201 \text{ V} \quad (15\text{-}25)$$

When we multiplied Reaction 15-23 by 2, *we did not multiply its potential by 2*. The reason why the potential is not multiplied can be seen in Equation 15-3. The potential is the work done *per coulomb* of charge carried through that potential (E = Work/q). The work *per coulomb* is the same whether 0.1, 2.3, or 10^4 coulombs have been transferred. The *total work* is different in each case, but the work per coulomb is constant. Therefore, we do not double E^0 when we multiply a reaction by 2.

The examples above illustrate how to find E^0 values for *balanced net reactions* in which no electrons appear. Box 15-1 tells you how to find E^0 for a *half-reaction* that is the sum of two or more half-reactions of known potential.

15-4 NERNST EQUATION

The standard potential is a measure of the inherent driving force of a chemical reaction under standard conditions. Changing the concentration of reactants and products influences the free energy change of the reaction. We know from Le Châtelier's principle that increasing reactant concentration drives a reaction to the right and increasing product concentration drives the reaction to the left. The net interaction between the inherent drive of a reaction and the effect of concentration is expressed in the **Nernst equation.**

Suppose that we add two half-reactions to obtain the balanced net reaction (15-26):

A reaction is spontaneous if ΔG is negative and E is positive. ΔG^0 and E^0 refer to the free energy change and potential when the activities of all reactants and products are unity. $\Delta G^0 = -nFE^0$

Cathode:	$aA + ne^- \rightleftharpoons cC$	E_1^0
Anode:	$bB \rightleftharpoons dD + ne^-$	E_2^0
Net reaction:	$aA + bB \rightleftharpoons cC = dD$	$E_3^0 = E_1^0 - E_2^0$ (15-26)

The Nernst equation for this reaction has the following form:

$$E(cell) = E^0(cell) - \frac{RT}{nF}\left(\ln \frac{\mathscr{A}_C^c \mathscr{A}_D^d}{\mathscr{A}_A^a \mathscr{A}_B^b}\right) \qquad (15\text{-}27)$$

Justify that Le Châtelier's principle requires a negative sign in front of the log term in the Nernst equation.

where $E^0(cell)$ = standard potential for the net cell reaction

R = gas constant (8.314 41 V C K^{-1} mol^{-1})

T = temperature (K)

n = number of electrons transferred in the balanced net reaction

Box 15-1 LATIMER DIAGRAMS

A **Latimer diagram** summarizes the standard reduction potentials (E^0) connecting various oxidation states of an element. For example, in acidic solution the following standard reduction potentials are observed:

$$
\begin{array}{c}
\overset{?}{\underset{}{\text{- -}}}\\
\overset{1.34}{\underset{}{\text{- - - - - - - - -}}}\\
IO_4^- \xrightarrow{1.653} IO_3^- \xrightarrow{1.134} HOI \xrightarrow{1.45} I_2(s) \xrightarrow{0.536} I^-\\
(+7)\qquad (+5)\qquad (+1)\qquad (0)\qquad (-1)\\
\underset{1.195}{\underline{}}
\end{array}
$$

Oxidation state of iodine

As an example of what each arrow means, let's write the balanced equation represented by the arrow connecting IO_3^- and HOI:

$$IO_3^- \xrightarrow{1.134} HOI$$

means

$$IO_3^- + 5H^+ + 4e^- \rightleftharpoons HOI + 2H_2O \qquad E^0 = +1.134 \text{ V}$$

It is possible to derive reduction potentials for arrows that are not shown in the diagram. Suppose we wish to determine E^0 for the reaction shown by the dashed line in the Latimer diagram, which is

$$IO_3^- + 6H^+ + 6e^- \rightleftharpoons I^- + 3H_2O$$

We can use a thermodynamic cycle to find E^0 for this reaction by expressing the desired reaction as a sum of reactions whose potentials are known.

The standard free energy change, ΔG^0, for a reaction is given by

$$\Delta G^0 = -nFE^0$$

When two reactions are added to give a third reaction, the sum of the individual ΔG^0 values must equal the overall value of ΔG^0.

To apply free energy to the problem above, we write two reactions whose sum is the desired reaction:

$$
\begin{aligned}
IO_3^- + 6H^+ + 5e^- &\xrightarrow{E_1^0 = 1.195} \tfrac{1}{2}I_2(s) + 3H_2O & \Delta G_1^0 &= -5F(1.195)\\
\tfrac{1}{2}I_2(s) + e^- &\xrightarrow{E_2^0 = 0.536} I^- & \Delta G_2^0 &= -1F(0.536)\\
\hline
IO_3^- + 6H^+ + 6e^- &\xrightarrow{E_3^0 = ?} I^- + 3H_2O & \Delta G_3^0 &= -6FE_3^0
\end{aligned}
$$

But since $\Delta G_1^0 + \Delta G_2^0 = \Delta G_3^0$, we may solve for E_3^0:

$$\Delta G_3^0 = \Delta G_1^0 + \Delta G_2^0$$

$$-6FE_3^0 = -5F(1.195) - 1F(0.536)$$

$$E_3^0 = \frac{5(1.195) + 1(0.536)}{6} = 1.085 \text{ V}$$

F = Faraday constant $(9.648\ 456 \times 10^4\ \text{C mol}^{-1})$

$\mathscr{A}_i$ = activity of species i

The logarithmic term in the Nernst equation is the **reaction quotient** (Q).

$$Q = \frac{\mathscr{A}_C^c \mathscr{A}_D^d}{\mathscr{A}_A^a \mathscr{A}_B^b} \qquad (15\text{-}28)$$

Q has the same form as the equilibrium constant, but the activities need not have their equilibrium values. That is, $Q \neq K$ unless the system happens to be at equilibrium. Recall that pure solids, pure liquids, and solvents are omitted from Q because their activities are unity (or close to unity), while concentrations of solutes are expressed as mol/L and pressures of gases in atm. When all activities are unity, $Q = 1$ and $\ln Q = 0$. Therefore, $E(\text{cell}) = E^0(\text{cell})$, when all activities are unity.

$E = E^0$, when all activities are unity.

Converting the natural logarithm in Equation 15-27 to the base 10 logarithm, and inserting $T = 298.15\ \text{K}$ (25.00°C) gives the most useful form of the Nernst equation.

Appendix A tells how to convert ln to log.

$$E(\text{cell}) = E^0(\text{cell}) - \frac{0.059\ 16\ \text{V}}{n} \log Q \qquad \text{(at 25°C)} \quad (15\text{-}29)$$

We see that the potential (E) changes by 59.16 mV for each factor-of-ten change in the value of Q.

Using the Nernst Equation

Let's apply the Nernst equation to calculating the potential of the cell in Figure 15-6 when the concentration of $CdCl_2$ is 0.010 M and the concentration of $AgNO_3$ is 0.50 M. As has been our custom, we will neglect activity coefficients. In the reaction quotient, we will replace the activities with concentrations. For the net cell reaction (15-14), we can write the following Nernst equation:

$$E(\text{cell}) = E^0(\text{cell}) - \frac{0.059\ 16}{2} \log \frac{[Cd^{2+}]}{[Ag^+]^2} \qquad (15\text{-}30)$$

Pure solids, liquids, and solvents are omitted from Q.

In Equation 15-25, we found that E^0 for this cell is $+1.201$ V. Putting all of the numerical values into Equation 15-30 gives

Box 15-2 shows that the cell potential does not depend on how many electrons you choose to use when balancing the cell reaction.

$$E(\text{cell}) = +1.201 - \frac{0.059\ 16}{2} \log \frac{(0.010)}{(0.50)^2} = 1.242\ \text{V} \qquad (15\text{-}31)$$

Box 15-2 THE CELL POTENTIAL DOES NOT DEPEND ON HOW YOU WRITE THE CELL REACTION

Suppose that the cell reaction for Figure 15-6 were written with just one electron being transferred, instead of two:

$$\frac{1}{2}Cd(s) \rightleftharpoons \frac{1}{2}Cd^{2+} + e^- \qquad E^0 = -0.402$$

$$Ag^+ + e^- \rightleftharpoons Ag(s) \qquad E^0 = +0.799$$

$$\overline{\frac{1}{2}Cd(s) + Ag^+ \rightleftharpoons \frac{1}{2}Cd^{2+} + Ag(s)} \qquad E^0 = +1.201 \ V$$

For 0.010 M $CdCl_2$ and 0.50 M $AgNO_3$ we calculate

$$E\,(\text{cell}) = E^0\,(\text{cell}) - \frac{0.059\ 16}{1} \log \frac{[Cd^{2+}]^{1/2}}{[Ag^+]} \qquad (15\text{-}32)$$

$$E\,(\text{cell}) = +1.201 - \frac{0.059\ 16}{1} \log \frac{\sqrt{(0.010)}}{(0.50)} = 1.242 \ V$$

This is the same potential calculated with Equation 15-30, which was based on a balanced reaction in which two electrons are transferred:

$$Cd(s) + 2Ag^+ \rightleftharpoons Cd^{2+} + 2Ag(s)$$

Why do Equations 15-30 and 15-32 give the same potential? The reason is that the factor $1/n$ in front of the log term is related to the exponents of the concentrations within the log term. Making use of the identity

$$\log a^b = b \log a$$

we can write

$$\underbrace{\frac{1}{2} \log \frac{[Cd^{2+}]}{[Ag^+]^2}}_{\substack{\text{The form in} \\ \text{Equation 15-30}}} = \frac{1}{2} \log \frac{[Cd^{2+}]^{(1/2)\cdot 2}}{[Ag^+]^{1\cdot 2}} = \underbrace{\frac{2}{2} \log \frac{[Cd^{2+}]^{1/2}}{[Ag^+]}}_{\substack{\text{The form in} \\ \text{Equation 15-32}}}$$

The cell potential does not depend on how you write the cell reaction.

The positive value of E(cell), not the positive value of E^0(cell), guarantees that the reaction will be spontaneous. In some cases, E^0(cell) could be positive and E(cell) negative.

EXAMPLE

For the cell in Figure 15-8, we can write

Anode:	$H_2(g) \rightleftharpoons 2H^+ + 2e^-$
Cathode:	$Cd^{2+} + 2e^- \rightleftharpoons Cd(s)$
Net reaction:	$Cd^{2+} + H_2(g) \rightleftharpoons Cd(s) + 2H^+$

$$(15\text{-}33)$$

$$E(\text{cell}) = E^0(\text{cell}) - \frac{0.059\ 16}{2} \log \frac{[H^+]^2}{[Cd^{2+}]P_{H_2}}$$

If $[H^+] = [Cd^{2+}] = 1$ M and $P_{H_2} = 1$ atm, then $E(\text{cell}) = E^0(\text{cell}) = -0.402$ V. That is, the net cell reaction proceeds spontaneously in the *reverse* direction. By Le Châtelier's principle, we should be able to force the reaction to go forward by decreasing the concentration of H^+. At what pH will the cell reaction be in equilibrium if $[Cd^{2+}]$ remains 1 M and P_{H_2} remains 1 atm?

The cell reaction goes forward if $E(\text{cell})$ is positive and backward if $E(\text{cell})$ is negative. If $E(\text{cell}) = 0$, the cell reaction is in equilibrium. If $E(\text{cell}) = 0$ in the Nernst equation, this allows us to solve for $[H^+]$:

$$0 = -0.402 - \frac{0.059\ 16}{2} \log \frac{[H^+]^2}{(1)(1)}$$

$$[H^+] = 1.6 \times 10^{-7}\ \text{M} \Rightarrow \text{pH} = 6.80$$

At a pH of 6.80 (*in the left half-cell*), the cell reaction will be in equilibrium.

Question: Why does this pH apply to the left half-cell only?

Now we turn our attention to the cell in Figure 15-3, whose half-reactions can be expressed as follows:

$$Cd(s) \rightleftharpoons Cd^{2+} + 2e^- \qquad\qquad E^0 = -0.402$$

$$\underline{2AgCl(s) + 2e^- \rightleftharpoons 2Ag(s) + 2Cl^- \qquad E^0 = +0.222}$$

$$Cd(s) + 2AgCl(s) \rightleftharpoons Cd^{2+} + 2Ag(s) + 2Cl^- \qquad E^0 = +0.624 \quad (15\text{-}33)$$

The Nernst equation for Reaction 15-33 looks like this:

$$E(\text{cell}) = +0.624 - \frac{0.059\ 16}{2} \log[Cl^-]^2[Cd^{2+}] \qquad (15\text{-}34)$$

The cell in Figure 15-3 contains $0.016\ 7$ M $CdCl_2$, so the Nernst equation becomes

$$E(\text{cell}) = +0.624 - \frac{0.059\ 16}{2} \log(0.033\ 4)^2(0.016\ 7) = 0.764\ \text{V} \quad (15\text{-}35)$$

We expect a potential of $+0.764$ V.

Suppose that a different author wrote this book and chose to describe the cell in Figure 15-3 in a different way:

312

Box 15-3 CONCENTRATIONS IN THE OPERATING CELL

Doesn't operation of a cell change the concentrations in the cell? Yes, but cell potentials are measured under conditions of *negligible current flow*. For example, the resistance of a high-quality pH meter is $10^{13}\ \Omega$. If you use this meter to measure a potential of 1 V, the current is

$$I = E/R = 1/10^{13} = 10^{-13}\ A$$

If the cell in Figure 15-6 produces 50 mV, the current through the circuit is $0.050\ V/10^{13}\ \Omega = 5 \times 10^{-15}\ A$. This corresponds to a flow of

$$\frac{5 \times 10^{-15}\ C/s}{9.648 \times 10^4\ C/mol} = 5 \times 10^{-20}\ \text{mol of } e^-/s$$

The rate at which Cd^{2+} is produced is only 2.5×10^{-20} mol/s. Clearly, this will not have any effect on the cadmium concentration in the cell. *The purpose of the potentiometer is to measure the potential of the cell without affecting the concentrations in the cell.*

If the salt bridge were left in a real cell for very long, the concentrations and ionic strength would change because of diffusion between each compartment and the salt bridge. We assume that the cells are set up for a short enough time that this does not happen.

Another description of Figure 15-3.

$$
\begin{aligned}
Cd(s) &\rightleftharpoons Cd^{2+} + 2e^- & E^0 &= -0.402 \\
2Ag^+ + 2e^- &\rightleftharpoons 2Ag(s) & E^0 &= +0.799 \\
\hline
Cd(s) + 2Ag^+ &\rightleftharpoons Cd^{2+} + 2Ag(s) & E^0 &= +1.201\ V
\end{aligned}
\qquad (15\text{-}36)
$$

Equation 15-36 is just as valid as Equation 15-33. One person chose to describe the silver half-reaction as $AgCl(s) + e^- \rightleftharpoons Ag(s) + Cl^-$, and the other chose to look at it as $Ag^+ + e^- \rightleftharpoons Ag(s)$. In both equations, Ag(I) is being reduced to Ag(0).

If the two descriptions are equally valid, they should predict the same cell potential. The Nernst equation for Reaction 15-36 is

$$E(\text{cell}) = +1.201 - \frac{0.059\,16}{2} \log \frac{[Cd^{2+}]}{[Ag^+]^2} \qquad (15\text{-}37)$$

To find the concentration of Ag^+, we must use the solubility product for AgCl. Since the cell contains $0.033\,4$ M Cl^- and solid AgCl, we can say

$K_{sp} = [Ag^+][Cl^-]$

$$[Ag^+] = \frac{K_{sp}\ (\text{for AgCl})}{[Cl^-]} = \frac{1.8 \times 10^{-10}}{0.033\,4} = 5.4 \times 10^{-9}\ \text{M} \qquad (15\text{-}38)$$

Putting this value into Equation 15-37 gives

$$E(\text{cell}) = +1.201 - \frac{0.059\,16}{2} \log \frac{0.016\,7}{(5.4 \times 10^{-9})^2} = 0.764 \text{ V} \quad (15\text{-}39)$$

Lo and behold! Equations 15-39 and 15-35 give the same potential. They certainly should, because they describe the same cell.

Some advice

When you are faced with a cell drawing or a line diagram, the first step toward understanding the cell is to write an anode reaction and a cathode reaction. To write these reactions, *look for an element in the cell in two oxidation states*. For the cell

$$Pb(s)\,|\,PbF_2(s)\,|\,F^-(aq)\,||\,Cu^{2+}(aq)\,|\,Cu(s)$$

we see Pb in two oxidation states, as $Pb(s)$ and $PbF_2(s)$, and Cu in two oxidation states, as Cu^{2+} and $Cu(s)$. Thus, the half-reactions are

$$\text{Anode:} \quad Pb(s) + 2F^- \rightleftharpoons PbF_2(s) + 2e^- \quad (15\text{-}40)$$

$$\text{Cathode:} \quad Cu^{2+} + 2e^- \rightleftharpoons Cu(s) \quad (15\text{-}41)$$

There is no need to think about which is the anode and which is the cathode. By convention we *always* choose the left-hand electrode as the anode in calculating the cell potential.

If, perhaps, the phase of the moon were different, you might choose to write the Pb half-reaction as

$$\text{Anode:} \quad Pb(s) \rightleftharpoons Pb^{2+} + 2e^- \quad (15\text{-}42)$$

since you know that if $PbF_2(s)$ is present, there *must* be some Pb^{2+} in the solution. As we saw in the $AgCl\,|\,Ag$ example above, Equations 15-40 and 15-42 are equally valid descriptions of the cell, and each should predict the same cell potential. The choice of using Equation 15-40 or 15-42 depends on whether the F^- or Pb^{2+} concentration is more easily available to you.

We described the left half-cell in terms of a redox reaction involving Pb, because Pb is the element that appears in two oxidation states. We would *not* write a reaction such as

$$2F^- \rightleftharpoons F_2(g) + 2e^- \quad (15\text{-}43)$$

because $F_2(g)$ is not shown in the cell.

Single-Electrode Potentials

It is convenient to be able to write the Nernst equation for a *half-reaction* (in which electrons appear) in the same way that we write a Nernst equation

How to figure out the cell reactions.

Don't invent species not shown in the cell. Use what is shown in the cell diagram to select the half-reactions.

314

15 / FUNDAMENTALS OF ELECTROCHEMISTRY

A single-electrode potential is what would be measured when the cell of interest is coupled to a standard hydrogen electrode, as in Figure 15-8.

for a whole reaction (in which electrons do not appear). For example, the Nernst equation for the reaction

$$Cd^{2+} + 2e^- \rightleftharpoons Cd(s)$$

can be written

$$E = E^0 - \frac{0.059\ 16}{2} \log \frac{1}{[Cd^{2+}]} \tag{15-44}$$

We will make extensive use of single-electrode potentials in the next chapter.

15-5 RELATION OF E^0 AND THE EQUILIBRIUM CONSTANT

A galvanic cell produces electricity because the net cell reaction is not at equilibrium. When we use a potentiometer to measure the cell potential, we allow negligible current flow so that the concentrations in each half-cell remain unchanged. If we replaced the potentiometer with a wire, much more current would flow, the concentrations of reactants would decrease, and the concentrations of products would increase. This process would continue until the cell reached equilibrium. At that point, there is no more force driving the reaction, and E(cell) is zero.

When E(cell) = 0, a cell is at equilibrium. Therefore, the reaction quotient, Q, is equal to the equilibrium constant, K, when E(cell) = 0. This allows us to derive a most important relation between K and E^0 for a chemical reaction:

When the cell is at equilibrium, E(cell) (not E^0) = 0.

$$E(\text{cell}) = E^0 - \frac{0.059\ 16}{n} \log Q \qquad \text{(at any time)} \tag{15-45}$$

$$0 = E^0 - \frac{0.059\ 16}{n} \log K \qquad \text{(at equilibrium)} \tag{15-46}$$

Rearranging Equation 15-46 gives

From K we can find E^0.

$$\boxed{\frac{0.059\ 16}{n} \log K = E^0} \qquad \text{at } 25°C \tag{15-47}$$

or

From E^0 we can find K.

$$\boxed{K = 10^{nE^0/0.059\ 16}} \qquad \text{at } 25°C \tag{15-48}$$

Equation 15-48 allows us to deduce the equilibrium constant for any reaction for which E^0 is known. Alternatively, knowing the equilibrium constant for a reaction lets us find E^0, using Equation 15-47.

The correct form of Equation 15-48 at any temperature is

$$K = 10^{nFE^0/RT \ln 10} \qquad (15\text{-}49)$$

If you want to determine K at some particular temperature, you must know E^0 at that temperature. The value of E^0 is temperature-dependent.

For the reaction

$$Cu(s) + 2Fe^{3+} \rightleftharpoons 2Fe^{2+} + Cu^{2+} \qquad E^0 = 0.433 \text{ V} \qquad (15\text{-}50)$$

we can use Equation 15-48 to evaluate K as follows:

$$K = 10^{(2)(0.433)/(0.059\,16)} = 4 \times 10^{14} \qquad (15\text{-}51)$$

Note that a very large value of E^0 is not required to produce a very large equilibrium constant. The value of K in Equation 15-51 is correctly expressed with one significant figure. The value of E^0 has three figures. Two are used for the exponent (14), and only one is left for the multiplier (4).

By judicious choice of half-reactions, we can evaluate equilibrium constants for reactions that need not be redox reactions. For example, the sum of the two half-reactions below gives the solubility equation for ferrous carbonate:

$$FeCO_3(s) + 2e^- \rightleftharpoons Fe(s) + CO_3^{2-} \qquad E^0 = -0.756$$

$$Fe(s) \rightleftharpoons Fe^{2+} + 2e^- \qquad E^0 = -0.440$$

$$FeCO_3(s) \rightleftharpoons Fe^{2+} + CO_3^{2-} \ (K \equiv K_{sp}) \qquad E^0 = -0.316 \text{ V} \qquad (15\text{-}52)$$

ferrous carbonate

$$K_{sp} = 10^{(2)(-0.316)/(0.059\,16)} = 2 \times 10^{-11} \qquad (15\text{-}53)$$

Significant figures for logs and exponents were discussed on page 30.

The net reaction need not be a redox reaction. We can still use E^0 to find K.

Potentiometric measurements provide one of the most useful means of measuring equilibrium constants that are too small or too large to measure by determining concentrations of reactants and products directly.

The general form of a problem involving the relation between E^0 values for half-reactions and K for a net reaction is

Cathode:	E_1^0
Anode:	E_2^0
Net reaction:	E_3^0, K

Any two pieces of information allow us to calculate the third piece.

If you know E_1^0 and E_2^0, you can find $E_3^0 (= E_1^0 - E_2^0)$ and $K (= 10^{nE_3^0/0.059\,16})$. Alternatively, if you know E_3^0 and either E_1^0 or E_2^0, you can find the missing E^0. If you know K, you can calculate E_3^0 and use it to find either E_1^0 or E_2^0.

15 / FUNDAMENTALS OF ELECTROCHEMISTRY

EXAMPLE

Suppose we have the following information:

$$Ni^{2+} + 2 \text{ glycine} \rightleftharpoons Ni(\text{glycine})_2^{2+} \qquad K = \beta_2 = 1.2 \times 10^{11}$$

$$Ni^{2+} + 2e^- \rightleftharpoons Ni(s) \qquad E^0 = -0.250 \text{ V}$$

From the value of the overall formation constant of $Ni(\text{glycine})_2^{2+}$ plus the value of E^0 for the $Ni^{2+}|Ni(s)$ couple, deduce the value of E^0 for the reaction

$$Ni(\text{glycine})_2^{2+} + 2e^- \rightleftharpoons Ni(s) + 2 \text{ glycine} \qquad E^0 = ?$$

To accomplish this task, we need to see the relation among the three reactions above.

$$Ni^{2+} + 2e^- \rightleftharpoons Ni(s) \qquad\qquad E_1^0 = -0.250$$

$$\underline{Ni(s) + 2 \text{ glycine} \rightleftharpoons Ni(\text{glycine})_2^{2+} + 2e^- \qquad E_2^0 = ?}$$

$$Ni^{2+} + 2 \text{ glycine} \rightleftharpoons Ni(\text{glycine})_2^{2+} \qquad E_3^0 = ? \qquad K = 1.2 \times 10^{11}$$

We know that $E_1^0 - E_2^0$ must equal E_3^0, We can therefore deduce the value of E_2^0 if we can find E_3^0. But E_3^0 can be determined from the equilibrium constant for the net reaction.

$$K = 10^{nE_3^0/0.059\ 16} \Rightarrow E_3^0 = \frac{0.059\ 16}{n} \log K$$

$$E_3^0 = \frac{0.059\ 16}{2} \log(1.2 \times 10^{11}) = 0.328 \text{ V}$$

Hence,

$$E_2^0 = E_1^0 - E_3^0 = -0.578 \text{ V}$$

15-6 USING CELLS AS CHEMICAL PROBES

It is essential to distinguish two classes of equilibria associated with galvanic cells:

1. Equilibrium *between* the two half-cells.
2. Equilibrium *within* each half-cell.

If a galvanic cell is producing a nonzero potential (either positive or negative), then the net cell reaction cannot be at equilibrium. We say that equilibrium *between* the two half-cells has not been established.

On the other hand, in any galvanic cell it is implicit that *equilibrium exists for any reaction that can occur* **within** *one half-cell*. For example, in the right-hand half-cell in Figure 15-9, the equilibrium (15-54)

A chemical reaction that can occur *within one half-cell* will reach equilibrium and is assumed to remain at equilibrium. Such a reaction is not the net cell reaction.

$$AgCl(s) \rightleftharpoons Ag^+(aq) + Cl^-(aq) \qquad\qquad (15\text{-}54)$$

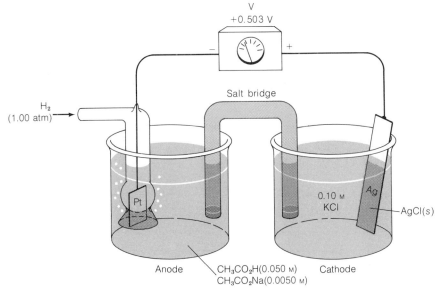

Figure 15-9
This galvanic cell can be used to measure the pH of the left half-cell.

exists with or without the presence of another half-cell. Reaction 15-54 is not part of the net cell reaction. It is simply a chemical reaction whose equilibrium will be established when AgCl(s) is in contact with an aqueous solution. In the left half-cell, the reaction

$$CH_3CO_2H \rightleftharpoons CH_3CO_2^- + H^+ \qquad (15\text{-}55)$$

has also come to equilibrium. Neither of these is a redox reaction involved in the net cell reaction.

The half-reaction for the right-hand cell is quite obviously

$$AgCl(s) + e^- \rightleftharpoons Ag(s) + Cl^-(aq) \qquad E^0 = 0.222 \text{ V} \qquad (15\text{-}56)$$

But what is the half-reaction in the left cell? The only element we find in two oxidation states is hydrogen. We see that $H_2(g)$ is being bubbled into the cell, and we also realize that every aqueous solution contains H^+. Therefore, hydrogen is present in two oxidation states, and the half-reaction can be written as

$$H_2(g) \rightleftharpoons 2H^+(aq) + 2e^- \qquad E^0 = 0 \qquad (15\text{-}57)$$

The net cell reaction must be

$$2AgCl(s) + H_2(g, 1.00 \text{ atm}) \rightleftharpoons 2H^+(x \text{ M}) + 2Cl^-(0.10 \text{ M}) + 2Ag(s) \qquad (15\text{-}58)$$

318

15 / FUNDAMENTALS OF ELECTROCHEMISTRY

The reaction cannot be at equilibrium, because the measured cell potential is 0.503 V.

When we write the Nernst equation for the net cell reaction

$$E(\text{cell}) = 0.222 - \frac{0.059\ 16}{2} \log[H^+]^2[Cl^-]^2 \tag{15-59}$$

we discover that all quantities except $[H^+]$ are known. ($[Cl^-] = 0.10$ M.) *The measured cell potential therefore allows us to find the concentration of H^+ in the left half-cell.*

$$0.503 = 0.222 - \frac{0.059\ 16}{2} \log[H^+]^2[0.10]^2$$

$$\Rightarrow [H^+] = 1.76 \times 10^{-4} \text{ M} \tag{15-60}$$

This, in turn, allows us to evaluate the equilibrium constant for the acid–base reaction that has come to equilibrium in the left half-cell:

> *Question:* Why can we assume that the concentrations of acetic acid and acetate ion are equal to their initial (formal) concentrations?

$$K_A = \frac{[CH_3CO_2^-][H^+]}{[CH_3CO_2H]} = \frac{(0.005\ 0)(1.8 \times 10^{-4})}{0.050} = 1.8 \times 10^{-5} \tag{15-61}$$

The potential of a galvanic cell may serve as a probe, providing information on the concentration of an unknown within the cell. This particular cell behaves as a pH meter, a probe for H^+.

The cell in Figure 15-9 may be thought of as a *probe* to measure the unknown H^+ concentration in the left half-cell. Using this type of cell, we could determine the equilibrium constant for dissociation of any acid or the hydrolysis of any base placed in the left half-cell. The use of electrochemical cells as probes will be explored further in the next chapter.

The problems at the end of this chapter include several brainbusters designed to bring together your knowledge of electrochemistry, chemical equilibrium, solubility, complex formation, and acid–base chemistry. They require you to find the equilibrium constant for a reaction that occurs in only one half-cell. The reaction of interest is *not* the net cell reaction and is not even a redox reaction. A good approach to such problems is outlined below:

The half-reactions *must* involve species that appear in two oxidation states in the cell.

Step 1. Write the anode and cathode half-reactions and their standard potentials. If you choose a half-reaction for which you cannot find E^0, find another way to write the reaction.

Step 2. Write a Nernst equation for the net reaction, and put in all of the known quantities. If all is well, there will be only one unknown in the equation.

Step 3. Solve for the unknown concentration, and use that concentration to solve the chemical equilibrium problem that was originally posed.

EXAMPLE

The cell in Figure 15-10 can be used to measure the formation constant (K_f) of $Hg(EDTA)^{2-}$. The solution in the cathode compartment is prepared from 0.500 mmol

15-6 USING CELLS AS CHEMICAL PROBES

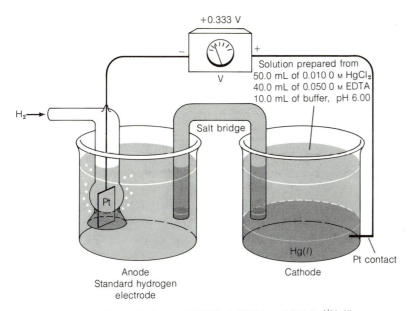

S.H.E.‖Hg(EDTA)$^{2-}$(aq, 0.005 00 M), EDTA(aq, 0.015 0 M)|Hg(l)

Figure 15-10
A galvanic cell that can be used to measure the formation constant for Hg(EDTA)$^{2-}$.

of Hg^{2+} and 2.00 mmol of EDTA in a volume of 0.100 L buffered to pH 6.00. If the measured cell potential is +0.333 V, find the value of K_f for Hg(EDTA)$^{2-}$.

Step 1. The left half-cell is a standard hydrogen electrode for which we can say

Anode: $H_2(g, 1.00 \text{ atm}) \rightleftharpoons 2H^+(1.00 \text{ M}) + 2e^-$ $E^0 = 0$

In the right half-cell, the reaction between Hg^{2+} and EDTA is

$$Hg^{2+} + Y^{4-} \xrightleftharpoons{K_f} HgY^{2-}$$

Since we expect K_f to be large, we will assume that virtually all of the Hg^{2+} has reacted to make HgY^{2-}. Therefore, the concentration of HgY^{2-} is 0.500 mmol/100 mL = 0.005 00 M. The remaining EDTA has a total concentration of (2.00 − 0.50) mmol/100 mL = 0.015 0 M. The cathode compartment therefore contains 0.005 00 M HgY^{2-}, 0.015 0 M EDTA, and a very small, unknown concentration of Hg^{2+}.

The formation constant for HgY^{2-} can be written

$$K_f = \frac{[HgY^{2-}]}{[Hg^{2+}][Y^{4-}]} = \frac{[HgY^{2-}]}{[Hg^{2+}]\alpha_{Y^{4-}}[EDTA]}$$

where [EDTA] is the formal concentration of EDTA not bound to metal. In this cell, [EDTA] = 0.015 0 M. Since we know that [HgY^{2-}] = 0.005 00 M, all we need to find is [Hg^{2+}] in order to evaluate K_f. Since we need to find [Hg^{2+}], let us write the cathode reaction as

Cathode: $Hg^{2+} + 2e^- \rightleftharpoons Hg(l)$ $E^0 = 0.854$ V

320

15 / FUNDAMENTALS OF ELECTROCHEMISTRY

Step 2. The net cell reaction is

$$Hg^{2+} + H_2(g) \rightleftharpoons Hg(l) + 2H^+ \qquad E^0 = 0.854$$

and the Nernst equation is

$$E(\text{cell}) = E^0 - \frac{0.059\ 16}{2} \log \frac{[H^+]^2}{[Hg^{2+}]P_{H_2}}$$

Step 3. Putting in all of the known values allows us to solve for $[Hg^{2+}]$:

$$0.333 = 0.854 - \frac{0.059\ 16}{2} \log \frac{(1.00)^2}{[Hg^{2+}](1.00)}$$

$$[Hg^{2+}] = 2.4 \times 10^{-18}\ M$$

This value of $[Hg^{2+}]$ can be used to evaluate the formation constant for HgY^{2-}:

$$K_f = \frac{[HgY^{2-}]}{[Hg^{2+}]\alpha_{Y^{4-}}[EDTA]} = \frac{(0.005\ 00)}{(2.4 \times 10^{-18})(2.3 \times 10^{-5})(0.015\ 0)} = 6 \times 10^{21}$$

The mixture of EDTA plus $Hg(EDTA)^{2-}$ in the cathode serves as a mercuric ion "buffer" that fixes the concentration of Hg^{2+}. This, in turn, determines the value of the cell potential.

15-7 BIOCHEMISTS USE $E^{0\prime}$

Many of the reactions in living organisms are redox reactions. Perhaps the most important redox reactions are involved in the respiratory process in which molecules of food are oxidized by O_2 to yield energy or metabolic intermediates. The standard reduction potentials that we have been using so far apply to systems in which all activities of reactants and products are unity. If H^+ is involved in the reaction, E^0 applies when pH $= 0$ ($\mathscr{A}_{H^+} = 1$). *Whenever H^+ appears in a redox reaction, or whenever reactants or products are acids or bases, reduction potentials are pH-dependent.*

Since the pH inside a cell is in the neighborhood of 7, the reduction potentials that apply at pH 0 are not particularly appropriate and may be misleading. For example, at pH 0, ascorbic acid (vitamin C) is a more powerful reducing agent than is succinic acid. However, at pH 7, this order is reversed. It is the reducing strength at pH 7, not at pH 0, that is relevant to the chemistry of a living cell.

The formal potential at pH $= 7$ is called $E^{0\prime}$.

The **standard potential** for a redox reaction is defined for a galvanic cell in which all activities are unity. The **formal potential** is the reduction potential that applies under a *specified* set of conditions (including pH, ionic strength, concentration of complexing agents, etc.). Biochemists call the formal potential at pH 7 $\mathbf{E^{0\prime}}$ (read "E zero prime"). Table 15-2 lists $E^{0\prime}$ values for a few biologically important redox couples.

Table 15-2
Reduction potentials of biological interest

Reaction	E^0 (V)	$E^{0\prime}$ (V)
$O_2 + 4H^+ + 4e^- \rightleftharpoons 2H_2O$	+1.229	+0.816
$Fe^{3+} + e^- \rightleftharpoons Fe^{2+}$	+0.770	+0.770
$I_2 + 2e^- \rightleftharpoons 2I^-$	+0.536	+0.536
$O_2(g) + 2H^2 + 2e^- \rightleftharpoons H_2O_2$	+0.69	+0.295
Cytochrome a $(Fe^{3+}) \rightleftharpoons$ cytochrome a (Fe^{2+})	+0.290	+0.290
Cytochrome c $(Fe^{3+}) \rightleftharpoons$ cytochrome c (Fe^{2+})	—	+0.254
2,6-dichlorophenolindophenol $+ 2H^+ + 2e^- \rightleftharpoons$ reduced 2,6-dichlorophenolindophenol	—	+0.22
Dehydroascorbate $+ 2H^+ + 2e^- \rightleftharpoons$ ascorbate	+0.390	+0.058
Fumarate $+ 2H^+ + 2e^- \rightleftharpoons$ succinate	+0.433	+0.031
Methylene blue $+ 2H^+ + 2e^- \rightleftharpoons$ reduced product	+0.532	+0.011
Glyoxylate $+ 2H^+ + 2e^- \rightleftharpoons$ glycolate	—	−0.090
Oxalacetate $+ 2H^+ + 2e^- \rightleftharpoons$ malate	+0.330	−0.102
Pyruvate $+ 2H^+ + 2e^- \rightleftharpoons$ lactate	+0.224	−0.190
Riboflavin $+ 2H^+ + 2e^- \rightleftharpoons$ reduced riboflavin	—	−0.208
FAD $+ 2H^+ + 2e^- \rightleftharpoons FADH_2$	—	−0.219
(Glutathione-S)$_2$ $+ 2H^+ + 2e^- \rightleftharpoons$ 2 glutathione-SH	—	−0.23
Safranine T $+ 2e^- \rightleftharpoons$ leucosafranine T	−0.235	−0.289
$(C_6H_5S)_2 + 2H^+ + 2e^- \rightleftharpoons 2C_6H_5SH$	—	−0.30
$NAD^+ + H^+ + 2e^- \rightleftharpoons NADH$	−0.105	−0.320
$NADP^+ + H^+ + 2e^- \rightleftharpoons NADPH$	—	−0.324
Cystine $+ 2H^+ + 2e^- \rightleftharpoons$ 2 cysteine	—	−0.340
Acetoacetate $+ 2H^+ + 2e^- \rightleftharpoons$ L-β-hydroxybutyrate	—	−0.346
Xanthine $+ 2H^+ + 2e^- \rightleftharpoons$ hypoxanthine $+ H_2O$	—	−0.371
$2H^+ + 2e^- \rightleftharpoons H_2$	0.000	−0.414
Gluconate $+ 2H^+ + 2e^- \rightleftharpoons$ glucose $+ H_2O$	—	−0.44
$SO_4^{2-} + 2e^- + 2H^+ \rightleftharpoons SO_3^{2-} + H_2O$	—	−0.454
$2SO_3^{2-} + 2e^- + 4H^+ \rightleftharpoons S_2O_4^{2-} + 2H_2O$	—	−0.527

Note: A more complete table can be found in the *Handbook of Biochemistry and Molecular Biology*, 3rd ed. (Cleveland: CRC Press, 1976), *Physical and Chemical Data*, Vol. 1, pp. 122–129.

Relation Between E^0 and $E^{0\prime}$

Consider the half-reaction

$$aA + ne^- \rightleftharpoons bB + mH^+ \qquad E^0 \qquad (15\text{-}62)$$

in which A is an oxidized species and B is a reduced species. Both A and B might be acids or bases, as well. The Nernst equation for the reduction is written

$$E = E^0 - \frac{0.059\,16}{n} \log \frac{[B]^b[H^+]^m}{[A]^a} \qquad (15\text{-}63)$$

322

15 / FUNDAMENTALS OF ELECTROCHEMISTRY

The recipe for finding $E^{0\prime}$.

Remember:

For a monoprotic acid:

$$F = [HA] + [A^-]$$

For a diprotic acid:

$$F = [H_2A] + [HA^-] + [A^{2-}]$$

To find $E^{0\prime}$, we must rearrange the Nernst equation to a form in which the log term contains only the *formal concentrations* of A and B raised to the powers a and b, respectively.

$$E = \underbrace{E^0 + \text{other terms}} - \frac{0.059\ 16}{n} \log \frac{F_B^b}{F_A^a} \tag{15-64}$$

All of this is
called $E^{0\prime}$
when pH $= 7$

The entire collection of terms over the brace, evaluated at pH $= 7$, is called $E^{0\prime}$.

To convert $[A]$ or $[B]$ to F_A or F_B, we make use of equations such as 13-6, 13-7, 13-19, 13-20, or 13-21, which relate the formal (i.e., total) concentration of *all* forms of an acid or a base to its concentration in a *particular* form. These useful equations are repeated below:

Monoprotic system: $\quad [HA] = \alpha_0 F = \dfrac{[H^+]F}{[H^+] + K_A} \tag{15-65}$

$$[A^-] = \alpha_1 F = \frac{K_A F}{[H^+] + K_A} \tag{15-66}$$

Diprotic system: $\quad [H_2A] = \alpha_0 F = \dfrac{[H^+]^2 F}{[H^+]^2 + [H^+]K_1 + K_1 K_2} \tag{15-67}$

$$[HA^-] = \alpha_1 F = \frac{K_1[H^+]F}{[H^+]^2 + [H^+]K_1 + K_1 K_2} \tag{15-68}$$

$$[A^{2-}] = \alpha_2 F = \frac{K_1 K_2 F}{[H^+]^2 + [H^+]K_1 + K_1 K_2} \tag{15-69}$$

where K_A is the acid dissociation constant for HA and K_1 and K_2 are the acid dissociation constants for H_2A.

EXAMPLE

Find $E^{0\prime}$ for the reaction

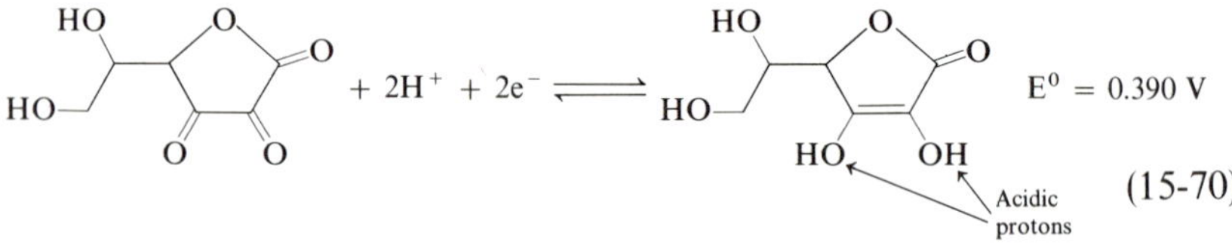

dehydroascorbic acid
(oxidized)

ascorbic acid
(vitamin C)
(reduced)

$E^0 = 0.390$ V

(15-70)

$$pK_1 = 4.10 \qquad pK_2 = 11.79$$

Abbreviating dehydroascorbic acid as D, and ascorbic acid as H_2A, we can rewrite the reduction as

$$D + 2H^+ + 2e^- \rightleftharpoons H_2A$$

for which the Nernst equation is

$$E = E^0 - \frac{0.05916}{2} \log \frac{[H_2A]}{[D][H^+]^2} \qquad (15\text{-}71)$$

D is not an acid or a base, so $F_D = [D]$. For the diprotic acid H_2A, we can use Equation 15-67 to express $[H_2A]$ in terms of F_{H_2A}:

$$[H_2A] = \frac{[H^+]^2 F_{H_2A}}{[H^+]^2 + [H^+]K_1 + K_1K_2}$$

Putting these values into Equation 15-71 gives

$$E = E^0 - \frac{0.05916}{2} \log \left(\frac{\dfrac{[H^+]^2 F_{H_2A}}{[H^+]^2 + [H^+]K_1 + K_1K_2}}{F_D[H^+]^2} \right)$$

which can be rearranged to the form

$$E = \underbrace{E^0 - \frac{0.05916}{2} \log \frac{1}{[H^+]^2 + [H^+]K_1 + K_1K_2}}_{E^{0\prime}} - \frac{0.05916}{2} \log \frac{F_{H_2A}}{F_D} \qquad (15\text{-}72)$$

Putting the values of E^0, K_1, and K_2 into Equation 15-72 and setting $[H^+] = 10^{-7.00}$, we find $E^{0\prime} = +0.062$ V.

Figure 15-11 shows how the formal potential for Reaction 15-70 depends on pH. The potential decreases as the pH increases, until pH $\approx pK_2$. Above pK_2, A^{2-} is the dominant form of ascorbic acid, and no protons are involved in the net redox reaction. Therefore, the potential becomes independent of pH.

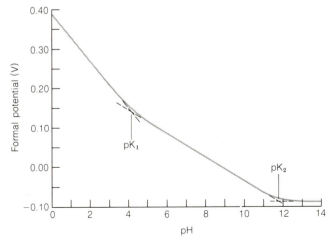

Figure 15-11
Formal reduction potential of ascorbic acid, showing its dependence on pH. This is a graph of the function labeled $E^{0\prime}$ in Equation 15-72.

Summary

The work done when a charge of q coulombs passes through a potential difference of E volts is E · q. The maximum work that can be done by a spontaneous chemical change is expressed as Work = $-\Delta G$. If that chemical change produces an electric potential, the relation between the free energy and the potential is $\Delta G = -nFE$. Ohm's law (I = E/R) describes the relation among current, voltage, and resistance in an electric circuit. It can be combined with the definitions of work and power (P = Work/s) to give $P = E \cdot I = I^2R$.

A galvanic cell is one that uses a spontaneous redox reaction to produce electricity. The electrode at which oxidation occurs is called the anode, and the one at which reduction occurs is the cathode. Our convention in calculating a cell potential assumes that the left half-cell is the anode and is connected to the negative terminal of a potentiometer. The two half-cells are occasionally in the same compartment, but more often they must be separated by a salt bridge that allows ionic diffusion but prevents rapid mixing of the two half-cells. If a half-cell is connected as a cathode to a standard hydrogen electrode, the measured potential is called the standard potential. The word "standard" means that the activities of all reactants and products are unity. The standard potential for any complete redox reaction is given by $E^0 = E^0$(cathode half-reaction) $- E^0$(anode half-reaction). If several half-reactions can be added to give another half-reaction, the standard potential of the net half-reaction can be found by equating the free energy of the net half-reaction to the sum of free energies of the other half-reactions.

When the activities of reactants and products are not unity, the potential of a cell is given by the Nernst equation: $E = E^0 - (0.059\ 16/n) \log Q$ (at 25°C), where Q is the reaction quotient. The reaction quotient has the same form as the equilibrium constant, but it is evaluated with the concentrations existing at the time of interest. When Q = K, the cell is at equilibrium and E = 0. Therefore, $E^0 = (0.059\ 16/n) \log K$, and the equilibrium constant for a chemical reaction may be calculated from the standard potential for that reaction.

Complex equilibria can be studied by making the system part of a galvanic cell. If we measure the cell potential and know the concentrations (activities) of all but one of the reactants and products, the Nernst equation allows us to compute the concentration of that one unknown species. In this way a galvanic cell serves as a probe for that one species.

The Nernst equation can be written for a half-reaction as well as for a complete reaction. Biochemists prefer to use the formal potential of a half-reaction at pH 7 ($E^{0\prime}$) instead of the standard potential (E^0), which applies at pH 0. The value of $E^{0\prime}$ is found by writing the Nernst equation for the desired half-reaction and grouping together all terms except the logarithm containing the formal concentrations of reactant and product. The combination of terms, evaluated at pH 7, is $E^{0\prime}$.

Terms to Understand

ampere	Ohm's law
anode	oxidant
cathode	oxidation
coulomb	oxidizing agent
current	power
$E^{0\prime}$	reaction quotient
electric potential	redox reaction
electrode	reducing agent
Faraday constant	reductant
formal potential	reduction
galvanic cell	salt bridge
half-reaction	single-electrode potential
joule	standard hydrogen electrode
Latimer diagram	standard reduction potential
Nernst equation	watt

Exercises

15-A. The mercury cell used to power heart pacemakers runs on the following reaction:

$$Zn(s) + HgO(s) \rightarrow ZnO(s) + Hg(l) \quad E^0 = 1.35 \text{ V}$$

If the power required to operate the pacemaker is 0.010 0 W, how many kilograms of HgO will be consumed in 365 days? How many pounds of HgO is this? (1 pound = 453.6 g)

15-B. Calculate E^0 and K for each of the following reactions.
(a) $I_2(s) + 5Br_2(aq) + 6H_2O$
    $\rightleftharpoons 2IO_3^- + 10Br^- + 12H^+$
(b) $Cr^{2+} + Fe(s) \rightleftharpoons Fe^{2+} + Cr(s)$
(c) $Mg(s) + Cl_2(g) \rightleftharpoons Mg^{2+} + 2Cl^-$
(d) $5MnO_2(s) + 4H^+ \rightleftharpoons 3Mn^{2+} + 2MnO_4^- + 2H_2O$
(e) $Ag^+ + 2S_2O_3^{2-} \rightleftharpoons Ag(S_2O_3)_2^{3-}$
(f) $CuI(s) \rightleftharpoons Cu^+ + I^-$

15-C. Calculate the potential of each cell below.
(a) $Fe(s)|FeBr_2(0.010 \text{ M})\|NaBr(0.050 \text{ M})$
    $|Br_2(l)|Pt(s)$
(b) $Cu(s)|Cu(NO_3)_2(0.020 \text{ M})$
    $\|Fe(NO_3)_2(0.050 \text{ M})|Fe(s)$
(c) $Hg(l)|Hg_2Cl_2(s)|KCl(0.060 \text{ M})\|KCl(0.040 \text{ M})$
    $|Cl_2(g, 0.50 \text{ atm})|Pt(s)$

15-D. Consider the cell below. The anode reaction can be written in *either* of two ways:

$$Ag(s) + I^- \rightarrow AgI(s) + e^- \quad (1)$$

or

$$Ag(s) \rightarrow Ag^+ + e^- \quad (2)$$

The cathode reaction is

$$H^+ + e^- \rightarrow \tfrac{1}{2}H_2(g) \quad (3)$$

(a) Using Reactions 2 and 3, calculate E^0 and write the Nernst equation for the cell.
(b) Use the value of K_{sp} for AgI to compute $[Ag^+]$ and find the cell potential.
(c) Suppose, instead, that you wish to describe the cell with Reactions 1 and 3. We know that the cell potential (E, not E^0) must be the same, no matter which description we use. Write the Nernst equation for Reactions 1 and 3 and use it to solve for E^0 in Reaction 1. Compare your answer with the value in Appendix H.

15-E. Calculate the potential of the cell

$$Cu(s)|Cu^{2+}(0.030 \text{ M})\|K^+Ag(CN)_2^-(0.010 \text{ M}),$$
$$HCN(0.10 \text{ F}), \text{ buffer to pH } 8.21|Ag(s)$$

You may wish to refer to the following reactions:

$$Ag(CN)_2^- + e^- \rightleftharpoons Ag(s) + 2CN^- \quad E^0 = -0.310 \text{ V}$$

$$HCN \rightleftharpoons H^+ + CN^- \quad pK_A = 9.21$$

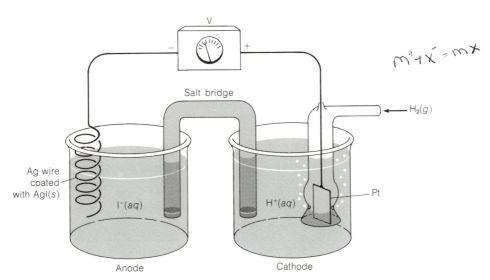

$Ag(s)|AgI(s)|NaI(0.10 \text{ M})\|HCl(0.10 \text{ M})|H_2(g, 0.20 \text{ atm})|Pt(s)$

15-F. (a) Write a balanced equation for the reaction $PuO_2^+ \rightarrow Pu^{4+}$ and calculate E^0 for the reaction.

$$PuO_2^{2+} \xrightarrow{+0.93} PuO_2^+ \xrightarrow{?} Pu^{4+} \xrightarrow{+0.97} Pu^{3+}$$
$$\underset{1.02}{\underline{\qquad\qquad\qquad\qquad}}$$

(b) Predict whether or not an equimolar mixture of PuO_2^{2+} and PuO_2^+ will oxidize H_2O to O_2 at a pH of 2.00. You may assume that $P_{O_2} = 0.20$ atm. Will O_2 be liberated at pH 7.00?

15-G. Calculate the potential of the cell below, in which KHP is potassium acid phthalate, the monopotassium salt of phthalic acid.

$Hg(l)|Hg_2Cl_2(s)|KCl(0.10\ M)\|KHP(0.050\ M)$
$|H_2(g, 1.00\ atm)|Pt(s)$

15-H. The cell below has a potential of -0.321 V:

$Hg(l)|Hg(NO_3)_2(0.001\ 0\ M), KI(0.010\ M)\|S.H.E.$

Calculate the equilibrium constant for the reaction

$$Hg^{2+} + 4I^- \rightleftharpoons HgI_4^{2-}$$

You may assume that the only forms of mercury in solution are Hg^{2+} and HgI_4^{2-}.

15-I. The formation constant for $Cu(EDTA)^{2-}$ is 6.3×10^{18}, and the value of E^0 for the reaction $Cu^{2+} + 2e^- \rightleftharpoons Cu(s)$ is $+0.337$ V. From this information, find E^0 for the reaction

$$CuY^{2-} + 2e^- \rightleftharpoons Cu(s) + Y^{4-}$$

15-J. Using the reaction below, state which compound, $H_2(g)$ or glucose, is the more powerful reducing agent at pH = 0.00.

$$\begin{array}{c}CO_2H\\|\\HCOH\\|\\HOCH\\|\\HCOH\\|\\HCOH\\|\\CH_2OH\end{array} + 2H^+ + 2e^- \rightleftharpoons \begin{array}{c}CHO\\|\\HCOH\\|\\HOCH\\|\\HCOH\\|\\HCOH\\|\\CH_2OH\end{array} + H_2O \quad E^{0\prime} = -0.45\ V$$

gluconic acid
$pK_A = 3.56$

glucose
(no acidic protons)

15-K. Living cells convert energy derived from sunlight or combustion of food into energy-rich ATP (adenosine triphosphate) molecules. For ATP synthesis, $\Delta G = +34.5$ kJ/mol. This energy is then made available to the cell when ATP is hydrolyzed to ADP (adenosine diphosphate). In animals, ATP is synthesized when protons pass through a complex enzyme in the mitochondrial membrane.[†] Two factors account for the movement of protons through this enzyme into the mitochondrion (see the figure below). One factor is the gradient of concentration of H^+, higher outside the mitochondrion than inside. This gradient arises because protons are *pumped* out of the mitochondrion by enzymes involved in the oxidation of food molecules. A second

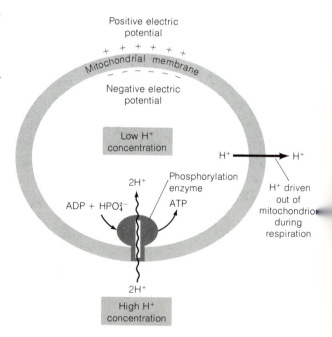

[†] For an interesting discussion of this topic, see "How Cells Make ATP," by P. C. Hinkle and R. E. McCarty, *Scientific American* (March 1978), p. 104.

factor is that the inside of the mitochondrion is negatively charged with respect to the outside. The synthesis of one ATP molecule requires two protons to pass through the phosphorylation enzyme.

(a) The difference in free energy when a molecule travels from a region of high activity to a region of low activity is given by

$$\Delta G = -RT \ln \frac{\mathscr{A}_{high}}{\mathscr{A}_{low}}$$

How big must the pH difference be if the passage of two protons is to provide enough energy to synthesize one ATP molecule? ·

(b) pH differences this large have not been observed in mitochondria. How great an electric potential difference between inside and outside is necessary for the movement of two protons to provide energy to synthesize ATP? In answering this question, neglect any contribution from the pH difference.

(c) It is thought that the energy for ATP synthesis is provided by *both* the pH difference and the electric potential. If the pH difference is 1.00 pH unit, how many millivolts must be the potential difference?

Problems

15-1. (a) How many electrons are in one coulomb?
(b) How many coulombs are there in one mole of charge?

15-2. Use the numerical values of R and F to derive Equation 15-29 from Equation 15-27. What would be the value of the numerical constant in front of the log term at $0°C$? At $37°C$?

15-3. A 6.00 V battery is connected across a 2.00 kΩ resistor.
(a) How many electrons per second flow through the circuit?
(b) How many joules of heat are produced for each electron?
(c) If the circuit operates for 30.0 min, how many moles of electrons will have flowed through the resistor?
(d) What voltage would the battery need to deliver for the power to be 100 watts?

15-4. A light-weight rechargable battery developed for electric automobiles uses the following cell:

$$Zn(s)|ZnCl_2(aq)||Cl^-(aq)|Cl_2(l)|C(s)$$

If the battery delivers a constant current of 1.00×10^3 A for 1.00 hr, how many kilograms of Cl_2 will be consumed?

15-5. Which will be the strongest oxidizing agent under standard conditions (all activities = 1): HNO_2, Se, UO_2^{2+}, Cl_2, H_2SO_3, or MnO_2?

15-6. Which will be the strongest reducing agent under standard conditions (all activities = 1): Se, Sn^{4+}, Cr^{2+}, Mg^{2+}, or $Fe(CN)_6^{4-}$?

15-7. Use Le Châtelier's principle and half-reactions from Appendix H to find which of the following become stronger oxidizing agents as the pH is lowered. Which are unchanged, and which become weaker?

$$Cl_2 \qquad Cr_2O_7^{2-} \quad Fe^{3+} \qquad MnO_4^- \qquad IO_3^-$$

chlorine dichromate ferric permanganate iodate

15-8. Calculate E^0, ΔG^0, and K for each of the following reactions.
(a) $4Co^{3+} + 2H_2O \rightleftharpoons 4Co^{2+} + O_2(g) + 4H^+$
(b) $Cu(s) + Cu^{2+} \rightleftharpoons 2Cu^+$
(c) $Ag(S_2O_3)_2^{3-} + Fe(CN)_6^{4-}$
$\rightleftharpoons Ag(s) + 2S_2O_3^{2-} + Fe(CN)_6^{3-}$

(d) $2Cu^{2+} + 2I^- + HO\!-\!\langle\bigcirc\rangle\!-\!OH$

hydroquinone

$\rightleftharpoons 2CuI(s) + O\!=\!\langle\bigcirc\rangle\!=\!O + 2H^+$

quinone

15-9. A half-cell was prepared by dipping a Pt wire in a beaker containing an equimolar mixture of Cr^{2+} and Cr^{3+}. Another half-cell contained a Tl rod immersed in 1.00 M $TlClO_4$.
(a) Use line notation to describe this cell.
(b) Calculate the cell potential.
(c) Write the spontaneous net cell reaction.
(d) When the two electrodes are connected by a salt bridge and a wire, which terminal (Pt or Tl) will be the anode?

15-10. A solution contains 0.100 M Ce^{3+}, 1.00×10^{-4} M Ce^{4+}, 1.00×10^{-4} M Mn^{2+}, 0.100 M MnO_4^-, and 1.00 M $HClO_4$.
  (a) Write a balanced net reaction that can occur between the species in this solution.
  (b) Calculate ΔG^0 and K for the reaction.
  (c) Calculate E for the conditions given above.
  (d) Calculate ΔG for the conditions given above.
  (e) At what pH would the concentrations of Ce^{4+}, Ce^{3+}, Mn^{2+}, and MnO_4^- listed above be in equilibrium at 298 K? That is, at what pH would there be no net reaction?

15-11. (a) Draw a picture of the following cell, showing the location of each species:

$$Pt(s)|Fe^{3+}(aq), Fe^{2+}(aq)\|Cr_2O_7^{2-}(aq),$$
$$Cr^{3+}(aq), HA(aq)|Pt(s)$$

  (b) Write a balanced equation for the cell reaction.

15-12. (a) In the presence of cyanide ion, the reduction potential of Fe(III) is decreased from 0.770 to 0.356 V.

$$Fe^{3+} + e^- \rightleftharpoons Fe^{2+} \qquad E^0 = 0.770 \text{ V}$$
ferric ferrous

$$Fe(CN)_6^{3-} + e^- \rightleftharpoons Fe(CN)_6^{4-} \qquad E^0 = 0.356 \text{ V}$$
ferricyanide ferrocyanide

Which ion, Fe^{3+} or Fe^{2+}, is stabilized more by complexing with CN^-?
  (b) Using Appendix H, answer the same question when the ligand is phenanthroline instead of cyanide.

phenanthroline

15-13. For each picture at the bottom of the page, write the line notation to describe the cell. Also write the anode and cathode reactions.

15-14. Draw a picture of each of the following cells, showing the location of each chemical species. For each cell, write the anode and cathode reactions.
  (a) $Au(s)|Fe(CN)_6^{4-}(aq), Fe(CN)_6^{3-}(aq)$
      $\|Ag(CN)_2^-(aq), KCN(aq)|Ag(s)$
  (b) $Pt(s)|Hg(l)|Hg_2Cl_2(s)|KCl(aq)$
      $\|ZnCl_2(aq)|Zn(s)$

15-15. Consider the cell

$$\text{S.H.E.}\|Ag(S_2O_3)_2^{3-}(aq, 0.010 \text{ M}),$$
$$S_2O_3^{2-}(aq, 0.050 \text{ M})|Ag(s)$$

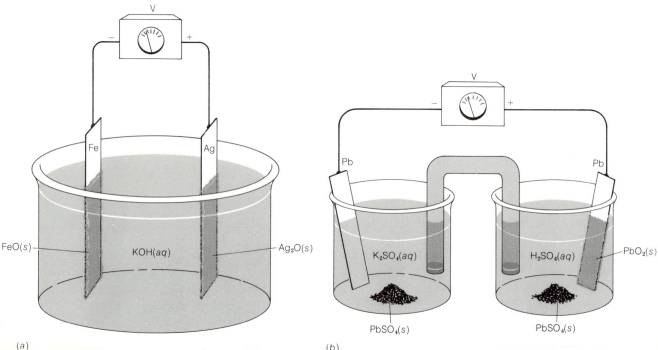

(a)

(b)

(a) Using the half-reaction $Ag(S_2O_3)_2^{3-} + e^- \rightleftharpoons Ag(s) + 2S_2O_3^{2-}$, calculate the cell potential (E, not E^0).

(b) Alternatively, the cell could have been described with the half-reaction $Ag^+ + e^- \rightleftharpoons Ag(s)$. Using the cell potential from part a, calculate $[Ag^+]$ in the right half-cell.

(c) Use the answer to part b to find the formation constant for the reaction

$$Ag^+ + 2S_2O_3^{2-} \xrightleftharpoons{K_f} Ag(S_2O_3)_2^{3-}$$
thiosulfate

15-16. (a) Write the line notation for the cell below.

(b) Calculate the cell potential (E, not E^0), and state the direction in which electrons will flow through the potentiometer.

(c) The left half-cell was loaded with 14.3 mL of $Br_2(l)$ (density = 3.12 g/mL). The aluminum electrode contains 12.0 g of Al. Which element, Br_2 or Al, is the limiting reagent in this cell? (That is, which reagent will be used up first?)

(d) If the cell is somehow operated under conditions in which it produces a constant voltage of 1.50 V, how much electrical work will have been done when 0.231 mL of $Br_2(l)$ has been consumed?

(e) If the potentiometer is replaced by a 1.20 kΩ resistor, and if the heat dissipated by the resistor is 1.00×10^{-4} J/s, at what rate (grams per second) is Al(s) dissolving? (In this question the voltage is not 1.50 V.)

15-17. Without neglecting activities, calculate the potential of the cell:

$$Ni(s)|NiSO_4(0.020 \text{ M})||CuCl_2(0.030 \text{ M})|Cu(s)$$

15-18. Suppose that the concentrations of NaF and KCl were each 0.10 M in the cell $Pb(s)|PbF_2(s)|F^-(aq)||Cl^-(aq)|AgCl(s)|Ag(s)$.

(a) Using the half-reactions $Pb(s) + 2F^- \rightleftharpoons PbF_2(s) + 2e^-$ and $AgCl(s) + e^- \rightleftharpoons Ag(s) + Cl^-$, calculate the cell potential.

(b) Now calculate the cell potential using the reactions $Pb(s) \rightleftharpoons Pb^{2+} + 2e^-$ and $Ag^+ + e^- \rightleftharpoons Ag(s)$. For this part, you will need the solubility products for PbF_2 and AgCl.

15-19. The quinhydrone electrode was introduced in 1921 as a means of measuring pH. A cell employing this electrode is shown on the following page. The solution whose pH is to be measured is placed in the left half-cell, which also contains a 1:1 mole ratio of quinone and hydroquinone. The cell reaction is

HO—⬡—OH + $Hg_2Cl_2(s) \rightleftharpoons$
hydroquinone

O=⬡=O + $2H^+ + 2Hg(l) + 2Cl^-$
quinone

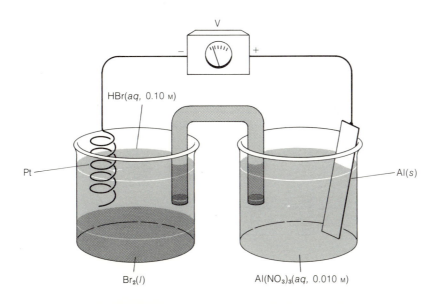

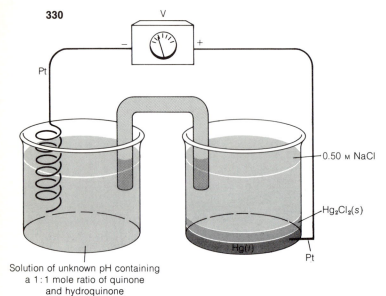

Solution of unknown pH containing a 1:1 mole ratio of quinone and hydroquinone

(a) Ignoring activities and using the relation pH = $-\log[H^+]$, the Nernst equation can be changed into the form

$$E(\text{cell}) = A + B \cdot pH$$

where A and B are constants. Find the numerical values of A and B at 25°C.

(b) If the pH were 4.50, in which direction would electrons flow through the potentiometer?

15-20. The potential for the cell below is 0.490 V. Find K_B for the organic base, RNH_2.

$$Pt(s)|H_2(1.00 \text{ atm})|RNH_2(aq, 0.10 \text{ M}),$$
$$RNH_3^+Cl^-(aq, 0.050 \text{ M})\|S.H.E.$$

15-21. Do not ignore activity coefficients in this problem. If the potential for the cell below is 0.482 V, find K_{sp} for $Cu(IO_3)_2$.

$$Ni(s)|NiSO_4(0.025 \text{ M})\|KIO_3(0.10 \text{ M})$$
$$|Cu(IO_3)_2(s)|Cu(s)$$

15-22. Calculate the standard potential for the half-reaction

$$Pd(OH)_2(s) + 2e^- \rightleftharpoons Pd(s) + 2OH^-$$

given that K_{sp} for $Pd(OH)_2$ is 3×10^{-28} and given that the standard potential for the reaction $Pd^{2+} + 2e^- \rightleftharpoons Pd(s)$ is 0.915 V.

15-23. The solubility of Br_2 (F.W. 159.808) in water is 29 g/L. Given

$$Br_2(aq) + 2e^- \rightleftharpoons 2Br^- \quad E^\circ = 1.087 \text{ V}$$

find the potential for the half-reaction

$$Br_2(l) + 2e^- \rightleftharpoons 2Br^- \quad E^\circ = ?$$

15-24. For the cell below, E (not E°) = -0.289 V. Write the net cell reaction and calculate its equilibrium constant. Do not use E° values from Appendix H to answer this question.

$$Pt(s)|VO^{2+}(0.116 \text{ M}), V^{3+}(0.116 \text{ M}),$$
$$H^+(1.57 \text{ M})\|Sn^{2+}(0.031\ 8 \text{ M}), Sn^{4+}(0.031\ 8 \text{ M})|Pt(s)$$

15-25 The following cell has a potential of 0.956 V. Find K_A for formic acid, HCO_2H.

$$Pt(s)|UO_2^{2+}(0.050 \text{ M}), U^{4+}(0.050 \text{ M}), HCO_2H(0.10 \text{ M}),$$
$$HCO_2Na(0.30 \text{ M})\|Fe^{3+}(0.050 \text{ M}), Fe^{2+}(0.025 \text{ M})|Pt(s)$$

15-26. Using the reaction

$$HPO_4^{2-} + 2H^+ + 2e^- \rightleftharpoons HPO_3^{2-} + H_2O$$
$$E^\circ = -0.234 \text{ V}$$

and any acid dissociation constants from Appendix G, calculate E° for the reaction

$$H_2PO_4^- + H^+ + 2e^- \rightleftharpoons HPO_3^{2-} + H_2O$$

15-27. Given the following information,

$$FeY^- + e^- \rightleftharpoons Fe^{2+} + Y^{4-} \quad E^\circ = -0.730 \text{ V}$$
$$FeY^{2-} \quad K_f = 2.1 \times 10^{14}$$
$$FeY^- \quad K_f = 1.3 \times 10^{25}$$

calculate the standard potential for the reaction

$$FeY^- + e^- \rightleftharpoons FeY^{2-}$$

where Y is EDTA.

15-28. Write a balanced chemical equation (in acidic solution) for the reaction represented by the question mark below. Calculate E° for the reaction.

$$BrO_3^- \xrightarrow{0.54} BrO^- \xrightarrow{0.45} Br_2 \xrightarrow{1.06} Br^-$$

with overall 0.61 above and ? below.

15-29. Write a balanced chemical equation (in acid solution) for the reaction represented by the question mark below. Calculate E^0 for the reaction.

$$HNO_3 \xrightarrow{0.775} NO_2 \xrightarrow{1.093} HNO_2 \xrightarrow{?} NO$$
$$\underset{0.957}{\xleftarrow{\hspace{4cm}}}$$

15-30. What must be the relationship between E_1^0 and E_2^0 if the species X^+ is to disproportionate spontaneously to X^{3+} and $X(s)$? Write a balanced equation for the disproportionation.

$$X^{3+} \xrightarrow{E_1^0} X^+ \xrightarrow{E_2^0} X(s)$$

15-31. Calculate $E^{0\prime}$ for the reaction

$$H_2C_2O_4 + 2H^+ + 2e^- \rightleftharpoons 2HCO_2H$$
$$E^0 = 0.074 \text{ V}$$

15-32. The standard reduction potential of anthraquinone-2,6-disulfonate is 0.229 V.

[structure: anthraquinone-2,6-disulfonate] $SO_3^- + 2H^+ + 2e^-$

anthraquinone-2,6-disulfonate

[structure: reduced form with OH groups and SO_3^-, Acidic protons]

The reduced product is a diprotic acid with $pK_1 = 8.10$ and $pK_2 = 10.52$. Calculate $E^{0\prime}$ (pH = 7) for anthraquinone-2,6-disulfonate.

15-33. Suppose that HOx is a monoprotic acid with dissociation constant 1.4×10^{-5} and H_2Red is a diprotic acid with dissociation constants of 3.6×10^{-4} and 8.1×10^{-8}. Calculate E^0 for the reaction

$$HOx + e^- \rightleftharpoons H_2Red \qquad E^{0\prime} = 0.062 \text{ V}$$

15-34. Given the information below, find K_A for nitrous acid, HNO_2.

$$NO_3^- + 3H^+ + 2e^- \rightleftharpoons HNO_2 + H_2O$$
$$E^0 = 0.940 \text{ V}$$
$$E^{0\prime} = 0.433 \text{ V}$$

15-35. The potential of the cell below is -0.246 V. The right half-cell contains the metal ion, M^{2+}, whose standard reduction potential is -0.266 V.

$$M^{2+} + 2e^- \rightleftharpoons M(s) \qquad E^0 = -0.266 \text{ V}$$

Calculate K_f for the metal–EDTA complex.

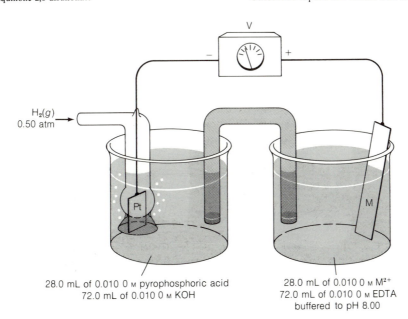

28.0 mL of 0.010 0 M pyrophosphoric acid
72.0 mL of 0.010 0 M KOH

28.0 mL of 0.010 0 M M^{2+}
72.0 mL of 0.010 0 M EDTA
buffered to pH 8.00

15-36. For the cell at the bottom of the page, the half-reactions can be written

Anode: $Pb(s) \rightleftharpoons Pb^{2+} + 2e^-$

Cathode: $Pb^{2+} + 2e^- \rightleftharpoons Pb(s)$

(a) Given that K_{sp} for $Pb(HPO_4)(s)$ is 2.0×10^{-10}, find $[HPO_4^{2-}]$ in the left half-cell.
(b) If the measured cell potential is 0.097 V, calculate K_{sp} for $PbF_2(s)$.

15-37. The cell below was constructed to find the difference in K_{sp} between two naturally occurring forms of $CaCO_3(s)$, called *calcite* and *aragonite*.[†]

$Pb(s)|PbCO_3(s)|CaCO_3(s, calcite)|$
    aqueous buffer(pH 7.00)$\|$aqueous buffer(pH 7.00)
    $|CaCO_3(s, aragonite)|PbCO_3(s)|Pb(s)$

Each compartment of the cell contains a mixture of solid $PbCO_3$ ($K_{sp} = 7.4 \times 10^{-14}$) and either calcite or aragonite, both of which have $K_{sp} \approx 5 \times 10^{-9}$. Each solution was buffered to pH 7.00 with an inert buffer, and the cell was completely isolated from atmospheric CO_2. The measured cell potential was -1.8 mV. Find the ratio of solubility products for calcite and aragonite.

$$\frac{K_{sp} \text{ (for calcite)}}{K_{sp} \text{ (for aragonite)}} = ?$$

15-38. The oxidized form (Ox) of a flavoprotein that functions as a one-electron reducing agent has a molar absorptivity (ε) of 1.12×10^4 M^{-1} cm^{-1} at 457 nm at pH = 7.00.[‡] For the reduced form (Red), $\varepsilon = 3.82 \times 10^3$ at 457 nm at pH 7.00.

$$Ox + e^- \rightleftharpoons Red \qquad E^{0\prime} = -0.128 \text{ V}$$

The substrate (S) is the molecule reduced by the protein.

$$Red + S \rightleftharpoons Ox + S^-$$

Both S and S$^-$ are colorless. A solution at pH 7.00 was prepared by mixing enough protein plus substrate (Red + S) so that each would have a concentration of 5.70×10^{-5} M if they did not react with each other. The absorbance at 457 mm was 0.500 in a 1.00 cm cell.

(a) Calculate the concentrations of Ox and Red from the absorbance data.
(b) Calculate the concentrations of S and S$^-$.
(c) Calculate the value of $E^{0\prime}$ for the reaction $S + e^- \rightleftharpoons S^-$.

[†] The cell in this problem would not give an accurate result because of the *junction potential* at each liquid junction (Section 16-3). A clever way around this problem, using a cell without any liquid junctions, is described by P. A. Rock, *J. Chem. Ed.*, **52**, 787 (1975).

[‡] This problem is an application of Beer's law, which you can read about in Sections 20-1 and 20-2.

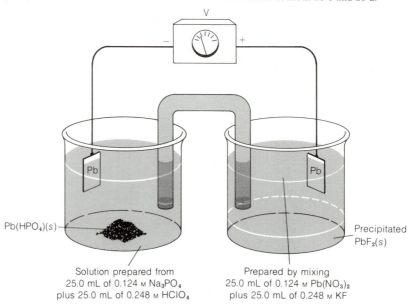

16 / Electrodes and Potentiometry

We saw in the last chapter that a galvanic cell can serve as a *probe* for one chemical species when the activities of all other species are known. That is, if we know the activites of all but one species in a cell, the cell potential tells us the activity of that one species.

Imagine a solution containing an **electroactive species** whose activity (concentration) we wish to measure. An electroactive species is one that can donate or accept electrons from an electrode. We can turn the unknown into a half-cell by inserting an electrode (such as a Pt wire) into the solution to transfer electrons to or from the species of interest. Since this electrode responds directly to the analyte, it is called the **indicator electrode.** We then connect this half-cell to a second half-cell via a salt bridge. The second half-cell should have a known, fixed composition so that it contributes a known, constant potential. Since the second half-cell contributes a constant potential, it is called a **reference electrode.** The net cell potential is composed of a constant potential from the reference electrode and a variable potential that reflects changes in the analyte activity.

Indicator electrode: responds to analyte activity

Reference electrode: maintains a fixed (reference) potential

16-1 REFERENCE ELECTRODES

Suppose that you have a solution with variable amounts of Fe^{2+} and Fe^{3+}. If you are clever, you can make this solution part of a cell in a way that the cell potential will tell you the ratio $[Fe^{2+}]/[Fe^{3+}]$. Figure 16-1 shows one way to do this. A Pt wire is inserted as an indicator electrode through which Fe^{3+} can receive electrons or Fe^{2+} can lose electrons. The left half-cell serves to complete the galvanic cell and contributes a known, constant potential.

333

16 / ELECTRODES AND POTENTIOMETRY

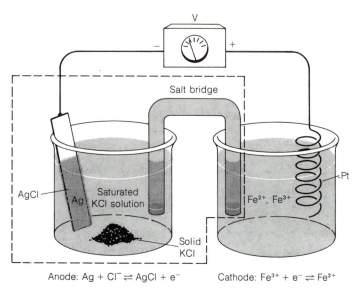

Figure 16-1
A galvanic cell that can be used to measure the ratio $[Fe^{2+}]/[Fe^{3+}]$ in the right half-cell. The Pt wire is the *indicator electrode*, and the entire left half-cell plus salt bridge (enclosed by the dashed line) can be considered a *reference electrode*.

The net cell reaction can be written as follows:

Anode: $\quad Ag(s) + Cl^- \rightleftharpoons AgCl(s) + e^- \quad E^0 = 0.222$ V (16-1)

Cathode: $\quad Fe^{3+} + e^- \rightleftharpoons Fe^{2+} \quad E^0 = 0.770$ V (16-2)

Net reaction: $\quad Fe^{3+} + Ag(s) + Cl^- \rightleftharpoons Fe^{2+} + AgCl(s) \quad E^0 = +0.548$ V (16-3)

and the Nernst equation has the form

$$E(\text{cell}) = +0.548 \text{ V} - \frac{0.059\,16}{1} \log \frac{[Fe^{2+}]}{[Fe^{3+}][Cl^-]} \quad (16\text{-}4)$$

But the concentration of Cl^- in the left half-cell is constant, fixed by the solubility of KCl, with which the solution is saturated. Since $[Cl^-]$ is constant, it can be taken out of the log term, and the Nernst equation can be rearranged to the form

$$E(\text{cell}) = \underbrace{0.548 - 0.059\,16 \log[Cl^-]}_{\text{A constant}} - 0.059\,16 \log \frac{[Fe^{2+}]}{[Fe^{3+}]} \quad (16\text{-}5)$$

Equation 16-5 tells us that the cell potential is proportional to the logarithm of the ratio $[Fe^{2+}]/[Fe^{3+}]$.

The half-cell on the left in Figure 16-1 can be thought of as a *reference electrode*. We can picture the cell and salt bridge enclosed by the dashed

---

Actually, the potential tells us the ratio of activities, $\mathscr{A}_{Fe^{2+}}/\mathscr{A}_{Fe^{3+}}$. However, we will neglect activity coefficients and continue to write the Nernst equation in terms of concentrations instead of activities.

line as a single unit dipped into the analyte solution, as shown in Figure 16-2. The Pt wire is the indicator electrode, whose potential responds to the ratio $[Fe^{2+}]/[Fe^{3+}]$. The reference electrode is necessary to complete the redox reaction and to provide a *constant* reference potential. Changes in the cell potential can therefore be unambiguously assigned to changes in the ratio $[Fe^{2+}]/[Fe^{3+}]$.

How Analytical Chemists Write the Nernst Equation

In dealing with electrodes, it is convenient to divide the Nernst equation into two parts, one from each half-cell. The conventional treatment is to write both the anode and cathode reactions as *reductions* and to write a single-electrode Nernst equation for each. We then calculate the cell potential by subtracting the single-electrode potential, E(anode), from the single-electrode potential, E(cathode):

$$\boxed{E(\text{cell}) = E(\text{cathode}) - E(\text{anode})} \quad (16\text{-}6)$$

for both reactions written as *reductions*.

Equation 16-6 applies to the cell in Figure 16-1 as follows:

Anode: $\quad AgCl(s) + e^- \rightleftharpoons Ag(s) + Cl^- \quad (16\text{-}7)$

$$E(\text{anode}) = 0.222 \text{ V} - \frac{0.059\,16}{1} \log[Cl^-] \quad (16\text{-}8)$$

Cathode: $\quad Fe^{3+} + e^- \rightleftharpoons Fe^{2+} \quad (16\text{-}9)$

$$E(\text{cathode}) = 0.770 \text{ V} - \frac{0.059\,16}{1} \log \frac{[Fe^{2+}]}{[Fe^{3+}]} \quad (16\text{-}10)$$

$E(\text{cell}) = E(\text{cathode}) - E(\text{anode})$

$$= \underbrace{\left(0.770 - 0.059\,16 \log \frac{[Fe^{2+}]}{[Fe^{3+}]}\right)}_{\text{Single-electrode potential for cathode}} - \underbrace{(0.222 - 0.059\,16 \log[Cl^-])}_{\text{Single-electrode potential for anode}}$$

$$= +0.548 - 0.059\,16 \log \frac{[Fe^{2+}]}{[Fe^{3+}][Cl^-]} \quad (16\text{-}11)$$

Equation 16-11 is exactly the same as Equation 16-4. Both ways of setting up the Nernst equation necessarily give the same result.

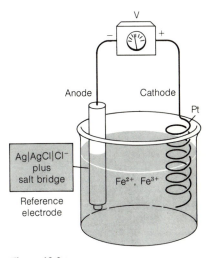

Figure 16-2
Another view of Figure 16-1. The contents of the dashed box in Figure 16-1 are now considered to be a reference electrode dipped into the analyte solution.

Note that the anode reaction is written as a *reduction*.

From now on we will write both the anode and cathode reactions as *reductions*. We then calculate the cell potential using the formula

$E(\text{cell}) = E(\text{cathode}) - E(\text{anode})$

16 / ELECTRODES AND POTENTIOMETRY

The two most common reference electrodes are the silver–silver chloride electrode and the calomel electrode.

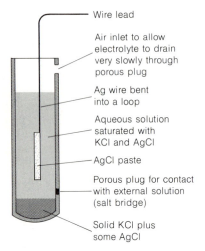

Figure 16-3
Diagram of a silver–silver chloride reference electrode.

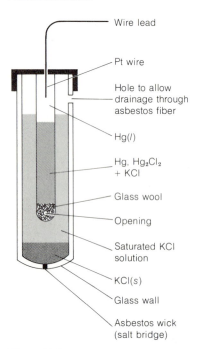

Figure 16-4
A saturated calomel electrode (S.C.E.).

Common Reference Electrodes

Silver–silver chloride electrode

The reference electrode enclosed by the dashed line in Figure 16-1 is called a **silver–silver chloride electrode.** Figure 16-3 shows how the half-cell in Figure 16-1 might be reconstructed as a thin, glass-enclosed electrode that can be dipped into the analyte solution in Figure 16-2.

The standard reduction potential for the AgCl|Ag couple is $+0.222$ V at $25°C$. This would be the potential of a silver–silver chloride electrode if $\mathscr{A}_{Cl^-}$ were unity. But the activity of Cl^- in a saturated solution of KCl at $25°C$ is not unity, and the potential of the electrode in Figure 16-3 is found to be $+0.197$ V with respect to a standard hydrogen electrode at $25°C$.

$$AgCl(s) + e^- \rightleftharpoons Ag(s) + Cl^- \qquad E^0 = +0.222 \text{ V} \quad (16\text{-}12)$$

$$E(\text{saturated KCl}) = +0.197 \text{ V}$$

Calomel electrode

The **calomel electrode** in Figure 16-4 is based on the reaction

$$Hg_2Cl_2(s) + 2e^- \rightleftharpoons 2Hg(l) + 2Cl^- \qquad (16\text{-}13)$$

mercurous chloride (calomel)

$$E = E^0 - \frac{0.05916}{2} \log[Cl^-]^2 \qquad (16\text{-}14)$$

The standard potential (E^0) for this reaction is $+0.268$ V. If the cell is saturated with KCl at $25°C$, the activity of Cl^- is such that the potential of the electrode is $+0.241$ V. The calomel electrode saturated with KCl is called the **saturated calomel electrode.** It is encountered so frequently that it is abbreviated **S.C.E.** and any chemist will immediately recognize what this means. The advantage in using a saturated KCl solution is that the concentration of chloride does not change if some of the liquid evaporates.

16-2 METALLIC INDICATOR ELECTRODES

The most common metallic indicator electrode is the platinum electrode that we have encountered several times already (in Figure 16-2, for example). Platinum is used because it is relatively *inert*—it does not participate in many chemical reactions. When used as an electrode, its purpose is simply to transmit electrons to or from a reactive species in solution.

In cases where platinum reacts with the electrolyte solution, a gold electrode can usually be used. Gold is more inert (and more expensive) than platinum. A metal electrode works best when its surface is large and clean.

Demonstration 16-1 POTENTIOMETRY WITH AN OSCILLATING REACTION

The principles of potentiometry can be illustrated in a fascinating manner using **oscillating reactions.** An oscillating reaction is one in which the concentrations of various species oscillate between high and low values, instead of monotonically approaching their equilibrium values. In spite of this oscillatory behavior, the free energy of the system decreases throughout the entire reaction.[†]

One interesting example is the Belousov–Zhabotinskii reaction, in which the net reactions are

$$CH_2(CO_2H)_2 + 6Ce^{4+} + 2H_2O \rightarrow 2CO_2 + HCO_2H + 6Ce^{3+} + 6H^+$$

malonic   ceric
acid

$$10Ce^{3+} + 2BrO_3^- + 12H^+ \rightarrow 10Ce^{4+} + Br_2 + 6H_2O$$

cerous   bromate

During this cerium-catalyzed oxidation of malonic acid by bromate, the ratio $[Ce^{3+}]/[Ce^{4+}]$ oscillates between a high and low value.[‡] When Ce^{4+} is the major species, the solution is yellow. When Ce^{3+} predominates, the solution becomes colorless. Using redox indicators (Section 17-3), this reaction can be made to oscillate through a spectacular sequence of colors.[§]

The simple oscillation between yellow and colorless is the basis for this demonstration. A 300 mL beaker is loaded with the following solutions:

    160 mL of 1.5 M H_2SO_4
      40 mL of 2 M malonic acid
      30 mL of 0.5 M $NaBrO_3$ (or saturated $KBrO_3$)
       4 mL of saturated ceric ammonium sulfate, $(Ce(SO_4)_2 \cdot 2(NH_4)_2SO_4 \cdot 2H_2O)$

After an induction period of five to ten minutes with magnetic stirring, oscillations can be initiated by adding 1 mL of ceric ammonium sulfate solution. The reaction is somewhat temperamental and may need to be treated once or twice more with Ce^{4+} over a five-minute period to initiate oscillations.

A galvanic cell is built around the reaction as shown at the top of the following page. The $[Ce^{3+}]/[Ce^{4+}]$ ratio is monitored by a Pt indicator electrode, with an S.C.E. reference electrode. You should be able to write the cell reactions.

In place of a potentiometer (a pH meter), we use a chart recorder to obtain a permanent record of the oscillations. Since the potential oscillates over a range of ~ 100 mV, but is centered near ~ 1.2 V, the cell voltage is offset by ~ 1.2 V with any available power supply.

[†] General discussions of oscillating reactions can be found in H. Degn, *J. Chem. Ed.*, **49**, 302 (1972); and J. Higgins, *Ind. Eng. Chem.*, **59**, 19 (1967).

[‡] For references on the rather complicated mechanism of this reaction, see E. J. Heilweil, M. J. Henchman, and I. R. Epstein, *J. Amer. Chem. Soc.*, **101**, 3698 (1979); Z. Noszticzius and J. Bódiss, *J. Amer. Chem. Soc.*, **101**, 3177 (1979); and Z. Noszticzius, *J. Amer. Chem. Soc.*, **101**, 3663 (1979).

[§] For some beautiful lecture demonstrations and a laboratory experiment, see R. J. Field, *J. Chem. Ed.*, **49**, 308 (1972); J. N. Demas and D. Diemente, *J. Chem. Ed.*, **50**, 357 (1973); and J. F. Lefelhocz, *J. Chem. Ed.*, **49**, 312 (1972).

(continued)

Demonstration 16-1 (*continued*)

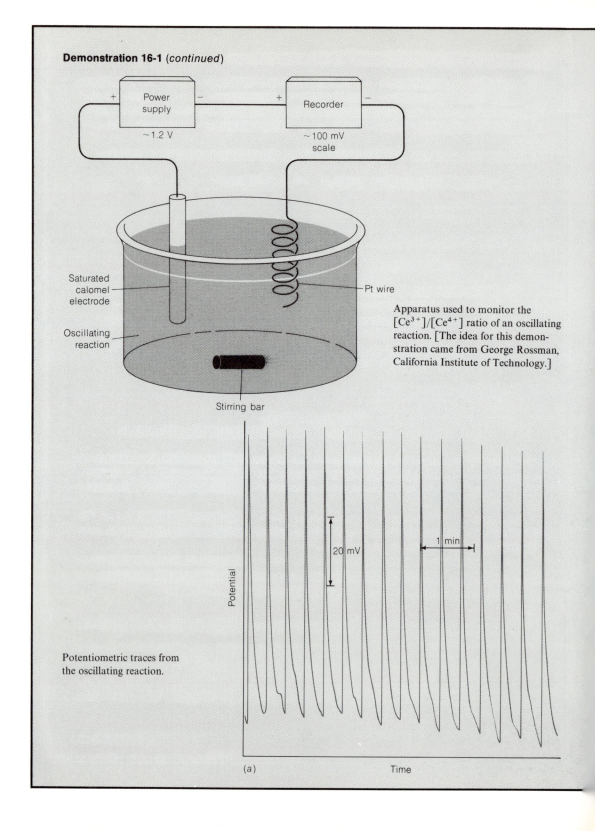

Apparatus used to monitor the $[Ce^{3+}]/[Ce^{4+}]$ ratio of an oscillating reaction. [The idea for this demonstration came from George Rossman, California Institute of Technology.]

Potentiometric traces from the oscillating reaction.

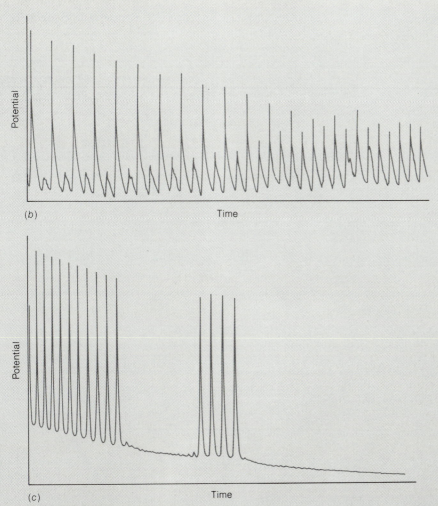

Using the setup shown, we recorded the traces. Trace a shows what is usually observed. The potential changes rapidly during the abrupt colorless-to-yellow color change and gradually during the gentle yellow-to-colorless change. Trace b shows two different cycles superimposed in the same solution. This unusual event happened spontaneously in a reaction that had been oscillating normally for about 30 minutes. Trace c represents a reaction that had been oscillating for several hours. It stopped for a while, began again spontaneously, and finally died.

This same experiment can be used to demonstrate a bromide ion-selective electrode, since the concentration of Br^- also oscillates in this solution. Simultaneous monitoring of the $[Ce^{3+}]/[Ce^{4+}]$ ratio and the Br^- concentration would be worthwhile. An even more striking oscillation, in which the concentration of I^- oscillates over four orders of magnitude, has been described.[†] Monitoring the oscillating I^- concentration with a silver–silver iodide electrode[†] is an informative classroom demonstration.

[†] T. S. Briggs and W. C. Rauscher, *J. Chem. Ed.*, **50**, 496 (1973).

A brief dip in concentrated nitric acid, followed by rinsing with distilled water, is often effective for cleaning an electrode surface.

Figure 16-5 shows how a silver electrode can be used in conjunction with a calomel reference electrode to measure silver ion concentration (actually activity). The cathode reaction in this cell is simply

$$Ag^+ + e^- \rightleftharpoons Ag(s) \qquad E^0 = 0.799 \text{ V} \qquad (16\text{-}15)$$

The anode reaction, written as a *reduction*, is that of the calomel electrode, Reaction 16-13:

$$Hg_2Cl_2(s) + 2e^- \rightleftharpoons 2Hg(l) + 2Cl^- \qquad E^0 = 0.268 \text{ V}$$

Using Equation 16-10, we can write a Nernst equation for this cell as follows:

$$E(\text{cell}) = E(\text{cathode}) - E(\text{anode})$$

$$E(\text{cell}) = \left(0.799 - 0.05916 \log \frac{1}{[Ag^+]}\right) - \left(0.268 - \frac{0.05916}{2}\log[Cl^-]^2\right)$$

In this equation E(cathode) and E(anode) are each a Nernst equation for a single-electrode reaction written as a reduction.

[Cl$^-$] refers to concentration of Cl$^-$ in calomel electrode

(16-16)

But the term E(anode) in Equation 16-16 is a constant, equal to 0.241 V. It is constant because [Cl$^-$] is constant inside the reference electrode. The cell potential can therefore be rewritten in a simpler form:

$$E(\text{cell}) = \left(0.799 - 0.05916 \log \frac{1}{[Ag^+]}\right) - (0.241)$$

$$E(\text{cell}) = 0.558 + 0.05916 \log[Ag^+] \qquad (16\text{-}17)$$

That is, the potential measured by the cell in Figure 16-5 is a direct measure of the concentration of Ag$^+$. Ideally, the potential changes by 59 mV (at 25°C) for each factor-of-ten change in [Ag$^+$].

In Figure 9-5 we used a silver electrode as the anode and a *glass electrode* as the cathode. The glass electrode responds to the pH of the solution, as will be discussed in Section 16-4. The cell shown in Figure 9-5 contains a buffer to maintain a constant pH. Therefore, the glass electrode maintains a constant potential throughout the experiment; it is being used in an unconventional way as a reference electrode.

You should memorize the calomel half-reaction:

$$Hg_2Cl_2(s) + 2e^- \rightleftharpoons 2Hg(l) + 2Cl^-$$

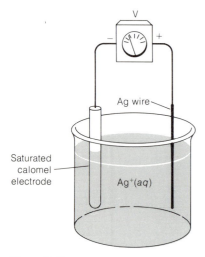

Figure 16-5
Use of silver and calomel electrodes to measure the concentration of Ag$^+$ in a solution.

Challenge: In Equation 16-16, we used *one* electron for the cathode half-reaction and *two* electrons for the anode half-reaction. Show that if you had used two electrons for each reaction, Equation 16-17 would still be the same.

EXAMPLE

A 100.0 mL solution containing 0.1000 M NaCl was titrated with 0.1000 M AgNO$_3$, and the potential was measured with the circuit shown in Figure 16-5. Calculate the cell potential after the addition of 65.0, 100.0, and 103.0 mL of AgNO$_3$.

The titration reaction is

$$Ag^+ + Cl^- \rightarrow AgCl(s)$$

for which V_e (the equivalence point) is 100.0 mL.

65.0 mL: 65.0% of the chloride has been precipitated and 35.0% remains in solution:

$$[Cl^-] = \underbrace{(0.350)}_{\substack{\text{Fraction} \\ \text{remaining}}} \underbrace{(0.100\,0)}_{\substack{\text{Initial} \\ \text{concentration} \\ \text{of } Cl^-}} \underbrace{\left(\frac{100.0}{165.0}\right)}_{\substack{\text{Dilution} \\ \text{factor}}} = 0.021\,2 \text{ M}$$

where the 100.0 is the initial volume of Cl^- and the 165.0 is the total volume of solution.

To find the cell potential, we need to know $[Ag^+]$:

$$[Ag^+] = K_{sp} \text{ (for AgCl)}/[Cl^-]$$
$$[Ag^+] = (1.8 \times 10^{-10})/0.021\,2 = 8.5 \times 10^{-9} \text{ M}$$

The cell potential is computed using Equation 16-17.

$$E(\text{cell}) = 0.558 + 0.059\,16 \log(8.5 \times 10^{-9}) = 0.081 \text{ V}$$

100.0 mL: This is the equivalence point, at which $[Ag^+] = [Cl^-]$:

$$[Ag^+][Cl^-] = [Ag^+]^2 = K_{sp}$$
$$[Ag^+] = \sqrt{K_{sp}} = 1.34 \times 10^{-5} \text{ M}$$
$$E(\text{cell}) = 0.558 + 0.059\,16 \log(1.34 \times 10^{-5}) = 0.270 \text{ V}$$

103.0 mL: Now there is an excess of 3.0 mL of 0.100 0 M $AgNO_3$ in the solution. Therefore,

$$[Ag^+] = \underbrace{(0.100\,0)}_{\substack{\text{Initial} \\ \text{concentration} \\ \text{of } Ag^+}} \underbrace{\left(\frac{3.0}{203.0}\right)}_{\substack{\text{Dilution} \\ \text{factor}}} = 1.48 \times 10^{-3} \text{ M}$$

where the 3.0 is the volume of excess Ag^+ and the 203.0 is the total volume of solution.

and we can write

$$E(\text{cell}) = 0.558 + 0.059\,16 \log(1.48 \times 10^{-3}) = 0.391 \text{ V}$$

Using a little imagination, *a silver electrode is also a halide electrode, if solid silver halide is present in the cell.* For example, if the solution contains AgCl(s), we can write

$$[Ag^+][Cl^-] = K_{sp}$$
$$[Ag^+] = K_{sp}/[Cl^-] \tag{16-18}$$

The cell responds to a change in Cl^- concentration because such a change necessarily changes the Ag^+ concentration, since $[Ag^+][Cl^-] = K_{sp}$.

16 / ELECTRODES AND POTENTIOMETRY

Putting this value of $[Ag^+]$ into Equation 16-17 gives

$$E(cell) = 0.558 + 0.059\ 16 \log \frac{K_{sp}}{[Cl^-]} \qquad (16\text{-}19)$$

That is, the potential responds to changes in Cl^- concentration.

Some metals, such as Ag, Cu, Zn, Cd, and Hg, can be used as indicator electrodes for their aqueous ions. Most metals, however, are unsuitable for this purpose because the equilibrium, $M \rightleftharpoons M^{n+} + ne^-$, is not readily established at the metal surface.

16-3   WHAT IS A JUNCTION POTENTIAL?

E(measured) = E(cell) + E(junction)

Since the junction potential is usually unknown, the value of E(cell) is uncertain.

There is a fundamental problem with most potentiometric measurements. It concerns a small potential, called the **junction potential,** that exists at the interface between the salt bridge and each half-cell. This potential is usually small, but it is almost always of unknown magnitude. *The junction potential puts a fundamental limitation on the accuracy of direct potentiometric measurements*, because we usually do not know the contribution of the junction to the measured potential.

Any time two dissimilar electrolyte solutions are placed in contact, an electric potential develops at their interface. To see why this junction potential occurs, consider a solution containing NaCl in contact with distilled water (Figure 16-6). The Na^+ and Cl^- ions will begin to diffuse from the NaCl solution into the water phase. However, Cl^- ion has a greater **mobility** than Na^+. That is, Cl^- diffuses faster than Na^+. As a result, a region rich in Cl^- develops at the front. This region has excess negative charge. Behind it is a region depleted of Cl^- and thus containing excess positive charge. The result is an electric potential at the junction of the NaCl and H_2O phases. This junction potential opposes the movement of Cl^- and accelerates the movement of Na^+. The steady-state junction potential represents a balance between the unequal mobilities that create a charge imbalance and the tendency of the resulting charge imbalance to retard the movement of Cl^-.

The mobilities of several ions are shown in Table 16-1. You can see that K^+ and Cl^- have similar mobilities. Therefore, the junction potential at the

Table 16-1

Mobilities of ions in water at 25°C

Ion	Mobility ($m^2\ s^{-1}\ V^{-1}$)[†]
H^+	36.30×10^{-8}
K^+	7.62×10^{-8}
Ba^{2+}	6.59×10^{-8}
Na^+	5.19×10^{-8}
Li^+	4.01×10^{-8}
OH^-	20.50×10^{-8}
SO_4^{2-}	8.27×10^{-8}
Cl^-	7.91×10^{-8}
NO_3^-	7.40×10^{-8}
HCO_3^-	4.61×10^{-8}

[†] The mobility of an ion is the terminal velocity that the particle achieves in an electric field of $1\ V\ m^{-1}$. Mobility = Velocity/Field. The dimensions of mobility are therefore $m\ s^{-1}/V\ m^{-1} = m^2 s^{-1}\ V^{-1}$.

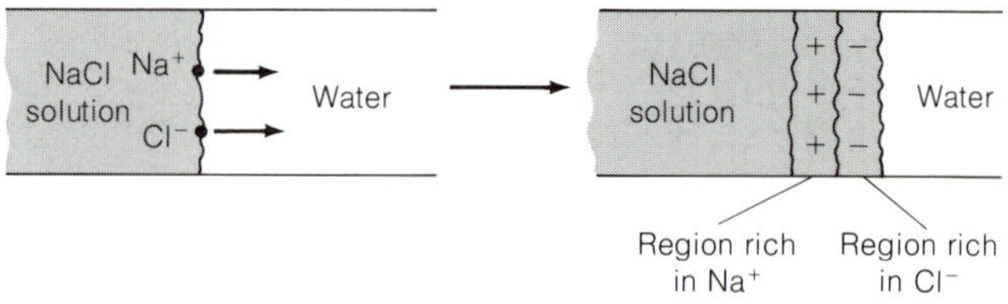

Figure 16-6

Development of the junction potential caused by unequal mobilities of Na^+ and Cl^-.

interface of a KCl salt bridge with another solution will be slight. Furthermore, the junction potentials at each end of a salt bridge often partially cancel each other. Although a salt bridge necessarily contributes some *unknown* net potential to a galvanic cell, the contribution is fairly small (a few millivolts).

Several liquid junction potentials are listed in Table 16-2. You can see that a high concentration of KCl in one solution reduces the magnitude of the potential. This is the reason why saturated KCl is used in salt bridges. Any time a liquid junction is present in a cell, there will be some (usually unknown) junction potential. Therefore, the measured cell potential cannot be attributed entirely to the two half-reactions.

16-4 pH MEASUREMENT WITH A GLASS ELECTRODE

A junction potential exists at each end of a salt bridge.

Table 16-2
Liquid junction potentials at 25°C

Junction	Potential (mV)
0.1 M NaCl│0.1 M KCl	−6.4
0.1 M NaCl│3.5 M KCl	−0.2
1 M NaCl│3.5 M KCl	−1.9
0.1 M HCl│0.1 M KCl	+27
0.1 M HCl│3.5 M KCl	+3.1

Note: A positive sign means that the right side of the junction becomes positive with respect to the left side.
SOURCE: Data are from G. D. Christian, *Analytical Chemistry*, 3rd ed. (New York: Wiley, 1980), p. 294.

EXAMPLE
A 0.1 M NaCl solution was placed in contact with a 0.1 M $NaNO_3$ solution. Which side of the junction will be positive and which will be negative?

Since $[Na^+]$ is equal on both sides, there will be no net diffusion of Na^+ across the junction. However, Cl^- will diffuse into the $NaNO_3$, and NO_3^- will diffuse into the NaCl. Since the mobility of Cl^- is greater than that of NO_3^-, the NaCl region will be depleted of Cl^- faster than the $NaNO_3$ region will be depleted of NO_3^-. The $NaNO_3$ side will become negative, and the NaCl side will become positive.

16-4 pH MEASUREMENT WITH A GLASS ELECTRODE

Probably the most widely employed electrode is the **glass electrode** for measuring pH. A diagram of a typical **combination electrode,** incorporating both the glass and the reference electrodes in one body, is shown in Figure 16-7. A line diagram of this cell can be written as follows:

$$\underbrace{Ag(s)\,|\,AgCl(s)\,|\,Cl^-(aq)}_{\substack{\text{Outer reference}\\\text{electrode}}}\,\|\,\underbrace{H^+(aq,\text{ outside})}_{\substack{H^+ \text{ outside}\\\text{glass electrode}\\\text{(analyte solution)}}}\!\!\overset{\substack{\text{Glass}\\\text{membrane}}}{\vdots}\!\underbrace{H^+(aq,\text{ inside}),\ Cl^-(aq)}_{\substack{H^+ \text{ inside}\\\text{glass}\\\text{electrode}}}\,|\,\underbrace{AgCl(s)\,|\,Ag(s)}_{\substack{\text{Inner reference}\\\text{electrode}}}$$

The pH-sensitive part of the electrode is the specially constructed thin glass membrane at the bottom of the electrode. Figure 16-8 shows the structure of the silicate lattice of which glass is made. The irregular silicate network contains negatively charged oxygen atoms available for coordination to metal cations of suitable size. The monovalent cations, particularly Na^+, are able to move somewhat through the silicate lattice.

A schematic cross section of the glass membrane of the pH electrode is shown in Figure 16-9. The two surfaces exposed to aqueous solution absorb some water and become swollen. Most of the metal cations in these

16 / ELECTRODES AND POTENTIOMETRY

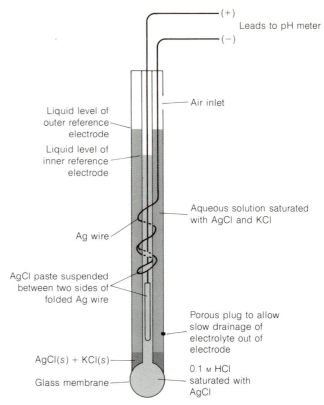

Figure 16-7
Diagram of a glass combination electrode having a silver–silver chloride reference electrode. The glass electrode is immersed in a solution of unknown pH to where the porous plug, on the lower right, is below the surface of the liquid. The two silver electrodes measure the potential across the glass membrane.

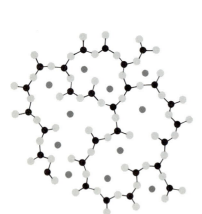

Figure 16-8
Schematic diagram of the structure of glass, which consists of an irregular network of SiO_4 tetrahedra connected through their oxygen atoms. ○ = O, ● = Si, ◐ = Cation. Cations such as Li^+, Na^+, K^+, and Ca^{2+} are coordinated to the oxygen atoms. The silicate network is not planar. This diagram is a projection of each tetrahedron onto the plane of the page. [Adapted from G. A. Perley, *Anal. Chem.*, **21**, 394 (1949).]

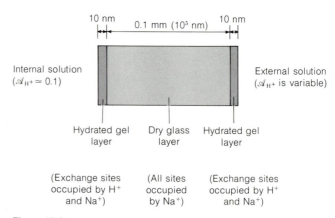

Figure 16-9
Schematic diagram showing a cross section of the glass membrane of a pH electrode.

hydrated gel regions of the membrane diffuse out of the glass and into solution. Concomitantly, H^+ from solution can diffuse into the swollen silicate lattice and occupy some of the cation binding sites. We refer to the equilibrium in which H^+ replaces metallic cations in the glass as an **ion-exchange equilibrium.**

The H^+ ion-exchange equilibrium is depicted schematically in Figure 16-10. Suppose that the internal solution (which is typically 0.1 M HCl) contains a higher concentration of H^+ than the external solution. In this case more H^+ ions will occupy ion-exchange sites of the inner hydrated gel layer than of the outer hydrated gel layer. The inside-facing surface of the glass membrane will be positive with respect to the outside-facing surface. If the H^+ in the external solution were changed from 10^{-2} M to 10^{-3} M, there would be less H^+ bound to the glass membrane's outer surface, which would thus become still more negative.

The potential difference between the inner and outer silver—silver chloride electrodes depends on the chloride concentration in each electrode compartment and on the potential across the glass membrane. Since the chloride concentration is fixed in each electrode compartment, and since the H^+ concentration is fixed on the inside of the glass membrane, the only factor causing a change in cell potential is a change in pH of the analyte solution outside the glass membrane.

From thermodynamics, we expect the potential difference across the glass membrane to obey the Nernst equation. To see why, we need to know that the free energy change when one mole of H^+ passes from a region where the activity of H^+ is $\mathcal{A}_1$ to a region in which the activity of H^+ is $\mathcal{A}_2$ is given by

$$\Delta G = -RT \ln \frac{\mathcal{A}_1}{\mathcal{A}_2} \qquad (16\text{-}20)$$

The electric potential difference (E) across the membrane must be such that

$$\Delta G = -n F E = -RT \ln \frac{\mathcal{A}_1}{\mathcal{A}_2}$$

$$E = \frac{RT}{nF} \ln \frac{\mathcal{A}_1}{\mathcal{A}_2} = \frac{RT \ln 10}{F} \log \frac{\mathcal{A}_1}{\mathcal{A}_2}$$

$$E = \frac{RT \ln 10}{F}[-\log \mathcal{A}_2 - (-\log \mathcal{A}_1)]$$

$$= 0.059\,16\ \Delta\text{pH} \quad \text{(volts, at 25°C)} \qquad (16\text{-}21)$$

Equation 16-21 predicts that at 25°C there should be a potential difference of 59.16 mV for every one-pH-unit difference between the solutions inside and outside the glass electrode. *Note also that the pH electrode responds to hydrogen ion activity, not concentration.*

16-4 pH MEASUREMENT WITH A GLASS ELECTRODE

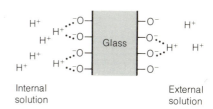

Figure 16-10
Ion-exchange equilibria on the inner and outer surfaces of the glass membrane.

$\ln x = (\ln 10)(\log x)$, and $n = 1$ for H^+.

The pH electrode measures H^+ activity, not H^+ concentration.

16 / ELECTRODES AND POTENTIOMETRY

Real electrodes come near to obeying the Nernst equation.

A pH electrode **must** be calibrated before it can be used. It should be calibrated about every two hours during sustained use.

The pH of the calibration standards should bracket the pH of the unknown.

Do not leave a glass electrode out of water (or in a nonaqueous solvent) any longer than necessary.

For real glass electrodes, the response to changes in pH is nearly Nernstian and can be described by the equation

$$E = \text{Constant} - \beta(0.059\ 16) \log \frac{\mathscr{A}_{H^+}(\text{inside})}{\mathscr{A}_{H^+}(\text{outside})} \quad (\text{at } 25°C) \quad (16\text{-}22)$$

The number β, called the **electromotive efficiency,** is close to 1.00 (typically > 0.98). It varies with each type of glass and with each individual electrode. The constant term would be zero if both sides of the glass were identical. That is, no potential would be observed if the activities of H^+ were the same inside and outside and the membrane were ideal. However, no two sides of any real object are identical, and a small potential exists even if $\mathscr{A}_{H^+}$ is the same on both sides of the membrane. We correct for this **asymmetry potential** by calibrating the electrode in solutions of known pH. Because the asymmetry potential varies with time and conditions (temperature, concentrations, phase of the moon, color of your socks), electrodes must be calibrated frequently.

In order for the pH meter to measure potential, a small current must flow through the instrument. The current in the glass membrane, whose resistance is typically $10^8\ \Omega$, is carried mainly by Na^+ ions, which diffuse through the silicate lattice better than the other ions in the glass. It has been shown by tritium tracer studies that H^+ does not cross the glass membrane and therefore cannot carry electric current.

Calibrating a Glass Electrode

Before using a pH electrode, you should calibrate it, using two (or more) buffers of known pH bracketing the pH of the unknown. The exact calibration procedure is different for each model of pH meter, so consult the manufacturer's instructions for your meter. Before calibrating the electrode, wash it with distilled water and gently *blot* it dry with a tissue. Do not *wipe* it, because this might produce a static electric charge on the glass.

For calibration, dip the electrode in a standard buffer and allow it to equilibrate for at least a minute. For best results, all solutions used for pH calibration and measurement should be stirred continuously before and during the measurement. Following the manufacturer's instructions, adjust the meter reading (usually with a knob labeled "Calibrate") to indicate the pH of the standard buffer. The electrode is then washed, blotted dry, and immersed in a second standard buffer. If the electrode response were perfectly Nernstian, the potential would change by 0.059 16 V/pH unit at 25°C. The actual change is slightly less, so these two measurements serve to establish the value of β in Equation 16-22. This value is told to the pH meter with a knob that may be labeled "Slope" or "Temperature" on different instruments. Finally, the electrode is dipped in a solution of unknown pH. The potential is translated directly into pH by the meter.

A glass electrode must be stored in aqueous solution so that the hydrated gel layer of the glass does not dry out. If the electrode has been allowed to dry out, it should be reconditioned by soaking in water for several hours.

If the electrode is to be used at a pH above 9, soak it in a high-pH buffer.

Many pH standards are commercially available. The pH values of the solutions in Table 16-3 were measured at the United States National Bureau of Standards and are considered to be accurate to ± 0.01 pH unit.

Glass electrodes slowly wear out, partly because the composition of the glass changes near the solution interface, as ions diffuse in and out. If

Table 16-3
pH values of National Bureau of Standards buffers

Temperature (°C)	Saturated (25°C) potassium hydrogen tartrate	0.05 m potassium dihydrogen citrate	0.05 m potassium hydrogen phthalate	0.025 m potassium dihydrogen phosphate 0.025 m disodium hydrogen phosphate	0.008 695 m potassium dihydrogen phosphate 0.030 43 m disodium hydrogen phosphate	0.01 m borax	0.025 m sodium bicarbonate 0.025 m sodium carbonate
0	—	3.863	4.003	6.984	7.534	9.464	10.317
5	—	3.840	3.999	6.951	7.500	9.395	10.245
10	—	3.820	3.998	6.923	7.472	9.332	10.179
15	—	3.802	3.999	6.900	7.448	9.276	10.118
20	—	3.788	4.002	6.881	7.429	9.225	10.062
25	3.557	3.776	4.008	6.865	7.413	9.180	10.012
30	3.552	3.766	4.015	6.853	7.400	9.139	9.966
35	3.549	3.759	4.024	6.844	7.389	9.102	9.925
38	3.548	—	4.030	6.840	7.384	9.081	—
40	3.547	3.753	4.035	6.838	7.380	9.068	9.889
45	3.547	3.750	4.047	6.834	7.373	9.038	9.856
50	3.549	3.749	4.060	6.833	7.367	9.011	9.828
55	3.554	—	4.075	6.834	—	8.985	—
60	3.560	—	4.091	6.836	—	8.962	—
70	3.580	—	4.126	6.845	—	8.921	—
80	3.609	—	4.164	6.859	—	8.885	—
90	3.650	—	4.205	6.877	—	8.850	—
95	3.674	—	4.227	6.886	—	8.833	—

Note: The designation m stands for molality.

In the buffer solution preparations, it is essential to use high-purity materials and to employ freshly distilled or deionized water of specific conductivity not greater than 5 micromhos. Solutions having pH 6 or above should be stored in plastic containers, and preferably with an NaOH trap to prevent ingress of atmospheric carbon dioxide. They can normally be kept for 2–3 weeks, or slightly longer in a refrigerator.

1. Saturated (25°C) potassium hydrogen tartrate, $KHC_4H_4O_6$. An excess of the salt is shaken with water, and it can be stored in this way. Before use, it should be filtered or decanted at a temperature between 22°C and 28°C.
2. 0.05 m potassium dihydrogen citrate, $KH_2C_6H_5O_7$. Dissolve 11.41 g of the salt in one liter of solution at 25°C.
3. 0.05 m potassium hydrogen phthalate. Although this is not usually essential, the crystals may be dried at 110°C for an hour, then cooled in a desiccator. At 25°C, 10.12 g $C_6H_4(CO_2H)(CO_2K)$ is dissolved in water, and the solution made up to one liter.
4. 0.025 m disodium hydrogen phosphate, 0.025 m potassium dihydrogen phosphate. The anhydrous salts are best used; each should be dried for two hours at 120°C and cooled in a desiccator, since they are slightly hygroscopic. Higher drying temperatures should be avoided to prevent formation of condensed phosphates. Dissolve 3.53 g Na_2HPO_4 and 3.39 g KH_2PO_4 in water to give one liter of solution at 25°C.
5. 0.008 695 m potassium dihydrogen phosphate, 0.030 43 m disodium hydrogen phosphate. Prepare as in Step 4 and dissolve 1.179 g KH_2PO_4 and 4.30 g Na_2HPO_4 in water to give one liter of solution at 25°C.
6. 0.01 m sodium tetraborate decahydrate. Dissolve 3.80 g $Na_2B_4O_7 \cdot 10H_2O$ in water to give one liter of solution. This borax solution is particularly susceptible to pH change from carbon dioxide absorption, and it should be correspondingly protected.
7. 0.025 m sodium bicarbonate, 0.025 m sodium carbonate. Primary standard grade Na_2CO_3 is dried at 250°C for 90 minutes and stored over $CaCl_2$ and Drierite. Reagent-grade $NaHCO_3$ is dried over molecular sieves and Drierite for two days at room temperature. Dissolve 2.092 g of $NaHCO_3$ and 2.640 g of Na_2CO_3 in one liter of solution at 25°C.

SOURCE: R. G. Bates, *J. Res. National Bureau of Standards*, **66A**, 179 (1962); and B. R. Staples and R. G. Bates, *J. Res. National Bureau of Standards*, **73A**, 37 (1969). The instructions for preparing these solutions are taken, in part, from G. Mattock in C. N. Reilley, ed., *Advances in Analytical Chemistry and Instrumentation* (New York: Wiley, 1963), Vol. 2, p. 45. See also R. G. Bates, *Determination of pH: Theory and Practice*, 2nd ed. (New York: Wiley, 1973), Chap. 4.

electrode response becomes sluggish or if the electrode cannot be calibrated properly, try washing it with 6 M HCl, followed by water. As a last resort, the electrode can be soaked in 20% (wt/wt) aqueous ammonium bifluoride, NH_4HF_2, for one minute in a plastic beaker. This reagent dissolves a little of the glass and exposes fresh surface. Wash the electrode with water and try calibrating it again.

Errors in pH Measurement

To make intelligent use of a glass electrode, it is important to understand its limitations:

1. Our knowledge of an analyte's pH cannot be any better than our knowledge of the pH of the buffers used to calibrate the meter and the electrode. This error is typically on the order of ±0.01 pH unit.

2. A *junction potential* exists across the porous plug near the bottom of the electrode in Figure 16-7. The porous plug is the salt bridge connecting the outer silver–silver chloride electrode with the analyte solution. If the ionic composition of the analyte solution is different from that of the standard buffer, the junction potential will change *even if the pH of the two solutions is the same*. This change appears as a change in pH and leads to an uncertainty in the pH of at least ~0.01 pH unit.

> The apparent pH will change if the ionic composition of the analyte changes, even when the actual pH is constant.

3. When the concentration of H^+ is very low and the concentration of Na^+ is high (a typical set of conditions in strongly basic solution), the electrode responds to Na^+ as well as to H^+. This arises because Na^+ can participate in ion exchange with the hydrated gel layer.

$$Na^+(\text{in aqueous phase}) \rightleftharpoons Na^+(\text{in hydrated gel layer})$$

This reaction is not nearly as important as H^+ ion exchange, but becomes significant when $\mathscr{A}_{H^+}$ is low and $\mathscr{A}_{Na^+}$ is high. Under these conditions, the electrode behaves as if Na^+ were H^+, and the apparent pH is lower than the true pH. This behavior leads to the **alkaline error,** or **sodium error,** and its magnitude varies with the composition of the glass (Figure 16-11). To reduce the alkaline error, lithium has largely replaced sodium in glass electrodes.

4. In very acidic solutions, the measured pH is higher than the actual pH. The reason for this is not clear, but it may be related to the decreased activity of water in concentrated acid solution.

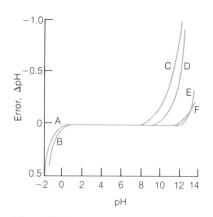

Figure 16-11
Acid and alkaline errors of some glass electrodes. A: Corning 015, H_2SO_4.
B: Corning 015, HCl. C: Corning 015, 1 M Na^+. D: Beckman-GP, 1 M Na^+.
E: L & N Black Dot, 1 M Na^+.
F: Beckman Type E, 1 M Na^+.
[R. G. Bates, *Determination of pH: Theory and Practice*, 2nd ed (New York: Wiley, 1973).]

5. Adequate time must be allowed for the glass membrane to equilibrate with a fresh solution. In a well-buffered solution, this takes just seconds with adequate stirring. In a poorly buffered solution (such as near the equivalence point of a titration), it could take many minutes.

6. An electrode that has not been stored in water becomes dehydrated and requires several hours of soaking before it responds to H^+ correctly.

Errors 1 and 2 above are unavoidable and effectively limit the accuracy of pH measurement with the glass electrode to ±0.02 pH units, at best. Measurement of pH differences *between* solutions can be accurate to about ±0.002 pH units with good equipment. But knowledge of the true pH will still be at least an order of magnitude more uncertain. It is instructive to realize that an uncertainty of ± 0.02 pH units corresponds to an uncertainty of $\pm 5\%$ in $\mathscr{A}_{H^+}$.

One final caution: A pH meter has a temperature knob that allows you to vary the slope (0.059 16) in Equation 16-22 for different temperatures. (That is, a change of 1.00 pH unit corresponds to 54 mV at 0°C, 59 mV at 25°C, and 64 mV at 50°C.) *A pH meter must be calibrated at the same temperature at which the measurement will be made.* You cannot calibrate your equipment at one temperature and then make an accurate measurement at a second temperature simply by changing the temperature adjustment. The meter must be calibrated at the same temperature as the unknown. Furthermore, the pH of the standard buffer changes with temperature, and this variation must be known in order to use the buffer for calibration purposes.

> Calibration must be performed at the same temperature as the analyte.

16-5 ION-SELECTIVE ELECTRODES

The glass pH electrode develops a potential dependent upon the activity of H^+ on each side of the glass membrane. Any electrode that preferentially responds to one species is called an **ion-selective electrode.** No electrode responds exclusively to one kind of ion, but the glass electrode is among the most selective. A high-pH glass electrode responds to Na^+ only when $[H^+] \lesssim 10^{-12}$ M and $[Na^+] \gtrsim 10^{-2}$ M (Figure 16-11).

By varying the composition of the glass, it is possible to alter its sensitivity to different ions. Corning 015 glass for pH electrodes contains 22% (wt/wt) Na_2O, 6% CaO, and 72% SiO_2. By contrast, a glass membrane made of 11% Na_2O, 18% Al_2O_3, and 71% SiO_2 is used in a **sodium ion electrode.** Such an electrode is 2800 times more responsive to Na^+ than to K^+ at pH 11. Using different glass compositions, Li^+, K^+, and Ag^+ ion-selective electrodes have been fabricated from glass membranes.

An electrode used for the measurement of the ion X may also respond to the ion Y. The sensitivity of the electrode to different species is given by the **selectivity coefficient,** defined as

$$ k_{XY} = \frac{\text{Response to Y}}{\text{Response to X}} \qquad (16\text{-}23) $$

Ideally, the selectivity coefficient should be very small ($k \ll 1$) or else significant interference occurs. The sodium ion-selective electrode described above has a selectivity coefficient $k_{Na^+, K^+} = 1/2\,800$ at pH 11, and $k_{Na^+, K^+} = 1/300$ at pH 7. The selectivity coefficient, k_{Na^+, H^+}, has a value of 36, which means that the electrode is more sensitive to H^+ than to Na^+, even though the electrode is used for sodium.

In general, the behavior of most ion-selective electrodes can be described by the equation

$$ E = \text{Constant} \pm \beta \frac{0.059\,16}{n_X} \log \left[\mathscr{A}_X + \sum_Y (k_{XY} \mathscr{A}_Y^{n_X/n_Y}) \right] \qquad (16\text{-}24) $$

> Equation 16-24 describes the response of an electrode to its primary ion, X, and to all interfering ions, Y.

Principle of operation of an ion-selective electrode:
1. Ion-exchange equilibrium on each surface of the membrane
2. Mobile ions to carry current within the membrane

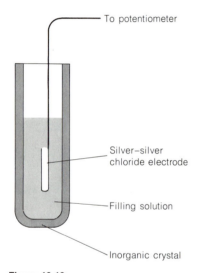

Figure 16-12
Schematic diagram of an ion-selective electrode employing an inorganic salt crystal as the ion-sensitive membrane.

The principle of operation is similar to that of the glass electrode. Ion-exchange equilibria are established on the surface of an inorganic crystal.

where $\mathscr{A}_X$ is the activity of the ion intended to be measured and $\mathscr{A}_Y$ is the activity of any interfering species (Y). The charge of each species is n_X or n_Y, and k_{XY} is the selectivity coefficient defined above. The sign before the log term is positive if X is a cation and negative if X is an anion. The value of β is near 1 for most electrodes.

The principle of most ion-selective electrodes appears to be the same as that of the glass pH electrode. There must be an ion-exchange equilibrium on each side of the membrane surface in order to establish an ion-dependent membrane potential. In order to measure this potential, there must also be some means by which a small current can flow through the membrane. This is accomplished in the glass electrode by slightly mobile Na^+ ions within the glass.

There are many imaginative strategies for producing an electrode that is selectively sensitive to one species. Most electrodes fall into the following classes:

1. *Glass membranes* for H^+ and certain monovalent cations.
2. *Solid-state electrodes* based on inorganic salt crystals.
3. *Liquid-based electrodes* using a hydrophobic membrane saturated with a hydrophobic liquid ion exchanger.
4. *Compound electrodes* involving a species-selective electrode enclosed by a region in which that species is separated from other species or produced by a chemical reaction.

We have already discussed glass membranes and will now examine the other three main types.

Solid-State Electrodes

A schematic diagram showing the construction of an ion-selective electrode based on an inorganic salt crystal is shown in Figure 16-12. One common electrode of this type is the fluoride electrode, employing a crystal of LaF_3 doped with Eu(II) to increase its conductivity. In this case, the filling solution contains 0.1 M NaF and 0.1 M NaCl.

Ion-exchange equilibria exist on each side of the crystal, in which F^- from solution becomes adsorbed on the crystal surface. Current is conducted through the crystal apparently by mobile F^- ions changing positions within the crystal lattice. The amount of F^- adsorbed on the upper surface of the crystal is constant, because the filling solution contains 0.1 M NaF. The amount of F^- adsorbed on the lower surface changes whenever the electrode is immersed in a new analyte solution. By analogy with the pH electrode, we can describe the response of the F^- electrode in the form

$$E = \text{Constant} - \beta(0.059\,16)\log[\mathscr{A}_{F^-}(\text{outside})] \quad (16\text{-}25)$$

where β is close to 1.00.

The LaF_3 electrode gives a nearly Nernstian response over a F^- concentration range from about 10^{-6} M to 1 M. The electrode is more responsive to F^- than to other ions by a factor greater than 1 000. The only interfering species is OH^-, which must be kept to less than one-tenth of the F^- concentration to avoid interference. At low pH, F^- is converted to HF ($pK_A =$ 3.17), to which the electrode is insensitive. The LaF_3 electrode is used to continuously monitor and control the fluoridation of municipal water supplies.

EXAMPLE

When a fluoride ion-selective electrode was immersed in standard fluoride solutions (maintained at a constant ionic strength of 0.1 M with $NaNO_3$), the following potentials (versus S.C.E.) were observed:

$[F^-]$ (M)	E (mV)	
1.00×10^{-5}	100.0	} -58.5 mV
1.00×10^{-4}	41.5	
1.00×10^{-3}	-17.0	} -58.5 mV

Since the ionic strength was held constant, the response should be proportional to the logarithm of the F^- *concentration*. What potential is expected if $[F^-] = 5.00 \times 10^{-5}$ M? What concentration of F^- will give a potential of 0.0 V?

We seek to fit the calibration data with an equation in the form of Equation 16-25:

$$E = m \underbrace{\log[F^-]}_{x} + b$$
$$ y$$

Plotting E versus $\log[F^-]$ gives a straight line with a slope of -58.5 mV and an intercept -192.5 mV. Setting $[F^-] = 5.00 \times 10^{-5}$ M gives

$$E = (-58.5) \log[5.00 \times 10^{-5}] - 192.5 = 59.1 \text{ mV}$$

If E = 0 mV, we can solve for the concentration of $[F^-]$:

$$0 = (-58.5) \log[F^-] - 192.5 \Rightarrow [F^-] = 5.12 \times 10^{-4} \text{ M}$$

Another common inorganic crystal electrode uses Ag_2S for the membrane. This electrode responds to Ag^+ and to S^{2-}. By doping the electrode with CuS, CdS, or PbS, it is possible to prepare electrodes sensitive to Cu^{2+}, Cd^{2+}, or Pb^{2+}, respectively. Table 16-4 lists some ion-selective electrodes based on inorganic crystals.

Liquid-Based Ion-Selective Electrodes

The principle of **liquid-based ion-selective electrodes** is the same as that of solid-state electrodes and the glass electrode. However, in liquid-based

The principle of operation is similar to that of the solid-state electrode. Ion-exchange equilibria are established on the surface of a membrane saturated with a liquid ion exchanger.

16 / ELECTRODES AND POTENTIOMETRY

Table 16-4

Properties of some solid-state ion-selective electrodes

Ion	Concentration range (M)	Membrane material	pH range	Some interfering species
F^-	$10^{-6}-1$	LaF_3	5–8	OH^-
Cl^-	$10^{-4}-1$	$AgCl$	2–11	$CN^-, S^{2-}, I^-, S_2O_3^{2-}, Br^-$
Br^-	$10^{-5}-1$	$AgBr$	2–12	CN^-, S^{2-}, I^-
I^-	$10^{-6}-1$	AgI	3–12	S^{2-}
SCN^-	$10^{-5}-1$	$AgSCN$	2–12	$S^{2-}, I^-, CN^-, Br^-, S_2O_3^{2-}$
CN^-	$10^{-6}-10^{-2}$	AgI	11–13	S^{2-}, I^-
S^{2-}	$10^{-5}-1$	Ag_2S	13–14	

SOURCE: P. L. Bailey, *Analysis with Ion-Selective Electrodes* (London: Heyden, 1976), pp. 95–99; and *Orion Research Analytical Methods Guide* (Cambridge, Mass.: Orion Research Inc., 1975).

Table 16-5

Properties of some liquid-based ion-selective electrodes

Ion	Concentration range (M)	Carrier	Solvent for carrier	pH range	Some interfering species
Ca^{2+}	$10^{-5}-1$	Calcium didecylphosphate	Dioctylphenyl phosphonate	6–10	$Zn^{2+}, Pb^{2+}, Fe^{2+}, Cu^{2+}$
NO_3^-	$10^{-5}-1$	Tridodecylhexadecylammonium nitrate	Octyl-2-nitrophenyl ether	3–8	$ClO_4^-, I^-, ClO_3^-, Br^-, HS^-, CN^-$
ClO_4^-	$10^{-5}-1$	*tris*(substituted 1,10-phenanthroline) Iron(II) perchlorate	*p*-Nitrocymene	4–10	I^-, NO_3^-, Br^-
BF_4^-	$10^{-5}-1$	*tris*(substituted 1,10-phenanthroline) Nickel(II) tetrafluoroborate	*p*-Nitrocymene	2–12	NO_3^-
Cl^-	$10^{-5}-1$	Dimethyldioctadecylammonium chloride		3–10	$ClO_4^-, I^-, NO_3^-, SO_4^{2-}, Br^-, OH^-, HCO_3^-, F^-,$ acetate

SOURCE: P. L. Bailey, *Analysis with Ion-Selective Electrodes* (London: Heyden, 1976), pp. 127–130; and *Orion Research Analytical Methods Guide* (Cambridge, Mass.: Orion Research Inc., 1975).

$$[(CH_3(CH_2)_8CH_2O)_2\overset{O}{\overset{\|}{P}O}]_2Ca$$

calcium didecylphosphate

$$(CH_3(CH_2)_6CH_2O)_2\overset{O}{\overset{\|}{P}}C_6H_5$$

dioctylphenylphosphonate

electrodes, the ion exchanger is a liquid soaked into a hydrophobic membrane. Several such electrodes are listed in Table 16-5.

Figure 16-13 shows a calcium ion-selective electrode, which is basically similar to the solid-state electrode in Figure 16-14. The difference is that the solid crystal is replaced by a membrane saturated with the liquid ion exchanger. A solution of the exchanger is housed in a reservoir surrounding the internal silver–silver chloride electrode.

In order to establish a Ca^{2+}-dependent potential, Ca^{2+} ions must interact with sites on each surface of the membrane. The hydrophobic membrane is not permeable to water, but is soaked with calcium didecylphosphate dissolved in dioctylphenylphosphonate. The didecylphosphate anion can react with calcium ions at each surface of the membrane:

$$[(RO)_2PO_2^-]_2Ca \rightleftharpoons 2(RO)_2PO_2^- + Ca^{2+} \qquad (16\text{-}26)$$

Since there is a different activity of Ca^{2+} on each side of the membrane, a different amount of Ca^{2+} becomes bound to ion exchanger at each surface, and a nearly Nernstian potential develops across the membrane:

$$E = \text{Constant} + \beta\left(\frac{0.059\,16}{2}\right) \log \mathscr{A}_{Ca^{2+}} \text{ (outside)} \qquad (16\text{-}27)$$

where β is close to 1.00.

Note that Equations 16-27 and 16-25 have different signs before the log term, because one involves a cation and the other involves an anion. Note also that the charge of the calcium ion requires a factor of 2 in the denominator before the logarithm.

The most serious interference with the Ca^{2+} electrode comes from Zn^{2+}, Fe^{2+}, Pb^{2+}, and Cu^{2+}; but high concentrations of Sr^{2+}, Mg^{2+}, Ba^{2+}, and Na^+ also interfere. Interference from H^+ is substantial below a pH of 4–5.

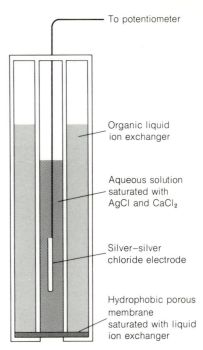

Figure 16-13
Schematic diagram of a calcium ion-selective electrode based on a liquid ion exchanger.

Box 16-1 MICROELECTRODES INSIDE LIVING CELLS

To understand how nerve and muscle cells work, an accurate knowledge of the intracellular ionic composition is necessary. The figure below shows the design of a *microelectrode* used to make intracellular measurements of K^+ and Cl^-. It is similar to the liquid-based electrode in Figure 16-13, but the glass tip is drawn out to microscopic dimensions. Inside this tip a droplet of liquid ion exchanger is held by capillary action. To make a measurement, the cell must be impaled with this electrode and also with a reference electrode of similar dimensions. The reference electrode is a silver–silver chloride electrode containing only 3 M KCl and no liquid ion exchanger.

A microelectrode for measuring ion activities inside living cells.
[J. L. Walker, Jr., *Anal. Chem.*, **43** (No. 3), 89A (1971).]

Compound Electrodes

Gas-sensing and enzyme-based electrodes are usually of a **compound** design, incorporating a conventional electrode and an additional membrane to isolate (or generate) the species that the electrode detects. A CO_2 gas-sensing electrode is shown in Figure 16-14. It consists of an ordinary glass pH electrode surrounded by an electrolyte solution and enclosed by a semi-permeable membrane made of rubber, Teflon, or polyethylene. Immersed in the electrolyte solution is a silver–silver chloride band serving as the reference electrode. When CO_2 diffuses through the semipermeable membrane, it lowers the pH in the electrolyte compartment. The response of the glass electrode to the change in pH is measured. Other acidic or basic gases, including NH_3, SO_2, H_2S, and NO_x (nitrogen oxides) can be detected in the same manner. These electrodes can be used to measure gases *in the gas phase* or dissolved in solution.

Numerous ingenious compound electrodes using *enzymes* have been built. These devices contain a conventional electrode coated with an enzyme that catalyzes a reaction of the analyte. The product of the reaction is detected by the electrode.

Acidic or basic gases are detected by a pH electrode surrounded by an electrolyte and enclosed in a gas-permeable electrode.

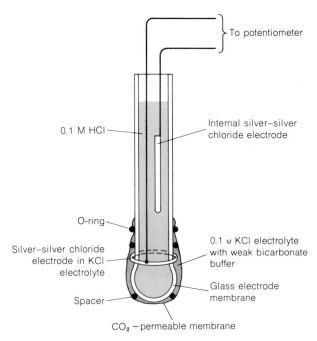

Figure 16-14
Schematic diagram of a CO_2 gas-sensing electrode.

Use and Abuse of Ion-Selective Electrodes

The advantages of ion-selective electrodes are many. Electrodes respond in a linear manner to most analyte species over a wide range of concentrations (four to six orders of magnitude). They do not destroy the unknown sample, and they introduce negligible contamination. Their response time is usually short (seconds to minutes), so they may be used to monitor flowing samples in industrial or clinical applications. Highly colored or turbid solutions are amenable to measurements with electrodes, but not to spectrophotometry. Finally, specially designed electrodes can be used in otherwise inaccessible environments, such as the interior of living cells.

Advantages of ion-selective electrodes:
1. Wide range of linear response
2. Nondestructive
3. Noncontaminating
4. Short response time
5. Unaffected by color or turbidity

On the other hand, special care is needed to obtain reliable results. The precision of ion-selective electrode measurements is rarely better than 1%, and is usually worse. Electrodes can be fouled by proteins or other organic solutes, leading to sluggish response and drifting potentials. Certain ionic species interfere with or poison particular electrodes.

A 1 mV error in potential corresponds to a 4% error in monovalent ion activity. A 5 mV error corresponds to a 22% error. The relative error *doubles* for divalent ions and *triples* for trivalent ions.

Extreme care in the preparation of samples and standards is critical to obtaining meaningful results. The electrode responds to the *activity* of *uncomplexed* analyte ion. Therefore, potential ligands must be absent or masked. Since we usually wish to know concentrations, not activities, an inert salt is often used to bring all of the standards and samples to a high and constant ionic strength. If the activity coefficients remain constant, the electrode potential gives concentrations directly.

Electrodes respond to the *activity* of *uncomplexed* ion. If the ionic strength is held constant, the concentration is proportional to the activity and the electrode measures concentrations.

Standard addition method

It is important that the composition of the standard solutions should closely approximate the composition of the unknown. The medium in which the analyte exists is called the **matrix.** In cases where the matrix is complex or unknown, the **standard addition method** can be used. In this technique, the electrode is immersed in the unknown and the potential is recorded. Then a known addition of standard solution is made to the unknown. The addition should be of a small volume so as not to change the ionic strength of the sample. The change in potential tells how the electrode responds to analyte and how much analyte was in the original solution.

The matrix (solution composition) of the standards should be the same as the matrix of the unknown.

The **standard addition method** is discussed further in Sections 19-3 and 21-4. A general treatment of the standard addition method can be found in an article by M. Bader, *J. Chem. Ed.*, **57**, 703 (1980).

EXAMPLE

A perchlorate ion-selective electrode immersed in 50.0 mL of unknown perchlorate solution gave a potential of 358.7 mV versus S.C.E. When 1.00 mL of 0.050 0 M $NaClO_4$ was added, the potential changed to 346.1 mV. Assuming that the electrode has a Nernstian response ($\beta = 1.00$), find the concentration of ClO_4^- in the unknown.

The first solution contains x moles of ClO_4^- in 0.050 0 L. The standard addition adds $(0.001\ 00\ \text{L})(0.050\ 0\ \text{M}) = 5.00 \times 10^{-5}$ moles of ClO_4^-. Therefore, the second solution contains $x + (5.00 \times 10^{-5})$ moles in 0.051 0 L. We can write a Nernst equation for the first solution:

$$E_1 = \text{Constant} - 0.059\ 16 \log[ClO_4^-]_1$$

356

16 / ELECTRODES AND POTENTIOMETRY

and another for the second solution:

$$E_2 = \text{Constant} - 0.059\,16 \log[\text{ClO}_4^-]_2$$

We set $E_1 = 0.358\,7$ V, $E_2 = 0.346\,1$ V, $[\text{ClO}_4^-]_1 = x/0.050\,0$, and $[\text{ClO}_4^-]_2 = (x + 5.00 \times 10^{-5})/0.051\,0$. This allows us to solve for x by subtracting one equation from the other:

$$\begin{array}{r} 0.358\,7 = \text{Constant} - 0.059\,16 \log[x/0.050\,0] \\ - \quad 0.346\,1 = \text{Constant} - 0.059\,16 \log[(x + 5.00 \times 10^{-5})/0.051\,0] \\ \hline 0.012\,6 = -0.059\,16 \log \dfrac{x/0.050\,0}{(x + 5.00 \times 10^{-5})/0.051\,0} \end{array}$$

$$\frac{x/0.050\,0}{(x + 5.00 \times 10^{-5})/0.051\,0} = 0.612 \Rightarrow x = 7.51 \times 10^{-5} \text{ moles}$$

The original perchlorate concentration was therefore

$$\frac{7.51 \times 10^{-5} \text{ moles}}{0.050\,0 \text{ L}} = 1.50 \text{ mM}$$

The standard addition method works best if the quantity of added analyte is approximately 50–200% of the original analyte. Results are more accurate if the effects of several additions are averaged.

Metal ion buffers

Glass vessels should not be used for low concentration standard solutions, because ions can be adsorbed on the glass surface. Bottles and beakers made of polyethylene or similar materials are much better than glass for handling dilute solutions.

$\text{HN(CH}_2\text{CO}_2\text{H)}_3^+$
$\equiv \text{H}_4\text{NTA}^+$

$\text{p}K_1 = 1.1$
$\text{p}K_2 = 1.650$
$\text{p}K_3 = 2.940$
$\text{p}K_4 = 10.334$

If you wanted to prepare a 10^{-6} M H^+ standard, you would never dream of doing it by diluting HCl until it was 10^{-6} M. (Why?) Similarly, it is poor form to prepare, say, a 10^{-6} M Ca^{2+} standard by diluting CaCl_2 to this low concentration. At such a low concentration, the Ca^{2+} ion could be lost by adsorption on a glass wall or reaction with some impurity.

A better idea is to prepare a **metal ion buffer.** Such a buffer is prepared from the metal and a suitable ligand. For example, consider the reaction of Ca^{2+} with nitrilotriacetic acid (NTA) at a pH high enough for NTA to be in its fully basic form (NTA^{3-}):

$$\text{Ca}^{2+} + \text{NTA}^{3-} \rightleftharpoons \text{CaNTA}^- \tag{16-28}$$

$$K_f = \frac{[\text{CaNTA}^-]}{[\text{Ca}^{2+}][\text{NTA}^{3-}]} = 10^{6.57} \text{ in } 0.1 \text{ M } \text{KNO}_3 \tag{16-29}$$

If equal concentrations of NTA^{3-} and CaNTA^- are present in a solution, the concentration of Ca^{2+} would be

$$[\text{Ca}^{2+}] = \frac{[\text{CaNTA}^-]}{K_f[\text{NTA}^{3-}]} = 10^{-6.57} \text{ M} \tag{16-30}$$

EXAMPLE

What concentration of NTA^{3-} should be added to 1.00×10^{-2} M $CaNTA^-$ in 0.1 M KNO_3 to give $[Ca^{2+}] = 1.00 \times 10^{-6}$ M?

Using Equation 16-29, we write

$$[NTA^{3-}] = \frac{[CaNTA^-]}{K_f[Ca^{2+}]} = \frac{1.00 \times 10^{-2}}{(10^{6.57})(1.00 \times 10^{-6})} = 2.69 \times 10^{-3} \text{ M}$$

These are practical concentrations of $CaNTA^-$ and of NTA^{3-}.

Summary

In potentiometric measurements, the indicator electrode responds to changes in the activity of analyte, and the reference electrode is a self-contained half-cell producing a constant reference potential. The most common reference electrodes are the calomel and silver–silver chloride electrodes. The indicator electrodes encountered in this chapter include (1) the inert Pt electrode, (2) a silver electrode responsive to Ag^+, halides, and other ions that react with Ag^+, and (3) ion-selective electrodes. Writing the Nernst equation in the form E(cell) = E(cathode) − E(anode), you should be able to analyze complex cells, such as those in which complexometric or precipitation titrations are occurring. The existence of small, unknown junction potentials at the interface of two electrolyte solutions sets a fundamental limit on the accuracy of potentiometric measurements. If small potentials must be measured accurately, it is necessary to construct a cell without a liquid junction.

The glass pH electrode and other ion-selective electrodes function by virtue of ion-exchange equilibria established on each surface of the electrode membrane. A gradient of activity in any species produces a gradient of free energy equal to $\Delta G = -RT \ln \mathscr{A}_1/\mathscr{A}_2$. The electric potential corresponding to this free energy difference, $E = -\Delta G/nF = (RT/nF) \ln \mathscr{A}_1/\mathscr{A}_2$, is the signal employed by an ion-selective electrode. Most electrodes are sensitive to many species, and their net response can be described by the equation $E = \text{Constant} \pm \beta(0.059\ 16/n_X)\log[\mathscr{A}_X + \sum_Y (k_{X,Y}\mathscr{A}_Y^{n_X/n_Y})]$, where $k_{X,Y}$ is the selectivity coefficient for each species. Common ion-selective electrodes can be classified as solid-state, liquid-based, and compound, depending on their construction. Metal ion buffers are especially appropriate for establishing and maintaining low concentrations of ions.

Terms to Understand

alkaline (sodium) error	matrix
asymmetry potential	metal ion buffer
calomel electrode	mobility
combination electrode	potentiometer
compound electrode	reference electrode
electroactive species	S.C.E. (saturated calomel electrode)
glass electrode	selectivity coefficient
indicator electrode	silver–silver chloride electrode
ion-selective electrode	solid-state electrode
junction potential	standard addition method
liquid-based ion-selective electrode	

Exercises

16-A. The apparatus in Figure 9-5 was used to monitor the titration of 100.0 mL of solution containing 50.0 mL of 0.100 M AgNO$_3$ and 50.0 mL of 0.100 M TlNO$_3$. The titrant contained 0.200 M NaBr. Suppose that the glass electrode (used as a *reference electrode* in this experiment) gives a constant potential of +0.200 V. The glass electrode is attached to the *positive* terminal of the pH meter, and the silver wire to the negative terminal. Calculate the cell potential at each of the following volumes of NaBr, and sketch the titration curve: 1.0, 15.0, 24.0, 24.9, 25.2, 35.0, 50.0, 60.0 mL.

16-B. The apparatus below can be used to follow the course of an EDTA titration and was used to generate the curves in Figure 14-6. The heart of the cell is a pool of liquid Hg in contact with the solution and with a Pt wire. A small amount of HgY^{2-} added to the analyte equilibrates with a very tiny amount of Hg^{2+}:

$$HgY^{2-} \rightleftharpoons Hg^{2+} + Y^{4-} \quad K = \frac{[Hg^{2+}][Y^{4-}]}{[HgY^{2-}]} = \frac{1}{K_f}$$

$$= 2.0 \times 10^{-22} \quad (A)$$

The redox equilibrium, Hg^{2+} + 2e$^-$ $\rightleftharpoons$ Hg(l), is established rapidly at the surface of the Hg electrode, so the Nernst equation for the cell can be written in the form

$$E(\text{cell}) = E(\text{cathode}) - E(\text{anode})$$

$$= \left(0.854 - \frac{0.059\ 16}{2} \log \frac{1}{[Hg^{2+}]}\right) - E(\text{anode})$$

(B)

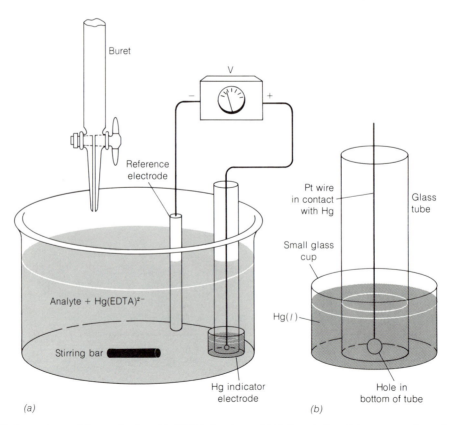

(a) Apparatus used for a potentiometric EDTA titration. (b) Enlarged view of the mercury electrode.

where E(anode) is the constant potential of the reference electrode. But from Equation A, we can write $[Hg^{2+}] = [HgY^{2-}]/K_f[Y^{4-}]$, and this can be substituted into Equation B to give

$$E(cell) = 0.854 - \frac{0.059\ 16}{2} \log \frac{[Y^{4-}]K_f}{[HgY^{2-}]} - E(anode)$$

$$E(cell) = 0.854 - E(anode) - \frac{0.059\ 16}{2} \log \frac{K_f}{[HgY^{2-}]}$$

$$- \frac{0.059\ 16}{2} \log[Y^{4-}] \qquad (C)$$

where K_f is the formation constant for HgY^{2-}. This apparatus thus responds to the changing EDTA concentration during an EDTA titration.

Suppose that 50.0 mL of 0.010 0 M $MgSO_4$ is titrated with 0.020 0 M EDTA at pH 10.0 using the apparatus above with an S.C.E. reference electrode. Assume that the analyte contains 1.0×10^{-4} M $Hg(EDTA)^{2-}$ added at the beginning of the titration. Calculate the cell potential at the following volumes of added EDTA, and draw a graph of millivolts versus milliliters: 0, 10.0, 20.0, 24.9, 25.0, and 26.0 mL.

16-C. A solid-state fluoride ion-selective electrode responds to F^-, but not to HF. It also responds to hydroxide ion at high concentration when $[OH^-] \approx [F^-]/10$. Suppose that such an electrode gave a potential of $+100$ mV (versus S.C.E.) in 10^{-5} M NaF and $+41$ mV in 10^{-4} M NaF. Sketch qualitatively how the potential would vary if the electrode were immersed in 10^{-5} M NaF and the pH varied from 1 to 13.

16-D. One commercial glass-membrane sodium ion-selective electrode has a selectivity coefficient $k_{Na^+, H^+} = 36$. When this electrode was immersed in 1.00 mM NaCl at pH 8.00, a potential of -38 mV (versus S.C.E.) was recorded.
   (a) Neglecting activity coefficients and assuming $\beta = 1$ in Equation 16-24 what would be the potential if the electrode were immersed in 5.00 mM NaCl at pH 8.00?
   (b) What would be the potential for 1.00 mM NaCl at pH 3.87?
   After answering this question, you should realize that pH is a critical variable in the use of a sodium ion-selective electrode.

16-E. An ammonia gas-sensing electrode gave the following calibration points when all solutions contained 1 M NaOH.

NH_3 (M)	E (mV)	NH_3 (M)	E (mV)
1.00×10^{-5}	268.0	5.00×10^{-4}	368.0
5.00×10^{-5}	310.0	1.00×10^{-3}	386.4
1.00×10^{-4}	326.8	5.00×10^{-3}	427.6

A dry food sample weighing 312.4 mg was digested by the Kjeldahl procedure (page 134) to convert all of the nitrogen to NH_4^+. The digestion solution was diluted to 1.00 L, and 20.0 mL was transferred to a 100 mL volumetric flask. The flask was treated with 10.0 mL of 10.0 M NaOH plus enough NaI to complex the Hg catalyst from the digestion, and diluted to 100.0 mL. When measured with the ammonia electrode, this solution gave a reading of 339.3 mV. Calculate the percent nitrogen in the food sample.

16-F. Cyanide ion was measured indirectly with a solid-state ion-specific electrode containing a silver sulfide membrane. Assuming that the electrode response is Nernstian and that the ionic strength of all solutions is constant, the electrode response can be written as follows:

$$E = Constant + 0.059\ 16 \log[Ag^+] \qquad (1)$$

To an unknown cyanide solution was added $Ag(CN)_2^-$, such that $[Ag(CN)_2^-] = 1.0 \times 10^{-5}$ M in the final solution. This complex ion behaves as a silver ion buffer in the presence of CN^-:

$$Ag^+ + 2CN^- \rightleftharpoons Ag(CN)_2^-$$

$$K = \beta_2 = \frac{[Ag(CN)_2^-]}{[Ag^+][CN^-]^2} = 10^{19.85} \qquad (2)$$

   (a) Suppose that the unknown contained 8.0×10^{-6} M CN^- and the potential was found to be $+206.3$ mV. Then a *standard addition* of CN^- was made to bring the CN^- concentration to 12.0×10^{-6} M. What will be the new potential?
   (b) Now consider a real experiment in which 50.0 mL of unknown gave a potential of 134.8 mV before a standard addition of CN^- was made. After adding 1.00 mL of 2.50×10^{-4} M KCN, the potential dropped to 118.6 mV. What was the concentration of CN^- in the 50.0 mL sample? For this part, we do not know the value of the constant in Equation 1 above. That is, you cannot use the value of the constant derived from part a of this question.

Problems

16-1. Suppose that the silver–silver chloride electrode in Figure 16-2 is replaced by a saturated calomel electrode. Calculate the cell potential if $[Fe^{2+}]/[Fe^{3+}] = 2.5 \times 10^{-3}$.

16-2. For a silver–silver chloride electrode, the following potentials are observed:

$$E^0 = 0.222 \text{ V} \qquad E(\text{saturated KCl}) = 0.197 \text{ V}$$

Predict the value of E for a calomel electrode saturated with KCl, given that E^0 for the calomel electrode is 0.268 V. (Your answer will not be exactly the value 0.241 used in this text.)

16-3. Based on the potentials below, calculate the *activity* of Cl^- in 1 M KCl.

$$E^0(\text{calomel electrode}) = 0.268 \text{ V}$$

$$E(\text{calomel electrode, 1 M KCl}) = 0.280 \text{ V}$$

16-4. Why is the 0.1 M HCl|0.1 M KCl junction potential of opposite sign and greater magnitude than the 0.1 M NaCl|0.1 M KCl potential in Table 16-2?

16-5. How many seconds will it take for (a) H^+ and (b) NO_3^- to migrate a distance of 12 cm in a field of 7800 V m^{-1}?

16-6. Suppose that the Ag|AgCl outer electrode in Figure 16-7 is filled with 0.1 M NaCl instead of saturated KCl. Suppose that the electrode is calibrated in a dilute buffer containing 0.1 M KCl at pH 6.54 at 25°C. The electrode is then dipped in a second buffer *at the same pH* and same temperature, but containing 3.5 M KCl. Use Table 16-2 to estimate how much the indicated pH will change.

16-7. Suppose that an ideal hypothetical cell such as that in Figure 15-7 were set up to measure E^0 for the half-reaction $Ag^+ + e^- \rightleftharpoons Ag(s)$.
(a) Calculate the equilibrium constant for the net cell reaction.
(b) If there were a junction potential of +2 mV (increasing E(cell) from 0.799 to 0.801 V), by what percent would the calculated equilibrium constant increase?
(c) Answer parts a and b with E^0 for the cathode reaction being 0.100 V instead of 0.799 V.

16-8. Describe how you would calibrate a pH electrode and measure the pH of blood (which is ~7.5) *at 37° C*. Use the standard buffers in Table 16-3.

16-9. (a) When the difference in pH across the membrane of a glass electrode at 25°C is 4.63 pH units, how much potential is generated by the pH gradient?
(b) What would be the potential for the same pH difference at 37°C?

16-10. List the sources of error associated with pH measurement using the glass electrode.

16-11. If electrode C in Figure 16-11 is placed in a solution of pH 11.0, what will the pH reading be?

16-12. Which National Bureau of Standards buffer(s) would you use to calibrate an electrode for pH measurements in the range 3–4?

16-13. A solution prepared by mixing 25.0 mL of 0.200 M KI with 25.0 mL of 0.200 M NaCl was titrated with 0.100 M AgNO$_3$ in the following cell:

$$\text{S.C.E.} \| \text{titration solution} | Ag(s)$$

Calling the solubility products of AgI and AgCl K_I and K_{Cl}, respectively, the answers to parts a and b should be expressions containing these constants.
(a) Calculate the concentration of $[Ag^+]$ in the solution when 25.0 mL of titrant has been added.
(b) Calculate the concentration of $[Ag^+]$ in the solution when 75.0 mL of titrant has been added.
(c) Write an expression showing how the cell potential depends on $[Ag^+]$.
(d) The titration curve is shown below. Calculate the numerical value of the ratio K_{Cl}/K_I.

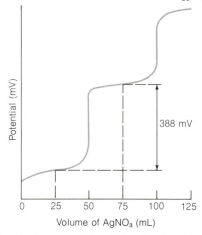

16-14. A solution containing 50.0 mL of 0.100 M EDTA buffered to pH 10.00 was titrated with 50.0 mL of

0.0200 M $Hg(ClO_4)_2$ in the cell below.

$$Hg(l)\,|\,\text{titration solution}\,\|\,\text{S.C.E.}$$

From the cell potential of $+0.034$ V, calculate the formation constant of $Hg(EDTA)^{2-}$. (*Hint:* See Exercise 16-B.)

16-15. The titration solution in the cell below had a total volume of 50.0 mL and contained 0.100 M Mg^{2+} and 1.00×10^{-5} M $Zn(EDTA)^{2-}$ at a pH of 10.00.

$$Zn(s)\,|\,\text{titration solution}\,\|\,\text{S.C.E.}$$

What will be the cell potential when 10.0 mL of 0.100 M EDTA has been added? (*Hint:* See Exercise 16-B.)

16-16. To determine the *concentration* of a dilute analyte with an ion-selective electrode, why is it desirable to use standards with a constant, high concentration of an inert salt?

16-17. The data below were obtained when a Ca^{2+} ion-selective electrode was immersed in a series of standard solutions whose ionic strength was constant at 2.0 M.

$[Ca^{2+}]$ (M)	E (mV)	$[Ca^{2+}]$ (M)	E (mV)
3.38×10^{-5}	-74.8	3.38×10^{-2}	$+10.0$
3.38×10^{-4}	-46.4	3.38×10^{-1}	$+37.7$
3.38×10^{-3}	-18.7		

(a) Make a graph of the data and find the concentration of Ca^{2+} in a sample that gave a potential of -22.5 mV.

(b) Calculate the value of β in Equation 16-27.

16-18. A calcium ion-selective electrode obeys Equation 16-24, in which $\beta = 0.970$ and $n_{Ca^{2+}} = 2$. The selectivity coefficients for several ions are listed below.

Interfering ion, Y	$k_{Ca^{2+},Y}$
Mg^{2+}	0.040
Ba^{2+}	0.021
Zn^{2+}	0.081
K^+	6.6×10^{-5}
Na^+	1.7×10^{-4}

In a pure 1.00×10^{-3} M calcium solution, the measured potential was $+300.0$ mV. What would be the potential if the solution had the same calcium concentration plus $[Mg^{2+}] = 1.00 \times 10^{-3}$ M, $[Ba^{2+}] = 1.00 \times 10^{-3}$ M, $[Zn^{2+}] = 5.00 \times 10^{-4}$ M,

$[K^+] = 0.100$ M, and $[Na^+] = 0.0500$ M. (Use concentrations instead of activities to answer this question.) If they are present at *equal* concentrations, which ion in the table above interferes the most with the Ca^{2+} electrode?

16-19. (a) Calculate the slope and intercept (and their standard deviations) of the best straight line through the points in Problem 16-17, using the method of least squares (Section 4-4).

(b) Calculate the concentration (and its associated uncertainty) of a sample that gave a potential of $-22.5\ (\pm 0.3)$ mV.

16-20. A lead ion buffer was prepared by mixing 0.100 mmol of $Pb(NO_3)_2$ with 2.00 mmol of $Na_2C_2O_4$ in a volume of 10.0 mL.

(a) Given the equilibrium below, find the concentration of free Pb^{2+} in this solution.

$$Pb^{2+} + 2C_2O_4^{2-} \rightleftharpoons Pb(C_2O_4)_2^{2-}$$
$$K = \beta_2 = 10^{6.54}$$

(b) How many mmol of $Na_2C_2O_4$ should be used to give $[Pb^{2+}] = 1.00 \times 10^{-7}$ M?

16-21. A magnesium ion buffer was made by mixing 10.0 mL of 1.00 mM $MgSO_4$, 10.0 mL of 1.30 mM EDTA, and 5.00 mL of buffer, pH 10.00. What is the concentration of free metal ion in this solution? Answer the same question for $MnSO_4$ used instead of $MgSO_4$.

16-22. The standard addition table below assumes that an electrode has a Nernstian response to analyte:

Standard addition table: 10 mL of standard added to 100 mL of sample (To obtain sample concentration, multiply standard concentration by Q.)

ΔE	Q	ΔE	Q	ΔE	Q
0 mv	1.00	10 mv	0.160	20 mv	0.0716
1	0.696	11	0.145	21	0.0671
2	0.529	12	0.133	22	0.0629
3	0.423	13	0.121	23	0.0591
4	0.351	14	0.112	24	0.0556
5	0.297	15	0.1030	25	0.0523
6	0.257	16	0.0952	26	0.0494
7	0.225	17	0.0884	27	0.0466
8	0.199	18	0.0822	28	0.0440
9	0.178	19	0.0767	29	0.0416

SOURCE: *Orion Research Analytical Methods Guide* (Cambridge, Mass.: Orion Research, Inc., 1975), p. 5.

As an example, suppose chloride ion is measured with an ion-selective electrode and a reference electrode. The electrodes are placed in 100.0 mL of sample, and a reading of 228.0 mV is obtained. Then 10.0 mL of standard containing 100.0 ppm Cl^- is added, and a new reading of 210.0 mV is observed. Since $|\Delta E| = 18.0$ mV, $Q = 0.082\ 2$ in the table above. Therefore, the original concentration of Cl^- was $(0.082\ 2)(100\ ppm) = 8.22$ ppm.

(a) What molarity of Cl^- is 8.22 ppm? Assume that the density of unknown is 1.00 g/mL.

(b) Use the original concentration of 8.22 ppm and the original potential of 228.0 mV to show that the second potential should be 210.0 mV if the electrode response obeys the equation

$$E = Constant - 0.059\ 16\ \log[Cl^-]$$

(c) How would you change the table above to use it for Ca^{2+} instead of Cl^-?

16-23. Do not ignore activities in this problem. Citric acid is a triprotic acid (H_3A) whose anion (A^{3-}) forms stable complexes with many metal ions.

(a) A calcium ion-selective electrode gave a calibration curve similar to Figure B-2 in Appendix B. The slope of the curve was 29.58 mV. When immersed in a solution having $\mathscr{A}_{Ca^{2+}} = 1.00 \times$

10^{-3}, the electrode potential was $+2.06$ mV. When immersed in the solution to be described in part b of this problem, the potential was -25.90 mV. Calculate the activity of Ca^{2+} in the solution in part b.

(b) Ca^{2+} forms a 1:1 complex with citrate under the conditions of this problem.

$$Ca^{2+} + A^{3-} \xrightleftharpoons{K_f} CaA^-$$

A solution was prepared by mixing equal volumes of solutions 1 and 2 below.

Solution 1: $[Ca^{2+}] = 1.00 \times 10^{-3}$ M,

$pH = 8.00, \mu = 0.10$ M

Solution 2: $[citrate]_{Total} = 1.00 \times 10^{-3}$ M,

$pH = 8.00, \mu = 0.10$ M

The activity of the calcium ion in the resulting solution was determined in part a of this problem. Calculate the formation constant, K_f, for CaA^-. For the sake of this calculation, you may assume that the size of CaA^- is 500 pm. At pH 8.00 and $\mu = 0.10$ M, the fraction of free citrate in the form A^{3-} is 0.998.

17 / Redox Titrations

A redox titration is based on an oxidation–reduction reaction between analyte and titrant. For example, hydroquinone can be analyzed by titration with standard dichromate solution:

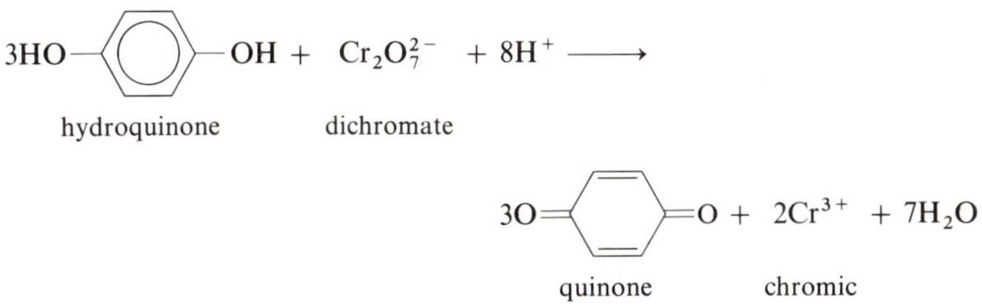

$$3HO-\langle\ \rangle-OH + Cr_2O_7^{2-} + 8H^+ \longrightarrow$$

hydroquinone   dichromate

$$3O=\langle\ \rangle=O + 2Cr^{3+} + 7H_2O \quad (17\text{-}1)$$

quinone   chromic

The equilibrium constant for the titration reaction is easy to calculate from standard reduction potentials:

$$Cr_2O_7^{2-} + 14H^+ + 6e^- \rightleftharpoons 2Cr^{3+} + 7H_2O \qquad E^0 = 1.33 \text{ V}$$

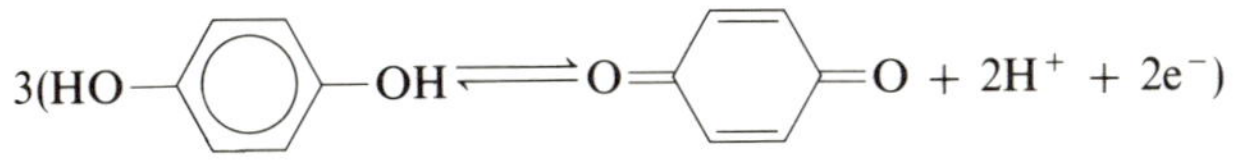

$$3(HO-\langle\ \rangle-OH \rightleftharpoons O=\langle\ \rangle=O + 2H^+ + 2e^-) \qquad E^0 = 0.700 \text{ V}$$

$$Cr_2O_7^{2-} + 3C_6H_6O_2 + 8H^+ \rightleftharpoons 2Cr^{3+} + 3C_6H_4O_2 + 7H_2O \qquad E^0 = 0.63 \text{ V}$$

$$K = 10^{nE^0/0.059\,16} = 10^{6(0.63)/0.059\,16} = 10^{64}$$

Relation between E^0 and the equilibrium constant:

$$K = 10^{nE^0/0.059\,16} \text{ at } 25°C$$

$$(17\text{-}2)$$

In practice, the reaction is too slow for a titration at room temperature, but it proceeds rapidly enough at 40–60°C to be a useful analytical procedure. The enormous equilibrium constant (typical of many redox reactions) assures us that the reaction will be quantitative (i.e., it will go to completion). The

363

endpoint is detected with the *redox indicator* diphenylamine, whose color changes from colorless to violet when the titration reaction is complete.

In this chapter we first deal with the theory of redox titrations and then discuss practical aspects of some common reagents. The theory is necessary to understand how redox indicators and potentiometric detection procedures work.

17-1 THE SHAPE OF A REDOX TITRATION CURVE

Consider the titration of Fe^{2+} with standard Ce^{4+}, the course of which could be monitored potentiometrically as shown in Figure 17-1. The titration reaction is

$$Ce^{4+} + Fe^{2+} \rightarrow Ce^{3+} + Fe^{3+} \qquad (17\text{-}3)$$

ceric   ferrous   cerous   ferric

The *titration reaction* is $Ce^{4+} + Fe^{2+} \rightarrow Ce^{3+} + Fe^{3+}$. It goes to completion after each addition of titrant.

for which $K \approx 10^{17}$ in 1 M $HClO_4$. Each mole of ceric ion oxidizes one mole of ferrous ion rapidly and quantitatively. The titration reaction thus creates a mixture of Ce^{4+}, Ce^{3+}, Fe^{2+}, and Fe^{3+} in the beaker in Figure 17-1.

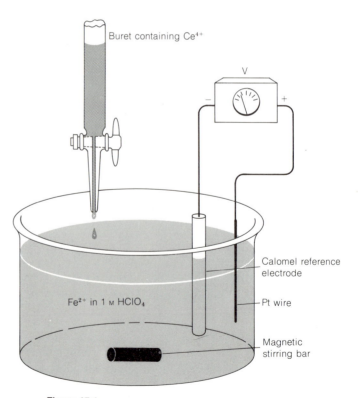

Figure 17-1
Apparatus for potentiometric titration of Fe^{2+} with Ce^{4+}.

To follow the course of the reaction, a pair of electrodes is inserted into the reaction mixture. At the *calomel* reference electrode, the reaction is

$$2Hg(l) + 2Cl^- \rightleftharpoons Hg_2Cl_2(s) + 2e^- \qquad (17\text{-}4)$$

At the *Pt indicator electrode*, there are *two* reactions that come to equilibrium:

$$Fe^{3+} + e^- \rightleftharpoons Fe^{2+} \qquad E^0 = 0.767 \text{ V} \qquad (17\text{-}5)$$

$$Ce^{4+} + e^- \rightleftharpoons Ce^{3+} \qquad E^0 = 1.70 \text{ V} \qquad (17\text{-}6)$$

These redox equilibria are established at the Pt electrode.

The potentials cited here are the formal potentials that apply in 1 M $HClO_4$. The cell reaction can be described in either of two ways:

$$2Fe^{3+} + 2Hg(l) + 2Cl^- \rightleftharpoons 2Fe^{2+} + Hg_2Cl_2(s) \qquad (17\text{-}7)$$

or

$$2Ce^{4+} + 2Hg(l) + 2Cl^- \rightleftharpoons 2Ce^{3+} + Hg_2Cl_2(s) \qquad (17\text{-}8)$$

The titration reaction goes to completion. The cell reactions proceed to a negligible extent. The potentiometric circuit *measures* the concentrations of species in solution, but does not *change* them.

The cell reactions are not the same as the titration reaction (17-3). The potentiometer has no interest in how the Ce^{4+}, Ce^{3+}, Fe^{3+}, and Fe^{2+} happened to get into the beaker. What it does care about is how hard electrons want to flow from the anode to the cathode through the meter. If the solution has come to equilibrium, the potential driving both Reaction 17-7 and Reaction 17-8 must be the same. **We may describe the cell potential using either Equation 17-7 or Equation 17-8, or both, however we please.**

To reiterate, the physical picture of the titration is this: Ce^{4+} is added from the buret to create a mixture of Ce^{4+}, Ce^{3+}, Fe^{3+}, and Fe^{2+} in the beaker. Since the equilibrium constant for Reaction 17-3 is so large, the titration reaction goes to "completion" after each addition of Ce^{4+}. The potentiometer measures the potential driving electrons from the reference electrode, through the meter, and out at the Pt electrode. That is, **the circuit measures the potential for reduction of Fe^{3+} or Ce^{4+} at the Pt surface provided by the $Hg|Hg_2Cl_2$ couple in the reference electrode.** The titration reaction, on the other hand, is an oxidation of Fe^{2+} and a reduction of Ce^{4+}. The titration reaction produces a certain mixture of Ce^{4+}, Ce^{3+}, Fe^{3+}, and Fe^{2+}. The circuit measures the potential for reduction of Ce^{4+} and Fe^{3+}. **The titration reaction goes to completion. The cell reaction is negligible. The cell is being used to measure activities, not to change them.**

We now set out to calculate how the cell potential will change as Fe^{2+} is titrated with Ce^{4+}. The calculations are of three types.

Region 1: Before the equivalence point

As each aliquot of Ce^{4+} is added, the titration reaction (17-3) goes to "completion," consuming the Ce^{4+} and making an equal number of moles of Ce^{3+}

366

17 / REDOX TITRATIONS

Either Reaction 17-5 or 17-6 can be used to describe the cell potential at any time. However, since we know the concentrations of Fe^{2+} and Fe^{3+}, it is more convenient for now to use Reaction 17-5.

and Fe^{3+}. Prior to the equivalence point, excess unreacted Fe^{2+} remains in the solution. Therefore, we can find the concentrations of Fe^{2+} and Fe^{3+} without any difficult calculations. On the other hand, we cannot find the concentration of Ce^{4+} without solving a fancy little equilibrium problem. Since the amounts of Fe^{2+} and Fe^{3+} are both known, it is *convenient* to calculate the cell potential using the cathode reaction (17-5) instead of Reaction 17-6.

$$E(\text{cell}) = E(\text{cathode}) - E(\text{anode})$$

$$E(\text{cell}) = \left(0.767 - 0.05916 \log \frac{[Fe^{2+}]}{[Fe^{3+}]}\right) - (0.241) \qquad (17\text{-}9)$$

Formal potential for Fe^{3+} reduction in $1\,M\,HClO_4$

potential of saturated calomel electrode

$$E(\text{cell}) = +0.526 - 0.05916 \log \frac{[Fe^{2+}]}{[Fe^{3+}]} \qquad (17\text{-}10)$$

For Reaction 17-5, $E = E^0(Fe^{3+}|Fe^{2+})$ when $V = \frac{1}{2}V_e$.

One special point is reached before the equivalence point. When the volume of titrant is one-half of the amount required to reach the equivalence point ($V = \frac{1}{2}V_e$), the concentrations of Fe^{3+} and Fe^{2+} are equal. In this case the log term of Equation 17-9 is zero, and $E(\text{cathode}) = E^0$ for the $Fe^{3+}|Fe^{2+}$ couple. *This is analogous to the point at which* $pH = pK_A$ *when* $V = \frac{1}{2}V_e$ *in an acid–base titration.*

Region 2: At the equivalence point

Exactly enough Ce^{4+} has been added to react with all of the Fe^{2+}. Virtually all of the cerium is in the form Ce^{3+}, and virtually all of the iron is in the form Fe^{3+}. Tiny amounts of Ce^{4+} and Fe^{2+} are present at equilibrium. From the stoichiometry of Reaction 17-3, we can say that

$$[Ce^{3+}] = [Fe^{3+}] \qquad (17\text{-}11)$$

and

$$[Ce^{4+}] = [Fe^{2+}] \qquad (17\text{-}12)$$

To understand why Equations 17-11 and 17-12 are true, imagine that *all* of the cerium and iron has been converted to Ce^{3+} and Fe^{3+}. Since we are at the equivalence point, $[Ce^{3+}] = [Fe^{3+}]$. Now let Reaction 17-3 come to equilibrium:

$$Fe^{3+} + Ce^{3+} \rightleftharpoons Fe^{2+} + Ce^{4+} \quad (\text{reverse of Reaction 17-3})$$

If a little bit of Fe^{3+} goes back to Fe^{2+}, an equal number of moles of Ce^{4+} must be made. So $[Ce^{4+}] = [Fe^{2+}]$.

Effect of Ionic Strength on Ionic Dissociation

Color Plate 1
Beakers containing same mixture of $FeSCN^{2+}$, Fe^{3+}, and SCN^-.

Color Plate 2
Change when KNO_3 is added to left beaker.

Colloids and Dialysis

Color Plate 3
Ordinary aqueous Fe(III) (left) and colloidal Fe(III) (right).

Color Plate 4
Result of 24-hour dialysis of colloidal Fe(III) (left) and solution of Cu(II) (right).

Fajans Titration of Cl⁻ with AgNO$_3$ Using Dichlorofluorescein

Color Plate 5
Indicator before beginning titration.

Color Plate 6
AgCl precipitate before endpoint.

Color Plate 7
Indicator adsorbed on precipitate after endpoint.

HCl Fountain

Color Plate 8
Basic indicator solution in beaker.

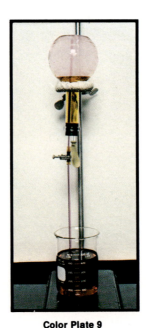

Color Plate 9
Indicator as it mixes with HCl in flask.

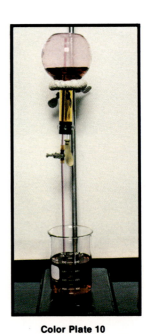

Color Plate 10
Later, mixing is not very efficient, and indicator retains red color.

Thymol Blue

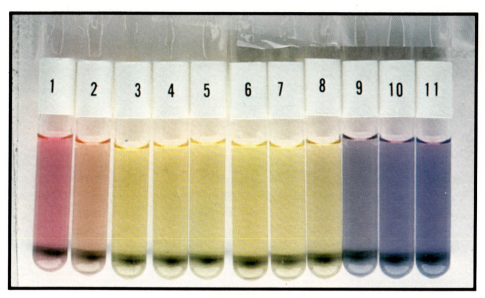

Color Plate 11

Acid–base indicator thymol blue as function of pH. The pK values are 1.7 and 8.9, and pH of each solution is marked on each test tube.

Indicators and Acidity of CO_2

Sequence of events when dry ice is added to cylinders containing basic solutions of phenolphthalein (left) and bromothymol blue (right).

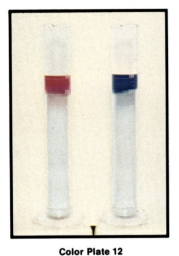

Color Plate 12

Cylinders before adding dry ice. Ethanol indicator solutions have not yet mixed with entire cylinder.

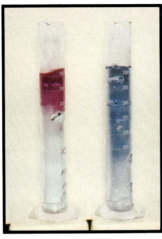

Color Plate 13

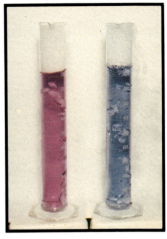

Color Plate 14

Adding dry ice causes vigorous bubbling, which mixes cylinder contents and lowers pH.

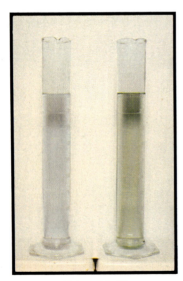

Color Plate 15

Phenolphthalein changes to its colorless acidic form, and bromothymol blue color is that of a mixture of its acidic and basic forms.

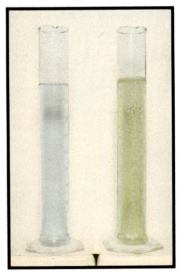

Color Plate 16

After addition of HCl and stirring of right-hand cylinder, bubbles of CO_2 can be seen leaving solution, and indicator changes completely to its acidic color.

Titration of Mg^{2+} by EDTA Using Eriochrome Black T

Color Plate 17

Before (left), near (center), and after (right) equivalence point.

Color Plate 18

Same titration with methyl red added as inert dye to alter colors.

Electrochemical Writing

Color Plate 19

Stylus used as cathode.

Color Plate 20

Stylus used as anode.

Color Plate 21

Foil backing has opposite polarity from stylus and produces reverse color on bottom sheet of paper.

Absorption Spectra

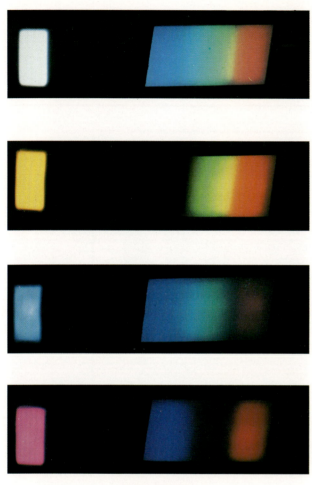

Color Plate 22

Projected visible spectra of (from top to bottom) white light, potassium dichromate, bromophenol blue, and phenolphthalein.

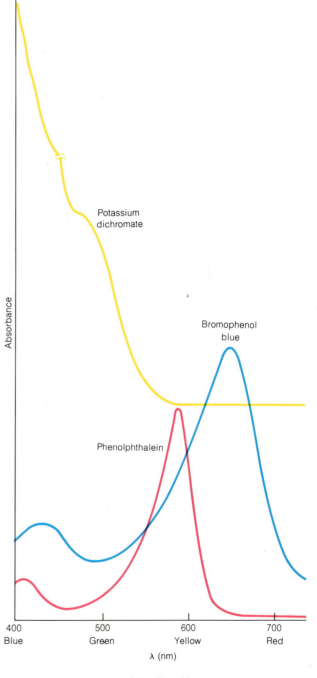

Color Plate 23

Spectrophotometric visible absorption spectra of colored compounds whose projected spectra are shown in Color Plate 22. Bromophenol blue and potassium dichromate spectra are displaced upward for clarity.

Countercurrent Distribution

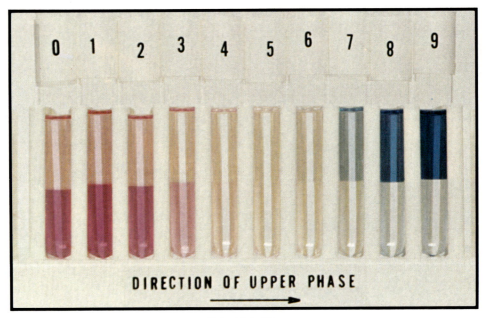

Color Plate 24

Countercurrent separation of phenol red and bromocresol green.
Upper phase is 1-butanol; lower phase is 0.1 M aqueous Na_2CO_3.

At any time, Reactions 17-5 and 17-6 are *both* in equilibrium at the cathode. At the equivalence point, it is *convenient* to use both reactions to describe the cell potential. The Nernst equations for the reactions are, respectively,

$$E(\text{cathode}) = 0.767 - 0.059\,16 \log \frac{[Fe^{2+}]}{[Fe^{3+}]} \qquad (17\text{-}13)$$

$$E(\text{cathode}) = 1.70 - 0.059\,16 \log \frac{[Ce^{3+}]}{[Ce^{4+}]} \qquad (17\text{-}14)$$

At the equivalence point, it is convenient to use both Reactions 17-5 and 17-6 to calculate the cell potential. This is strictly a matter of algebraic convenience.

Here is where we stand: Each equation above is a statement of algebraic truth. But neither one alone allows us to find E(cathode) because we do not know exactly what tiny concentrations of Fe^{2+} and Ce^{4+} are present. It is possible to solve the four simultaneous equations, 17-11, 17-12, 17-13, and and 17-14, by first *adding* Equations 17-13 and 17-14. This is strictly an algebraic manipulation, the logic and beauty of which will become apparent very soon.

Adding Equations 17-13 and 17-14 gives

$$2E(\text{cathode}) = 0.767 + 1.70 - 0.059\,16 \log \frac{[Fe^{2+}]}{[Fe^{3+}]} - 0.059\,16 \log \frac{[Ce^{3+}]}{[Ce^{4+}]}$$

$$2E(\text{cathode}) = 2.46_7 - 0.059\,16 \log \frac{[Fe^{2+}][Ce^{3+}]}{[Fe^{3+}][Ce^{4+}]} \qquad (17\text{-}15)$$

Remember that $\log a + \log b = \log ab$.

But since $[Ce^{3+}] = [Fe^{3+}]$ and $[Ce^{4+}] = [Fe^{2+}]$ at the equivalence point, the ratio of concentrations in the log term is unity. The logarithm is zero. Therefore,

$$2E(\text{cathode}) = 2.46_7 \text{ V}$$

$$E(\text{cathode}) = 1.23 \text{ V} \qquad (17\text{-}16)$$

and the cell potential is

$$E(\text{cell}) = E(\text{cathode}) - E(\text{anode})$$

$$= 1.23 - 0.241 = 0.99 \text{ V} \qquad (17\text{-}17)$$

In this particular titration, the equivalence-point potential is independent of the concentrations and volumes of the reactants.

Region 3: After the equivalence point

Now virtually all of the iron is Fe^{3+}. The moles of Ce^{3+} equal the moles of Fe^{3+}. There is also a known excess of unreacted Ce^{4+}. Since we know both

368

17 / REDOX TITRATIONS

After the equivalence point, it is convenient to use Equation 17-6 because we can easily calculate the concentrations of Ce^{3+} and Ce^{4+}. It is not convenient to use Equation 17-5 because we do not know the concentration of Fe^{2+}, which has been "used up."

The potential (versus S.H.E.) in a redox titration varies roughly between the standard potentials of the two couples involved in the titration reaction.

$[Ce^{3+}]$ and $[Ce^{4+}]$, it is *convenient* to use Equation 17-6 to describe the cathode reaction.

$$E(cell) = E(cathode) - E(anode)$$

$$= \left(1.70 - 0.059\ 16 \log \frac{[Ce^{3+}]}{[Ce^{4+}]}\right) - (0.241) \qquad (17\text{-}18)$$

At the special point when $V = 2V_e$, $[Ce^{3+}] = [Ce^{4+}]$ *and* $E(cathode) = E^0(Ce^{4+}|Ce^{3+}) = 1.70\ V$.

Roughly speaking, the cell potential will be fairly level before and after the equivalence point, with a rapid rise near the equivalence point. Before the equivalence point, the potential is roughly in the range $E(cathode) \approx E^0(Fe^{3+}|Fe^{2+})$ or $E(cell) = E(cathode) - E(anode) \approx E^0(Fe^{3+}|Fe^{2+}) - 0.241$ V. After the equivalence point, the potential levels off such that $E(cathode) \approx E^0(Ce^{4+}|Ce^{3+})$ and $E(cell) \approx E^0(Ce^{4+}|Ce^{3+}) - 0.241$ V.

EXAMPLE

Suppose that 100.0 mL of 0.050 0 M Fe^{2+} is titrated with 0.100 M Ce^{4+}, using the cell in Figure 17-1. The equivalence point occurs when $V_{Ce^{4+}} = 50.0$ mL, since the Ce^{4+} is twice as concentrated as the Fe^{2+}. Calculate the cell potential at the following points: 36.0, 50.0, and 63.0 mL.

36.0 mL: This is 36.0/50.0 of the way to the equivalence point. Therefore, 36.0/50.0 of the iron is in the form Fe^{3+}, and 14.0/50.0 is in the form Fe^{2+}. Putting the ratio $[Fe^{2+}]/[Fe^{3+}] = 14.0/36.0$ into Equation 17-10 gives a cell potential of 0.550 V.

50.0 mL: This is the equivalence point. Equation 17-17 tells us that $V_e = 0.99$ V, regardless of the concentrations of analyte and titrant for this particular redox titration. For many titrations the potential at the equivalence point is constant, but, as we will see in the next section, this is not always true.

63.0 mL: The first 50.0 mL of cerium has been converted to Ce^{3+}. Since 13.0 mL of excess Ce^{4+} has been added, the ratio $[Ce^{3+}]/[Ce^{4+}]$ in Equation 17-18 is 50.0/13.0, and the cell potential is 1.424 V.

The calculated titration curve for Reaction 17-3 is shown in Figure 17-2. As in all previous titrations of all types, the endpoint is marked by a steep rise in the curve. Because both reactants are one-electron redox reagents, the curve is symmetric near the equivalence point. As we shall see in the next section, this is not the case when the stoichiometry of reactants is not 1:1. The calculated cathode potential when $V_{Ce^{4+}} = \frac{1}{2}V_e$ is the formal potential of the $Fe^{3+}|Fe^{2+}$ couple, since the ratio $[Fe^{2+}]/[Fe^{3+}]$ in Equation 17-9 is unity at this point. The calculated potential at any point in this titration depends only on the *ratio* of reactants; their *concentrations*

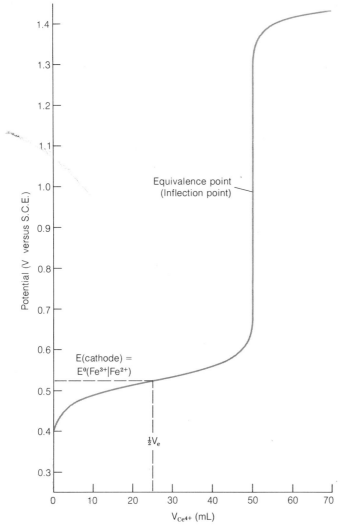

Figure 17-2
Theoretical curve for titration of 100.0 mL of 0.050 0 M Fe^{2+} with 0.100 M Ce^{4+} in 1 M HClO$_4$.

The shape of the curve in Figure 17-2 is essentially independent of the concentrations of analyte and titrant. The curve is symmetric near V$_e$ because the stoichiometry is 1:1.

do not figure in any calculations in this example. We expect, therefore, that the curve in Figure 17-2 will be independent of dilution. We should observe the same curve even if both reactants were diluted by a factor of ten.

The potential before Ce^{4+} is added to the reaction (V$_{Ce^{4+}}$ = 0) cannot be calculated because we do not know how much Fe^{3+} is present in solution. If no Fe^{3+} is present, the potential calculated with Equation 17-10 is $-\infty$. In fact, there must be some Fe^{3+} present in each reagent, either as an impurity or from oxidation of Fe^{2+} by atmospheric oxygen. In any case, the potential could never be lower than that needed to reduce the solvent (H$_2$O + e$^-$ → $\frac{1}{2}$H$_2$ + OH$^-$).

370

17 / REDOX TITRATIONS

A Slightly More Complicated Redox Calculation

Iodate can be used to titrate Tl^+ in a concentrated solution of HCl:

$$IO_3^- + 2Tl^+ + 2Cl^- + 6H^+ \rightarrow ICl_2^- + 2Tl^{3+} + 3H_2O \quad (17\text{-}19)$$

iodate   thallous   thallic

From the standard potentials below

$$IO_3^- + 2Cl^- + 6H^+ + 4e^- \rightarrow ICl_2^- + 3H_2O \qquad E^0 = 1.24 \text{ V} \quad (17\text{-}20)$$

$$Tl^{3+} + 2e^- \rightarrow Tl^+ \qquad E^0 = 0.77 \text{ V} \quad (17\text{-}21)$$

we expect E^0 for Reaction 17-19 to be $1.24 - 0.77 = 0.47$ V and $K = 10^{4E^0/0.059\ 16} = 10^{32}$. The equilibrium constant is very large, so the reaction goes to completion after each addition of IO_3^-.

Let's calculate the theoretical titration curve that results when 100.0 mL of 0.010 0 M Tl^+ is titrated with 0.010 0 M IO_3^-. We will assume that both solutions have a constant concentration of HCl equal to 1.00 M. Because one mole of IO_3^- consumes two moles of Tl^+, the equivalence point occurs at $V_{IO_3^-} = 50.0$ mL. Suppose that the reference electrode is a saturated calomel electrode, and the apparatus shown in Figure 17-1 is used for the titration. We will calculate the potential at one point in each region of the titration.

Region 1: Before the equivalence point

Before the equivalence point, we know the concentrations of Tl^+ and Tl^{3+}, so it is convenient to use Reaction 17-21 to calculate the cathode potential.

Suppose $V_{IO_3^-} = 10.0$ mL. Since the entire reaction requires 50.0 mL, the reaction is 10.0/50.0 complete. That is, one-fifth of the thallium is in the form Tl^{3+}, and four-fifths is in the form Tl^+. We may choose Reaction 17-21 to describe the cathode potential because we know the concentrations of both Tl^+ and Tl^{3+}:

$$E(\text{cell}) = E(\text{cathode}) - E(\text{anode})$$

$$= \underbrace{\left(0.77 - \frac{0.059\ 16}{2} \log \frac{[Tl^+]}{[Tl^{3+}]}\right)}_{\substack{\text{Nernst equation for} \\ \text{Reaction 17-21}}} - \underbrace{(0.241)}_{\substack{\text{Potential of} \\ \text{calomel electrode}}} \qquad (17\text{-}22)$$

$$= \left(0.77 - \frac{0.059\ 16}{2} \log \frac{4/5}{1/5}\right) - 0.241 = 0.511 \text{ V}$$

Region 2: At the equivalence point

Now virtually all of the reactants have been converted to products. We can say that

$$[Tl^{3+}] = 2[ICl_2^-] \qquad (17\text{-}23)$$

since two moles of Tl^{3+} are created for each mole of ICl_2^-. Although only tiny amounts of the reactants are present, we can say that

$$[Tl^+] = 2[IO_3^-] \qquad (17\text{-}24)$$

You can justify this by imagining that only products are present. When they react, they produce a tiny amount of Tl^+ and IO_3^- in a 2:1 mole ratio.

Once again it is convenient to use both half-reactions, 17-20 and 17-21, to describe the cathode potential at the equivalence point. From Reaction 17-20, we can say that

It is convenient to use Equations 17-20 and 17-21 *together* to find E(cathode) at the equivalence point. We do not have enough information to use either equation by itself.

$$E(\text{cathode}) = 1.24 - \frac{0.059\ 16}{4} \log \frac{[ICl_2^-]}{[IO_3^-][Cl^-]^2[H^+]^6} \qquad (17\text{-}25)$$

From Equation 17-21, we may write

$$E(\text{cathode}) = 0.77 - \frac{0.059\ 16}{2} \log \frac{[Tl^+]}{[Tl^{3+}]} \qquad (17\text{-}26)$$

To solve the four simultaneous equations (17-23 to 17-26), it is algebraically expedient to add Equations 17-25 and 17-26 together. However, Equation 17-25 should first be multiplied by 2 so that the coefficients of the log terms are equal. Multiplying Equation 17-25 by 2 gives

$$2E(\text{cathode}) = 2.48 - \frac{0.059\ 16}{2} \log \frac{[ICl_2^-]}{[IO_3^-][Cl^-]^2[H^+]^6} \qquad (17\text{-}27)$$

Adding Equations 17-26 and 17-27 gives

$$3E(\text{cathode}) = 3.25 - \frac{0.059\ 16}{2} \log \frac{[ICl_2^-][Tl^+]}{[IO_3^-][Cl^-]^2[H^+]^6[Tl^{3+}]} \qquad (17\text{-}28)$$

Substituting $[Tl^{3+}] = 2[ICl_2^-]$ and $[Tl^+] = 2[IO_3^-]$ into Equation 17-28 simplifies the log term:

$$3E(\text{cathode}) = 3.25 - \frac{0.059\ 16}{2} \log \frac{[I\cancel{Cl_2^-}]2[I\cancel{O_3^-}]}{[I\cancel{O_3^-}][Cl^-]^2[H^+]^6 2[I\cancel{Cl_2^-}]} \qquad (17\text{-}29)$$

Since we have stipulated that $[Cl^-] = [H^+] = 1.00\ \text{M}$ in this example, the log term is again zero. If the concentrations were not $1.00\ \text{M}$, we would simply put their values into Equation 17-29 and evaluate the log term. Putting $[H^+] = [Cl^-] = 1.00\ \text{M}$ into Equation 17-29 gives

$$3E(\text{cathode}) = 3.25\ \text{V} \Rightarrow E(\text{cathode}) = 1.08_3\ \text{V} \qquad (17\text{-}30)$$

$$E(\text{cell}) = E(\text{cathode}) - E(\text{anode}) = 1.08_3 - 0.241 = 0.84\ \text{V} \qquad (17\text{-}31)$$

In some other titrations, the concentrations of products do not entirely cancel in the log term. In such cases, we simply calculate the molarity of product in the solution at the equivalence point, assuming that all reactant has been converted to product.

Region 3: After the equivalence point

If $V_{IO_3^-} = 57.6$ mL, we can say that

$$\frac{[ICl_2^-]}{[IO_3^-]} = \frac{50.0}{7.6} \begin{array}{l} \leftarrow \text{Volume at equivalence point} \\ \leftarrow \text{Volume past equivalence point} \end{array} \quad (17\text{-}32)$$

Now we know both $[ICl_2^-]$ and $[IO_3^-]$, so we can use Equation 17-20 to find the cathode potential.

Using Equation 17-20 to describe the cathode reaction, we may write

$$E(\text{cell}) = E(\text{cathode}) - E(\text{anode})$$

$$= \left(1.24 - \frac{0.059\ 16}{4} \log \frac{[ICl_2^-]}{[IO_3^-][Cl^-]^2[H^+]^6}\right) - E(\text{anode}) \quad (17\text{-}33)$$

Substituting $[ICl_2^-]/[IO_3^-] = 50.0/7.6$ and $[H^+] = [Cl^-] = 1.00$ M, we obtain

$$E(\text{cell}) = 1.24 - \frac{0.059\ 16}{4} \log \frac{50.0}{(7.6)(1.00)^2(1.00)^6} - 0.241 \quad (17\text{-}34)$$

$$E(\text{cell}) = 0.987 \text{ V} \quad (17\text{-}35)$$

When the stoichiometry of the titration reaction is not 1:1, the curve is not symmetric around the equivalence point.

The theoretical titration curve is shown in Figure 17-3. *Note that the curve is not symmetric about the equivalence point.* This is true whenever

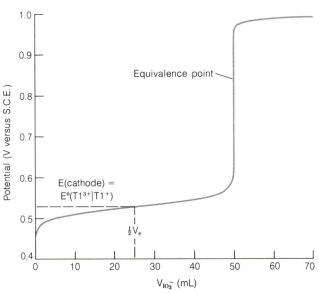

Figure 17-3
Theoretical curve for titration of 100.0 mL of 0.010 0 M Tl$^+$ with 0.010 0 M IO$_3^-$ in 1.00 M HCl.

Demonstration 17-1 POTENTIOMETRIC TITRATION OF Fe²⁺ WITH MnO₄⁻

The reaction of Fe²⁺ with KMnO₄ is an excellent illustration of the principles of potentiometric titrations. Dissolve 0.60 g of Fe(NH₄)₂(SO₄)₂·6H₂O (F.W. 392.13, 1.5 mmol) in 400 mL of 1 M H₂SO₄. Titrate the well-stirred solution with 0.02 M KMnO₄ (∼15 mL will be required to reach the equivalence point), using Pt and calomel electrodes with a pH meter as a potentiometer. The reference socket of the pH meter is the negative input terminal. Before starting the titration, set the millivolt scale of the meter to zero while connecting the two input sockets directly to each other through a wire.

Calculate some points on the theoretical titration curve before performing the experiment. The formal potential of the Fe³⁺|Fe²⁺ couple in 1 M H₂SO₄ is 0.68 V. Compare the theoretical and experimental results. Also note the coincidence of the potentiometric and visual endpoints.

> *Question:* Potassium permanganate is purple, and all of the other species in this titration are colorless (or very faintly colored). What color change is expected at the equivalence point?

the stoichiometry of reactants is not 1:1. Still, the curve is so steep near the equivalence point that negligible error is introduced if the center of the steepest portion is taken as the endpoint. Demonstration 17-1 illustrates a titration curve with an asymmetric endpoint.

The magnitude of the change in potential near the equivalence point is smaller in Figure 17-3 than in Figure 17-2 because E^0 for Reaction 17-3 is greater than E^0 for Reaction 17-19. In general, the clearest results are achieved with the strongest oxidizing and reducing agents. This is analogous to acid–base titrations, where we normally use a strong acid or strong base as titrant, to get the sharpest break at the equivalence point.

17-2 TITRATION OF A MIXTURE

The titration of two species will exhibit two breaks if the standard potentials of the redox couples are sufficiently different. Figure 17-4 shows the theoretical titration curve for an equimolar mixture of Tl⁺ and Sn²⁺ titrated with IO₃⁻. The two **titration reactions** are

First: $IO_3^- + 2Sn^{2+} + 2Cl^- + 6H^+ \rightarrow ICl_2^- + 2Sn^{4+} + 3H_2O$ (17-36)

Second: $IO_3^- + 2Tl^+ + 2Cl^- + 6H^+ \rightarrow ICl_2^- + 2Tl^{3+} + 3H_2O$ (17-37)

The relevant half-reactions are 17-20, 17-21, and 17-38:

$$Sn^{4+} + 2e^- \rightleftharpoons Sn^{2+} \qquad E^0 = 0.139 \text{ V} \qquad (17\text{-}38)$$

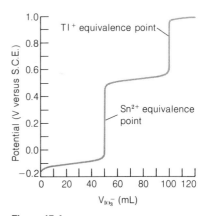

Figure 17-4
Theoretical curve for 0.0100 M Tl⁺ plus 0.0100 M Sn²⁺ titrated with 0.0100 M IO₃⁻. The initial volume of analyte is 100.0 mL, and all solutions contain 1.00 M HCl.

Sn²⁺ reacts before Tl⁺ because Sn²⁺ is a stronger reducing agent than Tl⁺. The strongest reagents react first, just as the strongest acids or bases react first in acid–base reactions.

374

17 / REDOX TITRATIONS

Question: What is the value of E(cathode) at $V = 25.0$ and $V = 75.0$ mL in the titration described above?

Because the $Sn^{4+}|Sn^{2+}$ couple has a lower reduction potential, Sn^{2+} will be oxidized before Tl^+. That is, the equilibrium constant for Reaction 17-36 is larger than for Reaction 17-37. This is another way of saying that Sn^{2+} is a stronger reducing agent than Tl^+.

The first part of the titration curve in Figure 17-4 is calculated using Reaction 17-36, as if no Tl^+ were present. After the Sn^{2+} equivalence point, the remainder is calculated using Reaction 17-37, as if no Sn^{2+} were present.

17-3 REDOX INDICATORS

A chemical indicator may be used to detect the endpoint of a redox titration, just as an indicator may be used in an acid–base titration. A **redox indicator** is a compound that changes color when it goes from its oxidized to its reduced state. One common indicator is ferroin, whose color change is from pale blue (almost colorless) to red.

$$\text{oxidized ferroin} \quad + e^- \rightleftharpoons \quad \text{reduced ferroin} \tag{17-39}$$

oxidized ferroin
(pale blue)
In(oxidized)

reduced ferroin
(red)
In(reduced)

To predict the potential range over which the indicator color will change, we first write a Nernst equation for the indicator.

$$\text{In(oxidized)} + ne^- \rightleftharpoons \text{In(reduced)} \tag{17-40}$$

$$E = E^0 - \frac{0.059\ 16}{n} \log \frac{[\text{In(reduced)}]}{[\text{In(oxidized)}]} \tag{17-41}$$

As with acid–base indicators, the color of In(reduced) will be observed when

$$\frac{[\text{In(reduced)}]}{[\text{In(oxidized)}]} \gtrsim \frac{10}{1} \tag{17-42}$$

and the color of In(oxidized) will be observed when

$$\frac{\text{In(reduced)}}{\text{In(oxidized)}} \lesssim \frac{1}{10} \tag{17-43}$$

Putting these ratios into Equation 17-41 tells us that the color change will occur over the range

$$E = \left(E^0 \pm \frac{0.059}{n} \right) \text{ volts} \qquad (17\text{-}44)$$

For ferroin, with $E^0 = 1.147$ V (Table 17-1), we expect the color change to occur in the approximate range 1.088 V to 1.206 V with respect to the standard hydrogen electrode. If a saturated calomel electrode is used as the reference instead, the indicator transition range will be

A redox indicator changes color over a range of $\pm(59/n)$ mV, centered at E^0 for the indicator.

$$\begin{pmatrix} \text{Indicator transition} \\ \text{range versus calomel} \\ \text{electrode} \end{pmatrix} = \begin{pmatrix} \text{Indicator transition} \\ \text{range versus standard} \\ \text{hydrogen electrode} \end{pmatrix} - E(\text{calomel}) \quad (17\text{-}45)$$

$$= (1.088 \text{ to } 1.206) - (0.241)$$

$$= 0.847 \text{ V to } 0.965 \text{ V (versus S.C.E.)}$$

Ferroin would therefore be a useful indicator for the titrations in Figures 17-2 and 17-3.

The indicator's transition range should match the steep part of the titration curve.

We have seen that the larger the difference in standard potential (or formal potential) between titrant and analyte, the sharper the break in the titration curve at the equivalence point. A redox titration is usually feasible if the difference in potentials between analyte and titrant is $\gtrsim 0.2$ V. However, the endpoint of such a titration is not very sharp, and it is probably best detected potentiometrically. If the difference in formal potentials is $\gtrsim 0.4$ V, then a redox indicator will usually give a satisfactorily sharp endpoint.

Challenge: Select suitable indicators from Table 17-1 for the two endpoints of the titration in Figure 17-4. What color change might you expect at each of the two endpoints? Remember that *both* indicators will be in the solution together.

Table 17-1
Some redox indicators

	Color		
Indicator	Oxidized	Reduced	E°
Phenosafranine	Red	Colorless	0.28
Indigo tetrasulfonate	Blue	Colorless	0.36
Methylene blue	Blue	Colorless	0.53
Diphenylamine	Violet	Colorless	0.75
4'-ethoxy-2,4-diaminoazobenzene	Yellow	Red	0.76
Diphenylamine sulfonic acid	Red-violet	Colorless	0.85
Diphenylbenzidine sulfonic acid	Violet	Colorless	0.87
tris(2,2'-bipyridine)iron	Pale blue	Red	1.120
tris(1,10-phenanthroline)iron (ferroin)	Pale blue	Red	1.147
tris(5-nitro-1,10-phenanthroline)iron	Pale blue	Red-violet	1.25
tris(2,2'-bipyridine)ruthenium	Pale blue	Yellow	1.29

The Starch–Iodine Complex

Numerous analytical procedures are based on titrations involving iodine. Starch is the indicator of choice for these procedures because it forms an intense blue complex with iodine. Starch is not a redox indicator because it responds specifically to the presence of I_2, and not to a change in redox potential.

The active fraction of starch is amylose, a polymer of the sugar α-D-glucose, with the repeating unit shown in Figure 17-5. The polymer exists as a coiled helix into which small molecules can fit. In the presence of starch and I^-, iodine molecules form long chains of I_5^- ions that occupy the center of the amylose helix (Figure 17-6).

$$\cdot\cdot[I-I-I-I-I]^- \cdot\cdot [I-I-I-I-I]^- \cdot\cdot$$

It is a visible absorption band of this I_5^- chain bound within the helix that gives rise to the characteristic starch–iodine color.

Starch is readily biodegraded, so either it should be freshly dissolved, or the solution should contain a preservative, such as HgI_2 or thymol. One of the hydrolysis products of starch is glucose, which is a reducing agent. A partially hydrolyzed solution of starch could thus be a source of error in a redox titration.

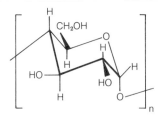

Figure 17-5
Structure of the repeating unit of the sugar amylose.

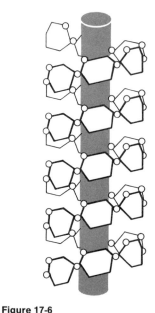

Figure 17-6
Schematic structure of the starch–iodine complex. The amylose sugar chain forms a helix around the nearly linear iodine chain. [R. C. Teitelbaum, S. L. Ruby, and T. J. Marks, *J. Amer. Chem. Soc.*, **102**, 3322 (1980).]

It is necessary to remove the excess preadjustment reagent so that it will not interfere in the subsequent titration.

17-4 COMMON REDOX REAGENTS

In this section we will discuss some of the more common redox agents used in volumetric analysis.[†] Table 17-2 lists some frequently encountered oxidizing and reducing agents. Most of the oxidizing agents can be used as titrants. However, most of the reducing agents react with oxygen and are therefore less suitable as titrants. Only a few of the reagents in Table 17-2 can be used as primary standards.

Adjustment of Analyte Oxidation State

For many analyses it is necessary to adjust the oxidation state of the analyte to one that can be titrated with an oxidizing or a reducing agent. For example, Mn(II) might be quantitatively **preoxidized** to MnO_4^- and then titrated with standard Fe^{2+}. The preadjustment reaction must be quantitative, and it must be possible to remove or destroy the excess preadjustment reagent.

[†] Some sources of information on redox titrations include H. A. Laitinen and W. E. Harris, *Chemical Analysis*, 2nd ed. (New York: McGraw-Hill, 1975); I. M. Kolthoff, R. Belcher, V. A. Stenger, and G. Matsuyama, *Volumetric Analysis*, Vol. 3 (New York: Wiley, 1957); A. Berka, J. Vulterin, and J. Zýka, *Newer Redox Titrants* (H. Weisz, trans.), (Oxford: Pergamon, 1965); A. I. Vogel, *Quantitative Inorganic Analysis* (New York: Wiley, 1961).

Table 17-2
Some common oxidizing and reducing agents

Oxidants		Reductants	
BiO_3^-	bismuthate		
BrO_3^-	bromate	AsO_3^{3-}	arsenite
Br_2	bromine	(structure)	ascorbic acid
Ce^{4+}	ceric	Cr^{2+}	chromous
$CH_3-\bigcirc-SO_2NCl^-$	chloramine T	$S_2O_4^{2-}$	dithionite
$Cr_2O_7^{2-}$	dichromate	$Fe(CN)_6^{4-}$	ferrocyanide
H_2O_2	hydrogen peroxide	Fe^{2+}	ferrous
OCl^-	hypochlorite	N_2H_4	hydrazine
IO_3^-	iodate	$HO-\bigcirc-OH$	hydroquinone
I_2	iodine	NH_2OH	hydroxylamine
$Pb(acetate)_4$	lead(IV) acetate	Hg_2^{2+}	mercurous
$HClO_4$	perchloric acid	Sn^{2+}	stannous
IO_4^-	periodate	SO_3^{2-}	sulfite
MnO_4^-	permanganate	SO_2	sulfur dioxide
$S_2O_8^{2-}$	peroxydisulfate	$S_2O_3^{2-}$	thiosulfate

Preoxidation

There are several powerful oxidants available that can be easily removed after preoxidation. *Peroxydisulfate* ($S_2O_8^{2-}$, also called *persulfate*) is a strong oxidant that requires Ag^+ as a catalyst.

$$S_2O_8^{2-} + Ag^+ \rightarrow SO_4^{2-} + \underbrace{SO_4^- + Ag^{2+}}_{\text{Two powerful oxidants}} \qquad (17\text{-}46)$$

The excess reagent is destroyed by boiling the solution after the oxidation of analyte is complete.

$$2S_2O_8^{2-} + 2H_2O \xrightarrow{\text{boiling}} 4SO_4^{2-} + O_2 + 4H^+ \qquad (17\text{-}47)$$

378

17 / REDOX TITRATIONS

The $S_2O_8^{2-}/Ag^+$ mixture is able to oxidize Mn(II) to MnO_4^-, Ce(III) to Ce(IV), Cr(III) to $Cr_2O_7^{2-}$, and V(IV) to V(V).

Silver (II) oxide (AgO) dissolves in concentrated mineral acids to give Ag^{2+}, with oxidizing power similar to the $S_2O_8^{2-}/Ag^+$ combination. Excess Ag^{2+} can be removed by boiling:

$$4Ag^{2+} + 2H_2O \xrightarrow{\text{boiling}} 4Ag^+ + O_2 + 4H^+ \qquad (17\text{-}48)$$

Solid *sodium bismuthate* ($NaBiO_3$) is of similar oxidizing strength to Ag^{2+} and $S_2O_8^{2-}$. The excess solid oxidant is removed by filtration.

Hydrogen peroxide is a good oxidant in basic solution. It can transform Co(II) to Co(III), Fe(II) to Fe(III), and Mn(II) to Mn(IV). In acidic solution it can *reduce* $Cr_2O_7^{2-}$ to Cr^{3+} and MnO_4^- to Mn^{2+}. Excess H_2O_2 spontaneously *disproportionates* in boiling water.

$$2H_2O_2 \xrightarrow{\text{boiling}} O_2 + 2H_2O \qquad (17\text{-}49)$$

> Do you remember what "disproportionation" means? Look in the Glossary if you don't.

Boiling, concentrated (72% wt/wt), aqueous *perchloric acid* is a common, but dangerous, oxidant. Cold or dilute $HClO_4$ has essentially no oxidizing abilities, but the hot, concentrated acid can be used to destroy organic matter (oxidizing it to CO_2) and leave inorganic components in high oxidastates. Hot, concentrated $HClO_4$ has been the cause of many serious explosions, especially in the presence of ethanol, cellulose, or other polyhydroxy alcohols.

> *Challenge:* Write one half-reaction in which H_2O_2 behaves as an oxidant and one half-reaction in which it behaves as a reductant.

The process in which the organic portion of a sample is destroyed prior to analyzing an inorganic constituent is called **wet ashing.** The sample is heated with nitric acid prior to adding perchloric acid. The nitric acid oxidizes much of the sample and lowers the risk of the $HClO_4$ contacting any material that is too easily oxidized (and therefore explosive). You should *never* use boiling perchloric acid without a strong barrier between you and the sample.

> A sample should be wet-ashed with HNO_3 prior to $HClO_4$ to minimize the explosion hazard.

Prereduction

Stannous chloride ($SnCl_2$) has been used to reduce Fe(III) to Fe(II) in hot HCl. The excess reducing agent is destroyed by adding excess $HgCl_2$:

$$Sn(II) + 2HgCl_2 \rightarrow Sn(IV) + Hg_2Cl_2 + 2Cl^- \qquad (17\text{-}50)$$

Chromous chloride is a very powerful reductant sometimes used for prereduction. Any excess Cr^{2+} is oxidized by atmospheric oxygen. *Sulfur dioxide* and *hydrogen sulfide* are mild reducing agents that can be expelled by boiling an acidic solution after the reduction is complete.

An important prereduction technique uses a column packed with a solid reducing agent. Figure 17-7 shows the *Jones reductor*, which is packed with zinc coated with zinc amalgam. The amalgam is prepared by mixing granular zinc with 2% (wt/wt) aqueous $HgCl_2$ for 10 minutes, then washing

> An *amalgam* is a solution of anything in mercury.

with water. For example, a sample of Fe(III) can be reduced to Fe(II) by being passed through a Jones reductor using 1 M H_2SO_4 as solvent. The column is washed well with water, and the combined washings can be titrated with standard MnO_4^-, Ce(IV), or $Cr_2O_7^{2-}$. It is necessary to do a blank determination on a solution passed through the reductor in the same manner as the unknown.

Most reduced analytes are easily reoxidized by atmospheric oxygen. To avoid air oxidation, the reduced analyte may be collected in a solution containing excess Fe(III). The ferric ion is immediately reduced to Fe(II), which is stable in acid. The Fe(II) is then titrated with an oxidant. By this means, such elements as Cr, Ti, V, and Mo can be analyzed.

Zinc is such a powerful reducing agent that the Jones reductor is not very selective.

$$Zn^{2+} + 2e^- \rightleftharpoons Zn(s) \qquad E^0 = -0.764 \text{ V} \qquad (17\text{-}51)$$

More selective is the *Walden reductor*, filled with solid Ag and 1 M HCl. The reduction potential for the silver–silver chloride couple (0.222 V) is high enough that such species as Cr^{3+} and TiO^{2+} are not reduced and therefore do not interfere in the analysis of a metal such as Fe^{3+}. Another more selective reductor uses a filling of granular Cd metal. In determining levels of nitrogen oxides for air-pollution monitoring,[†] the gases are first converted to NO_3^-, which is not easy to analyze. Passing the nitrate through a Cd-filled column reduces NO_3^- to NO_2^-, for which a convenient spectrophotometric analysis is available. Significant disadvantages of all reductors are the time, labor, and expense associated with their use.

Once the analyte has been adjusted to the desired oxidation state, it may be titrated with an appropriate oxidant or reductant. We will now examine a few of the most common titrants.

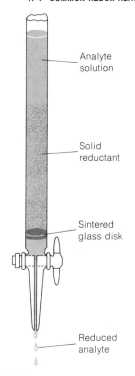

Figure 17-7
A column filled with a solid reagent used for prereduction of analyte is called a *reductor*. Often the analyte is drawn through by suction.

Oxidation with Potassium Permanganate

Potassium permanganate ($KMnO_4$) is an oxidizing agent of an intense violet color. In strongly acidic solutions (pH ≲ 1), it is reduced to colorless Mn^{2+}.

$$\underset{\text{permanganate}}{MnO_4^-} + 8H^+ + 5e^- \rightleftharpoons \underset{\text{manganous}}{Mn^{2+}} + 4H_2O \qquad E^0 = 1.51 \text{ V} \qquad (17\text{-}52)$$

In neutral or alkaline solution, the product is the brown solid, MnO_2.

$$MnO_4^- + 4H^+ + 3e^- \rightleftharpoons \underset{\text{manganese dioxide}}{MnO_2(s)} + 2H_2O \qquad E^0 = 1.695 \text{ V} \qquad (17\text{-}53)$$

[†] J. H. Margeson, J. C. Suggs, and M. R. Midgett, *Anal. Chem.*, **52**, 1955 (1980).

380

17 / REDOX TITRATIONS

In very strongly alkaline solution (2 M NaOH), green manganate ion is produced.

$$MnO_4^- + e^- \rightleftharpoons MnO_4^{2-} \qquad E^0 = 0.558 \text{ V} \qquad (17\text{-}54)$$

$$\text{manganate}$$

Some representative permanganate titrations are listed in Table 17-3. For titrations in strongly acidic solution, $KMnO_4$ serves as its own indicator. The product, Mn^{2+}, is colorless. The endpoint is taken as the first persistent appearance of pale pink MnO_4^-. If the titrant is too dilute to be seen, an indicator such as ferroin may be used.

KMnO₄ serves as its own indicator in acidic solution.

Preparation and standardization

KMnO₄ is not a primary standard.

Potassium permanganate is not pure enough to be a primary standard, since traces of MnO_2 are invariably present. In addition, distilled water usually contains enough organic impurities to reduce some freshly dissolved MnO_4^- to MnO_2. To prepare a stable solution, $KMnO_4$ is dissolved in distilled water, boiled for an hour to hasten the reaction between MnO_4^- and organic impurities, and filtered through a clear, sintered glass filter to remove precipitated MnO_2. Filter paper (organic matter!) should never be used. The reagent is stored in a dark glass bottle. Thermodynamically, aqueous $KMnO_4$ is unstable by virtue of the reaction

A properly stored solution of 0.02 M KMnO₄ decomposes ≤0.1% in several months. More dilute solutions should be prepared fresh from this stock solution and standardized.

$$4MnO_4^- + 2H_2O \rightarrow 4MnO_2(s) + 3O_2 + 4OH^- \qquad (17\text{-}55)$$

But the oxidation of water by MnO_4^- is very slow in the absence of such agents as MnO_2, Mn^{2+}, heat, light, acids, and bases.

Potassium permanganate can be standardized by titration of sodium oxalate ($Na_2C_2O_4$), pure electrolytic iron wire, or arsenious oxide (As_4O_6). Dry (105°C, two hours) sodium oxalate (available in a 99.9–99.95% pure form) is dissolved in 1 M H_2SO_4 and treated with 90–95% of the required $KMnO_4$ solution at room temperature. The solution is then warmed to 55–60°C, and the titration is completed by slow addition of $KMnO_4$. A blank value is subtracted to account for the quantity of titrant (usually one drop) needed to impart a pink color to the solution.

$$2MnO_4^- + 5H_2C_2O_4 + 6H^+ \rightarrow 2Mn^{2+} + 10CO_2 + 8H_2O \quad (17\text{-}56)$$

If pure Fe wire is used as a standard, it is dissolved in warm 1.5 M H_2SO_4 under nitrogen. The product is Fe(II), and the cooled solution can be used to standardize $KMnO_4$ (or other oxidants) with no special precautions. Addition of 5 mL of 86% (wt/wt) phosphoric acid per 100 mL of solution masks the yellow color of Fe^{3+} and makes the endpoint easier to see. Alternatively, metallic iron can be dissolved less cautiously, and Fe^{3+} reduced to Fe^{2+} with $SnCl_2$. Ferrous ammonium sulfate—$Fe(NH_4)_2(SO_4)_2 \cdot 6H_2O$—and ferrous ethylenediammonium sulfate—$Fe(H_3NCH_2CH_2NH_3)(SO_4)_2 \cdot 2H_2O$—are available in sufficiently pure forms to be used as primary standards for most purposes.

Table 17-3
Some analytical applications of permanganate titrations

Species analyzed	Oxidation reaction	Notes
Fe^{2+}	$Fe^{2+} \rightleftharpoons Fe^{3+} + e^-$	Fe^{3+} is reduced to Fe^{2+} with Sn^{2+} or a Jones reductor. Titration is carried out in 1 M H_2SO_4 or 1 M HCl containing Mn(II), H_3PO_4, and H_2SO_4. Mn(II) inhibits oxidation of Cl^- by MnO_4^-. H_3PO_4 complexes Fe(III) to prevent formation of yellow Fe(III)–chloride complexes.
$H_2C_2O_4$	$H_2C_2O_4 \rightleftharpoons 2CO_2 + 2H^+ + 2e^-$	Add 95% of titrant at 25°C, then complete titration at 55–60°C.
Br^-	$Br^- \rightleftharpoons \frac{1}{2}Br_2(g) + e^-$	Titrate in boiling 2 M H_2SO_4 to remove $Br_2(g)$.
H_2O_2	$H_2O_2 \rightleftharpoons O_2(g) + 2H^+ + 2e^-$	Titrate in 1 M H_2SO_4.
HNO_2	$HNO_2 + H_2O \rightleftharpoons NO_3^- + 3H^+ + 2e^-$	Add excess standard $KMnO_4$ and back-titrate after 15 minutes at 40°C with Fe(II).
As(III)	$H_3AsO_3 + H_2O \rightleftharpoons H_3AsO_4 + 2H^+ + 2e^-$	Titrate in 1 M HCl with KI or ICl catalyst.
Sb(III)	$H_3SbO_3 + H_2O \rightleftharpoons H_3SbO_4 + 2H^+ + 2e^-$	Titrate in 2 M HCl.
Mo(III)	$Mo^{3+} + 2H_2O \rightleftharpoons MoO_2^{2+} + 4H^+ + 3e^-$	Reduce Mo in a Jones reductor, and run the Mo^{3+} into excess Fe^{3+} in 1 M H_2SO_4. Titrate the Fe^{2+} formed.
W(III)	$W^{3+} + 2H_2O \rightleftharpoons WO_2^{2+} + 4H^+ + 3e^-$	Reduce W with Pb(Hg) at 50°C and titrate in 1 M HCl.
U(IV)	$U^{4+} + 2H_2O \rightleftharpoons UO_2^{2+} + 4H^+ + 3e^-$	Reduce U to U^{3+} with a Jones reductor. Expose to air to produce U^{4+}, which is titrated in 1 M H_2SO_4.
Ti(III)	$Ti^{3+} + H_2O \rightleftharpoons TiO^{2+} + 2H^+ + e^-$	Reduce Ti to Ti^{3+} with a Jones reductor, and run the Ti^{3+} into excess Fe^{3+} in 1 M H_2SO_4. Titrate the Fe^{2+} that is formed.
$Mg^{2+}, Ca^{2+}, Sr^{2+}, Ba^{2+}, Zn^{2+},$ $Co^{2+}, La^{3+}, Th^{4+}, Pb^{2+},$ Ce^{3+}, BiO^+, Ag^+	$H_2C_2O_4 \rightleftharpoons 2CO_2 + 2H^+ + 2e^-$	Precipitate the metal oxalate. Dissolve in acid and titrate the $H_2C_2O_4$.
K^+	$K_2NaCo(NO_2)_6 + 6H_2O \rightleftharpoons Co^{2+} + 6NO_3^-$ $+ 12H^+ + 2K^+ + Na^+ + 11e^-$	Precipitate potassium sodium cobaltinitrite. Dissolve in acid and titrate. Co^{3+} is *reduced* to Co^{2+}, and HNO_2 is oxidized to NO_3^-.
Na^+	$U^{4+} + 2H_2O \rightleftharpoons UO_2^{2+} + 4H^+ + 2e^-$	Precipitate $NaZn(UO_2)_3(acetate)_9$. Dissolve in acid, reduce the UO_2^{2+} (as above), and titrate.
$S_2O_8^{2-}$	$S_2O_8^{2-} + 2Fe^{2+} + 2H^+ \rightleftharpoons 2Fe^{3+} + 2HSO_4^-$	Peroxydisulfate is added to excess standard Fe^{2+} containing H_3PO_4. Unreacted Fe^{2+} is titrated with MnO_4^-.
PO_4^{3-}	$Mo^{3+} + 2H_2O \rightleftharpoons MoO_2^{2+} + 4H^+ + 3e^-$	$(NH_4)_3PO_4 \cdot 12MoO_3$ is precipitated and dissolved in H_2SO_4. The Mo(VI) is reduced (as above) and titrated.

382

17 / REDOX TITRATIONS

Oxidation with Cerium(IV)

The reduction of Ce(IV) to Ce(III) proceeds cleanly in acidic solution. The aquo ion—$Ce(H_2O)_n^{4+}$—probably does not exist in any of these solutions, as the cerium ion binds the acid counterion (ClO_4^-, SO_4^{2-}, NO_3^-, Cl^-) very strongly to give a variety of complexes. The variation of the $Ce^{4+}|Ce^{3+}$ formal potential with the medium is indicative of these interactions:

> The variation of potential in each solvent implies that different species of cerium are present in each solvent.

$$Ce(IV) + e^- \rightleftharpoons Ce^{3+} \tag{17-57}$$

$$\text{Formal potential} = 1.70 \text{ V in 1 F } HClO_4$$

$$= 1.61 \text{ V in 1 F } HNO_3$$

$$= 1.44 \text{ V in 1 F } H_2SO_4$$

$$= 1.28 \text{ V in 1 F } HCl$$

Ce(IV) is yellow and Ce(III) is colorless, but the color change is not distinct enough for cerium to be its own indicator. Ferroin and other substituted phenanthroline redox indicators (Table 17-1) are well suited to titrations with Ce(IV).

Preparation and standardization

> $(NH_4)_2Ce(NO_3)_6$ is a primary standard.

Primary-standard-grade ammonium hexanitratocerate(IV)—$(NH_4)_2$-$Ce(NO_3)_6$—can be dissolved in 1 M H_2SO_4 and used directly. Although the oxidizing strength of Ce(IV) is greater in $HClO_4$ or HNO_3, solutions in these acids undergo slow photochemical decomposition with concomitant oxidation of water. Solutions of Ce(IV) in H_2SO_4 are indefinitely stable. Solutions in HCl are unstable because Cl^- is oxidized to Cl_2. Sulfuric acid solutions of Ce(IV) can be used to titrate unknowns (such as Fe^{2+}) in HCl solution because the reaction with analyte is favored over the slow reaction with Cl^-.

Less expensive, less pure salts of Ce(IV), including $Ce(HSO_4)_4$, $(NH_4)_4Ce(SO_4)_4 \cdot 2H_2O$, and $CeO_2 \cdot xH_2O$—also called $Ce(OH)_4$—are perfectly adequate for preparing titrants that are subsequently standardized. The procedures for Ce(IV) standardization are similar to those for MnO_4^-, with As_4O_6, $Na_2C_2O_4$, and Fe being useful primary standards.

Analytical applications

Ce(IV) can be used in place of $KMnO_4$ in most of the procedures mentioned in the previous section. In addition, ceric ion finds applications in analysis of many organic compounds. In the oscillating reaction in Demonstration 16-1, Ce(IV) oxidizes malonic acid to CO_2 and formic acid:

$$CH_2(CO_2H)_2 + 2H_2O + 6Ce(IV) \rightarrow 2CO_2 + HCO_2H + 6Ce(III) + 6H^+$$

malonic acid   formic acid

$$\tag{17-58}$$

This reaction can be used for quantitative analysis of malonic acid by heating a sample in 4 M $HClO_4$ with excess standard Ce(IV) and back-titrating the unreacted Ce(IV) with Fe^{2+}. Analogous procedures are available for many alcohols, aldehydes, ketones, and carboxylic acids.

Oxidation with Potassium Dichromate

In acidic solution, the orange dichromate ion is a powerful oxidant that is reduced to chromic ion (Cr^{3+}):

$$Cr_2O_7^{2-} + 14H^+ + 6e^- \rightleftharpoons 2Cr^{3+} + 7H_2O \qquad E^0 = 1.33 \text{ V} \quad (17\text{-}59)$$

In 1 M HCl, the formal potential is just 1.00 V; and in 2 M H_2SO_4, it is 1.11 V, so dichromate is a less powerful oxidizing agent than MnO_4^- or Ce(IV). In basic solution, $Cr_2O_7^{2-}$ is converted to the yellow chromate ion (CrO_4^{2-}), whose oxidizing ability is nil:

$$CrO_4^{2-} + 4H_2O + 3e^- \rightleftharpoons Cr(OH)_3(s, \text{hydrated}) + 5OH^-$$
$$E^0 = -0.13 \text{ V} \qquad (17\text{-}60)$$

The advantages of $K_2Cr_2O_7$ are that it is pure enough to be a primary standard, its solutions are stable, and it is cheap. Because $Cr_2O_7^{2-}$ is orange and complexes of Cr^{3+} range from green to violet, indicators with very distinctive color changes must be used to find a dichromate endpoint. Indicators such as diphenylamine sulfonic acid or diphenylbenzidine sulfonic acid are suitable. Alternatively, the reaction can be monitored with Pt and calomel electrodes.

Since potassium dichromate is not as strong an oxidant as $KMnO_4$ or Ce(IV), it is not used as widely. It is employed chiefly for the determination of Fe^{2+} and, indirectly, for many species that will oxidize Fe^{2+} to Fe^{3+}. For indirect analyses, the unknown is treated with a measured excess of Fe^{2+}, and the unreacted Fe^{2+} then titrated with $K_2Cr_2O_7$. Among the species that can be analyzed in this way are ClO_3^-, NO_3^-, MnO_4^-, and organic peroxides.

diphenylbenzidine sulfonate(reduced) (colorless)

diphenylbenzidine sulfonate(oxidized) (violet)
$+ 2H^+ + 2e^-$

Methods Involving Iodine

A great many analytical procedures are based on reactions of iodine. When a reducing analyte is titrated directly with iodine (to produce I^-), the method is called **iodimetry.** In **iodometry,** an oxidizing analyte is added to excess I^- to produce iodine, which is then titrated with standard thiosulfate solution.

Molecular iodine is only slightly soluble in water (1.33×10^{-3} M at 20°C), but its solubility is greatly enhanced by complexation with iodide.

$$I_2(aq) + I^- \rightleftharpoons I_3^- \qquad K = 7.0 \times 10^2 \qquad (17\text{-}61)$$

$$\text{iodine} \qquad \text{iodide} \qquad \text{triiodide}$$

Iodimetry—titration *with* iodine.

Iodometry—titration *of* iodine produced by a chemical reaction.

The form of iodine used as a titrant or produced by reaction with analyte is almost always I_3^-, not I_2.

384

17 / REDOX TITRATIONS

A typical 0.05 M solution of I_3^- for titrations is prepared by dissolving 0.12 mol of KI plus 0.05 mol of I_2 in one liter of water. When we speak of using iodine as a titrant, we almost always mean that we are using a solution of I_2 plus excess I^-.

Use of starch indicator

As described in Section 17-3, starch is used as an indicator for iodine. In a solution with no other colored species, it is possible to see the color of $\sim 5 \times 10^{-6}$ M I_3^-. With a starch indicator, the limit of detection is extended by about a factor of ten.

In iodimetry (titration *with* I_3^-), starch can be added at the beginning of the titration. The first drop of excess I_3^- after the equivalence point causes the solution to turn dark blue. In iodometry (titration *of* I_3^-), I_3^- is present throughout the reaction up to the equivalence point. *Starch should not be added to such a reaction until immediately before the equivalence point* (as detected visually, by fading of the I_3^- color). Otherwise some iodine tends to remain bound to starch particles after the equivalence point is reached.

An alternative to using starch is to add a few milliliters of chloroform to the vigorously stirred titration vessel. After each addition of reagent near the endpoint, stirring is stopped long enough to examine the color of the chloroform phase at the bottom of the flask. I_2 is much more soluble in $CHCl_3$ than in water, and its color is readily detected in the organic phase.

The starch–iodine complex is very temperature-sensitive. At 50°C, the color is only one-tenth as intense as at 25°C. Organic solvents also decrease the affinity of iodine for starch and markedly reduce the utility of the indicator.

Preparation and standardization of I_3^- solutions

There is a significant vapor pressure of toxic I_2 above solid I_2 and aqueous I_3^-. Vessels containing I_2 or I_3^- should be sealed or, better, kept in a fume hood. Waste solutions of I_3^- should not be dumped into a sink in the open lab.

As described above, I_3^- is prepared by dissolving solid I_2 in excess KI. Sublimed I_2 is pure enough to be a primary standard, but it is seldom used as a standard due to significant vaporization of the solid during the weighing procedure. Instead, the approximate amount is rapidly weighed, and the solution of I_3^- is standardized with a pure sample of the intended analyte or with As_4O_6 or $Na_2S_2O_3$ as described below.

Acidic solutions of I_3^- are unstable because the excess I^- is slowly oxidized by air:

$$6I^- + O_2 + 4H^+ \rightarrow 2I_3^- + 2H_2O \qquad (17\text{-}62)$$

In neutral solutions, Reaction 17-62 is insignificant in the absence of heat, light, and metal ions.

Triiodide can be standardized by reaction with primary standard arsenious oxide, As_4O_6, the structure of which is shown in Figure 17-8. When As_4O_6 is dissolved in acidic solution, arsenious acid is formed:

$$As_4O_6(s) + 6H_2O \rightleftharpoons 4H_3AsO_3 \qquad (17\text{-}63)$$

arsenious oxide   arsenious acid

Challenge: Use standard potentials from Appendix H to show that the equilibrium constant for Reaction 17-64 is 0.17.

The latter reacts with I_3^- as follows:

$$H_3AsO_3 + I_3^- + H_2O \rightleftharpoons H_3AsO_4 + 3I^- + 2H^+ \qquad (17\text{-}64)$$

Because the equilibrium constant for Reaction 17-64 is small, the concentration of H^+ must be kept low to ensure complete reaction. If the pH is *too* high (pH $\gtrsim$ 11), triiodide disproportionates to hypoiodous acid, iodate, and iodide. For best results, the standardization is carried out at pH 7–8 with bicarbonate as a buffer.

An alternative and excellent way to make a standard solution of I_3^- is to add a weighed quantity of pure potassium iodate to a small excess of KI. Addition of excess strong acid (to give pH $\approx$ 1) gives a quantitative reverse disproportionation reaction in which I_3^- is formed:

$$IO_3^- + 8I^- + 6H^+ \rightleftharpoons 3I_3^- + 3H_2O \qquad (17\text{-}65)$$
iodate

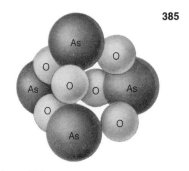

Figure 17-8
As_4O_6 consists of an As_4 pyramid with a bridging oxygen atom on each edge.

A freshly acidified solution of iodate plus iodide can be used to standardize thiosulfate or any other reagent that reacts with I_3^-. The reagent must be used immediately, or else air oxidation of I^- takes place. The only disadvantage of KIO_3 is its low molecular weight relative to the number of electrons it accepts. This leads to a larger-than-desirable relative weighing error in preparing solutions.

HOI—hypoiodous acid.

IO_3^-—iodate.

KIO_3 is a good primary standard for the generation of I_3^-.

Use of sodium thiosulfate

Sodium thiosulfate is the almost universal titrant for triiodide. In neutral or acidic solution, triiodide oxidizes thiosulfate to tetrathionate:

$$I_3^- + 2S_2O_3^{2-} \rightleftharpoons 3I^- + {}^-O-\underset{O^-}{\underset{|}{S}}(=O)-S-S-\underset{O^-}{\underset{|}{S}}(=O) \qquad (17\text{-}66)$$

thiosulfate   tetrathionate

Reaction of iodine with thiosulfate.

(In basic solution, I_3^- disproportionates to I^- and HOI. Because hypoiodite oxidizes thiosulfate to sulfate, the stoichiometry of Reaction 17-66 changes and the titration of I_3^- with thiosulfate is not carried out in base.) The common form of thiosulfate, $Na_2S_2O_3 \cdot 5H_2O$, is not pure enough to be a primary standard. Instead, thiosulfate is usually standardized by reaction with a fresh solution of I_3^- prepared from KIO_3 plus KI or a solution of I_3^- standardized with As_4O_6.

A stable solution of $Na_2S_2O_3$ can be prepared by dissolving the reagent in high-quality, freshly boiled distilled water. The quality of the water is important because dissolved CO_2 promotes disproportionation of $S_2O_3^{2-}$:

$$S_2O_3^{2-} + H^+ \rightleftharpoons HSO_3^- + S(s) \qquad (17\text{-}67)$$
bisulfite   sulfur

and metal ions catalyze the atmospheric oxidation of thiosulfate:

$$2Cu^{2+} + 2S_2O_3^{2-} \rightarrow 2Cu^+ + S_4O_6^{2-} \qquad (17\text{-}68)$$

$$2Cu^+ + \tfrac{1}{2}O_2 + 2H^+ \rightarrow 2Cu^{2+} + H_2O \qquad (17\text{-}69)$$

A solution of thiosulfate should be stored in the dark. Addition of 0.1 g of sodium carbonate per liter maintains the pH in an optimum range for stability of the solution. An acidic solution of thiosulfate is unstable, but the reagent can be used to titrate I_3^- in acidic solution because the reaction with triiodide is faster than Reaction 17-67.

Analytical applications of iodine

Reducing agent $+ I_3^- \rightarrow 3I^-$.

Reducing agents can be titrated directly with standard I_3^- in the presence of starch. The endpoint is marked by the appearance of the intense blue starch–iodine complex. An example is the iodimetric determination of vitamin C:

ascorbic acid (vitamin C) $+ I_3^- \longrightarrow$ dehydroascorbic acid $+ 2H^+ + 3I^-$ (17-70)

In some cases, excess standard I_3^- is used to drive the reaction to completion, as in the analysis of glucose or other reducing sugars (sugars with an aldehyde group).

$$\text{glucose} + 3OH^- + I_3^- \longrightarrow \text{gluconate} + 3I^- + 2H_2O$$

(17-71)

Reaction 17-71 is carried out in a basic solution, which is then acidified and the excess I_3^- back-titrated with standard thiosulfate. Examples of iodimetric analyses are given in Table 17-4.

Oxidizing agent $+ 3I^- \rightarrow I_3^-$.

Oxidizing agents can be treated with excess I^- to produce I_3^- (Table 17-5). The iodometric analysis is completed by titrating the liberated I_3^- with standard thiosulfate. Starch is not added until just before the endpoint.

Table 17-4
Titrations with standard triiodide (iodimetric titrations)

Species analyzed	Oxidation reaction	Notes
As(III)	$H_3AsO_3 + H_2O \rightleftharpoons H_3AsO_4 + 2H^+ + 2e^-$	Direct titration in $NaHCO_3$ solution with I_3^-.
Sb(III)	$SbO(O_2CCHOHCHOHCO_2)^- + H_2O \rightleftharpoons$ $SbO_2(O_2CCHOHCHOHCO_2)^- + 2H^+ + 2e^-$	Direct titration in $NaHCO_3$ solution, using tartarate to mask As(III).
Sn(II)	$SnCl_4^{2-} + 2Cl^- \rightleftharpoons SnCl_6^{2-} + 2e^-$	Sn(IV) is reduced to Sn(II) with granular Pb or Ni in 1 M HCl and titrated in the absence of oxygen.
N_2H_4	$N_2H_4 \rightleftharpoons N_2 + 4H^+ + 4e^-$	Titrate in $NaHCO_3$ solution.
SO_2	$SO_2 + H_2O \rightleftharpoons H_2SO_3$ $H_2SO_3 + H_2O \rightleftharpoons SO_4^{2-} + 4H^+ + 2e^-$	Add SO_2 (or H_2SO_3 or HSO_3^- or SO_3^{2-}) to excess standard I_3^- in dilute acid and back-titrate unreacted I_3^- with standard thiosulfate.
H_2S	$H_2S \rightleftharpoons S(s) + 2H^+ + 2e^-$	Add H_2S to excess I_3^- in 1 M HCl and back-titrate with thiosulfate.
$Zn^{2+}, Cd^{2+}, Hg^{2+}, Pb^{2+}$, etc.	$M^{2+} + H_2S \rightarrow MS(s) + 2H^+$ $MS(s) \rightleftharpoons M^{2+} + S + 2e^-$	Precipitate and wash metal sulfide. Dissolve in 3 M HCl with excess standard I_3^- and back-titrate with thiosulfate.
Cysteine, glutathione, thioglycolic acid, mercaptoethanol	$2 RSH \rightleftharpoons RSSR + 2H^+ + 2e^-$	Titrate the sulfhydryl compound at pH 4–5 with I_3^-.
HCN	$I_2 + HCN^- \rightleftharpoons ICN + I^- + H^+$	Titrate in carbonate–bicarbonate buffer, using $CHCl_3$ as an extraction indicator.
$H_2C{=}O$	$H_2CO + 3OH^- \rightleftharpoons HCO_2^- + 2H_2O + 2e^-$	Add excess I_3^- plus NaOH to the unknown. After five minutes, add HCl and back-titrate with thiosulfate.
Glucose (and other reducing sugars)	$\underset{\displaystyle RCH}{\overset{\displaystyle O \atop \|}{}} + 3OH^- \rightleftharpoons RCO_2^- + 2H_2O + 2e^-$	Add excess I_3^- plus NaOH to the sample. After five minutes add HCl and back-titrate with thiosulfate.
Ascorbic acid (vitamin C)		Direct titration with I_3^-.
H_3PO_3	$H_3PO_3 + H_2O \rightleftharpoons H_3PO_4 + 2H^+ + 2e^-$	Titrate in $NaHCO_3$ solution.

Table 17-5
Titration of I_3^- produced by analyte (iodometric titrations)

Species analyzed	Reaction	Notes
Cl_2	$Cl_2 + 3I^- \rightleftharpoons 2Cl^- + I_3^-$	Reaction in dilute acid.
$HOCl$	$HOCl + H^+ + 3I^- \rightleftharpoons Cl^- + I_3^- + H_2O$	Reaction in 0.5 M H_2SO_4.
Br_2	$Br_2 + 3I^- \rightleftharpoons 2Br^- + I_3^-$	Reaction in dilute acid.
BrO_3^-	$BrO_3^- + 6H^+ + 9I^- \rightleftharpoons Br^- + 3I_3^- + 3H_2O$	Reaction in 0.5 M H_2SO_4.
IO_3^-	$2IO_3^- + 16I^- + 12H^+ \rightleftharpoons 6I_3^- + 6H_2O$	Reaction in 0.5 M HCl.
IO_4^-	$2IO_4^- + 22I^- + 16H^+ \rightleftharpoons 8I_3^- + 8H_2O$	Reaction in 0.5 M HCl.
O_2	$O_2 + 4Mn(OH)_2 + 2H_2O \rightleftharpoons 4Mn(OH)_3$ $2Mn(OH)_3 + 6H^+ + 6I^- \rightleftharpoons 2Mn^{2+} + 2I_3^- + 6H_2O$	The sample is treated with Mn^{2+}, NaOH, and KI. After one minute, it is acidified with H_2SO_4, and the I_3^- is titrated.
H_2O_2	$H_2O_2 + 3I^- + 2H^+ \rightleftharpoons I_3^- + 2H_2O$	Reaction in 1 M H_2SO_4 with NH_4MoO_3 catalyst.
O_3	$O_3 + 3I^- + 2H^+ \rightleftharpoons O_2 + I_3^- + H_2O$	O_3 is passed through neutral 2% (wt/wt) KI solution. Add H_2SO_4 and titrate.
NO_2^-	$2HNO_2 + 2H^+ + 3I^- \rightleftharpoons 2NO + I_3^- + 2H_2O$	The nitric oxide is removed (by bubbling CO_2 generated *in situ*) prior to titration of I_3^-.
As(V)	$H_3AsO_4 + 2H^+ + 3I^- \rightleftharpoons H_3AsO_3 + I_3^- + H_2O$	Reaction in 5 M HCl.
Sb(V)	$SbCl_6^- + 3I^- \rightleftharpoons SbCl_4^- + I_3^- + 2Cl^-$	Reaction in 5 M HCl.
$S_2O_8^{2-}$	$S_2O_8^{2-} + 3I^- \rightleftharpoons 2SO_4^{2-} + I_3^-$	Reaction in neutral solution. Then acidify and titrate.
Cu^{2+}	$2Cu^{2+} + 5I^- \rightleftharpoons 2CuI(s) + I_3^-$	NH_4HF_2 is used as a buffer.
$Fe(CN)_6^{3-}$	$2Fe(CN)_6^{3-} + 3I^- \rightleftharpoons 2Fe(CN)_6^{4-} + I_3^-$	Reaction in 1 M HCl.
MnO_4^-	$2MnO_4^- + 8H^+ + 15I^- \rightleftharpoons 2Mn^{2+} + 5I_3^- + 8H_2O$	Reaction in 0.1 M HCl.
MnO_2	$MnO_2(s) + 4H^+ + 3I^- \rightleftharpoons Mn^{2+} + I_3^- + 2H_2O$	Reaction in 0.5 M H_3PO_4 or HCl.
$Cr_2O_7^{2-}$	$Cr_2O_7^{2-} + 14H^+ + 9I^- \rightleftharpoons 2Cr^{3+} + 3I_3^- + 7H_2O$	Reaction in 0.4 M HCl requires five minutes for completion and is particularly sensitive to air oxidation.
Ce(IV)	$2Ce(IV) + 3I^- \rightleftharpoons 2Ce(III) + I_3^-$	Reaction in 1 M H_2SO_4.

Analysis of Organic Compounds with Periodic Acid

H_5IO_6—paraperiodic acid.

IO_4^-—metaperiodate.

Periodic acid is a powerful oxidizing agent that exists under different conditions as *paraperiodic acid* (H_5IO_6), *metaperiodic acid* (HIO_4), and various depronated forms of the acids. Sodium metaperiodate ($NaIO_4$) is usually used to prepare a solution that is standardized by addition to excess KI in bicarbonate solution at pH 8–9:

In acidic solution, the reaction goes further:

$H_5IO_6 + 11I^- + 7H^+ \rightleftharpoons 4I_3^- + 6H_2O$

$$IO_4^- + 3I^- + H_2O \rightleftharpoons IO_3^- + I_3^- + 2OH^- \tag{17-72}$$

The I_3^- released is titrated with thiosulfate to complete the standardization.

Periodate solutions are especially useful for the analysis of organic compounds (such as carbohydrates) containing the following functional groups:

hydroxyl adjacent to hydroxyl
carbonyl adjacent to hydroxyl
carbonyl adjacent to carbonyl

amine adjacent to hydroxyl
amine adjacent to carbonyl

In this oxidation, known as the *Malaprade reaction*, the carbon–carbon bond between the two functional groups is broken and the following changes occur:

1. A hydroxyl group is oxidized to an aldehyde or a ketone.
2. A carbonyl group is oxidized to a carboxylic acid.
3. An amine is converted to an aldehyde plus ammonia (or a substituted amine if the original compound was a secondary amine).

When there are three or more adjacent functional groups, oxidation begins near one end of the molecule.

The reactions are performed at room temperature for about one hour with a known excess of periodate. At higher temperatures, further nonspecific oxidations occur. Solvents such as methanol, ethanol, dioxane, or acetic acid may be added to the aqueous solution to increase the solubility of the organic reactant. After the reaction is complete, the unreacted periodate is analyzed using Reaction 17-72, followed by thiosulfate titration of the liberated I_3^-.

Some examples of the Malaprade reaction are given below:

(1) 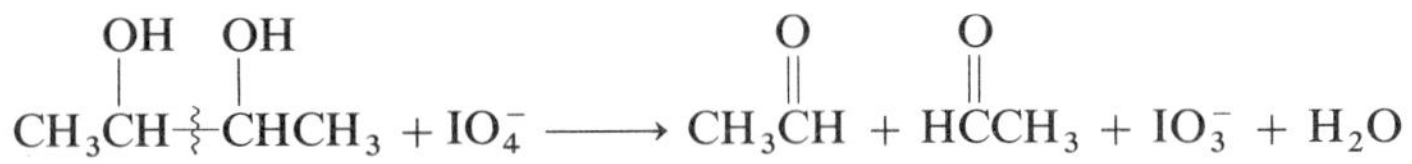
$$\overset{\overset{\displaystyle OH}{|}}{CH_3CH}\text{---}\overset{\overset{\displaystyle OH}{|}}{CHCH_3} + IO_4^- \longrightarrow CH_3\overset{\overset{\displaystyle O}{\|}}{CH} + H\overset{\overset{\displaystyle O}{\|}}{C}CH_3 + IO_3^- + H_2O$$
2,3-dihydroxybutane

(2) 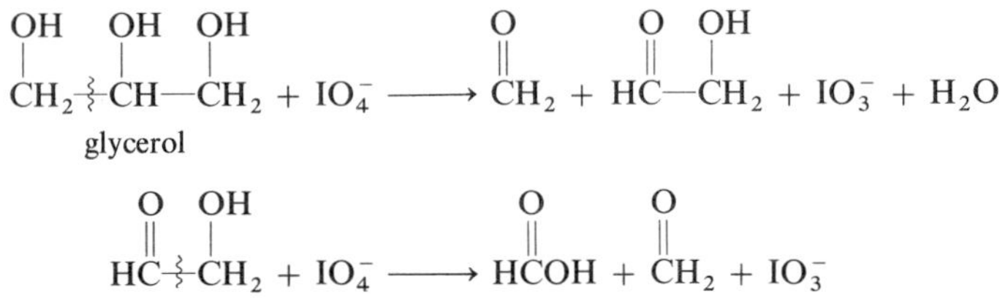
$$\overset{\overset{\displaystyle OH}{|}}{CH_2}\text{---}\overset{\overset{\displaystyle OH}{|}}{CH}\text{---}\overset{\overset{\displaystyle OH}{|}}{CH_2} + IO_4^- \longrightarrow \overset{\overset{\displaystyle O}{\|}}{CH_2} + H\overset{\overset{\displaystyle O}{\|}}{C}\text{---}\overset{\overset{\displaystyle OH}{|}}{CH_2} + IO_3^- + H_2O$$
glycerol

$$\overset{\overset{\displaystyle O}{\|}}{HC}\text{---}\overset{\overset{\displaystyle OH}{|}}{CH_2} + IO_4^- \longrightarrow H\overset{\overset{\displaystyle O}{\|}}{C}OH + \overset{\overset{\displaystyle O}{\|}}{CH_2} + IO_3^-$$

(3)
$$\overset{\overset{\displaystyle OH}{|}}{CH_2}\text{---}\overset{\overset{\displaystyle NH_3^+}{|}}{CH}\text{---}CO_2^- + IO_4^- \longrightarrow \overset{\overset{\displaystyle O}{\|}}{CH_2} + \overset{\overset{\displaystyle O}{\|}}{CH}\text{---}CO_2^- + NH_4^+ + IO_3^-$$ 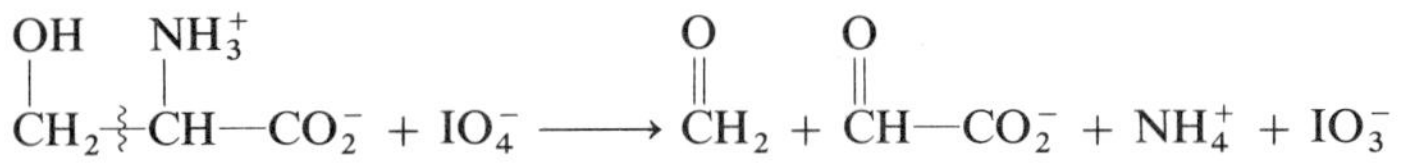
serine

Titrations with Reducing Agents

Most analytical redox titrations involve an oxidizing titrant. Reducing titrants are less common because they are generally unstable in the presence of oxygen and must therefore be stored and used under an inert atmosphere.

17 / REDOX TITRATIONS

390

Fe(NH$_4$)$_2$(SO$_4$)$_2$·6H$_2$O is usually used as a primary standard for preparing solutions of Fe(II).

Fewer indicators are suitable for reductive titrations, so potentiometry is usually used to find the endpoint. In Section 18-4, we will see how reducing agents can be conveniently generated *in situ* by electrolytic reduction of an appropriate precursor. Some reagents used for reductive titrations include Fe(II), Cr(II), Ti(III), Hg$_2$(NO$_3$)$_2$, and ascorbic acid. Solutions of Fe(II) in 1 M H$_2$SO$_4$ are stable to oxygen and are used to standardize strong oxidants such as MnO$_4^-$, Cr$_2$O$_7^{2-}$, Ce(IV), Au(III), and V(V).

Chromous ion, prepared by reducing K$_2$Cr$_2$O$_7$ with H$_2$O$_2$, followed by zinc amalgam, is the most powerful reducing agent commonly used in volumetric analysis:

$$Cr_2O_7^{2-} + 3H_2O_2 + 8H^+ \rightarrow 2Cr^{3+} + 3O_2(g) + 7H_2O \quad (17\text{-}73)$$

$$2Cr^{3+} + Zn(Hg) \rightarrow 2Cr^{2+} + Zn^{2+} \quad (17\text{-}74)$$

The chromous solution must be completely protected from oxygen in storage and use. It can be standardized by titrating Cu(II) in 6 M HCl, giving Cu(I) and Cr^{3+}.

R—NO$_2$—a nitro compound.

R—NO—a nitroso compound.

R—N=N—R—an azo compound.

Species that can be analyzed with Cr(II) titrant include Fe(CN)$_6^{3-}$, NO$_3^-$, CN$^-$, Fe(III), Ti(IV), V(V), Cr(VI), Mo(VI), W(VI), Ag(I), Au(III), Hg(II), Sn(IV), Sb(V), Bi(III), and Se(IV). Organic nitro, nitroso, and azo compounds are reduced rapidly at room temperature to their corresponding amines by excess Cr(II).

$$\text{azobenzene} + 4Cr^{2+} + 4H^+ \longrightarrow 2 \ \text{aniline} - NH_2 + 4Cr^{3+}$$

$$(17\text{-}75)$$

The excess Cr(II) is back-titrated with standard Fe(III). Ti(III) is not as strong a reducing agent as Cr(II), but has similar analytical applications, including the reactions with nitro, nitroso, and azo compounds.

Summary

Redox titrations are based on an oxidation–reduction reaction between analyte and titrant. Sometimes a quantitative chemical preoxidation (with such reagents as S$_2$O$_8^{2-}$, AgO, NaBiO$_3$, H$_2$O$_2$, or HClO$_4$) or prereduction (with such reagents as SnCl$_2$, CrCl$_2$, SO$_2$, H$_2$S, or a metallic reductor column) is necessary to adjust the oxidation state of the analyte prior to analysis. The endpoint of a redox titration is commonly detected with a redox indicator or by potentiometry. A useful indicator must have a transition range (= E^0 (indicator) $\pm$ 0.059/n V) that overlaps the abrupt change in potential of the titration curve.

The greater the difference in reduction potential between the analyte and titrant, the sharper will be the endpoint of a titration. The plateaus before and after the equivalence point are centered near E^0(analyte) and E^0(titrant). Calculations regarding the shape of the titration curve fall into three natural categories. Prior to the equivalence point, the half-reaction involving analyte is used because the concentration of both the oxidized and the reduced forms of analyte can be readily calculated. After the equivalence point, the half-reaction involving titrant is employed for the same reason. At the equivalence point, both half-reactions must be used simultaneously to find the potential.

Common oxidizing titrants discussed in this chapter include $KMnO_4$, Ce(IV), and $K_2Cr_2O_7$. Periodic acid is an oxidant used specifically to analyze organic reagents with certain adjacent functional groups (Malaprade re- action). A very large number of procedures is based on oxidation with I_3^- or titration of I_3^- liberated in a chemical reaction. Titrations with reducing agents such as Fe(II), Cr(II), or Ti(III) are not so common because the reductant must be protected from the air.

Terms to Understand

amalgam
disproportionation
iodimetry
iodometry

preoxidation
prereduction
redox indicator
wet ashing

Exercises

17-A. Consider the titration of 120.0 mL of 0.010 0 M Fe^{2+} (buffered to pH 1.00) with 0.020 0 M $Cr_2O_7^{2-}$. Calculate the potential (versus a silver–silver chloride anode saturated with KCl) at the following volumes of $Cr_2O_7^{2-}$: 0.100, 2.00, 4.00, 6.00, 8.00, 9.00, 9.90, 10.00, 10.10, 11.00, 12.00 mL. Sketch the titration curve.

17-B. Vanadium (II) undergoes three stepwise oxidation reactions:

$$V^{2+} \rightarrow V^{3+} \rightarrow VO^{2+} \rightarrow VO_2^+$$

Calculate the potential (versus S.C.E. anode) at each of the following volumes when 10.0 mL of 0.010 0 M V^{2+} is titrated with 0.010 0 M Ce(IV) in 1 M $HClO_4$: 5.0, 15.0, 25.0, and 35.0 mL. Sketch the titration curve.

17-C. Select an indicator from Table 17-1 that would be useful for finding (a) the second endpoint and (b) the third endpoint of the titration in Problem 17-B. What color change would you see at each point?

17-D. A 128.6 mg sample of a protein (M.W. 58 600) was treated with 2.000 mL of 0.048 7 M $NaIO_4$ to react with all of the serine and threonine residues.

$$
\begin{array}{cc}
\text{OH} & \text{CH}_3 \\
| & | \\
\text{CH}_2 & \text{HOCH} \\
| & | \\
\text{H}_3\overset{+}{\text{N}}\text{CHCO}_2^- & \text{H}_3\overset{+}{\text{N}}\text{CHCO}_2^- \\
\text{serine} & \text{threonine}
\end{array}
$$

The solution was then treated with excess iodide to convert the unreacted periodate to iodine (Equa- tion 17-72). Titration of the iodine required 823 μL of 0.098 8 M thiosulfate.

(a) Write balanced equations for the reaction of IO_4^- with serine and threonine.

(b) Calculate the number of serine plus threonine residues per molecule of protein. Answer to the nearest integer.

(c) How many milligrams of As_4O_6 (F.W. 395.68) would be required to react with the I_3^- liberated in this experiment?

Problems

17-1. What is the difference between the *titration reaction* and the *cell reaction(s)* in a potentiometric titra- tion?

17-2. Calculate the potential (versus S.C.E. anode) at each point listed for the titration of 25.00 mL of 0.020 00 M Cr^{2+} with 0.010 00 M Fe^{3+}: 5.00, 25.00, 50.00, 100.0 mL.

17-3. Suppose 50 mL of 0.020 8 M Fe^{3+} was titrated with 0.017 3 M ascorbic acid at pH 1.00.

dehydroascorbic acid

$E^0 = +0.390$ V

ascorbic acid

The potential was measured with a Pt cathode and an S.C.E. anode.

(a) Write a balanced equation for the titration reaction.

(b) Write two balanced equations to describe the cell reactions. (This question is asking for two complete reactions, not two half-reactions.)

(c) Calculate the potential after 50.0 mL of titrant has been added.

(d) Calculate the potential at the equivalence point.

17-4. Suppose that 50.0 mL of 0.050 0 M Fe^{2+} was titrated with 0.050 0 M MnO_4^- in a solution buffered to pH 1.00. Calculate the potential (versus S.C.E. anode) at each volume of MnO_4^-: 2.00, 4.00, 5.00, 6.00, 8.00, 9.00, 9.90, 10.00, 10.10, 11.00, 13.00 mL. Sketch the titration curve.

17-5. A solution containing 50.0 mL of 0.050 M UO_2^{2+} buffered to pH 0.00 was titrated with 0.100 M Sn^{2+}, and the potential was measured with a Pt electrode relative to a saturated silver–silver chloride anode. The *unbalanced* reaction is

$$UO_2^{2+} + Sn^{2+} \rightarrow U^{4+} + Sn^{4+}$$

(a) Calculate the potential when 35.0 mL of Sn^{2+} has been added.

(b) Calculate the potential at the equivalence point.

17-6. Suppose that the 0.010 0 M IO_3^- reagent in Figure 17-4 were replaced by 0.010 0 M $Cr_2O_7^{2-}$. Calculate the potential at the following volumes: $\frac{1}{2}V_e$, $3/2V_e$, $2V_e$, and $3V_e$. (V_e is the volume at the first equivalence point.)

17-7. A 25.0 mL solution containing a mixture of UO_2^+ and Fe^{2+} in 1 M $HClO_4$ was titrated with 0.009 87 M $KMnO_4$.

(a) Write two balanced titration reactions in the order in which they occur.

(b) The two potentiometric endpoints were observed at 12.73 and 31.21 mL. Calculate the molarities of UO_2^+ and Fe^{2+} in the unknown.

(c) Calculate the potential (versus S.C.E. anode) at $\frac{1}{2}V_1$, $V_1 + \frac{1}{2}V_2$, V_2, and $V_2 + 1.00$ mL, where V_1 and V_2 are the two equivalence points. Assume that $[H^+]$ is constant at 1.00 M.

17-8. Suppose 25 mL of 0.010 0 M U^{4+} at pH 1.00 was oxidized to UO_2^{2+} with 0.005 00 M $Br_2(aq)$. Calculate the potential (versus S.C.E. anode) after addition of 10.0, 50.0, and 51.0 mL of Br_2.

17-9. Would indigo tetrasulfonate be a suitable redox indicator for the titration of $Fe(CN)_6^{4-}$ with Tl^{3+}?

17-10. Would *tris*(2,2'-bipyridine) iron be a useful indicator for the titration of Sn^{2+} with $Mn(EDTA)^-$?

17-11. Suppose that 100.0 mL of solution containing 0.100 M Fe^{2+} and 0.005 00 M *tris*(1,10-phenanthroline)Fe(II) (ferroin) is titrated with 0.050 0 M Ce^{4+}. Calculate the potential (versus S.C.E. anode) at the following volumes of Ce^{4+}: 1.0, 10.0, 100.0, 190.0, 199.0, 201.0, 205.0, 209.0, 210.0, 211.0, 220.0 mL. Assume that the reaction contains 1 M $HClO_4$. Sketch the titration curve. Is ferroin a suitable indicator for the titration of Fe^{2+} by Ce^{4+}?

17-12. An enzyme involved in the catalysis of redox reactions has an oxidized and a reduced form differing by two electrons. The oxidized form of this enzyme was mixed at pH 7 with the oxidized form of a redox indicator whose two forms differ by one electron. The mixture was protected by an inert atmosphere and partially reduced with sodium dithionite. The equilibrium composition of the mixture was determined by spectroscopic analysis:

$$\text{Enzyme(oxidized)} = 4.2 \times 10^{-5} \text{ M}$$

$$\text{Indicator(oxidized)} = 3.9 \times 10^{-5} \text{ M}$$

$$\text{Enzyme(reduced)} = 1.8 \times 10^{-5} \text{ M}$$

$$\text{Indicator(reduced)} = 5.5 \times 10^{-5} \text{ M}$$

Given that $E^{0'} = -0.187$ V for the indicator, find $E^{0'}$ for the enzyme.

17-13. From the reduction potentials below

$$I_2(s) + 2e^- \rightleftharpoons 2I^- \qquad E^0 = 0.534\ 5 \text{ V}$$

$$I_2(aq) + 2e^- \rightleftharpoons 2I^- \qquad E^0 = 0.619\ 7 \text{ V}$$

$$I_3^- + 2e^- \rightleftharpoons 3I^- \qquad E^0 = 0.535\ 5 \text{ V}$$

(a) Calculate the equilibrium constant for the reaction

$$I_2(aq) + I^- \rightleftharpoons I_3^-$$

(b) Calculate the equilibrium constant for the reaction

$$I_2(s) + I^- \rightleftharpoons I_3^-$$

(c) Calculate the solubility (g/L) of $I_2(s)$ in water.

17-14. When 25.00 mL of unknown was passed through a Jones reductor, molybdate ion (MoO_4^{2-}) was converted to Mo(III). The filtrate required 16.43 mL of 0.010 33 M $KMnO_4$ to reach the purple endpoint.

$$MnO_4^- + Mo^{3+} \rightarrow Mn^{2+} + MoO_2^{2+}$$

A blank required 0.04 mL. Find the molarity of molybdate in the unknown.

17-15. A solution of I_3^- was standardized by titrating freshly dissolved arsenious oxide (As_4O_6, F.W. 395.683). The titration of 25.00 mL of a solution prepared by dissolving 0.366 3 g of As_4O_6 in a volume of 100.0 mL required 31.77 mL of I_3^-.
(a) Calculate the molarity of the I_3^- solution.
(b) Does it matter whether starch indicator is added at the beginning or near the endpoint in this titration?

17-16. An aqueous glycerol solution weighing 100.0 mg was treated with 50.0 mL of 0.083 7 M Ce(IV) in 4 M $HClO_4$ at 60°C for 15 minutes to oxidize the glycerol to formic acid:

$$\underset{\underset{\text{glycerol}}{HO \quad HO \quad HO}}{CH_2-CH-CH_2} \qquad \underset{\text{formic acid}}{3HCO_2H}$$

The excess Ce(IV) required 12.11 mL of 0.044 8 M Fe(II) to reach a ferroin endpoint. What is the weight percent of glycerol in the unknown?

17-17. A mixture of nitrobenzene and nitrosobenzene weighing 24.43 mg was titrated with Cr^{2+} to give aniline:

nitrosobenzene
F.W. 107.112

nitrobenzene
F.W. 123.111

aniline

The titration required 21.57 mL of 0.050 00 M Cr^{2+} to reach a potentiometric endpoint. Find the weight percent of nitrobenzene in the mixture.

17-18. Potassium bromate ($KBrO_3$) is a good primary standard for the generation of Br_2 in acidic solution:

$$BrO_3^- + 5Br^- + 6H^+ \rightleftharpoons 3Br_2(aq) + 3H_2O$$

The Br_2 can be used to analyze many unsaturated organic compounds. Al^{3+} was analyzed as follows: An unknown was treated with 8-hydroxyquinoline (oxine) at pH 5 to precipitate aluminum oxinate, $Al(C_9H_6ON)_3$. The precipitate was washed, dissolved in warm HCl containing excess KBr, and treated with 25.00 mL of 0.020 00 M $KBrO_3$.

The excess Br_2 was reduced with KI, which was converted to I_3^-. The I_3^- required 8.83 mL of 0.051 13 M $Na_2S_2O_3$ to reach a starch endpoint. How many milligrams of Al were in the unknown?

17-19. Nitrite (NO_2^-) can be determined by oxidation with excess Ce^{4+}, followed by back titration of the unreacted Ce^{4+}. A 4.030 g sample of solid containing only $NaNO_2$ and $NaNO_3$ was dissolved in 500.0 mL. A 25 mL sample of this solution was treated with 50.00 mL of 0.118 6 M Ce^{4+} in strong acid for five minutes, and the excess Ce^{4+} was back-titrated with 31.13 mL of 0.042 89 M ferrous ammonium sulfate.

$$2Ce^{4+} + NO_2^- + H_2O \rightarrow 2Ce^{3+} + NO_3^- + 2H^+$$

$$Ce^{4+} + Fe^{2+} \rightarrow Ce^{3+} + Fe^{3+}$$

Calculate the weight percent of $NaNO_2$ in the solid.

17-20. A 50.00 mL sample containing La^{3+} was treated with sodium oxalate to precipitate $La_2(C_2O_4)_3$, which was washed, dissolved in acid, and titrated with 18.04 mL of 0.006 363 M $KMnO_4$. Calculate the molarity of La^{3+} in the unknown.

17-21. Write a balanced equation for the reaction of periodate (IO_4^-) with

(a)

	CHO	
H	—	OH
OH	—	H
H	—	OH
H	—	OH
	CH_2OH	

glucose

(b)

	CHO	
H	—	NH_2
HO	—	H
H	—	OH
H	—	OH
	CH_2OH	

2-amino-2-deoxyglucose

(c)

dihydroxyacetone

17-22. A solution containing Be and several other metals was treated with excess EDTA to *mask* the other metals. Then excess acetoacetanilide was added at 50°C at pH 7.5 to precipitate beryllium ion:

$$Be^{2+} + 2$$

(s)

The precipitate was dissolved in 6 M HCl and treated with 50.0 mL of solution containing 0.139 2 g of $KBrO_3$ (F.W. 167.00) plus 0.5 g of KBr. Bromine produced by these reagents reacts with acetoacetanilide as follows:

$$+ Br_2 \longrightarrow$$

$$\longrightarrow \quad Br + HBr$$

After five minutes, the excess Br_2 was destroyed by adding 2 g of KI (F.W. 166.00). The I_2 released

by the Br_2 required 19.18 mL of 0.050 00 M $Na_2S_2O_3$. Calculate the number of milligrams of Be in the original solution.

17-23. A sensitive titration of Bi^{3+} is based on the following sequence:[†]

1. Bi^{3+} is precipitated by $Cr(SCN)_6^{3-}$:

$$Bi^{3+} + Cr(SCN)_6^{3-} \rightarrow Bi[Cr(SCN)_6](s) \quad (a)$$

2. The precipitate is filtered, washed, and treated with bicarbonate to release $Cr(SCN)_6^{3-}$:

$$Bi[Cr(SCN)_6] + HCO_3^- + H_2O \rightarrow$$
$$(BiO)_2CO_3(s) + Cr(SCN)_6^{3-} + H^+ \quad (b)$$

3. I_2 is added to the filtrate after removal of the $(BiO)_2CO_3(s)$:

$$Cr(SCN)_6^{3-} + I_2 + H_2O \rightarrow$$
$$SO_4^{2-} + ICN + I^- + H^+ + Cr^{3+} \quad (c)$$

4. Upon acidification to pH 2.5, HCN is created:

$$ICN + I^- + H^+ \rightarrow I_2 + HCN \quad (d)$$

5. The I_2 in this mixture is removed by extraction with chloroform. Excess bromine water is then added to the aqueous phase to convert iodide to iodate and HCN to BrCN:

$$Br_2 + I^- + H_2O \rightarrow IO_3^- + H^+ + Br^- \quad (e)$$
$$Br_2 + HCN \rightarrow BrCN + H^+ + Br^- \quad (f)$$

6. Excess Br_2 is destroyed with formic acid:

$$Br_2 + HCO_2H \rightarrow Br^- + CO_2 + H^+ \quad (g)$$

7. Addition of excess I^- produces I_2:

$$IO_3^- + I^- + H^+ \rightarrow I_2 + H_2O \quad (h)$$
$$BrCN + I^- + H^+ \rightarrow I_2 + HCN + Br^- \quad (i)$$

8. Finally, the iodine is titrated with a standard solution of sodium thiosulfate.

Show that 228 moles of thiosulfate will be required for each mole of Bi^{3+} that is analyzed.

[†] G. A. Parker, *J. Chem. Ed.*, **57**, 721 (1980).

18 / Electrogravimetric and Coulometric Analysis

So far we have seen how *spontaneous* electrochemical reactions can be adapted for analytical purposes. Potentiometric measurements with galvanic cells (in which negligible current is permitted to flow) can be used to measure concentrations of species in the cell. In redox titrations a spontaneous reaction between analyte and titrant is employed. In Chapters 18 and 19 we will see how *nonspontaneous* redox reactions, driven by an external source of electricity, are used in analytical chemistry.

This chapter deals with electrogravimetric and coulometric techniques. In **electrogravimetric analysis** the analyte is quantitatively deposited as a solid on the cathode or anode. The increase in mass of the electrode is a direct measure of the amount of analyte. **Coulometric analyses** are based on the measurement of current and time needed to complete a chemical reaction. Before discussing these methods, we must examine how the potential of an electrochemical cell changes when a significant current flows.

18-1 ELECTROLYSIS: PUTTING ELECTRONS TO WORK

Electrolysis is the process in which a reaction is driven in its nonspontaneous (endergonic) direction by the application of an electric current. For example, Reaction 18-1 has a standard potential of -1.455 V.

Exergonic means that ΔG is negative.

Endergonic means that ΔG is positive.

$$Pb^{2+} + 2H_2O \rightleftharpoons PbO_2(s) + H_2(g) + 2H^+ \qquad E^0 = -1.455 \text{ V} \quad (18\text{-}1)$$

Suppose you wish to drive this reaction to the right in a solution containing 5.00 mM Pb^{2+}, 2.00 M HNO_3, and solid PbO_2 under 1.00 atm of $H_2(g)$. Under these conditions, we calculate the potential to be

18 / ELECTROGRAVIMETRIC
AND COULOMETRIC ANALYSIS

ΔG (for Reaction 18-1) = $-nFE$ = $-nF(-1.541 \text{ V})$ = $+297 \text{ kJ/mol}$.

$$E = E^0 - \frac{0.059\ 16}{2} \log \frac{P_{H_2}[H^+]^2}{[Pb^{2+}]}$$

$$E = -1.455 - \frac{0.059\ 16}{2} \log \frac{(1.00)(2.00)^2}{(5.00 \times 10^{-3})} = -1.541 \text{ V} \quad (18\text{-}2)$$

We predict that a potential greater than 1.541 V must be *applied* to the solution to force Reaction 18-1 to proceed as written.

Figure 18-1 shows how this might be done. A pair of Pt electrodes is immersed in the solution, and a potential more positive than 1.541 V is applied with an external power supply. At the cathode (where reduction takes place), the reaction is

$$2H^+ + 2e^- \rightarrow H_2(g) \quad (18\text{-}3)$$

and at the anode (where oxidation occurs), the reaction is

$$Pb^{2+} + 2H_2O \rightarrow PbO_2(s) + 4H^+ + 2e^- \quad (18\text{-}4)$$

The net reaction is 18-1.

If a current I flows for a time t, the amount of charge that has passed any point in the circuit is

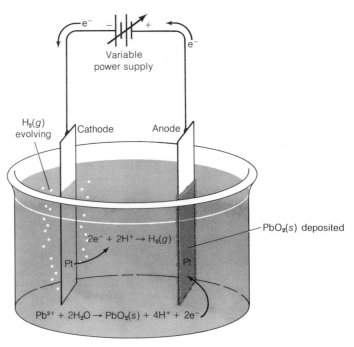

Figure 18-1
Electrolysis of Pb^{2+} in acidic solution. The symbol ⇥|⊢ will be used to represent a variable source of direct current.

$$q \quad = \quad I \cdot t \qquad\qquad (18\text{-}5)$$

$$\text{coulombs} \qquad \text{amps} \cdot \text{seconds}$$

The number of moles of electrons is

$$\text{moles of e}^- = \frac{\text{Coulombs}}{\text{Coulombs/mol}} = \frac{I \cdot t}{F} \qquad\qquad (18\text{-}6)$$

If a species requires n electrons per molecule in its redox half-reaction, the moles of that species that will have reacted in a time t is simply

$$\text{Moles reacted} = \frac{I \cdot t}{nF} \qquad\qquad (18\text{-}7)$$

Amperes = Coulombs/Second

$\quad$ F = 96 484.56 C/mol

Moles of electrons = $I \cdot t/F$

EXAMPLE

If a current of 0.17 A flows for 16 minutes through the cell in Figure 18-1, how many grams of PbO_2 will have been deposited?

To answer this, we first calculate the moles of e$^-$ that have flowed through the cell:

$$\text{Moles of e}^- = \frac{I \cdot t}{F} = \frac{\left(0.17\,\dfrac{C}{s}\right)(16\ \text{min})\left(60\,\dfrac{s}{\text{min}}\right)}{96\,485\left(\dfrac{C}{\text{mol}}\right)} = 1.69 \times 10^{-3}\ \text{mol}$$

For the half-reaction

$$Pb^{2+} + 2H_2O \rightleftharpoons PbO_2(s) + 4H^+ + 2e^-$$

two moles of electrons are required for each mole of PbO_2 deposited. Therefore,

$$\text{Moles of } PbO_2 = \tfrac{1}{2}(\text{moles of e}^-) = 8.45 \times 10^{-4}\ \text{mol}$$

The mass of PbO_2 will be

$$\text{g } PbO_2 = (8.45 \times 10^{-4}\ \text{mol})(239.2\ \text{g mol}^{-1}) = 0.20\ \text{g}$$

Demonstration 18-1 is an impressive classroom illustration of electrolysis. Box 18-1 describes one approach to the conversion of solar energy into storable fuel by means of a photo-assisted electrolysis.

18-2 WHY THE VOLTAGE CHANGES WHEN CURRENT FLOWS THROUGH A CELL

In previous chapters we considered the potential of galvanic cells only under conditions of negligible current flow. For a cell to do any useful work or for an electrolysis to occur, a significant current must flow. Whenever current

When current flows through a galvanic cell, the observed voltage is *less* than the equilibrium voltage.

Demonstration 18-1 ELECTROCHEMICAL WRITING[†]

Electrolysis is dramatically demonstrated in this experiment. The apparatus consists of a sheet of aluminum foil taped or cemented to a glass or wood surface. Any size will work, but an area about fifteen centimeters on a side is convenient for a classroom demonstration. On the metal foil is taped (at one edge only) a sandwich consisting of filter paper, typing paper, and another sheet of filter paper. A stylus is prepared from a length of copper wire (18 gauge or thicker) looped at the end and passed through a length of glass tubing.

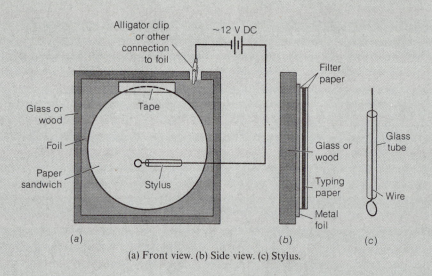

(a) Front view. (b) Side view. (c) Stylus.

A fresh solution is prepared from 1.6 g of KI, 20 mL of water, 5 mL of 1% (wt/wt) starch solution, and 5 mL of phenolphthalein indicator solution. (If the solution darkens after standing for several days, it can be decolorized by adding drops of dilute $Na_2S_2O_3$.) The three layers of paper are soaked with the KI/starch/phenolphthalein solution. Connect the stylus and foil to a 12 V DC power source, and write on the paper with the stylus.

When the stylus is the cathode, pink color appears from the reaction of OH^- with phenolphthalein:

$$\text{Cathode:} \quad H_2O + e^- \rightarrow \tfrac{1}{2}H_2(g) + OH^-$$

When the polarity is reversed and the stylus is the anode, a black (very dark blue) color appears from the reaction of I_2 with starch:

$$\text{Anode:} \quad I^- \rightarrow \tfrac{1}{2}I_2 + e^-$$

Pick up the top sheet of filter paper and the typing paper, and you will discover that the writing appears in the opposite color on the bottom sheet of filter paper. This sequence is shown in Color Plates 19–21.

[†] E. C. Gilbert in H. N. Alyea and F. B. Dutton, eds., *Tested Demonstrations in Chemistry* (Easton, Penn.: Journal of Chemical Education, 1965), p. 145.

flows, three factors act to decrease the output voltage of a galvanic cell or to increase the applied potential needed for electrolysis. These factors are called the *ohmic potential, concentration polarization,* and *overpotential.*

Ohmic Potential

Any cell possesses some electrical resistance. The voltage needed to force current (ions) to flow through the cell is called the **ohmic potential** and is given by Ohm's law:

$$E_{ohmic} = IR \qquad (18\text{-}8)$$

where I is the current and R is the resistance of the cell.

In a galvanic cell at equilibrium, there is no ohmic potential because $I = 0$. If a current is drawn from the cell, the cell potential decreases because part of the free energy released by the chemical reaction is needed to overcome the resistance of the cell itself. The voltage applied to an electrolysis cell must be great enough to provide the free energy for the chemical reaction and to overcome the cell resistance.

In the absence of any other effects, *the potential of a galvanic cell is decreased by IR, and the magnitude of the applied potential in an electrolysis must be increased by IR* in order for current to flow.

> **18-2 WHY THE VOLTAGE CHANGES WHEN CURRENT FLOWS THROUGH A CELL**
>
> Performing an electrolysis requires *more* than the calculated equilibrium voltage for the reaction.

> The output potential of a galvanic cell is decreased by $I \cdot R$.

> The magnitude of the input potential for an electrolysis cell must be increased by $I \cdot R$.

EXAMPLE

Consider the cell

$$Cd(s)\,|\,CdCl_2(aq, 0.167\ \text{M})\,|\,AgCl(s)\,|\,Ag(s)$$

in which the spontaneous chemical reaction is

$$Cd(s) + 2AgCl(s) \rightarrow Cd^{2+} + 2Ag(s) + 2Cl^-$$

In Equation 15-35, we calculated that the cell potential should be 0.764 V. (a) If the cell has a resistance of 6.42 Ω and a current of 28.3 mA is drawn, what will be the cell potential? (b) Suppose that the same cell is operated in reverse as an electrolysis. What voltage must be applied to reverse the reaction above?

(a) In the absence of electron flow, the potential ($E_{equilibrium}$) is 0.764 V. With a current of 28.3 mA, the potential will *decrease* to

$$E = E_{equilibrium} - IR = 0.764 - (0.028\ 3\ \text{A})(6.42\ \Omega) = 0.582\ \text{V}$$

(b) By convention, the potential applied to an electrolysis cell is given a negative sign. The potential needed to reverse the spontaneous reaction will be

$$E = -E_{equilibrium} - IR = -0.764 - (0.028\ 3\ \text{A})(6.42\ \Omega) = -0.946\ \text{V}$$

Notice that the magnitude of **any galvanic cell potential is *decreased*** by the ohmic potential. The magnitude of the potential that must be applied to any electrolysis cell is *increased* by the ohmic potential.

Box 18-1   PHOTO-ASSISTED ELECTROLYSIS†

A major goal of solar energy research is to devise a means of converting solar energy into a fuel that can be stored and used later. Mother Nature accomplishes this in photosynthesis. The green leaves of plants use the energy of sunlight to reduce atmospheric CO_2 to carbohydrates, which can be "burned" by living organisms to provide chemical energy.

Man-made systems for converting solar energy (visible light) to fuels are not yet economically practical, but are the subject of much research. A system using ultraviolet light to reduce water to H_2 (which is probably the fuel of the future) provides a model for devices that must eventually operate with visible light. In a practical system, the H_2 would be stored and burned later in an engine.

A remarkably simple photoelectrolysis cell is shown below. In this cell, an electric power supply provides part of the energy needed to electrolyze water, and ultraviolet light provides the remainder. The energy content of the liberated hydrogen is greater than the input of electrical energy, but smaller than the total input of electrical plus radiant energy.

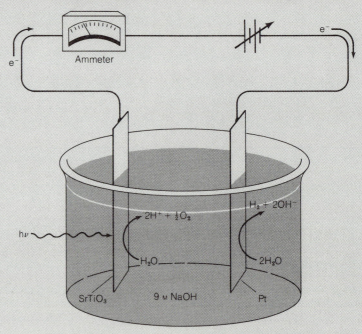

Photo-assisted electrolysis cell for the production of H_2 from ultraviolet light.

The key element of the cell is a $SrTiO_3$ semiconducting photo-anode. The valence electrons of a semiconductor lie in a band of energy levels called the *valence band*. At higher energy lies the *conduction band*, whose electrons are free to roam through the bulk material. In a conductor, the conduction band lies right above the valence band and is populated by electrons with thermal energy. In an insulator, the conduction band is at such a high energy that it is almost never populated. A semiconductor represents a case intermediate between insulators and conductors.

---

† M. S. Wrighton, A. B. Ellis, P. T. Wolczanski, D. L. Morse, H. B. Abrahamson, and D. S. Ginley, *J. Amer. Chem. Soc.*, **98**, 2774 (1976).

When the SrTiO$_3$ electrode is irradiated with ultraviolet light, electrons are promoted from the valence band to the conduction band. The electrons can either flow out of the anode and through the circuit or drop back down to the valence band, liberating heat. The secret is to find a semiconductor in which electrons flow through the circuit more readily than they drop back to the valence band.

The photos below show what happens when the SrTiO$_3$ anode is irradiated with 330 nm light. Electrons flow out of the SrTiO$_3$ electrode, through the circuit, and into the Pt cathode. At the cathode, H$_2$(g) is liberated:

$$2H_2O + 2e^- \rightarrow H_2(g) + 2OH^-$$

At the anode, H$_2$O is oxidized by the now electron-deficient semiconductor:

$$H_2O \rightarrow 2H^+ + \tfrac{1}{2}O_2 + 2e^-$$

This process occurs at low efficiency when the anode is irradiated. The efficiency can be increased by applying a slight positive potential (~ 0.3 V) to the anode. This apparently helps draw conduction-band electrons from the SrTiO$_3$ and decreases the rate at which they fall back to the valence band. Conversion of light to useful chemical energy (burnable H$_2$) occurs with 20% efficiency under optimum conditions in this system. Major research challenges include finding systems that can operate with visible sunlight and devising electrodes that do not deteriorate as the cell operates.

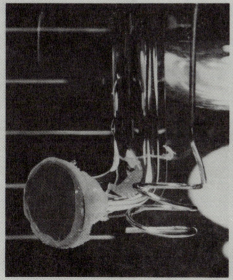 

(a) (b)

Photographs showing the SrTiO$_3$ (flat plate) and Pt (coiled wire) electrodes of the electrolysis cell (a) in the absence of ultraviolet light and (b) in the presence of ultraviolet light. Upon irradiation, gas is evolved at both electrodes. [Courtesy Mark Wrighton, Massachusetts Institute of Technology.]

Concentration Polarization

Consider the cadmium anode in Figure 18-2, for which the reaction is

$$Cd(s) \rightarrow Cd^{2+} + 2e^- \qquad (18\text{-}9)$$

This reaction creates Cd^{2+} ions in the layer of solution immediately surrounding the Cd electrode. If these ions diffuse rapidly or are transported rapidly away from the electrode by stirring, the concentration of Cd^{2+} will be essentially constant in the entire solution. We will call the concentration of Cd^{2+} in the bulk solution $[Cd^{2+}]_0$. We will call the concentration of Cd^{2+} in the immediate vicinity of the electrode surface $[Cd^{2+}]_s$. This concentration will equal $[Cd^{2+}]_0$ if diffusion or stirring is sufficiently rapid. Otherwise, $[Cd^{2+}]_s$ will be greater than $[Cd^{2+}]_0$, since Cd^{2+} ions are created at the electrode.

The anode potential depends on $[Cd^{2+}]_s$, not $[Cd^{2+}]_0$, because $[Cd^{2+}]_s$ is the actual concentration at the electrode surface. Reversing Reaction 18-9 to write it as a reduction (our continuing custom), the anode potential is given by the equation

The electrode potential depends on the concentration of species in the region immediately surrounding the electrode.

$$E(\text{anode}) = E^0(\text{anode}) - \frac{0.059\,16}{2} \log \frac{1}{[Cd^{2+}]_s} \qquad (18\text{-}10)$$

If $[Cd^{2+}]_s = [Cd^{2+}]_0$, the anode potential will be that expected from the bulk Cd^{2+} concentration.

When ions are not transported to or from an electrode as rapidly as they are consumed or created, we say that concentration polarization exists. That is, concentration polarization means that $[X]_s \neq [X]_0$.

If the current is flowing so fast that Cd^{2+} cannot escape from the region around the electrode as fast as it is made, $[Cd^{2+}]_s$ will be greater than

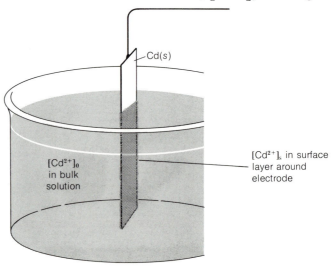

Partial view of cell

Figure 18-2
The potential of the $Cd^{2+}|Cd$ couple depends on the concentration of Cd^{2+} in the layer surrounding the electrode. The layer is greatly exaggerated in this drawing; it is only a few molecules thick.

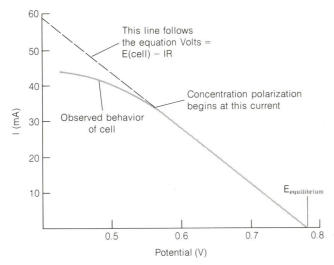

Figure 18-3
The hypothetical behavior of a galvanic cell, illustrating concentration polarization that occurs when $[Cd^{2+}]_s > [Cd^{2+}]_0$. The resistance of the cell is 6.42 Ω.

$[Cd^{2+}]_0$. When $[Cd^{2+}]_s \neq [Cd^{2+}]_0$, we say that **concentration polarization** exists. The anode potential in Equation 18-10 will increase and E(cell) (= E(cathode) − E(anode)) will decrease. This behavior is shown in Figure 18-3, where the straight line shows the behavior expected if only the ohmic potential (IR) affects the net cell potential. The deviation of the curve from the straight line at high currents is due to concentration polarization. In a galvanic cell, concentration polarization *decreases* the potential below the value expected in the absence of polarization.

In electrolysis the situation is reversed; reactant is depleted and product accumulates. Therefore, *concentration polarization requires us to apply a voltage of greater magnitude (more negative) than that expected in the absence of polarization.* By custom, the applied potential is given a negative sign.

Among the factors causing ions to move toward or away from the electrode are diffusion, convection, and electrostatic attraction or repulsion. Raising the temperature increases the rate of diffusion and thereby decreases concentration polarization. Mechanical stirring is very effective in transporting species through the cell. Increasing ionic strength decreases the electrostatic forces between ions and the electrode. These factors can all be used to affect the degree of polarization. Also, the greater the electrode surface area, the more current can be passed without polarization.

Concentration polarization *decreases* the voltage of a galvanic cell and *increases* the magnitude of the voltage necessary for electrolysis.

To decrease concentration polarization:
1. Raise the temperature.
2. Increase stirring.
3. Increase electrode surface area.
4. Change ionic strength to increase or decrease electrostatic interaction between the electrode and the reactive ion.

Overpotential

Even when concentration polarization is absent and ohmic potential is taken into account, the potentials of some electrochemical cells still show unexpected values. Some electrolyses require a greater-than-expected applied potential, and some galvanic cells produce less voltage than we anticipate.

Overpotential increases as the rate of reaction increases.

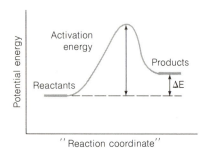

Figure 18-4
Schematic energy profile of a chemical reaction.

Overpotential is a kinetic phenomenon, needed to overcome the activation energy barrier for a reaction.

The difference between the expected potential (after accounting for IR drop and concentration polarization) and the applied potential is called the **overpotential**. *The overpotential is found to increase as the rate of the electrochemical reaction increases.*

The overpotential can be traced to the activation energy barrier for the electrode reaction.[†] Figure 18-4 shows a schematic "reaction coordinate" diagram for a chemical reaction. The activation energy for any chemical reaction is the barrier that must be overcome before reactants can be converted to products. The higher the temperature, the more molecules have sufficient energy to overcome the barrier, and the faster the reaction proceeds.

Now consider the reaction in which an electron from a metal electrode is transferred to H_3O^+ to initiate the reduction to H_2:

$$H_3O^+ + e^- \rightarrow \tfrac{1}{2}H_2 + H_2O \tag{18-11}$$

Figure 18-5a shows a schematic energy level diagram for this process. A certain energy barrier must be overcome for the electron transfer to occur, and the rate of reduction may be very slow. If an electric potential (the overpotential) is applied to the electrode, the energy of the electrons *in the metal electrode* is increased (Figure 18-5b). This decreases the height of the activation energy barrier and increases the rate of electron transfer. *The overpotential is that voltage needed to sustain a particular rate of electron transfer.* The greater the desired rate, the higher must be the overpotential. Thus, the overpotential increases as the current density $(A \cdot m^{-2})$ increases

---

[†] J. O'M. Bockris, *J. Chem. Ed.*, **48**, 352 (1971).

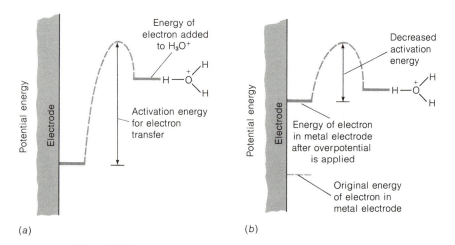

Figure 18-5
Schematic energy profile for electron transfer from a metal to H_3O^+ (a) With no applied potential. (b) After a potential is applied to the electrode. The overpotential increases the energy of the electrons in the electrode.

Table 18-1
Overpotential (V) for gas evolution at 25°C

| Electrode | 10 A m^{-2} | | 100 A m^{-2} | | 1 000 A m^{-2} | | 10 000 A m^{-2} | |
	H$_2$	O$_2$	H$_2$	O$_2$	H$_2$	O$_2$	H$_2$	O$_2$
Platinized Pt	0.015 4	0.398	0.030 0	0.521	0.040 5	0.638	0.048 3	0.766
Smooth Pt	0.024	0.721	0.068	0.85	0.288	1.28	0.676	1.49
Cu	0.479	0.422	0.584	0.580	0.801	0.660	1.254	0.793
Ag	0.475 1	0.580	0.761 8	0.729	0.874 9	0.984	1.089 0	1.131
Au	0.241	0.673	0.390	0.963	0.588	1.244	0.798	1.63
Graphite	0.599 5		0.778 8		0.977 4		1.220 0	
Sn	0.856 1		1.076 7		1.223 0		1.230 6	
Pb	0.52		1.090		1.179		1.262	
Zn	0.716		0.746		1.064		1.229	
Cd	0.981		1.134		1.216		1.254	
Hg	0.9		1.0		1.1		1.1	
Fe	0.403 6		0.557 1		0.818 4		1.291 5	
Ni	0.563	0.353	0.747	0.519	1.048	0.726	1.241	0.853

SOURCE: *International Critical Tables*, **6**, 339 (1929). This reference also gives overpotentials for Cl$_2$, Br$_2$, and I$_2$.

(Table 18-1). Because overpotential is a kinetic phenomenon, it is also called **kinetic polarization.**

The overpotential depends on the composition of the electrode, since the potential energy of an electron depends on what metal it is in. The overpotential also depends on how a given reactant interacts with the metal surface. Overpotential tends to decrease with increasing temperature.

18-3 ELECTROGRAVIMETRIC ANALYSIS

In *electrogravimetric analysis* the analyte is electrolytically deposited as a solid on an electrode. The increase in the mass of the electrode tells us how much analyte was present.

Electrogravimetric analysis is a combination of electrolysis and gravimetric analysis.

EXAMPLE

A solution containing 0.402 49 g of $CoCl_2 \cdot xH_2O$ was exhaustively electrolyzed to deposit 0.099 37 g of metallic cobalt on a platinum cathode.

$$Co^{2+} + 2e^- \rightarrow Co(s)$$

Calculate the number of moles of water per mole of cobalt in the reagent.

Assuming that the reagent contains only $CoCl_2$ and H_2O, we can write

$$g \text{ of } CoCl_2 = \left(\frac{g \text{ of } Co}{\text{atomic wt of } Co} \right)(\text{F.W. of } CoCl_2)$$

$$= 0.218\,93 \text{ g}$$

$$\text{g of } H_2O = 0.402\,49 - 0.218\,93 = 0.183\,56 \text{ g}$$

$$\frac{\text{mol of } H_2O}{\text{mol of Co}} = \frac{0.183\,56/\text{F.W. of } H_2O}{0.099\,37/\text{atomic wt of Co}} = 6.043$$

The reagent is apparently close to the composition $CoCl_2 \cdot 6H_2O$.

The basic requirements of an apparatus for electrogravimetric analysis are shown in Figure 18-6. The analyte is typically deposited on a carefully cleaned Pt gauze cathode, used because of its large surface area and chemical inertness. A simple power supply is adequate, but more elaborate apparatus, such as that in Figure 18-7, is available.

How do you know when an electrolysis is complete? One way is to observe the disappearance of color in a solution from which a colored ion (such as Co^{2+} or Cu^{2+}) is removed. Another way is to expose most, but not all, of the surface of the cathode to the solution during electrolysis. When you believe the reaction is complete, raise the level of the beaker or add some water so that some fresh surface of the cathode is exposed to the solution. After an additional period of electrolysis (fifteen minutes, say), see if the

Tests for completion of the deposition:
1. Disappearance of color.
2. Deposition on freshly exposed electrode surface.
3. Qualitative test for analyte in solution.

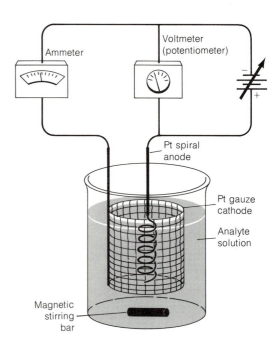

Figure 18-6
Simple apparatus for electrogravimetric analysis. The analyte is deposited on the large Pt gauze electrode. If the electrolytic deposition involves oxidation of the analyte, rather than reduction, the polarity of the power supply is simply reversed so that the deposition always occurs on the large electrode.

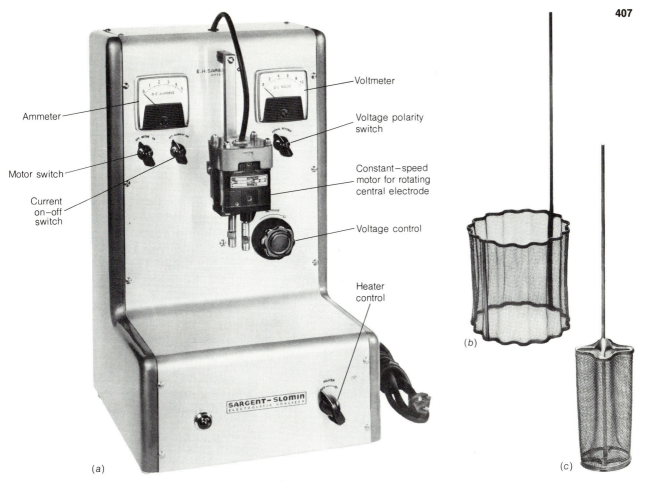

Figure 18-7
(a) Apparatus for electrogravimetric analysis. Spinning of the inner electrode serves to stir the solution. Heating is available to facilitate some electrolyses. (b) Outer electrode, usually the cathode. (c) Inner, rotating electrode. [Courtesy Sargent–Welch Co., Skokie, Ill.]

newly exposed electrode surface has a deposit. If it does, repeat the procedure. If not, the electrolysis is done. A third method is to remove a small sample of solution and perform a qualitative test for the analyte species.

Electrogravimetric analysis would be a simple matter if it merely involved electrolysis of a single analyte species from an otherwise inert solution. In practice, there may be other *electroactive* species that interfere by codeposition with the desired analyte. Even the solvent (water) is electroactive, since it decomposes to $H_2 + \frac{1}{2}O_2$ at a sufficiently high applied potential. Although these gases are liberated from the solution, their presence at the electrode surface interferes with deposition of solids. Because of these complications, control of the applied potential is an important feature of a successful gravimetric analysis.

Electroactive species are those that can be oxidized or reduced at an electrode.

Current–Voltage Behavior During Electrolysis

One of the most common applications of electrogravimetry is the analysis of copper. Suppose that a solution containing 0.20 M Cu^{2+} and 1.0 M H^+ is electrolyzed to deposit $Cu(s)$ on a Pt cathode and to liberate O_2 at a Pt anode.

Cathode: $\quad\quad Cu^{2+} + 2e^- \rightleftharpoons Cu(s)$ $\quad\quad\quad\quad\quad$ $E^0 = \quad 0.337$

Anode: $\quad\quad\quad\quad\quad H_2O \rightleftharpoons \frac{1}{2}O_2(g) + 2H^+ + 2e^-$ $\quad\quad$ $E^0 = \quad 1.229$

Net reaction: $\quad H_2O + Cu^{2+} \rightleftharpoons Cu(s) + \frac{1}{2}O_2(g) + 2H^+$ $\quad\quad$ $E^0 = -0.892$

$$(18\text{-}12)$$

Assuming that O_2 is liberated at a pressure of 0.20 atm, we naively calculate the potential needed for the electrolysis as follows:

$$E = E^0 - \frac{0.059\ 16}{n} \log \frac{P_{O_2}^{1/2}[H^+]^2}{[Cu^{2+}]}$$

$$E = -0.892 - \frac{0.059\ 16}{2} \log \frac{(0.20)^{1/2}(1.0)^2}{(0.20)} = -0.902\ V \quad (18\text{-}13)$$

In the absence of any polarization effects, we expect that no reaction should occur if the applied potential is more positive than -0.902 V. When the applied potential is more negative than -0.902 V, we expect deposition of Cu and liberation of O_2 to occur.

A small *residual current* is observed at all times. Substantial reaction begins at the *decomposition potential*.

The actual behavior of the electrolysis (using a pair of Pt electrodes) is shown in Figure 18-8. At low applied potential, a small current called the **residual current** is observed, even though no current was expected. At -0.902 V, nothing special happens. At a potential around one volt more negative, the **decomposition potential,** the reaction appears to take off in earnest.

It requires about 1 V of extra potential to overcome the barrier to O_2 formation at the anode.

Why is it necessary to apply a potential considerably more negative than -0.902 V to drive the reaction at an appreciable rate? The principal reason is the overpotential for oxygen formation at a Pt surface. In Table 18-1, we see that an extra potential of around one volt is needed to overcome the activation energy for oxidation of water at a smooth platinum anode, giving $O_2(g)$ as the product.

Why is a residual current observed at a low potential? There must be some oxidation occurring at the anode and some (equal amount) of reduction occurring at the cathode to support this residual current. Probably, these reactions involve mainly impurities. For example, dissolved oxygen can be reduced to H_2O_2, and Fe^{3+} could be reduced to Fe^{2+} at the cathode. Electrode surface oxide impurities could be reduced as well. At the anode, some small amount of water can be oxidized, or some oxidizable impurity in the solution might react.

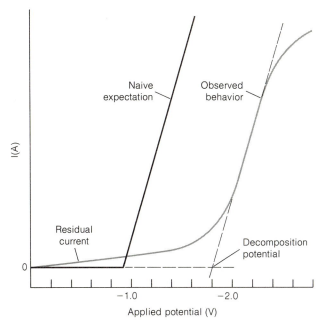

Figure 18-8
Schematic current-voltage relationship for the electrolysis of Cu^{2+}.

Another reason why the calculation in Equation 18-13 is naive is that when copper is deposited on a platinum electrode, the initial activity of Cu(s) is infinitely small. The calculation in Equation 18-13 assumes that copper is being deposited on a copper surface, for which the activity of copper is unity. Suppose that at some time early in the deposition of Cu(s) on Pt, the activity of Cu(s) on the Pt surface is 10^{-6}. Then a more realistic estimate of the potential for Reaction 18-12 is the following:

$$E = E^0 - \frac{0.059\ 16}{2} \log \frac{P_{O_2}^{1/2}[H^+]^2[Cu(s)]}{[Cu^{2+}]}$$

$$E = -0.892 - \frac{0.059\ 16}{2} \log \frac{(0.20)^{1/2}(1.0)^2(10^{-6})}{(0.20)} = -0.724\ V \quad (18\text{-}14)$$

Note the value of [Cu(s)] in this equation.

That is, when very little Cu(s) has been deposited, the activity of Cu(s) is much less than unity. It is not necessary to apply as much potential to electrolyze Cu^{2+} under these circumstances.

The curve in Figure 18-8 can therefore be explained in the following way:

1. Before reaching the decomposition potential, a small residual current is present, due both to oxidation and reduction of impurities and to a small amount of the intended electrolysis.

Residual current.

410

Overpotential.

Ohmic potential.

Concentration polarization

2. By the time the decomposition potential is applied, the desired electrolysis is the main reaction. The potential is shifted from the expected value by the overpotential for oxygen formation.

3. At potentials more negative than the decomposition potential, the current is essentially linear with respect to voltage, according to Ohm's law. The ratio between current and voltage tells us the resistance of the cell.

4. At a sufficiently high current, the curve deviates from Ohm's law. The Cu^{2+} ions cannot diffuse to the cathode rapidly enough to support the current predicted by Ohm's law. Concentration polarization has set in. It requires a progressively greater potential to increase the current.

EXAMPLE

Suppose that the resistance of the cell described by Figure 18-8 is 0.20 Ω. Estimate the potential needed to maintain a current of 0.40 A. Assume that the smooth Pt anode is supporting a current density of 1 000 A m^{-2} and that there is no concentration polarization.

Our estimate includes contributions from the naively calculated cell potential, the ohmic potential, and the overpotential.

$$E(\text{applied}) = \underbrace{E(\text{cathode}) - E(\text{anode})}_{\substack{E(\text{cell}) \\ (= E_{\text{equilibrium}}) \\ = 0.902 \text{ V} \\ (\text{from Equation 18-13})}} - IR - \underbrace{\text{Overpotential}}_{\substack{1.28 \text{ V} \\ \\ (\text{from Table 18-1})}}$$

$$E(\text{applied}) = -0.902 - (0.40 \text{ A})(0.20 \ \Omega) - 1.28 \text{ V}$$

$$= -0.902 - 0.080 - 1.28$$

$$= -2.26 \text{ V}$$

In this example the overpotential is large and the ohmic potential is fairly small.

Question: If the cell in Figure 18-8 has a resistance of 0.20 Ω, what will be the slope (A/V) of the linear portion of the curve between about -2.0 and -2.3 V?

Electrolysis at Constant Applied Potential

We now know that the potential calculated for a simple reversible cell must be modified to include the contributions of the ohmic potential and overpotential.

$$E(\text{applied}) = \underbrace{E(\text{cathode}) - E(\text{anode})}_{\substack{\text{Value of } E(\text{cell}) \\ \text{calculated for} \\ \text{reversible cell with} \\ \text{negligible current flow}}} - IR - \text{Overpotential} \qquad (18\text{-}15)$$

Suppose that the applied potential for an electrolysis is held at a constant value. For example, we might electrolyze 0.10 M Cu^{2+} in 1.0 M HNO_3 at a constant potential of -2.0 V. As Cu(s) is deposited, the concentration of Cu^{2+} in solution decreases. Eventually, there is too little Cu^{2+} in solution, and Cu^{2+} cannot be transported to the cathode rapidly enough to maintain the initial current of the cell. Since the current decreases, the ohmic potential and overpotential in Equation 18-15 decrease. The value of E(anode) is fairly constant due to the high concentration of solvent being oxidized at the anode. (Recall that the anode reaction is $H_2O \rightarrow \frac{1}{2}O_2 + 2H^+ + 2e^-$.)

Now think about Equation 18-15: If E(applied) and E(anode) are constant and if IR and overpotential decrease in magnitude, E(cathode) must become more negative to maintain the algebraic equality. This is shown schematically in Figure 18-9. The value of E(cathode) becomes more negative with time because of the concentration polarization that occurs as Cu^{2+} is consumed.

The value of E(cathode) continues to drop until it is negative enough to reduce H^+ to H_2:

$$H^+ + e^- \rightarrow \tfrac{1}{2}H_2(g) \qquad (18\text{-}16)$$

The overpotential of Reaction 18-16 determines where the curve in Figure 18-9 will level off. At a potential near -0.4 V, steady reduction of H^+ ensues. As E(cathode) falls from its initial value of $+0.3$ V to its steady value near -0.4 V, other ions in the solution might react. For example, Co^{2+}, Sn^{2+}, and Ni^{2+}, if present, would be reduced. In general, then, electrolysis at constant applied potential is not selective. Any solute more easily reduced than H^+ will be electrolyzed.

Formation of H_2 sometimes weakens the cathode deposit, causing it to crumble and fall off the electrode. To avoid this and to prevent the potential from becoming so negative that unintended ions are reduced, a **cathodic depolarizer** such as NO_3^-, is added to the solution. The cathodic depolarizer is more easily reduced than H^+ and yields a harmless product.

$$NO_3^- + 10H^+ + 8e^- \rightarrow NH_4^+ + 3H_2O \qquad (18\text{-}17)$$

From inspection of reduction potentials, we would expect that NO_3^- would be reduced *before* Cu^{2+}, since E^0 for Reaction 18-17 is 0.88 V while E^0 for the reduction of Cu^{2+} to Cu(s) is 0.337 V. Apparently, the overpotential for reduction of NO_3^- is high enough that reduction of Cu^{2+} happens first.

Electrolysis can be made somewhat selective through control of pH. In strongly acidic solution, Cu^{2+} can be reduced without concomitant reduction of such ions as Zn^{2+}, Ni^{2+}, or Cd^{2+}, because H^+ is more readily reduced than Zn^{2+}, Ni^{2+}, or Cd^{2+}. If the concentration of H^+ is not great enough, these metal ions will be reduced ahead of H^+. The redox behavior of various analytes is also affected in a predictable way by chelating agents, which tend to stabilize the cations and thereby make them harder to reduce.

Question: Why do the ohmic potential and overpotential decrease as the current decreases?

E(cathode) becomes more negative as Cu^{2+} is consumed because of concentration polarization.

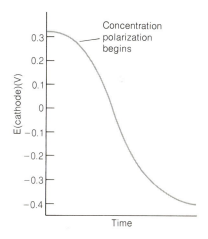

Figure 18-9
Schematic drawing showing the change of E(cathode) with time for an electrolysis conducted at a constant applied potential, ΔE.

A *cathodic depolarizer* is reduced in preference to solvent. It prevents E(cathode) from becoming so negative that water and impurities are reduced.

The selectivity of electrolysis can be affected by pH and by chelate concentrations.

18 / ELECTROGRAVIMETRIC AND COULOMETRIC ANALYSIS

Question: How would the curve in Figure 18-9 be different if the electrolysis were conducted at a constant current?

Working electrode—where the analytical reaction occurs.

Auxiliary electrode—the other electrode needed for current flow.

Reference electrode—the third electrode, used to measure the potential of the working electrode.

In controlled-potential electrolysis, the potential of the working electrode is kept constant with respect to the reference electrode. The applied potential (with respect to the auxiliary electrode) does change.

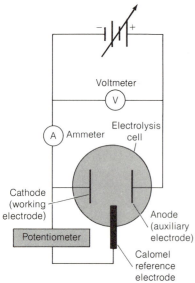

Figure 18-10
Circuit used for electrolysis using constant cathode potential.

Constant-Current Electrolysis

This mode of electrolysis is least selective. Concentration polarization reduces the current if the applied potential is constant. To maintain constant current, the magnitude of the applied potential must be progressively increased. The cathode potential rapidly becomes more negative until it levels off at a value fixed by the reduction of H^+ or a cathodic depolarizer. Reduction of the desired analyte continues as well.

Controlled-Potential Electrolysis

We have seen that electrolysis using a constant applied potential or a constant current is rather unselective. In both techniques the cathode potential becomes more and more negative, leading to reduction of species other than the intended analyte. A *three-electrode cell* (as shown in Figure 18-10) can be used to maintain a *constant cathode potential* and thereby greatly increase the selectivity of the electrolysis.

The electrode at which the reaction of interest occurs is called the **working electrode.** The calomel electrode serves as a **reference electrode** against which the potential of the working electrode can be measured. The third electrode (the current-supporting partner of the working electrode) is called the **auxiliary electrode.** A significant current flows between the working and auxiliary electrodes. Negligible current flows between the working and reference electrodes.

In a **controlled-potential electrolysis,** the potential applied between the working and auxiliary electrodes is varied in such a way that the potential between the working and reference electrodes is *constant*. The cathode potential can be maintained at this constant value using an electronic device called a **potentiostat.**

To see why a constant cathode potential permits selective deposition of analyte, consider a solution containing 0.1 M Cu^{2+} and 0.1 M Sn^{2+}. From the standard potentials below, we expect Cu^{2+} to be reduced more easily than Sn^{2+}:

$$Cu^{2+} + 2e^- \rightleftharpoons Cu(s) \qquad E^0 = 0.337 \text{ V}$$

$$Sn^{2+} + 2e^- \rightleftharpoons Sn(s) \qquad E^0 = -0.136 \text{ V}$$

The cathode potential at which Cu^{2+} ought to be reduced is calculated as

$$E(\text{cathode}) = 0.337 - \frac{0.059\,16}{2} \log \frac{1}{0.1} = 0.31 \text{ V} \qquad (18\text{-}18)$$

If 99.99% of the Cu^{2+} were deposited, the concentration of Cu^{2+} remaining in solution would be 10^{-5} M, and the cathode potential required to continue reduction would be

$$E(\text{cathode}) = 0.337 - \frac{0.059\ 16}{2} \log \frac{1}{10^{-5}} = 0.19\ \text{V} \qquad (18\text{-}19)$$

At a cathode potential of 0.19 V, then, rather complete deposition of copper is expected. Would Sn^{2+} be reduced at this potential? To deposit $Sn(s)$ from a solution containing 0.1 M Sn^{2+}, a cathode potential of -0.17 V is required:

$$E(\text{cathode, for reduction of } Sn^{2+}) = -0.136 - \frac{0.059\ 16}{2} \log \frac{1}{[Sn^{2+}]}$$

$$= -0.136 - \frac{0.059\ 16}{2} \log \frac{1}{0.1}$$

$$= -0.17\ \text{V} \qquad (18\text{-}20)$$

We do not expect significant reduction of Sn^{2+} at a cathode potential more positive than -0.17 V.

If the *cathode* potential is kept constant at a value near 0.19 V, we predict that 99.99% of the Cu^{2+} will react without deposition of Sn^{2+}. On the other hand, if the cell were run with a constant potential between the working and auxiliary electrodes, the cathode potential would behave as in Figure 18-9 and Sn^{2+} would be reduced. The price of achieving selective reduction with a constant cathode potential is that the current decreases and electrolysis is slower than for a constant applied potential.

The cathode potentials calculated in Equations 18-18 through 18-20 are implicitly stated with respect to the standard hydrogen electrode (since E^0 is taken from a table of standard potentials). To calculate the potential that would be measured with respect to a saturated calomel electrode in Figure 18-10, we use the relation

Controlled potential means that a constant potential difference is maintained between the working and *reference* electrodes. *Constant potential* means that a constant potential difference is maintained between the working and *auxiliary* electrodes. Controlled potential affords high selectivity, but the procedure is slower than constant potential electrolysis.

$$E\ (\text{versus S.C.E.}) = E\ (\text{versus S.H.E.}) - E(\text{calomel electrode})$$

$$E\ (\text{versus S.C.E.}) = E\ (\text{versus S.H.E.}) - 0.241 \qquad (18\text{-}21)$$

To obtain a cathode potential of 0.19 V (versus S.H.E.), we must maintain a cathode potential of $0.19 - 0.241 = -0.05$ V with respect to the saturated calomel reference electrode in the three-electrode cell of Figure 18-10.

18-4 COULOMETRIC ANALYSIS

Coulometry is based on counting the number of electrons used in a chemical reaction. For example, cyclohexene may be titrated with Br_2 generated by electrolytic oxidation of Br^-:

Coulometric methods are based on measuring the number of electrons that participate in a chemical reaction.

$$2Br^- \rightarrow Br_2 + 2e^- \qquad (18\text{-}22)$$

$$Br_2 + \text{cyclohexene} \longrightarrow \text{trans-1,2-dibromocyclohexene} \quad (18\text{-}23)$$

In this sequence of reactions, just enough Br_2 is generated to react with all of the cyclohexene. The moles of electrons liberated in Reaction 18-22 will be equal to twice the moles of Br_2 and therefore twice the moles of cyclohexene.

The reaction is conveniently carried out using the apparatus in Figure 18-11. Br_2 is generated by the Pt anode at the left. As the Br_2 is formed, it reacts with cyclohexene. When all of the cyclohexene has been consumed, the concentration of Br_2 in the solution suddenly rises, signaling the end of the reaction.

The rise in Br_2 concentration is detected **amperometrically,** using the circuit shown on the right in Figure 18-11. A small voltage (~ 0.25 V) is applied between the two electrodes on the right. This voltage is not great enough to electrolyze any of the solutes, so only a small residual current of <1 μA flows through the sensitive ammeter. When $[Br_2]$ suddenly increases, a current can flow by virtue of the following reactions:

Amperometric methods are based on the measurement of electric current. Amperometry is discussed further in Chapter 19.

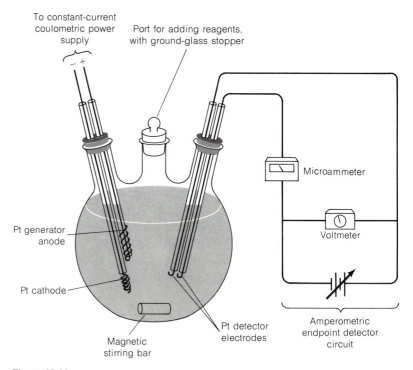

Figure 18-11
Apparatus for coulometric titration of cyclohexene with Br_2. The solution contains cyclohexene, 0.15 M KBr, and 3 mM mercuric acetate in a mixed solvent of acetic acid, methanol, and water. The mercuric acetate catalyzes the addition of Br_2 to the olefin. [Adapted from D. H. Evans, *J. Chem. Ed.*, **45**, 88 (1968).]

$$\text{Detector anode:} \quad 2Br^- \rightarrow Br_2 + 2e^- \qquad (18\text{-}24)$$

$$\text{Detector cathode:} \quad Br_2 + 2e^- \rightarrow 2Br^- \qquad (18\text{-}25)$$

The sudden increase of current is taken as the endpoint of the cyclohexene titration.

In practice, enough Br_2 is first generated in the absence of cyclohexene to give a detector current of 20.0 μA. When cyclohexene is added, the current decreases to a very small value because bromine is consumed. Bromine is then generated by the coulometric circuit, and the endpoint is taken when the detector again indicates 20.0 μA. This endpoint is easier to reproduce than "a sudden increase of current." Further, since the reaction is begun with enough Br_2 to carry 20.0 μA, any impurities that can react with Br_2 are eliminated.

The current for the bromine-generating electrodes is controlled by a hand-operated switch. As the detector current approaches 20.0 μA, the operator closes the switch for shorter and shorter intervals. This is closely analogous to adding titrant dropwise from a buret near the end of a titration. *The switch in the coulometer circuit serves as a "stopcock" for addition of Br_2 to the reaction.*

EXAMPLE

A 2.000 mL volume of solution containing 0.611 3 mg of cyclohexene/mL is to be titrated in the apparatus shown in Figure 18-11. If the coulometer is operated at a constant current of 4.825 mA, how much time will be required for complete titration?

The quantity of cyclohexene is

$$\frac{(2.000 \text{ mL})(0.611\ 3 \text{ mg/mL})}{(82.146 \text{ mg/mmol})} = 0.014\ 88 \text{ mmol}$$

In Reactions 18-22 and 18-23, each mole of cyclohexene requires one mole of Br_2, which in turn requires two moles of electrons to flow through the circuit. For 0.014 88 mmol of cyclohexene to react, 0.029 76 mmol of electrons must flow. From Equation 18-6, we can write

$$\text{Moles of } e^- = \frac{I \cdot t}{F} \Rightarrow t = \frac{(\text{mol } e^-)F}{I}$$

$$t = \frac{(0.029\ 76 \times 10^{-3} \text{ mol})(96\ 485 \text{ C/mol})}{(4.825 \times 10^{-3} \text{ C/s})} = 595.1 \text{ s}$$

It will require just under ten minutes to complete the reaction if the current is constant.

The example above illustrates the accuracy and sensitivity afforded by coulometric titrations. A common commercial coulometric power supply delivers current with an accuracy of $\sim 0.1\%$. With extreme care, the value of the Faraday constant has been determined to seven significant digits by a

Advantages of coulometry:
1. Accuracy.
2. Sensitivity.
3. Unstable reagents are generated *in situ*.

18 / ELECTROGRAVIMETRIC AND COULOMETRIC ANALYSIS

coulometric procedure (see Box 18-2). With careful attention to electronic circuitry, coulometric titration of Cl^- at a concentration of 10^{-5} M (0.3 ppm) is possible.[†] Another advantage of coulometric titrations is that such unstable reagents as Ag(II), Cu(I), Mn(III), and Ti(III) are generated and used in the same vessel. There is no need to handle reagents in the air or transfer them between vessels.

Some Details of Coulometry

Type of coulometry

The two common coulometric methods employ either a *constant current* or a *controlled potential*. Constant-current methods, as in the Br_2/cyclohexene example above, are called **coulometric titrations.** Knowing the current, it is only necessary to measure the time needed for complete reaction, in order to count the coulombs (Equation 18-5):

$$q = I \cdot t$$

> *Question:* What was that reason?

Controlled-potential coulometry is inherently more selective than constant-current coulometry. This is for the same reason that controlled-potential electrogravimetric analysis is more selective than analysis at constant applied potential.

In controlled-potential coulometry, the initial current is high, but decreases exponentially as the analyte concentration decreases. Since the current is not constant, the coulombs must be measured by integrating the current over the time of the reaction:

Chemical methods for performing this integration (measuring coulombs) are discussed later in this section.

$$q = \int_0^t I \, dt \qquad (18\text{-}26)$$

Endpoint detection

Methods of endpoint detection used for other types of titrations apply to coulometric titrations as well. These include the use of indicators and spectrophotometric methods. Potentiometric methods are useful for acid–base reactions (glass electrode), redox reactions (Pt electrode), EDTA titrations (mercury electrode, as in Exercise 16-B), and reactions for which ion-selective electrodes are available. For halide precipitations, a silver electrode might be used. The amperometric endpoint detector in Figure 18-11 is useful for a wide variety of coulometric titrations.

In controlled-potential coulometry, the equivalence point is never reached because the current decays exponentially. However, you can approach the equivalence point by letting the current decay to an arbitrarily

---

[†] M. J. Zetlmeisl and D. F. Lawrence, *Anal. Chem.*, **49,** 1557 (1977).

set value. For example, the current will ideally be 1% of its initial value when 99% of the analyte has been consumed, and it will be 0.1% of its initial value when 99.9% of the analyte has been consumed. (These figures refer to the current *above* the residual current.)

Mediators

A necessary condition for coulometric analysis is that the analytical reaction(s) must proceed with 100% electrochemical efficiency. Electrons cannot be siphoned off into side reactions, or the measurement of total coulombs becomes meaningless. While a few reactions meet this requirement directly, most analyses incorporate a **mediator** to improve the efficiency.

For example, Fe^{2+} might be analyzed coulometrically by titration to Fe^{3+}:

$$Fe^{2+} \rightarrow Fe^{3+} + e^- \qquad (18\text{-}27)$$

Initially, this reaction accounts for all of the current. When the concentration of Fe^{2+} decreases sufficiently, concentration polarization develops. The anode potential might increase to the point where water could be oxidized to O_2 and current would be carried mainly by the reaction

$$H_2O \rightarrow \tfrac{1}{2}O_2 + 2H^+ + 2e^- \qquad (18\text{-}28)$$

To avoid the side reaction 18-28, excess Ce^{3+} can be added to the solution. As the electrode becomes polarized, the reaction

$$Ce^{3+} \rightarrow Ce^{4+} + e^- \qquad (18\text{-}29)$$

begins at a lower potential than that needed for oxidation of water. The Ce^{4+} diffuses into the solution and oxidizes any Fe^{2+} it meets:

$$Ce^{4+} + Fe^{2+} \rightarrow Ce^{3+} + Fe^{3+} \qquad (18\text{-}30)$$

As long as Reaction 18-30 is rapid, the net reaction is the oxidation of Fe^{2+} to Fe^{3+}. The Ce^{3+} ion, which undergoes no net change, acts as a *mediator*.

Proteins normally do not react directly at electrode surfaces. They diffuse too slowly in solution, and their redox active sites may be too deep within the molecule to exchange electrons with the electrode surface. Because of this, mediators are often used to study the electrochemistry of proteins. Superoxide dismutase is an enzyme (a protein) that catalyzes the disproportionation of superoxide ion

$$2O_2^- + 2H^+ \rightarrow H_2O_2 + O_2 \qquad (18\text{-}31)$$
superoxide

Question: Why is Reaction 18-28 undesirable? Will it lead to an estimate of Fe^{2+} concentration that is higher or lower than the true value?

A *mediator* transports electrons quantitatively between the analyte and the working electrode. The mediator undergoes no net reaction itself.

Box 18-2 MEASURING THE FARADAY CONSTANT

The most accurate measurement of the Faraday constant comes from a very careful coulometric experiment conducted at the U.S. National Bureau of Standards.[†] A schematic diagram of the experiment is shown in part a of the figure on the facing page. Part b is a photograph of the actual equipment. The method consists of the electrolytic dissolution of a highly purified metallic Ag anode in an aqueous solution of 20% (wt/wt) $HClO_4$ containing 0.5% (wt/wt) $AgClO_4$.

$$Ag \text{ anode:} \qquad Ag(s) \rightarrow Ag^+ + e^-$$

$$Pt \text{ cathode:} \quad Ag^+ + e^- \rightarrow Ag(s)$$

This electrolyte was chosen because metallic Ag is very stable in this solution and does not spontaneously dissolve.

In a typical experiment, electrolysis was conducted using a potential of 1.018 209 8 V and a current of 0.203 639 0 A for 18 000.075 s. The loss of mass at the anode amounted to 4.097 900 g. The number of coulombs passed through the cell is therefore

$$q = I \cdot t = (0.203\ 639\ 0 \text{ A})(18\ 000.075 \text{ s}) = 3\ 665.517\ 3 \text{ C}$$

The silver lost from the anode amounts to

$$\text{Moles of Ag} = \frac{4.097\ 900 \text{ g}}{107.868 \text{ g/mole}} = 3.798\ 995 \times 10^{-2} \text{ mol}$$

The coulombs and moles can be combined to find the Faraday constant:

$$F = \frac{\text{Coulombs}}{\text{Moles}} = \frac{3\ 665.517\ 3}{3.798\ 995 \times 10^{-2}} = 96\ 486.5 \text{ C mol}^{-1}$$

Not all of the silver lost from the anode was oxidized. A fraction (ranging from 0.01% to 15%) of the Ag simply fell off as contiguous parts of the electrode were electrolyzed. This sediment was collected and weighed at the end of the experiment, so that the true mass of oxidized silver could be calculated. The intermediate beakers and siphons in the apparatus shown above are used to physically separate the anode and cathode compartments. This prevents Ag deposited on the Pt cathode from falling off and being weighed with the anode sediment.

Very great care was taken to purify the Ag anode. Needless to say, dust had to be excluded from all phases of the electrolysis and from the purification of the metallic silver. The electrode was prepared from "pure" electrolytic silver, obtained from the U.S. Mint, dissolved in HNO_3, and crystallized as $AgNO_3$. The silver was then deposited in metallic form on the purest available Ag electrode. The fresh deposit was scraped off the electrode, and the scrapings were washed for two weeks with HF (to dissolve any silica from the glass apparatus) and for two weeks with very pure water. The dry scrapings were fused in silica tubes from which impurities had been leached with hot, concentrated HNO_3. The fused Ag was again etched with HF until a constant mass was achieved. Under vacuum, the metal was melted to remove any oxide, then etched with 10% aqueous NH_3, and washed with very pure water. Spectrochemical analysis indicated less than 1 ppm of impurities. The Faraday constant was corrected for this tiny quantity of impurities. Also, the ratio $^{107}Ag/^{109}Ag$ was monitored throughout the purification and electrolysis procedures. No variation was found.

[†] D. N. Craig, J. I. Hoffman, C. A. Law, and W. J. Hamer, *J. Res. National Bureau of Standards*, **64A**, 381 (1960).

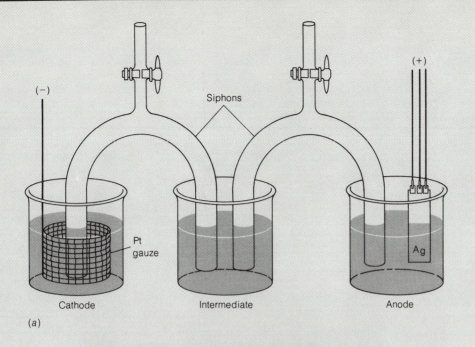

(a) Schematic view of the coulometer. (b) Photograph of experimental apparatus with two intermediate beakers.

Question: Ordinary Ag consists of 51.82% ^{107}Ag and 48.18% ^{109}Ag. How would selective electrolysis of ^{107}Ag affect the value of F determined in this experiment?

One form of this enzyme contains Mn^{3+}. It can be analyzed by coulometric titration using the dye methyl viologen as a mediator:

$$CH_3-{}^+N\!\!\underset{\text{methyl viologen cation}\atop(MV^{2+})}{\bigcirc\!\!-\!\!\bigcirc}\!\!N^+\!-CH_3 + e^- \longrightarrow CH_3-{}^+\dot{N}\!\!\underset{\text{methyl viologen radical cation}\atop(MV^+)}{\bigcirc\!\!=\!\!\bigcirc}\!\!N-CH_3$$

(18-32)

$$MV^+ + Mn(III)(\text{in enzyme}) \rightarrow MV^{2+} + Mn(II)(\text{in enzyme}) \quad (18\text{-}33)$$

Separation of anode and cathode reactions

In Figure 18-11, the reactive species (Br_2) is generated at the anode (the working electrode). The cathode products (H_2 from solvent and Hg from the catalyst) do not interfere with the reaction of Br_2 and cyclohexene. Therefore, the cathode can be in the same compartment with the analyte. In some cases, the H_2 or Hg might react with the analyte. Frequently, then, it is desirable to separate the electrolysis product of the auxiliary electrode from the bulk solution. This can be done with the cell in Figure 18-12. An oxidizing agent, for instance, might be generated at the left-hand electrode operating as an anode. Gaseous H_2 produced at the cathode bubbles innocuously out of the cathode chamber without mixing with the bulk solution.

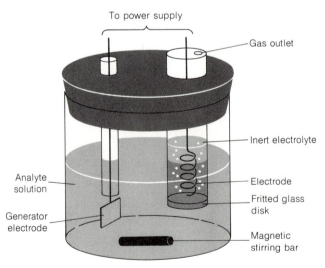

Figure 18-12
Cell showing how one electrode can be isolated from the analyte. Electrical contact is made through the porous fritted glass disk.

Counting Coulombs

In coulometric titrations, a constant current is delivered whenever the power supply is connected to the electrodes. The power supply is built to automatically measure the time of operation. Multiplying current by the length of time gives the number of coulombs used in the titration. In controlled-potential coulometry, the current decreases with time. The power supply contains a circuit that automatically integrates Equation 18-26 to calculate the number of coulombs delivered.

In the absence of a fancy power supply, you can use any common power supply (even a dry cell) and integrate Equation 18-26 chemically. A **chemical coulometer** uses a chemical reaction to measure the total number of electrons passing through the coulometer. The device is attached in series with the analytical experiment. For every electron passing through the analytical apparatus, one electron must pass through the coulometer.

Figure 18-13 shows a chemical coulometer in which the quantity of gas collected in the buret is directly related to the number of coulombs that has

A chemical coulometer is attached in series with the coulometric analysis. For every electron passing through the analytical system, one electron passes through the coulometer.

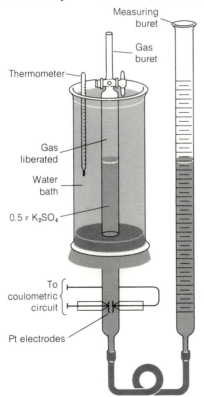

Figure 18-13
A gas coulometer. The tubing at the bottom allows the level of electrolyte in both burets to be kept the same.
[Adapted from J. J. Lingane, *J. Amer. Chem. Soc.*, **67**, 1916 (1945).]

18 / ELECTROGRAVIMETRIC
AND COULOMETRIC ANALYSIS

passed through the electrodes at the base of the buret. The electrolyte in the gas buret is 0.5 M K_2SO_4, and the electrochemical reactions are

$$\text{Coulometer anode:} \qquad \tfrac{1}{2}H_2O \rightarrow \tfrac{1}{4}O_2(g) + H^+ + e^- \qquad (18\text{-}34)$$

$$\text{Coulometer cathode:} \quad H_2O + e^- \rightarrow \tfrac{1}{2}H_2(g) + OH^- \qquad (18\text{-}35)$$

For every mole of electrons flowing through the coulometer, three-fourths of a mole of gas ($\tfrac{1}{2}$ mole of H_2 plus $\tfrac{1}{4}$ mole of O_2) is liberated. The gas is collected in a temperature-controlled buret and displaces an equal volume of water into the measuring buret on the right in Figure 18-13.

Summary

In electrolysis, an endergonic chemical reaction is forced to occur by the flow of electricity through a cell. The moles of electrons flowing through the cell are simply It/F, where I is current, t is time, and F is the Faraday constant. The potential that must be applied to an electrolysis cell is greater than predicted by the Nernst equation because of three factors:

1. Ohmic potential ($=$ IR) is that voltage needed to overcome the resistance of the cell itself.

2. Concentration polarization describes the condition in which the concentration of electroactive species near an electrode is not the same as the concentration in bulk solution. It always opposes the desired reaction and requires that a greater potential be applied.

3. Overpotential describes that voltage required, in addition to those above, to overcome the kinetic barrier (activation energy) of an electrode reaction. Overpotential increases as the rate of reaction increases.

In electrogravimetric analysis, the analyte is electrolytically deposited on an electrode, whose increase in mass is then measured. In a typical electrolysis, a small residual current flows at a level below that of the decomposition potential; above this, rapid reaction occurs. With a constant potential applied between the working and auxiliary electrodes, electrolysis is not very selective. Greater selectivity results from a constant potential being maintained between the working and reference electrodes. The most rapid and least selective electrolysis results from a constant current between the working and auxiliary electrodes.

Coulometry comprises a series of analytical techniques in which the quantity of electricity needed to carry out a chemical reaction is used to measure the quantity of analyte present. These techniques are particularly well suited to automation. Coulometric titrations are performed with a constant current. Measuring the time needed for complete reaction thus directly measures the number of electrons consumed. The endpoint can be found by any conventional means, but amperometry is especially appropriate in many cases. Reactive mediators are often required to transfer electrons between the electrode and the analyte. Controlled-potential coulometry is inherently more selective than constant-current coulometry, but the reaction never reaches completion. Measuring the electrons consumed in the reaction is accomplished by electronic integration of the current-versus-time curve or by a chemical coulometer connected in series with the analytical cell.

Terms to Understand

ampere
amperometry
auxiliary electrode
chemical coulometer
concentration polarization
controlled-potential electrolysis
coulomb

coulometric titration
coulometry
decomposition potential
depolarizer
electroactive species
electrogravimetric analysis
electrolysis

mediator
ohmic potential
overpotential
potentiostat

reference electrode
residual current
working electrode

Exercises

18-A. Suppose that the cell below delivers a constant potential of 1.02 V and is used to operate a light bulb with a resistance of 2.8 Ω.

$$Zn(s)|Zn^{2+}(aq)\|Cu^{2+}(aq)|Cu(s)$$

How many hours are required for 5.0 g of Zn to be consumed?

18-B. A dilute Na_2SO_4 solution is to be electrolyzed with a pair of smooth Pt electrodes at a current density of 100 A m^{-2} and a current of 0.100 A. Calculate the potential that must be applied to cause electrolysis if the cell resistance is 2.00 Ω and there is no concentration polarization. Assume that $H_2(g)$ and $O_2(g)$ are both produced at 1.00 atm. What would your answer be if the Pt electrodes were replaced by Au electrodes?

18-C. (a) What percent of 0.10 M Cu^{2+} could be reduced electrolytically before 0.010 M SbO^+ in the same solution begins to be reduced at pH 0.00? Consider the reaction

$$SbO^+ + 2H^+ + 3e^- \rightleftharpoons Sb(s) + H_2O$$

$$E^\circ = 0.212 \text{ V}$$

(b) What will be the cathode potential (measured with respect to a silver–silver chloride electrode saturated with KCl) when Sb(s) deposition commences?

18-D. Calculate the cathode potential (versus S.C.E.) needed to reduce the total concentration of cobalt(II) to 1.0 μM in each solution below. In each case, Co(s) is the product of reaction.
(a) A solution containing 0.10 M $HClO_4$.
(b) A solution containing 0.10 M $C_2O_4^{2-}$. Use the reaction

$$Co(C_2O_4)_2^{2-} + 2e^- \rightleftharpoons Co(s) + 2C_2O_4^{2-}$$

$$E^\circ = -0.474 \text{ V}$$

This question is asking you to find the potential at which $[Co(C_2O_4)_2^{2-}]$ will be reduced to 1.0 μM.
(c) A solution containing 0.10 M EDTA at pH 7.00.

18-E. Ions that react with Ag^+ can be determined electrogravimetrically by deposition on a silver anode:

$$Ag(s) + X^- \rightarrow AgX(s) + e^-$$

(a) What will be the final mass of a silver anode used to electrolyze 75.00 mL of 0.023 80 M KSCN if the initial mass of the anode is 12.463 8 g?
(b) At what anode potential (versus S.C.E. cathode) will 0.10 M Br^- be deposited as AgBr(s)?
(c) Is it theoretically possible to separate 99.99% of 0.10 M KI from 0.10 M KBr by controlled anode potential electrolysis?

18-F. The galvanic cell below can be used to measure the concentration of O_2 in gases.[†] Oxygen is quantitatively reduced as it passes through the porous Ag bubbler, and Cd is oxidized to Cd^{2+} to complete

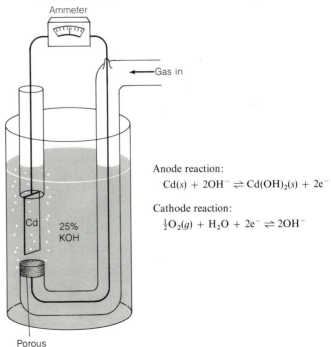

Anode reaction:
$$Cd(s) + 2OH^- \rightleftharpoons Cd(OH)_2(s) + 2e^-$$

Cathode reaction:
$$\tfrac{1}{2}O_2(g) + H_2O + 2e^- \rightleftharpoons 2OH^-$$

[†] F. A. Keidel, *Ind. Eng. Chem.*, **52**, 491 (1960).

the cell. Suppose that gas at 293 K and 1.00 atm is bubbled through the cell at a constant rate of 30.0 mL/min. Calculate the current that would be measured if the gas contains 1.00 ppt or 1.00 ppm (vol/vol) O_2.

18-G. An interesting ligand for coulometric complexometric titrations is R-(−)-1,2-propylenediaminetetraacetic acid (PDTA), an optically active relative of EDTA.[†]

$$\begin{array}{c} N(CH_2CO_2H)_2 \\ | \\ C \text{—} H \\ / \quad \backslash \\ H_3C \quad CH_2N(CH_2CO_2H)_2 \end{array}$$

R-(−)-1,2-propylenediaminetetraacetic acid (PDTA)

This chelate has a negative optical rotation at 365 nm and gives metal complexes with positive optical rotations. The reaction of PDTA with a metal can be studied by observing changes in the optical rotation of the solution. A schematic coulometric titration curve is shown at right.

The initial solution contains PDTA and excess Hg^{2+}. Between points A and B, the cathode current reduces excess Hg^{2+} to Hg(l), which has no effect on the optical rotation. Between points B and C,

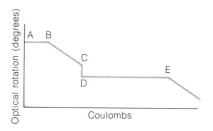

[†] R. A. Gibbs and R. J. Palma, Sr., *Anal. Chem.*, **48**, 1983 (1976).

$Hg(PDTA)^{2-}$ is reduced to Hg(l) plus $PDTA^{4-}$. This decreases the optical rotation. A large amount of $Hg(PDTA)^{2-}$ is still in solution at point C. When analyte solution containing, say, Zn^{2+} is added, the Zn^{2+} displaces Hg^{2+} from PDTA, changing the optical rotation from C to D. Then coulometric reduction of the liberated Hg^{2+} is continued beyond point D. At point E, the free Hg^{2+} is used up and $Hg(PDTA)^{2-}$ begins to be reduced, lowering the optical rotation once again. In a typical experiment, 2.000 mL of Zn^{2+} was added to the cell at point C. The coulombs measured in each region are

AB	2.60 C
BC	3.89 C
DE	14.47 C

Calculate the molarity of Zn^{2+} in the unknown.

Problems

18-1. What is the difference between a galvanic cell and an electrolysis cell?

18-2. Why does overpotential increase with current density?

18-3. State the difference between an electrolysis conducted with a constant *applied* potential and one conducted with a constant *cathode* (or *anode*) potential.

18-4. Explain how the amperometric endpoint detector operates in Figure 18-11.

18-5. How many hours are required for 0.100 mol of electrons to flow through a circuit if the current is 1.00 A?

18-6. The two reactions occurring in the electrolysis cell at right are

$$Mn(s) \rightarrow Mn^{2+} + 2e^-$$
$$M^{3+} + 3e^- \rightarrow M(s)$$

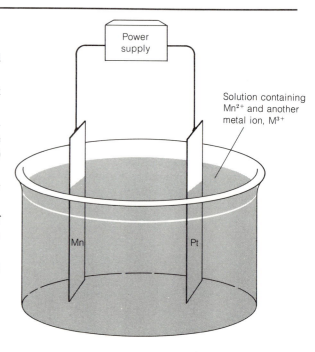

The initial volume of the cell is 1.00 L, and the initial concentration of Mn^{2+} is 0.025 0 M.
(a) Is the Mn electrode the anode or the cathode?
(b) A constant current of 2.60 A was passed through the cell for 18.0 minutes, causing 0.504 g of the metal M to plate out on the Pt electrode. What is the atomic weight of M?
(c) What will be the concentration of Mn^{2+} in the cell at the end of the experiment?

18-7. The Weston cell shown below is a very stable source of potential used as a voltage standard in potentiometers. (The potentiometer compares the unknown input potential to the potential of the

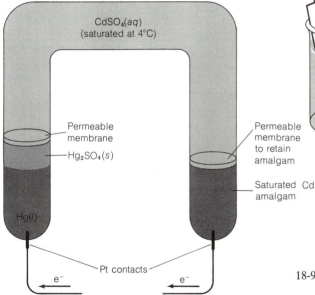

A saturated Weston cell contains solid $CdSO_4$ and is therefore saturated with this salt at room temperature. The saturated cell is a more precise voltage standard than the unsaturated cell, but it is more sensitive to temperature and mechanical shock and cannot be easily incorporated into portable electronic equipment.

standard. In contrast to the conditions of this problem, very little current may be drawn from the cell if it is to be an accurate voltage standard.)
(a) How much work (J) can be done by the Weston cell if the potential is 1.02 V and 1.00 mL of Hg (density = 13.53 g/mL) is deposited?
(b) Suppose that the cell is used to pass current through a 100 Ω resistor. If the heat dissipated by the resistor is 0.209 J/min, how many grams of cadmium are oxidized each hour? Part b is not meant to be consistent with part a. That is, the voltage is no longer 1.02 volts, and you do not know what the voltage is.

18-8. The electrolysis cell below was run at a constant current of 0.021 96 A. On one side, 49.22 mL of H_2 was produced (at 303 K and 0.983 atm); on the other side, Cu metal was oxidized to Cu^{2+}.

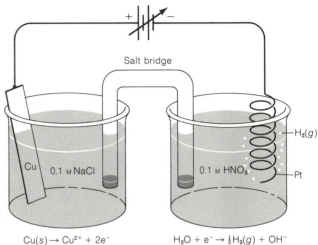

(a) How many moles of H_2 were produced?
(b) If 47.36 mL of EDTA were required to titrate the Cu^{2+} produced by the electrolysis, what was the molarity of the EDTA?
(c) For how many hours was the electrolysis run?

18-9. Consider the cell below, whose resistance is 3.50 Ω.

$Pt(s)|Fe^{2+}(0.10\ M),\ Fe^{3+}(0.10\ M),\ HClO_4(1\ M)\|$
$Ce^{3+}(0.050\ M),\ Ce^{4+}(0.10\ M),\ HClO_4(1\ M)|Pt(s)$

Suppose that there is no concentration polarization or overpotential.
(a) Calculate the potential of the galvanic cell if it produces 30.0 mA.
(b) Calculate the potential that must be applied to run the reaction in reverse, as an electrolysis, at 30.0 mA.

18-10. Suppose that the galvanic cell in Problem 18-9 delivers 100 mA under the following conditions: $[Fe^{2+}]_s = 0.050\ M$, $[Fe^{3+}]_s = 0.160\ M$, $[Ce^{3+}]_s = 0.180\ M$, and $[Ce^{4+}]_s = 0.070\ M$. Considering the ohmic potential and concentration polarization, calculate the cell potential.

18-11. An unknown weighing 0.326 8 g and containing lead lactate $[Pb(CH_3CHOHCO_2)_2]$ plus inert material was electrolyzed to produce 0.111 1 g of PbO_2. Was the PbO_2 deposited at the anode or the cathode? Find the weight percent of lead lactate in the unknown.

18-12. A solution of Sn^{2+} is to be electrolyzed to reduce the Sn^{2+} to $Sn(s)$. Calculate the cathode potential (versus S.C.E.) needed to reduce the Sn^{2+} concentration to 1.0×10^{-8} M if no concentration polarization occurs. Would the potential be more positive or more negative if concentration polarization occurred?

18-13. Calculate the initial potential that should be applied to electrolyze 0.010 M $Zn(OH)_4^{2-}$ in 0.10 M NaOH using Ni electrodes. Assume that the current is 0.20 A, the anode current density is 100 A m^{-2}, the cell resistance is 0.35 Ω, and O_2 is evolved at 0.20 atm. The reactions are

Cathode: $\quad Zn(OH)_4^{2-} + 2e^- \rightleftharpoons Zn(s) + 4OH^-$

$$E^0 = -1.214 \text{ V}$$

Anode: $\qquad\qquad H_2O \rightleftharpoons \tfrac{1}{2}O_2 + 2H^+ + 2e^-$

18-14. The free energy change for the formation of $H_2(g) + \tfrac{1}{2}O_2(g)$ from $H_2O(l)$ is $\Delta G^0 = +237.19$ kJ.
 (a) Calculate the standard potential needed to decompose water into its elements by electrolysis.
 (b) Explain why the cell in Box 18-1 is able to electrolyze water with an applied potential of just -0.3 V.

18-15. The sensitivity of a coulometer is governed by the delivery of its minimum current for its minimum time. Suppose that 5 mA can be delivered for 0.1 s.
 (a) To how many moles of electrons does this correspond?

(b) How many milliliters of a 0.01 M solution of a two-electron reducing agent are required to deliver the same number of electrons?

18-16. A mixture of trichloroacetate and dichloroacetate can be analyzed by selective reduction in a solution containing 2 M KCl, 2.5 M NH_3, and 1 M NH_4Cl. At a mercury cathode potential of -0.90 V (versus S.C.E.), only trichloroacetate is reduced:

$$Cl_3CCO_2^- + H_2O + 2e^- \rightarrow$$
$$Cl_2CHCO_2^- + OH^- + Cl^-$$

At a potential of -1.65 V, dichloroacetate will react:

$$Cl_2CHCO_2^- + H_2O + 2e^- \rightarrow$$
$$ClCH_2CO_2^- + OH^- + Cl^-$$

A hygroscopic mixture of trichloro- and dichloroacetic acid containing an unknown quantity of water weighed 0.721 g. Upon controlled potential electrolysis, 224 C passed at -0.90 V, and 758 C was required to complete the electrolysis at -1.65 V. Calculate the weight percent of each acid in the mixture.

18-17. H_2S in aqueous solution can be analyzed by titration with coulometrically generated I_2:

$$H_2S + I_2 \rightarrow S(s) + 2H^+ + 2I^-$$

To 50.00 mL of sample was added 4 g of KI. Electrolysis required 812 s at a constant current of 52.6 mA. Calculate the concentration of H_2S (μg/mL) in the sample.

19 / Voltammetry

A sophisticated collection of analytical techniques is based on **voltammetry.** These methods are based on observing the relationship between voltage and current during electrochemical processes. The major subdivision of voltammetry is *polarography*, a highly sensitive electroanalytical technique that is especially useful for trace analysis. A second major classification within voltammetry is *amperometry*, which was introduced in conjunction with coulometric endpoint detection in Chapter 18.

19-1 POLAROGRAPHY

Consider an experiment in which analyte is reduced or oxidized at the working electrode of an electrolysis cell. In **polarography** the current flowing through the cell is measured as a function of the potential of the working electrode. Usually, this current is proportional to the concentration of analyte. The most sensitive polarographic procedures have a detection limit near 10^{-9} M and a precision around 5%. Less sensitive polarographic methods operating with $\sim 10^{-3}$ M analyte are capable of a precision of a few tenths of a percent, though 2–3% is most common.

In *polarography*, we measure the current as a function of the potential of the working electrode.

Apparatus for a direct-current polarographic experiment is shown in Figure 19-1. The working electrode is a mercury droplet suspended from the bottom of a glass capillary tube. Analyte is either reduced or oxidized at the surface of the mercury drop. The current-carrying auxiliary electrode is a platinum wire, and for reference a saturated calomel electrode is used. The potential of the mercury drop is measured with respect to the calomel electrode, through which negligible current flows.

19 / VOLTAMMETRY

The experiment in Figure 19-1 was first done by J. Heyrovsky in 1922. For his pioneering work in polarography, Heyrovsky received the Nobel Prize in 1959.

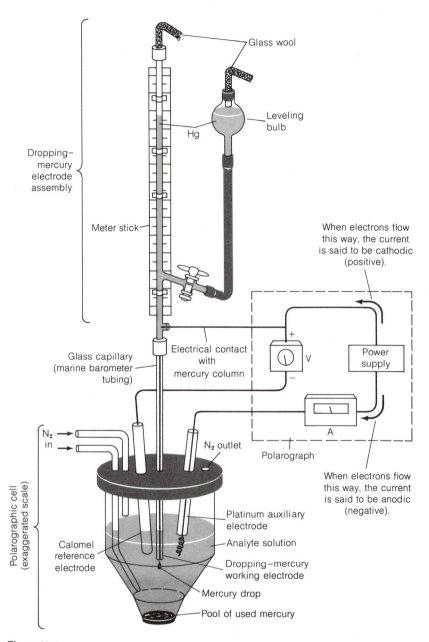

Figure 19-1
Apparatus for polarography. Note the definitions of the directions of positive and negative current flow.

Why We Use the Dropping-Mercury Electrode

At first sight, the **dropping-mercury electrode** in Figure 19-1 seems a little strange. It consists of a very small diameter capillary through which mercury drips from an adjustable reservoir. The height of mercury, measured from the outlet of the capillary tube, is typically ~ 30 cm. With a capillary length of 10–20 cm and a capillary diameter of 0.05 mm, drops of mercury with a diameter of 0.5 mm are formed at a rate of 10–20 min^{-1}. The drop interval is therefore 3–6 seconds. The drop rate is controlled by raising or lowering the leveling bulb at the top right of Figure 19–1.

The reason for using the dropping-mercury electrode is its ability to yield reproducible current-potential data. This reproducibility can be attributed to the continuous exposure of fresh surface on the growing mercury drop. With any other electrode (such as Pt in various forms), the potential depends on its surface condition and therefore on its previous treatment.

> A *dropping-mercury electrode* is used because fresh Hg is continuously exposed to the analyte. This gives more reproducible behavior than does a static surface, whose characteristics change with use.

The vast majority of reactions studied with the mercury electrode are reductions. At a Pt surface, reduction of solvent is expected to compete with reduction of many analyte species, especially in acidic solutions:

$$2H^+ + 2e^- \rightarrow H_2(g) \qquad E^0 = 0 \qquad (19\text{-}1)$$

But Table 18-1 shows that there is a rather large *overpotential* for reduction of H^+ at the Hg surface. Therefore, thermodynamically more difficult reactions than 19–1 can be carried out without competitive reduction of H^+. In neutral or basic solutions, even alkali metal cations can be reduced more easily than H^+, despite their lower standard potentials. This is partly because the potential for reduction of a metal forming a mercury amalgam is higher than its potential for reduction to the solid state:

> Another reason for using a mercury electrode is that there is a high overpotential for H^+ reduction at the mercury surface. Therefore, H^+ reduction does not interfere with many reductions.

> It is easier to reduce most metals to their amalgam than to the solid metal.

$$K^+ + e^- \rightarrow K(s) \qquad E^0 = -2.925 \text{ V} \qquad (19\text{-}2)$$

$$K^+ + e^- + Hg \rightarrow K(\text{in Hg}) \qquad E^0 = -1.971 \text{ V} \qquad (19\text{-}3)$$

A mercury electrode is not very useful for performing oxidations, because Hg is too easily oxidized. For most oxidations, some other working electrode must be employed. In a noncomplexing medium, Hg is oxidized near $+0.25$ V (versus S.C.E.). If the concentration of the complexing ion, Cl^-, is 1 M, the Hg oxidation potential is near 0.0 V. The oxidation is made easier by the stabilization of the Hg(II) product:

> *Question:* What do these potentials imply about the relative stabilities of K(s) and K(in Hg)?

$$Hg(l) + 4Cl^- \rightleftharpoons HgCl_4^{2-} + 2e^- \qquad (19\text{-}4)$$

The capillary of the dropping-mercury electrode will function for years if it is not allowed to clog. A flow of mercury should always be started prior to immersing the capillary in any solution. At the conclusion of an experiment, the electrode is raised and rinsed with distilled water with mercury still flowing. The clean electrode is stored either dry or in a pool of liquid mercury.

> Respect your local electrode!

19-2 SHAPE OF THE POLAROGRAM

A graph of current versus potential in a polarographic experiment is called a **polarogram.** In Figure 19-2, we see the result of the polarographic reduction of Cd^{2+} in HCl solution.

$$Cd^{2+} + 2e^- \rightleftharpoons Cd(s, \text{ in Hg}) \qquad (19\text{-}5)$$

When the potential is only slightly negative with respect to the calomel electrode, essentially no reduction of Cd^{2+} occurs. Only a small **residual current** flows. At a sufficiently negative potential (the *decomposition potential*), reduction of Cd^{2+} commences and the current increases. The reduced Cd dissolves in the Hg to form an amalgam. After a steep increase in current, **concentration polarization** sets in: The rate of electron transfer becomes limited by the rate at which Cd^{2+} can diffuse from bulk solution to the surface of the electrode. The magnitude of this **diffusion current** (I_d) is proportional to Cd^{2+} concentration and is used for quantitative analysis. The upper trace in Figure 19-2 is called a **polarographic wave.**

When the potential is sufficiently negative, around -1.2 V, reduction of H^+ begins and the curve rises steeply. This can be seen at the right-hand edge of Figure 19-2. At positive potentials (near the left side of the polarogram), oxidation of the Hg electrode produces a negative current. *By convention, a negative current means that the working electrode is behaving as the*

Residual current and decomposition potential appear in Section 18-3 in connection with electrogravimetric analysis.

The magnitude of the diffusion current is proportional to the concentration of analyte. This is why polarography is used for quantitative analysis.

Speaking of "reduction of H^+" is equivalent to speaking of "reduction of H_2O."

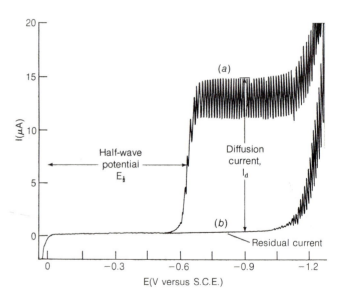

Figure 19-2
Polarograms. (a) 5×10^{-4} M Cd^{2+} in 1 M HCl. (b) 1 M HCl alone. Note that the scale of current is μA. [D. T. Sawyer and J. L. Roberts, Jr., *Experimental Electrochemistry for Chemists* (New York: Wiley, 1974).]

anode with respect to the auxiliary electrode. A positive current means that the working electrode is behaving as the cathode.

The oscillating current in Figure 19-2 is due to the growth and fall of the Hg drops. As each drop begins to form, there is very little Hg surface and correspondingly little current can flow. As the drop grows, its area increases, more solute can reach the surface in a given time, and more current flows. The current increases as the drop grows until, finally, the drop falls off and the current decreases sharply.

Diffusion Current

When the potential of the working electrode is sufficiently negative, the rate of reduction of Cd^{2+} ions in Equation 19-5 is governed by the rate at which Cd^{2+} can reach the electrode. In Figure 19-2, this occurs at potentials more negative than -0.7 V. When the rate of reduction is controlled by the rate of diffusion of analyte to the electrode, the limiting current is called the **diffusion current.**

The rate of diffusion of a solute from bulk solution to the surface of the electrode is proportional to the concentration difference between the two regions:

> The diffusion current is the limiting current when the rate of electrolysis is controlled by the rate of diffusion of species to the electrode.

$$\text{Current} \propto \text{Rate of diffusion} \propto [C]_0 - [C]_s \qquad (19\text{-}6)$$

> The symbol $\propto$ is read "is proportional to."

where $[C]_0$ is the concentration in bulk solution and $[C]_s$ is the concentration at the surface of the electrode (See Figure 18-2). The greater the difference in concentrations, the more rapid will be the diffusion. At a sufficiently negative potential, the reduction is so fast that $[C]_s \ll [C]_0$ and Equation 19-6 reduces to the form

$$\text{Limiting current} \equiv \text{Diffusion current} \propto [C]_0 \qquad (19\text{-}7)$$

The ratio of the diffusion current to the bulk-solute concentration is the basis for the use of polarography in analytical chemistry.

The magnitude of the diffusion current, *measured at the top of each oscillation* in Figure 19-2, is given (to an accuracy of a few percent) by the Ilkovič equation:[†]

> Equations 19-6 and 19-7 should actually be formulated in terms of activities instead of concentrations. However, in the presence of a large quantity of supporting electrolyte, the ionic strength is constant, so the activity coefficients are also constant. Under these conditions, both equations are valid in terms of concentrations.

$$I_d = (7.08 \times 10^4)n\text{C}D^{1/2}m^{2/3}t^{1/6} \qquad (19\text{-}8)$$

where

I_d = diffusion current, measured at the top of the oscillations in Figure 19-2, with the dimensions μA

[†] In the older literature, current was measured at the center of each oscillation, using a damped ballistic galvanometer. In that case, the constant in Equation 19-8 should be 6.07×10^4. A. J. Bard and L. R. Faulkner, *Electrochemical Methods* (New York: Wiley, 1980), pp. 147–150.

19 / VOLTAMMETRY

Table 19-1

Influence of supporting electrolyte on the limiting current for reduction of Pb^{2+}

Electrolyte concentration (M)	Limiting current (μA)	
	KCl	KNO$_3$
0	17.6	17.6
0.000 1	16.3	16.2
0.000 2	15.9	15.0
0.000 5	13.3	13.4
0.001	11.8	12.0
0.005	9.8	9.8
0.1	8.35	8.45
1.0	8.00	8.45

Note: 50 mL of 9.5×10^{-4} M PbCl$_2$ was analyzed at 25°C with 0.2 mL of 0.1% (wt/wt) sodium methyl red present as a maximum suppressor.

SOURCE: Data from I. M. Kolthoff and J. J. Lingane, *Polarography*, Vol. I (New York: Wiley, 1952), p. 123.

Under proper, analytically useful conditions, the limiting current should be proportional to $\sqrt{h}$. Otherwise, effects other than diffusion are controlling the rate of reaction. See page 446.

Three ways that electrolytes get to the electrode:
1. Diffusion (desired in polarography).
2. Mechanical transport (not desired).
3. Electrostatic attraction (not desired).

Electrostatic attraction is minimized by using a high concentration of supporting electrolyte.

n = number of electrons per molecule involved in the oxidation or reduction of the electroactive species

C = concentration of electroactive species, with the dimensions mmol/L

D = diffusion coefficient of electroactive species, with the dimensions $m^2\ s^{-1\dagger}$

m = rate of flow of Hg, in mg s^{-1}

t = drop interval, in s

The number 7.08×10^4 is a combination of several constants whose dimensions are such that I_d will be given in μA.

Clearly, the magnitude of the diffusion current depends on several factors in addition to analyte concentration. The quantity $m^{2/3}t^{1/6}$ in Equation 19-8 is called the **capillary constant**, and is very nearly proportional to the square root of the Hg column height (h), measured from the top of the Hg meniscus to the bottom of the Hg electrode in Figure 19-1. To demonstrate that the limiting current is indeed *diffusion*-controlled, you can measure the current at various heights of the Hg column and see if the current is proportional to $\sqrt{h}$.

In quantitative polarography, it is important to control the temperature within a few tenths of a degree. This is because the diffusion coefficient for most species increases by about 2% per degree. If the capillary constant is to be used, it should be measured at the same potential used to measure I_d, since t depends on the applied potential.

The transport of solute to the electrode depends on three factors: diffusion, mechanical transport (stirring and convection), and electrostatic attraction. In polarography, we try to minimize the latter two mechanisms. We want the limiting current to be controlled solely by the rate of diffusion of Cd^{2+} to the electrode.

To minimize mechanical transport, the solution is *not* stirred, and effort is made to reduce vibrations. Setting the apparatus on a heavy base (such as the marble slab used for sensitive analytical balances) helps to reduce the effect of vibrations of the building.

Electrostatic attraction (or repulsion) of analyte ions by the electrode is reduced to a negligible level by the presence of a high concentration of **supporting electrolyte** (1 M HCl in Figure 19-2). As shown in Table 19-1, increasing concentrations of electrolyte reduce the net current flow, since the rate of arrival of cationic analyte at the negative Hg surface is decreased.

[†] The diffusion coefficient is defined from Fick's first law of diffusion: The rate (J) at which molecules diffuse across a plane of unit area is given by

$$J = -D\frac{dc}{dx}$$

where D is the diffusion coefficient and dc/dx is the gradient of concentration in the direction of diffusion. The larger the diffusion coefficient, the more rapidly the molecules diffuse.

Typically, a supporting electrolyte concentration 50–100 times greater than the analyte concentration will reduce electrostatic transport of the analyte to a negligible level.

Residual Current

To measure the value of the diffusion current in a polarogram, we subtract the value of the **residual current** from the observed limiting current in the diffusion-controlled region. This difference is labeled I_d in Figure 19-2, where the magnitude of the residual current is quite small. At lower analyte concentration, the proportion of residual current will be greater, since residual current will be the same, while diffusion current decreases. For accurate work, you should always measure the residual current of a solution containing the same supporting electrolyte as your sample. The reagents should come from the same stock solutions, since different batches will have different impurities contributing to the residual current.

An enlarged plot of the residual current of a 0.1 M HCl solution is shown in Figure 19-3. The dashed lines show that the residual current has an

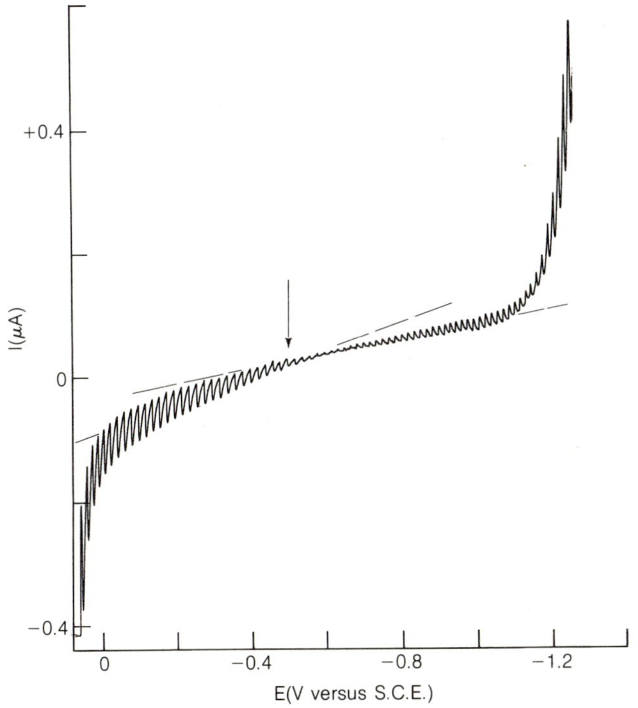

> *Challenge:* See if you can understand why the presence of supporting electrolyte will *increase* the limiting current for reduction of an anion, such as IO_3^-.

I_d = Limiting current $-$ Residual current

Figure 19-3
Residual current of 0.1 M HCl. The arrow marks the inflection point that occurs at the electrocapillary maximum. This occurs when the charge of the Hg drop is zero with respect to the solution. If there were no faradaic current, the net current would be zero at the electrocapillary maximum. [L. Meites, *Polarographic Techniques*, 2nd ed. (New York: Wiley, 1965).]

434

19 / VOLTAMMETRY

Residual current has two components:
1. Condenser current (due to charging of Hg drops).
2. Faradaic current (due to redox reactions).

inflection point near -0.5 V in this case. At this point, called the **electrocapillary maximum,** the charge on the drop of mercury is zero. At more positive potentials, the charge on the drop is positive with respect to the solution. At more negative potentials, the charge is negative. Because the mercury drop and the solution can have different charges, the mercury–solution interface behaves as a capacitor.

The residual current has two components. One is the current needed to charge or discharge the capacitor formed by the mercury–solution interface. This is called the **condenser current** or **charging current.** It is present in all polarographic experiments, regardless of the purity of reagents. As each drop of mercury falls, it carries its charge with it to the bottom of the cell. The new drop requires more current for charging.

Any current that flows as a result of reduction or oxidation of a species in solution is called a **faradaic current.** The second component of the residual current is a small faradaic current due mainly to the reduction (or oxidation) of impurities in the supporting electrolyte. Since the concentration of supporting electrolyte is very high, the concentration of trace impurities can be significant.

Shape of the Polarographic Wave

The two most common reactions at the dropping Hg electrode involve reduction of an ion to an amalgam or reduction of a soluble ion to another soluble ion:

$$M^{n+} + ne^- + Hg \rightleftharpoons M(Hg) \tag{19-9}$$

$$X^{a+} + ne^- \rightleftharpoons X^{(a-n)+} \tag{19-10}$$

It can be shown for both cases that if the reactions are reversible, the equation relating the current and the potential in the polarographic wave is

$$E = E_{1/2} - \frac{0.059\ 16}{n} \log \frac{I}{I_d - I} \tag{19-11}$$

Recall that a *cathodic current* represents a flow of electrons from the Hg electrode to the analyte. An *anodic current* represents a flow of electrons from the analyte to the electrode.

$E_{1/2}$ is the **half-wave potential,** drawn in Figure 19-2. It is the potential at which $I = \frac{1}{2}I_d$ (both corrected for residual current).

Reversible anodic and composite anodic–cathodic waves have the same shape defined by Equation 19-11. Figure 19-4 shows polarograms of (a) ferric, (b) ferric plus ferrous, and (c) ferrous ions. In curve a, Fe(III) is reduced to Fe(II) at a half-wave potential of $+0.05$ V (versus S.C.E.). Only a residual current is observed at $+0.15$ V, since Fe(III) is not reduced at this potential. Curve c shows an anodic current at $+0.15$ V because Fe(II) is being oxidized. At -0.05 V, only a residual current is observed because no Fe(III) is present to be reduced. Curve b shows an anodic diffusion current at $+0.15$ V due to oxidation of Fe(II) and a cathodic current at -0.05 V due to reduction of

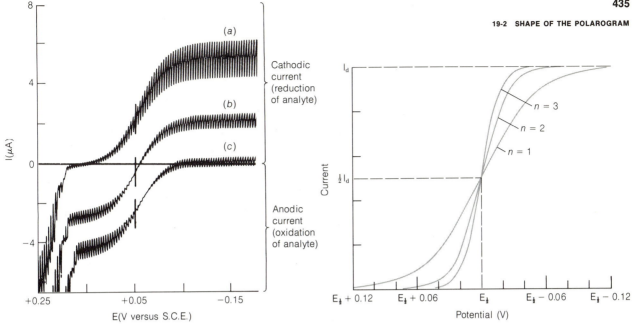

Figure 19-4
Polarograms. (a) 1.4 mM Fe(III). (b) 0.7 mM Fe(III) plus 0.7 mM Fe(II). (c) 1.4 mM Fe(II). Each solution contains saturated oxalic acid as supporting electrolyte and 0.000 2% methyl red as maximum suppressor. Vertical lines are drawn at the observed half-wave potentials. [L. Meites, *Polarographic Techniques*, 2nd ed. (New York: Wiley, 1965).]

Figure 19-5
Graph of Equation 19-11 for $n = 1$, 2, and 3. The greater the value of n, the steeper is the polarographic wave.

Fe(III). All three curves have the same value of $E_{1/2}$ and the same shape.

The graph of Equation 19-11 in Figure 19-5 shows how the shape of a reversible polarographic wave depends on n, the number of electrons in the half-reaction. The larger the value of n, the steeper the polarographic wave. The function in Equation 19-11 predicts that a graph of E versus $\log(I/I_d - I)$ should be linear, with a slope of $-0.059\,16/n$. Such a graph allows us to verify that the reaction is reversible and to find n, the number of electrons involved. The best results are obtained if the effect of cell resistance on the current is also considered.

Relation Between $E_{1/2}$ and E^0

For reversible electrochemical reactions (those with negligible overpotential), there is a relatively straightforward thermodynamic interpretation of the half-wave potential, $E_{1/2}$. Consider Reaction 19-9, in which the product dissolves in mercury. If the reduction product were not stabilized by dissolution in mercury, then we would expect $E_{1/2}$ to be equal to E^0. In fact, the product is stabilized when it dissolves in the mercury, so $E_{1/2} \neq E^0$.

19 / VOLTAMMETRY

Table 19-2
Test of Equation 19-12 for several metals at 25°C

Ion	E^0 (V)	E_s (V)	C (M)	γ	$E_{1/2}$ [Eq. 19-12]	$E_{1/2}$ (observed)
Pb^{2+}	-0.372	0.006	0.96	0.72	-0.371	-0.388
Tl^+	-0.582	0.003	27.4	8.3	-0.440	-0.459
Cd^{2+}	-0.647	0.051	6.40	1.15	-0.570	-0.578
Zn^{2+}	-1.008	0	4.37	0.74	-0.993	-0.997
Na^+	-2.961	0.780	3.52	1.3	-2.14	-2.12
K^+	-3.170	1.001	1.69	5.6	-2.11	-2.14

Note: E^0 and E_s are with respect to S.C.E.
SOURCE: Data from I. M. Kolthoff and J. L. Lingane, *Polarography* (New York: Wiley, 1952), p. 198.

Equation 19-12 tells us that the reaction $M^{n+} + ne^- \rightleftharpoons M(Hg)$ is facilitated if
1. Electrons flow spontaneously from M(s) to M(Hg).
2. M is very soluble in liquid mercury.

The relation between the half-wave potential and E^0 is[†]

$$E_{1/2} \approx E^0 + E_s + \frac{0.059\ 16}{n} \log C\gamma \qquad (19\text{-}12)$$

The symbols in Equation 19-12 have the following meanings: E^0 is the ordinary standard reduction potential for M^{n+} with respect to the reference electrode used in the polarographic cell. For a calomel reference electrode, $E^0 = E^0$ (versus S.H.E.) $- 0.241$ V. E_s is the standard potential for the cell $M(s)|M^{n+}|M(Hg)_{saturated}$, in which the left electrode is solid metal and the right electrode is a saturated amalgam of the same metal. C is the concentration of M in the saturated amalgam, and γ is the activity coefficient of M in the saturated amalgam; therefore, $C\gamma$ is the activity of metal M in the saturated amalgam.

The term E_s in Equation 19-12 measures the tendency of electrons to flow between solid analyte and analyte amalgam. A positive value means that electrons flow spontaneously from pure reduced analyte to reduced analyte dissolved in mercury. The more positive the value of E_s, the easier it will be to reduce the species in the presence of mercury (compared to reduction in the absence of mercury). The log term in Equation 19-12 simply gives the free energy change for dissolving reduced analyte in mercury. The more soluble in mercury the reduced product is, the easier it will be to carry out the reduction.

Table 19–2 gives some data that demonstrate these effects. In the case of Pb^{2+}, both E_s and the log term are small and $E_{1/2}$ is close to E^0. For the other ions in the table, either E_s or the log term is sufficiently large so that $E_{1/2}$ is not very close to E^0.

Now consider the reversible Reaction 19-10, in which the oxidized and reduced species remain in solution. In this case the relation between $E_{1/2}$ and E^0 is given by[‡]

[†] I. M. Kolthoff and J. L. Lingane, *Polarography* (New York: Wiley, 1952), p. 201. Equation 19-12 ignores small terms involving diffusion and activity coefficients.

[‡] L. Meites, *Polarographic Techniques*, 2nd ed. (New York: Wiley, 1965), p. 278.

$$E_{1/2} = E^0 - \frac{0.059\,16}{n} \log \frac{\gamma_{red} D_{ox}^{1/2}}{\gamma_{ox} D_{red}^{1/2}} \tag{19-13}$$

where E^0 is the standard reduction potential for Reaction 19-10, with respect to S.C.E. The activity coefficients of the reduced and oxidized species are γ_{red} and γ_{ox}. The corresponding diffusion coefficients are D_{red} and D_{ox}. The quotient in the log term is usually close to unity, so the logarithm is usually small. Therefore, $E_{1/2}$ for reaction of soluble reactants and products is usually close to E^0 for the redox couple.

For the reaction $X^{a+} + ne^- \rightleftharpoons X^{(a-n)+}$, $E_{1/2}$ is usually close to E^0.

Other Factors Affecting the Shape of the Polarogram

Current maxima

Figure 19-6 shows the polarogram of a mixture of Pb(II) and Zn(II) in 2 M NaOH. The wave near -0.8 V is due to reduction of Pb(II), and the smaller wave near -1.5 V is due to reduction of Zn(II). In trace a, the current "overshoots" the value of I_d, then settles back down. These current maxima are common and have been attributed to convection currents near the surface of the electrode. Adding traces of certain agents, such as gelatin, Triton X-100, or methyl red, generally eliminates this behavior, as shown in trace b of Figure 19-6. These agents are called **maximum suppressors** and apparently alter the convective behavior of the solution.

Small quantities of certain surface-active agents are routinely used to suppress current maxima.

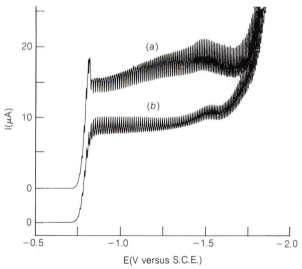

Figure 19-6
Polarograms of 3 mM Pb(II) and 0.25 mM Zn(II) in 2 M NaOH. (a) In the absence of a suppressor. (b) In the presence of 0.002% Triton X-100. [L. Meites, *Polarographic Techniques*, 2nd ed. (New York: Wiley, 1965).]

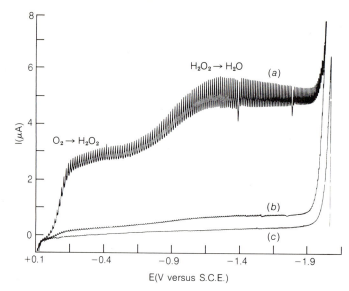

Figure 19-7

Polarogram of 0.1 M KCl. (a) Saturated with air. (b) After partial deaeration. (c) After further deaeration. [L. Meites, *Polarographic Techniques*, 2nd ed. (New York: Wiley, 1965).]

Oxygen

Figure 19-7 shows that oxygen gives rise to a pair of intense polarographic waves. The first wave is due to its reduction to H_2O_2, and the second is from further reduction to H_2O.

$$O_2 + 2H^+ + 2e^- \rightleftharpoons H_2O_2 \quad E_{1/2} \approx -0.1 \text{ V (versus S.C.E.)} \quad (19\text{-}14)$$

$$H_2O_2 + 2H^+ + 2e^- \rightleftharpoons 2H_2O \quad E_{1/2} \approx -0.9 \text{ V (versus S.C.E.)} \quad (19\text{-}15)$$

Since oxygen is dissolved in any solution exposed to air, these waves would be superimposed on the polarogram of the analyte if the oxygen were not removed. Bubbling nitrogen gas through the solution removes enough oxygen to eliminate these waves for ordinary polarographic work. The cell in Figure 19-1 has ports for purging the solution with nitrogen and for maintaining a blanket of nitrogen over the solution during the experiment.

19-3 APPLICATIONS OF POLAROGRAPHY

Polarography serves many purposes in analytical chemistry. One of the most straightforward is in the qualitative identification of an unknown. For a reversible redox process, the half-wave potential is characteristic of the electroactive species (the analyte) and the medium in which it is analyzed. Such factors as analyte concentration and the electrode capillary constant have no effect on $E_{1/2}$. Extensive tables of half-wave potentials exist,[†] so $E_{1/2}$ for an unknown can be compared with known values to try to identify the species by polarography. For irreversible processes, the waves are broader and $E_{1/2}$ is not independent of concentration or capillary characteristics. Nonetheless, the variation in half-wave potential is not large enough to preclude the use of polarography in qualitative analysis.

A unique identification of an unknown cannot be made with a single half-wave potential. However, once a list of suspect ions is compiled, varying the medium usually permits further shortening of the list of possibilities. For example, two ions that have similar reduction potentials in 0.1 M tartrate solution are not likely to have similar potentials in ammonia solution. If a species undergoes more than one redox process, the positions of successive waves (as well as their relative heights) can be very diagnostic.

Many organic functional groups give rise to polarographic waves (Table 19-3). Once again, the half-wave potentials can help to distinguish one possible functional group from another. Table 19-3 might lead you to ask, "Why not do synthetic, organic, redox chemistry by controlled-potential electrolysis?" In fact, many organic reactions can be conducted electrochemically, and a large body of literature based on this technique does exist.[‡]

> For qualitative analysis, the half-wave potential of an unknown is measured in several different complexing media. Comparison with a table of half-wave potentials allows the unknown to be identified.

> Oxidations and reductions can be carried out on a synthetically useful scale by controlled-potential electrolysis, as well as by using oxidizing or reducing agents.

Quantitative Analysis

The principal use of polarography is in quantitative analysis. Since the magnitude of the diffusion current is proportional to the concentration of analyte, the height of a polarographic wave tells how much analyte is present. In the following sections, we will describe the use of *standard curves*, *standard additions*, and *internal standards* for quantitative analysis. These methods are completely general and by no means restricted to polarography; they can be used in conjunction with any quantitative technique, such as potentiometry, spectrophotometry, or chromatography.

Standard curves

The most reliable, but tedious, method of quantitative analysis is to prepare a series of known concentrations of analyte in otherwise identical solutions. A polarogram of each solution is recorded, and a graph of the diffusion current versus analyte concentration is prepared. Finally, a polarogram of the unknown is recorded, using the same conditions. From the measured diffusion current and the standard curve, the concentration of analyte can be determined. Figure 19-8 shows an example of the linear relationship between diffusion current and concentration.

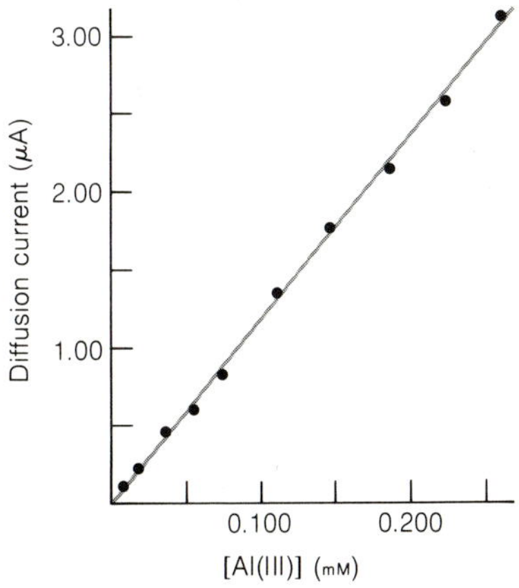

Figure 19-8
Standard curve for polarographic analysis of Al(III) in 0.2 M sodium acetate, pH 4.7, with 0.6 mM pontachrome violet SW used as a maximum suppressor. I_d is corrected for the residual current and for the diffusion current due to reaction of the maximum suppressor. [Data from H. H. Willard and J. A. Dean, *Anal. Chem.*, **22**, 1264 (1950).]

[†] L. Meites, *Handbook of Analytical Chemistry* (New York: McGraw-Hill, 1963), pp. 5-53 to 5-103.

[‡] See N. L. Weinberg, ed., *Technique of Electroorganic Synthesis* (New York: Wiley, 1974); J. Chang, R. F. Large, and G. Popp in A. Weissberger and B. W. Rossiter, eds., *Physical Methods of Chemistry*, Vol. I, Part IIB (New York: Wiley, 1971).

Table 19-3
Polarographic behavior of some functional groups

Group	Reaction

C—C RO_2C—⟨C₆H₄⟩—$CN + H^+ + 2e^- \longrightarrow RO_2C$—⟨C₆H₄⟩—$H + CN^-$

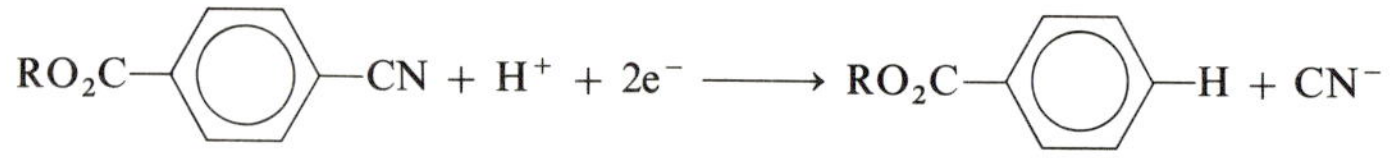

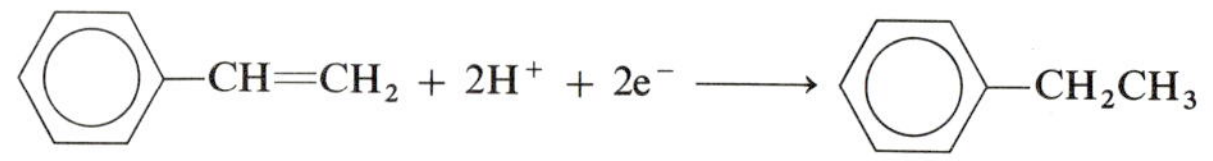

C=C ⟨C₆H₅⟩—$CH=CH_2 + 2H^+ + 2e^- \longrightarrow$ ⟨C₆H₅⟩—CH_2CH_3

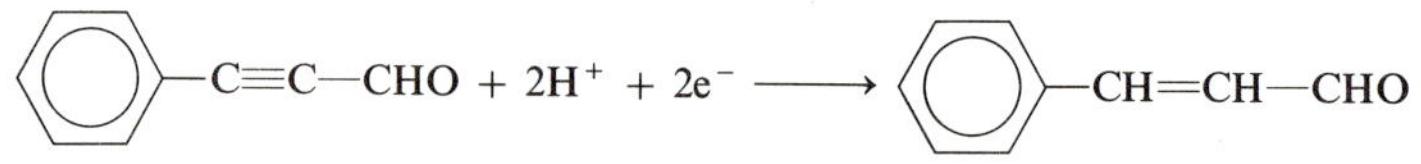

C≡C ⟨C₆H₅⟩—$C\equiv C$—$CHO + 2H^+ + 2e^- \longrightarrow$ ⟨C₆H₅⟩—$CH=CH$—CHO

C—X $RCH_2Br + H^+ + e^- \rightarrow RCH_3 + Br^-$

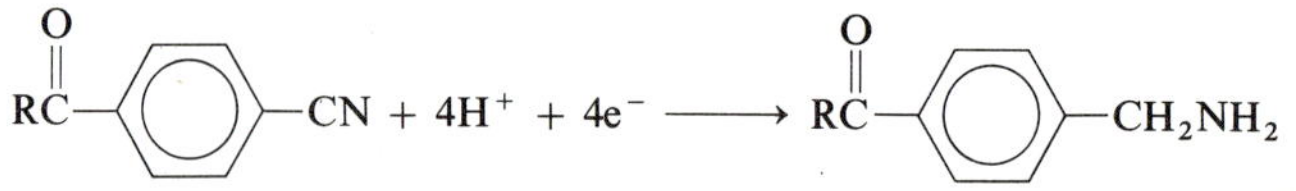

C=O ⟨C₆H₅⟩—$\overset{\overset{\displaystyle O}{\|}}{C}R + 2H^+ + 2e^- \longrightarrow$ ⟨C₆H₅⟩—$\overset{\overset{\displaystyle OH}{|}}{\underset{\underset{\displaystyle H}{|}}{C}}R$

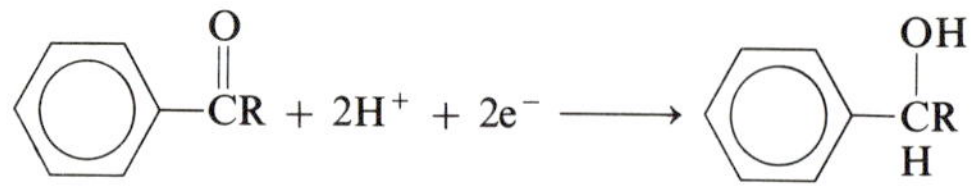

C—N $R\overset{\overset{\displaystyle O}{\|}}{C}CH_2NR_2 + 2H^+ + 2e^- \longrightarrow R\overset{\overset{\displaystyle O}{\|}}{C}CH_3 + HNR_2$

C=N ⟨C₆H₅⟩—$CH=CH$—$C\overset{\displaystyle \diagup NH}{\diagdown R} + 2H^+ + 2e^- \rightarrow$ ⟨C₆H₅⟩—$CH=CH$—$CHRNH_2$

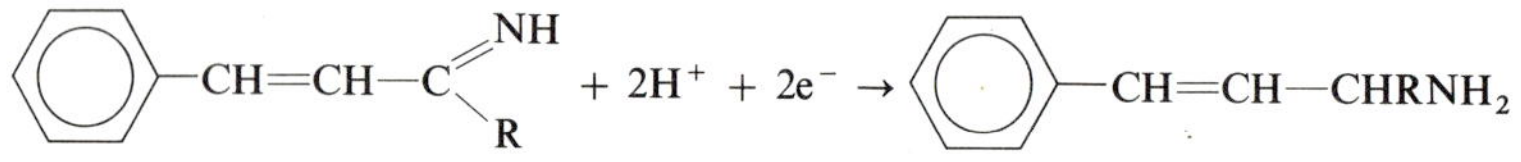

C≡N $R\overset{\overset{\displaystyle O}{\|}}{C}$—⟨C₆H₄⟩—$CN + 4H^+ + 4e^- \longrightarrow R\overset{\overset{\displaystyle O}{\|}}{C}$—⟨C₆H₄⟩—$CH_2NH_2$

N=N ⟨C₆H₅⟩—$N=N$—⟨C₆H₅⟩ $+ 2H^+ + 2e^- \longrightarrow$ ⟨C₆H₅⟩—NH—NH—⟨C₆H₅⟩

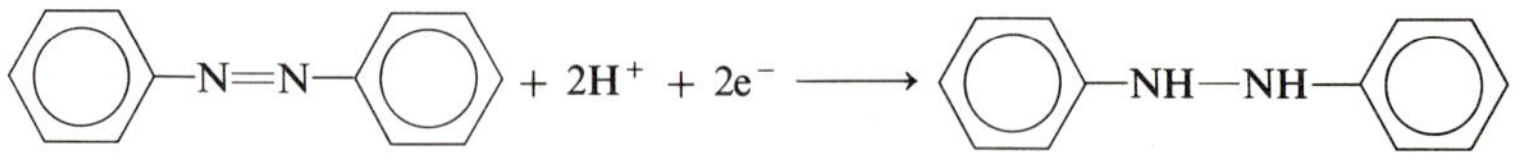

N=O R—$NO + 2H^+ + 2e^- \rightarrow RNHOH$

NO_2 $RNO_2 + 4H^+ + 4e^- \rightarrow RNHOH + H_2O$

O—O $ROOR + 2H^+ + 2e^- \rightarrow 2ROH$

S—S $RSSR + 2H^+ + 2e^- \rightarrow 2RSH$

S=O $R_2S=O + 2H^+ + 2e^- \rightarrow R_2S + H_2O$

EXAMPLE

Suppose that 5.00 mL of an unknown sample of Al(III) was placed in a 100 mL volumetric flask containing 25.00 mL of 0.8 M sodium acetate (pH 4.7) and 2.4 mM ponta-chrome violet SW (a maximum suppressor). After dilution to 100 mL, an aliquot of the solution was analyzed by polarography. The height of the polarographic wave was 1.53 μA, and the residual current—measured at the same potential with a similar solution containing no Al(III)—was 0.12 μA. Find the concentration of Al(III) in the unknown.

The corrected diffusion current is $1.53 - 0.12 = 1.41 \ \mu A$. In Figure 19-8, this current corresponds to $[Al(III)] = 0.126$ mM. Since the unknown was diluted by a factor of 20.0 (from 5.00 mL to 100 mL) for analysis, the original concentration of unknown must have been $(20.0)(0.126) = 2.46$ mM.

Standard addition method

The standard addition method is most useful when the sample composition is unknown or difficult to duplicate in synthetic standard solutions. This method is faster, but usually not as reliable as the method employing a standard curve.

First, a polarogram of the unknown is recorded. Then, a solution containing a known quantity of the analyte is added to the sample. *Assuming that the response is linear*, the increase in diffusion current of this new solution can be used to estimate the amount of unknown in the original solution. For greatest accuracy, several standard additions are made. Section 21-4 describes how a sequence of several consecutive standard additions can be used.

To see how the wave height is affected by the standard addition, consider the first sample containing just the unknown in supporting electrolyte. The diffusion current of the unknown will be proportional to the concentration of unknown, C_x:

$$I_d \text{ (unknown)} = kC_x \tag{19-16}$$

where k is a constant of proportionality. Let the concentration of standard solution be C_s. When V_s mL of standard solution is added to V_x mL of unknown, the diffusion current is the sum of diffusion currents due to the unknown and the standard.

$$I_d \text{ (unknown + standard)} = kC_x\left(\frac{V_x}{V_x + V_s}\right) + kC_s\left(\frac{V_s}{V_x + V_s}\right) \tag{19-17}$$

where the quantities in parentheses are dilution factors. Dividing the left side of Equation 19-17 by I_d (unknown), and dividing the right side by kC_x, allows us to rearrange and solve for C_x:

$$C_x = \frac{C_s V_s}{R(V_x + V_s) - V_x} \tag{19-18}$$

where R is the ratio I_d (unknown + standard)/I_d (unknown). All of the quantities on the right side of Equation 19-18 are known, so C_x can be calculated. The standard addition method is most accurate when the wave height of the combined solution is about twice that of the unknown solution.

In the method of *standard additions*, a known amount of analyte is added to the unknown. The increase in signal intensity tells us how much analyte was present prior to the standard addition.

The method of standard additions relies on a linear relation between signal and concentration. If the polarographic response is not proportional to analyte concentration over the whole range of concentrations employed, there will be an error in the calculated concentration of unknown. It is impossible to detect this with a single standard addition, but several standard additions *might* indicate that the relation is not linear.

To apply the method of standard additions to any other technique, substitute the ratio of whatever signals are measured (such as absorbance or the height of chromatographic peaks) for the ratio I_d(unknown + standard)/I_d(unknown).

442

19 / VOLTAMMETRY

EXAMPLE

A 25.0 mL sample of Ni^{2+} gave a wave height of 2.36 μA (corrected for residual current) in a polarographic analysis. When 0.500 mL of solution containing 28.7 mM Ni^{2+} was added, the wave height increased to 3.79 μA. Find the concentration of Ni^{2+} in the unknown.

Using Equation 19-18, we can write

$$C_x = \frac{(28.7 \text{ mM})(0.500 \text{ mL})}{\left(\dfrac{3.79 \ \mu A}{2.36 \ \mu A}\right)(25.0 + 0.500 \text{ mL}) - 25.0 \text{ mL}} = 0.900 \text{ mM}$$

Use of an internal standard

An *internal standard* is a known amount of a second compound added to the analyte. The ratio of signals due to known and unknown is compared with the ratio of signals from a solution in which both concentrations are known.

This procedure is based on the fact that in a particular medium the diffusion currents due to two different species will have a ratio in direct proportion to their concentration ratio. The half-wave potentials of the two species must be sufficiently far apart ($\gtrsim 0.2$ V) that the limiting current of the first can be measured before the onset of the second wave.

To use this method, a polarogram of a known mixture of analyte (say, Tl^+) plus internal standard (say, Cd^{2+}) must be recorded. This polarogram establishes the relative response to the two species. Next, the unknown (containing an unknown concentration of Tl^+) is mixed with a known quantity of internal standard (Cd^{2+}), and the polarogram is recorded again. Comparing the ratios of diffusion currents in the two experiments tells us the concentration of unknown.

EXAMPLE

Chloroform and the pesticide DDT exhibit the following half-wave potentials in a medium consisting of 0.05 M $(CH_3)_4NBr$ in 3:1 (vol/vol) dioxane/water:

$$CHCl_3 \qquad Cl-\underset{\text{DDT}}{\bigcirc}-\overset{\overset{\displaystyle CCl_3}{|}}{CH}-\bigcirc-Cl$$

chloroform
$E_{1/2} = -1.6$ V

DDT
$E_{1/2} = -0.8$ V

Suppose that a polarogram of a mixture containing 0.500 mM chloroform and 0.800 mM DDT gives the following relative wave heights:

$$\frac{\text{Wave height of } CHCl_3}{\text{Wave height of DDT}} = 1.53 \quad \text{when} \quad \frac{[CHCl_3]}{[DDT]} = \frac{0.500 \text{ mM}}{0.800 \text{ mM}} = 0.625$$

To use chloroform as an *internal standard* for the analysis of DDT, a known concentration of chloroform is added to an unknown solution of DDT. Suppose that when the concentration of internal standard is 0.462 mM, the relative wave heights are

$$\frac{\text{Wave height of } CHCl_3}{\text{Wave height of DDT}} = 1.11 \quad \text{when} \quad \frac{[CHCl_3]}{[DDT]} = \frac{0.462 \text{ mM}}{\text{Unknown}} \equiv x$$

Find the concentration of DDT in the unknown.

To do this, we can set up a ratio. Let x be the quotient $[CHCl_3]/[DDT]$ in the mixture of unknown plus internal standard. When the relative wave heights were 1.53, the relative concentrations were 0.625. When the relative wave heights are 1.11, the relative concentration, x, must be given by

$$\frac{\text{Relative concentrations in unknown}}{\text{Relative wave heights in unknown}} = \frac{\text{Relative concentrations in known}}{\text{Relative wave heights in known}}$$

$$\frac{x}{1.11} = \frac{0.625}{1.53} \Rightarrow x = 0.453$$

But the internal standard, $CHCl_3$, has a concentration of 0.462 mM (after it has been mixed with unknown). Since

$$\frac{[CHCl_3]}{[DDT]} = x = 0.453$$

and since $[CHCl_3] = 0.462$ mM, we find

$$[DDT] = \frac{[CHCl_3]}{x} = \frac{0.462}{0.453} = 1.02 \text{ mM}$$

Internal standards are most useful when loss of sample is unavoidable in the course of a chemical procedure. Suppose that a known amount of internal standard is added to an unknown. During various subsequent operations, much of the sample might be lost or diluted. However, the *ratio* of concentrations of internal standard and unknown remains constant throughout. An analysis performed at the end of the various operations will give the correct ratio of unknown to standard, regardless of any losses that occurred. As with the method of standard additions, the use of internal standards relies on a linear response to both analyte and standard over the entire range of concentrations employed.

> In the older literature, an internal standard in polarography was called a "pilot ion."

Polarographic Study of Chemical Equilibrium

Polarography is one of the more common electrochemical techniques used in measuring the equilibrium constants of reactions. To see why, consider a solution containing 1.00 mM Fe^{3+} and 1.00 mM Fe^{2+}. The cathode potential for reduction of Fe^{3+} is given by

$$E = 0.770 - 0.059\,16 \log \frac{[Fe^{2+}]}{[Fe^{3+}]} = 0.770 \qquad (19\text{-}19)$$

> Any equilibrium that affects either $[Fe^{3+}]$ or $[Fe^{2+}]$ will alter the reduction potential of the solution.

where 0.770 V is the standard reduction potential of Fe^{3+}. Suppose that a ligand that binds only to Fe^{3+} is added to the solution. The cathode potential needed for reduction of Fe^{3+} will no longer be 0.770 V, because the ratio $[Fe^{2+}]/[Fe^{3+}]$ is no longer 1.00. Since the ligand decreases the con-

19 / VOLTAMMETRY

centration of Fe^{3+}, E will become more negative. It will be harder to reduce Fe^{3+} after the ligand has been added than in its absence.

The change in reduction potential can be predicted if we know the stability constant of the metal–ligand complex. Alternatively, measurement of the reduction potential tells us the magnitude of the stability constant. In polarography, the half-wave potential is related to the reduction potential of analyte and will therefore be sensitive to any equilibria involving the analyte. We now present results for two common classes of reactions whose equilibria are amenable to polarographic analysis.

Reversible reduction of one soluble complex to another

We can formulate a general reduction involving two soluble species as

$$ML_p^{(a-pb)} + ne^- \rightleftharpoons ML_q^{(a-n-qb)} + (p-q)L^{-b} \qquad (19\text{-}20)$$

where M = metal

    L = ligand

    a = charge of free metal

    b = charge of free ligand

    n = number of electrons in half-reaction

    p and q are stoichiometry coefficients

For the overall formation constants of ML_p and ML_q, we can write

$$M^a + pL^{-b} \rightleftharpoons ML_p^{(a-pb)} \qquad \beta_p = \frac{[ML_p^{(a-pb)}]}{[M^a][L^{-b}]^p} \qquad (19\text{-}21)$$

$$M^{(a-n)} + qL^{-b} \rightleftharpoons ML_q^{(a-n-qb)} \qquad \beta_q = \frac{[ML_q^{(a-n-qb)}]}{[M^{a-n}][L^{-b}]^q} \qquad (19\text{-}22)$$

In these terms, it can be shown that $E_{1/2}$ for Reaction 19-20 is given approximately by

A graph of $E_{1/2}$ versus $\log[L^{-b}]$ will have

$$\text{Slope} = \frac{-(p-q)(0.059\,16)}{n}$$

$$\text{Intercept} = E^0 - \frac{0.059\,16}{n}\log\frac{\beta_p}{\beta_q}$$

$$E_{1/2} \approx E^0 - \frac{0.059\,16}{n}\log\frac{\beta_p}{\beta_q} - \frac{(p-q)0.059\,16}{n}\log[L^{-b}] \quad (19\text{-}23)$$

where E^0 applies to the reaction

$$M^a + ne^- \rightleftharpoons M^{a-n} \qquad (19\text{-}24)$$

measured in a noncomplexing medium with respect to the same reference electrode used for the polarographic experiment. By measuring $E_{1/2}$ as a function of the ligand concentration, it is possible to find $(p-q)$ and the ratio β_p/β_q.

Reversible reduction of a soluble complex to an amalgam

Now consider the reduction of a complex ion to yield an amalgam plus free ligand:

$$ML_p^{n-pb} + ne^- + Hg \rightleftharpoons M(Hg) + pL^{-b} \qquad (19\text{-}25)$$

For such a reaction, it can be shown that $E_{1/2}$ is given by

$$E_{1/2} = E_{1/2} \text{ (for free } M^{n+}) - \frac{0.059\ 16}{n} \log \beta_p - \frac{0.059\ 16\ p}{n} \log[L^{-b}] \qquad (19\text{-}26)$$

For Reaction 19-25, a graph of $E_{1/2}$ versus $\log[L^{-b}]$ will give the values of p and the equilibrium constant.

The value of $E_{1/2}$ (for free M^{n+}) refers to $E_{1/2}$ in a noncomplexing medium. A graph of $E_{1/2}$ versus $\log[L^{-b}]$ should have a slope of $-0.059\ 16\ p/n$ and an intercept of $[E_{1/2} \text{ (for free } M^{n+}) - (0.059\ 16/n) \log \beta_p]$.

The equations we have just given apply to reversible electrode reactions. It is possible to extend the treatment to estimate consecutive equilibrium constants for multiple equilibria and to find equilibrium constants when the electrode reactions are not reversible.[†]

Polarographic Study of Chemical Kinetics

Polarographic techniques can be used to study the rates of chemical reactions that compete with the electrode reactions. Consider a solution of Ti(IV) containing both oxalate and sulfate. The composition of the individual complexes is not known, but let us write

$$Ti(IV)(oxalate) \underset{k_r}{\overset{k_f}{\rightleftharpoons}} Ti(IV)(sulfate) \qquad (19\text{-}27)$$

where k_f and k_r are forward and reverse rate constants for equilibrium between the two species.

When a mixture of Ti(IV), oxalate, and sulfate is subjected to polarographic analysis, two waves are observed for the reduction of Ti(IV) to Ti(III). The more easily reduced species is Ti(IV)(oxalate), and the less easily reduced species is Ti(IV)(sulfate). If Reaction 19-27 were rapid compared to the rate of diffusion of Ti(IV)(oxalate) to the electrode, there would be only one wave, since Ti(IV)(sulfate) would produce Ti(IV)(oxalate) as fast as the latter is reduced. If Reaction 19-27 were infinitely slow compared to the rate of diffusion, there would be two waves of fixed height. First, Ti(IV)(oxalate) would be reduced, then Ti(IV)(sulfate) would be reduced at more negative potential.

If the rate of Reaction 19-27 is intermediate, then as Ti(IV)(oxalate) is reduced, more is created from Ti(IV)(sulfate). At this intermediate rate, the magnitude of the limiting current for Ti(IV)(oxalate) will depend on the rate constant k_r in Reaction 19-27.

[†] D. R. Crow, *Polarography of Metal Complexes* (London: Academic Press, 1969).

19 / VOLTAMMETRY

A polarographic wave whose height depends on the rate constant of a chemical reaction is called a *kinetic wave.*

Polarographic waves that depend on the rate constants of chemical reactions are called **kinetic waves.** Detailed analysis of the polarogram provides information on the rate constants for Reaction 19-27.

Now consider what happens when chlorate, ClO_3^-, is added to a solution of Ti(IV) and a polarographic measurement is made. Although ClO_3^- is a strong oxidizing agent, it does not give a polarographic wave because of the large overpotential for its reduction. However, as Ti(IV) is reduced to Ti(III), ClO_3^- reoxidizes the Ti(III) back to Ti(IV):

$$Ti(IV) + e^- \rightarrow Ti(III)(at\ Hg\ electrode) \qquad (19\text{-}28)$$

$$Ti(III) + ClO_3^- \rightarrow Ti(IV) + Cl^- \qquad (19\text{-}29)$$

Sometimes the polarographic reaction serves to catalyze another reaction in the solution. If this second reaction alters the concentration of polarographic reactant, the height of the wave will depend on the rate of the second reaction. Such a polarographic wave is called a *catalytic wave.*

The magnitude of the limiting current depends on the rate at which Ti(IV) diffuses to the electrode and also on the rate at which Reaction 19-29 creates Ti(IV) near the electrode surface. Because Reaction 19-29 is rapid, the limiting current is much larger than that expected for diffusion alone. Since the electrogenerated Ti(III) is acting as a catalyst for the net consumption of ClO_3^-, we call the increase in current a **catalytic current.** Analysis of the current can yield information on the rate constant for Reaction 19-29.

Beyond its use in finding rate constants, the information in this section has another practical value. In Section 19-2, it was stated that the *capillary constant* of a dropping-mercury electrode is proportional to the square root of the height of the mercury column. For quantitative analysis, the height of a polarographic wave must be diffusion-controlled. A graph showing that the wave height is proportional to $\sqrt{h}$ is good evidence that the electrode reaction rate is diffusion-controlled. If kinetic or catalytic currents are present, a graph of I_d versus $\sqrt{h}$ will *not* be linear.

19-4 POWERFUL VARIATIONS OF POLAROGRAPHY

Many variations on the classical polarographic experiment have been developed to increase the sensitivity and resolving power of this analytical technique. The *direct current polarography* that we have discussed so far can detect concentrations as low as 10^{-5} M in favorable cases and can resolve species with half-wave potentials differing by $\gtrsim 0.2$ V. One of the most effective variations of polarography in common use at this time is *differential pulse polarography.* This technique provides a detection limit as low as 10^{-8} M and a resolution of $\gtrsim 0.05$ V. Differential pulse polarography is mainly an electronic variation of direct current polarography. The reaction cell and electrodes are similar for both techniques. The detection limit of polarography can be extended to $\sim 10^{-9}$ M by a method called **stripping analysis,** in which the analyte is first concentrated from a dilute solution into a single drop of mercury. This is one of the most sensitive of all analytical techniques. Both differential pulse and stripping analysis will be discussed in this section. Finally, we will mention *cyclic voltammetry,* which finds much of its use in the study of electrochemical reaction mechanisms.

Differential Pulse Polarography

In direct current polarography, the voltage applied to the working electrode increases linearly with time, as shown in Figure 19-9a. The current is recorded continuously, and a polarogram such as that in Figure 19-2 results. The shape of the plot in Figure 19-9a is called a **linear voltage ramp.** In differential pulse polarography, pulses are superimposed on the linear voltage ramp, as in Figure 19-9b. The height of a pulse is called its **modulation amplitude.** Each pulse of magnitude 5–100 mV is applied during the last 60 ms of the life of each mercury drop. The drop is then mechanically dislodged. The current is not measured continuously. Rather, it is measured once before the pulse and again for the last 17 ms of the pulse. The polarograph subtracts the first current from the second and plots this difference

> Differential pulse polarography produces a curve that is nearly the derivative of an ordinary direct current polarogram.

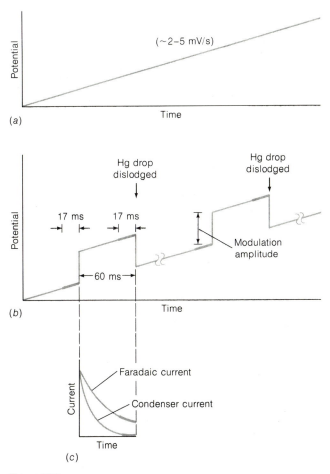

Figure 19-9
(a) Linear voltage ramp used in direct current polarography.
(b) Pulsed ramp of differential pulse polarography. Current is measured only during the intervals shown by heavy lines.
(c) Behavior of faradaic and condenser currents during each pulse.

versus the applied potential. The resulting differential pulse polarogram is nearly the *derivative* of a direct current polarogram, as shown in Figure 19-10.

To see why a derivative is produced, consider the idealized polarographic wave in Figure 19-11. Periodically, the polarograph steps up the voltage and measures the resulting current increase. At the potential V_1 in Figure 19-11, the current increase (ΔI_1) is rather small. At the potential V_2, ΔI_2 is larger, and at the potential V_3, ΔI_3 is still larger. By plotting ΔI versus V, the differential pulse polarograph produces a curve that is nearly the derivative of the direct current polarogram.

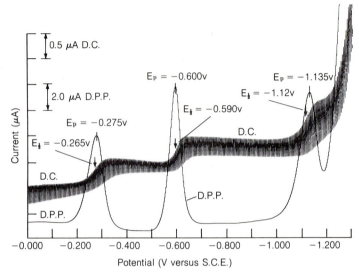

Figure 19-10
Comparison of direct current (D.C.) and differential pulsed polarography (D.P.P.) of 1.2×10^{-4} M chlordiazepoxide (the drug Librium) in 3 mL of 0.05 M H_2SO_4. Modulation amplitude = 50 mV. Note the different current scales for each curve. [M. R. Hackman, M. A. Brooks, J. A. F. de Silva, and T. S. Ma, *Anal. Chem.*, **46**, 1075 (1974).] The chemistry responsible for each wave is believed to be as shown below. [E. Jacobsen and T. V. Jacobsen, *Anal. Chim. Acta*, **55**, 293 (1971).]

19-4 POWERFUL VARIATIONS
OF POLAROGRAPHY

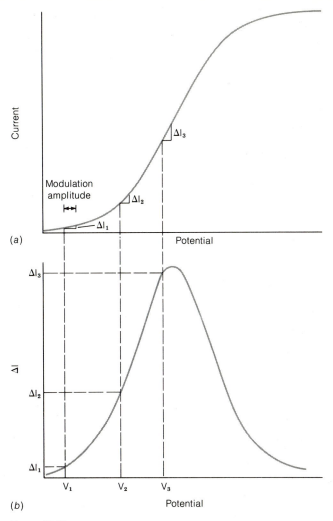

Figure 19-11
Illustration showing why the differential pulse polarogram (b) is very nearly the derivative of the direct current polarogram (a).

Figure 19-12 shows the effect of increasing modulation amplitude in the analysis of a mixture of Fe(III) and Mn(II). As the modulation amplitude is increased, the signal increases, but the resolution (separation of neighboring peaks) decreases. If the modulation amplitude is made too great, the differential pulse polarogram no longer gives a faithful derivative shape.

The enhanced sensitivity of pulsed polarography compared with direct current polarography is due mainly to an increase in the *faradaic current* and a decrease in the *condenser current*. Consider the sample solution when the cathode potential is -0.200 V. The surface concentration of electroactive species is that which is consistent with -0.200 V. Suddenly, a pulse is applied and the voltage is changed to -0.250 V. If the voltage had been *slowly* changed from -0.200 to -0.250 V, the concentration of analyte near the

Increasing the modulation amplitude increases the signal height, but decreases resolution of neighboring peaks.

Faradaic current is due to redox reactions.
Condenser current is due to charging of the Hg drop.

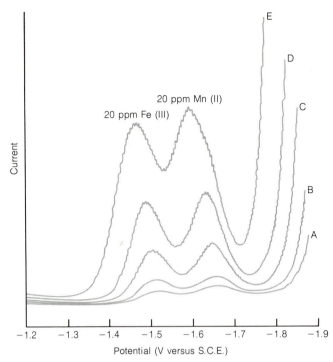

Figure 19-12
Effect of modulation amplitude on peak height and resolution in differential pulse polarography. Modulation amplitude: A = 5 mV, B = 10 mV, C = 25 mV, D = 50 mV, E = 100 mV. [Courtesy Princeton Applied Research Corp., Application Note 151.]

electrode surface would have decreased. Instead, when the pulse is applied, the concentration of analyte is greater than expected for -0.250 V. At the instant that each pulse is applied, there is a surge of electroactive species toward the working electrode. The faradaic current suddenly increases as the species reacts, and its concentration approaches a new steady-state value consistent with the new applied voltage. As the concentration of electroactive species is reduced, the faradaic current decays, as shown schematically in Figure 19-9c. At the instant the pulse is applied, the condenser current also increases to charge the drop up to its new equilibrium charge. As the drop becomes charged, the condenser current also decays, as shown in Figure 19-9c. The condenser current decays faster than the faradaic current. By the time the total current is measured (~ 40 ms after applying the pulse), the condenser current has decayed to near zero, but the faradaic current is substantial.

As compared with direct current polarography, pulsed measurement increases the faradaic current and almost eliminates the condenser current. Both of these effects enhance the sensitivity of the pulse technique. Another reason for increased sensitivity is that the current is measured only during the last 17 ms of the drop life, so the maximum area of the drop is available for electron transfer. Differential pulse polarography also provides better

Differential pulse polarography provides greater sensitivity and better resolution than does direct current polarography.

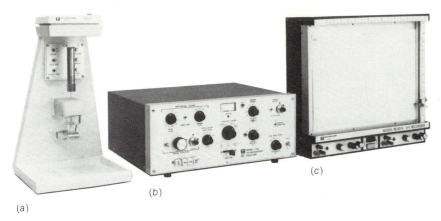

Figure 19-13
Modern polarographic apparatus. (a) This unit includes the dropping mercury electrode and polarographic cell. (b) The analyzer can perform direct current or pulsed experiments. (c) The results of the experiments are displayed on the X–Y recorder. [Courtesy EG&G Princeton Applied Research Corp., Princeton, N.J.]

resolution of adjacent waves because it is easier to distinguish partially overlapping derivative maxima than partially overlapping polarographic waves.

The classical polarographic apparatus of Figure 19-1 is being replaced by more sophisticated equipment, such as that in Figure 19-13. The cell and electrode are contained in the unit at the left. This device uses an electrically controlled, mechanical mechanism to suspend a static drop of mercury at the base of the electrode. After recording the current and the potential, the drop is mechanically dislodged and a fresh, identical drop is created. A new current is recorded at the new potential of the new drop. Because the mercury drop does not change its size during the polarographic measurement, no oscillations are observed in the polarogram. Further, by waiting for a short time after the creation of each new drop, the condenser current decays to near zero (Figure 19-9c), and the signal-to-noise ratio is increased.

Stripping Analysis

In **stripping analysis,** the analyte from a dilute solution is first concentrated in a single drop of Hg by electroreduction. The electroactive species is then *stripped* from the electrode by making the potential more *positive* and *oxidizing* the species back into solution. The current measured during the oxidation is related to the quantity of analyte that was initially deposited. Stripping analysis can be done by conventional direct current polarography or by any variation, such as differential pulse polarography.

The customary setup for stripping analysis involves a **hanging-drop electrode.** Apparatus such as that in Figure 19-13 can create and suspend a single drop of Hg. Analyte is reduced and dissolved in the hanging drop by

In *stripping analysis*, analyte is first concentrated into a drop of Hg by reduction. The concentrated analyte is then permitted to be oxidized by making the potential more positive. The polarographic signal is recorded during the oxidation process.

452

19 / VOLTAMMETRY

Cyclic voltammetry uses a periodic, tri-angular wave form.

The current decreases after the cathodic peak because of concentration polarization.

applying a constant potential more negative than the half-wave potential for the analyte. Because only a fraction of analyte from the solution is deposited, the deposition must be done for a reproducible time (such as 5 minutes) and with reproducible stirring. After concentrating analyte in the drop for a certain time, stirring is stopped and the potential is made more positive at a constant rate to reoxidize the analyte from the drop. The current measured during this oxidation reaches a maximum value that is proportional to the quantity of analyte that was deposited. A fresh drop of mercury is created for each analysis.

Stripping analysis is the most sensitive of polarographic techniques, because analyte is first concentrated from a dilute solution into a single drop of mercury. The longer the period of concentration, the greater will be the signal. A particularly sensitive application of stripping analysis is described in Box 19-1.

Cyclic Voltammetry

Cyclic voltammetry is a form of polarography that is used principally to characterize the redox properties of compounds and to study the mechanisms of redox reactions. Although it is not usually a quantitative tool, we introduce it here because it appears widely in the chemical literature.

In cyclic voltammetry, the triangular wave form in Figure 19-14a is applied to the working electrode. The portion between times t_0 and t_1 is a linear voltage ramp. However, unlike ordinary polarography, in which the ramp is applied over a period of a few minutes, in cyclic voltammetry the time is just a few seconds. Furthermore, in cyclic voltammetry the ramp is then reversed to bring the potential back to its initial value at time t_2. The cycle may be repeated many more times.

The polarogram in Figure 19-14b is recorded either with an oscilloscope or a fast $X-Y$ recorder. The initial portion of the current–potential curve looks like an ordinary polarogram, with a residual current followed by a **cathodic wave.** Instead of leveling off at the top of the wave, the current decreases as the potential is increased further. This happens because the electroactive species becomes depleted in the region around the electrode surface. In this example, at the time of the maximum voltage (t_1), the cathodic current has decayed to a fairly small value. After t_1, the potential change is reversed, but a cathodic current continues to flow because the potential is still negative enough to reduce the analyte. When the potential becomes sufficiently less negative, the reduced analyte in the layer around the electrode surface begins to be oxidized. This gives rise to an **anodic** current peak. Finally, as the reduced species is depleted, the anodic current decays back toward the initial value of the residual current.

Figure 19-14b illustrates the behavior of a reversible electrode reaction. The peak anodic and peak cathodic currents have equal magnitudes in a reversible process. For a reversible process, the difference in potential between the anodic and cathodic peaks is

Box 19-1 THIN-LAYER DIFFERENTIAL PULSE POLAROGRAPHY[†]

The high sensitivity of differential pulse polarographic stripping analysis can be combined with the very small sample volume required by a *thin-layer cell* as shown in the figure below. Analyte is first reduced into a thin layer of Hg on the surface of the graphite working electrode at the bottom of the cell. Salt bridges leading to reference and auxiliary electrodes are at the top of the cell. Although the entire volume of sample in the cell is 70 μL, only 6 μL of solution in a 0.30 mm layer above the graphite surface is electrolyzed.

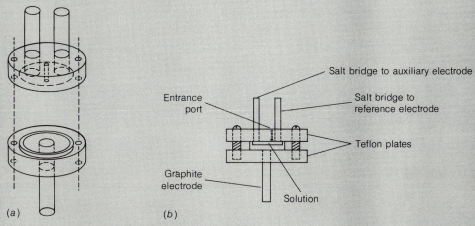

Thin-layer cell used for stripping analysis of very small volume samples. (a) Exploded view of cell. (b) Side view of assembled cell. [T. P. DeAngelis and W. R. Heineman, *Anal. Chem.*, **48**, 2262 (1976).]

The thin layer of Hg on the working electrode is actually deposited along with analyte by electrolysis of the sample solution, which contains 10^{-4} M Hg^{2+}. The electrolysis also reduces O_2 to H_2O, eliminating the need to degas the cell.

In one application, the cell was used to measure the Pb content of human blood, using the standard addition method. Standard Pb^{2+} at a concentration of 10^{-7} M (25 ppb) gave a signal height of 3 μA in the polarogram. Deposition and analysis each required just 60 seconds. Replicate 100 μL samples of blood gave a precision of 4% and an accuracy estimated at 9%.

[†] T. P. DeAngelis, R. E. Bond, E. E. Brooks, and W. R. Heineman, *Anal. Chem.*, **49**, 1792 (1977); T. P. DeAngelis and W. R. Heineman, *Anal. Chem.*, **48**, 2262 (1976).

$$E_a - E_c = \frac{2.22\,RT}{nF} = \frac{57.0}{n}\,(mV) \qquad (19\text{-}30)$$

where n is the number of electrons in the half-reaction. The half-wave potential, $E_{1/2}$, lies midway between the two peak potentials. With an irreversible reaction, the cathodic and anodic peaks become more drawn out and more separated. At the limit of irreversibility, where the oxidation is very slow, no anodic peak would be seen.

For a reversible reaction, $E_{1/2}$ lies midway between the cathodic and anodic peaks.

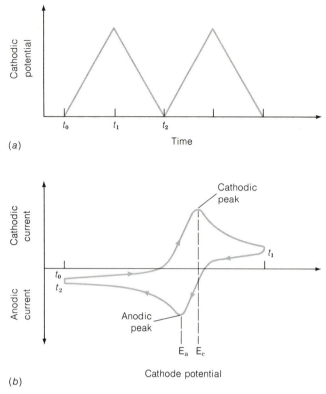

Figure 19-14
(a) Waveform used in cyclic voltammetry. (b) A schematic cyclic voltammogram for a reversible process.

A study of the peak current as a function of the rate of change of applied potential permits an evaluation of the rate constant for the electrochemical reaction. If there are chemical reactions competing for the electrochemical reactants or products, the shape of the voltammogram will reflect the rates of these competing reactions.

Cyclic voltammetry is widely used to characterize the redox behavior of compounds and to elucidate the kinetics of electrode reactions and competing chemical reactions. Common electrodes of Pt, C, Au, or Hg permit the study of either oxidations or reductions. Nonaqueous solvents such as alcohols, dioxane, acetonitrile, dimethylsulfoxide, and dimethylformamide are routinely employed with supporting electrolytes such as LiCl, LiClO$_4$, or tetraalkylammonium salts.

Identifying the species in a complex sequence of redox reactions is not a trivial matter. Cleverly designed cells permit us to measure optical and magnetic resonance spectra of some intermediates, thereby helping to establish structures. Box 19-2 describes how the optical absorbance spectrum can be measured concurrently with a voltammetric experiment.

Box 19-2 AN OPTICALLY TRANSPARENT THIN-LAYER ELECTRODE

A very clever cell design permits the simultaneous measurement of the optical and electrochemical properties of a solution.[†] The key element is a thin gold screen, which is the working electrode. As shown below, the screen is sandwiched between two microscope slides, which form a thin-layer cell. The screen has an optical transmittance of 82%, which means that a visible spectrum of the solution can be recorded by placing the cell in a spectrophotometer. The working volume of the cell is only 30–50 µL, and complete electrolysis of the solute requires only 30–60 s.

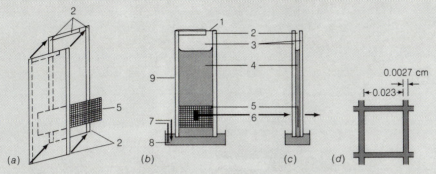

Diagram of an optically transparent thin-layer electrode. (a) Assembly of the cell. (b) Front View. (c) Side View. (d) Dimensions of gold grid. *Key:* (1) Point where suction is applied to change solution. (2) Teflon tape spacers. (3) Microscope slides. (4) Solution. (5) Gold grid electrode. (6) Optical path of spectrometer. (7) Reference and auxiliary electrodes. (8) Solution cup. (9) Epoxy cement. [T. P. DeAngelis and W. R. Heineman, *J. Chem. Ed.*, **53**, 594 (1976).]

The cell can be used for such purposes as characterization of polarographic products. It has also been used to measure the reduction potentials of colored redox enzymes[‡] and the lifetime of reduced chlorophyll.[§] The figure below shows a series of spectra recorded during the reduction of *o*-tolidine. From these spectra the fraction of *o*-tolidine reduced at each potential can be measured.

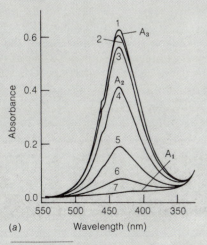

(a) Thin-layer spectra of 0.97 mM *o*-tolidine, 0.5 M acetic acid, 1.0 M HClO$_4$ for different values of applied potential (V versus S.C.E.). (1) 0.800. (2) 0.660. (3) 0.640. (4) 0.620. (5) 0.600. (6) 0.580. (7) 0.400. Cell thickness is 0.17 mm. (b) *o*-Tolidine (reduced form). [T. P. DeAngelis and W. R. Heineman, *J. Chem. Ed.*, **53**, 594 (1976).]

[†] T. P. DeAngelis and W. R. Heineman, *J. Chem. Ed.*, **53**, 594 (1976).
[‡] W. R. Heineman, B. J. Norris, and J. F. Goelz, *Anal. Chem.*, **47**, 79 (1975).
[§] T. Watanabe and K. Honda, *J. Amer. Chem. Soc.*, **102**, 370 (1980).

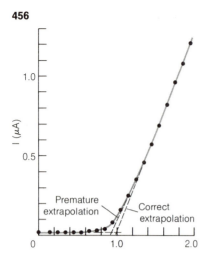

Figure 19-15
Argentometric titration of 100 mL of 10^{-5} M Br^- in 0.01 M HNO_3 with 0.005% gelatin. The potential of the rotating Pt working electrode was +0.15 V versus S.C.E. [J. T. Stock, *Amperometric Titrations* (New York: Wiley, 1965).]

The potential of a *polarizable electrode* is easily changed. The potential of a *nonpolarizable electrode* is hard to change.

Challenge Write the titration reaction and the working electrode half-reaction for the titration in Figure 19-15.

Box 19-3 describes one of the most important applications of amperometry: the measurement of dissolved oxygen.

19-5 AMPEROMETRIC TITRATIONS

Amperometry refers to the measurement of electric current. An **amperometric titration** employs a current measurement to detect the endpoint of a titration. In Figure 18-11, an amperometric circuit was used to detect excess Br_2 at the end of a coulometric titration of cyclohexene. The detection system was simply a pair of platinum electrodes with a potential of 0.2 V applied between them. Only a small residual current is observed until the endpoint is reached; beyond the endpoint, the current increases markedly. The reason for this behavior is that prior to the endpoint, only Br^- is present; while beyond it, both Br_2 and Br^- are present. In the presence of both species, the two reactions below can carry substantial current between the electrodes.

$$\text{Cathode:} \quad Br_2 + 2e^- \rightarrow 2Br^- \tag{19-31}$$

$$\text{Anode:} \quad 2Br^- \rightarrow Br_2 + 2e^- \tag{19-32}$$

A platinum electrode is said to be **polarizable** because its potential is easily changed when only a small current flows. In contrast, a calomel electrode is said to be **nonpolarizable** because its potential remains very nearly constant unless a large current is flowing. Amperometric titrations utilize either one or two polarizable electrodes.

Systems with One Polarizable Electrode

Conventional polarography, with a polarizable working electrode and a nonpolarizable reference electrode, can be used to follow the progress of a titration. Figure 19-15 shows polarographic results for the titration of 10^{-5} M Br^- with Ag^+. For each drop, before the equivalence point, almost all of the Ag^+ precipitates, so only a tiny residual current is observed for the polarographic reduction of Ag^+.

The endpoint in Figure 19-15 is taken as the intersection of the two linear portions. Note that it is necessary to record points out as far as twice the equivalence volume in order to define the truly linear portion of the curve beyond the endpoint. Premature extrapolation of the data just beyond the equivalence volume gives an endpoint that is 8% low.

The working electrode of the polarograph for the titration in Figure 19-15 is the **rotating platinum electrode** shown in Figure 19-16. This electrode spins at a constant rate of ~600 rpm, bringing analyte to the surface by convection as well as by diffusion. The resulting current can be as much as twenty times greater than the diffusion current alone. The rotating Pt electrode is preferred over the dropping Hg electrode for very easily reduced species—e.g., Ag^+, Br_2, and Fe(III)—and for anodic reactions. The Hg electrode cannot be used for many anodic reactions because the electrode itself is easily oxidized. The Pt electrode has a less useful cathodic range because of the low overpotential for H^+ reduction at the Pt surface. For anodic reactions with the Pt electrode, the solution need not be deoxygenated, since the potential is too positive to reduce O_2. The current produced by the rotating Pt electrode is not as reproducible as that of the Hg electrode, but this is not critical for finding the endpoint of a titration.

Systems with Two Polarizable Electrodes

In ordinary polarography, we use one polarizable (working) electrode and one nonpolarizable (reference) electrode. For the amperometric detection of a titration endpoint, it is often desirable to have two polarizable electrodes. An amperometric titration employing two polarizable electrodes is called a **biamperometric titration.**

Figure 19-17 illustrates the shape of the curve obtained for the biamperometric titration of I_2 with $S_2O_3^{2-}$. The titration reaction is

$$2S_2O_3^{2-} + I_2 \rightarrow 2I^- + S_4O_6^{2-} \tag{19-33}$$

The $I_2|I^-$ couple reacts reversibly at a Pt electrode, but the $S_2O_3^{2-}|S_4O_6^{2-}$ couple does not. That is, Reaction 19-34 occurs, but Reaction 19-35 does not.

$$I_2 + 2e^- \rightleftharpoons 2I^- \tag{19-34}$$

$$S_4O_6^{2-} + 2e^- \rightleftharpoons 2S_2O_3^{2-} \tag{19-35}$$

Now consider what happens when $S_2O_3^{2-}$ is added to I_2. At the beginning of the titration, no I^- is present and only residual current is observed. As the reaction proceeds, I^- is created and the reactions below conduct current between the two polarizable electrodes:

$$\text{Cathode:} \quad I_2 + 2e^- \rightarrow 2I^- \tag{19-36}$$

$$\text{Anode:} \quad 2I^- \rightarrow I_2 + 2e^- \tag{19-37}$$

The current reaches a maximum near the middle of the titration, when

In a *biamperometric* titration, the current between two electrodes that are held at a constant potential difference is measured.

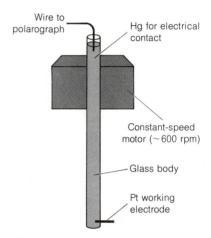

Figure 19-16
A rotating Pt electrode.

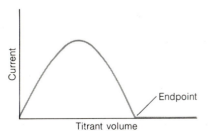

Figure 19-17
Schematic biamperometric titration curve for the addition of $S_2O_3^{2-}$ to I_2.

Box 19-3 THE CLARK OXYGEN ELECTRODE

The simplest means of measuring dissolved O_2 is with a Clark electrode. This combination electrode consists of a Pt cathode held at a potential of -0.6 V with respect to a silver–silver chloride anode. The cell is covered by a semipermeable Teflon membrane, across which oxygen can diffuse in a few seconds. The electrode is dipped into a sample solution, and a short time is allowed for oxygen to equilibrate between the sample and the electrode solution. The current that flows between the two electrodes is proportional to the dissolved oxygen concentration:

$$\text{Cathode:} \quad O_2 + 2H^+ + 2e^- \rightleftharpoons H_2O_2$$

$$\text{Anode:} \quad 2Ag + 2Cl^- \rightleftharpoons 2AgCl + 2e^-$$

The electrode must first be calibrated in solutions of known oxygen concentration.

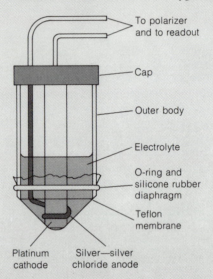

Clark oxygen electrode. [D. T. Sawyer and J. L. Roberts, Jr., *Experimental Electrochemistry for Chemists* (New York: Wiley, 1974).]

maximal quantities of both I_2 and I^- are present. The current then decreases as more I_2 is consumed by Reaction 19-33. Beyond the endpoint, there is no I_2 present and only residual current is observed.

The endpoint of a biamperometric titration is sometimes called a *dead stop* endpoint.

Challenge: A schematic biamperometric titration curve for the addition of Ce^{4+} to Fe^{2+} is shown in Figure 19-18. The titration reaction is

$$Ce^{4+} + Fe^{2+} \rightarrow Ce^{3+} + Fe^{3+}$$

and both couples ($Ce^{4+}|Ce^{3+}$ and $Fe^{3+}|Fe^{2+}$) react reversibly at the Pt electrodes. Explain the shape of the titration curve.

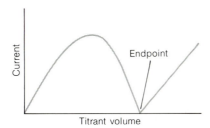

Figure 19-18
Schematic biamperometric titration curve for the addition of Ce^{4+} to Fe^{2+}.

Karl Fischer titration of H₂O

A very important technique in analytical chemistry is the analysis of water by the **Karl Fischer titration.** This sensitive method can be used to measure residual water in purified solvents and water of hydration in crystals. The Karl Fischer reagent consists of I_2, pyridine, and SO_2 in a 1:10:3 mole ratio, dissolved in methanol or ethylene glycol monomethyl ether (CH_3-OCH_2CH_2OH). Addition of water from the analyte begins the sequence of reactions below:

$$\text{Py} \cdot I_2 + \text{Py} \cdot SO_2 + \text{Py} + H_2O \longrightarrow$$
$$2\,\text{Py}NH^+I^- + \text{Py}\overset{+}{N}{-}SO_3^- \quad (19\text{-}38)$$

$$\text{Py}\overset{+}{N}{-}SO_3^- + CH_3OH \longrightarrow \text{Py}NH^+CH_3OSO_3^- \quad (19\text{-}39)$$

The alcoholic solvent is needed to drive Reaction 19-39 (and therefore 19-38) to the right. The most common procedure is to titrate a known volume of

Demonstration 19-1   THE KARL FISCHER JACKS OF A pH METER

Most pH meters contain a pair of sockets at the back that are labeled "K–F" or "Karl Fischer." When the manufacturer's instructions are followed, a constant current (usually around 10 μA) is applied across these terminals. To perform a bipotentiometric titration, a pair of Pt electrodes is connected to the K–F sockets. The meter is set to the millivolt scale, which indicates the potential needed to maintain the constant current between the electrodes immersed in a solution.

The figure below shows the results of a bipotentiometric titration of ascorbic acid with I_3^-. Ascorbic acid (146 mg) was dissolved in 200 mL of water in a 400 mL beaker. Two Pt electrodes spaced about 4 cm apart were immersed in the solution, which was magnetically stirred. Each electrode was attached to one of the K–F outlets of the pH meter. The solution was titrated with 0.04 M I_3^- (prepared by dissolving 2.4 g of KI plus 1.2 g of I_2 in 100 mL of water), and the potential was recorded after each addition.

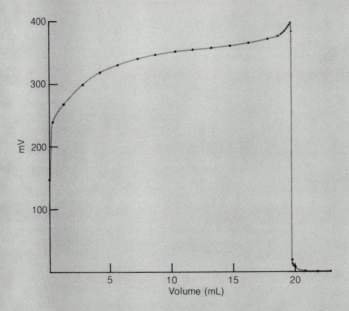

Prior to the equivalence point, all of the I_3^- is reduced to I^- by the excess ascorbic acid. Reaction 19-37 can take place at the anode, but Reaction 19-36 cannot occur at the cathode. A potential of about 300 mV is required to support a constant current of 10 μA. (The ascorbate|dehydroascorbate couple does not react at a Pt electrode and cannot carry current.) After the equivalence point, excess I_3^- is present, so Reactions 19-36 and 19-37 can both occur, and the potential drops precipitously.

standardized Karl Fischer reagent with a solution of the unknown in methanol. The reagent is first standardized with a methanol solution containing a known amount of water.

A *bipotentiometric* measurement is the most common way to detect the endpoint of a Karl Fischer titration. Most pH meters have a pair of electrode receptacles for a Karl Fischer (or other bipotentiometric) titration. The meter maintains a *constant current* (usually 5 or 10 μA) between electrodes plugged into these sockets. When immersed in a solution containing I_2 and I^-, this current flows at a very low applied potential. Prior to the equivalence point, the solution contains I_2 from the Karl Fischer reagent and I^- produced by Reaction 19-38. At the endpoint, the I_2 has been consumed. In order to maintain a current of 10 μA, the cathode potential must increase dramatically to carry current by means of the reduction of solvent:

> In a *bipotentiometric titration*, the potential between two electrodes is measured while a constant current is forced to flow between the electrodes.

$$CH_3OH + e^- \rightleftharpoons CH_3O^- + \tfrac{1}{2}H_2 \qquad (19\text{-}40)$$

The sudden increase of potential (read directly on the pH meter) marks the endpoint. Demonstration 19-1 shows how the Karl Fischer jacks of a pH meter are used.

Summary

In direct current polarography, we observe the current flowing through an electrochemical cell as a function of the potential applied to a dropping-mercury working electrode. This electrode gives reproducible polarograms because fresh surface is continuously exposed. The overpotential for H^+ reduction allows the observation of cathodic processes that are inherently less favorable than H^+ reduction. Most anodic processes must be observed with other electrodes because Hg is too readily oxidized. Polarography is useful in quantitative analysis because the diffusion current (= limiting current − residual current) is generally proportional to analyte concentration. Methods employing standard curves, standard additions, or internal standards are most common in quantitative analysis. Polarography is also useful in qualitative analysis, since $E_{1/2}$ varies characteristically with each electroactive species. The shape of a polarographic wave can be used to find the number of electrons participating in a reversible electrode reaction. The dependence of $E_{1/2}$ on ligand concentration can be used to measure metal–ligand stability constants.

In differential pulse polarography, the voltage ramp is periodically stepped up, and current is measured during the last part of the life of each mercury drop. This process produces a derivative-shaped polarogram. Residual current is diminished because the condenser current for each mercury drop decreases more rapidly than the faradaic current. The signal height can be increased (up to a point) by increasing the modulation amplitude, and the resolution of neighboring waves is better than in direct current polarography. In anodic stripping voltammetry, the most sensitive form of polarography, analyte is concentrated into a single drop of mercury by reduction at a fixed voltage for a fixed time. The potential is then made more positive, and current is measured as the analyte is spontaneously oxidized and leaves the mercury drop. In cyclic voltammetry, a triangular wave form is applied, and cathodic and anodic processes are observed in succession.

In amperometric titrations, the current flowing between a pair of electrodes is used to locate the endpoint. Systems with one polarizable electrode typically use ordinary polarographic apparatus or a rotating Pt electrode to measure the concentration of one species in the reaction. A system with two polarizable electrodes can be used for two types of measurements. In a biamperometric titration, there is a fixed potential difference between the electrodes, and the current needed to sustain it is recorded. In a bipotentiometric titration, the potential needed to maintain a constant current is measured. The current or the potential changes abruptly at the endpoint in these titrations, because at the endpoint one member of an electroactive redox couple is either created or destroyed.

462

Terms to Understand

amperometric titration	cyclic voltammetry	maximum suppressor
amperometry	differential pulse polarography	modulation amplitude
anodic wave	diffusion current	nonpolarizable electrode
biamperometric titration	direct current polarography	polarizable electrode
bipotentiometric titration	dropping-mercury electrode	polarographic wave
capillary constant	electrocapillary maximum	polarography
catalytic current	faradaic current	residual current
cathodic wave	half-wave potential	rotating platinum electrode
charging current	hanging-drop electrode	standard addition method
Clark electrode	internal standard	standard curve
concentration polarization	Karl Fischer titration	stripping analysis
condenser current	kinetic wave	supporting electrolyte
current maximum	linear voltage ramp	voltammetry

Exercises

19-A. The analysis of Ni(II) at the nanogram level is possible using differential pulse polarography.[†] Addition of dimethylglyoxime to an ammonium citrate buffer enhances the response to Ni(II) by a factor of 15. Some representative data are given below.

Ni(II) (ppb)	Peak current (μA)
19.1	0.095
38.2	0.173
57.2	0.258
76.1	0.346
95.0	0.429
114	0.500
132	0.581
151	0.650
170	0.721

Construct a standard curve from these data and use it to answer the following question. What current is expected if 54.0 μL of solution containing 10.0 ppm Ni(II) is added to 5.00 mL of buffer for polarographic analysis?

19-B. A mixture containing Tl^+, Cd^{2+}, and Zn^{2+} exhibited the following diffusion currents in two different experiments (A and B) run with the same electrolyte on different occasions:

	Tl^+ (mM)	Tl^+ (I_d, μA)
A	1.15	6.38
B	1.21	6.11

	Cd^{2+} (mM)	Cd^{2+} (I_d, μA)
A	1.02	6.48
B	?	4.76

	Zn^{2+} (mM)	Zn^{2+} (I_d, μA)
A	1.23	6.93
B	?	8.54

Calculate the Cd^{2+} and Zn^{2+} concentrations in Experiment B.

19-C. A large quantity of Fe(III) interferes with the polarographic analysis of Cu(II) because the Fe(III) is reduced at less negative potentials than Cu(II) in most supporting electrolytes. The Fe(III) reduction wave can be eliminated by addition of hydroxylamine (NH_2OH), which reduces Fe(III), but not Cu(II). The figure on the following page shows the differential pulse polarogram of Cu(II) in the presence of 1000 ppm of Fe(III) with saturated

---

[†] C. J. Flora and E. Nieboer, *Anal. Chem.*, **52**, 1013 (1980).

$NH_3OH^+Cl^-$ as supporting electrolyte. Each sample in this analysis was made up to the *same final volume*. Averaging the response for the two standard additions, calculate the concentration of the sample solution.

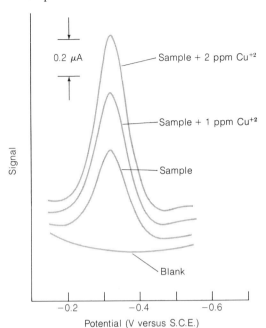

Differential pulse polarograms. Modulation amplitude: 25 mV. Drop interval: 1 second. Scan rate: 2 mV/s. Note that each scan is offset (displaced vertically) from the previous one. When you measure the peak heights relative to the blank, subtract the vertical displacement at the left edge of the scan. [Courtesy EG&G Princeton Applied Research Corp., Application Note 151.]

19-D. Polarographic data for the reaction

$$HPbO_2^- + H_2O + 2e^- + Hg \rightleftharpoons Pb(Hg) + 3OH^-$$

are given below.[†]

$E_{1/2}$ (V, versus S.C.E.)	[OH$^-$] (M, free OH$^-$)
−0.603	0.011
−0.649	0.038
−0.666	0.060
−0.681	0.099
−0.708	0.201
−0.734	0.448
−0.747	0.702
−0.755	1.09

[†] J. J. Lingane, *Chem. Rev.*, **29**, 1 (1941).

The species HPO_2^- can be treated as $Pb(OH)_3^-$ by virtue of the equilibrium

$$Pb(OH)_3^- \rightleftharpoons HPbO_2^- + H_2O$$

Use the data above to show that $p = 3$ in Equation 19-25, and find the value of β_3 for $Pb(OH)_3^-$. The value of $E_{1/2}$ (for free Pb^{2+}) in Equation 19-26 is −0.41 V. This is the half-wave potential for the reaction $Pb^{2+} + 2e^- \rightleftharpoons Pb(Hg)$ in 1M HNO_3.

19-E. Shown below is a cyclic voltammogram of Co(III)-$(B_9C_2H_{11})_2^-$ in 1,2-dimethoxyethane solution.

$E_{1/2}$ (V, versus S.C.E.)	$\dfrac{I_{anodic\ peak}}{I_{cathodic\ peak}}$	$E_a - E_c$ (mV)
−1.38	1.01	60
−2.38	1.00	60

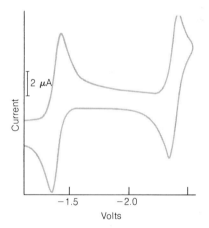

Cyclic voltammogram of Co(III)$(B_9C_2H_{11})_2^-$. [W. E. Geiger, Jr., W. L. Bowden, and N. El Murr, *Inorg. Chem.*, **18**, 2358 (1979).]

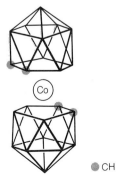

Suggest a chemical reaction to account for each wave. Are the reactions reversible? How many electrons are involved in each step? Sketch the direct current and differential pulse polarograms expected for this compound.

19-F. Amperometric titration curves for the precipitation of Pb(II) with $Cr_2O_7^{2-}$ are shown below.

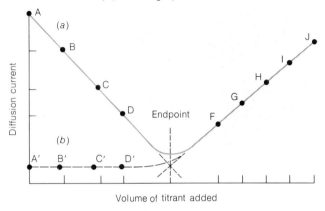

Amperometric titration curves. [D. T. Sawyer and J. L. Roberts, Jr., *Experimental Electrochemistry for Chemists* (New York: Wiley, 1974).]

The potential of the dropping-mercury electrode was -0.8 V versus S.C.E. for curve a and 0.0 V versus S.C.E. for curve b. Explain the shapes of the two curves.

19-G. Sketch the shape of a bipotentiometric titration curve (E versus volume added) for the addition of Ce(IV) to Fe(II). Both the Ce(IV)|Ce(III) and Fe(III)|Fe(II) couples react reversibly at Pt electrodes.

19-H. The Karl Fischer reagent is usually standardized by titration with H_2O dissolved in methanol. A 25.00 mL aliquot of Karl Fischer reagent reacted with 34.61 mL of methanol to which was added 4.163 mg of H_2O per mL. When pure "dry" methanol was titrated, 25.00 mL of methanol reacted with 3.18 mL of the same Karl Fischer reagent. A suspension of 1.000 g of a hydrated crystalline salt in 25.00 mL of methanol consumed a total of 38.12 mL of Karl Fischer reagent. Calculate the weight percent of water in the crystal.

Problems

19-1. Why does the half-wave potential for polarographic reduction of a metal ion to the zero oxidation state depend on the composition of the working electrode? That is, why would a reaction such as Cu(I) → Cu(0) have different half-wave potentials with Hg and Pt electrodes? Will the half-wave potential for Reaction 19-10 be dependent on the composition of the electrode?

19-2. Why is a dropping-mercury electrode preferred for cathodic reactions in amperometric titrations, while a rotating platinum electrode is preferred for anodic reactions?

19-3. For the polarogram in Figure 19-2, the diffusion current is 14 μA.
(a) If the solution contains 25 mL of 0.50 mM Cd^{2+}, calculate the fraction of Cd^{2+} reduced per minute of current passage.
(b) If the drop interval is 4 s, how many minutes were required to scan from -0.6 V to -1.2 V in Figure 19-2?
(c) By what percent has the concentration of Cd^{2+} changed during the scan from -0.6 V to -1.2 V?

19-4. In 1 M NH_3–1 M NH_4Cl solution, Cu(II) is reduced to Cu(I) near -0.3 V (vs S.C.E.), and Cu(I) is reduced to Cu(Hg) near -0.6 V.
(a) Sketch a qualitative polarogram for a solution of Cu(I).
(b) Sketch a qualitative polarogram for a solution of Cu(II).
(c) Suppose that Pt, instead of Hg, were used as the working electrode. Which, if any, reduction potential would you expect to change?

19-5. Use Equation 19-11 to derive an expression for $E_{3/4} - E_{1/4}$, where $E_{3/4}$ is the potential when $I = \frac{3}{4}I_d$ and $E_{1/4}$ applies when $I = \frac{1}{4}I_d$. The value of $E_{3/4} - E_{1/4}$ is used as a criterion for reversibility of a polarographic wave. If $E_{3/4} - E_{1/4}$ is close to the value you have calculated, the reaction is probably reversible.

19-6. For the data in Table 19-1, $m^{2/3}t^{1/6} = 2.28$ $mg^{2/3} \cdot s^{-1/2}$. Calculate the diffusion coefficient for Pb^{2+} in 1.0 M KNO_3. The limiting current in Table 19-1 is given by Equation 9-8, using the constant 6.07×10^4 for the reason described in the footnote on page 431.

19-7. Polarographic data for the reduction of A1(III) in 0.2 M sodium acetate, pH 4.7, is given below.[†] Construct a standard curve and determine the best straight line by the method of least squares. Calculate the standard deviation for the slope and the intercept. If an unknown solution gives $I_d = 0.904\ \mu A$, calculate the concentration of A1(III) and estimate the uncertainty in concentration, using the procedure of Section 4-4.

[A1(III)] (mM)	I_d(corrected for residual current) (μA)
0.009 25	0.115
0.018 5	0.216
0.037 0	0.445
0.055 0	0.610
0.074 0	0.842
0.111	1.34
0.148	1.77
0.185	2.16
0.222	2.59
0.259	3.12

19-8. The diffusion currents below were measured at -0.6 V for $CuSO_4$ in 2 M NH_4Cl/2 M NH_3.[‡] Use the method of least squares in Section 4-4 to estimate the molarity and uncertainty in molarity of an unknown solution giving $I_d = 15.6\ \mu A$.

[Cu(II)] (mM)	I_d (μA)	[Cu(II)] (mM)	I_d (μA)
0.039 3	0.256	0.990	6.37
0.078 0	0.520	1.97	13.00
0.158 5	1.058	3.83	25.0
0.489	3.06	8.43	55.8

19-9. The differential pulse polarogram of 3.00 mL of solution containing the antibiotic tetracycline in 0.1 M acetate, pH 4, gives a maximum current of 152 nA at a half-wave potential of -1.05 V (versus S.C.E.). When 0.500 mL containing 2.65 ppm of tetracycline was added, the current increased to 206 nA. Calculate the parts per million of tetracycline in the original solution.

19-10. Br_2 can be generated for quantitative analysis by addition of standard BrO_3^- to excess Br^- in acidic solution:

$$BrO_3^- + 5Br^- + 6H^+ \rightleftharpoons 3H_2O + 3Br_2$$

[†] H. H. Willard and J. A. Dean, *Anal. Chem.*, **22**, 1264 (1950).
[‡] I. M. Kolthoff and J. J. Lingane, *Polarography*, Vol. I (New York: Wiley, 1952), p. 378.

Consider the titration of As(III) with Br_2, using biamperometric endpoint detection:

$$Br_2 + H_3AsO_3 + 1H_2O \rightleftharpoons 2Br^- + H_3AsO_4 + 2H^+$$

A solution containing As(III) and Br^- is titrated with standard BrO_3^-. Given that the $H_3AsO_4|H_3AsO_3$ couple does not react at a Pt electrode, predict the shape of the titration curve. The titration curve is a graph of I versus volume of titrant.

19-11. Predict the shape of a bipotentiometric titration curve for the titration of H_3AsO_3 with I_2. You should sketch a graph of E versus volume of I_2 added.

19-12. Ammonia can be titrated with hypobromite, but the reaction is somewhat slow:

$$2NH_3 + 3OBr^- \rightleftharpoons N_2 + 3Br^- + 3H_2O$$

The titration can be performed with a rotating Pt electrode held at $+0.20$ V (versus S.C.E.) to monitor the concentration of OBr^-:

$$\text{Cathode:}\quad OBr^- + 2H^+ + 2e^- \rightleftharpoons Br^- + H_2O$$

The current would be near zero before the equivalence point if the OBr^- were consumed quickly in the titration. However, the sluggish titration reaction does not consume all of the OBr^- after each addition, and some current is observed. The increasing current beyond the equivalence point can be extrapolated back to the residual current to find the endpoint. A solution containing 30.0 mL of 4.43×10^{-5} M NH_4Cl was titrated with NaOBr in 0.2 M $NaHCO_3$ with the following results:

OBr^- (mL)	I (μA)	OBr^- (mL)	I (μA)
0.000 0	0.03	0.700	1.63
0.100	1.42	0.720	2.89
0.200	2.61	0.740	4.17
0.300	3.26	0.760	5.53
0.400	3.74	0.780	6.84
0.500	3.79	0.800	8.08
0.600	3.20	0.820	9.36
0.650	2.09	0.840	10.75

Prepare a graph of current versus volume of OBr^-, and find the molarity of the NaOBr solution.

19-13. A series of spectra for *o*-tolidine in an optically transparent, thin-layer electrode was shown in Box 19-2. The Nernst equation for the reaction

$$o\text{-tolidine(oxidized)} + ne^- \rightleftharpoons o\text{-tolidine(reduced)}$$

can be written

$$E_{applied} = E' - \frac{0.059\ 16}{n} \log \frac{[\text{tolidine(reduced)}]}{[\text{tolidine(oxidized)}]}$$

where E′ is the formal reduction potential in the medium used for the experiment. The spectrum labeled 1 is that of the oxidized form of *o*-tolidine. The spectrum labeled 7 is that of the reduced form. Spectra 2 through 6 represent mixtures of both forms. The ratio of species can be calculated from Beer's law. For example, for spectrum 4,

$$\frac{[\text{tolidine(reduced)}]}{[\text{tolidine(oxidized)}]} = \frac{A_3 - A_2}{A_2 - A_1}$$

(a) Measure the ratio [tolidine(reduced)]/[tolidine(oxidized)] using the absorption maxima for curves 2 through 6.

(b) Prepare a graph of $E_{applied}$ versus log([tolidine(reduced)]/[tolidine(oxidized)]).

(c) Use this graph to find E′(versus S.C.E.). Convert your value to E′(versus S.H.E.), which is given by

$$E'(\text{versus S.H.E.}) = E'(\text{versus S.C.E.}) + 0.241\ V$$

(d) Use the slope of the graph to calculate the number of electrons in the half-reaction. Propose a structure for the oxidized form of *o*-tolidine.

20 / Spectrophotometry

A large fraction of all analytical procedures is spectrophotometric in nature. By **spectrophotometry,** we refer broadly to the use of light to measure chemical concentrations. In this chapter, we discuss the fundamental principles of absorption and emission of radiation by molecules and how these processes are used in quantitative analysis.

20-1 PROPERTIES OF LIGHT

It is convenient to describe light in terms of both particles and waves. Light waves consist of perpendicular, oscillating electric and magnetic fields. For simplicity, a *plane-polarized* wave is shown in Figure 20-1. In this figure, the electric field is confined to the xy plane, and the magnetic field is confined to the xz plane. The **wavelength,** λ, is the crest-to-crest distance between waves. The **frequency,** ν, is the number of complete oscillations that the wave makes each second. The unit of frequency is s^{-1}. One oscillation per second is also called one **hertz** (Hz). A frequency of 10^6 s^{-1} is therefore said to be 10^6 Hz, or one *megahertz* (MHz).

The electric field of *plane-polarized* light is confined to a single plane. Ordinary, unpolarized light has electric field components in all planes.

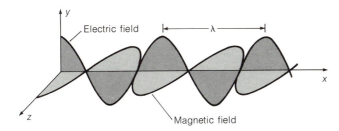

Figure 20-1
Plane-polarized electromagnetic radiation of wavelength λ, propagating along the x axis.

468

20 / SPECTROPHOTOMETRY

The relation between frequency and wavelength is

$$\lambda v = c \qquad (20\text{-}1)$$

where c is the speed of light ($2.997\,924\,58 \times 10^8$ m s^{-1} in vacuum). In any medium other than vacuum, the speed of light is c/n, where n is the **refractive index** of that medium. Since n is always ≥ 1, light travels more slowly through any medium than through vacuum. When light enters a medium of higher refractive index, its frequency remains unchanged but its wavelength decreases.

With regard to its energy, it is more convenient to think of light as particles called **photons.** Each photon carries the energy, E, which is given by

Fundamental equations:

$$\lambda v = c$$

$$E = hv$$

$$E = hv \qquad (20\text{-}2)$$

where h is **Planck's constant** ($6.626\,176 \times 10^{-34}$ J s). One mole of photons is called one *einstein.*

Equation 20-2 states that energy is proportional to frequency. Combining Equations 20-1 and 20-2, we can write

$$E = hc/\lambda \equiv hc\bar{v} \qquad (20\text{-}3)$$

where $\bar{v}$, equal to $1/\lambda$, is called the **wavenumber.** We see that energy is inversely proportional to wavelength and directly proportional to wavenumber. Red light, with a longer wavelength than blue light, is thus less energetic than blue light. The SI unit for wavenumber is m^{-1}. However, the most common unit of wavenumber in the chemical literature is cm^{-1}, read "reciprocal centimeters" or "wavenumbers."

The major regions of the **electromagnetic spectrum** are labeled in Figure 20-2. The various names reflect the history of physical science. There are no discontinuities in the properties of radiation as we pass from one region of the spectrum to another. Note that visible light, which is the kind our eyes detect, represents only a very small fraction of the electromagnetic spectrum.

20-2 ABSORPTION OF LIGHT

When a molecule absorbs a photon, the energy of the molecule is increased. We say that the molecule is promoted to an **excited state** (Figure 20-3). If a molecule emits a photon, its energy is lowered. The lowest energy state of a molecule is called the **ground state.**

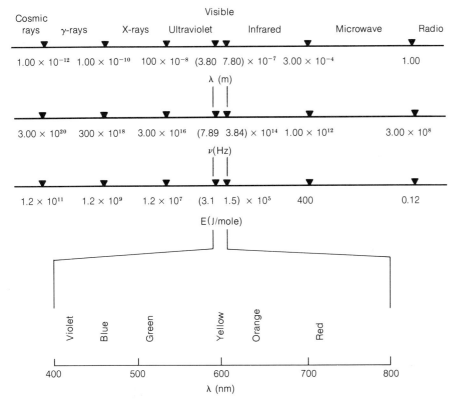

Figure 20-2
The electromagnetic spectrum. The units of wavelength in the visible portion of the spectrum are nanometers (nm, 10^{-9} m).

When light is absorbed by a sample, the **radiant power** of the beam of light is decreased. Radiant power, P, refers to the energy per second per unit area of the light beam. A rudimentary spectrophotometric experiment is illustrated in Figure 20-4. Light is passed through a **monochromator** (a prism, grating, or even a filter) to select one wavelength. Light of this wavelength, with radiant power P_0, strikes a sample of length b. The radiant power of the beam emerging from the other side of the sample is P. Some of the light may be absorbed by the sample, so $P \leq P_0$.

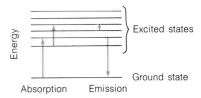

Figure 20-3
Absorption of light increases the energy of a molecule. Emission of light decreases the energy.

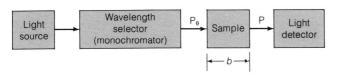

Figure 20-4
Schematic diagram of a spectrophotometric experiment.

470

20 / SPECTROPHOTOMETRY

The **transmittance,** T, is defined as the fraction of the original light that passes through the sample.

$$T = P/P_0 \qquad (20\text{-}4)$$

Relation between transmittance and absorbance:

P/P_0	$\%T$	A
1	100	0
0.1	10	1
0.01	1	2

Therefore, T has the range zero to one. The **percent transmittance** is simply $100 \cdot T$ and varies between zero and one hundred percent. A more useful quantity is the **absorbance,** defined as

$$A = \log_{10}(P_0/P) = -\log T \qquad (20\text{-}5)$$

When no light is absorbed, $P = P_0$ and $A = 0$. If 90% of the light is absorbed, 10% is transmitted and $P = P_0/10$. This gives $A = 1$. If only 1% of the light is transmitted, $A = 2$. The absorbance is sometimes called *optical density*, abbreviated OD or, occasionally, E.

The reason absorbance is so important is that it is directly proportional to the concentration of light-absorbing species in the sample:

Beer's law: $A = \varepsilon bc$.

Box 20-1 explains why absorbance, not transmittance, is proportional to concentration.

$$\boxed{A = \varepsilon bc} \qquad (20\text{-}6)$$

Equation 20-6, which is the heart of spectrophotometry as applied to analytical chemistry, is called the Beer–Lambert law, or simply **Beer's law.** The absorbance, A, is a dimensionless ratio. The concentration of the sample, c, is usually given in units of moles/liter (M). The pathlength, b, is commonly expressed in centimeters. The quantity ε (epsilon) is called the **molar absorptivity** (or *extinction coefficient*, in the older literature) and has the units $M^{-1}\ cm^{-1}$ (since the product εbc must be dimensionless). The molar absorptivity, ε, is often expressed without dimensions, in which case $M^{-1}\ cm^{-1}$ should be assumed.

Equation 20-6 could be written

$$A_\lambda = \varepsilon_\lambda bc \qquad (20\text{-}7)$$

because the values of A and ε depend on the wavelength of light. The quantity ε is simply a coefficient of proportionality between absorbance and the product bc. The larger the value of ε, the greater is A. An **absorption spectrum** is a graph showing how A (or ε) varies with wavelength. Demonstration 20-1 illustrates the meaning of an absorption spectrum and shows several examples of spectra.

The color of a solution is the complement of the color of light that it absorbs.

The part of a molecule responsible for light absorption is called a **chromophore.** Any substance that absorbs visible light will appear colored. A rough guide to colors is given in Table 20-1. The observed color is said to be the *complement* of the absorbed color. As an example, bromophenol blue has a visible absorbance maximum at 614 nm, and its observed color is blue.

Box 20-1 WHY IS THERE A LOGARITHMIC RELATION BETWEEN TRANSMITTANCE AND CONCENTRATION?

Beer's law, Equation 20-6, states that the *absorbance* of a sample is directly proportional to the concentration of the absorbing species. The fraction of light passing through the sample (the *transmittance*) is related logarithmically, not linearly, to the sample concentration. Why should this be?

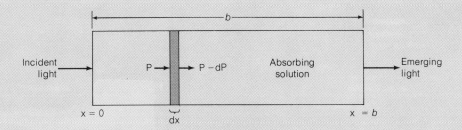

Imagine light of radiant power P passing through an infinitesimally thin layer of solution whose thickness is dx. The decrease in power (dP) is proportional to the incident power (P), to the concentration of absorbing species (c), and to the thickness of the section (dx):

$$dP = -\alpha P c \, dx$$

where α is a constant of proportionality. The equation above can be rearranged and integrated quite simply:

$$-\frac{dP}{P} = \alpha c \, dx$$

$$-\int_{P_0}^{P} \frac{dP}{P} = \alpha c \int_0^b dx$$

The limits of integration are $P = P_0$ at $x = 0$ and $P = P$ at $x = b$. Evaluating the integrals gives us

$$-\ln P - (-\ln P_0) = \alpha c b$$

$$\ln(P_0/P) = \alpha c b$$

Finally, converting ln to log, using the relation $\ln z = (\ln 10)(\log z)$, gives

$$\log\left(\frac{P_0}{P}\right) = \underbrace{\left(\frac{\alpha}{\ln 10}\right)}_{\text{a constant}} cb$$

or

$$A = \varepsilon c b$$

which is Beer's law!

The logarithmic relation of P_0/P and concentration arises from the fact that in each infinitesimal portion of the total volume, *the decrease in power is proportional to the power incident upon that section.* As light travels through the sample, the drop in power in each succeeding layer, in absolute terms, decreases, because the magnitude of the incident power that reaches each layer is decreasing.

Demonstration 20-1 ABSORPTION SPECTRA[†]

The spectrum of visible light can be projected on a screen in a darkened room in the following manner: Four layers of plastic diffraction grating[‡] are mounted on a cardboard frame having a square hole large enough to cover the lens of an overhead projector. This assembly is taped over the projector lens facing the screen. An opaque cardboard surface with two 1 × 3 cm slits is placed on the working surface of the projector.

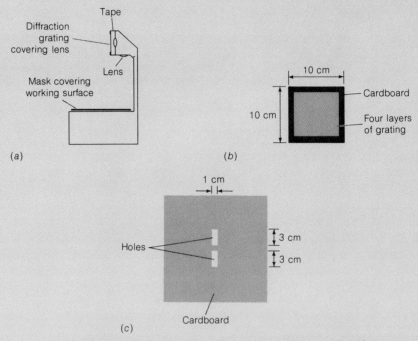

(a) Overhead projector. (b) Diffraction grating mounted on cardboard.
(c) Mask for working surface.

When the lamp is turned on, the white image of each slit is projected on the center of the screen. A visible spectrum appears on either side of each image. By placing a beaker of colored solution over one slit, you can see its color projected on the screen where the white image previously appeared. The spectrum beside the colored image loses its intensity in regions where the colored species absorbs light.

Shown in Color Plate 22 are the spectrum of white light and the absorption spectra of three different colored solutions. You can see that potassium dichromate, which appears orange or yellow, absorbs blue wavelengths. Bromophenol blue absorbs red wavelengths and appears blue to our eyes. The absorption of phenolphthalein is located near the center of the visible spectrum. For comparison, the spectra of these three solutions as recorded with a spectrophotometer are shown in Color Plate 23.

This same setup can be used to demonstrate fluorescence and the properties of colors.[†]

[†] D. H. Alman and F. W. Billmeyer, Jr., *J. Chem. Ed.*, **53**, 166 (1976).
[‡] One inexpensive $8\frac{1}{2}$ × 11 inch sheet of plastic diffraction grating is all that is required. It is available from Edmund Scientific Co., 5975 Edscorp Building, Barrington, N.J. 08007, catalog no. 40,267.

Table 20-1
Colors of visible light

Wavelength of maximum absorption (nm)	Color absorbed	Color observed
380–420	Violet	Green–yellow
420–440	Violet–blue	Yellow
440–470	Blue	Orange
470–500	Blue–green	Red
500–520	Green	Purple
520–550	Yellow–green	Violet
550–580	Yellow	Violet–blue
580–620	Orange	Blue
620–680	Red	Blue–green
680–780	Purple	Green

20-3 WHAT HAPPENS WHEN A MOLECULE ABSORBS LIGHT?

When a molecule absorbs a photon, the molecule is necessarily promoted to a more energetic *excited state* (Figure 20-3). Conversely, when a molecule emits a photon, the energy of the molecule necessarily drops by an amount equal to the energy of the photon. We will now consider the physical processes associated with absorption and emission of radiation.

Excited States of Molecules

As a concrete example, let's consider the molecule formaldehyde, whose structure is shown in Figure 20-5a. In its ground state, the molecule is planar with a double bond between carbon and oxygen. From the simplest electron-dot description of formaldehyde, we expect two pairs of nonbonding electrons to be localized on the oxygen atom. The double bond consists of a sigma bond between carbon and oxygen and a pi bond made from the $2p_y$ (out-of-plane) atomic orbitals of carbon and oxygen.

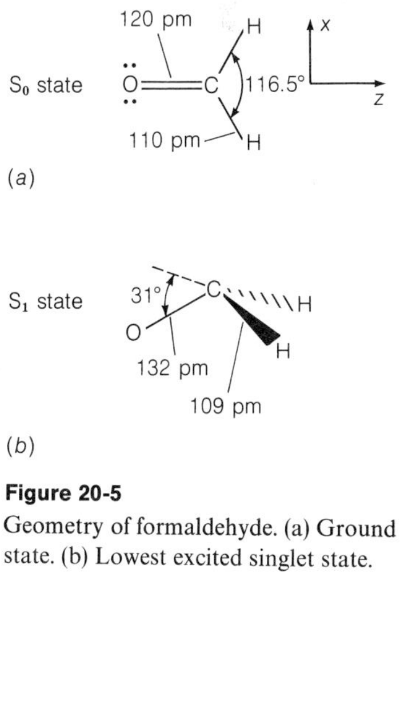

Electronic states of formaldehyde

A molecular orbital description of the valence shell of formaldehyde is given in Figure 20-6. The contours denote the electron density in the various molecular orbitals. The **molecular orbitals** describe the distribution of electrons in a molecule, just as *atomic orbitals* describe the distribution of electrons in an atom. In the molecular orbital description of formaldehyde, one of the nonbonding orbitals of oxygen is thoroughly mixed with the three sigma-bonding orbitals. These four orbitals are labeled σ_1 through σ_4 in Figure 20-6, and each is occupied by a pair of electrons with opposite spin (spin quantum numbers $= +1/2$ and $-1/2$). At higher energy is an occupied

(a)

(b)

Figure 20-5
Geometry of formaldehyde. (a) Ground state. (b) Lowest excited singlet state.

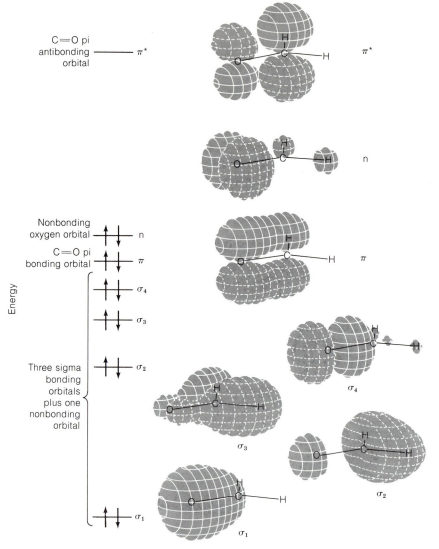

Figure 20-6
Molecular orbital diagram of formaldehyde, showing energy levels and orbital drawings. [Orbital drawings from W. L. Jorgensen and L. Salem, *The Organic Chemist's Book of Orbitals* (New York: Academic Press, 1973).]

We are labeling the plane of the molecule *xz*, as in Figure 20-5.

pi-bonding orbital (π), made of the p_y atomic orbitals of carbon and oxygen. The highest-energy occupied orbital is the nonbonding orbital (n), composed principally of the oxygen $2p_x$ atomic orbital. The lowest-energy unoccupied orbital is the pi-antibonding orbital (π^*). An electron in this orbital produces a repulsion, rather than an attraction, between the carbon and oxygen atoms.

In an **electronic transition,** an electron from one molecular orbital moves to another orbital, with a concomitant increase or decrease in the energy of the molecule. The lowest-energy electronic transition of formaldehyde involves the promotion of a nonbonding (n) electron to the antibonding pi orbital (π*). There are actually two possible transitions, depending on the spin quantum numbers in the excited state (Figure 20-7). The state in which the spins are opposed ($+1/2, -1/2$) is called a **singlet state.** If both electrons have the same spin quantum number ($+1/2, +1/2$), we call the excited state a **triplet state.**

The lowest-energy excited singlet and triplet states are called S_1 and T_1, respectively. In general, T_1 is of lower energy than S_1. In formaldehyde, the transition n → π*(T_1) requires the absorption of visible light with a wavelength of 397 nm. The n → π*(S_1) transition ocurrs when ultraviolet radiation with a wavelength of 355 nm is absorbed. In general, higher excited states of the molecule are designated $S_2, S_3, \ldots,$ and $T_2, T_3, \ldots$.

With an electronic transition near 397 nm, you might expect solutions of formaldehyde to be green–yellow (Table 20-1) in appearance. In fact, formaldehyde is colorless, because the probability of undergoing the n → π*(T_1) transition is exceedingly small. The solution absorbs so little light at 397 nm that our eyes do not detect any absorbance at all. The reason for the low probability is that in the ground state the two spin quantum numbers are $+1/2$ and $-1/2$. In the T_1 excited state, the quantum numbers are $+1/2$ and $+1/2$. The probability of simultaneously changing orbitals and spin quantum numbers is very low. Therefore, very little visible light is absorbed by formaldehyde. The n → π*(S_1) transition is much more probable, and the ultraviolet absorption is more intense.

Although formaldehyde is planar in its ground state, it has a pyramidal structure in both the T_1 and S_1 excited states (Figure 20-5). The excited-state electron distribution actually leads to a change in the geometry of the molecule. Promotion of a nonbonding electron to an antibonding C–O orbital also leads to considerable lengthening of the C–O bond.

Vibrational and rotational states of formaldehyde

We have seen that absorption of visible or ultraviolet radiation can promote electrons to higher-energy orbitals in formaldehyde. Infrared and microwave radiation is not energetic enough to induce electronic transitions, but it can cause changes in the vibrational or rotational motion of the molecule.

Consider formaldehyde as a collection of four atoms. Each atom can move along three axes in space, so the entire molecule can move in $4 \times 3 = 12$ different ways. We say that formaldehyde has twelve *degrees of freedom*. Three degrees of freedom correspond simply to translation of the entire molecule in the x, y, and z directions. Another three degrees of freedom correspond to rotation about the x, y, and z axes of the molecule. The remaining six degrees of freedom represent vibrations of the molecule. The six possible kinds of vibrations of formaldehyde are shown in Figure 20-8.

When a molecule of formaldehyde absorbs an infrared photon with a wavenumber of, say, 1 251 cm^{-1}, the asymmetric bending vibration depicted in Figure 20-8 is stimulated. This means that the oscillations of the atoms are increased in amplitude, and the energy of the molecule increases.

20-3 WHAT HAPPENS WHEN A MOLECULE ABSORBS LIGHT?

The terms *singlet* and *triplet* are used because a triplet state is split into three slightly different energy levels in the presence of a magnetic field, but a singlet state is not split.

The shorter the wavelength of light, the greater the energy.

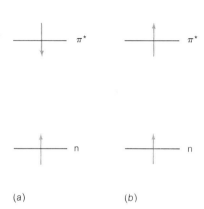

Figure 20-7
Diagram showing the two possible electronic states arising from an n → π* transition. (a) Excited singlet state, S_1. (b) Excited triplet state, T_1.

A nonlinear molecule with *n* atoms has $3n - 6$ vibrational modes and 3 possible rotations. A linear molecule can rotate about only two axes; it therefore has $3n - 5$ vibrational modes and 2 rotations.

Just as the geometry of the excited electronic state is not the same as the geometry of the ground state, the vibrational and rotational energy levels of the excited states are not the same as in the ground state. For example, the C–O stretching vibration of formaldehyde is reduced from 1 746 cm^{-1} in the S_0 state to 1 183 cm^{-1} in the S_1 state. This is reasonable because the strength of the C–O bond is lessened when the antibonding π orbital is populated.

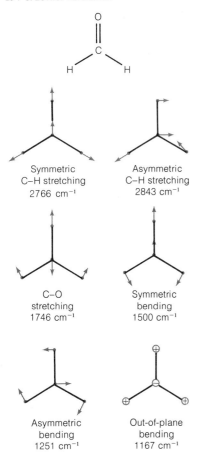

Figure 20-8
The six kinds of vibrations of formaldehyde. The wavenumber of each vibration is given in units of cm^{-1}. These are the wavenumbers of infrared radiation needed to stimulate each kind of motion.

Vibrational transitions usually involve simultaneous rotational transitions. Electronic transitions usually involve simultaneous vibrational and rotational transitions.

EXAMPLE
By how many kilojoules per mole is the energy of formaldehyde increased when 1 251 cm^{-1} radiation is absorbed?
The energy is increased by

$$\Delta E = h\nu = h\frac{c}{\lambda} = hc\bar{\nu} \left(\text{since } \bar{\nu} = \frac{1}{\lambda} \right)$$

$$\Delta E = (6.626\ 2 \times 10^{-34}\ \text{Js})(2.997\ 9 \times 10^{8}\ \text{ms}^{-1})(1\ 251\ \text{cm}^{-1})(100\ \text{cm/m})$$

$$\Delta E = 2.485 \times 10^{-20}\ \text{J/molecule} = 14.97\ \text{kJ/mol}$$

The rotational energy levels of a molecule lie at even lower energy than do the vibrational energy levels. The three lowest rotational energy levels of formaldehyde lie at 0, 0.029 07, and 0.087 16 kJ/mol. A molecule in the rotational ground state could absorb microwave photons with energies of 0.029 07 or 0.087 16 kJ/mol (wavelengths of 4.115 or 1.372 mm) to be promoted to the two lowest excited states. A molecule in the lowest excited state could absorb a photon whose energy is (0.087 16 − 0.029 07) = 0.058 09 kJ/mol and be promoted to the second excited state. In general, absorption of microwave radiation leads to rotational excitation of molecules. In a rotationally excited state, the molecule rotates faster than it does in its ground state.

Combined electronic, vibrational, and rotational transitions

In general, when a molecule absorbs light having sufficient energy to cause an electronic transition, the vibrational and rotational states can change as well. Thus, for example, formaldehyde can absorb one photon with just the right energy to promote simultaneously an electron from the S_0 to the S_1 electronic state, from the ground vibrational state of S_0 to an excited vibrational state of S_1, and from one rotational state of S_0 to a different rotational state of S_1.

The reason why electronic absorption bands are usually very broad (as in Color Plate 22) is that many different vibrational and rotational levels are available at slightly different energies. Therefore, a molecule could absorb photons with a fairly wide range of energies and still be promoted from the ground electronic state to one particular excited electronic state.

What Happens to Absorbed Energy?

We are now in a position to discuss what can happen to a molecule when it absorbs a photon. Many of the possibilities are shown schematically in Figure 20-9. Suppose that the absorption promotes the molecule from the ground electronic state, S_0, to a vibrationally and rotationally excited level of the excited electronic state S_1. Usually, the first process following this

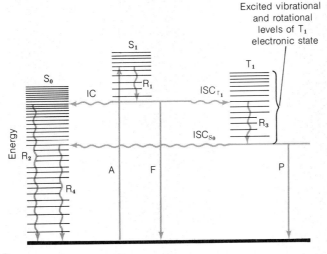

Figure 20-9
Diagram illustrating some of the physical processes that can occur after a molecule absorbs a photon. S_0 is the ground electronic state of the molecule. S_1 and T_1 are the lowest excited singlet and triplet states, respectively. Straight arrows represent processes involving photons, and wavy arrows represent radiation-less transitions. A = absorption, F = fluorescence, P = phosphorescence, IC = internal conversion, ISC = intersystem crossing, R = vibrational relaxation.

absorption is *vibrational relaxation* to the ground vibrational level of S_1. This radiationless transition is labeled R_1 in Figure 20-9. The vibrational energy lost in this relaxation is transferred to other molecules (solvent, for example) through collisions. The net effect is to convert part of the energy of the absorbed photon into heat spread throughout the entire medium.

From the S_1 level, many things can happen. The molecule could enter a very highly excited vibrational state of S_0 having the same energy as S_1. This is called **internal conversion**. From this excited state, the molecule can relax back to the ground vibrational state and transfer its energy to neighboring molecules through collisions. This radiationless process is labeled R_2. If a molecule follows the path A–R_1–IC–R_2 in Figure 20-9, the entire energy of the photon will have been converted to heat.

Alternatively, the molecule could cross from S_1 into an excited vibrational level of T_1. Such an event is known as **intersystem crossing**. Following the radiationless vibrational relaxation R_3, the molecule finds itself at the lowest vibrational level of T_1. From here, the molecule might undergo a second intersystem crossing to S_0, followed by the radiationless relaxation R_4. All of the processes mentioned so far have the net effect of converting light to heat.

In contrast, from S_1 or T_1, a molecule could relax to S_0 by emitting a photon. The transition $S_1 \rightarrow S_0$ is called **fluorescence,** and the transition $T_1 \rightarrow S_0$ is called **phosphorescence.** The relative rates of internal conversion, intersystem crossing, fluorescence, and phosphorescence depend on the molecule, the solvent, and physical conditions such as temperature and pressure. Box 20-2 describes some occasions when light is emitted by excited molecules.

Internal conversion is a radiationless transition between states with the same spin quantum numbers (e.g., $S_1 \rightarrow S_0$).

Intersystem crossing is a radiationless transition between states with different spin quantum numbers (e.g., $T_1 \rightarrow S_0$).

Fluorescence is a radiational transition between states with the same spin quantum numbers (e.g. $S_1 \rightarrow S_0$). *Phosphorescence* is a radiational transition between states with different quantum numbers (e.g., $T_1 \rightarrow S_0$).

Box 20-2 FLUORESCENT LAMPS AND LITTLE-KNOWN FLUORESCENT OBJECTS

A fluorescent lamp is a glass tube filled with mercury vapor; the inner walls are coated with a *phosphor* (luminescent substance) consisting of a calcium halophosphate ($Ca_5(PO_4)_3F_{1-x}Cl_x$) doped with Mn^{2+} and Sb^{3+}. The mercury atoms, promoted to an excited state by electric current passing through the lamp, emit mostly ultraviolet radiation at 254 and 185 nm. This radiation is absorbed by the Sb^{3+} dopant, and some of the energy is passed on to Mn^{2+}. The Sb^{3+} emits blue light, and the Mn^{2+} emits yellow light, with the combined emission appearing white. The emission spectrum is shown below. Fluorescent lamps are important energy-saving devices because they are more efficient than incandescent lamps in the conversion of electricity to light.

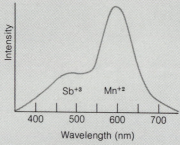

Emission spectrum of the phosphor used in fluorescent lamps. [J. A. DeLuca, *J. Chem. Ed.*, **57**, 541 (1980).]

A little-known occurrence of photoemission is that from most white fabrics. Just for fun, turn on an ultraviolet lamp in a darkened room containing several people. You will discover a surprising amount of emission from white fabrics (shirts, pants, shoelaces, and lots more) that have been treated with fluorescent compounds to enhance their whiteness. You may also be surprised to see fluorescence from teeth and from recently bruised areas of skin that show no surface damage.

The *lifetime* of a state is the time needed for the population of that state to decay to $1/e$ of its initial value, where e is the base of natural logarithms.

Most molecules under ordinary conditions return to the ground state through radiationless processes. Fluorescence and phosphorescence are relatively rare. The *lifetime* of fluorescence is always very short (10^{-4}–10^{-8} s). If a molecule is *not* to fluoresce, internal conversion or intersystem crossing must be even more rapid. The lifetime of phosphorescence is much longer, being in the range 10^{-4}–10^2 s. This means that phosphorescence is even rarer than fluorescence, since a molecule in the T_0 state has a good chance of undergoing intersystem crossing to S_1 before phosphorescence can occur. In molecules containing transition metals, electronic states other than singlets and triplets are possible. Emission from a transition metal complex is usually called simply **luminescence,** which makes no distinction between fluorescence and phosphorescence.

A molecule can dissipate the energy of an absorbed photon by emitting a photon or by creating heat throughout the medium. Alternatively, a bond may break, and *photochemistry* may occur. That is, the energy of the excited molecule could overcome the activation energy of a chemical reaction. In some chemical reactions (not necessarily stimulated by light), part of the energy released appears in the form of light. Light emitted during a chemical reaction is called **chemiluminescence.**

20-4 THE SPECTROPHOTOMETER

The minimum requirements for a spectrophotometer—a device to measure absorbance of light—were shown in Figure 20-4. Light from a continuous source is passed through a monochromator, which selects a narrow band of wavelengths from the incident beam. This "monochromatic" light travels through a sample of pathlength b, and the radiant power of the emergent light is measured.

> Monochromatic radiation is radiation of a single wavelength. Of course, it is impossible to produce truly monochromatic light. However, the better the monochromator, the narrower is the range of wavelengths.

One of the most common ways to measure absorbance of visible light in the undergraduate laboratory is with the Bausch and Lomb Spectronic 20 spectrophotometer, shown in Figure 20-10. In this instrument, the light source is an ordinary tungsten lamp whose emission covers the entire visible spectrum, extending somewhat into the ultraviolet and infrared regions. The light is dispersed into its component wavelengths by a grating, and only one small band of wavelengths is passed through the sample. The detector is a phototube that creates an electric current proportional to the radiant power of light striking the tube. The output is expressed on a meter that reports both transmittance and absorbance (Figure 3-1). It is important to be aware of which scale you are reading on such an instrument.

> If you record transmittance when you really want absorbance, you can use Equation 20-5 to convert one to the other.

The sample is introduced in a cell called a **cuvette.** We do not measure the incident radiant power, P_0, directly. Rather, the radiant power of light passing through a cuvette containing pure solvent is *defined* as P_0. This cuvette is then removed and replaced by an identical one containing sample. The radiant power of light striking the detector is then taken as P, permitting T or A to be determined. The **reference cuvette,** containing pure solvent, compensates for light scattering or absorption due to the cuvette and solvent. The radiant power of light striking the detector would not be the same if the reference cuvette were removed from the beam.

We will use the symbol P_r for the radiant power measured when a blank (the reference) is placed in the spectrophotometer. This is effectively the value of P_0 in Equation 20-5, because it is the maximum power that can reach the detector in the absence of an absorbing species in the solution. We denote by the symbol P_s the power measured at the detector with the sample in the beam. The absorbance is therefore

$$A = \log \frac{P_r}{P_s} \qquad (20\text{-}8)$$

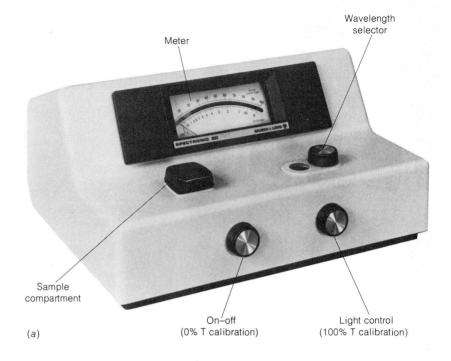

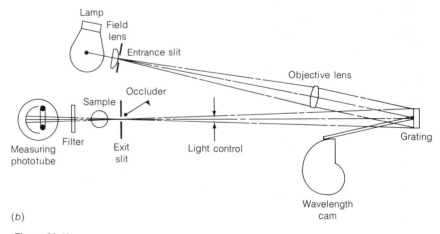

Figure 20-10
Bausch and Lomb Spectronic 20 spectrophotometer. [Courtesy Bausch and Lomb, Analytical Systems Division, Rochester, N.Y.]

Double-Beam Strategy

The Spectronic 20 is an example of a **single-beam spectrophotometer,** because the beam of light follows a single path through one sample at a time. To measure the absorbance of a sample, we must first measure the value of P_0 with a separate blank sample. This process is inconvenient because two different samples must be placed alternately in the beam. It is inaccurate because both

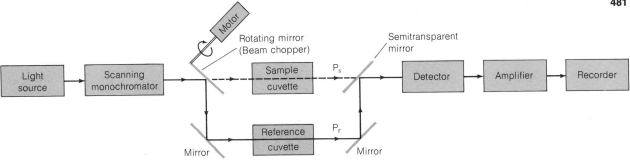

Figure 20-11
Schematic diagram of a double-beam scanning spectrophotometer. The incident beam is passed alternately through the sample and reference cuvettes by the rotating beam chopper.

the output of the source and the response of the detector fluctuate. If there is a change in either one between the measurement of the reference solution and that of the sample solution, the apparent absorbance will be in error. A single-beam instrument is poorly suited to continuous measurements of absorbance, as in a kinetics experiment, because both the source intensity and the detector response drift.

In a **double-beam spectrophotometer** (Figure 20-11), light passes alternately through the sample and the reference cuvettes. This is accomplished by a motor that rotates a mirror into and out of the light path. When the *chopper* is not diverting the beam, the light passes through the sample, and the detector measures the radiant power P_s. When the chopper diverts the beam through the reference cuvette, the detector measures P_r. The beam is chopped several times per second, and the circuitry automatically compares P_r and P_s to obtain absorbance (using Equation 20-8). This procedure provides automatic correction for drift of the source intensity or detector response, since the power emerging from the two samples is compared so frequently. Most research-quality spectrophotometers also provide for automatic wavelength scanning and continuous recording of the absorbance.

When recording an absorbance spectrum, it is a routine procedure to first record the baseline spectrum with blank solutions (pure solvent or a reagent blank) in both cuvettes. The absorbance of the blank is then subtracted from the absorbance of the sample to obtain the true absorbance of the sample at any wavelength.

> The double-beam instrument makes an essentially continuous measurement of the light emerging from the sample and the reference cells.

Major Components

As outlined in Figure 20-4, the essential components of any spectrophotometer are the source, the monochromator, the sample cell, and the detector. We will briefly examine the most commonly encountered forms of these components.

Sources

A tungsten lamp is an excellent source of visible and near-infrared light. A typical tungsten filament operates at a temperature of 2 900 K and produces useful radiation in the range 320–2 500 nm. This covers the entire visible region and parts of the infrared and ultraviolet regions as well. The lamp does not have enough output in the mid-infrared region (4 000–200 cm^{-1}, 2 500–50 000 nm) to be used for vibrational spectroscopy. Mid-infrared radiation is commonly obtained from a silicon carbide rod called a globar, heated to 1 500–2 000 K.

Ultraviolet spectroscopy normally employs a deuterium arc lamp. In such a lamp, an electric discharge causes D_2 to dissociate and emit ultraviolet light over the range 160–375 nm.

$$D_2 \xrightarrow{\text{electric arc}} 2D + h\nu \qquad (20\text{-}9)$$

A whole spectrum of radiant energy is emitted because a wide range of deuterium kinetic energies is created. The only constraint is that the sum of the kinetic energy of the D atoms plus the energy of the photon must equal the energy absorbed by the D_2 molecule.

Ultraviolet light is harmful to the naked eye and should not be viewed without protection.

Monochromators

Light of one wavelength (or color) is said to be **monochromatic.** Gratings and prisms are the two most common devices used to disperse light into its component wavelengths.

A typical design of a grating monochromator is shown in Figure 20-12. *Polychromatic* radiation from the entrance slit is collimated into a beam of parallel rays by a concave mirror. These rays fall on a reflection **grating,**

Monochromatic—one wavelength.

Polychromatic—many wavelengths.

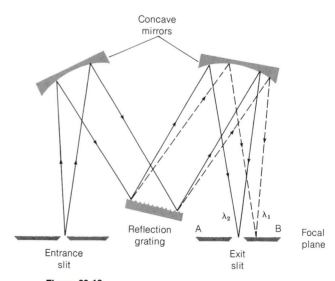

Figure 20-12
Diagram of a Czerney–Turner grating monochromator.

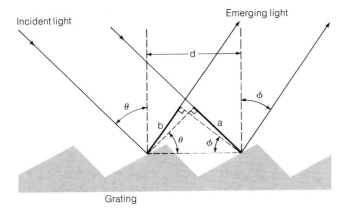

Figure 20-13
Diagram illustrating the principle of a reflection grating.

whereupon different wavelengths are *diffracted* at different angles. The light strikes a second concave mirror, which focuses each wavelength at a different point on the focal plane. The orientation of the reflection grating directs only one narrow band of wavelengths to the exit slit of the monochromator. By rotation of the grating, different wavelengths are allowed to pass through the exit slit.

The principle of the **diffraction** grating is shown in Figure 20-13. The grating is ruled with a series of closely spaced parallel grooves. Light is reflected from the grating, with each groove behaving as a source of radiation. When adjacent light rays are in phase, they reinforce each other. When they are not in phase, they partially or completely cancel (Figure 20-14).

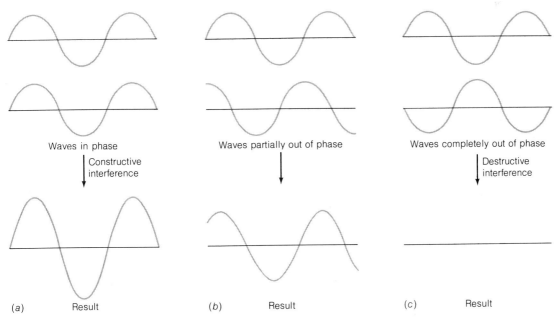

Figure 20-14
Interference of adjacent waves that are (a) 0°, (b) 90°, and (c) 180° out of phase.

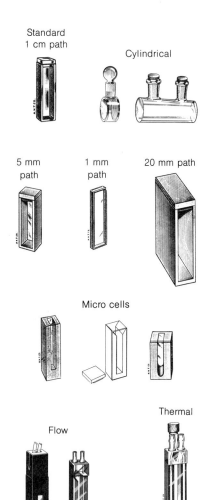

Figure 20-15
Common spectrophotometer cuvettes. The flow cells allow for continuous flow of solution through the cell. They are especially useful for measuring the absorbance of a solution flowing out of a chromatography column. The thermal cell permits liquid from a constant-temperature bath to flow through the cell jacket and thereby maintain the contents of the cell at a desired temperature. [Courtesy A. H. Thomas Co., Philadelphia, Pa.]

Consider the two rays shown in Figure 20-13. Fully constructive interference will occur only if the difference in length of the two paths is exactly equal to the wavelength of light. The difference in path is equal to the distance $a - b$ in Figure 20-13. Constructive interference occurs if

$$n\lambda = a - b \qquad (20\text{-}10)$$

where $n = 1, 2, 3, 4, \ldots$. The interference maximum for which $n = 1$ is called *first-order diffraction*. When $n = 2$, we have *second-order diffraction*, etc. The most intense component of diffracted light occurs when $n = 1$.

From the geometry in Figure 20-13, we see that $a = d \sin \theta$ and $b = d \sin \phi$. Therefore, the condition for constructive interference is

$$n\lambda = d(\sin \theta - \sin \phi) \qquad (20\text{-}11)$$

For each angle, θ, there is a series of angles, ϕ, at which a given wavelength will produce maximum constructive interference. Many sophisticated spectrophotometers contain two monochromators in series (a *double monochromator*) to produce radiation of an even narrower bandwidth.

In any type of monochromator, the width of the exit slit determines what range of wavelengths is passed on to the sample. The narrower the slit width, the narrower is the **bandwidth** (the range of wavelengths) emerging from the monochromator. For a spectrum with very sharp peaks, a narrow slit is needed to resolve closely spaced peaks. However, the narrower the slit, the less light is passed through the sample, and the less signal is measured. Thus, **resolution** is achieved at the expense of increased noise. For quantitative analysis involving broad peaks, it is better to use a broad bandwidth, so that there is more signal to measure.

Sample cells

Cells come in all sizes and shapes appropriate for various experiments (Figure 20-15). The most common cuvettes for measuring visible and ultraviolet spectra have quartz windows, have a 1.000 cm pathlength, and are sold in matched sets (for sample and reference). Glass cells are suitable for visible measurements, but not for ultraviolet spectroscopy (because they absorb the ultraviolet light). Quartz is transparent through the normally accessible visible and ultraviolet regions. Cells for infrared measurements are commonly constructed of NaCl, KBr, or AgCl, which transmit infrared radiation.

Detectors

The general property of a detector is the ability to produce an electrical signal when struck by photons. Many devices are currently employed for this purpose, including several solid-state devices. We will describe only two vacuum-tube detectors.

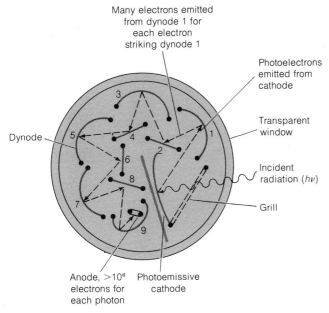

Figure 20-16
Diagram of a photomultiplier tube with nine dynodes. Amplification of the signal occurs at each dynode, which is approximately 90 volts more positive than the previous dynode.

A simple **phototube** emits electrons from a photosensitive, negatively charged surface when struck by light. The electrons flow through a vacuum to a positively charged collector whose current is essentially proportional to the radiation intensity.

A more sophisticated and very sensitive device is the **photomultiplier tube** (Figure 20-16). In this device, the electrons emitted from the photosensitive surface strike a second surface, called a *dynode*, which is kept positive with respect to the photosensitive emitter. The electrons from the emitter are therefore accelerated and strike the dynode with more than their original kinetic energy. These energetic electrons cause more electrons to be emitted from the dynode than strike the dynode. These new electrons are accelerated toward a second dynode, which is more positive than the first dynode. Upon striking the second dynode, even more electrons are knocked off and accelerated toward a third dynode. This process is repeated several times, with the result that more than 10^6 electrons are finally collected for each photon striking the first surface. By this means, extremely low light intensities may be translated into measurable electric signals.

Detectors that respond to a *single photon* are available.

20-5 ERRORS IN SPECTROPHOTOMETRY

For a given spectrophotometer, the manufacturer normally provides some specifications for the instrumental errors to be expected. Additional errors associated with sample preparation and cell positioning will increase the error in any measurement.

486

20 / SPECTROPHOTOMETRY

For greatest precision, samples should be free of dust, and cuvettes must be free of fingerprints or other contamination.

Errors in sample preparation and handling can be minimized by suitable care and common sense. For example, samples must be dust-free because small particles scatter light and increase the apparent "absorbance" of the sample. Keeping all containers covered will lower the concentration of dust in solutions. Filtering the final solution through a very fine filter may be necessary in critical work. Cuvettes should not be handled with fingers and must be kept scrupulously clean to avoid surface contamination, which leads to scattering.

Another Look at Beer's Law

Beer's law can be expected to apply to dilute solutions only.

Beer's law states that absorbance is proportional to the concentration of the absorbing species. It works very well for dilute solutions ($\lesssim 0.01$ M) of most substances. Apparent deviations from Beer's law at higher concentrations can be traced to changes in the absorbing species or in the properties of the bulk solution.

As a solution becomes more concentrated, solute molecules begin to influence each other due to their proximity. When one solute molecule interacts with another, the electrical properties of each (including absorption of light) are likely to change. The apparent result is that a graph of absorbance versus concentration is no longer a straight line. In the extreme case, at very high concentration, the solute *becomes* the solvent. Clearly, you cannot expect the electrical properties of a molecule to be the same in different solvents.[†] Sometimes, nonabsorbing solutes in a solution may interact with the absorbing species and alter the apparent absorptivity.

The physical interaction of two solutes is just one example of a chemical equilibrium (the association of two solutes) affecting the apparent absorbance. An even simpler example is that of a weak electrolyte, such as a weak acid. In concentrated solution, the predominant form of the acid will be the undissociated form, HA. As the solution is diluted, more dissociation occurs. If the absorptivity of A^- is not the same as that of HA, the solution will appear not to obey Beer's law as it is diluted.

Beer's law can be applied strictly only to monochromatic radiation. However, deviations from Beer's law are not serious unless the spectral bandwidth is very broad and the variation of ε with wavelength is substantial.

Choosing the wavelength and the bandwidth

Consider the absorption spectrum of the analytically important complex (ferrozine)$_3$Fe(II) in Figure 20-17. If this species is to be used for the spectrophotometric analysis of iron, it would be sensible to choose the wavelength of maximum absorbance (562 nm) because this gives the greatest absorbance for a given concentration of iron. Also, the variation of absorbance with

---

[†] The apparent absorbance also depends on the refractive index, n, of the solution. At sufficiently high concentrations of solute, the refractive index will change and the absorbance will appear to deviate from Beer's law. The dependence on refractive index is given by $A = \varepsilon bcn/(n^2 + 2)^2$. See G. Kortum and M. Seiler, *Angew. Chem.*, **52**, 687 (1939).

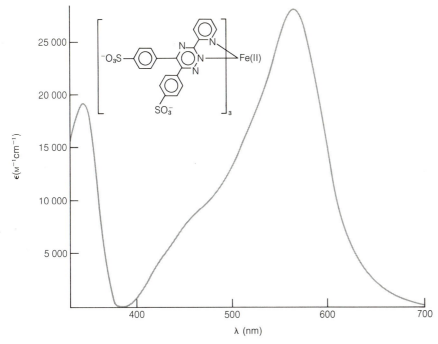

Figure 20-17
Visible absorption spectrum of the complex (ferrozine)$_3$Fe(II) used in the colorimetric analysis of iron.

wavelength (dA/dλ) is a minimum at the peak. Therefore, the effect of not using perfectly monochromatic radiation will be insignificant, since ε is fairly constant over a small range of wavelengths.

The breadth of the peak at 562 nm determines how great a monochromator bandwidth can be used. A bandwidth of 4 nm would lead to negligible deviation from Beer's law due to polychromaticity. In solutions, vibrational structure of an electronic absorption band is usually broadened beyond recognition. Under the broad band of Figure 20-17 are myriad vibrational and rotational transitions. They are so broadened that only a single, nearly featureless envelope is seen.

For quantitative analysis:
1. Monochromator bandwidth should be small compared to the band being measured.
2. Absorbance should be measured at a peak or shoulder where dA/dλ is small.

Instrument Errors

Most spectrophotometers exhibit their minimum error at intermediate levels of absorbance (say, A ≈ 0.2–0.9). If too little light gets through the sample (high absorbance), the intensity is hard to measure. If too much light gets through (low absorbance), then it is hard to detect the difference between the sample and the reference cuvettes. It is therefore desirable to adjust the concentration of the sample so that its absorbance falls in this intermediate range.

Figure 20-18 shows some measured errors for a research-quality spectrophotometer. The curve labeled "dark current noise" gives the imprecision

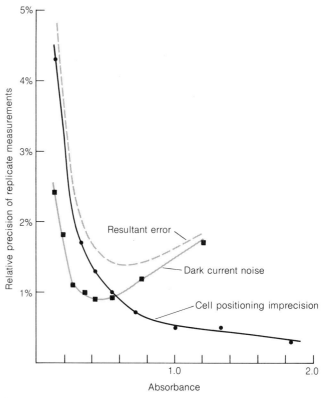

Figure 20-18
Errors in spectrophotometric measurements due to dark current noise and cell positioning imprecision. [Data from L. D. Rothman, S. R. Crouch, and J. D. Ingle, Jr., *Anal. Chem.*, **47**, 1226 (1975).]

due to instrument noise that was essentially independent of the sample absorbance. This noise is caused by such factors as thermal motion of electrons in electronic components and noise in the readout device. By comparison, noise due to flicker of the light source and photomultiplier tube was negligible.

The largest source of imprecision for absorbances less than 0.6 was nonreproducible positioning of the cuvette in the sample holder. This error was large despite great care in placement and cleaning of the cuvette. The resulting error curve in Figure 20-18 reaches a minimum near A = 0.6. At lower absorbances the relative error due to cell positioning is limiting, and at higher absorbances instrumental noise is most important.

20-6 TYPICAL ANALYTICAL PROCEDURES

For a compound to be analyzed by spectrophotometry, it must absorb light, and this absorption should be distinguishable from that due to other species in the sample. Since most compounds absorb ultraviolet radiation, those results tend to be inconclusive, and analysis is usually restricted to the visible

Spectrophotometric analyses employing visible radiation are called *colorimetric* analyses.

spectrum. If there are no interfering species, however, ultraviolet absorbance can be used as well. For example, solutions of proteins are normally assayed in the ultraviolet region at 280 nm because the aromatic groups present in virtually every protein have an absorbance maximum at 280 nm. In this section, we will describe a few typical ways that absorption spectrophotometry is used in quantitative analysis.

Serum Iron Determination

Iron for biosynthesis is transported through the bloodstream attached to the protein transferrin. The procedure described below is used to measure the iron content of transferrin in the blood.[†] This analysis is quite sensitive, with only about 1 μg (μg = microgram = 10^{-6} g) of iron needed to provide an accuracy of ~2–5%. Human blood usually contains about 45% (vol/vol) cells and 55% plasma (liquid). If blood is collected without an anticoagulant, the blood clots, and the liquid that remains is called *serum*. Serum normally contains about 1 μg of Fe/mL attached to transferrin.

To measure the serum iron content three steps are needed:

1. Fe(III) in transferrin is reduced to Fe(II) and thereby released from the protein. Commonly employed reducing agents are hydroxylamine hydrochloride ($NH_3OH^+Cl^-$), thioglycolic acid, or ascorbic acid.

2. Trichloroacetic acid (Cl_3CCO_2H) is added to precipitate all of the proteins, leaving Fe(II) in solution. The proteins are removed by centrifugation. If protein is left in the solution, it will partially precipitate in the final solution. Light scattering by particles of precipitate would be mistaken for absorbance of light in the spectrophotometer.

3. A measured volume of supernatant liquid from Step 2 is transferred to a fresh vessel and treated with excess ferrozine to form the purple complex, whose absorbance is measured. A buffer is also added to keep the pH in a range in which the formation of the ferrozine–iron complex is complete. The structure and spectrum of this complex were given in Figure 20-17.

$HSCH_2CO_2H$

thioglycolic acid

ascorbic acid (vitamin C)

In most spectrophotometric analyses, it is important to prepare a **reagent blank** containing all reagents, but with analyte replaced by distilled water. Any absorbance of the blank is due to the color of uncomplexed ferrozine plus the color caused by the iron impurities in the reagents and glassware. *The absorbance of the blank is subtracted from all other absorbances before any calculations are done.*

The blank should contain all sources of absorbance other than the analyte.

It is also important to use a series of iron standards to establish a calibration curve. Figure 20-19 is a typical calibration curve for this analysis. Beer's law is clearly obeyed over the concentration range in this illustration. The standards should always be prepared using the same procedure as for the unknowns. The absorbance of the unknown should always fall within the region covered by the standards, so that there is no question about the validity of the calibration curve.

[†] D. C. Harris, *J. Chem. Ed.*, **55**, 539 (1978).

Spectrophotometry is well suited to automation. Box 20-3 describes how a *continuous-flow* spectrophotometric analysis operates.

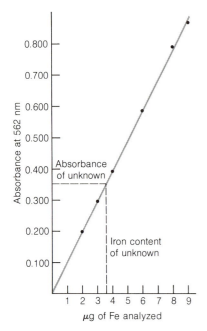

Figure 20-19
Calibration curve showing the validity of Beer's law for the (ferrozine)$_3$Fe(II) complex used in the serum iron determination. Each sample was diluted to a final volume of 5.00 mL. Therefore, 1.00 µg of iron gives a concentration of 3.58 × 10^{-6} M.

If all samples and standards are prepared in the same way and with identical volumes, then the quantity of iron in the unknown can be read directly from the calibration curve. For example, if the unknown has an absorbance of 0.357 (after subtracting the absorbance of the blank), Figure 20-19 tells us that it contains 3.59 µg of iron. To appreciate the uncertainty of the result, the method of least squares in Section 4-4 should be used.

In the serum iron determination just described, the values obtained would be about 10% high due to reaction of serum copper with the ferrozine. This interference can be eliminated if neocuproine or thiourea is added.[†] These reagents form strong complexes with copper, thereby **masking** it.

$$2 \text{ neocuproine} + \text{Cu(I)} \rightarrow [\text{complex}] \text{Cu(I)} \quad (20\text{-}12)$$

EXAMPLE

Serum iron and standard iron solutions were analyzed according to the following procedure:

1. To 1.00 mL of sample is added 2.00 mL of reducing agent and 2.00 mL of acid to reduce and release Fe from transferrin.

2. The serum proteins are precipitated with 1.00 mL of 30% (wt/wt) trichloroacetic acid. The volume change of the solution is negligible when the protein precipitates and can be said to remain 1.00 + 2.00 + 2.00 + 1.00 = 6.00 mL (assuming no changes in volume due to mixing). The mixture is centrifuged to remove protein.

3. A 4.00 mL aliquot of solution is transferred to a fresh test tube and treated with 1.00 mL of solution containing ferrozine and buffer. The absorbance of this solution is measured after a ten-minute waiting period.

The following data were obtained:

Sample	A (at 562 nm in 1.000 cm cell)
Blank	0.038
3.00 µg Fe standard	0.239
Serum sample	0.129

Assuming that Beer's law has been shown to be valid in control experiments, use the data above to find the concentration of Fe in the serum. Also, calculate the molar absorptivity of (ferrozine)$_3$Fe(II).

[†] J. R. Duffy and J. Gaudin, *Clin. Biochem.*, **10**, 122 (1977).

The calculation of the serum iron content is trivial. Since the sample and standard were prepared in an identical manner, their Fe ratio must be equal to their absorbance ratio (corrected for the blank absorbance).

$$\frac{\text{Fe in sample}}{\text{Fe in standard}} = \frac{\text{Corrected absorbance of sample}}{\text{Corrected absorbance of standard}} = \frac{0.129 - 0.038}{0.239 - 0.038} = 0.453$$

Since the standard contained $3.00\ \mu g$ of Fe, the sample must have contained $(0.453)(3.00\ \mu g) = 1.359\ \mu g$ of Fe. The concentration of Fe in the serum is

$$[\text{Fe}] = \text{Moles of Fe/Liters of serum}$$

$$[\text{Fe}] = \left(\frac{1.359 \times 10^{-6}\ \text{g Fe}}{55.847\ \text{g Fe/mol Fe}}\right)\bigg/(1.00 \times 10^{-3}\ \text{L}) = 2.43 \times 10^{-5}\ \text{M}$$

To find ε for $(\text{ferrozine})_3\text{Fe(II)}$, we can use the absorbance of the standard. In the procedure above, a volume of $1.00\ \text{mL}$ of standard containing $3.00\ \mu g$ of Fe was diluted to $6.00\ \text{mL}$ with other reagents. Then $4.00\ \text{mL}$ (containing $4.00/6.00 \times 3.00\ \mu g = 2.00\ \mu g$ of Fe) was transferred to a new vessel and diluted with $1.00\ \text{mL}$ of reagent. The final concentration of Fe is

$$[\text{Fe}] = \left(\frac{2.00 \times 10^{-6}\ \text{g Fe}}{55.847\ \text{g Fe/mol Fe}}\right)\bigg/(5.00 \times 10^{-3}\ \text{L}) = 7.16 \times 10^{-6}\ \text{M}$$

All of this Fe is in the form $(\text{ferrozine})_3\text{Fe(II)}$. The molar absorptivity is

$$\varepsilon = \frac{A}{bc} = \frac{0.239 - 0.038}{(1.00\ \text{cm})(7.16 \times 10^{-6}\ \text{M})} = 2.81 \times 10^4\ \text{M}^{-1}\ \text{cm}^{-1}$$

Analysis of a Mixture

The absorbance of a solution at a particular wavelength is the sum of the absorbances at that wavelength of each species in the solution.

Absorbance is additive.

$$A' = \varepsilon_X' b[\text{X}] + \varepsilon_Y' b[\text{Y}] + \varepsilon_Z' b[\text{Z}] + \cdots \qquad (20\text{-}13)$$

where the prime refers to measurements made at wavelength λ' and ε_i' is the molar absorptivity of species i at wavelength λ'. Let's apply Equation 20-13 to the analysis of a mixture containing two components.

The problem is illustrated schematically in Figure 20-20. Pure compound X has a maximum at λ', and pure compound Y exhibits a maximum at λ''. A mixture of X and Y produces the spectrum shown by the dashed line. The absorbance at any wavelength is the sum of the absorbances at that wavelength of each component. For the absorbance at λ', we can write

$$A' = \varepsilon_X' b[\text{X}] + \varepsilon_Y' b[\text{Y}] \qquad (20\text{-}14)$$

Similarly, at λ'', the absorbance is

BOX 20-3 CONTINUOUS-FLOW SPECTROPHOTOMETRIC ANALYSIS

In many industrial and clinical applications, it is valuable to be able to perform automatic or semiautomatic analyses on an essentially continuous basis. For this purpose, a flow system is often employed.

One typical system is designed to analyze ten samples of oxalate per hour. Sample containers are held in a circular rack that automatically changes position after each sample has been analyzed. With each change of position, a plastic tube is removed from the old sample and inserted into the new one. A peristaltic pump draws liquid from the tube and feeds it into the analyzer.

Once in the analyzer, the sample is diluted with an appropriate buffer drawn continuously from a stock reservoir. In this particular analytical system, the diluted oxalate is passed through a chamber containing the enzyme oxalate oxidase, covalently attached to inert beads inside the chamber. The enzyme catalyzes the oxidation of oxalic acid to CO_2:

$$HO_2CCO_2H + O_2 \xrightarrow[\text{oxidase}]{\text{oxalate}} 2CO_2 + \underbrace{H_2O_2}_{\text{hydrogen peroxide}}$$

This chemical reaction produces one mole of H_2O_2 for each mole of oxalate that is consumed.

After passing through the enzyme chamber, the stream is mixed with a new solution containing reagents that react with H_2O_2 to form a colored product. After passing through a mixing coil that allows the color-forming reaction to go to completion, the solution is directed through a flow cell in which the absorbance is continuously monitored. Traces such as those in the figure below show that the absorbance is proportional to the original concentration of oxalate; the precision is about 2%.

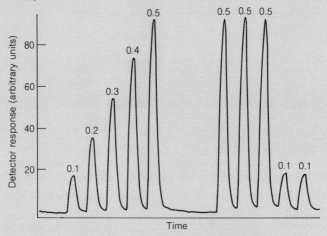

Representative recording of the oxalate determination using standard solutions. Numbers refer to oxalate concentration (mM). [R. Bais, N. Potezny, J. Edwards, A. M. Rofe, and R. A. J. Conyers, *Anal. Chem.* **52**, 508 (1980).]

Between samples, the instrument is programmed to conduct a wash cycle, in which diluent solution is used to flush the previous sample out of the line. Air bubbles are used to separate consecutive samples from each other and to aid in mixing reagents. Such automatic analyzers are finding steadily increasing use in clinical and industrial settings.

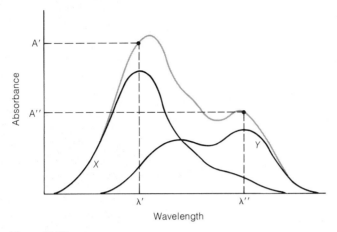

Figure 20-20
Absorption spectra of two pure compounds (black lines) and a mixture of the two (gray line).

$$A'' = \varepsilon_X'' b[X] + \varepsilon_Y'' b[Y] \qquad (20\text{-}15)$$

The absorptivities of X and Y at each wavelength must each be measured in a separate experiment.

Combining Equations 20-14 and 20-15, we can solve for the concentrations of each species in the mixture that produced the dashed-line spectrum in Figure 20-20. The final result is

$$[X] = \frac{\begin{vmatrix} A' & \varepsilon_Y' b \\ A'' & \varepsilon_Y'' b \end{vmatrix}}{\begin{vmatrix} \varepsilon_X' b & \varepsilon_Y' b \\ \varepsilon_X'' b & \varepsilon_Y'' b \end{vmatrix}} \qquad [Y] = \frac{\begin{vmatrix} \varepsilon_X' b & A' \\ \varepsilon_X'' b & A'' \end{vmatrix}}{\begin{vmatrix} \varepsilon_X' b & \varepsilon_Y' b \\ \varepsilon_X'' b & \varepsilon_Y'' b \end{vmatrix}} \qquad (20\text{-}16)$$

Recall that the determinant $\begin{vmatrix} a & b \\ c & d \end{vmatrix}$ means $ad - bc$.

To analyze a mixture of two compounds, therefore, it is necessary to measure the absorbances at two wavelengths and to know ε at each wavelength for each compound. Similarly, a mixture of n components may be analyzed by making n absorbance measurements at n wavelengths. Usually, the best accuracy is obtained when absorbance is measured at λ_{max} for each component. The smaller the overlap between the spectra of individual compounds, the more accurate the analysis will be.

EXAMPLE
The molar absorptivities of compounds X and Y were measured with pure samples of each:

	ε at 272 nm (M^{-1} cm^{-1})	ε at 327 nm (M^{-1} cm^{-1})
X	16 400	3 990
Y	3 870	6 420

A mixture of compounds X and Y in a 1.00 cm cell had an absorbance of 0.957 at 272 nm and 0.559 at 327 nm. Find the concentrations of X and Y in the mixture.

Using Equations 20-16 and setting $b = 1.00$, we find

$$[X] = \frac{\begin{vmatrix} 0.957 & 3\,870 \\ 0.559 & 6\,420 \end{vmatrix}}{\begin{vmatrix} 16\,400 & 3\,870 \\ 3\,990 & 6\,420 \end{vmatrix}} = \frac{(0.957)(6\,420) - (3\,870)(0.559)}{(16\,400)(6\,420) - (3\,870)(3\,990)} = 4.43 \times 10^{-5}\ \text{M}$$

$$[Y] = \frac{\begin{vmatrix} 16\,400 & 0.957 \\ 3\,990 & 0.559 \end{vmatrix}}{\begin{vmatrix} 16\,400 & 3\,870 \\ 3\,990 & 6\,420 \end{vmatrix}} = 5.95 \times 10^{-5}\ \text{M}$$

Isosbestic points

Often one absorbing species, X, is converted to another absorbing species, Y, during the course of a chemical reaction. This sort of transformation leads to a very obvious and characteristic behavior, shown in Figure 20-21. If the spectra of pure X and pure Y cross each other at any wavelength, then every spectrum recorded during this chemical reaction will cross at that same point, called an **isosbestic point.**

Methyl red is an acid–base indicator that has two pK_A values:

$$(20\text{-}17)$$

In a solution of methyl red at pH 4.5, the predominant species is the red compound HIn. As the pH is raised, yellow In^- is formed. The spectra of these two species (at the same concentration) happen to cross at 465 nm. Since the solution contains essentially HIn and In^- in the pH range 4.5–7.1, all spectra in Figure 20-21 cross at one point.

To see why this is so, we write an equation for the absorbance of the solution at 465 nm:

$$A^{465} = \varepsilon_{HIn}^{465}b[HIn] + \varepsilon_{In^-}^{465}b[In^-] \qquad (20\text{-}18)$$

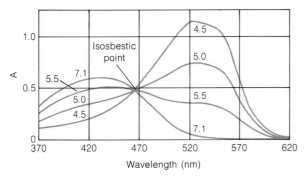

Figure 20-21
Absoprtion spectrum of 3.7×10^{-4} M methyl red as a function of pH between pH 4.5 and 7.1. [E. J. King, *Acid–Base Equilibria* (Oxford: Pergamon Press, 1965).]

But since the spectra of pure HIn and pure In⁻ (at the same concentration) cross at 465 nm, ε_{HIn}^{465} must be equal to $\varepsilon_{In^-}^{465}$. Setting $\varepsilon_{HIn}^{465} = \varepsilon_{In^-}^{465} \equiv \varepsilon^{465}$, Equation 20-18 can be factored as follows:

$$A^{465} = \varepsilon^{465} b([HIn] + [In^-]) \qquad (20\text{-}19)$$

In Figure 20-21, all solutions contain the same total concentration of methyl red (= [HIn] + [In⁻]). Only the pH varies. Therefore, the sum of concentrations in Equation 20-19 is constant, and A^{465} is constant. *The existence of an isosbestic point during a chemical reaction is good evidence that only two principal species are present.*†

An isosbestic point occurs when $\varepsilon_X = \varepsilon_Y$ and $[X] + [Y]$ is constant.

Spectrophotometric Titrations

Absorption of light is one of many common physical properties whose change may be used to monitor the progress of a titration. For example, a solution of the iron-transport protein, transferrin, may be titrated with iron to measure the transferrin content. Transferrin without iron, called apotransferrin, is colorless. Each molecule binds two atoms of Fe(III) and has a molecular weight of 81 000. When the iron binds to the protein, a red color (λ_{max} = 465 nm) develops. The appearance of the red color may be used to follow the course of a titration of an unknown amount of transferrin with a standard solution of Fe(III).

$$\text{apotransferrin} + 2\text{Fe(III)} \rightarrow [\text{Fe(III)}]_2\text{transferrin} \qquad (20\text{-}20)$$
$$\text{(colorless)} \qquad\qquad\qquad\qquad \text{(red)}$$

---

† Under certain conditions it is possible for a solution with more than two principal species to exhibit an isosbestic point. See D. V. Stynes, *Inorg. Chem.*, **14**, 453 (1975).

Ferric nitrilotriacetate is used because uncomplexed ferric ion precipitates in neutral solution. Nitrilotriacetate is $N(CH_2CO_2^-)_3$.

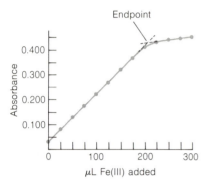

Figure 20-22
Spectrophotometric titration of transferrin with Fe(III). The initial absorbance of the solution, before iron is added, is due to a colored impurity.

Since absorbance is proportional to *concentration* (not activity), concentrations must be converted to activities to get true equilibrium constants. Sometimes this is easy to do, and often it is not. We will neglect activity coefficients in this section.

Figure 20-22 shows the results of a titration of 2.000 mL of a solution of apotransferrin with 1.79×10^{-3} M ferric nitrilotriacetate solution. As Fe(III) is added to the protein, the red color develops and the absorbance increases. When the protein is saturated with iron, no further color-forming reaction can occur, and the curve in Figure 20-22 levels off abruptly. The extrapolated intersection of the two straight portions of the titration curve is taken as the endpoint—203 μL of Fe(III) in Figure 20-22. The absorbance continues to rise slowly after the equivalence point because the ferric nitrilotriacetate solution has some absorbance at 465 nm.

In constructing the graph in Figure 20-22, the effect of dilution must be considered, because the volume is different at each point. Each point plotted on the graph represents the absorbance that would be observed *if the solution had not been diluted from its original volume of 2.000 mL*. For example, the observed absorbance after adding 125 μL (= 0.125 mL) of Fe(III) was 0.260. The solution volume was 2.000 + 0.125 = 2.125 mL. If the volume had been 2.000 mL, the absorbance would have been greater than 0.260 by a factor of (2.125)/(2.000).

$$\text{Corrected absorbance} = \left(\frac{\text{Total volume}}{\text{Initial volume}}\right)(\text{Observed absorbance})$$

$$= \left(\frac{2.125}{2.000}\right)(0.260) = 0.276 \qquad (20\text{-}21)$$

The absorbance plotted in the figure is 0.276, the corrected absorbance.

Measuring an Equilibrium Constant: The Scatchard Plot

Measuring an equilibrium constant requires that we measure the concentration (actually activity) of the species involved in the equilibrium. It is usually not necessary to measure all concentrations, because some are related to others through various mass balance equations. In general, any physical property related to concentration or activity can be useful in measuring the equilibrium constant. We have seen how pH, other potentiometric measurements, and polarographic measurements can be used to find equilibrium constants.[†] In this section, we will see how absorbance can be used to measure an equilibrium constant.[‡]

We confine our attention to the simplest equilibrium, in which the species P and X react to form PX.

$$P + X \rightleftharpoons PX \qquad (20\text{-}22)$$

[†] One of the best sources regarding the determination of equilibrium constants is F. J. C. Rossotti and H. Rossotti, *The Determination of Stability Constants* (New York: McGraw-Hill, 1961). This book is out of print, but available in many libraries.

[‡] An excellent article dealing with practical aspects of this subject has been written by R. W. Ramette, *J. Chem. Ed.*, **44**, 647 (1967).

Neglecting activity coefficients, we can write

$$K = \frac{[PX]}{[P][X]} \tag{20-23}$$

Consider a series of solutions in which increments of X are added to a constant amount of P. Letting the total concentration of P (in the form P or PX) be called P_0, we can write

$$P_0 = [P] + [PX] \tag{20-24}$$

or

$$[P] = P_0 - [PX] \tag{20-25}$$

Clearing the cobwebs from your brain, you should realize that Equation 20-24 is a mass balance equation.

Now the equilibrium expression, 20-23, can be rearranged as follows:

$$\frac{[PX]}{[X]} = K[P]$$

$$\frac{[PX]}{[X]} = K(P_0 - [PX]) \tag{20-26}$$

A *Scatchard plot* is a graph of $[PX]/[X]$ versus $[PX]$. The slope is $-K$.

A graph of $[PX]/[X]$ versus $[PX]$ will have a slope of $-K$ and is called a **Scatchard plot.**[†] It is widely used (in various forms) to measure equilibrium constants, especially in biochemistry.

If we can find $[PX]$, we can find $[X]$ with the mass balance:

$$X_0 = [\text{total X}] = [PX] + [X] \tag{20-27}$$

To measure $[PX]$, we might use the absorbance of the solution. Suppose that P and PX each have some absorbance at a certain wavelength, but X has no absorbance at this wavelength. Suppose, for simplicity, that all measurements are made in a cell of pathlength 1.00 cm. This will allow us to omit $b\ (= 1.00$ cm) when writing Beer's law.

The absorbance of the solution at some wavelength is the sum of absorbances of PX and P:

$$A = \varepsilon_{PX}[PX] + \varepsilon_P[P] \tag{20-28}$$

Substituting $[P] = P_0 - [PX]$, we can write

$$A = \varepsilon_{PX}[PX] + \underbrace{\varepsilon_P P_0}_{A_0} - \varepsilon_P[PX] \tag{20-29}$$

[†] G. Scatchard, *Ann. N.Y. Acad. Sci.*, **51,** 660 (1949).

20 / SPECTROPHOTOMETRY

But $\varepsilon_P P_0$ is A_0, the initial absorbance before any X is added. Regrouping Equation 20-29 gives

$$A = [PX](\varepsilon_{PX} - \varepsilon_P) + \varepsilon_P P_0$$

$$A = [PX]\Delta\varepsilon + A_0 \qquad (20\text{-}30)$$

where $\Delta\varepsilon = \varepsilon_{PX} - \varepsilon_P$. Solving Equation 20-30 for $[PX]$ gives

$$[PX] = \frac{A - A_0}{\Delta\varepsilon} = \frac{\Delta A}{\Delta\varepsilon} \qquad (20\text{-}31)$$

where ΔA is the observed absorbance minus the initial absorbance for each point in the titration.

Substituting the value of $[PX]$ from Equation 20-31 into Equation 20-26 gives a useful result:

$$\frac{\Delta A}{\Delta\varepsilon[X]} = K\left(P_0 - \frac{\Delta A}{\Delta\varepsilon}\right)$$

A useful result.

$$\frac{\Delta A}{[X]} = K\,\Delta\varepsilon P_0 - K\,\Delta A \qquad (20\text{-}32)$$

That is, a graph of $\Delta A/[X]$ versus ΔA should be a straight line with a slope of $-K$. In this way, absorbances measured while P is titrated with X can be used to find the equilibrium constant for the reaction of X with P.

Two cases commonly arise in the application of Equation 20-32. If the binding constant is small, then large concentrations of X are needed to observe the formation of PX. Therefore, $X_0 \gg P_0$, and the concentration of unbound X in Equation 20-32 can be set equal to the total concentration, X_0. Alternatively, if K is not small, then $[X]$ is not equal to X_0, and $[X]$ must be measured somehow. The best approach is to have an independent measurement of $[X]$, either by measurement at another wavelength or by measurement of a different physical property.

Challenge: Use the substitution $[X] = X_0 - [PX]$ to show that Equation 20-32 can be rewritten in the form

$$\frac{\Delta A}{X_0 - \dfrac{\Delta A}{\Delta\varepsilon}} = KP_0\,\Delta\varepsilon - K\,\Delta A \qquad (20\text{-}33)$$

If $\Delta\varepsilon \,(= \varepsilon_{PX} - \varepsilon_P)$ is known, Equation 20-33 allows us to use only the measured absorbance to make the Scatchard plot: $\Delta A/(X_0 - (\Delta A/\Delta\varepsilon))$ versus ΔA.

In practice, the errors inherent in a Scatchard plot are often substantial and sometimes overlooked. Defining the fraction of saturation of P as

$$\text{Fraction of saturation} = S = \frac{[PX]}{P_0} \quad (20\text{-}34)$$

it can be shown that the most accurate data are obtained for $0.2 \lesssim S \lesssim 0.8$.[†] Furthermore, data should be obtained throughout a range representing about 75% of the total saturation curve before it can be verified that the equilibrium (Equation 20-22) is obeyed. Many people have made mistakes by exploring too little of the binding curve and by not including the region $0.2 \lesssim S \lesssim 0.8$.

20-7 LUMINESCENCE

Although absorption measurements account for the majority of analytical spectrophotometric methods at present, luminescence measurements will probably find increasing use in the future. This is because luminescence measurements are inherently more sensitive than absorption measurements and instruments designed to exploit this advantage are becoming more common. Luminescence measurements are not universally applicable, because many molecules produce weak or negligible emission when they are irradiated.

Relation Between Absorption and Emission Spectra

In general, molecular fluorescence or phosphorescence is observed at a *lower* energy than that of the absorbed radiation (the *excitation* energy). That is, molecules emit longer-wavelength radiation than they absorb. A typical example is shown in Figure 20-23. Let's try to understand why emission occurs at lower energy, why there is so much structure in Figure 20-23, and why the emission spectrum is the approximate mirror image of the absorption spectrum.

A molecule absorbing radiation is initially in its electronic ground state, S_0. This molecule possesses a certain geometry and solvation. Suppose that the excited state is S_1. When radiation is first absorbed, the excited molecule still possesses its S_0 geometry and solvation (Figure 20-24). Very shortly after the excitation, the geometry and solvation revert to their most favorable values for the S_1 state. This must lower the energy of the excited molecule. When an S_1 molecule fluoresces, it returns to the S_0 state, but retains the S_1 geometry and solvation. This unstable configuration must have a higher energy than an S_0 molecule with S_0 geometry and solvation. As

[†] D. A. Deranleau, *J. Amer. Chem. Soc.*, **91**, 4044 (1969).

Luminescence refers to any emission of radiation, including both fluorescence (singlet → singlet emission) and phosphorescence (triplet → singlet emission).

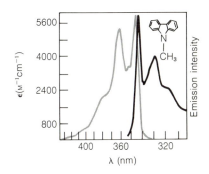

Figure 20-23
Absorption (black line) and emission (gray line) spectra of N-methylcarbazole in cylohexane solution, illustrating the approximate mirror-image relationship between absorption and emission. [I. B. Berlman, *Handbook of Fluorescence Spectra of Aromatic Molecules* (New York: Academic Press, 1971).]

Electronic transitions are so fast, relative to nuclear motion, that each atom has nearly the same position and momentum before and after a transition. This is called the *Franck–Condon principle*.

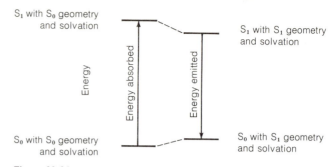

Figure 20-24
Diagram showing why the absorbed energy is greater than the emitted energy.

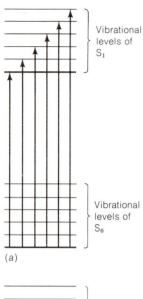

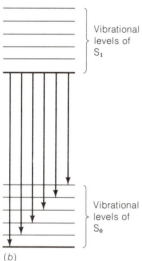

Figure 20-25
Energy-level diagram showing why structure is seen in the (a) absorption and (b) emission spectra, and why the spectra seem roughly mirror images of each other.

Emission spectrum—constant λ_{ex} and variable λ_{em}.

Excitation spectrum—variable λ_{ex} and constant λ_{em}.

shown in Figure 20-24, the net effect is that the emission energy is less than the excitation energy.

Figure 20-25 explains the structure in the spectra and shows why the emission spectrum is roughly the mirror image of the absorption spectrum. The structure in the absorption spectrum is due to absorption of zero or more quanta of vibrational energy in addition to one quantum of electronic energy. In polar solvents, the vibrational structure is often broadened beyond recognition, and only a broad envelope of absorption is observed. In Figure 20-23, the solvent is cyclohexane, and the vibrational structure is very evident. Following absorption, the vibrationally excited S_1 molecule relaxes back to the ground vibrational level of S_1 prior to emitting any radiation. As shown in Figure 20-25, emission from S_1 can occur to any of the vibrational levels of S_0. This gives rise to a series of peaks in the emission spectrum. The absorption and emission spectra will have an approximate mirror-image relationship if the spacings between vibrational levels are roughly equal and if the transition probabilities are similar.

Emission Intensity

The general outline of an emission experiment is shown in Figure 20-26. An excitation wavelength is selected by one monochromator, and the luminescence is examined with a second monochromator, usually positioned at a 90° angle to the incident light. By holding the excitation wavelength (λ_{ex}) fixed and scanning through the emitted radiation, an **emission spectrum** is produced. An emission spectrum is a graph of emission intensity versus emission wavelength. If the emission wavelength (λ_{em}) is held constant and the excitation wavelength is varied, an **excitation spectrum** is produced. An excitation spectrum is a graph of emission intensity versus excitation wavelength.

In emission spectroscopy, we are measuring the absolute intensity of the emission, rather than the fraction of radiant power striking the detector. Since all photomultipliers (or other detectors) have different responses for

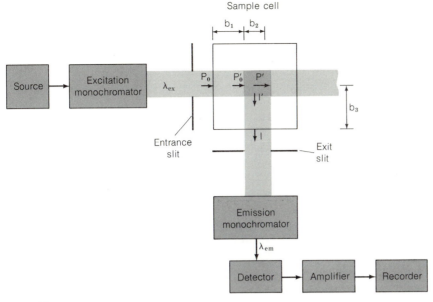

Figure 20-26
Block diagram of a fluorescence spectrophotometer, with the sample cell magnified to show various pathlengths.

different wavelengths, the recorded emission spectrum is usually not a true profile of emission intensity versus emission wavelength. For analytical measurements employing a single emission wavelength, this effect is inconsequential. If a true profile is required (which is rare), it is necessary to calibrate the detector.

To derive a relation between the incident radiant power and the emission intensity, consider the sample cell in Figure 20-26. We expect the emission intensity to be proportional to the radiant power absorbed by the sample. That is, a certain proportion of the absorbed radiation will appear as emission under a given set of conditions (solvent, temperature, etc.). The exit slit in Figure 20-26 is set to observe emission from a region whose width is b_2.

Let the incident radiant power striking the cell be called P_0. Some of this is absorbed by the sample over the pathlength b_1 in Figure 20-26. The radiant power striking the central region of the cell is diminished by absorbance over the pathlength b_1:

$$\text{Power striking central region} \equiv P_0' = P_0 10^{-\varepsilon_{ex} b_1 c} \qquad (20\text{-}35)$$

where ε_{ex} is the molar absorptivity for the wavelength λ_{ex}. The radiant power of the beam when it has traveled the additional distance b_2 is

$$P' = P_0' 10^{-\varepsilon_{ex} b_2 c} \qquad (20\text{-}36)$$

The emission intensity ought to be proportional to the radiant power absorbed in the central region of the cell:

Challenge: Explain why an excitation spectrum will bear a very close resemblance to the absorption spectrum.

Equation 20-35 follows from Beer's law:
$\varepsilon_{ex} b_1 c = A = \log(P_0/P_0') \Rightarrow P_0' = P_0 10^{-\varepsilon_{ex} b_1 c}$

$$\text{Emission intensity} \equiv I' = k'(P_0' - P') \qquad (20\text{-}37)$$

where k' is a constant of proportionality dependent on the emitting molecule and the conditions. Not all of the radiation emitted from the center of the cell and directed at the exit slit is observed. Part of it is absorbed by the solution between the center and the edge of the cell. The emission intensity, I, emerging from the cell is given by Beer's law:

$$I = I'10^{-\varepsilon_{em}b_3c} \qquad (20\text{-}38)$$

where ε_{em} is the molar absorptivity at the emission wavelength and b_3 is the distance from the center to the side of the cell (Figure 20-26).

Combining Equations 20-37 and 20-38, we obtain an expression for the observed emission intensity:

$$I = k'(P_0' - P')10^{-\varepsilon_{em}b_3c} \qquad (20\text{-}39)$$

Substituting values of P_0' and P' from Equations 20-35 and 20-36, we obtain a relation between the incident radiant power and the emission intensity:

$$I = k'(P_010^{-\varepsilon_{ex}b_1c} - P_010^{-\varepsilon_{ex}b_1c}10^{-\varepsilon_{ex}b_2c})10^{-\varepsilon_{em}b_3c}$$

$$I = k'P_010^{-\varepsilon_{ex}b_1c}(1 - 10^{-\varepsilon_{ex}b_2c})10^{-\varepsilon_{em}b_2c} \qquad (20\text{-}40)$$

Equation 20-40 allows us to calculate the emission intensity as a function of solute concentration. When the concentration is sufficiently high, Equation 20-40 has some interesting properties that are explored in Box 20-4.

For quantitative analysis, it is helpful to have a simple, monotonic, preferably linear relation between emission intensity and solute concentration. When the concentration, c, in Equation 20-40 is sufficiently low, the equation can be greatly simplified. Analytical emission experiments are ordinarily performed with solutions so dilute that their absorbance is negligible. This means that the exponents $\varepsilon_{ex}b_1c$, $\varepsilon_{ex}b_2c$, and $\varepsilon_{em}b_3c$ are all very small, and the terms $10^{-\varepsilon_{ex}b_1c}$, $10^{-\varepsilon_{ex}b_2c}$, and $10^{-\varepsilon_{em}b_3c}$ are all very close to unity. We can replace $10^{-\varepsilon_{ex}b_1c}$ and $10^{-\varepsilon_{em}b_3c}$, by unity in Equation 20-40 whenever the absorbance is negligible. We cannot replace $10^{-\varepsilon_{ex}b_2c}$ by unity, because it appears in the term $1 - 10^{-\varepsilon_{ex}b_2c}$, which would become zero.

To find the value of $1 - 10^{-\varepsilon_{ex}b_2c}$ when $10^{-\varepsilon_{ex}b_2c}$ is close to unity, we can expand $10^{-\varepsilon_{ex}b_2c}$ in power series:

The series 20-41 follows from the relation $10^{-A} = (e^{\ln 10})^{-A} = e^{-A\ln 10}$ and the power series expansion of e^x:

$$e^x = 1 + \frac{x}{1!} + \frac{x^2}{2!} + \frac{x^3}{3!} + \cdots$$

$$10^{-\varepsilon_{ex}b_2c} = 1 - \varepsilon_{ex}b_2c \ln 10 + \frac{(\varepsilon_{ex}b_2c \ln 10)^2}{2!} - \frac{(\varepsilon_{ex}b_2c \ln 10)^3}{3!} + \cdots \qquad (20\text{-}41)$$

The term $1 - 10^{-\varepsilon_{ex}b_2c}$ in Equation 20-40 becomes

$$1 - 10^{-\varepsilon_{ex}b_2c} = \varepsilon_{ex}b_2c \ln 10 - \frac{(\varepsilon_{ex}b_2c \ln 10)^2}{2!} + \frac{(\varepsilon_{ex}b_2c \ln 10)^3}{3!} - \cdots \qquad (20\text{-}42)$$

Box 20-4 ABSORPTION-QUENCHING OF EMISSION SPECTRA†

When the absorbance of a sample is not negligible, Equation 20-44 is not an adequate description of emission behavior and Equation 20-40 must be used. At low concentrations, the emission intensity increases with increasing concentration of analyte, because absorption is small and emission is proportional to emitter concentration. At high concentration, the observed emission actually *decreases*, because the absorption increases more rapidly than the emission. We say that emission is *quenched* by self-absorption.

The concentration-dependence of the fluorescence intensity of biacetyl (CH$_3$CCCH$_3$, with two C=O groups) is shown below. Not only does self-absorption cause the observed emission to decrease at high concentration, it also causes the *shape* of the spectrum to change. The reason the shape changes is that absorption of the incident and emitted radiation is dependent on wavelength.

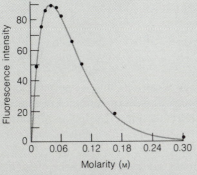

Concentration-dependence of the observed fluorescence intensity of biacetyl in CCl$_4$. Circles are experimental points, and the line was calculated with Equation 20-40 using $\lambda_{ex} = 422$ nm and $\lambda_{em} = 464$ nm.

Besides absorption-quenching, *self-quenching* can also occur in some cases. In this process, a molecule of analyte absorbs the energy from a neighboring excited analyte molecule by a radiationless process that eventually dissipates the excitation energy as heat. Self-quenching occurs when the concentration is high and analyte molecules are therefore close together in the solution.

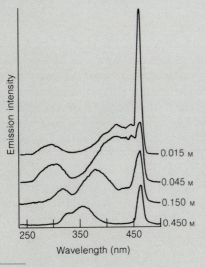

Concentration-dependence of the fluorescence excitation spectrum of biacetyl in CCl$_4$. The apparent change in the spectrum is not due to concentration-dependent energy levels, but to absorption quenching described by Equation 20-40.

† G. Henderson, *J. Chem. Ed.*, **54**, 57 (1977).

504

20 / SPECTROPHOTOMETRY

But when the absorbance of the solution is very small, $\varepsilon_{ex}b_2c$ is very small, and it is a good approximation to neglect all but the first term in the power series:

$$1 - 10^{-\varepsilon_{ex}b_2c} \approx \varepsilon_{ex}b_2c \ln 10 \quad \text{(when } \varepsilon_{ex}b_2c \text{ is small)} \qquad (20\text{-}43)$$

Substituting Equation 20-43 into Equation 20-40 and setting the other exponential terms equal to unity, gives

$$I = k'P_0\varepsilon_{ex}b_2c \ln 10$$

or

In analytical experiments, the absorbance is low, and the emission intensity is given simply by $I = kP_0c$.

$$\boxed{I = kP_0c} \qquad (20\text{-}44)$$

where $k = k'\varepsilon_{ex}b_2 \ln 10$. That is, *when the absorbance is small, the emission intensity is directly proportional to the sample concentration, c, and to the incident radiant power, P_0.*

For most analytical applications, the concentration of analyte is sufficiently small that Equation 20-44 is obeyed and the emission intensity is directly proportional to concentration. The linear relationship between I and P_0 does not extend to arbitrarily high power levels, but it provides substantially increased sensitivity in emission measurements as compared with absorbance measurements. That is, doubling the incident radiant power will double the emission intensity, whereas doubling P_0 has no effect whatsoever on the absorbance. Another advantage of fluorescence is that sensitivity can be increased simply by using a more sensitive detector.

This is the same reason why nephelometry is more sensitive than turbidimetry. See Box 9-1 if you have forgotten about these techniques.

Luminescence in Analytical Chemistry

Some analytes are naturally fluorescent and can be analyzed directly. A typical procedure involves establishing a working curve of luminescence intensity versus analyte concentration (Blank samples invariably scatter light and must be run in every analysis.) Among the more important naturally fluorescent compounds are riboflavin (vitamin B_2), many drugs, polycyclic aromatic compounds (an important class of carcinogens), and proteins.

Most compounds are not naturally luminescent enough to be analyzed direct. However, coupling to a fluorescent moiety provides an easy route to sensitive fluorimetric analyses. For example, airborne aliphatic isocyanates (RNCO) found in workplaces employing polyurethane foam are a significant health hazard. In one sensitive analytical procedure, a mixture of ioscyanates collected from an air sample is treated with 1-naphthyl-methylamine to form fluorescent derivatives that can be separated by liquid chromatography.[†]

riboflavin
(vitamin B_2)

[†] S. P. Levine, J. H. Hoggott, E. Chladek, G. Jungclaus, and J. L. Gerlock, *Anal. Chem.*, **51**, 1106 (1979).

505

20-7 LUMINESCENCE

If the analyte is not fluorescent, it may be coupled to something that is, and then analyzed.

The fluorescence intensities of fractions eluted from the chromatography column are used to quantify the various isocyanates in the air.

Metal ions can be analyzed following their reaction with a fluorescent chelating agent. For example, calcein forms a fluorescent complex with calcium. Fluoride ion can be analyzed because of its ability to *quench* (decrease) the fluorescence of the Al^{3+} complex of alizarin garnet R. A working curve of fluorescence intensity versus F^- concentration decreases as $[F^-]$ increases.

With sophisticated laser fluorometers, it is possible to measure the time-dependence of fluorescence. Each member of a multicomponent mixture is likely to have a different fluorescence lifetime, even if the emission wavelengths are similar. By measuring intensity as a function of time and wavelength, complex mixtures with overlapping spectra may be analyzed.

Summary

Light can be thought of as waves whose frequency (λ) and wavelength (v) have the important relation $\lambda v = c$, where c is the speed of light. Alternatively, light may be viewed as consisting of photons whose energy (E) is given by $E = hv = hc/\lambda = hc\bar{v}$, where h is Planck's constant and $\bar{v} = (1/\lambda)$ is the wavenumber. Absorption of light is commonly measured by absorbance (A) or transmittance (T), defined as $A = \log(P_0/P)$ and $T = P/P_0$, where P_0 is the radiant power of light incident on a sample and P is the power emerging from the other side. The major analytical utility of absorption spectroscopy is derived from the fact that absorbance is proportional to the concentration of the absorbing species (Beer's law): $A = \varepsilon bc$. In this equation, b is pathlength, c is concentration, and the constant of proportionality, ε, is the molar absorptivity. Beer's law works very well for dilute solutions and reasonably monochromatic radiation. If a sample does not obey Beer's law, it is likely that a chemical reaction is altering the concentration of chromophore when the sample concentration is changed.

The basic components of a spectrophotometer include a radiation source (such as a tungsten or deuterium lamp), a monochromator (such as a grating or prism), a sample cell, and a detector (such as a photomultiplier tube). In double-beam spectrophotometry, light is passed alternately through the sample and the reference cuvettes by a rotating beam chopper, and the light beams emerging from each are continually compared to measure absorbance. To minimize errors in spectrophotometry, samples should be free of particles, and cuvettes should be clean. The spectral bandwidth should be small compared to the absorption band, and measurements should be made at a wavelength of maximum absorbance. Instrument errors tend to be minimized if the absorbance falls in the approximate range $A \approx 0.2\text{–}0.9$.

The most common analytical application of spectrophotometry makes use of the proportionality between absorbance and concentration. If the absorbance of a series of standards is measured, the concentration of an unknown treated in the same way can be obtained by

506

direct comparison to the standards. In such analyses, a suitable reagent blank should be prepared, and interfering species should be removed, masked, or otherwise accounted for. A series of n measurements of absorbance at n different wavelengths is, in principle, sufficient to find the concentrations of n absorbing components in a mixture. Spectrophotometry can also be used to follow the course of a titration reaction and to measure equilibrium constants (using a Scatchard plot).

When a molecule absorbs light, it is promoted to an excited state from which it may return to the ground state by radiationless processes or by fluorescence (singlet →

singlet emission) or phosphorescence (triplet → singlet emission). Any form of luminescence is potentially useful for quantitative analysis, because emission intensity is proportional to sample concentration at low concentrations. At higher concentration, self-absorption and self-quenching distort the emission. An excitation spectrum (a graph of emission intensity versus excitation wavelength) is very similar to an absorption spectrum (a graph of absorbance versus wavelength). An emission spectrum (a graph of emission intensity versus emission wavelength) comes at lower energy and tends to be the mirror image of the absorption spectrum.

Terms to Understand

absorbance	luminescence
absorption spectrum	masking agent
bandwidth	molar absorptivity
Beer's law	molecular orbital
chemiluminescence	monochromatic light
chromophore	monochromator
cuvette	phosphorescence
diffraction	photomultiplier tube
electromagnetic spectrum	phototube
electronic transition	radiant power
emission spectrum	reagent blank
excitation spectrum	refractive index
excited state	resolution
extinction coefficient	rotational transition
fluorescence	Scatchard plot
frequency	singlet state
grating	spectrophotometry
ground state	transmittance
hertz	triplet state
internal conversion	vibrational transition
intersystem crossing	wavelength
isosbestic point	wavenumber

Exercises

20-A. (a) What value of absorbance corresponds to 45.0% T?

(b) If a 0.010 0 M solution exhibits 45.0% T at some wavelength, what will be the percent transmittance for a 0.020 0 M solution of the same substance?

20-B. Ammonia can be determined spectrophotometrically by reaction with phenol in the presence of hypochlorite (OCl^-):

A 4.37 mg sample of protein was chemically digested to convert all of its nitrogen to ammonia. After this treatment, the volume of the sample was 100.0 mL. Then 10.0 mL of the solution was placed in a 50 mL volumetric flask and treated with 5 mL of phenol solution plus 2 mL of sodium hypochlorite solution. The sample was diluted to 50.0 mL, and the absorbance at 625 nm was measured in a 1.00 cm cuvette after 30 minutes. For reference, a standard solution was prepared from 1.00×10^{-2} g of NH_4Cl dissolved in 1.00 L of water. Then 10.0 mL of this standard was placed in a 50 mL volumetric flask and analyzed in the same manner as the unknown. A reagent blank was prepared using distilled water in place of unknown.

Sample	Observed absorbance at 625 nm
Blank	0.140
Reference	0.308
Unknown	0.592

(a) Calculate the molar absorptivity of the blue product.

(b) Calculate the weight percent of nitrogen in the protein.

20-C. Cu(I) reacts with neocuproine to form a colored complex with an absorption maximum at 454 nm (Equation 20-12). Neocuproine is particularly useful because it reacts with few other metals. The copper complex is soluble in 3-methy-1-butanol (isoamyl alcohol), an organic solvent that does not dissolve appreciably in water. This means that if isoamyl alcohol is added to water, a two-layered mixture results, with the denser water layer at the bottom. If Cu(I)–neocuproine is present, virtually all of it goes into the organic phase. For the purpose of this problem, assume that the isoamyl alcohol does not dissolve in the water at all and that all of the colored complex will be in the organic phase. Suppose that the following procedure is carried out:

1. A rock containing copper is pulverized, and all metals are extracted from it with strong acid. The acidic solution is neutralized with base and made up to 250.0 mL in flask A.

2. Next 10.00 mL of the solution is transferred to flask B and treated with 10.00 mL of a reducing agent to reduce all Cu to Cu(I). Then 10.00 mL of buffer is added to bring the pH to a value suitable for complex formation with neocuproine.

3. After that, 15.00 mL of this solution is withdrawn and placed in flask C. To the flask is added 10.00 mL of an aqueous solution containing neocuproine and 20.00 mL of isoamyl alcohol. After shaking well and allowing the phases to separate, all Cu(I)–neocuproine is in the organic phase.

4. A few milliliters of the upper layer are withdrawn, and the absorbance at 454 nm is measured in a 1.00 cm tube. A blank carried through the same procedure gave an absorbance of 0.056.

(a) Suppose that the rock contained 1.00 mg of Cu. What will be the concentration of Cu (moles per liter) in the isoamyl alcohol phase?

(b) If the molar absorptivity of Cu(I)–neocuproine is 7.90×10^3 M^{-1} cm^{-1}, what will be the observed absorbance? Remember that a blank carried through the same procedure gave an absorbance of 0.056.

(c) A rock is analyzed and found to give a final absorbance of 0.874 (uncorrected for the blank). How many milligrams of Cu are in the rock?

20-D. Transferrin is the iron-transport protein found in blood. It has a molecular weight of 81 000 and carries two Fe(III) ions. Desferrioxamine B is a potent iron chelator used to treat patients with iron overload. It has a molecular weight of about 650 and can bind one iron atom as Fe(III). Desferrioxamine can take iron from many sites within the body and is excreted (with its iron) through the kidneys. The molar absorptivities of these compounds (saturated with iron) at two wavelengths are given below. Both compounds are colorless (no visible absorption) in the absence of iron.

λ (nm)	Transferrin	Desferrioxamine	λ_{max}
428	3 540	2 730	Transferrin—470 nm
470	4 170	2 290	Desferrioxamine—428 nm

(a) A solution of transferrin exhibits an absorbance of 0.463 at 470 nm in a 1.000 cm cell. Calculate the concentration of transferrin in milligrams per milliliter and the concentration of iron in micrograms per milliliter.

(b) A short time after adding some desferrioxamine (which dilutes the sample), the absorbance at 470 nm was 0.424, and the absorbance at

428 nm was 0.401. Calculate the fraction of iron in transferrin and the fraction in desferrioxamine. Remember that transferrin binds two iron atoms and desferrioxamine binds only one.

20-E. The metal chelator semi-xylenol orange is yellow at pH 5.9, but turns red (λ_{max} = 490 nm) when it reacts with Pb^{2+}. A 2.025 mL sample of semi-xylenol orange at pH 5.9 was titrated with 7.515×10^{-4} M $Pb(NO_3)_2$, with the following results:

Total μL Pb^{2+} added	$A^{1\,cm}_{490\,nm}$
0.0	0.227
6.0	0.256
12.0	0.286
18.0	0.316
24.0	0.345
30.0	0.370
36.0	0.399
42.0	0.425
48.0	0.445
54.0	0.448
60.0	0.449
70.0	0.450
80.0	0.447

Make a graph of A versus μL of Pb^{2+} added. Be sure to correct the absorbances in the table for the effect of dilution. That is, the corrected absorbance is what would be observed if the volume were not changed from its initial value of 2.025 mL. Assuming that the reaction of semi-xylenol orange with Pb^{2+} has a 1:1 stoichiometry, find the molarity of semi-xylenol orange in the original solution.

20-F. The compound P, which absorbs light at 305 nm, was titrated with X, which does not absorb at this wavelength. The product, PX, also absorbs at 305 nm. The absorbance of each solution was measured in a 1.000 cm cell, and the concentration of free X was determined by an independent method. The results are shown below.

Experiment	P_0 (M)	X_0 (M)	A	[X] (M)
0	0.0100	0	0.213	0
1	0.0100	0.00100	0.303	4.42×10^{-6}
2	0.0100	0.00200	0.394	9.10×10^{-6}
3	0.0100	0.00300	0.484	1.60×10^{-5}
4	0.0100	0.00400	0.574	2.47×10^{-5}
5	0.0100	0.00500	0.663	3.57×10^{-5}
6	0.0100	0.00600	0.752	5.52×10^{-5}
7	0.0100	0.00700	0.840	8.20×10^{-5}
8	0.0100	0.00800	0.926	1.42×10^{-4}
9	0.0100	0.00900	1.006	2.69×10^{-4}
10	0.0100	0.0100	1.066	5.87×10^{-4}
11	0.0100	0.0200	1.117	9.66×10^{-3}

Prepare a Scatchard plot and find the equilibrium constant for the reaction $X + P \rightleftharpoons PX$.

20-G. Consider a fluorescence experiment in which the cell in Figure 20-26 is arranged so that b_1 and b_3 are negligible and, therefore, self-absorption can be neglected. To a first approximation, the emission intensity will simply be proportional to solute concentration. At what absorbance ($= \varepsilon_{ex} b_2 c$) will the emission be 5% below the value expected if emission is proportional to concentration?

Problems

20-1. Fill in the blanks:
(a) If you double the frequency of electromagnetic radiation, you _____ the energy.
(b) If you double the wavelength, you _____ the energy.
(c) If you double the wavenumber, you _____ the energy.

20-2. (a) How much energy (in kilojoules) is carried by one einstein of red light with λ = 650 nm?
(b) How much is carried by one einstein of blue light with λ = 400 nm?

20-3. Calculate the frequency (in hertz), wavenumber (in cm^{-1}), and energy (in joules per photon and joules per einstein) of visible light with a wavelength of 562 nm.

20-4. The absorbance of a 2.31×10^{-5} M solution of a compound is 0.822 at a wavelength of 266 nm in a 1.00 cm cell. Calculate the molar absorptivity at 266 nm.

20-5. What color would you expect to observe for a solution of the ion $Fe(ferrozine)_3^{4-}$, which has a visible absorbance maximum at 562 nm?

20-6. The characteristic orange light produced by sodium in a flame is due to an intense emission called the sodium "D" line. This "line" is actually a doublet, with wavelengths (measured in vacuum) of 589.157 88 and 589.755 37 nm. The index of refraction of air at a wavelength near 589 nm is 1.000 292 6. Calculate the frequency, wavelength, and wavenumber of each component of the "D" line, measured in air.

20-7. A compound with a molecular weight of 292.16 was dissolved in a 5 mL volumetric flask. A 1.000 mL aliquot was withdrawn, placed in a 10 mL volumetric flask, and diluted to the mark. If ε_{340} for this compound is 6 130 and the absorbance of the dilute solution at 340 nm in a 1.000 cm cell is 0.427, how many milligrams were dissolved in the 5 mL flask?

20-8. While assaying the thiamine (vitamin B_1) content of a pharmaceutical preparation, the percent transmittance scale was accidentally read, instead of the absorbance scale of the spectrophotometer. One sample gave a reading of 82.2% T, and a second sample gave a reading of 50.7% T at a wavelength of maximum absorbance. What is the ratio of concentrations of thiamine in the two samples?

20-9. Nitrite ion (NO_2^-) is used as a preservative for bacon and other foods. It has been the center of controversy because it is potentially carcinogenic. A spectrophotometric determination of NO_2^- makes use of the following reactions:

$$HO_3S-\bigcirc-NH_2 + NO_2^- + 2H^+ \longrightarrow$$

sulfanilic acid

$$HO_3S-\bigcirc-\overset{+}{N}\equiv N + 2H_2O$$

$$HO_3S-\bigcirc-\overset{+}{N}\equiv N + \bigcirc\bigcirc-NH_2 \longrightarrow$$

1-aminonaphthalene

$$HO_3S-\bigcirc-N=N-\bigcirc\bigcirc-NH_2 + H^+$$

(colored product)
(λ_{max} = 520 nm)

An abbreviated procedure for the determination is given below:

1. To 50.0 mL of unknown solution containing nitrite is added 1.00 mL of sulfanilic acid solution.

2. After ten minutes, 2.00 mL of 1-aminonaphthalene solution and 1.00 mL of buffer is added.

3. After fifteen minutes, the absorbance is read at 520 nm in a 5.00 cm cell.

The following solutions were analyzed.

A. 50.0 mL of food extract known to contain no nitrite (that is, a negligible amount). Final absorbance = 0.153.

B. 50.0 mL of food extract suspected of containing nitrite. Final absorbance = 0.622.

C. Same as B, but with 10.0μL of 7.50×10^{-3} M $NaNO_3$ added to the 50.0 mL sample. Final absorbance = 0.967.

(a) Calculate the molar absorptivity, ε, of the colored product. Remember that a 5.00 cm cell was used.

(b) How many micrograms of NO_2^- were present in 50.0 mL of food extract?

20-10. Spectrophotometric analysis of phosphate can be performed by the following procedure:

Standard solutions
A. KH_2PO_4 (F.W. 136.09)—81.37 mg dissolved in 500.0 mL of water.

B. $Na_2MoO_4 \cdot 2H_2O$ (sodium molybdate)—1.25 g in 50 mL of 5 M H_2SO_4.

C. $H_3NNH_3^{2+}SO_4^{2-}$ (hydrazine sulfate)—0.15 g in 100 mL of H_2O.

Procedure
Place the sample (either an unknown or the standard phosphate solution, A) in a 5 mL volumetric flask, and add 0.500 mL B plus 0.200 mL C. Dilute to almost 5 mL with water, and heat at 100°C for ten minutes to form a blue product ("molybdenum blue"). Cool the flask to room temperature, dilute to the mark with water, mix well, and measure the absorbance at 830 nm in a 1.00 cm cell.

(a) When 0.140 mL of solution A was analyzed, an absorbance of 0.829 was recorded. A blank carried through the same procedure gave an absorbance of 0.017. Find the molar absorptivity of blue product, assuming that it contains one mole of phosphorus per mole of molybdenum blue.

(b) A solution of the phosphate-containing iron-storage protein ferritin was analyzed by this procedure. The unknown contained 1.35 mg of ferritin, which was digested in a total volume of 1.00 mL to release phosphate from the protein. Then 0.300 mL of this solution was analyzed by the procedure above and found to give an absorbance of 0.836. A blank carried through this procedure gave an absorbance of 0.038. Find the weight percent of phosphorus in the ferritin.

20-11. If a sample for spectrophotometric analysis is placed in a 10 cm cell, the absorbance will be ten times greater than the absorbance in a 1 cm cell. Will the absorbance of the reagent-blank solution also be increased by a factor of ten?

20-12. When I was a boy, Uncle Wilbur let me watch as he analyzed the iron content of runoff from his banana ranch. A 25.0 mL sample was acidified with nitric acid and treated with excess KSCN to form a red complex. (KSCN itself is colorless.) The solution was then diluted to 100.0 mL and put in a variable-pathlength cell. For comparison, a 10.0 mL reference sample of 6.80×10^{-4} M Fe^{3+} was treated with HNO_3 and KSCN and diluted to 50.0 mL. The reference was placed in a cell with a 1.00 cm light path. The runoff sample exhibited the same absorbance as the reference when the pathlength of the runoff cell was 2.48 cm. What was the concentration of iron in Uncle Wilbur's runoff?

20-13. Infrared spectra are customarily recorded on a %T scale so that weak and strong bands can be displayed on the same scale. The region near 2 000 cm^{-1} in the infrared spectra of compounds A and B is shown at the bottom of the page. Note that absorption corresponds to a downward peak on this scale. The spectra were recorded using a 0.010 0 M solution of each, in 0.005 00 cm pathlength cells. A mixture of A and B in a 0.005 00 cm cell gave a transmittance of 34.0% at 2 022 and 38.3% at 1 993 cm^{-1}. Find the concentrations of A and B.

	2 022	1 993 (cm^{-1})
pure A	31.0% T	79.7% T
pure B	97.4% T	20.0% T

20-14. The coenzyme NADP$^+$ can be assayed by a titration in which it is converted to the fluorescent product NADPH by the action of adenosine triphosphate (ATP) plus several enzymes. A hypothetical titration curve is shown below.

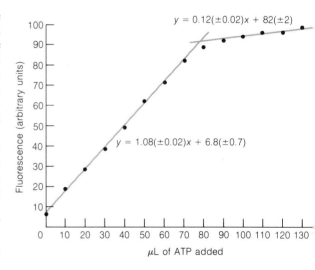

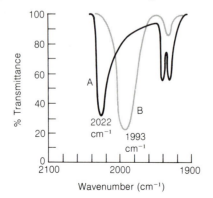

In this titration, the fluorescence intensity is plotted versus microliters of added ATP. The first eight points lie on one line, and the points from 90 to 140 μL lie on a second line. The uncertainties in slope and intercept of these lines are one standard deviation, as determined by the method of least squares. The endpoint of the titration lies at the intersection of the two lines. Using the equations for the two lines, *determine the volume of ATP (in microliters) at the equivalence point*. Also, use the standard deviations of the slopes and intercepts to *estimate the standard deviation of the volume of ATP at the equivalence point*. Express your answer (μL $\pm$ standard deviation) with an appropriate number of significant figures.

20-15. The metal ion indicator xylenol orange (Table 14-3) is yellow at pH 6 (λ_{max} = 439 nm). The spectral changes that occur as the indicator is titrated with VO^{2+} ion at pH 6 are shown below. The molar ratio VO^{2+}/xylenol orange at each point is

Trace	Ratio	Trace	Ratio
0	0		
1	0.10	9	0.90
2	0.20	10	1.0
3	0.30	11	1.1
4	0.40	12	1.3
5	0.50	13	1.5
6	0.60	14	2.0
7	0.70	15	3.1
8	0.80	16	4.1

Suggest a sequence of chemical reactions to explain the spectral changes, especially the isosbestic points at 457 and 528 nm.

20-16. How can the sample cell in Figure 20-26 be positioned to minimize the self-absorption expressed in the terms $10^{-\varepsilon_{ex}b_1 c}$ and $10^{-\varepsilon_{em}b_3 c}$ in Equation 20-40.

20-17. Consider a fluorescence experiment in which ε_{ex} = 1 530 M^{-1} cm^{-1}, ε_{em} = 495 M^{-1} cm^{-1}, b_1 = 0.400 cm, b_2 = 0.200 cm, and b_3 = 0.500 cm in Equation 20-40. Make a graph of relative fluorescence intensity versus concentration for the following concentration of solute: 1.00×10^{-7}, 1.00×10^{-6}, 1.00×10^{-5}, 1.00×10^{-4}, 1.00×10^{-3}, and 1.00×10^{-2} M.

20-18. Consider a reflection grating operating with an incident angle of 40° in Figure 20-13.
   (a) How many lines per centimeter should be etched in the grating if the first-order diffraction angle for 600 nm (visible) light is to be 30°?
   (b) Answer the same question for 1 000 cm^{-1} (infrared) light.

20-19. A sensitive assay for ATP is based on its participation in the light-producing reaction of the firefly.[†] The reaction catalyzed by the enzyme luciferase is

[†] J. J. Lemasters and C. R. Hackenbrock, *Methods of Enzymology*, **57**, 36 (1978).

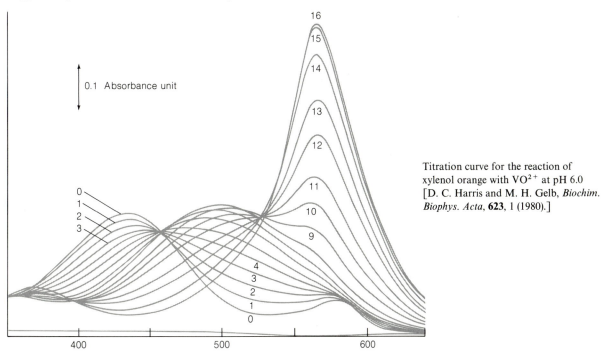

Titration curve for the reaction of xylenol orange with VO^{2+} at pH 6.0 [D. C. Harris and M. H. Gelb, *Biochim. Biophys. Acta*, **623**, 1 (1980).]

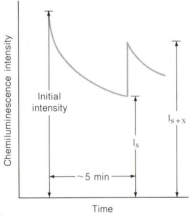

When the reactants are mixed, the solution gives off light. The light intensity decays slowly due to product inhibition of the reaction. Otherwise, the light would have a steady intensity because *the rate at which reactants are consumed is negligible.* That is, ATP and luciferin maintain their original concentrations throughout the few minutes that the reaction might be monitored. Some typical experimental results are shown below.

Let the initial concentration of ATP in the reaction be [S]. Suppose that additional ATP is added, increasing the concentration in the reaction to [S] + [X]. The kinetic description of the reaction predicts that the increase in light intensity after the addition will be given by

$$\frac{I_S}{I_{S+X}} = \frac{1}{[S] + [X]}\left(\frac{K[S]}{K + [S]}\right) + \frac{[S]}{K + [S]}$$

where K is a constant.

(a) Suppose that [S] = 250 μM and after five minutes I_S = 58.7 arbitrary intensity units. Then a standard addition of [X] = 200 μM is made, and I_{S+X} is found to be 74.5 units. Use these data to find the value of K in the equation above.

(b) When the intensity had decayed to 63.5 units, an unknown aliquot of ATP was added to the reaction, and the intensity increased to 74.6 units. How much was the increase in concentration caused by the unknown aliquot?

20-20. Compound P was titrated with X to form the complex PX. A series of solutions was prepared with the total concentration of P remaining fixed at 1.00×10^{-5} M. Both P and X have no visible absorbance, but PX has an absorption maximum at 437 nm. From the data below, calculate K for the equilibrium P + X ⇌ PX, and calculate ε for PX at 437 nm. The concentrations of X refer to the total molarity of X in both forms (X and PX). The absorbance was measured at 437 nm in a 5.00 cm cell.

[X]	A
0.002 00	0.125
0.004 00	0.213
0.006 00	0.286
0.008 00	0.342
0.010 0	0.406
0.020 0	0.535
0.040 0	0.631
0.060 0	0.700
0.080 0	0.708
0.100	0.765

21/ Atomic Spectroscopy

When heated to a sufficiently high temperature, most compounds break apart into atoms in the gaseous phase. Unlike the optical spectra of condensed phases, the spectra of these atoms consist of very sharp lines. For example, the spectrum of an iron complex in solution typically has broad bands 100 nm in width, but the spectrum of gaseous Fe is a series of sharp lines, whose natural width is <0.01 nm (Figure 21-1). This spectrum arises from transitions between the electronic states of the Fe atom. Each element has its own characteristic spectrum. Because the lines are so sharp, there is usually little overlap between the spectra of different elements in the same sample.

In *atomic spectroscopy*, samples are vaporized at very high temperatures and the concentrations of selected atoms are determined by measuring absorption or emission at their characteristic wavelengths. Because of its high sensitivity and the ease with which many samples can be examined, atomic spectroscopy has become one of the principal tools of analytical chemistry, especially in industrial settings. Analyte concentrations at the parts-per-million level are routine, and parts-per-billion levels are amenable to analysis in some cases. In analyzing the major constituents of an unknown, the sample is usually diluted to reduce concentrations to the parts-per-million level. Atomic spectroscopy is not as accurate as many wet chemical methods, since its precision is rarely better than $1-2\%$. The equipment is expensive, but widely available.

> The unit ppm (parts per million) refers to micrograms of solute per gram of solution. Since the density of dilute aqueous solutions is close to 1 g/mL, ppm is often used to mean μg/mL. A concentration of 1.00 ppm of Fe corresponds to 1.00×10^{-6} g Fe/mL $= 1.79 \times 10^{-5}$ M.

21-1 ABSORPTION, EMISSION, AND FLUORESCENCE

In conventional molecular spectroscopy, the absorbance of a sample placed in the beam of light is measured. Alternatively, the sample is irradiated, and its luminescence (fluorescence or phosphorescence) is measured in a direction perpendicular to the incident beam. Both of these experiments

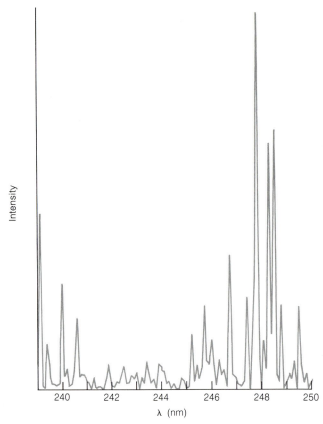

Figure 21-1
A small portion of the spectrum of a hollow-cathode Fe lamp, showing the series of sharp lines characteristic of gaseous atoms. The linewidths in this spectrum are artificially broadened by the monochromator, whose bandwidth is 0.08 nm.

Atomic spectroscopy:
1. Absorption.
2. Fluorescence (luminescence following absorption of radiation).
3. Emission (luminescence from a thermally populated excited state).

can also be done with an atomic vapor. In addition, at the high temperature of the vapor, many atoms are already in thermally populated, excited electronic states. They can spontaneously emit photons and return to a lower state. Therefore, atomic spectroscopy falls into three classes commonly designated as **absorption, fluorescence,** and **emission.** Routinely available instruments can perform absorption and emission experiments with equal ease. Equipment for **atomic fluorescence,** a technique potentially a thousand times more sensitive than absorption or emission, is currently under development and is not yet in general use.

Apparatus for an **atomic absorption** experiment is shown in Figure 21-2. The liquid sample is aspirated (sucked) into a flame whose temperature is 2 000–3 000 K. The sample is **atomized** (broken into atoms) in the flame, which takes the place of the cuvette in conventional spectrophotometry. The pathlength of the flame is typically 10 cm. To measure the light absorbance of Fe atoms in the flame, the light source uses a cathode made of Fe.

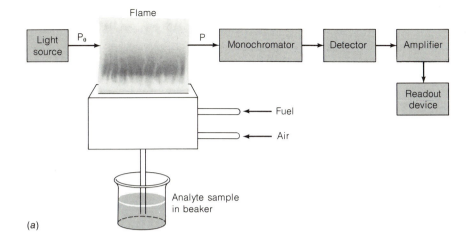

Figure 21-2
(a) Outline of an atomic absorption spectrometer. (b) Photograph of a research-quality instrument for atomic absorption and emission. The sample in the flask is being aspirated into the burner, which is behind the metal cage. Valves on the left control the flow rate of fuel and oxidizer. Dials on the right are used to select wavelengths, monochromator bandwidth, and observation modes. Results are displayed on the video tube or a printer. [Courtesy Instrumentation Laboratory, Wilmington, Mass.]

It emits the characteristic frequencies of Fe atoms. The remainder of the apparatus is not very different from an ordinary spectrophotometer.

Atomic emission spectroscopy is basically the same as atomic absorption spectroscopy. The difference is that no light source is needed. Some of the atoms in the flame are promoted to excited electronic states by collision with other atoms. The excited atoms emit their characteristic radiation as they return to their ground state. In atomic emission spectroscopy, the emission

Atomic emission requires the same equipment as atomic absorption, but the lamp is not used.

21 / ATOMIC SPECTROSCOPY

intensity at a characteristic wavelength of an element is nearly proportional to the concentration of the element in the sample. For both absorption and emission, standard curves are usually used to establish the relation between signal and concentration.

21-2 ATOMIZATION: FLAMES, FURNACES, AND PLASMAS

The essential feature that distinguishes atomic spectroscopy from ordinary spectroscopy is that the sample must be atomized. This is usually accomplished with a flame, less frequently with an electrically heated oven or a radio-frequency plasma. The sensitivity and interfering effects observed in atomic spectroscopy depend on the details of the heating process.

Premix Burner

Nebulization

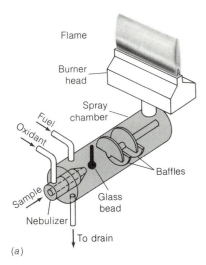

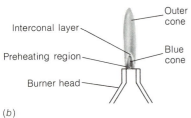

Figure 21-3
(a) Schematic diagram of a premix burner. (b) End view of flame.

Organic solvents having less surface tension than water are excellent for atomic spectroscopy because they form smaller droplets, leading to more efficient sample atomization.

Most modern flame atomic spectrometers use a **premix burner,** such as that in Figure 21-3, in which the sample, oxidant, and fuel are mixed before introduction into the flame. The sample solution (which need not be aqueous) is drawn into the **nebulizer** by the rapid flow of oxidant (usually air) past the tip of the sample capillary. The liquid breaks into a fine mist as it leaves the tip of the nebulizer. The spray is directed at high speed against a glass bead, upon which the droplets are broken into even smaller particles. The mist, oxidant, and fuel flow past a series of baffles that promotes further mixing and blocks large droplets of liquid. Liquid that collects at the bottom of the spray chamber flows out to a drain. Only a very fine mist of droplets reaches the flame.

The flame

The most common fuel–oxidizer combination is acetylene and air, which produces a flame temperature of $\sim 2\,400$–$2\,700$ K. Other fuels and oxidizers are listed in Table 21-1. When a hotter flame is required, the acetylene–nitrous

Table 21-1
Maximum flame temperatures

Fuel	Oxidant	Temperature (K)
Acetylene	Air	2 400–2 700
Acetylene	Nitrous oxide	2 900–3 100
Acetylene	Oxygen	3 300–3 400
Hydrogen	Air	2 300–2 400
Hydrogen	Oxygen	2 800–3 000
Cyanogen	Oxygen	4 800

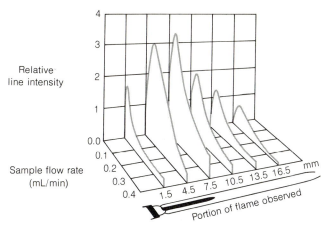

Figure 21-4
Profile of Ca emission line in a cyanogen/oxygen flame. [K. Fuwa, R. E. Thiers, B. L. Vallee, and M. R. Baker, *Anal. Chem.*, **31**, 2039 (1959).]

oxide combination is usually used. A flame profile is shown in Figure 21-3. Gas that enters the preheating region from the burner head is heated by downward conduction and radiation from the primary reaction zone (the blue cone). Combustion is completed in the outer cone, where surrounding air is drawn into the flame. Each flame has its own emission spectrum and therefore obscures the spectrum of analyte in certain regions.

Droplets entering the flame first lose their water through evaporation; then the remaining sample must vaporize and decompose into atoms. Many elements form oxides as they rise through the outer cone. Oxides do not have the same spectra as their free elements, and this lowers the atomic signal. If the flame is kept relatively rich in fuel (a "rich" flame), there is an excess of carbon species, which might reduce the metal oxides and thereby increase sensitivity. Opposite to a rich flame is a "lean" flame, which has excess oxidant and is hotter than a rich flame. Lean or rich flames are recommended in the analysis of different elements.

The position in the flame at which maximum atomic absorption or emission is observed depends on the element being measured as well as the flow rate of sample, fuel, and oxidizer. A profile of emission from Ca atoms in a cyanogen (N≡C—C≡N)/oxygen flame is shown in Figure 21-4. The decreasing intensity at higher flow rates is attributed to cooling of the flame by water from the sample. Different elements have different absorption and emission profiles. The sample flow rate, fuel and oxidant flow rates, and the level at which the flame is observed can be optimized for each element.

Hotter flames are needed for *refractory* (high vaporization temperature) elements or to decompose such species as metal oxides formed during passage through the flame.

Furnaces

The electrically heated furnace is an increasingly popular means of atomization in atomic absorption spectroscopy because it offers greater sensitivity than flames and requires a smaller volume of sample. Figure 21-5 shows a graphite-rod furnace mounted in the beam of a spectrometer. From 1 μL

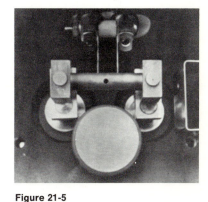

Figure 21-5
Photograph of an electrically heated graphite-rod furnace used for flameless atomic spectroscopy. Light travels the length of the furnace (~38 mm in this case), and sample is injected through the hole at the top. [Courtesy Instrumentation Laboratory, Wilmington, Mass.]

518

21 / ATOMIC SPECTROSCOPY

Furnaces offer increased sensitivity and require less sample than a flame, but usually give poorer precision.

The operator must determine reasonable time and temperature for each stage of the analysis. Once a program is established, it can be applied to a larger number of similar samples.

to 100 μL of sample is injected into the oven through the hole at the center. Each end of the oven is a window through which the light beam travels.

A **graphite furnace** provides higher sensitivity because the entire sample is confined in the light path for a few seconds. In flame spectroscopy, the sample is highly diluted by the time it has been nebulized, and it spends only a fraction of a second in the light path as it rises through the flame. Flames also require a much larger volume of sample, since sample is constantly flowing into the flame. Whereas 1–2 mL is needed for flame analysis, as little as 1 μL is adequate for a furnace.

Furnaces yield less precision than do flames. Reproducibility is rarely better than 5–10%. They also require more operator skill and greater effort to determine the proper conditions for each type of sample. The reason for the increased effort is that the furnace must be heated in three or more steps to properly atomize the sample. As an example, to analyze iron in the iron-storage protein ferritin, 10 μL of sample containing $\sim$0.1 ppm Fe was injected into the cold oven. The furnace was programmed to dry the sample at 125°C for 20 s to remove solvent. This was followed by 60 s of charring at 1 200°C to destroy organic matter, which would otherwise create a great deal of smoke and interfere with the Fe determination. Finally, atomization is accomplished by heating to 2 700°C for 10 s. During this 10 s heating, the absorbance reaches a maximum and then decreases as the Fe evaporates from the oven. Either the maximum absorbance measured on a recorder or the time-integrated absorbance is taken as the analytical signal. It is important to record these events with a recorder or an oscilloscope, because signals are also observed from smoke produced during charring and from the glow of the red-hot oven in the latter part of atomization. A skilled operator must interpret which signal was due to sample and which to other effects.

Inductively Coupled Plasma

The **inductively coupled plasma** is a type of flame that reaches a much higher temperature than ordinary combustion flames and is useful for emission spectroscopy. Its high temperature and stability eliminate many of the interferences and sources of error encountered with conventional flames. Because of these desirable features, the inductively coupled plasma is beginning to replace conventional flame burners. The plasma's principal disadvantage is its expense to purchase and operate.

A cross-sectional view of an inductively coupled plasma burner head is shown in Figure 21-6. Two turns of a radio-frequency induction coil are wrapped around the upper opening of the quartz apparatus. High-purity argon gas is fed through the plasma gas inlet. A spark from a Tesla coil is used to ionize the Ar gas. The Ar^+ ions are immediately accelerated by the powerful radio-frequency field that oscillates about the load coil at a frequency of 27 MHz. The accelerated ions transfer energy to the entire gas by collisions between atoms. Once the process is begun, the ions absorb enough energy from the electric field to maintain a temperature of 6 000 to

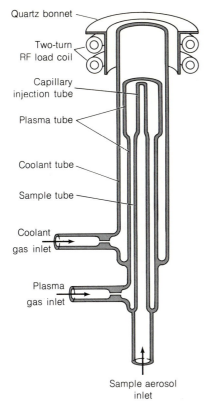

Figure 21-6
Diagram of an inductively coupled plasma burner head. [R. N. Savage and G. M. Hieftje, *Anal. Chem.*, **51**, 408 (1979).]

10 000 K in the flame (Figure 21-7). The plasma is so hot (especially near the coils) that the quartz burner must be protected by argon coolant gas flowing around the outer edge of the apparatus. Sample can be introduced into the flame by a conventional nebulizer.

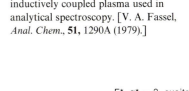

Figure 21-7
Temperature profile of a typical inductively coupled plasma used in analytical spectroscopy. [V. A. Fassel, *Anal. Chem.*, **51**, 1290A (1979).]

Effect of Temperature in Atomic Spectroscopy

Temperature is a critical factor in determining the degree to which a given sample breaks down to atoms. In addition, temperature determines the extent to which a given atom is found in its ground, excited, or ionized states.

Boltzmann distribution

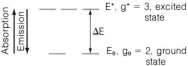

Figure 21-8
Two energy levels with different degeneracies. Ground-state atoms can absorb light to be promoted to the excited state. Excited-state atoms can emit light to return to the ground state.

Consider a molecule with two available energy levels separated by energy ΔE (Figure 21-8). Call the lower level E_0 and the upper level E^*. In general, an atom (or molecule) may have more than one state available at a given energy

520

21 / ATOMIC SPECTROSCOPY

level. In Figure 21-8, we show three states available at E* and two available at E_0. The number of states available at each energy level is called the *degeneracy* of the level. We will call the degeneracies g_0 and g^*.

If thermal equilibrium exists (which is not true in the blue cone of a flame, but is probably true above the blue cone), the relative populations of any two energy levels are given by

The Boltzmann distribution applies to a system at thermal equilibrium.

$$\frac{N^*}{N_0} = \frac{g^*}{g_0} e^{-\Delta E/kT} \tag{21-1}$$

where N is the population of each state, T is kelvins, and k is Boltzmann's constant ($1.380\,662 \times 10^{-23}$ J K^{-1}).

Effect of temperature on excited-state population

The lowest excited state of a sodium atom lies 3.371×10^{-19} J/atom above the ground state. The degeneracy of the excited state is 2, while that of the ground state is 1. Let's calculate the fraction of sodium atoms in the excited state in an acetylene–air flame at 2 600 K. Using Equation 21-1, we find

$$\frac{N^*}{N_0} = \left(\frac{2}{1}\right) e^{-3.371 \times 10^{-19}/(1.381 \times 10^{-23} \cdot 2\,600)} = 1.67 \times 10^{-4} \tag{21-2}$$

That is, less than 0.02% of the atoms are in the excited state.

How would the fraction of atoms in the excited state change if the temperature were 2 610 K instead?

A 10° temperature rise changes the excited-state population by 4% in this example.

$$\frac{N^*}{N_0} = \left(\frac{2}{1}\right) e^{-3.371 \times 10^{-19}/(1.381 \times 10^{-23} \cdot 2\,610)} = 1.74 \times 10^{-4} \tag{21-3}$$

The fraction of atoms in the excited state is still less than 0.02%, but that fraction has changed by $100(1.74 - 1.67)/1.67 = 4\%$.

Effect of temperature on absorption and emission

In the preceding section, we saw that more than 99.98% of the sodium atoms are in their ground state at 2 600 K. *Varying the temperature by 10° hardly affects the ground-state population and would not noticeably affect the signal in an atomic absorption experiment.*

It turns out that the emission spectrum of sodium is much more intense than the absorption spectrum, because the efficiency of emission is very much greater than the efficiency of absorption for this element. How would the emission intensity be affected by a 10° rise in temperature?

In Figure 21-8, it can be seen that absorption arises from ground-state atoms, but emission arises from excited-state atoms. The emission intensity should be proportional to the population of the excited state. *Since the excited-state population changes by 4% when the temperature rises 10°, the emission intensity will also rise 4%.* It is critical in atomic *emission* spectroscopy that flame conditions be very stable, or the emission intensity will vary significantly. In atomic *absorption* spectroscopy, flame temperature variation is not as critical.

The inductively coupled plasma is so hot that a substantial population of excited-state atoms and ions exists. The plasma is therefore almost always used for emission, not absorption, measurements. Relative to a flame, the plasma has a more uniform temperature profile and therefore gives more reproducible emission intensities.

> The intensity of atomic absorption is not very sensitive to temperature. The intensity of atomic emission is very sensitive to temperature.

21-3 INSTRUMENTATION

The fundamental requirements for an atomic absorption experiment are shown in Figure 21-2. The principal differences between atomic spectroscopy and ordinary molecular spectroscopy lie in the light source, the sample container (the flame), and the need to subtract the flame emission spectrum from the observed signal.

Source of Radiation

The linewidth problem

Beer's law applies to monochromatic radiation. In practical terms, this means that the linewidth of the radiation being measured should be substantially narrower than the bandwidth of the absorbing sample. Otherwise, the measured absorbance will not be proportional to the sample concentration.

Atomic absorption lines are very sharp, with an inherent width of only $\sim 10^{-4}$ nm. Two mechanisms serve to broaden the lines in atomic spectroscopy. One is the **Doppler effect.** An atom moving toward the radiation source samples the oscillating electromagnetic wave more frequently than one moving away from the source (Figure 21-9). That is, an atom moving toward the source sees higher-frequency light than one moving away. In the laboratory frame of reference, the atom moving toward the source absorbs a lower frequency of light than the one moving away. The linewidth, Δv, due to the Doppler effect, is given approximately by

$$\Delta v \approx v(7 \times 10^{-7})\sqrt{\frac{T}{M}} \qquad (21\text{-}4)$$

where v is the frequency of the peak maximum, T is kelvins, and M is the mass of the atom in atomic mass units. Δv is the width of the absorption line measured at half the height of the peak.

Another factor that affects atomic absorption linewidths is called **pressure broadening.** This arises because an atom does not behave as an isolated system during a collision with another atom. Its energy levels are

> The bandwidth of the source must be narrower than the bandwidth of the atomic vapor for Beer's law to be obeyed.

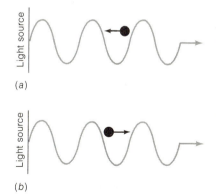

Figure 21-9
The Doppler effect. A molecule moving (a) toward the radiation source "feels" the electromagnetic field oscillate more often than one moving (b) away from the source.

Doppler and pressure effects broaden the atomic lines by 1–2 orders of magnitude as compared with their inherent linewidths.

perturbed, and it will not absorb the same frequency of radiation as an isolated atom. The pressure broadening, Δv_p, is roughly equal to the collision frequency and is proportional to pressure.

The Doppler effect and pressure broadening are of a similar order of magnitude. Together they yield linewidths of 10^{-3}–10^{-2} nm in atomic spectroscopy.

Hollow-cathode lamp

There is no monochromator that can produce linewidths smaller than 10^{-3}–10^{-2} nm. Lasers can do this, but they are not yet in general use. To produce such narrow lines of the correct frequency, we use a **hollow-cathode lamp** containing the same element as that being analyzed.

A hollow-cathode lamp, such as that shown in Figure 21-10, is filled with Ne or Ar at a pressure of ~ 130–700 Pa (1–5 torr). When a high enough voltage is applied between the anode and cathode, the filler gas becomes ionized and positive ions are accelerated toward the cathode. They strike the cathode with enough energy to "sputter" metal atoms from the cathode into the gas phase. Many of the sputtered atoms are in excited states; they emit photons and then return to the ground state. This atomic radiation is of exactly the same frequency as that which can be absorbed by the atoms of analyte. The linewidth is sufficiently sharp (narrow), with respect to that of the high-temperature analyte, to be nearly "monochromatic" (Figure 21-11).

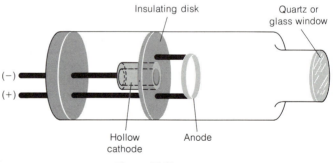

Figure 21-10
A hollow-cathode lamp.

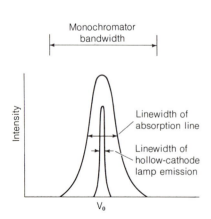

Figure 21-11
Relative linewidths of hollow-cathode emission, atomic absorption, and monochromator. Linewidths are measured at half the signal height.

Most of the sputtered atoms condense back on the cathode, but some are deposited on the glass walls of the lamp, eventually ruining the lamp. A different lamp is usually required for each element, although some lamps are made with more than one element in the cathode.

The emission from a lamp having a hollow iron cathode is shown in Figure 21-1. The linewidths are actually much narrower than the figure indicates, because the monochromator used to record the spectrum had a bandwidth of 0.08 nm. Some lines produced by the lamp arise from gaseous ions, such as Fe^+, Ne^+, or Ar^+.

The Spectrophotometer

We have already discussed the two most unusual features of an atomic absorption spectrophotometer. One is that the sample "container" is a flame or furnace. The other is that the lamp puts out just a few sharp lines at precisely those frequencies absorbed by the analyte. The remainder of the spectrophotometer is not very different from one used for ordinary absorption spectroscopy. The biggest difference is that some means must be provided in atomic spectroscopy to distinguish the analyte signal from the spectrum of the flame, which has its own emission.

Background correction

A **beam chopper** is usually used to distinguish the signal due to the flame from the desired atomic line at the same wavelength. As shown in Figure 21-12, the beam from the lamp is periodically blocked by the rotating chopper. The signal that reaches the detector while the beam is blocked must be due to flame emission. The signal reaching the detector when the beam is not blocked is the sum of the signals from the lamp and the flame. The difference between these two signals is the desired analytical signal, which is displayed by the spectrometer.

An alternative means of measuring the flame background involves modulating the lamp with an alternating current. The signal reaching the detector will therefore contain a constant contribution from the flame and a modulated contribution from the lamp. By electronically decomposing the detected signal into these two components, a flame emission correction is provided.

Some samples produce particulate matter (smoke or unvaporized particles), which scatters a significant fraction of light from the hollow-cathode lamp. Smoke is particularly troublesome with graphite furnaces. The detector cannot distinguish scattering from absorption, so this can lead to a systematic error in the measurement of analyte.

Many spectrometers provide a means to correct for background scattering and broad-band background absorption. Emission from a continuous radiation source (usually a D_2 lamp) is passed through the flame in alternation with that from the hollow cathode. The monochromator bandwidth is sufficiently wide that a negligible fraction of the D_2 lamp radiation is absorbed by the analyte's atomic absorption line. The attenuation of this continuous radiation is due to background scattering or broad-band absorption. The spectrometer corrects the analytical signal for the attenuation suffered by the background correction beam.†

Sensitivity and detection limit

The **sensitivity** of an atomic absorption spectrometer for a given element is defined as the concentration of that element needed to produce 99% trans-

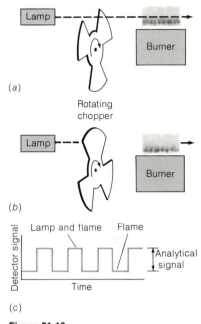

Figure 21-12
Operation of a beam chopper for subtracting the signal due to flame background emission. (a) Lamp and flame emission reach detector. (b) Only flame emission reaches detector. (c) Resulting signal.

Beam chopping and source modulation can correct for flame emission, but not for scattering.

---

† Another background correction technique coming into use is based on the *Zeeman effect*—the splitting of atomic states in a magnetic field. For an explanation of how this works, see S. D. Brown, *Anal. Chem.*, **49,** 1269A (1977).

524

21 / ATOMIC SPECTROSCOPY

Sensitivity—concentration producing 99%T.

Detection limit—concentration giving signal-to-noise ratio of 2.

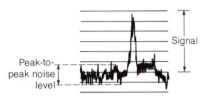

Figure 21-13
Illustration of the measurement of peak-to-peak noise level and signal level. The signal is measured from its base at the mid-point of the noise component along the slightly slanted baseline. This sample exhibits a signal-to-noise ratio of 3.2.

mittance (which corresponds to an absorbance of 0.004 36). The **detection limit** is that concentration of an element which gives a signal equal to twice the peak-to-peak noise level of the baseline (Figure 21-13). The baseline noise level should be measured while aspirating a blank sample into the flame.

Figure 21-14 compares the detection limits of flame and flameless (furnace) operation for a particular instrument. You can see that most elements can be determined by atomic absorption and that there are wide variations in the detection limits for different elements. This is attributable to the varying efficiencies of atomization and the differing absorptivities (ε) of different elements. The detection limit for flameless operation is typically two orders of magnitude lower than that observed with a flame. The main reason for this is that the sample is confined in a small volume for a relatively long time in the furnace, compared to its fleeting moment in a flame.

21-4 ANALYTICAL METHODS

Most elements of the periodic table can be analyzed by either or both atomic absorption and atomic emission. Figure 21-14 shows the most suitable method for each element. The equipment for both techniques is the same, except

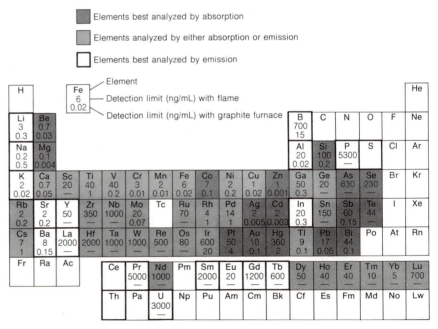

Figure 21-14
Atomic absorption detection limits (ng/mL) with the Perkin–Elmer 703 spectrophotometer. Adequate signals for quantitative analysis usually require a concentration about 100 times greater than the detection limit.

that a lamp is required for atomic absorption measurements. The elements most commonly analyzed by atomic emission are Li, Na, and K (Box 21-1). The choice of which method to use for other elements is often dictated by the availability of a lamp.

Standard Curve

The most common technique for quantitative analysis is to construct a standard curve, such as that in Figure 21-15, using known amounts of the desired element in a solution with a similar composition to the unknown. The standard curve is then used to find the concentration of unknown from

---

Box 21-1 THE FLAME PHOTOMETER IN CLINICAL CHEMISTRY

Measurement of sodium and potassium in serum and urine is a routine procedure in diagnostic clinical analysis. This was not true before the advent of atomic spectroscopy because laborious gravimetric techniques were required.

The flame photometer shown below is typical of clinical instruments. It operates on oxygen and natural gas. As many as eighteen samples containing 20 μL of serum plus 2 mL of water are placed in the rack on the right. Each sample is automatically aspirated into the flame in the housing above the sample rack. Simple optical filters are sufficient to isolate the strong emission lines of sodium, potassium, and lithium. The concentrations are read out directly on the meter, which is calibrated by aspirating standard samples. Lithium is often used as an internal standard for sodium and potassium.

(a)

(b)

(a) The Coleman Model 51 Flame Photometer used in clinical analysis.

(b) Meter scale. [Courtesy A. H. Thomas Co., Philadelphia, Pa.]

21 / ATOMIC SPECTROSCOPY

The composition of the standards should be as close as possible to the composition of the unknown.

its absorbance. The medium in which the analyte is contained is called the **matrix**. It is *critical* that the composition of the standards be as close as possible to that of the unknown, because different solutions have different types of interferences that affect the signal.

Standard Addition Method

In the **standard addition method**, known quantities of the desired element are added to the analyte, and the increase in signal is measured.[†] Each solution is diluted to the same total volume and should have the same final composition (except for analyte concentration). If the concentration of unknown is [X] and the concentration of added standard is [S], we can say that

$$\frac{[X]}{[X] + [S]} = \frac{A_X}{A_{S+X}} \tag{21-5}$$

where A_X is the absorbance (or emission intensity) of unknown and A_{S+X} is the absorbance (or emission intensity) of unknown plus standard. Equation 21-5 applies only if the absorbance or emission is linearly related to concentration. Most elements exhibit some concentration range where this is true.

Equation 21-5 can be solved directly for [X]. Alternatively, a series of standard additions can be made, and a graph such as Figure 21-16 can be used to find the concentration of unknown. In this graph, the *x* axis is the concentration of added analyte *after* it has been mixed with sample. The

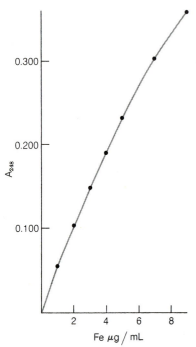

Figure 21-15
Atomic absorption calibration curve for Fe.

[†] For a complete discussion of standard addition methods, see M. Bader, *J. Chem. Ed.*, **57**, 703 (1980).

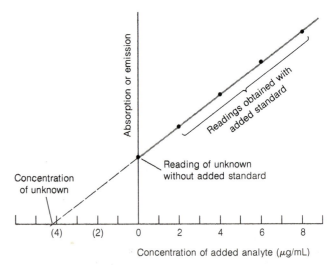

Figure 21-16
Graphical treatment of the method of standard additions.

x intercept of the extrapolated line is equal to the concentration of unknown in the original sample. In Figure 21-16, this value is near 4.2 µg/mL. Statistically, the most useful range of standard additions should increase the analyte concentration to between 1.5 and 3 times its original value. The main advantage of the standard addition method is that the matrix remains constant for all samples.

The graphic treatment in Figure 21-16 requires that the response be linear.

Internal Standard Method

An **internal standard** is a known amount of an element, not already present, which is added to a sample. To use an internal standard, *known* mixtures of standard (S) and analyte (X) are used to construct a standard curve, such as that in Figure 21-17. When a known amount of standard is added to an unknown sample, the calibration curve can be used to find the concentration of unknown.

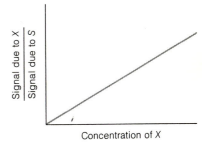

Figure 21-17
Calibration curve for an internal standard. To construct such a curve, the concentration of X would be varied while the concentration of S is held fixed.

EXAMPLE
A solution was prepared by mixing 5.00 mL of unknown (X) with 2.00 mL of solution containing 4.13 µg of standard (S) per milliliter, and diluting to 10.0 mL. The measured ratio of signals was

$$\frac{\text{Signal due to X}}{\text{Signal due to S}} = 0.808$$

In a separate experiment, it was found that for equal concentrations of X and S, the signal due to X was 1.31 times as intense as the signal due to S. Find the concentration of X in the unknown.

The concentration of internal standard in the unknown mixture was

$$[S] = (4.13 \text{ µg/mL}) \underbrace{\left(\frac{2.00}{10.0}\right)}_{\text{Dilution factor}} = 0.826 \text{ µg/mL}$$

Since the measured signal ratio is 1.31 when the concentration ratio is unity, we can say that

$$\frac{\text{Absorbance ratio in standard mixture}}{\text{Concentration ratio in standard mixture}} = \frac{\text{Absorbance ratio in unknown}}{\text{Concentration ratio in unknown}}$$

$$\frac{1.31}{1} = \frac{0.808}{[X]/[S]}$$

$$[X]/[S] = 0.617$$

Since $[X]/[S] = 0.617$ and since $[S] = 0.826$ µg/mL, $[X] = (0.617)(0.826) = 0.510$ µg/mL. But X was diluted by a factor of 2.00 when it was mixed with the standard. Therefore, the original concentration of X was 1.02 µg/mL.

528

21 / ATOMIC SPECTROSCOPY

We assume that changing the sample composition affects the signal from both elements equally.

Internal standards are most useful when unavoidable sample losses are expected. If a standard is added to a sample prior to any losses, then the fraction of standard lost is the same as the fraction of sample lost, and the ratio [Standard]/[Sample] remains constant. Internal standards are also useful when the sample matrix cannot be controlled. For example, the apparent concentration of Fe goes down by 20% as the NaCl concentration of an aqueous solution is increased from 0 to 10%. The reason is probably that the solution becomes more viscous and flows into the nebulizer more slowly. Decreased flow means that fewer atoms of iron reach the flame. The absorbance decreases by 20%. A way to circumvent this problem is to inject Mn as an internal standard. If the viscosity of the sample increased such that the flow rate decreased by 6%, *both* absorbances would decrease by 6%, but their *ratio* would remain the same. Many modern spectrometers are able to use two lamps at a time and thus detect absorption and emission at two wavelengths simultaneously. Such an instrument is especially convenient for use with an internal standard.

21-5 INTERFERENCE

Types of interference:
1. Spectral—unwanted signals overlapping analyte signal.
2. Chemical—chemical reactions decreasing the concentration of analyte atoms.
3. Ionization—ionization of analyte atoms decreases the concentration of neutral analyte atoms in the flame.

By *interference*, we refer to any effect that changes the signal when analyte concentration remains unchanged. In the measurement of atomic absorption or emission signals, interference is widespread and easy to overlook. If you are clever enough to discern that interference is occurring, it may be corrected by counteracting the source of interference or by preparing standards that exhibit the same interference.

Spectral interference refers to the overlap of analyte signal with signals due to other elements or molecules in the sample or with signals due to the flame or furnace. Interference from the flame can be subtracted using such means as D_2 lamp background correction. The best means of dealing with overlap between lines of different elements in the sample is to choose another wavelength for analysis.

Chemical interference is caused by any component of the sample that decreases the extent of atomization of analyte. For example, SO_4^{2-} and PO_4^{3-} hinder the atomization of Ca^{2+}, perhaps by forming involatile salts. **Releasing agents** are chemicals that can be added to a sample to decrease chemical interference. EDTA and 8-hydroxyquinoline protect Ca^{2+} from the interfering effects of SO_4^{2-} and PO_4^{3-}. Lanthanum(III) can also be used as a releasing agent, apparently because it reacts with PO_4^{3-} in place of Ca^{2+}. Use of a fuel-rich flame is recommended to reduce certain oxidized analyte species that would otherwise hinder atomization. Using a higher flame temperature eliminates many kinds of chemical interference.

Ionization interference can be a problem in the analysis of alkali metals at relatively low flame temperature, and for other elements at higher temperature. For any element, we can write a gas-phase ionization reaction:

$$M(g) \rightarrow M^+(g) + e^-(g) \qquad (21\text{-}6)$$

$$K = \frac{[M^+][e^-]}{[M]} \qquad (21\text{-}7)$$

Since the alkali metals have the lowest ionization potentials, they are most extensively ionized in a flame. At 2 450 K and a pressure of 0.1 Pa, sodium is expected to be 5% ionized. Potassium, with a lower ionization potential, is expected to be 33% ionized under the same conditions. Since ionized atoms have different energy levels from neutral atoms, the desired atomic signal is decreased.

An **ionization suppressor** is an element added to a sample to decrease the extent of ionization of analyte. For example, in the analysis of potassium, it is recommended that solutions contain 1 000 ppm of CsCl, since cesium is more easily ionized than potassium. By producing a high concentration of electrons in the flame, ionization of Cs suppresses ionization of K.

This is an application of Le Châtelier's principle to Reaction 21-6.

Virtues of the inductively coupled plasma

Many common interferences in atomic spectroscopy are eliminated by using an inductively coupled argon plasma for emission measurements. The plasma is twice as hot as a conventional flame, and the residence time of analyte in the flame is about twice as great. Therefore, atomization is more complete than in a flame, and the signal is correspondingly enhanced. Since the Ar gas is chemically inert, reaction of the analyte with the flame (which forms metal oxide molecules in conventional flames) is eliminated. The plasma is also remarkably free of background radiation in the region where sample emission is observed (15–35 mm above the load coil). The background concentration of electrons due to plasma formation is fairly high and uniform. The temperature is so high that many elements are observed as ions, and the concentrations of these ions appear to be fairly insensitive to the presence of potential suppressors.

A common problem in flame emission spectroscopy is that there is a lower concentration of electronically excited atoms in the cooler outer part of the flame than in the warmer central part of the flame. As a result, emission from the central region is absorbed in the outer region. This **self-absorption** increases with increasing concentration of analyte and leads to nonlinear calibration curves. In a plasma, the flame temperature is more uniform, and self-absorption is not nearly as important. Figure 21-18 shows calibration curves for plasma emission that are linear over nearly five orders of magnitude. In conventional flames, the linear range is much more restricted.

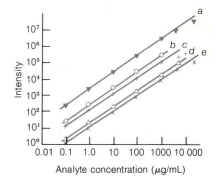

Figure 21-18
Analytical calibration curves for emission from (*a*) Ba^{2+}, (*b*) Cu^+, (*c*) Na^+, (*d*) Fe^+, and (*e*) Ba^+ in an inductively coupled plasma. [R. N. Savage and G. M. Hieftje, *Anal. Chem.*, **51**, 408 (1979).]

Summary

In atomic spectroscopy, the absorption, emission, or fluorescence from gaseous atoms is measured. Liquid samples may be atomized by a flame, furnace, or plasma. In a premix burner, the sample is mixed with fuel and oxidant before it flows into the flame, whose temperature is usually in the range 2 300–3 400 K. The fuel and oxidant chosen determine the temperature of the flame and affect the extent of spectral, chemical, or ionization interference that will be encountered. Flame temperature instability has little effect on atomic absorption, but a large effect on atomic emission, because the excited-state population is very temperature-dependent. An electrically heated furnace requires much less sample than a flame, has a lower detection limit, but is less precise. In an inductively coupled plasma, a radio-frequency induction coil is used to heat Ar^+ ions in an Ar gas stream to 6 000–10 000 K. At this high temperature, emission from excited atomic ions is the dominant signal. There is little chemical or spectral interference in an inductively coupled plasma, the flame temperature is very stable, and little self-absorption is observed.

Instrumentation for atomic spectroscopy is similar to that for ordinary spectroscopy, but the sample must be atomized, the radiation source must be highly monochromatic, and the background signal must be subtracted. Lamps with a hollow cathode made of the analyte element are usually used to obtain atomic lines sharper than those of the atomic vapor (whose lines are broadened by collisions and by the Doppler effect). Correction for background emission due to the flame usually involves a mechanical beam chopper. Correction for inadvertent light scattering can be made by measuring absorption using a broad-band deuterium lamp. Chemical interference can sometimes be reduced by appropriate releasing agents, which protect the analyte from reacting with interfering species. Ionization interference in flames may be suppressed by adding to the sample such easily ionized elements as Cs. The usual analytical methods employing standard curves, standard additions, or internal standards can be applied to quantitative analysis by atomic spectroscopy.

Terms to Understand

atomic absorption spectroscopy	internal standard
atomic emission spectroscopy	ionization interference
atomic fluorescence spectroscopy	ionization suppressor
atomization	matrix
beam chopper	nebulization
Boltzmann distribution	premix burner
chemical interference	pressure broadening
detection limit	releasing agent
Doppler effect	self-absorption
graphite furnace	sensitivity
hollow-cathode lamp	spectral interference
inductively coupled plasma	standard addition method

Exercises

21-A. Li was determined by the method of standard additions using atomic emission. From the results below, calculate the concentration of Li in pure unknown. The Li standard contained $1.62\ \mu g$ Li/mL.

Unknown (mL)	Standard (mL)	Final volume (mL)	Emission intensity (arbitrary units)
10.00	0.00	100.0	309
10.00	5.00	100.0	452
10.00	10.00	100.0	600
10.00	15.00	100.0	765
10.00	20.00	100.0	906

21-B. Mn was used as an internal standard for measuring Fe by atomic absorption. A standard mixture containing $2.00\ \mu g$ Mn/mL and $2.50\ \mu g$ Fe/mL gave a signal ratio (Fe signal/Mn signal) = 1.05. A mixture of volume 6.00 mL was prepared by mixing 5.00 mL of unknown Fe solution with 1.00 mL containing $13.5\ \mu g$ Mn/mL solution. The absorbance of this mixture at the Mn wavelength was 0.128, and the absorbance at the Fe wavelength was 0.185. Calculate the molarity of the unknown Fe solution.

21-C. The atomic absorption signal shown below was obtained with 0.048 5 μg Fe/mL in a graphite furnace. Estimate the detection limit for Fe.

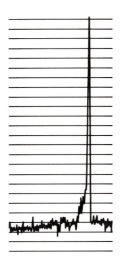

21-D. The laser atomic fluorescence excitation and emission spectra of sodium in an air–acetylene flame are shown at the right. In the *excitation* spectrum, the laser (bandwidth = 0.03 nm) was scanned through various wavelengths, while the detector monochromator (bandwidth = 1.6 nm) was held fixed near 589 nm. In the *emission* spectrum, the laser was fixed at 589.0 nm, and the detector monochromator wavelength was varied. Explain why the emission spectrum gives one broad line, while the excitation spectrum gives two sharp lines. How can the excitation linewidths be much narrower than the detector bandwidth?

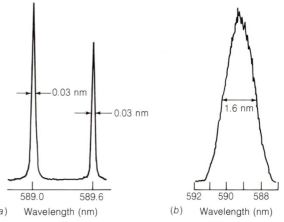

Fluorescence excitation and emission spectra of the two sodium D lines in an air–acetylene flame. (a) In the excitation spectrum the laser was scanned. (b) In the emission spectrum the monochromator was scanned. The monochromator slit width was the same for both spectra. [S. J. Weeks, H. Haraguchi, and J. D. Winefordner, *Anal. Chem.*, **50**, 360 (1978).]

Problems

21-1. In which technique, atomic absorption or atomic emission, is flame-temperature stability more critical? Why?

21-2. State the advantages and disadvantages of furnaces compared to flames in atomic absorption spectroscopy.

21-3. Calculate the wavelength (nanometers) of emission of excited atoms that lie 3.371×10^{-19} J/molecule above the ground state.

21-4. For Ca atoms, the first excited state is reached by absorption of light with $\lambda = 422.7$ nm.
(a) What is the energy difference (kilojoules per mole) between the ground state and the excited state?
(b) The relative degeneracies are $g^*/g_0 = 3$ for Ca. What is the ratio N^*/N_0 at 2 500 K?
(c) By what percentage will the fraction in part b be changed by a 15° rise in temperature?
(d) What will be the ratio N^*/N_0 at 6 000 K?

21-5. A series of potassium standards gave the following emission intensities at 404.3 nm. Find the concentration of potassium in the unknown.

Sample (μg K/mL)	Relative emission
Blank	≡0
5.00	124
10.00	243
20.0	486
30.0	712
Unknown	417

21-6. Calculate the Doppler linewidth in Hz($= s^{-1}$) for the 589 nm line of Na and for the 254 nm line of Hg, both at 2 000 K.

21-7. A series of Ca and Cu samples was run to determine the atomic absorbance of each element.

Ca (µg/mL)	$A_{422.7}$	Cu (µg/mL)	$A_{324.7}$
1.00	0.086	1.00	0.142
2.00	0.177	2.00	0.292
3.00	0.259	3.00	0.438
4.00	0.350	4.00	0.576

(a) Determine the average relative absorbance ($A_{324.7}/A_{422.7}$) produced by equal concentrations (µg/mL) of Ca and Cu.

(b) Copper was used as an internal standard in a Ca determination. A sample known to contain 2.47 µg Cu/mL gave $A_{324.7} = 0.269$ and $A_{422.7} = 0.218$. Calculate the concentration of Ca in micrograms per milliliter.

21-8. In Figure 21-15, a sample containing 1.00 µg Fe/mL gives an absorbance of 0.055. Estimate the sensitivity of this spectrometer for Fe.

21-9. Fluorescence excitation spectra of Mn and Mn + Ga solutions are shown below. The Ga line is within 0.2 nm of all three Mn lines. Explain why there is no spectral interference by Ga in the laser atomic fluorescence excitation analysis of Mn, even though the detector monochromator bandwidth is set to 1.0 nm. You may wish to refer to Exercise 21-D.

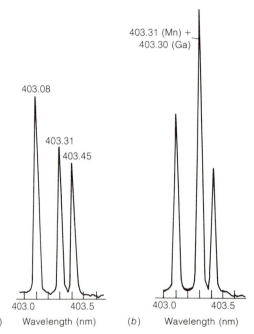

Fluorescence excitation spectrum for (a) 1 µg/mL Mn solution and (b) 1 µg/mL Mn plus 5 µg/mL Ga solution. [S. J. Weeks, H. Haraguchi, and J. D. Winefordner, *Anal. Chem.*, **50**, 360 (1978).]

22 / Introduction to Analytical Separations

Most chemical analyses of real samples require that the analyte be partially or completely separated from interfering components of the sample. For example, a sensitive but unselective colorimetric analysis of In(III) uses the reagent 4-(2-pyridylazo)-resorcinol, which forms a pink 2:1 complex with In(III) ($\varepsilon = 4.3 \times 10^4$ at 510 nm).

4-(2-pyridylazo)-resorcinol

The elements Zn, Pb, Cr, Al, Sn, Cd, Cu, Mn, Fe, Co, Ni, V, Zr, and Bi all interfere in this determination. To analyze a mineral containing just 0.1% In requires separation of In from all of these elements. One procedure involves dissolution in HCl plus HNO_3, removal of Sn^{4+} as the volatile $SnBr_4$, and precipitation of Pb^{2+} with SO_4^{2-}. $In(OH)_3$ and the hydroxides of several other metals are precipitated with NH_3 and redissolved in HCl. Addition of KI gives an indium–iodide complex, which can be extracted into diethyl ether, leaving all the other metals behind. Finally, the indium is extracted back into a fresh aqueous solution and analyzed colorimetrically.

This separation makes use of volatilization, precipitation, and solvent extraction.

Isolating one component of a mixture for analytical or preparative purposes is an important and challenging aspect of the practice of chemistry. In this and the next chapter, we discuss some of the more common methods of separation.

22-1 SOLVENT EXTRACTION

In its simplest form, **extraction** refers to the transfer of a solute from one liquid phase to another. The most common case is the extraction of an aqueous solution with an organic solvent. Diethyl ether, benzene, and other hydrocarbons are common solvents that are less dense than water and form a phase that sits on top of the aqueous phase. Chloroform, dichloromethane, and carbon tetrachloride are common solvents that are *immiscible* with and denser than water.† In a two-phase mixture, some of each solvent is found in *both* phases, but one phase is predominantly water and the other phase is predominantly organic. The volumes of each phase after mixing are not exactly equal to the volumes that were mixed. For simplicity, however, we will assume that the volumes of each phase are not changed by mixing.

Suppose that solute S is partitioned between phases 1 and 2 (Figure 22-1). The **partition coefficient**, K, is the equilibrium constant for the reaction

$$S \text{ (in phase 1)} \rightleftharpoons S \text{ (in phase 2)} \tag{22-1}$$

$$K = \frac{\mathscr{A}_{S_2}}{\mathscr{A}_{S_1}} \approx \frac{[S]_2}{[S]_1} \tag{22-2}$$

where $\mathscr{A}_{S_1}$ refers to the activity of solute in phase 1. In dilute solution, the ratio of activities can be replaced with a ratio of concentrations.

To be *miscible* means that the two liquid form a single phase when they are mixed in any ratio. To be *immiscible* means that each liquid remains in a separate phase.

In more concentrated solutions, which are often encountered, the concentration could deviate markedly from the activity. Nonetheless, lacking knowledge of the activity coefficients, we will write the partition coefficient in terms of concentrations.

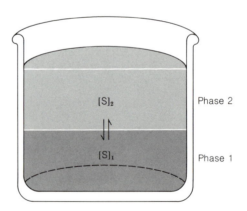

Figure 22-1
Partitioning of a solute between two liquid phases.

Suppose that a volume V_1 of solvent 1 containing m moles of solute S is extracted with volume V_2 of solvent 2. Let q be the fraction of S remaining in phase 1 when equilibrium is achieved. The molarity in phase 1 will therefore be qm/V_2. The fraction of total solute transferred to phase 2 will be $(1 - q)$, and the molarity in phase 2 is $(1 - q)m/V_2$. Putting these values of

---
† Whenever a choice exists between CHCl$_3$ and CCl$_4$, the less toxic CHCl$_3$ should be tried.

molarity into Equation 22-2 gives

$$K = \frac{(1 - q)m/V_2}{qm/V_1} \qquad (22\text{-}3)$$

from which we can solve for q:

$$q = \frac{V_1}{V_1 + KV_2} \qquad (22\text{-}4)$$

Equation 22-4 says that the fraction of solute remaining in phase 1 depends on the partition coefficient and the two volumes. If the phases are separated and a new quantity of fresh solvent 2 is mixed with phase 1, the fraction of solute remaining in phase 1 at equilibrium will be

$$\text{Fraction remaining after two extractions} = q \cdot q = \left(\frac{V_1}{V_1 + KV_2}\right)^2 \qquad (22\text{-}5)$$

After n extractions with volume V_2, the fraction remaining in phase 1 is

$$\text{Fraction remaining after } n \text{ extractions} = q^n = \left(\frac{V_1}{V_1 + KV_2}\right)^n \qquad (22\text{-}6)$$

Example of Equation 22-5: If $q = 1/4$, then 1/4 of the solute remains in phase 1 after one extraction. A second extraction will reduce the concentration to 1/4 of the value after the first extraction = $(1/4)(1/4) = 1/16$ of the initial concentration.

EXAMPLE

Solute A has a partition coefficient of 3 between benzene and water (with 3 times as much in the benzene phase). Suppose that 100 mL of a 0.01 M aqueous solution of A is extracted with benzene. What fraction of A remains in the aqueous phase (a) if one extraction with 500 mL is performed and (b) if five extractions with 100 mL are performed?

(a) Taking water as phase 1 and benzene as phase 2, Equation 22-4 says that after a 500 mL extraction, the fraction remaining in the aqueous phase is

$$q = \frac{100}{100 + (3)(500)} = 0.062 \approx 6\%$$

Many small extractions are much more effective than a few large extractions.

(b) With five 100 mL extractions, the fraction remaining is given by Equation 22-6:

$$\text{Fraction remaining} = \left(\frac{100}{100 + (3)(100)}\right)^5 = 0.000\ 98 \approx 0.1\%$$

It is much more efficient to do several small extractions than one big extraction.

---

pH Effects

Suppose that the solute being partitioned between phases 1 and 2 is an amine with a base constant K_B. Let's also assume that BH^+ is soluble *only* in the aqueous phase (1). Suppose that the neutral form, B, has partition coefficient, K, between the phases. The **distribution ratio, D,** is defined as

$$D = \frac{\text{Total concentration in phase 2}}{\text{Total concentration in phase 1}} \quad (22\text{-}7)$$

which becomes

$$D = \frac{[B]_2}{[B]_1 + [BH^+]_1} \quad (22\text{-}8)$$

Combining the relationships $K = [B]_2/[B]_1$ and $K_A = [H^+][B]/[BH^+] = K_w/K_B$ with Equation 22-8 leads to

$$D = \frac{K \cdot K_A}{K_A + [H^+]} \quad (22\text{-}9)$$

The distribution of solute between the two phases will be pH-dependent.

EXAMPLE
Suppose that $K = 3.0$ and $K_A = 1.0 \times 10^{-9}$. If 50 mL of 0.010 M aqueous amine is extracted with 100 mL of solvent 2, what will be the formal concentration remaining in the aqueous phase (a) at pH 10.00 and (b) at pH 8.00?

(a) At pH 10.00, $D = (3.0)(1.0 \times 10^{-9})/(1.0 \times 10^{-9} + 1.0 \times 10^{-10}) = 2.73$ (from Equation 22-9). Equation 22-4 says that the fraction remaining in the aqueous phase will be

$$q = \frac{50}{50 + (2.73)(100)} = 0.15 \Rightarrow 15\% \text{ left in water}$$

The concentration of amine in the aqueous phase will be 15% of 0.010 M = 0.001 5 M. *In the preceding equation, we have used the distribution ratio, D, in place of the partition coefficient, K, in Equation 22-4.*

(b) At pH 8.00, $D = (3.0)(1.0 \times 10^{-9})/(1.0 \times 10^{-9} + 1.0 \times 10^{-8}) = 0.273$. This time we find

$$q = \frac{50}{50 + (0.273)(100)} = 0.65 = 65\% \text{ left in water}$$

The concentration in the aqueous phase is 0.006 5 M. At pH 10, the base is predominantly in the form B and is extracted into the organic solvent. At pH 8, it is in the form BH^+ and remains in the water.

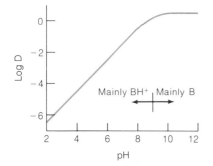

Figure 22-2
Effect of pH on the distribution coefficient for the extraction of a base into an organic solvent. In this example, $K = 3.0$ and $K_B = 1.0 \times 10^{-5}$.

Figure 22-2 shows the effect of pH on the distribution ratio in the above example. When you want to extract a base into water, it is clearly desirable to use a low enough pH to convert it to BH^+. By the same reasoning, to extract an acid (HA) into water, you should use a high enough pH to convert the acid to A^-.

Challenge: Suppose that the acid HA (with dissociation constant K_A) is partitioned between aqueous phase 1 and organic phase 2. Calling the partition coefficient K for HA and assuming that A^- is not soluble in the organic phase, show that the distribution coefficient is given by

$$D = \frac{K[H^+]}{[H^+] + K_A} \qquad (22\text{-}10)$$

Equation for the distribution of HA between two phases.

Extraction with a Metal Chelator

One scheme for separating metal ions from each other is to selectively complex one ion using an organic ligand and extract it into an organic solvent. Three ligands commonly employed for this purpose are shown below.

dithizone (diphenylthiocarbazone)   8-hydroxyquinoline (oxine)   cupferron

Each ligand can be represented as a weak acid, HL, which loses one proton when it binds to a metal ion through the atoms shown in bold type.

$$HL(aq) \rightleftharpoons H^+(aq) + L^-(aq) \quad K_A = \frac{[H^+]_{aq}[L^-]_{aq}}{[HL]_{aq}} \qquad (22\text{-}11)$$

$$nL^-(aq) + M^{n+}(aq) \rightleftharpoons ML_n(aq) \qquad \beta = \frac{[ML_n]_{aq}}{[M^{n+}]_{aq}[L^-]_{aq}^n} \qquad (22\text{-}12)$$

Each of these ligands can react with many different metal ions, but some selectivity is achieved by controlling the pH. Most complexes that can be extracted into organic solvents must be neutral. Charged complexes, such as $Fe(EDTA)^-$ or $Fe(2\text{-phenanthroline})_3^{2+}$, are not very soluble in organic solvents.

Let's derive an equation for the distribution coefficient of a metal between the two phases under a particular (but common) set of circumstances. We will assume that essentially all of the metal in the aqueous phase is in the form M^{n+} and that all of the metal in the organic phase is in the form ML_n (Figure 22-3). We define the partition coefficients for ligand and complex

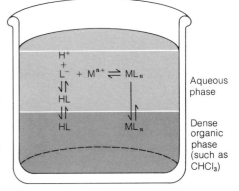

Figure 22-3
Extraction of a metal ion with a chelator. It is assumed that the predominant form of metal in the aqueous phase is M^{n+}, and the predominant form in the organic phase is ML_n.

We assume that M^{n+} is in the aqueous phase and that ML_n is in the organic phase.

22 / INTRODUCTION TO ANALYTICAL
SEPARATIONS

as follows:

$$HL(aq) \rightleftharpoons HL(org) \qquad K_L = \frac{[HL]_{org}}{[HL]_{aq}} \qquad (22\text{-}13)$$

$$ML_n(aq) \rightleftharpoons ML_n(org) \qquad K_M = \frac{[ML_n]_{org}}{[ML_n]_{aq}} \qquad (22\text{-}14)$$

where *org* refers to the organic phase.

The distribution coefficient we are seeking is

$$D = \frac{[\text{Total metal}]_{org}}{[\text{Total metal}]_{aq}} \approx \frac{[ML_n]_{org}}{[M^{n+}]_{aq}} \qquad (22\text{-}15)$$

From Equations 22-14 and 22-12, we can write

$$[ML_n]_{org} = K_M[ML_n]_{aq} = K_M\beta[M^{n+}]_{aq}[L^-]_{aq}^n \qquad (22\text{-}16)$$

Using the value of $[L^-]_{aq}$ from Equation 22-11 gives

$$[ML_n]_{org} = K_M\beta[M^{n+}]_{aq}K_A^n[HL]_{aq}^n/[H^+]_{aq}^n \qquad (22\text{-}17)$$

Putting this value of $[ML_n]_{org}$ into Equation 22-15 gives

$$D \approx K_M\beta K_A^n[HL]_{aq}^n/[H^+]_{aq}^n \qquad (22\text{-}18)$$

Since most of the HL is usually in the organic phase, we can use Equation 22-13 to rearrange Equation 22-18 to its most useful forms:

Distribution of the metal between the two phases is pH-dependent. You can select a pH to bring the metal into either phase.

$$D \approx \frac{K_M\beta K_A^n}{K_L^n} \frac{[HL]_{org}^n}{[H^+]_{aq}^n} \qquad (22\text{-}19)$$

Equation 22-19 says that the distribution coefficient for metal ion extraction depends on the pH and the ligand concentration. Since the various equilibrium constants are different for each metal, it is often possible to select a pH where D is large for one metal and small for another. For example, Figure 22-4 shows that Cu^{2+} could be separated from Pb^{2+} and Zn^{2+} by extraction with dithizone at pH 5. Demonstration 22-1 illustrates the pH-dependence of an extraction with dithizone.

Some Extraction Strategies

Some extraction techniques:
1. Extraction of ion pairs.
2. Masking interfering ions.
3. Use of hydrophobic counterion.
4. Use of hydrophobic ligand.

Separation by extraction is very much an art, with clever ideas and trial-and-error discovery of conditions (solvents, pH, chelators, etc.) being necessary. We will simply mention a few strategies that have been used in successful separations.

Sometimes **ion pairs** can be extracted from an aqueous phase into an

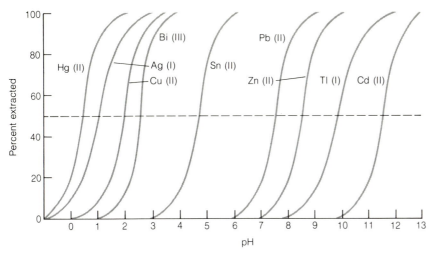

Figure 22-4
Extraction of metal ions by dithizone into CCl$_4$. [Adapted from G. H. Morrison and H. Freiser in C. L. Wilson and D. Wilson, eds., *Comprehensive Analytical Chemistry*, Vol. IA (New York: Elsevier, 1959).]

organic phase. For example, FeCl$_4^-$ ion can be extracted from 6 M HCl into diethyl ether as a tightly associated unit that can be written FeCl$_4^-\cdot$H$^+$, with no net charge. The distribution coefficient is a delicate function of conditions, with too much or too little HCl leading to incomplete extraction of FeCl$_4^-\cdot$H$^+$. The optimum conditions can be found only by trial and error. The indium–iodide system mentioned at the beginning of this chapter behaves similarly.

An ion that forms an extractable complex with a nonselective chelator might be isolated by masking other ions in the sample. For example, citrate and tartrate form strong polar complexes with many ions. These complexes would remain in the aqueous phase, while the desired ion might be extracted with another chelator. Clearly, the relative stability constants are all-important in this scheme.

Sometimes a relatively polar ion can be extracted into an organic phase in the presence of a hydrophobic counterion. The tetrabutylammonium cation [(C$_4$H$_9$)$_4$N$^+$] is commonly used for this purpose. Another way to extract a metal ion is to use a nonpolar molecule with one good ligand atom that may occupy one coordination site of a metal ion, thus pulling it into the organic phase. Trioctylphosphine oxide and dioctylsulfoxide are examples of hydrophobic ligands that can interact through their oxygen atoms with cations:

$$[CH_3(CH_2)_7]_3\overset{+}{P}-\overset{\:\:\:-}{\ddot{\underset{..}{O}}}: \quad\quad [CH_3(CH_2)_7]_2\overset{+}{\ddot{S}}-\overset{\:\:\:-}{\ddot{\underset{..}{O}}}:$$
$$\updownarrow \quad\quad\quad\quad\quad\quad\quad \updownarrow$$
$$[CH_3(CH_2)_7]_3P=\ddot{\underset{..}{O}} \quad\quad [CH_3(CH_2)_7]_2\ddot{S}=\ddot{\underset{..}{O}}$$

trioctylphosphine oxide   dioctylsulfoxide

Crown ethers and ionophores, described in Box 22-1, are specifically designed to extract metal ions into nonpolar solvents.

An ion pair is a pair of ions held closely together by electrostatic attraction. In solvents with a low dielectric constant, ions tend to aggregate in pairs.

Demonstration 22-1 EXTRACTION WITH DITHIZONE

Dithizone (diphenylthiocarbazone) is a green compound, soluble in nonpolar organic solvents and insoluble in water below pH 7. It forms red, hydrophobic complexes with most di- and trivalent metal ions.

$$2\,C_6H_5\text{–NH–N=C(SH)–N=N–}C_6H_5 + Pb^{2+} \underset{}{\overset{\beta_2 \approx 10^{16}}{\rightleftharpoons}} [\text{Pb(dithizonate)}_2] + 2H^+$$

dithizone (green)   (colorless)   (red)

Dithizone is widely used for analytical extractions, for colorimetric determinations of metal ions, and for removing traces of metals from aqueous buffers.

Box 22-1 CROWN ETHERS AND IONOPHORES

The problem of extracting metal ions into a nonpolar phase is important to synthetic chemists and living cells, as well as to analytical chemists. A class of synthetic compounds called *crown ethers* is able to envelop metal ions (especially alkali metal cations) in a pocket of oxygen ligands. The outside of the crown ether is very hydrophobic and soluble in nonpolar solvents. Crown ethers are used as **phase transfer catalysts.** They can extract a water-soluble ionic reagent into a nonpolar solvent, where reaction with a hydrophobic compound can occur. The structure of the potassium complex of dibenzo-30-crown-10 below shows that the K^+ ion is engulfed by ten

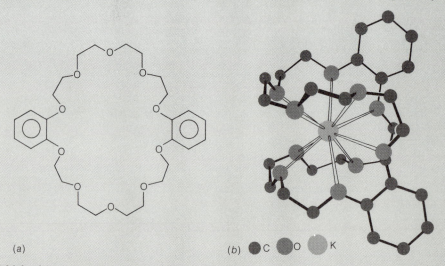

(a) Molecular structure of the crown ether designated dibenzo-30-crown-10. (b) The three-dimensional structure of its K^+ complex. [Adapted from M. A. Bush and M. R. Truter, *J. Chem. Soc. Chem. Commun.*, 1439 (1970).]

In the latter application, an aqueous buffer is extracted repeatedly with a green solution of dithizone in CHCl$_3$ or CCl$_4$. As long as the organic phase turns red, metal ions are being extracted from the buffer. When the extracts are green, the last traces of metal ions have been removed. Elements that react with dithizone include Mn, Fe, Co, Ni, Pd, Pt, Cu, Ag, Au, Zn, Cd, Hg, Ga, In, Tl, Sn, Pb, Bi, Se, Te, and Po. Figure 22-4 tells us that only certain metal ions can be extracted at a given pH.

The equilibrium between the green ligand and red complex is nicely demonstrated using three large test tubes sealed with tightly fitting rubber stoppers. In each tube is placed some hexane plus a few milliliters of dithizone solution (prepared by dissolving 1 mg of dithizone in 100 mL of CHCl$_3$). To tube A is added distilled water; to tube B is added tap water; and to tube C is added 2 mM Pb(NO$_3$)$_2$. After shaking and settling, tubes B and C contain a red upper phase, while A remains green.

The proton equilibrium indicated by the reaction above is shown by adding a few drops of 1 M HCl to tube C. After shaking, the dithizone turns green again. Competition with a stronger ligand is shown by adding a few drops of 0.05 M EDTA solution to tube B. Again, shaking causes a reversion to the green color.

oxygen atoms with K–O distances averaging 288 pm. Only the hydrophobic outside of the complex is exposed to the solvent.

Living cells must transport hydrophilic molecules through very hydrophobic membranes. For example, a nerve impulse requires that K$^+$ and Na$^+$ ions flow across the cell membrane of a neuron. The carriers of these ions appear to be an integral part of the membrane.

A class of antibiotics called **ionophores** works the same way as the crown ethers. Such compounds as nonactin, valinomycin, gramicidin, and nigericin alter the permeability of bacterial cells to metal ions and thus disrupt their metabolism. The structure of nonactin is shown below.

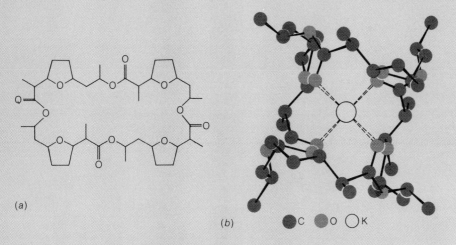

(a) Molecular structure of the antibiotic nonactin. (b) The three-dimensional structure of its K$^+$ complex, which can penetrate hydrophobic barriers. [Adapted from *Biochemistry* by Lubert Stryer. W. H. Freeman and Company. Copyright © 1975.

22-2 COUNTERCURRENT DISTRIBUTION

Countercurrent distribution is a serial extraction process devised by L. C. Craig in 1949. The process is a powerful improvement of liquid–liquid extraction. Although countercurrent distribution has been almost totally supplanted by chromatographic methods of separation, its theory is worth studying because it provides a basis for understanding chromatography.

The object of countercurrent distribution is to separate two or more solutes from each other by a series of partitions between two liquid phases. The scheme is shown in Figure 22-5. In Step 0, a simple extraction is performed. Some of each solute will appear in each phase. A necessary condition for separation is that the distribution coefficients for the two solutes be different. For example, suppose that the two solutes distribute themselves in the following fashion:

$$D_A = \frac{[A]_{\text{upper phase}}}{[A]_{\text{lower phase}}} = 4 \qquad D_B = \frac{[B]_{\text{upper phase}}}{[B]_{\text{lower phase}}} = 1 \qquad (22\text{-}20)$$

If 1 mmol of each solute is present, the quantities present in each phase upon equilibration in Step 0 of Figure 22-5 will be

A: upper phase (U0)—0.8 mmol
    lower phase (L0)—0.2 mmol

B: upper phase (U0)—0.5 mmol
    lower phase (L0)—0.5 mmol

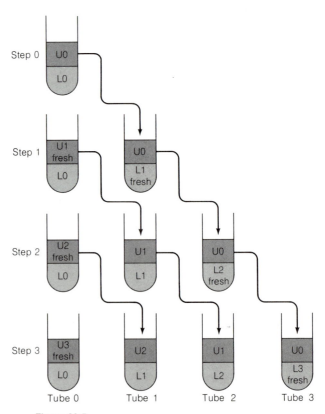

Figure 22-5
The scheme of extractions in countercurrent distribution.

To separate solutes from each other, they must have different distribution coefficients.

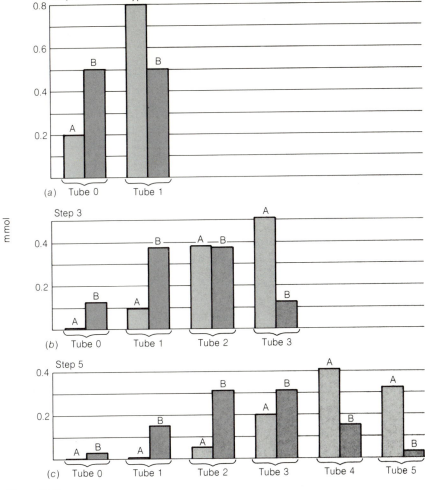

Figure 22-6
Total of each solute in each tube after (a) one, (b) three, and (c) five steps of countercurrent distribution.

In Step 1, phase U0 is transferred to a second tube containing *fresh* lower phase, L1. Likewise, L0 is placed in contact with *fresh* upper phase, U1. When both tubes are shaken, the following equilibria are established:

A: U1—0.16 mmol ⎫ total =   B: U1—0.25 mmol ⎫ total =
    L0—0.04 mmol ⎭ 0.2 mmol   L0—0.25 mmol ⎭ 0.5 mmol

    U0—0.64 mmol ⎫ total =   U0—0.25 mmol ⎫ total =
    L1—0.16 mmol ⎭ 0.8 mmol   L1—0.25 mmol ⎭ 0.5 mmol

In Step 2, phase U0 is transferred to a tube containing *fresh* L2. Phase U1 is transferred to the tube containing *old* L1. *Fresh* phase U2 is placed in the tube containing *old* L0. After equilibration, solute A will have a ratio 4:1 (upper:lower) in each tube, and solute B will have a ratio 1:1 (upper:lower) in each tube. The process is followed for five steps in Table 22-1. The percent of each solute in each tube (total of upper + lower phase) after one, three, and five steps is shown in Figure 22-6.

What Figure 22-6 shows is that both solutes, A and B, are carried to the

The ratios $\dfrac{[A]_{\text{upper phase}}}{[A]_{\text{lower phase}}}$ and $\dfrac{[B]_{\text{upper phase}}}{[B]_{\text{lower phase}}}$ are the same in every tube.

Challenge: Verify the numbers for Step 2 in Table 22-1.

Table 22-1
Countercurrent distribution of two solutes with distribution coefficients $D_A = 4$ and $D_B = 1$

Step number, n	Tube number, r: 0	1	2	3	4	5
0	A UPPER PHASE 0 / A LOWER PHASE 1; B UPPER PHASE 0 / B LOWER PHASE 1 A 0.8 / 0.2; B 0.5 / 0.5					
1	A 0.16 / 0.04; B 0.25 / 0.25	A 0.64 / 0.16; B 0.25 / 0.25				
2	A 0.032 / 0.008; B 0.125 / 0.125	A 0.256 / 0.064; B 0.25 / 0.25	A 0.512 / 0.128; B 0.125 / 0.125			
3	A 0.0064 / 0.0016; B 0.0625 / 0.0625	A 0.0768 / 0.0192; B 0.1875 / 0.1875	A 0.3072 / 0.0768; B 0.1875 / 0.1875	A 0.4096 / 0.1024; B 0.0625 / 0.0625		
4	A 0.00128 / 0.00032; B 0.03125 / 0.03125	A 0.02048 / 0.00512; B 0.125 / 0.125	A 0.12288 / 0.03072; B 0.1875 / 0.1875	A 0.32768 / 0.08192; B 0.125 / 0.125	A 0.32768 / 0.08192; B 0.03125 / 0.03125	
5	A 0.000256 / 0.000064; B 0.015625 / 0.015625	A 0.00512 / 0.00128; B 0.078125 / 0.078125	A 0.04096 / 0.01024; B 0.15625 / 0.15625	A 0.16384 / 0.04096; B 0.15625 / 0.15625	A 0.32768 / 0.08192; B 0.078125 / 0.078125	A 0.262144 / 0.065536; B 0.015625 / 0.015625
Total in each tube after five steps	A 0.00032; B 0.03125	A 0.0064; B 0.15625	A 0.0512; B 0.3125	A 0.2048; B 0.3125	A 0.4096; B 0.15625	A 0.32768; B 0.03125

right in successive steps. However, solute A moves faster than B because of the larger distribution coefficient of A. The net result is that after several steps A and B begin to separate from each other. For example, the mixture initially contained 1 mmol each of A and B. After five steps of countercurrent distribution, we might pool tubes 0, 1, and 2 in one flask, and tubes 4 and 5 in a second flask. The first flask would contain 0.5 mmol of B and 0.057 92 mmol of A. We would have a 50% recovery of 89% pure B. Tubes 4 and 5 would contain 0.737 28 mmol of A plus 0.187 5 mmol of B. This fraction gives a 74% recovery of 80% pure A. Instead of stopping the procedure after five steps, it could be continued to obtain further separation of A from B.

Color Plate 24 shows an experiment very much like that in Figure 22-6. A countercurrent extraction was performed beginning with 2 mL of 0.1 M aqueous Na_2CO_3 (saturated with 1-butanol) containing 40 μg of phenol red and 50 μg of bromocresol green. This lower phase was extracted with an upper phase of butanol (saturated with 0.1 M Na_2CO_3) to produce the 9-tube distribution shown. The bromocresol green is extracted readily by the butanol and moves rapidly to the right in the upper phase. The phenol red is more soluble in the aqueous phase and moves very slowly to the right.[†]

> Purity is expressed as $\dfrac{[B]}{[A] + [B]}$.

> The solute with the larger distribution coefficient moves faster.

Theoretical Distribution

Let's now develop a few equations describing the distribution of solute in the fractions of a countercurrent disbribution. The pattern of countercurrent extractions is shown below.

U9 fresh	U8 fresh	U7 fresh	U6 fresh	U5	U4	U3	U2	U1	U0	→				Mobile phase
			L0	L1	L2	L3	L4	L5	L6 fresh	L7 fresh	L8 fresh	L9 fresh		Stationary phase

The upper phase, which can be thought of as sliding along a fixed lower phase, will be called the **mobile phase.** The lower phase is called the **stationary phase.**

Consider a *single solute* in phase L0. Before equilibration with U0, the concentration in L0 will be called 1, and the concentration in U0 is 0.

Before zeroth equilibration

0	0	0	0	0				
				1	0	0	0	0

Tube number, $r = 0 \qquad 1 \qquad 2 \qquad 3 \qquad 4$

[†] This extraction can be used for a nice experiment on countercurrent distribution. See B. Arreguín, J. Padilla, and J. Herrán, *J. Chem. Ed.*, **39**, 539 (1962).

22 / INTRODUCTION TO ANALYTICAL SEPARATIONS

After the zeroth equilibration, we will call the *fraction of solute in the upper phase p* and the *fraction in the lower phase q*. Obviously, $p + q = 1$. The situation after the zeroth equilibration looks like this:

From Equation 22-20, we can say that $p = D/(D + 1)$.

After zeroth equilibration

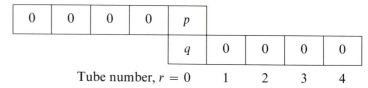

0	0	0	0	p				
				q	0	0	0	0

Tube number, $r = 0$ 1   2   3   4

Now we advance the mobile phase by one tube. Calling this first advance $n = 1$, we find p in tube 1 and q in tube 0:

Before first equilibration

0	0	0	0	p				
				q	0	0	0	0

($n = 1$ in Table 22-2)

Tube number, $r = 0$ 1   2   3   4

After equilibration of the phases, a fraction p must be in each upper phase, and a fraction q must be in each lower phase. Since tube 0 has a total quantity q, the quantities in the upper and lower phases are $p \cdot q$ and $q \cdot q$, respectively. The situation looks like this:

After first equilibration

0	0	0	pq	p^2			
			q^2	pq	0	0	0

Tube number, $r = 0$ 1   2   3   4

Upon making a second advance (defined as $n = 2$), the distributions before and after equilibration are as follows:

Before second equilibration

0	0	0	pq	p^2			
			q^2	pq	0	0	0

($n = 2$ in Table 22-2)

Tube number, $r = 0$ 1   2   3   4

After second equilibration

0	0	pq^2	$2p^2q$	p^3	
	q^3	$2pq^2$	p^2q		

Tube number, $r = 0 \quad 1 \quad 2 \quad 3 \quad 4$

One more advance gives the following:

Before third equilibration

0	0	pq^2	$2p^2q$	p^3	$(n = 3$ in Table 22-2)
	q^3	$2pq^2$	p^2q	0	0

Tube number, $r = 0 \quad 1 \quad 2 \quad 3 \quad 4$

After third equilibration

0	pq^3	$3p^2q^2$	$3p^3q$	p^4	
	q^4	$3pq^3$	$3p^2q^2$	p^3q	0

Tube number, $r = 0 \quad 1 \quad 2 \quad 3 \quad 4$

The fraction of solute in each tube upon each advance is shown in Table 22-2. You may recognize the coefficients in Table 22-2 as the binomial expansion $(q + p)^n$. The fraction of solute, f, in each tube, r, at each step, n, is given by

$$f = \frac{n!}{(n - r)!\, r!}\, p^r q^{n-r} \qquad (22\text{-}21)$$

The symbol $n!$ is read "n factorial" and means the product $n(n - 1)(n - 2) \cdots (1)$. For example, $6! = 6 \cdot 5 \cdot 4 \cdot 3 \cdot 2 \cdot 1 = 720$. The quantity $0!$ is defined as 1.

Table 22-2
Fraction of solute in each tube after each advance

Step number, n	Tube number, r					
	0	1	2	3	4	$\cdots$
1	q	p				
2	q^2	$2pq$	p^2			
3	q^3	$3pq^2$	$3p^2q$	p^3		
4	q^4	$4pq^3$	$6p^2q^2$	$4p^3q$	p^4	
$\vdots$						

22 / INTRODUCTION TO ANALYTICAL SEPARATIONS

EXAMPLE

Use Equation 22-21 to verify the quantity of each solute in tube 4 ($r = 4$) after five steps ($n = 5$) in Table 22-1.

For solute A, the distribution coefficient is 4 (Equation 22-20). This means that $p = \frac{4}{5}$ and $q = \frac{1}{5}$. Putting these values into Equation 22-21 gives:

$$f_A = \frac{5!}{(5-4)!4!}\left(\frac{4}{5}\right)^4\left(\frac{1}{5}\right)^{5-1} = \frac{(5 \cdot 4 \cdot 3 \cdot 2 \cdot 1)}{(1)(4 \cdot 3 \cdot 2 \cdot 1)}\left(\frac{4}{5}\right)^4\left(\frac{1}{5}\right) = 0.409\,6$$

which is the value in Table 22-1.

For solute B, the distribution coefficient is 1 (Equation 22-20). Setting $p = q = \frac{1}{2}$ gives

$$f_B = \frac{5!}{(5-4)!4!}\left(\frac{1}{2}\right)^4\left(\frac{1}{2}\right)^{5-1} = 0.156\,25$$

For a large number of countercurrent distribution steps, n, it can be shown[†] that the tube, r_{max}, containing the maximum amount of a solute with a fraction p in the mobile phase is given approximately by

$r_{max} \approx np$ means that the solute peak has always traveled a constant fraction of the distance that the mobile phase has traveled.

$$r_{max} \approx np \tag{22-22}$$

Thus in the example above, p for solute A is 4/5. We expect the peak fraction of A to be in tube $(100)(\frac{4}{5}) = 80$ after 100 steps. We expect the peak fraction of B to be in tube 50, since $p = \frac{1}{2}$ for solute B.

Bandwidth and Resolution

If solute A is maximal in tube 80 and solute B is maximal in tube 50 after 100 steps of countercurrent distribution, we might expect that the two bands are well separated from each other and that most of each solute can be isolated with a high degree of purity. The equations needed to calculate the degree of separation will now be discussed.

The binomial distribution in Equation 22-21 closely approximates a Gaussian distribution when n and r are large. Equation 22-21 can be expressed as[‡]

$$f \approx \frac{1}{\sqrt{npq}\sqrt{2\pi}}\,e^{-(r-np)^2/2npq} \tag{22-23}$$

At this point, it is worth recalling the Gaussian distribution, Equation 4-4:

$$y = \frac{1}{\sigma\sqrt{2\pi}}\,e^{-(x-\mu)^2/2\sigma^2}$$

[†] P. L. Nichols, Jr., *Anal. Chem.*, **22,** 915 (1950).
[‡] E. Nelson, *Anal. Chem.*, **28,** 1998 (1956).

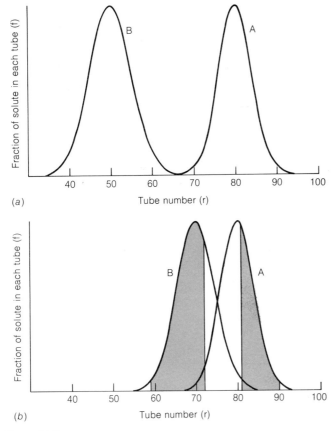

Figure 22-7
Countercurrent distribution separation of solutes A and B using 100 steps. (a) The distribution coefficients are $D_A = 4$ and $D_B = 1$. (b) $D_A = 4$ and $D_B = \frac{7}{3}$. Each Gaussian curve is normalized to the same maximum height.

The standard deviation is σ, and the mean is μ. By analogy between Equations 22-23 and 4-4, we see that the "standard deviation" of a solute band in countercurrent distribution is

$$"\sigma" \approx \sqrt{npq} \qquad (22\text{-}24)$$

The greater the number of separation steps, the broader is the band of solute.

We are now in a position to calculate the degree of overlap of two bands. If the partition coefficients are $D_A = 4$ and $D_B = 1$, the following parameters apply when $n = 100$ steps:

Solute A: $\quad p = \frac{4}{5}$ $\qquad\qquad$ Solute B: $\quad p = \frac{1}{2}$

$\qquad\qquad q = \frac{1}{5}$ $\qquad\qquad\qquad\qquad\qquad q = \frac{1}{2}$

$\qquad\qquad \sigma_A = \sqrt{npq} = 4$ $\qquad\qquad\quad \sigma_B = \sqrt{npq} = 5$

$\qquad\qquad r_{max} = np = 80$ $\qquad\qquad\quad r_{max} = np = 50$

A graph of the Gaussian curves describing these two bands is shown in the upper part of Figure 22-7. In a real experiment, each band would undoubtedly

550

22 / INTRODUCTION TO ANALYTICAL SEPARATIONS

In real separations, bands tend to be more spread at the extremes (especially at the rear of the band) than an ideal calculation would predict.

be less pure than we have calculated, because we are likely to leave some of each phase behind every time a phase is transferred from one tube to the next.

EXAMPLE

The lower portion of Figure 22-7 shows the separation of solutes A and B after 100 steps when $D_A = 4$ and $D_B = \frac{7}{3}$. If fractions 59–72 and 81–90 are each pooled, find the percent recovery and purity of each component.

For these parameters, we calculate the following:

$$\text{Solute A:} \quad p = 0.80 \qquad \text{Solute B:} \quad p = 0.70$$
$$q = 0.20 \qquad\qquad\qquad q = 0.30$$
$$\sigma_A = 4 \qquad\qquad\qquad \sigma_B = 4.58$$

Tube 70 is r_{max} for solute B. To find the fraction in tubes 59–72, we divide the curve into two portions and add their areas:

Tubes:	59–70	70–72	
Width:	$11/4.58 = 2.40\ \sigma_B$	$2/4.58 = 0.437\ \sigma_B$	
Area:	0.491 8	0.168 8	$\begin{cases}\text{From interpolation} \\ \text{in Table 4-1}\end{cases}$

$$\text{Total area} = 0.491\ 8 + 0.168\ 8 = 0.660\ 6$$

That is, tubes 59–72 contain 66% of B. The fraction of B to the right of tube 81 is 0.008 2, because tube 81 is $11/4.58 = 2.40\sigma_B$ from r_{max}.

The calculations for band A look like this:

Tubes:	81–90	80–81
Width:	$9/4 = 2.25\sigma_A$	$1/4 = 0.25\sigma_A$
Area:	0.487 6	0.098 6

$$\text{Total area} = 0.487\ 6 - 0.098\ 6 = 0.389\ 0$$
$$\uparrow$$
$$\text{Note the subtraction}$$

Since tube 72 is $8/4 = 2\sigma_A$ from the peak of A, the fraction of A lying on the left of tube 72 is 0.022 7.

The final results for this separation are

Tubes 59–72		Tubes 81–90	
Fraction of B	66.06%	Fraction of A	38.90 %
Fraction of A	2.27%	Fraction of B	0.82%
Purity	96.68%	Purity	97.94%

Band spreading

Important insight concerning these separations can be gained from Equations 22-22 and 22-24:

1. Since $r_{max} \approx np$, each solute migrates a distance equal to a constant fraction of the solvent "front" (the position to which fresh mobile phase moves).

2. The bandwidth increases with the square root of the number of separation steps, because $\sigma \approx \sqrt{npq}$. The farther the front has moved, the larger is n and the broader will be each solute band.

3. Increased separation is afforded when n is increased, because the distance that each band travels is proportional to n, but the band spreading is only proportional to $\sqrt{n}$.

Purification is possible because separation is proportional to n but spreading is only proportional to $\sqrt{n}$.

These physical properties apply to all forms of partition chromatography, as well as to countercurrent distribution.

22-3 CHROMATOGRAPHY

Chromatography is a logical extension of countercurrent distribution. Instead of performing a series of discrete extractions, a continuous equilibration of solute between two phases occurs. In chromatography, the mobile phase is either a liquid or a gas. The stationary phase is most commonly a liquid coated on the surface of solid particles. Alternatively, the solid particles themselves may serve as a stationary phase. In any case, the partitioning of solutes between the mobile and stationary phases accounts for the separation of solutes, just as it does in countercurrent distribution. This is shown schematically in Figure 22-8. The solute that has a greater affinity for the stationary phase will move through the column more slowly.

As it is currently practiced, chromatography is divided into several classes (Figure 22-9):

1. **Adsorption chromatography.** The oldest form of chromatography, this makes use of a solid stationary phase and a liquid or gaseous mobile phase. Solute can be adsorbed on the surface of the solid particles. Equilibration between the adsorbed state and the solution accounts for the separation of solute molecules.

2. **Partition chromatography.** In this technique, a stationary phase forms a thin film on the surface of a solid support. Solute equilibrates between this stationary liquid and a liquid or gaseous mobile phase.

3. **Ion-exchange chromatography.** Anions (such as $-SO_3^-$) or cations (such as $-N(CH_3)_3^+$) are covalently attached to the stationary solid phase, usually called a *resin*, in this type of chromatography. Solute ions of the opposite charge are attracted to the stationary phase by electrostatic force. The mobile phase is a liquid.

4. **Molecular exclusion chromatography.** Unlike other forms of chromatography, there is no attractive interaction between the "stationary phase" and the solute in the ideal case of molecular exclusion chromatography. Rather, the liquid or gaseous mobile phase passes through a porous gel. The pore sizes are small enough to exclude large solute molecules, but not small ones. The large molecules stream past without entering the gel. The small molecules take longer to pass through the column because they enter the gel and therefore must flow through a larger volume before leaving the column. Also called **gel filtration** or **gel permeation** chromatog-

In 1903, M. Tswett first applied adsorption chromatography to the separation of plant pigments, using a hydrocarbon solvent and $CaCO_3$ stationary phase. The separation of colored bands led to the name *chromatography*, from the Greek word *chromatos*, meaning color.

For their pioneering work on liquid–liquid partition chromatography in 1941, A. J. P. Martin and R. L. M. Synge received the Nobel Prize in 1954.

B. A. Adams and E. L. Holmes developed the first synthetic ion-exchange resins in 1935.

Large molecules pass through the column *faster* than small molecules

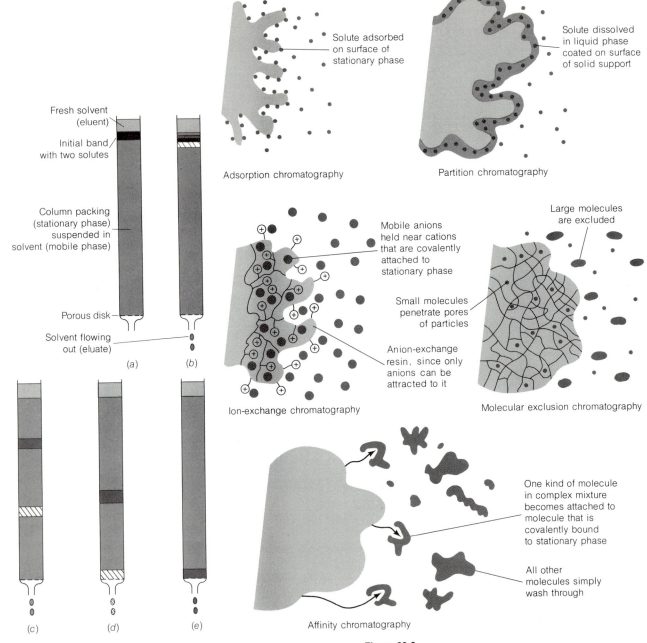

Figure 22-8
Schematic representation of a chromatographic separation. The solute with a greater affinity for the stationary phase remains on the column longer.

Figure 22-9
Major types of chromatography.

raphy, this technique separates molecules by size, with the larger solutes passing through most quickly.

5. **Affinity chromatography.** This newest and most selective kind of chromatography utilizes highly specific interactions between one kind of solute molecule and a second molecule covalently attached (immobilized) to the stationary phase. For example, the immobilized molecule might be an antibody to a particular protein. When a crude mixture containing a thousand proteins is passed through the column, only the one protein that reacts with the antibody is bound to the column. After washing all the other solutes off the column, the desired protein is dislodged from the antibody by changing the pH or ionic strength.

Some Terminology

The solvent (or gas) of the mobile phase is called the **eluent.** When the solvent emerges from the end of the column it is called the **eluate:**

$$\text{Eluent in} \longrightarrow \boxed{\text{COLUMN}} \longrightarrow \text{Eluate out}$$

Elu*ent*—in.

Elu*ate*—out.

The process of passing liquid (or gas) through a chromatography column is called **elution.**

The degree to which a solute is retained by the column is measured by the **retention ratio,** R, defined as

R is also called the *retardation factor.*

$$R = \frac{\text{Time required for solvent to pass through column}}{\text{Time required for solute to pass through column}} \quad (22\text{-}25)$$

and the denominator of Equation 22-25 is called the **retention time.** The fastest a solute can be eluted from a column is the time required for solvent to pass through the column. Therefore, R is always ≤ 1. For example, if solvent is eluted in 30 minutes and a solute requires 100 minutes for elution, $R = 30/100 = 0.30$.

The retention ratio gives the fraction of time that a solute molecule spends in the mobile phase:

$$R = \frac{t_m}{t_m + t_s} \quad (22\text{-}26)$$

where t_m is the time in the mobile phase and t_s is the time in the stationary phase. If a molecule spent 70% of the time in the stationary phase and 30% of the time in the mobile phase, it would have a retention ratio of $30/(30 + 70) = 0.30$. It would move at $3/10$ of the average speed of solvent and it would require $\frac{10}{3}$ times longer to be eluted.

Equation 22-26 can be rewritten as

$$R = \frac{1}{1 + \dfrac{t_s}{t_m}} \quad (22\text{-}27)$$

554

22 / INTRODUCTION TO ANALYTICAL SEPARATIONS

But t_s/t_m can be expressed as the ratio of the moles of solute in the stationary phase to the moles in the mobile phase. The moles are given by the concentration (C) times the volume (V) of each respective phase.

$$\frac{t_s}{t_m} = \frac{C_s V_s}{C_m V_m} = K\frac{V_s}{V_m} \tag{22-28}$$

K in chromatography is, unfortunately, defined as the inverse of K (or D) used in countercurrent distribution.

The ratio C_s/C_m is defined as the **partition coefficient,** K. Plugging this back into Equation 22-27 gives:

$$R = \frac{1}{1 + K\dfrac{V_s}{V_m}} = \frac{V_m}{V_m + KV_s} \tag{22-29}$$

The **retention volume,** V_r, is the volume of solvent required to elute a solute from the column. The relation between retention volume and volume of the mobile phase is simply

$$V_r = V_m/R \tag{22-30}$$

Solving for R in Equation 22-30 and equating it to R in Equation 22-29 gives a useful equation for chromatography:

Equation 22-31 will be used in Chapter 23 in connection with molecular exclusion chromatography.

$$R = \frac{V_m}{V_r} = \frac{V_m}{V_m + KV_s} \Rightarrow V_r = V_m + KV_s \tag{22-31}$$

Plate Theory of Chromatography

The simplest way to think about chromatography is to imagine that a very large number of countercurrent distribution equilibrations occur between the mobile and stationary phases as the solute travels through the column. Although chromatography is a continuous process, it can be imagined that the column is divided into N segments, in each of which one equilibration occurs. Each of these imaginary segments is called a **theoretical plate.** If the total length of the column is L, the **height equivalent to a theoretical plate** (H.E.T.P.) is

$$H.E.T.P. = L/N \tag{22-32}$$

It is not unusual for a chromatography column to possess several thousand theoretical plates. If a 1 m column possesses 10^4 theoretical plates, H.E.T.P. $= 1$ m$/10^4 = 10^{-4}$ m.

Figure 22-10 shows an ideal chromatogram with a Gaussian band shape. By measuring the retention time (or retention volume) and the bandwidth, the number of theoretical plates can be calculated from the equation

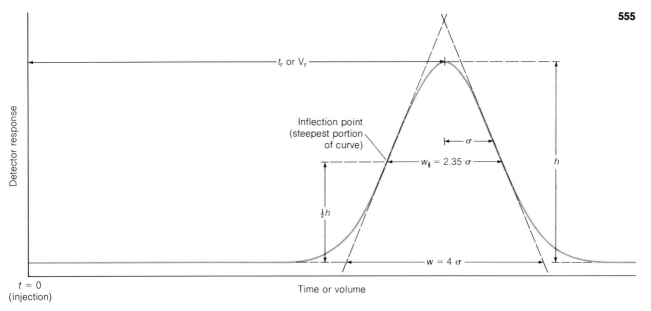

Figure 22-10
Idealized Gaussian chromatogram showing how w and $w_{1/2}$ are measured. The value w is obtained by extrapolating the tangents to the inflection points down to the baseline.

$$N = \frac{t_r^2}{\sigma^2} = \frac{16 t_r^2}{w^2} \qquad (22\text{-}33)$$

If you want to measure the number of theoretical plates from a real chromatogram, choose a peak with a capacity factor (defined on page 570) greater than 5.

where t_r is the retention time of the peak, σ is its standard deviation, and w is the width of the band (in units of time) measured at the base of the peak (as shown in Figure 22-10). This width is equal to four standard deviations for a Gaussian peak. Alternatively, if the width of the band is measured at a height equal to half of the peak height, then

$$N = \frac{5.55 t_r^2}{w_{1/2}^2} \qquad (22\text{-}34)$$

where $w_{1/2}$ is the width at half-height. In Equations 22-33 and 22-34, retention volume may be used in place of retention time. In this case, the width is measured in units of volume instead of time.

Equation 22-33 can be derived from the Gaussian distribution function for countercurrent distribution, using several approximations. Clearly, it is a gross simplification to consider a chromatography column as divided into N theoretical plates. Since a theoretical plate is an imaginary construct, Equation 22-33 can be considered to *define* a theoretical plate. Columns often behave as if they have different numbers of plates for different solutes in a mixture.

Challenge: If N is constant, show that the width of a chromatographic peak increases with increasing retention time. That is, successive peaks on a chromatogram should be increasingly broad.

22 / INTRODUCTION TO ANALYTICAL
SEPARATIONS

Rate Theory of Chromatography

A more realistic description of the movement of solute through a chromatography column takes into account the finite rate at which solute can equilibrate between the mobile and stationary phases. That is, equilibration is not infinitely fast (as plate theory assumes), and the resulting band shape depends on the rate of elution. The band shape is also affected by diffusion of solute along the length of the column and by the availability of different paths for different solute molecules to follow as they travel between particles of the stationary phase.

All of the effects just mentioned depend on the rate, v, at which the mobile phase passes through the column. Detailed consideration of the various mechanisms by which the solute band is broadened leads to the **van Deemter equation** for plate height:

The van Deemter equation.

$$H.E.T.P. = Av^{1/3} + \frac{B}{v} + Cv \qquad (22\text{-}35)$$

where A, B, and C are constants characteristic of a given column and solvent system. Equation 22-35 says that there will be an optimal velocity for the operation of any column, at which the plate height (H.E.T.P.) reaches its minimum value.

Let's examine each term of the van Deemter equation to understand the velocity-dependence of H.E.T.P. First, it should be realized that any mechanism that causes a band of solute to broaden will increase the height of a theoretical plate. Each term of Equation 22-35 therefore represents a zone-broadening mechanism.

A broad band equilibrates with a longer section of column than does a narrow band. The "section of column" required for equilibration is the theoretical plate.

Consider **longitudinal diffusion** of solute as it travels through the column (Figure 22-11 top). *Since the concentration of solute is lower at the edges of the solute zone than at the center, solute is always diffusing toward the edges of the zone.* The more time the solute spends on the column, the greater will be its diffusive spreading. This gives the B/v term of the van Deemter equation. The greater the mobile phase velocity, the less time is spent on the column and the less diffusion occurs.

Longitudinal diffusion means diffusion parallel to the direction of flow.

The term Cv arises from the finite time required for solute to equilibrate between the mobile and stationary phases. Consider an instant at which some solute is in each phase (Figure 22-11 center). If the mobile phase is moving rapidly and if solute cannot "escape" from the stationary phase rapidly, then the zone of solute in the mobile phase moves ahead of the zone in the stationary phase. The effect is to broaden the total solute band. The greater the velocity, v, the worse this broadening becomes. The term Cv expresses this effect.

The term $Av^{1/3}$ in the van Deemter equation is called the **eddy diffusion** term. As the mobile phase moves through the column packed with myriad tiny, stationary particles, different solute molecules follow different paths in a random manner (Figure 22-11 bottom). Some molecules take longer routes and some take shorter routes; some become trapped in swirling pools (eddies) of solvent whose forward progress has been stopped by a solid wall of par-

Terms in the van Deemter equation:

B/v—longitudinal diffusion.
Cv—speed of equilibration.
$Av^{1/3}$—eddy diffusion.

ticles. The existence of multiple paths of random length tends to broaden the zone of solute, since different molecules follow different paths. Zone broadening due to irregular flow patterns shows an empirical dependence on velocity, given approximately by $v^{1/3}$.[†]

Different columns can be conveniently compared through two parameters called the **reduced plate height,** h, and the *reduced velocity*, v

$$h = \text{H.E.T.P.}/d \tag{22-36}$$

$$v = vd/\text{D} \tag{22-37}$$

where d is the average diameter of the stationary-phase particles and D is the diffusion coefficient of solute in the mobile phase. The reduced plate height is an expression of the thickness of a theoretical plate in multiples of the size of stationary-phase particles. If $h = 3$, then H.E.T.P. equals three particle diameters.

In terms of reduced plate height and velocity, the van Deemter equation can be written

$$h = A'v^{1/3} + \frac{B'}{v} + C'v \tag{22-38}$$

where A', B', and C' are constants. A graph of log h versus log v is shown in Figure 22-12. The optimum condition occurs when h is a minimum. At velocities below the optimal velocity, longitudinal diffusion causes the zone to broaden and increases the plate height. At velocities above the optimum,

---

[†] J. N. Done, G. J. Kennedy, and J. H. Knox in S. G. Perry, ed., *Gas Chromatography* (Barking, Essex; U.K.: Applied Science Publishers, 1973), p. 145. In most texts, the slight velocity-dependence of the eddy diffusion term is ignored, and the first term of the van Deemter equation is considered to be simply the constant, A.

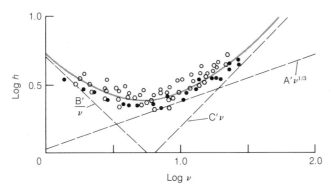

Figure 22-12
Graph of reduced plate height versus reduced velocity for high performance liquid chromatography. Particle diameters are d = 6 μm (○), 7.5 μm (◯), and 10 μm (●). Broken lines indicate approximate resolution of the curve into the three terms of the van Deemter equation (22-38). [G. R. Laird, J. Jurand, and J. H. Knox, *Proc. Soc. Anal. Chem.*, 310 (1974).]

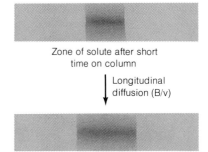

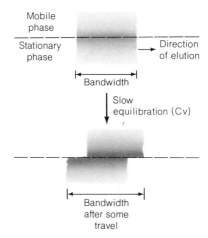

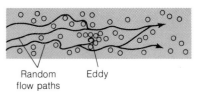

Figure 22-11
Mechanisms of band-broadening in chromatography.

558

22 / INTRODUCTION TO ANALYTICAL SEPARATIONS

H.E.T.P. is proportional to particle size. Smaller particles give more theoretical plates per length of column.

Equilibration with small particles is faster, so small particles allow higher flow rates.

Because gases diffuse into the stationary phase faster than liquids, the optimum flow rate in gas chromatography is greater than that for liquid chromatography.

slow equilibration of solute between the phases (and, to a lesser extent, eddy diffusion) causes the solute to spread out.

In Figure 22-12, the optimum reduced plate height is about 2.5. For stationary-phase particles of diameter $d = 10 \ \mu m$, H.E.T.P. $= 2.5 \ d = 25 \ \mu m$, giving 40 000 theoretical plates per meter. You can see the advantage of using the smallest possible particle size. For a given type of stationary phase, *the smaller the particle size the more efficient is the column*. Columns packed with such tiny particles provide a great resistance to solvent flow. Consequently, high pressures are needed to operate the most efficient liquid chromatography columns.

The optimum reduced velocity in Figure 22-12 is 5. This means that v(optimum) $= 5D/d$ for the columns to which Figure 22-12 applies. Since diffusion coefficients, D, are of the order of $10^{-9} \ m^2 \ s^{-1}$, the optimum velocity is around $5 \times 10^{-4} \ m \ s^{-1}$ when the particle diameter, d, is 10 μm. *The smaller the particle diameter, the higher is the optimal flow rate.* Remember, however, that the smaller the particles, the more resistance there is to flow and the more pressure is needed to obtain a given flow rate.

Because the diffusion coefficients of gases are so much greater than those of liquids, the optimum flow rate in gas chromatography is higher than that in liquid chromatography. While it is possible to run a gas chromatograph below its optimum flow rate, it is rare for the flow rate in liquid chromatography to be below the optimal value. In most cases, lowering the liquid flow rate raises the efficiency (decreasing H.E.T.P.) of a liquid chromatographic column.

Resolution

The **resolution** of two peaks from each other is defined as

$$\text{Resolution} = \frac{\Delta t_r}{w} = \frac{\Delta V_r}{w} \tag{22-39}$$

where Δt_r or ΔV_r is the separation between peaks (in units of time or volume) and w is the width of each peak, in corresponding units ($w = 4\sigma$). For simplicity, we assume that w is the same for both peaks. Figure 22-13 shows how well separated two peaks are at various values of resolution. When resolution $= 0.5$, the overlap of the two peaks is 16%. When resolution $= 1.0$, the overlap is 2.3%, and the overlap is reduced to 0.1% at a resolution of 1.5.

It can be shown[†] that the resolution is given by

$$\text{Resolution} = \frac{\sqrt{N}}{4} \left(\frac{\alpha - 1}{\alpha} \right) f_2 \tag{22-40}$$

where

N = number of theoretical plates in the column

[†] C. F. Simpson, *Practical High Performance Liquid Chromatography* (London: Heyden & Son, 1976), p. 11.

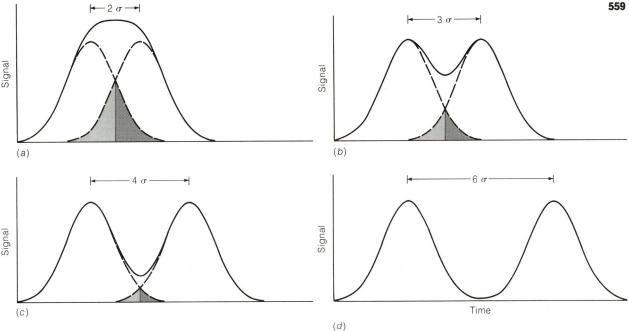

Figure 22-13
Resolution of two Gaussian peaks of equal area and amplitude. Dashed lines show individual peaks, and solid lines are the sum of two peaks. Overlapping area is shaded. (a) Resolution = 0.50. (b) Resolution = 0.75. (c) Resolution = 1.00. (d) Resolution = 1.50.

$\alpha = K_2/K_1$, where K_i is the partition coefficient for each solute. Component 1 is eluted before component 2

f_2 = fraction of time spent by component 2 in the stationary phase

The value of f_2 is

$$f_2 = \frac{K_2 V_s}{V_m + K_2 V_s} \qquad (22\text{-}41)$$

where V_s is the volume of the stationary phase and V_m is the volume of the mobile phase.

The most important feature of Equation 22-40 is that the resolution is proportional to $\sqrt{N}$. *Doubling the column length thus increases the resolution by $\sqrt{2}$.* Equation 22-40 also tells us that the resolution is proportional to the fraction of time spent by solute in the stationary phase. As a practical matter, however, very large values of f_2 (greater than ~0.85) are not useful, because the retention time becomes unreasonably long and the peaks become too broad.

Resolution $\propto \sqrt{N} \propto \sqrt{L}$

560

22 / INTRODUCTION TO ANALYTICAL
SEPARATIONS

A Touch of Reality

Band spreading before and after chromatography

In discussing the van Deemter equation, we only mentioned ways that the solute bands are broadened in the chromatographic process. However, the solute cannot be applied to the column in an infinitesimally thin zone, so the band has a finite width even before it begins spreading on the column. After elution, further broadening can occur in a poorly designed column outlet or in the detector used to measure the solute emerging from the column.

In well-designed outlets and detectors, laminar flow is achieved. This means that in the center of the tube, the velocity of the fluid is greatest and diminishes gradually until, at the walls of the tube, the velocity of fluid is zero. Because fluid in the center of the chamber moves more rapidly than fluid at the edges, the band spreads out. However, the spreading would be much worse in an outlet or detector where each new drop entering the chamber mixed totally with the contents of the chamber. In the large dead space beneath some crude chromatography columns, this does happen.

To minimize band spreading outside the chromatography column, all dead spaces and tubing lengths should be minimized. The sample should be applied in a narrow zone and allowed to enter the column before mixing with the eluent.[†]

Band shapes

A Gaussian band shape results when the partition coefficient, K ($= C_s/C_m$), is constant, independent of the quantity of solute on the column. In real columns, the ratio C_s/C_m changes somewhat as the total quantity of solute increases, and the resulting band shapes are skewed. A graph of C_s/C_m (at a given temperature) is called an *isotherm*. Three common isotherms and their resulting band shapes are shown in Figure 22-14. The center isotherm is the ideal one, leading to a symmetric peak.

The upper isotherm is typical of an *overloaded* column. This is one to which so much solute has been applied that the solute affects the physical properties of the stationary phase. The stationary phase can no longer be considered a dilute solution of solute. An overloaded stationary phase behaves more like a liquid phase made of the solute. Since solute is most soluble in itself ("like dissolves like"), the value C_s/C_m increases with increased solute loading.

The front of an overloaded peak exhibits a normal, gradually increasing concentration. As the concentration increases, the band becomes overloaded and the partition coefficient ($= C_s/C_m$) increases. The solute becomes so soluble in the zone on the stationary phase that there is little solute trailing

Overloading produces a gradual rise and an abrupt fall of the chromatographic peak.

[†] A quantitative treatment of the effect of mixing on bandwidths appears in D. G. Peters, J. M. Hayes, and G. M. Hieftje, *Chemical Separations and Measurements* (New York: Saunders, 1974), pp. 538–541.

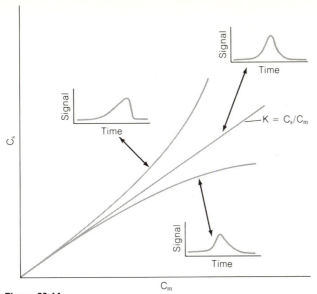

Figure 22-14
Common isotherms and their resulting chromatographic band shapes.

behind the peak. The band emerges gradually but ends abruptly (Figure 22-14, upper left inset).

The lower isotherm in Figure 22-14 results when small quantities of solute are retained more strongly than large quantities. This is exactly the opposite situation to overloading. It leads to a relatively abrupt rise of concentration at the beginning of the band and a long "tail" of gradually decreasing concentration after the peak.

One cause of tailing is the presence of sites at which the solute is held strongly by the stationary phase. When these sites are saturated with solute, the partition coefficient decreases because no more strong sites are available to hold fresh solute. In gas chromatography, a common solid phase support is made of diatomite, the hydrated silica skeletons of algae. Hydroxyl groups at the surface of the particles can form strong hydrogen bonds with polar solutes, leading to serious tailing. One strategy to reduce tailing is to cover these sites with a polar compound, such as acetic acid, which remains bound indefinitely. Another technique, called **silanization,** is to covalently attach dimethyldichlorosilane—$(CH_3)_2SiCl_2$—or hexamethyldisilazine to the silanol hydroxyl groups:

A long tail occurs when some sites retain solute more strongly than other sites.

```
   OH    OH                                          (CH₃)₃Si   Si(CH₃)₃
   |     |                                                |        |
                                                          O        O
   |     |                                                |        |
 —Si—O—Si—  + (CH₃)₃SiNHSi(CH₃)₃  ⟶               —Si—O—Si—   + NH₃
   |     |                                                |        |

 Solid phase    hexamethyldisilazine                  Protected surface
 with exposed
 —OH groups
```

Diatomite so treated produces less tailing than does untreated material. Glass columns used for gas and liquid chromatography are also often silanized to minimize interaction of the solute with active sites on the glass walls.

Summary

A solute may be extracted from one phase into a second in which it is more soluble. The ratio of solute concentrations in each phase at equilibrium is called the partition coefficient. If more than one form of the solute exists in the solution, we use a distribution coefficient instead of a partition coefficient. Simple equations relating the fraction of solute extracted to the partition coefficient, volumes of the phases, and pH were derived in this chapter. A series of small extractions is more effective than a few large extractions. A metal chelator, soluble only in organic solvents, may be used to extract metal ions from aqueous solution, with selectivity achieved by controlling pH. Hydrophobic counterions, ion-pair formation, or crown ethers may also be used to extract ions from water into a hydrophobic phase.

Countercurrent distribution is a serial extraction procedure in which solutes with differing partition coefficients may be separated. The distribution of solutes may be described by a binomial distribution in which the band position is proportional to the number of steps and the bandwidth is proportional to the square root of the number of steps.

In adsorption and partition chromatography, an essentially continuous equilibration of solute between the mobile and stationary phases occurs. In ion-exchange chromatography, the solute is attracted to the stationary phase by coulombic forces. In molecular exclusion chromatography, the fraction of stationary-phase volume available to solute decreases as the size of the solute molecules increases. Affinity chromatography relies on specific, noncovalent interactions between the stationary phase and one solute in a complex mixture.

We can speak of a theoretical plate as the length of column needed for one imaginary equilibration step to occur. The number of theoretical plates in a column producing Gaussian peaks is defined as $16t_r^2/w^2$, where t_r is the retention time of a peak and w is its width at the base. The van Deemter equation states that the height equivalent to a theoretical plate is given by $Av^{1/3} + B/v + Cv$, where v is the solvent velocity and A, B, and C are constants. The first term represents irregular flow paths and eddy diffusion, the second longitudinal diffusion, and the third the finite rate of equilibration. An optimum flow rate is one that minimizes the plate height. It is faster for gas chromatography than for liquid chromatography. Both the number of plates and the optimal flow rate increase as the stationary-phase particle size decreases. The resolution of a column is proportional to the square root of its length. Bands spread not only on the column, but during injection and detection. Overloading and tailing can be corrected by using smaller samples and by masking strong adsorption sites on the stationary phase.

Terms to Understand

adsorption chromatography
affinity chromatography
countercurrent distribution
distribution coefficient
eddy diffusion
eluate
eluent
elution
extraction
gel filtration
gel permeation chromatography
H.E.T.P.
ion-exchange chromatography
ionophore
ion pair

longitudinal diffusion
miscible
mobile phase
molecular exclusion chromatography
partition chromatography
partition coefficient
phase transfer catalysis
reduced plate height
resolution
retention ratio
retention time
retention volume
silanization
stationary phase
van Deemter equation

Exercises

22-A. A solute with a partition coefficient of 4.0 is extracted from 10 mL of phase 1 into phase 2.
  (a) What volume of phase 2 is needed to extract 99% of the solute in one extraction?
  (b) What will be the total volume of solvent 2 needed to remove 99% of the solute in three equal extractions instead?

22-B. Consider the 100-step countercurrent distribution of solutes A and B, with distribution coefficients $D_A = 1$ and $D_B = 0.6$. Suppose that the volumes of mobile and stationary phases are equal.
  (a) Calculate r_{max} and σ for each band.
  (b) If the fractions 28–44 and 44–60 are each pooled, find the percent recovery and purity of each solute.

22-C. A gas chromatogram of a mixture of toluene and ethyl acetate is shown below.

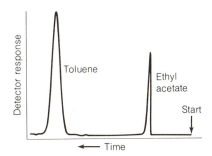

  (a) Use the width of each peak (measured at the base) to calculate the number of theoretical plates in the column. Estimate all lengths to the nearest 0.1 mm.
  (b) Using the width of the toluene peak at its base, calculate the width expected at half-height. Compare the measured and calculated values. When the thickness of the pen trace is significant compared to the length being measured, it is important to take the pen width into account. It is best to measure from the edge of one trace to the corresponding edge of the other trace, as shown below.

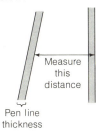

22-D. The three chromatograms below were obtained with 2.5, 1.0, and 0.4 μL of ethyl acetate injected on the same column under the same conditions. Explain why the asymmetry of the peak decreases as the sample size decreases.

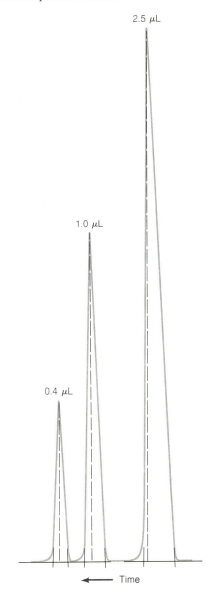

Problems

22-1. Give a physical interpretation of Equations 22-9 and 22-10 in terms of the fractional composition equations for a monoprotic acid in Section 13-2.

22-2. Why is the extraction of a metal ion into an organic solvent with 8-hydroxyquinoline more complete at higher pH?

22-3. Explain why one term of the van Deemter equation depends on v and why one term depends on $1/v$.

22-4. Why is the optimal flow rate greater for a chromatographic column if the stationary-phase particle size is smaller?

22-5. Describe how nonlinear partition isotherms lead to non-Gaussian band shapes. Draw the band shape produced by an overloaded column and a column with tailing.

22-6. Butanoic acid has a partition coefficient of 3 (favoring benzene) when distributed between water and benzene. Find the formal concentration of butanoic acid in each phase when 100 mL of 0.10 M aqueous butanoic acid is extracted with 25 mL of benzene (a) at pH 4.00 and (b) at pH 10.00.

22-7. For a given value of $[HL]_{org}$ in Equation 22-19, over what pH range (how many pH units) will D change from 0.01 to 100 if $n = 2$?

22-8. For the extraction of Cu(II) by dithizone in CCl_4, $K_L = 1.1 \times 10^4, K_M = 7 \times 10^4, K_A = 3 \times 10^{-5}$, $\beta = 5 \times 10^{22}$, and $n = 2$.
   (a) Calculate the distribution coefficient for extraction of 0.1 μM Cu(II) into CCl_4 by 0.1 mM dithizone at pH 1.0 and pH 4.0.
   (b) If 100 mL of 0.1 μM aqueous Cu(II) is extracted once with 10 mL of 0.1 mM dithizone at pH 4.0, what fraction of Cu(II) remains in the aqueous phase?

22-9. Calculate the fraction of each solute in Table 22-1 in each phase after seven steps. Make a graph similar to Figure 22-6, showing the relative proportions of A and B in each tube after seven steps.

22-10. Consider a countercurrent distribution in which substances A and B have the following distribution coefficients:

$$\frac{[A]_{upper}}{[A]_{lower}} = 3 \qquad \frac{[B]_{upper}}{[B]_{lower}} = \frac{1}{3}$$

   (a) Suppose that 1 mmol of A and 1 mmol of B are to be separated from each other. Calculate the number of millimoles of each substance in each tube after six steps of countercurrent distribution.
   (b) If tubes 5 and 6 are combined, what would be the percent recovery and purity of solute A?

22-11. Suppose that a solute with $D = 1.5$ is subjected to 100 steps of countercurrent distribution, using 4 mL of solvent (2 mL in each phase) in each tube.
   (a) Calculate the bandwidth (expressed as σ for the band) in milliliters and in tubes.
   (b) Find the bandwidth in milliliters and in tubes if the same countercurrent distribution is carried out with 200 tubes, each containing 1 mL of each phase.
   (c) What can you conclude about the utility of using more and smaller steps to effect a separation?

22-12. Derive a formula giving the number of countercurrent distribution steps (n) needed such that the separation of peaks A and B is $2\sigma_A + 2\sigma_B$. Call the fractional distribution of each solute p_A, q_A, p_B, and q_B. Use the Gaussian distribution as a good approximation for the shape of each band.
   (a) Calculate the value of n if $p_A = 0.40$ and $p_B = 0.30$.
   (b) Find n if $p_A = 0.36$ and $p_B = 0.34$.

22-13. If the volumes of mobile and stationary phases are not equal in countercurrent distribution, show that

$$p = \frac{DV_m}{DV_m + V_s} \qquad q = \frac{V_s}{DV_m + V_s}$$

where D is the distribution coefficient for solute between the two phases: $D = C_m/C_s$. (C is the concentration of solute in each phase.)

22-14. A band having a width of 4.0 mL and a retention volume of 49 mL was eluted from a chromatography column. What width is expected for a band with a retention volume of 127 mL? Assume that the only band spreading occurs on the column itself.

22-15. Two compounds with partition coefficients of 0.15 and 0.18 are to be separated on a column with

$V_m/V_s = 3.0$. Calculate the number of theoretical plates needed to produce a resolution of 1.5.

22-16. The retention volume of a solute was 76.2 mL for a column with $V_m = 16.6$ mL and $V_s = 12.7$ mL. Calculate the partition coefficient, K, for this solute.

22-17. Calculate the number of theoretical plates needed to achieve a resolution of 1.0 if
(a) $\alpha = 1.05$ and $f_2 = 0.40$
(b) $\alpha = 1.10$ and $f_2 = 0.40$
(c) $\alpha = 1.05$ and $f_2 = 0.80$
(d) How can you increase the value of f_2?

23 / Chromatographic Methods

Separation of compounds for the purpose of identification, quantification, or purification is one of the broadest areas of chemistry. In this chapter we discuss the major types of chromatography, which are among our most powerful methods of separation.

23-1 GAS CHROMATOGRAPHY

In **gas chromatography,** a volatile liquid or gaseous solute is carried by a gaseous mobile phase. The stationary phase is usually a relatively nonvolatile liquid coated on a solid support (Figure 22-9). This most common form of gas chromatography is called *gas-liquid partition chromatography*. Occasionally, solid particles on which solute can be adsorbed serve as the stationary phase (Figure 22-9). In this case, the technique can be called *gas-solid adsorption chromatography*.

Gas chromatography:
1. Mobile phase—gaseous.
2. Stationary phase—solid or nonvolatile liquid coated on solid support.
3. Solute—gas or volatile liquid.

A Gas Chromatograph

A schematic diagram of a gas chromatograph is shown in Figure 23-1. A volatile liquid sample is injected through a rubber **septum** (a thin disk) into a hot glass- or metal-lined injector port, which vaporizes the sample. Gaseous samples can be injected through a gas sampling valve. The sample is swept into the column by an inert carrier gas (usually He or N_2), which serves as the mobile phase. After passing through the column containing the stationary phase, the solutes are separated from each other. Finally, the gas stream

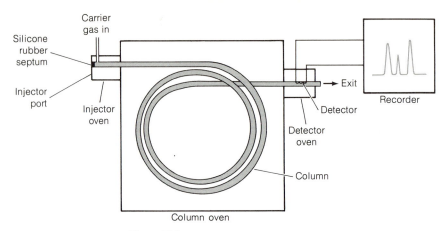

Figure 23-1
Schematic diagram of a gas chromatograph.

flows through a detector, which sends a signal to a recorder as solutes emerge from the column. The detector and injector are maintained at a higher temperature than the column, so that all solutes are gaseous in these chambers. The column temperature need not be above the boiling point of all solutes. It must only be hot enough for each solute to have sufficient vapor pressure to be eluted in a reasonable time.

To grasp the details of gas chromatography a little better, we will examine a simple but representative gas chromatograph used in research laboratories. The heart of the chromatograph in Figure 23-2 is a coiled column, which is housed in an insulated, thermostatically controlled oven. Typical columns are 2–5 m in length, with 2 mm inner diameter, but longer and/or wider columns will fit in this oven.

Helium carrier gas supplied from a tank passes first through a filter that removes traces of H_2O and enters the chromatograph at the inlet on the lower right of the side panel shown in Figure 23-2. This instrument uses a thermal conductivity detector, which requires that the carrier gas be split into two streams, one of which is used for reference. The silicone rubber septum in the injector port is easily removed and replaced after repeated puncturing by the injector syringe. The temperature of the injector, column, or detector can be selected for display on the temperature meter at the lower left. The injector is usually maintained about 50° above the column temperature, which is determined by the volatility of the solutes. The remaining controls on the front panel are associated with the detector.

The solutes emerging from the gas chromatograph can be collected for identification or can be sent directly into an infrared or mass spectrometer for immediate analysis of each chromatographic peak. To collect small quantities (a few microliters) of solute, a U-shaped glass tube is inserted in the carrier gas exit. The bottom of the U is cooled with dry ice or liquid nitrogen to condense the solute as it passes through.

The choice of carrier gas is dictated by the nature of the detector being used.

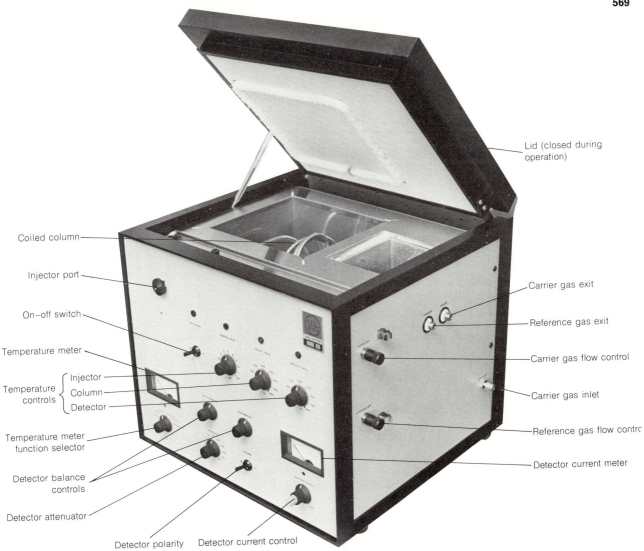

Figure 23-2
Photograph of the Varian Aerograph Model 920 gas chromatograph. The instrument measures about 40 cm on each edge. [Courtesy Varian Associates, Palo Alto, Calif.]

The Chromatogram

Retention time

A **chromatogram** is a graph showing the detector response as a function of elution time. In gas chromatography, the detector response is plotted by a strip chart recorder, as shown in Figure 23-3. The **retention time**, t_r, for each peak is the time needed after injection for the peak to reach the detector. The **adjusted retention time**, t'_r, is defined as

$$t'_r = t_r - t_m \tag{23-1}$$

where t_m is the time needed for the mobile phase to travel the length of the column. An unretained solute would be eluted in the time t_m. When a thermal conductivity detector is used, t_m is taken as the time needed for air to travel through the column. This can be determined by injecting a few microliters of air along with the sample (which is typically 0.1–10 μL).

The **capacity factor,** k', is defined by the equation

$$k' = \frac{t_r - t_m}{t_m} \tag{23-2}$$

The longer a component is retained by the column, the greater is the capacity factor. Large capacity factors favor good separation but also increase the time required for elution. It can be shown that

$$k' = \frac{KV_s}{V_m} \tag{23-3}$$

where K is the partition coefficient ($= C_s/C_m$), V_s is the volume of the stationary phase, and V_m is the volume of the mobile phase. The capacity

In calculating H.E.T.P., it is a good idea to choose a peak whose capacity factor is greater than 5.

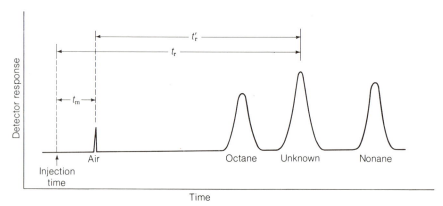

Figure 23-3
Schematic gas chromatogram showing measurement of retention times.

factor can be increased by increasing the volume of the stationary phase. It is a good practice to measure the capacity factor of a column periodically with a standard solution. Changes in the capacity factor reflect degradation of the column.

Retention index

One measure of relative retention time is the **Kovats index**, a logarithmic scale on which the adjusted retention time of a peak is compared with those of linear alkanes. An example is shown in Figure 23-3. The unknown happens to have a retention time between that of octane and nonane. The Kovats index, I, for the unknown is calculated from the formula

> The Kovats index relates the retention time of a solute to the retention times of linear alkanes.

$$I = 100\left[n + \frac{\log t_r'(\text{unknown}) - \log t_r'(n)}{\log t_r'(N) - \log t_r'(n)}\right] \tag{23-4}$$

where n is the number of carbon atoms in the *smaller* alkane

$t_r'(n)$ is the adjusted retention time of the *smaller* alkane

$t_r'(N)$ is the adjusted retention time of the *larger* alkane

This definition makes the Kovats index for a linear alkane equal to 100 times the number of carbon atoms. For octane $I = 800$, and for nonane $I = 900$.

EXAMPLE

Suppose that the unadjusted retention times in Figure 23-3 are $t_r(\text{air}) = 0.5$ min, $t_r(\text{octane}) = 14.3$ min, $t_r(\text{unknown}) = 15.7$ min, and $t_r(\text{nonane}) = 18.5$ min. The Kovats index for the unknown is

$$I = 100\left[8 + \frac{\log 15.2 - \log 13.8}{\log 18.0 - \log 13.8}\right] = 836$$

The retention index of an unknown measured on several different columns is useful for identifying the unknown by comparison with tabulated retention indices. For a homologous series of compounds (those with similar structures, but differing by the number of CH_2 groups in a chain, such as $CH_3CO(CH_2)_nCO_2CH_2CH_3$, $\log t_r'$ is usually a linear function of the number of carbon atoms.

> A plot of log t_r' versus (number of carbon atoms) is a fairly straight line for a homologous series of compounds.

Temperature programming

When separating a mixture of compounds with a wide range of boiling points or polarities, it is very useful to be able to change the column temperature *during* the separation. An example is shown in Figure 23-4. At a

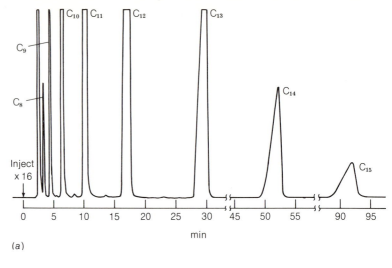

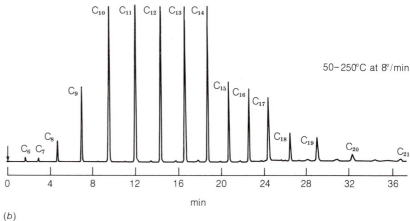

Figure 23-4
Comparison of (a) isothermal and (b) programmed temperature chromatography. Each sample contains linear alkanes run on a 1.6 mm × 6 m column containing 3% Apiezon L (liquid phase) on 100/120 mesh VarAport 30 solid support with an He flow rate of 10 mL/min. [H. M. McNair and E. J. Bonelli, *Basic Gas Chromatography* (Palo Alto, Calif.: Varian Instrument Division, 1968).]

constant temperature of 150°, the more volatile compounds emerge very close together and the less volatile compounds may not even be eluted from the column. If the temperature is increased from 50° to 250° at a rate of 8°/min, all of the compounds are eluted and the separation between peaks is fairly uniform. For rapid heating, it is desirable to use a small-diameter column. The equation describing the dependence of retention time on temperature is

If the retention time of a compound at two temperatures is known, Equation 23-5 lets us predict the retention time at a third temperature.

$$\log t'_r = \frac{a}{T} + b \qquad (23\text{-}5)$$

where T is kelvins and a and b are constants.

Columns

Columns for gas chromatography, usually made of stainless steel, aluminum, or glass, come in a variety of lengths and diameters. For analytical purposes, a small diameter is desirable. For preparative purposes, a fatter column capable of handling more sample is required.

Solid support

Most columns contain a nonvolatile liquid coated on fine particles of an "inert" support. The support should consist of strong, uniformly shaped, small particles with a proportionally large surface area. Most supports are made of diatomaceous earth, the silica skeletons of algae. Ideally, the support should not interact with the solutes, but no solid support is totally inert toward all solutes. **Silanization** (page 561) is one widely used means of reducing the interaction of silica particles with polar solutes. For some compounds that bind tenaciously to most solid supports, Teflon particles are a useful support.

Teflon is a polymer with the structure $-CF_2-CF_2-CF_2-CF_2-$. It is one of the more inert materials known.

Uniform particle size decreases the eddy diffusion term in the van Deemter equation (22-35), thereby reducing the H.E.T.P. and increasing the resolving power of the column. Small particle size decreases the time required for solute equilibration, thereby improving column efficiency. However, if too small a particle size is used, the column resistance reduces the flow rate below a useful value. Chromatographic particle sizes are given in **mesh size,** which is the number of openings per linear inch of screen. A 100/200 mesh particle will fall through a 100 mesh screen but not through a 200 mesh screen. Box 23-1 describes some impressive separations achieved with glass capillary columns.

Stationary phase

A bewildering assortment of liquid phases is available for gas-liquid partition chromatography. A few are listed in Table 23-1. The choice of liquid phase for a given problem is mostly empirical, with the rule that "like dissolves like" being most useful. The classes of solutes most strongly retained by each liquid phase are given in Table 23-2. In general, a variety of liquid phases might be suitable for a given separation problem.

Tables 23-1 and 23-2 tell us, for example, that the silicone gum rubber SE-30 would be a useful liquid phase for the separation of a mixture of olefins. In separating compounds of different classes, but with similar boiling points, the tables let us predict the order of elution. For example, suppose we wish to separate 1-propanol (b.p. 97°) from 2-chloropentane (b.p. 97°). Propanol is a polar compound in class III, while 2-chloropentane is a relatively nonpolar compound in class I. If the polar stationary phase Zonyl E-7 were used, propanol would have a longer retention time than 2-chloropentane. If the nonpolar liquid phase SE-30 were used, propanol would be eluted before 2-chloropentane.

Each stationary phase retains solutes in its own class best.

Box 23-1   GLASS CAPILLARY COLUMNS

Among the most powerful methods of separation is gas chromatography with long capillary columns. Glass columns 30–150 m in length, with inner diameters of 0.2–0.3 mm (outer diameter 0.8–0.9 mm) can be coiled to fit inside an ordinary gas chromatograph. The stationary phase is coated on the inner wall of the glass. The plate height of such columns is smaller than that of ordinary columns because there is no zone broadening from eddy diffusion. The decreased resistance to gas flow allows the use of much longer capillary columns than ordinary columns. Therefore, more plates and greater separation are possible. The capacity of a capillary column is very small, so only tiny quantities can be analyzed. The sample is injected through a *splitter*, which allows only 0.1–5% of the vaporized sample to enter the column.

Two very impressive examples of capillary gas chromatography are shown here. In the one below, deuterated and nondeuterated cyclohexanes, with a relative retention of 1.07, are separated with a resolution of 1.45.

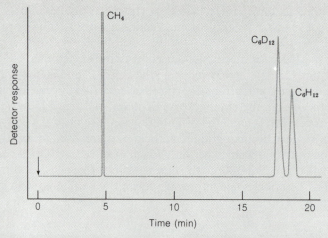

Separation of C_6H_{12} and C_6D_{12} at 40°C by a 30 m glass capillary column with a 0.22 mm inner diameter and squalane as the stationary phase. N = 15 000 plates. [F. A. Bruner and G. P. Cartoni, *Anal. Chem.* **36**, 1522 (1964).]

In the chromatogram on the facing page, the D and L enantiomers of amino acids are separated from each other on a glass column to which is covalently attached an optically active stationary phase:

Liquid phase containing L-valine covalently attached to glass surface

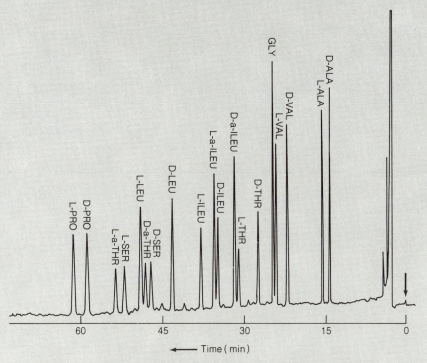

Separation of amino acid enantiomers at 110°C on a 0.30 mm (inner diameter) × 39 m glass capillary column with N = 100 000 plates. [W. A. Koenig and G. J. Nicholson, *Anal. Chem.*, **47**, 951 (1975).]

Amino acids are not volatile enough to be chromatographed directly. Instead, the more volatile N-trifluoroacetyl isopropyl esters are prepared. The use of volatile derivatives is commonplace in gas chromatography.

Table 23-1
Some common liquid phases used in gas chromatography

Liquid phase	Structure	Class of solutes retained most strongly	Maximum temperature (°C)
Squalane	CH_3—$CH(CH_2)_3$—CH—$(CH_2)_3$—CH—$(CH_2)_4$—CH—$(CH_2)_3$—CH—$(CH_2)_3$—CH with CH_3 branches	I	100
SE-30	$\left[\begin{array}{c} CH_3 \\ \mid \\ Si-O-Si-O \\ \mid \\ CH_3 \end{array} \right]_r$ (silicone)	I	350
Apiezon	(mixed hydrocarbons)	I	275–300 (depending on type of Apiezon)
Dibutyl tetrachlorophthalate	tetrachlorobenzene with $CO_2(CH_2)_3CH_3$ and $CO_2(CH_2)_3CH_3$, Cl substituents	II	150
Dinonyl phthalate	benzene with $CO_2(CH_2)_8CH_3$ and $CO_2(CH_2)_3CH_3$	II	175
QF-1	$Si(CH_3)_3$—$\left[O-\begin{array}{c} CF_3 \\ \mid \\ CH_2 \\ \mid \\ CH_2 \\ \mid \\ Si \\ \mid \\ CH_3 \end{array} \right]_x \left[O-\begin{array}{c} CH_3 \\ \mid \\ Si \\ \mid \\ CH_3 \end{array} \right]_y$ O—$Si(CH_3)_3$	II	250
OV-17	methyl phenyl silicone	II	300

DEGS	$\{CH_2-CH_2-O-CH_2-CH_2-O-\overset{\overset{O}{\|}}{C}-CH_2-CH_2-\overset{\overset{O}{\|}}{C}-O\}_n$	II	190
Tetracyanoethyl pentaerythritol	$N\equiv C-C_2H_4-O-CH_2-\overset{\overset{O-C_2H_4-C\equiv N}{\|}}{\underset{\underset{O-C_2H_4-C\equiv N}{\|}}{\overset{\overset{CH_2}{\|}}{C}}}-CH_2-O-C_2H_4-C\equiv N$	III	180
Zonyl E-7	$H\{CF_2\}_n CH_2-O-\overset{O}{C}$ ... $C-O-CH_2\{CF_2\}_n H$	III	200
XE-60	$Si(CH_3)_3-O\{Si-O\}Si(CH_3)_3$ (with CH_3, CH_2, CH_2, $C\equiv N$ groups)$_n$	III	275
Carbowax 20M	$OH\{CH_2-CH_2-O\}_n H$	IV	250
Versamid 900	$HO\{\overset{\overset{O}{\|}}{C}-R-\overset{\overset{O}{\|}}{C}-NH-R'-NH\}_n H$	IV	275
Tetrahydroxyethylethylenediamine	$HO-CH_2-H_2C$ and $HO-CH_2-H_2C$ $\rangle N-CH_2-CH_2-N\langle$ CH_2-CH_2-OH and CH_2-CH_2-OH	IV	150

Source: Adapted from H. M. McNair and E. J. Bonelli, *Basic Gas Chromatography* (Palo Alto, Calif.: Varian Instrument Division, 1968).

Raising percentage of stationary phase leads to:
1. Greater capacity for solute.
2. Longer retention time.
3. Increased H.E.T.P.

Lowering column temperature leads to:
1. Increased resolution.
2. Longer retention time.

Table 23-2
Solute classes for Table 23-1

I: Low polarity
Saturated hydrocarbons
Olefinic hydrocarbons
Aromatic hydrocarbons
Halocarbons
Mercaptans
Sulfides
CS_2

II: Intermediate polarity
Ethers
Ketones
Aldehydes
Esters
Tertiary amines
Nitro compounds (without α-H atoms)
Nitriles (without α-H atoms)

III: Polar
Alcohols
Carboxylic acids
Phenols
Primary and secondary amines
Oximes
Nitro compounds (with α-H atoms)
Nitriles (with α-H atoms)

IV: Very polar
Polyhydroxyalcohols
Amino alcohols
Hydroxy acids
Polyprotic acids
Polyphenols

SOURCE: Adapted from H. M. McNair and E. J. Bonelli, *Basic Gas Chromatography* (Palo Alto, Calif. Varian Instrument Division, 1968).

The amount of liquid phase used is expressed as a weight percent of the solid support. Loadings in the range 2–10% are most common. Beyond 30%, column efficiency decreases as pools of liquid phase are formed. Below 1%, the surface of the support may not be entirely covered. In general, a large percent of liquid phase leads to greater capacity for solute, longer retention times, and more separation between peaks. A lower percentage gives more rapid analysis and smaller values of H.E.T.P. Lowering column temperature increases resolution and retention time. Temperature control is used in conjunction with liquid-phase loading and sample volume to find the best compromise between resolution and speed.

The stationary phase is coated on the solid support by suspending the solid support in a solution of the liquid phase dissolved in a low-boiling solvent. After the solvent has been gently evaporated, the particles are poured into the column, being tapped continually to promote uniform packing. The ends of the column are plugged with glass wool. Finally, the packed column is conditioned for several hours at a temperature 25° above that intended for use (but below the maximum allowable temperature for the liquid phase). During conditioning, a low flow rate of carrier gas (~ 5 mL/min) is maintained, and the volatile compounds baked off the column are *not* passed through the detector (to prevent detector contamination).

Detectors

Thermal conductivity detector

As shown in Figure 23-5, this common detector usually consists of a hot tungsten–rhenium filament over which is directed the gas emerging from the column. The electrical resistance of the filament increases as the temperature of the filament increases. As long as carrier gas is flowing at a constant rate over the filament, the resistance is constant and a constant signal is sent to the recorder. When a solute emerges from the column, the thermal conductivity of the gas stream decreases, so the rate at which the filament is cooled by the gas stream decreases. The filament becomes hotter, its resistance increases, and a change in the signal sent to the recorder is observed.

Since the detector responds to *changes* in thermal conductivity of the gas stream, it is desirable that the conductivities of solute and carrier be as different as possible. Helium is almost always used as the carrier gas for a thermal conductivity detector because it has the second-highest thermal conductivity of any gas. (Only H_2 has a higher conductivity, but it would not be safe for routine use.)

It is common practice to split the carrier gas into two streams, sending part of it through the column and part over a reference filament. The resistance of the working filament is measured with respect to that of the reference filament. The sensitivity increases roughly with the square of the filament current. However, the maximum current recommended by the manufacturer should not be exceeded, to avoid burning out the filament. The filament should never be left on without carrier gas flow, since it can overheat and be ruined.

The sensitivity of a thermal conductivity detector (but *not* the flame ionization detector described in the next section) is inversely proportional to flow rate. This means that the detector is more sensitive at a lower flow rate and that its sensitivity will change whenever flow rate changes. The sensitivity also increases with greater temperature differences between the filament and the surrounding block in Figure 23-5. The detector block should therefore be maintained at the lowest temperature that will allow all solutes to remain gaseous.

To increase sensitivity with thermal conductivity detection:
1. Increase filament current (but do not exceed maximum).
2. Decrease flow rate.
3. Lower detector block temperature.

Flame ionization detector

A schematic diagram of this popular detector is shown in Figure 23-6. The column eluate is mixed with H_2 and air and burned in a flame inside the detector. Carbon atoms of organic compounds (with the notable exceptions of carbonyl and carboxyl carbons) can produce CH radicals, which go on to produce CHO^+ ions in the hydrogen–oxygen flame.

$$CH + O \rightarrow CHO^+ + e^-$$

Only about one in 10^5 carbon atoms produces an ion. However, this production is steady and is strictly proportional to the number of susceptible

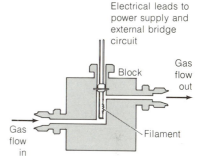

Figure 23-5
Schematic diagram of a thermal conductivity detector. [Courtesy Varian Associates, Palo Alto, Calif.]

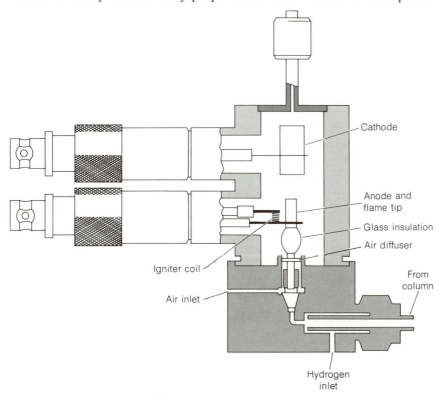

Figure 23-6
Schematic diagram of a flame ionization detector. [Courtesy Varian Associates, Palo Alto, Calif.]

23 / CHROMATOGRAPHIC METHODS

Features of flame ionization detector:
1. Signal is proportional to number of susceptible carbon atoms.
2. Extremely sensitive.
3. Wide range of linear response.
4. Very stable background (baseline).
5. N_2 can be used as carrier gas.

Cochromatography of an unknown and an authentic sample of the suspected compound on several different columns is a good way to identify a peak.

Peak heights can be used instead of areas if the peaks are very sharp.

carbon atoms entering the flame. The flame ionization detector is insensitive to inorganic compounds such as O_2, CO_2, H_2O, and NH_3.

The CHO^+ produced in the flame carries current to the cathodic collector above the flame. The current flowing between the anode at the base of the flame and the cathodic collector is measured and translated into a signal on a recorder. In the absence of organic solutes, the current is almost zero, and the detector response to organic compounds is directly proportional to solute mass over seven orders of magnitude. This sensitivity is about three orders of magnitude greater than that of the thermal conductivity detector. For alkanes, the flame ionization sensitivity is $\sim 10^{-12}$ g/s. Since there is no special requirement for the carrier gas, N_2 is used instead of the more expensive He.

Quantitative Analysis

For *qualitative* analysis, the peaks of a chromatogram can be identified by their retention times and by collecting samples for spectral analysis. A useful way to compare retention times is to add a known pure compound suspected of being in the sample. When this **cochromatography** is performed, one peak in the sample should increase in height if that component is part of the sample. The increase of one peak does not guarantee that the pure compound is the same as the unknown, since they could accidentally have the same retention times. However, equal retention times on several different columns would provide strong evidence of identity.

For *quantitative* analysis, the area of each peak of interest should be measured, since *the area is proportional to the quantity of each component* in the mixture. The common ways of measuring area, in order of preference, are:

1. Advanced chromatographs are coupled to a computer, which calculates the areas automatically.
2. Many recorders have an automatic integrator, which will be discussed below.
3. Peak areas can be measured by tracing over the edges with a mechanical device called a *planimeter*.
4. For a Gaussian peak, the product of peak height times width at half-height equals 84% of the total area.
5. A triangle can be drawn with two sides tangent to the inflection points on each side of the peak. When the third side is drawn along the baseline, the area of the triangle ($= \frac{1}{2}$(base $\times$ height)) is 96% of the area of a Gaussian peak.
6. Peaks can be cut out and weighed on a sensitive balance. The area is proportional to the mass of chart paper.

If the detector responded equally well to all solutes, the relative peak areas would be equal to the relative amounts of each. Thermal conductivity detectors have different sensitivities for different compounds, depending on their thermal conductivity. Flame ionization detectors also give varying response, depending mainly on the number of carbon atoms. An empirical

response factor (F) for each compound must therefore be determined for quantitative analysis by gas chromatography.

Figure 23-7 shows a chromatogram with an integrator trace at the bottom. The integrator describes a horizontal line when the detector signal is at the baseline. When a peak begins in the chromatogram, the integrator moves vertically in proportion to the area under the peak. The integrator scale has ten divisions, the total length of which we will call 100 units. The integration of the 1-butanol peak indicates 785 units of area:

First downstroke	58 units
7 full traverses	700 units
Last downstroke	27 units
Total area	785 units

If the baseline is slanted or not zeroed, you must estimate how much of the peak area is due to baseline and subtract it from the total area. For 1-pentanol,

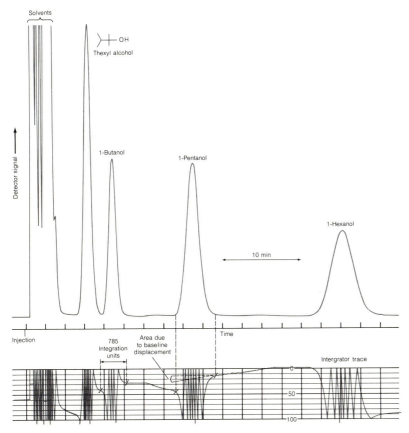

Figure 23-7
Chromatogram of 2 μL of a reaction mixture at 40°C on a 3 mm (outer diameter) × 0.76 m column containing 20% Carbowax 20M on Gas-Chrom R using a flame ionization detector. [Courtesy Norman Pearson.]

23 / CHROMATOGRAPHIC METHODS

the total area is 1 341 units. Extrapolation of the sloping integral following the peak suggests that 10 units be subtracted, giving a peak area of 1 331 units.

In Figure 23-7, a known amount of 1-pentanol was added to the reaction mixture as an **internal standard.** *The amounts of other solutes are then measured relative to that of the internal standard.* The response factor for each compound has to be measured in a separate experiment. We will define the response factor, F, with the relation

> The response factor measures the relative response of the detector to different compounds.

$$\frac{\text{Concentration of solute}}{\text{Concentration of standard}} = F\left(\frac{\text{Area of solute}}{\text{Area of standard}}\right) \qquad (23\text{-}6)$$

EXAMPLE

When a mixture containing 1.21 mmol of 1-butanol and 0.95 mmol of 1-pentanol was chromatographed under the same conditions used for Figure 23-7, the ratio of peak areas was butanol/pentanol = 1.02. Calculate the response factor for butanol.

Using Equation 23-6 with pentanol as the standard gives

$$F = \left(\frac{\text{Concentration of butanol}}{\text{Concentration of pentanol}}\right)\left(\frac{\text{Area of pentanol}}{\text{Area of butanol}}\right)$$

$$F = \left(\frac{1.21}{0.95}\right)\left(\frac{1}{1.02}\right) = 1.25$$

That is, the detector responds 1.25 times more to pentanol than to an equal number of moles of butanol. This is exactly what is expected for a flame ionization detector, since one compound has 5 carbon atoms and one has 4 carbons.

EXAMPLE

The mixture in Figure 23-7 contained 0.57 mmol of 1-pentanol added as an internal standard at the end of the reaction. How many millimoles of 1-butanol are present?

We calculated areas of 785 and 1 331 from the integration curves for these two peaks. With a response factor of 1.25, we find

$$\text{Millimoles of butanol} = F\left(\frac{\text{Area of butanol}}{\text{Area of pentanol}}\right)(\text{Millimoles of pentanol})$$

$$= 1.25\left(\frac{785}{1\ 331}\right)(0.57) = 0.42 \text{ mmol}$$

For each type of detector, there is a range of solute concentrations in which the detector response is linear (i.e., proportional to solute concentration). For quantitative analysis, it is important to be aware of this range and to stay within it.

23-2 LIQUID CHROMATOGRAPHY

Classical liquid chromatography utilizes gravity-fed columns such as that in Figure 22-8, with silica gel or alumina being the most common stationary phases. During the last decade, the practice of chromatography has been revolutionized by the introduction of high-performance liquid chromatography (**HPLC**). This technique makes use of high pressure to force liquid through a column packed with very efficient particles of very small diameter. The equipment is analogous to gas chromatography in many ways. In this section, we discuss both classical chromatography and HPLC. Ion-exchange, molecular-exclusion, and affinity chromatography—all forms of liquid chromatography—will be discussed in later sections.

Classical Liquid Chromatography

Columns

Columns as primitive as that in Figure 22-8 are still in widespread use. They are adequate when the elution bands are broad or the separation between bands is large. For more exacting separations in which the resolution is not as great, modern columns are designed to eliminate the spaces that may function as mixing chambers before and after the column bed.

In the column in Figure 23-8, the solid is supported by a porous nylon net beneath which is a very small dead space leading to the exit tube. An adjustable **flow adaptor** is pressed tightly against the top of the solid phase so that there is no space available for the sample to mix with a layer of solvent at the top of the column. The inlet and outlet tubing have a 1 mm diameter to further prevent mixing of liquids before and after passage through the column. Each of these measures decreases band spreading and increases resolution.

Stationary phases

The most common stationary phase is silica gel ($SiO_2 \cdot xH_2O$, also called silicic acid), obtained by acid precipitation of silicate solutions. The pore size, surface area, and surface pH depend on the conditions of preparation. The active adsorption sites are surface Si–O–H groups, which can adsorb water from the air and thereby slowly become deactivated. The gel can be activated by heating to 200°C to drive off the water. Heating to higher temperatures causes an irreversible dehydration with loss of surface area. Silica is weakly acidic and interacts most strongly with basic solutes.

Alumina ($Al_2O_3 \cdot xH_2O$), the other most common adsorbent, displays increased activity after heating to 1 100°C. Overnight heating to 400°C in air produces highly adsorbent *activity grade I* alumina. Equilibration with various quantities of water (for 24 hours with occasional shaking) reduces

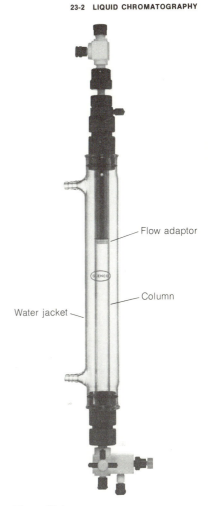

Figure 23-8
Glass chromatography column with a flow adaptor at the top. The column is enclosed by a jacket through which water from a constant-temperature bath may be circulated to keep delicate biological samples refrigerated. [Courtesy Glenco Scientific Co., Houston, Tex.]

Alumina grades:

I—anhydrous
II—3% H_2O
III—6% H_2O
IV—10% H_2O
V—15% H_2O

584

23 / CHROMATOGRAPHIC METHODS

Table 23-3
Elutropic series

Solvent	ε^0 (for alumina)
Fluoroalkanes	-0.25
n-Pentane	0.00
i-Octane	0.01
n-Heptane	0.01
n-Decane	0.04
Cyclohexane	0.04
Cyclopentane	0.05
Carbon disulfide	0.15
Carbon tetrachloride	0.18
1-Chloropentane	0.26
i-Propyl ether	0.28
i-Propyl chloride	0.29
Toluene	0.29
1-Chloropropane	0.30
Chlorobenzene	0.30
Benzene	0.32
Bromoethane	0.37
Diethyl ether	0.38
Chloroform	0.40
Dichloromethane	0.42
Tetrahydrofuran	0.45
1,2-Dichloroethane	0.49
2-Butanone	0.51
Acetone	0.56
Dioxane	0.56
Ethyl acetate	0.58
Methyl acetate	0.60
1-Pentanol	0.61
Dimethyl sulfoxide	0.62
Aniline	0.62
Nitromethane	0.64
Acetonitrile	0.65
Pyridine	0.71
2-Propanol	0.82
Ethanol	0.88
Methanol	0.95
1,2-Ethanediol	1.11
Acetic acid	Large

SOURCE: S. G. Perry, R. Amos, and P. I. Brewer, *Practical Liquid Chromatography* (New York; Plenum Press, 1972); and L. R. Snyder, *Principles of Adsorption Chromatography* (New York: Marcel Dekker, 1968).

the activity of the alumina. Alumina is slightly basic and behaves as if it has three types of active sites. The basic sites *strongly* adsorb acidic solutes. Electrophilic sites (Al^{3+}?) adsorb unsaturated compounds quite well. Aromatic molecules interact with the alumina by an apparent charge-transfer mechanism, in which the solute acts as an electron donor. Activated alumina contains sites that may react with esters, anhydrides, aldehydes, and ketones and may catalyze elimination or isomerization reactions.

Cellulose is used as an adsorbent for compounds that are too polar to be eluted from silica gel or alumina. Other adsorbents include activated charcoal, Florisil (a coprecipitate of silica and magnesium oxide), and magnesia ($MgO \cdot xH_2O$). The latter is particularly useful for olefins and aromatic compounds.

Solvents

In adsorption chromatography, the solvent competes with the solute for active adsorption sites on the stationary phase. The relative abilities of different solvents to elute a given solute from the column are nearly independent of the nature of the solute. That is, the elution can be described more as a displacement of solute from the adsorbent, rather than partitioning of the solute between two phases.

An **elutropic series** (Table 23-3) lists solvents by their relative abilities to displace solutes from a given adsorbent. The **eluent strength** (ε^0) is a measure of the solvent adsorption energy, with the value for pentane defined as zero. The eluent strengths in Table 23-3 apply to alumina, but a similar relative order is observed for silica gel. In general, the greater the eluent strength, the more rapidly will solutes be eluted from the column.

In practice, a *gradient* of eluent strength is used for many separations. First, the less highly retained solutes are eluted with a solvent of low eluent strength. Then a second solvent is mixed with the first, either in discrete steps or continuously, increasing eluent strength. By this means, more strongly adsorbed solutes can be eluted from the column. Eluent strength is not a linear function of the relative proportions of each solvent. A small amount of polar solvent will markedly increase the eluent strength of a non-polar solvent.

Detection

The most common detector in liquid chromatography is an ultraviolet spectrophotometer equipped with a flow cell (Figure 23-9). Since most compounds have ultraviolet absorbance, this detector is almost universal.

Choosing conditions

To select conditions for a chromatographic separation, two procedures are common. One is to use **thin-layer chromatography** (TLC) with different solvents and stationary phases to find the best conditions. In TLC, a layer of solid phase is coated on a glass or inert plastic surface. A spot of sample solution is applied near the bottom, and the plate is placed in a pool of solvent in a closed chamber. Liquid slowly creeps up the plate and performs an ascending chromatographic separation. When the solvent nears the top of the plate, the plate is removed from the solvent and dried. Spots are made visible by placing the plate in a warm chamber containing a crystal of I_2. The I_2 vapor is reversibly adsorbed on most substances and creates a dark spot wherever a compound appears on the plate. By this means, you can experiment rapidly with many different solvents to find the best one for a separation.

Alternatively, a disposable Pasteur pipet with a plug of glass wool makes a fine chromatography column with a total volume of about 1 mL. It is possible to run several such columns in a short time to select conditions for a larger-scale separation.

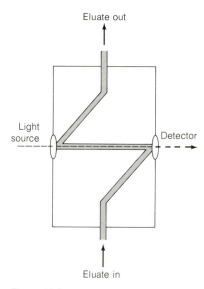

Figure 23-9
Light path in a micro flow cell of a spectrophotometric detector. Cells are available having a 0.5 cm pathlength and containing only 10 μL of liquid.

High-Performance Liquid Chromatography (HPLC)

The theoretical plate height of a chromatography column decreases as the size of the stationary-phase particles decreases. One reason is that the solute can equilibrate more rapidly between the mobile and stationary phases. Smaller particles reduce the depth (or thickness) of each phase and, with less distance to cover, the solute can diffuse between the two phases more quickly. Also, the flow paths followed by a molecule traveling through the column are more uniform with smaller particles. The eddy diffusion term in the van Deemter equation (22-35) is therefore smaller and the plate height is less. It can be seen in Figure 23-10 that the peaks become sharper and resolution increases with smaller particles.

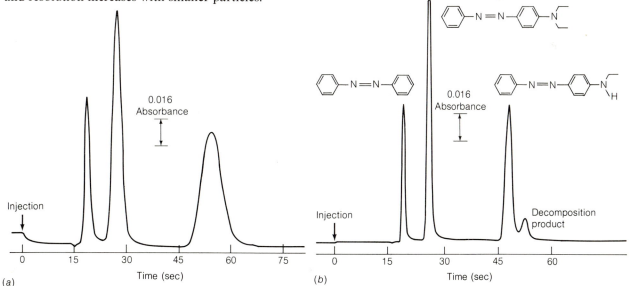

Figure 23-10
Chromatograms of the same sample using (a) 10 μm and (b) 5 μm particle diameter silica. [From R. E. Majors. *J. Chromatogr. Sci.*, **11**, 88 (1973).]

A programmed solvent gradient has the same effect in liquid chromatography that a programmed temperature gradient has in gas chromatography.

A column packed with fine particles offers much more resistance to flow than does a column packed with coarse particles. To obtain a useful flow rate, it is necessary to use high pressure to force the liquid through the column. Pressures of ~7–35 MPa (70–350 atm) are routinely used to yield flow rates of ~0.5–5 mL/min. Needless to say, the column and its associated plumbing must be made of a strong material—usually stainless steel.

The solvent must be of high quality and is routinely filtered through a fine membrane that rejects dust and other particulates. It is desirable to pass the solvent through a **precolumn** containing the same packing as the working column. Irreversibly adsorbed solvent impurities will be collected in the precolumn and will not degrade the working column. The sample should also be filtered before injection, or it can foul the equipment.

Elution with a single solvent is termed **isocratic.** Just as a programmed temperature change is useful in gas chromatography, a programmed increase of eluent strength is often desirable in liquid chromatography. Automatic gradient programmers will pump solvents from two or more reservoirs in a manner that continuously increases the eluent strength (Figure 23-11). By this means, strongly retained solutes can be eluted from the column in a reasonable length of time.

The eluate from the column flows through an absorbance monitor whose output is displayed on a recorder. Another nearly universal detector monitors the refractive index of the eluate. For appropriate samples, a polarographic detector is more sensitive than absorbance or refractive index monitors. The eluate is directed at a mercury drop in a specially designed polarographic cell. This detector responds to many organic functional groups and metal ions present at the nanogram level.

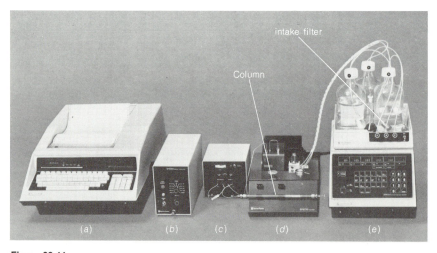

Figure 23-11
Equipment for high performance liquid chromatography (HPLC). (a) Computing integrator and plotter. (b) Detector control unit. (c) UV detector. (d) Organizer module, including pump, column, and injector. (e) Programmable three-reservoir delivery system. [Courtesy Spectra Physics, Santa Clara, Calif.]

Stationary phase

Three common types of stationary-phase particles are shown in Figure 23-12:

1. **Microporous particles** of silica or other materials are small (5–10 μm) particles with a very large surface area. They can be used for adsorption chromatography or may be coated with a stationary liquid phase for partition chromatography.
2. **Pellicular particles** consist of a large-diameter (50 μm) glass bead, coated with a thin layer (2 μm) of liquid phase. The thinness of the liquid layer is conducive to rapid equilibration of solute between the mobile and stationary phases. However, the small surface area of the spherical particle gives it very little capacity for solute. Pellicular particles thus offer high efficiency (small plate height) but very little capacity to retain solute. Small quantities of solute are eluted with good resolution in short times. By comparison, microporous particles have far more surface area, with greater capacity and longer retention times.
3. **Bonded-phase particles** are the most common type. They are usually microporous silica particles, to which a liquid phase is *covalently* attached. These particles afford excellent separations, and there is no bleeding of the liquid phase from the particles. The liquid phase can be attached to the silica surface by reactions such as

$$\text{Particle—Si—OH} + (CH_3CH_2O)_3\text{SiR} \xrightarrow{-CH_3CH_2OH} \text{Particle—Si—O—Si—R}\begin{array}{c}OCH_2CH_3\\|\\|\\OCH_2CH_3\end{array}$$

Common polar phases	Common nonpolar phases
$R = (CH_2)_3NH_2$	$R = (CH_2)_{17}CH_3$
$R = (CH_2)_3CN$	$R = (CH_2)_2$—⬡

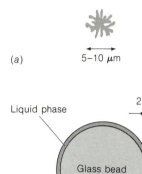

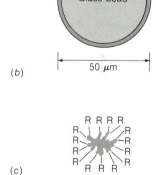

Figure 23-12
Three types of stationary phase particles used in HPLC. (a) Microporous. (b) Pellicular. (c) Bonded phase.

We refer to the case where a polar phase is attached to the solid support as **normal-phase chromatography**. A mobile phase less polar than the bonded phase is used for elution. The eluent strength of the mobile phase is increased by addition of more polar solvents. Chromatography with a nonpolar bonded phase and a polar solvent is called **reverse-phase chromatography**.

Reverse-phase chromatography has become the more important of the two methods. It eliminates tailing problems associated with adsorption of polar compounds by polar packings. It is also less sensitive to polar impurities (such as water) in the eluent.

Question: Why would the eluent strength of solvent in reverse-phase chromatography increase as the solvent is made *less* polar?

23-3 ION-EXCHANGE CHROMATOGRAPHY

Ion-exchange chromatography is based on the equilibration of solute ions between the solvent and charged sites fixed on the stationary phase (Figure 22-9). In **anion exchangers,** positively charged groups are covalently bound to the packing. Solute anions are attracted to these sites. **Cation exchangers** contain covalently bound, negatively charged sites that bind solute cations.

Anion exchangers contain bound *positive* groups.

Cation exchangers contain bound *negative* groups.

588

23 / CHROMATOGRAPHIC METHODS

Ion Exchangers

Polystyrene and polyacrylic acid resins

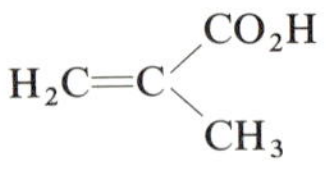

Polystyrene resins for ion exchange are made by the copolymerization of styrene and divinylbenzene (Figure 23-13). The divinylbenzene content is varied from 1% to 16% to increase the extent of crosslinking of the insoluble hydrocarbon polymer. The benzene rings can be modified to produce a cation-exchange resin, containing sulfonate groups ($—SO_3^-$), or an anion-exchange resin, containing ammonium groups ($—NR_3^+$). If methacrylic acid is used in place of styrene, a polymer with carboxyl groups results.

Ion exchangers are commonly classified as being strongly or weakly acidic or basic, as indicated in Table 23-4. The sulfonate groups ($—SO_3^-$) of strongly acidic resins remain ionized even in strongly acidic solutions. The carboxyl groups ($—CO_2^-$) of the weakly acidic resins are protonated at around pH 4 and lose their cation-exchange capacity. "Strongly basic" quaternary ammonium groups ($—CH_2NR_3^+$) remain cationic at all values of

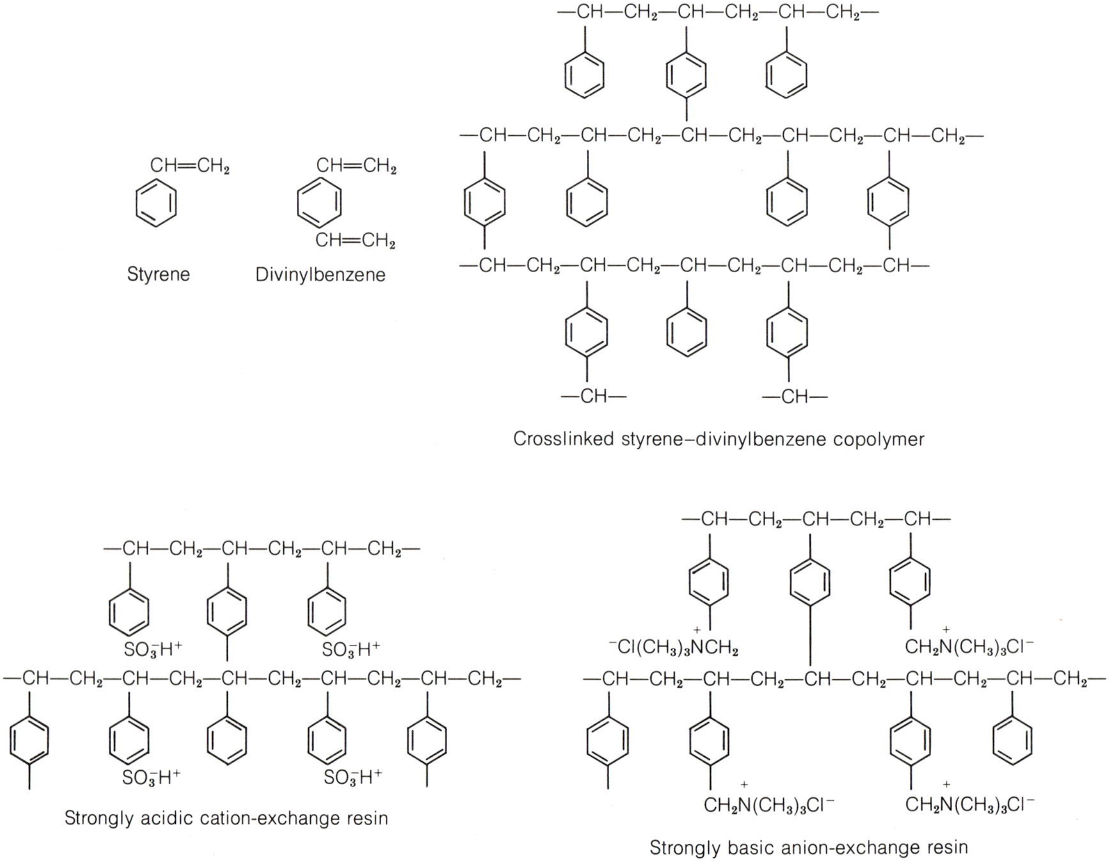

methacrylic
acid

Strongly acidic cation exchangers: RSO_3^-
Weakly acidic cation exchangers: RCO_2^-
"Strongly basic" anion exchangers: $RNR_3'^+$
Weakly basic anion exchangers: $RNR_2'H^+$

Styrene   Divinylbenzene

Crosslinked styrene–divinylbenzene copolymer

Strongly acidic cation-exchange resin

Strongly basic anion-exchange resin

Figure 23-13
Structures of styrene–divinylbenzene crosslinked ion-exchange resins.

Table 23-4
Ion-exchange resins

Resin type	Chemical constitution	Usual form as purchased	Common trade names		Selectivity	Thermal stability
			Rohm & Haas	Dow Chemical		
Strongly acidic cation exchanger	Sulfonic acid groups attached to styrene and divinyl-benzene copolymer	$\phi—SO_3^- H^+$	Amberlite IR-120	Dowex 50W	$Ag^+ > Rb^+ > Cs^+ > K^+ >$ $NH_4^+ > Na^+ > H^+ >$ Li^+ $Zn^{2+} > Cu^{2+} > Ni^{2+} >$ Co^{2+}	Good up 150°C
Weakly acidic cation exchanger	Carboxylic acid groups attached to acrylic and divinylbenzene copolymer	$R—COO^- Na^+$	Amberlite IRC-50	—	$H^+ \gg Ag^+ > K^+ >$ $Na^+ > Li^+$ $H^+ \gg Fe^{2+} > Ba^{2+}$ $Sr^{2+} > Ca^{2+} > Mg^{2+}$	Good up to 100°C
Strongly basic anion exchanger	Quaternary ammonium groups attached to styrene and divinylbenzene copolymer	$[\phi—CH_2N(CH_3)_3]^+ Cl^-$	Amberlite IRA-400	Dowex 1	$I^- >$ phenolate$^- >$ $HSO_4^- > ClO_3^- >$ $NO_3^- > Br^- > CN^- >$ $HSO_3^- > NO_2^- > Cl^- >$ $HCO_3^- > IO_3^- >$ $HCOO^- >$ Acetate$^- >$ $OH^- > F^-$	OH^- form fair up to 50°C Cl^- and other forms good up to 150°C
Weakly basic anion exchanger	Polyalkylamine groups attached to styrene and divinylbenzene copolymer	$[\phi—NH(R)_2]^+ Cl^-$	Amberlite IR-45	Dowex 3	$\phi SO_3H >$ Citric $> CrO_3 >$ $H_2SO_4 >$ tartaric $>$ oxalic $> H_3PO_4 >$ $H_3AsO_4 > HNO_3 >$ $HI > HBr > HCI >$ $HF > HCO_2H >$ $CH_3CO_2H > H_2CO_3$	Extensive information not available; tentatively limited to 65°C

SOURCE: Adapted from J. X. Khym, *Analytical Ion-Exchange Procedures in Chemistry and Biology* (Englewood Cliffs, N.J.: Prentice-Hall, 1974).

23 / CHROMATOGRAPHIC METHODS

pH and function as anion exchangers. The "weakly basic" tertiary ammonium ($-CH_2NHR_2^+$) anion exchangers are deprotonated in moderately basic solution and lose their ability to bind anions.

The extent of crosslinking is indicated by the notation "-XN" after the name of the resin. For example, Dowex 1-X4 contains 4% divinylbenzene, and Bio-Rad AG 50W-X12 contains 12% divinylbenzene. The resin becomes more rigid and less porous as the extent of **crosslinking** increases. Lightly crosslinked resins permit rapid equilibration of solute between the inside and outside of the particle. On the other hand, resins with little crosslinking swell in water. This decreases both the density of ion-exchange sites and the selectivity of the resin for different ions. More heavily crosslinked resins exhibit less swelling, higher exchange capacity and selectivity, but longer equilibration times.

The pore size of polystyrene and polyacrylic acid resins effectively limits their use to molecules weighing less than 500. Larger molecules cannot penetrate these resins. Also, the charge density of these exchangers is so great that large molecules with many charged groups can be irreversibly bound to the resin.

Cellulose, dextran, and related ion exchangers

Because they are much softer than polystyrene resins*, dextran and its relatives are called* gels.

These derivatized polysaccharides possess much larger pore sizes and lower density of charged groups. They are well suited to ion exchange of macromolecules, such as proteins. Cellulose and dextran are polymers of the sugar glucose. Dextran, crosslinked by glycerin, is sold under the name Sephadex. (Figure 23-14). Other macroporous ion exchangers are based on the polysaccharide agarose and on polyacrylamide.

The common charged functional groups that are synthetically bound to occasional hydroxyl groups of the polysaccharides are listed in Table 23-5. DEAE-Sephadex, for example, refers to an anion-exchange Sephadex containing diethylaminoethyl groups.

Inorganic ion exchangers

The resins and gels mentioned above are suitable for most applications of ion exchange. However, they are unstable at high temperature or in the presence of intense radiation. A number of inorganic substances are stable under these conditions and exhibit useful ion-exchange properties.

Hydrous oxides of Zr, Th, Ti, Sn, and W exhibit both anion- and cation-exchange properties. These same metals form insoluble phosphate, molybdate, tungstate, and arsenate salts with cation-exchange properties. A number of metal sulfides, notably CdS, exhibit ion-exchange behavior toward transition metal ions.

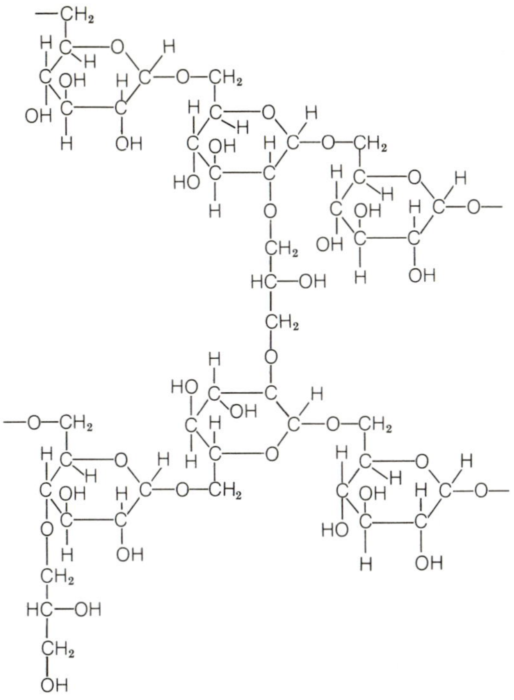

Figure 23-14

Structure of Sephadex, a crosslinked dextran sold by
Pharmacia Fine Chemicals, Piscataway, New Jersey.

Table 23-5

Some active groups of ion-exchange gels

Type	Abbreviation	Name	Structure
Cation exchangers			
Strong acid	SP	Sulfopropyl	$-OCH_2CH_2CH_2SO_3H$
	SE	Sulfoethyl	$-OCH_2CH_2SO_3H$
Intermediate acid	P	Phosphate	$-OPO_3H_2$
Weak acid	CM	Carboxymethyl	$-OCH_2CO_2H$
Anion exchangers			
Strong base	TEAE	Triethylaminoethyl	$-OCH_2CH_2\overset{+}{N}(CH_2CH_3)_3$
	QAE	Diethyl(2-hydroxypropyl) quaternary amino	$-OCH_2CH_2\overset{+}{N}(CH_2CH_3)_2$ $\qquad\qquad\quad\; CH_2CHOHCH_3$
Intermediate base	DEAE	Diethylaminoethyl	$-OCH_2CH_2N(CH_2CH_3)_2$
	ECTEOLA	Triethanolamine coupled to cellulose through gyceryl chains	
	BD	Benzoylated DEAE groups	
Weak base	PAB	*p*-Aminobenzyl	$-O-CH_2-\bigcirc-NH_2$

Ion-Exchange Equilibria

Selectivity

Consider the competition of Na^+ and Li^+ for sites on the cation-exchange resin, R^-:

$$R^-Na^+ + Li^+ \rightleftharpoons R^-Li^+ + Na^+ \qquad K = \frac{[R^-Li^+][Na^+]}{[R^-Na^+][Li^+]} \quad (23\text{-}7)$$

The equilibrium constant for Reaction 23-7 is called the **selectivity coefficient,** because it describes the relative selectivity of the resin for Li^+ and Na^+.

The relative selectivities of some strongly acidic and basic polystyrene resins are shown in Table 23-6. Note that the relative selectivities for certain ions increase with the extent of crosslinking. This is because the pore size of the resin shrinks as crosslinking increases. Ions such as Li^+, with a large hydrated radius (Table 6-1), do not have as much access to the resin as smaller ions, such as Cs^+.

In general, ion exchangers favor the binding of ions of higher charge, decreased hydrated radius, and increased polarizability. A fairly general order of selectivity for cations is the following:

$$Pu^{4+} \gg La^{3+} > Ce^{3+} > Pr^{3+} > Eu^{3+} > Y^{3+} > Sc^{3+} > Al^{3+} \gg$$

$$Ba^{2+} > Pb^{2+} > Sr^{2+} > Ca^{2+} > Ni^{2+} > Cd^{2+} > Cu^{2+} >$$

$$Co^{2+} > Zn^{2+} > Mg^{2+} > UO_2^{2+} \gg Tl^+ > Ag^+ > Rb^+ > K^+ >$$

$$NH_4^+ > Na^+ > H^+ > Li^+$$

Polarizability refers to the ability of an ion's electron cloud to be deformed by nearby charges. Deformation of the electron cloud induces a dipole in the ion. The attraction between the induced dipole and the nearby charge increases the binding of the ion to the resin.

Table 23-6
Relative selectivity coefficients of ion-exchange resins

	Sulfonic acid cation-exchange resin			Quaternary ammonium anion-exchange resin	
	% Divinylbenzene				Relative
Cation	4	8	10	Anion	selectivity
Li^+	1.00	1.00	1.00	F^-	0.09
H^+	1.30	1.26	1.45	OH^-	0.09
Na^+	1.49	1.88	2.23	Cl^-	1.0
NH_4^+	1.75	2.22	3.07	Br^-	2.8
K^+	2.09	2.63	4.15	NO_3^-	3.8
Rb^+	2.22	2.89	4.19	I^-	8.7
Cs^+	2.37	2.91	4.15	ClO_4^-	10.0
Ag^+	4.00	7.36	19.4		
Tl^+	5.20	9.66	22.2		

SOURCE: *Amberlite Ion Exchange Resins—Laboratory Guide* (Rohm & Haas Co., 1979).

Reaction 23-7 can be driven in either direction, even though Na^+ is bound more tightly than Li^+. Washing a column containing Na^+ with a substantial excess of a Li^+ salt will displace Na^+ and replace it with Li^+. Washing a column in the Li^+ form with Na^+ will convert it to the Na^+ form.

This is simply an application of Le Châtelier's principle.

Ion exchangers loaded with one kind of ion will generally bind small amounts of a different ion nearly quantitatively. Thus an Na^+-loaded resin will bind small amounts of Li^+ nearly quantitatively, even though the selectivity is greater for Na^+. The same column will tightly bind large quantities of, for example, Ni^{2+} or Fe^{3+}, because the resin has greater selectivity for these ions than for Na^+. Even though Fe^{3+} is bound more tightly than H^+, Fe^{3+} can be quantitatively removed from the resin by washing with a large excess of acid.

Donnan equilibrium

When an ion exchanger is placed in an electrolyte solution, *the concentration of electrolyte at equilibrium is higher outside the resin than inside it.* The Donnan membrane theory, which describes the distribution of charges about a semipermeable membrane, also accounts nicely for this property of ion exchangers.

A phase containing bound charges tends to exclude electrolyte.

Consider a quaternary ammonium anion-exchange resin (R^+) in its Cl^- form immersed in a solution of KCl. Let the concentration of an ion inside the membrane be $[X]_i$ and the concentration outside the membrane be $[X]_0$. It can be shown[†] from thermodynamics that the ion product inside the resin is approximately equal to the product outside the resin. In this example, we can write

$$[K^+]_i[Cl^-]_i = [K^+]_0[Cl^-]_0 \qquad (23\text{-}8)$$

From considerations of charge balance, we know that

$$[K^+]_0 = [Cl^-]_0 \qquad (23\text{-}9)$$

We are ignoring H^+ and OH^-, which are assumed to be negligible.

Inside the resin, there are three charged species, and the charge balance is

$$[R^+]_i + [K^+]_i = [Cl^-]_i \qquad (23\text{-}10)$$

where $[R^+]$ is the concentration of quaternary ammonium ions attached to the resin. Substituting Equations 23-9 and 23-10 into Equation 23-8 gives

$$[K^+]_i([K^+]_i + [R^+]_i) = [K^+]_0^2 \qquad (23\text{-}11)$$

Equation 23-11 says that $[K^+]_0$ must be greater than $[K^+]_i$.

[†] See, for example, W. J. Moore, *Physical Chemistry*, 4th ed. (Englewood Cliffs, N. J.: Prentice-Hall, 1972), p. 544.

23 / CHROMATOGRAPHIC METHODS

EXAMPLE

Suppose that the concentration of cationic sites in the resin is 6.0 M. When the Cl^- form of this resin is immersed in 0.050 M KCl, what will be the ratio $[K^+]_0/[K^+]_i$?

Let us assume that $[K^+]_0$ remains 0.050 M. Putting values into Equation 23-11 gives

$$[K^+]_i([K^+]_i + 6.0) = (0.050)^2$$

from which we find $[K^+]_i = 0.000\ 42$ M. The ratio $[K^+]_0/[K^+]_i$ is 120. The concentration of electrolyte inside the resin is less than 1% of that outside the resin.

The high concentration of positive charges within the resin repels cations from the resin.

Note that the Donnan theory predicts that ions with the *same* charge as the resin are excluded. (The quaternary ammonium resin excludes K^+.) The counterion, Cl^- in the above example, is *not excluded* from the resin. There is no electrostatic barrier to penetration of any anion into the resin. Anion exchange takes place freely in the quaternary ammonium resin, even though cations are repelled from the resin.

The volume V_m is available to an electrolyte. The volume $V_m + V_s$ is available to a nonelectrolyte.

The Donnan equilibrium is the basis of **ion-exclusion chromatography.** Since dilute electrolytes are effectively excluded from the resin, they will pass through a column when the volume of mobile phase (V_m) has been eluted. Nonelectrolytes, such as sugar, freely penetrate the resin. They will not be eluted until a volume $V_m + V_s$ (where V_s is the volume of liquid inside the resin) has passed. Thus, if a solution of NaCl and sugar is applied to an an ion-exchange column, the NaCl will emerge from the column *before* the sugar will.

Conducting Ion-Exchange Chromatography

Choosing an ion exchanger

The choice of an exchanger for a given application can usually be quickly narrowed to a particular class of resins, but further selection within that class is usually a trial-and-error procedure, with several acceptable solutions. We will consider three general classes of ion exchangers:

1. Ion-exchange **resins** of the styrene–divinylbenzene and methacrylic acid–divinylbenzene type.
2. Ion-exchange celluloses and **gels** of the dextran, polyacrylamide, and agarose types.
3. **Inorganic ion exchangers** such as hydrous metal hydroxides and sulfides.

It should also be noted that a variety of ion exchangers is available for HPLC separations.

In general, ion-exchange *resins* are used for applications involving small molecules (MW $\lesssim 500$), which can penetrate the small pores of the resin. Ion-exchange *gels* are used for large molecules (such as proteins and nucleic acids), which could not penetrate the pores of resins. The large molecules often have such great charge that if they could penetrate the resin they might be held too tightly to be eluted. The density of ion-exchange sites in gels is much lower than it is inside resins. For separations involving harsh chemical conditions (high temperature, high radiation levels, strongly basic solution, powerful oxidizing agents), the resins and gels are unsuitable. *Inorganic exchangers* should be tried in such cases.

For resins, mesh sizes of 100–200 or 200–400 are suitable for most work. The higher mesh number (smaller particle size) leads to finer separations but slower column operation. Very coarse particles are useful for gross separations or batch processes where speed or rapid settling of the resin from a suspension is desired. The selectivity of a resin increases with the degree of crosslinking, but the speed of equilibration decreases. For gels, the size of a macromolecule dictates the minimum pore size that can be used.

> The selectivity of a resin increases with crosslinking.

The choice between strong and weak ion exchangers depends on the operating pH and relative selectivity needed for a separation. A weak acid (RCO_2^-) cation exchanger becomes protonated below pH $\approx$ 4 and loses its ion-exchange capacity. Clearly, it would not be useful for a column eluted with a medium containing 1 M HCl. Similarly, a weak base anion exchanger (R_3NH^+) loses its charge in strongly basic solution. The strongly acidic (RSO_3^-) or strongly basic (R_4N^+) ion exchangers are useful over a greater range of pH. The order of ion selectivities of different ion exchangers is not the same. Manufacturers' catalogs usually contain some selectivity information for each resin.

Gradients

Elution of a column with a **gradient** of ionic strength or pH is extremely valuable in ion-exchange chromatography. The mechanics of creating a gradient are described in Section 23-6. Consider a column to which the anions A^- and B^- are bound. Suppose the binding of A^- is stronger than that of B^-. One good way to separate A^- from B^- is to elute the column with solution containing the anion C^- (which is less tightly bound than A^- or B^-). At low concentrations of C^-, neither A^- nor B^- is displaced. As the concentration of C^- is increased, B^- will eventually be displaced and move down the column. At a still higher concentration of C^-, the anion A^- will also be eluted. A nonlinear gradient may be used to increase the separation of some components (with a shallow gradient) or to decrease the separation of well-resolved components (with a steep gradient).

> An ionic-strength gradient is analogous to a solvent or temperature gradient.

Column dimensions

A column length-to-diameter ratio of 10 or 20 is adequate for most purposes. Difficult separations might require a greater length. In scaling up a procedure for preparative purposes, the *cross-sectional area* of the column should be increased in proportion to sample size. The column length need not be increased. For preparative separations, it is not uncommon for the sample to occupy 10–20% of the exchanger volume when the sample is first applied to the column.

23 / CHROMATOGRAPHIC METHODS

Water softeners use ion exchange to remove Ca^{2+} and Mg^{2+} from "hard" water.

Applications

In addition to its considerable value as a chromatographic separation technique, ion exchange has many other applications.

Deionization

Ionic impurities in water can be removed by passing the liquid through an anion-exchange resin in its OH^- form and a cation-exchange resin in its H^+ form. Suppose, for example, that $Cu(NO_3)_2$ is present in the solution. The cation-exchange resin binds Cu^{2+} and replaces it with $2H^+$. The anion-exchange resin binds NO_3^- and replaces it with OH^-. The effluent will contain just pure water:

$$\left. \begin{array}{l} Cu^{2+} \xrightarrow{\ H^+\ \text{ion exchange}\ } 2H^+ \\[4pt] 2NO_3^- \xrightarrow{\ OH^-\ \text{ion exchange}\ } 2OH^- \end{array} \right\} \longrightarrow \text{pure } H_2O$$

Water so treated is called *deionized water*.

Interconversion of salts

One salt can be converted to another by appropriate ion exchange. For example, tetrapropylammonium hydroxide solution can be prepared if a tetrapropylammonium salt of some other anion is available:

$$(CH_3CH_2CH_2)_4N^+I^- \xrightarrow[\text{OH}^-\ \text{form}]{\text{Anion exchanger}} (CH_3CH_2CH_2)_4N^+OH^-$$

tetrapropylammonium
iodide   tetrapropylammonium
hydroxide

A neutral organic acid can be prepared from its sodium salt as follows:

$$N(CH_2CO_2^-Na^+)_3 \xrightarrow[\text{H}^+\ \text{form}]{\text{Cation exchanger}} N(CH_2CO_2H)_3$$

trisodium
nitrilotriacetate   nitrilotriacetic
acid

If, instead, the trisodium salt had been titrated with HCl, the solution would contain nitrilotriacetic acid plus NaCl.

Concentration of trace species

It is sometimes necessary to concentrate trace components of a solution in order to obtain enough for analysis. For example, a large volume of fresh lake water could be passed through a cation-exchange resin in the H^+ form to concentrate metal ions from the water onto the resin. Chelex 100, a styrene–divinylbenzene resin containing iminodiacetic acid groups, is noteworthy for its ability to bind transition metal ions.

Resin—N$\overset{\displaystyle CH_2CO_2H}{\underset{\displaystyle CH_2CO_2H}{\diagdown}}$ Chelex 100

$\underbrace{}_{\text{iminodiacetic acid}}$

The metals can be eluted in a small volume by 2 M HNO_3, which protonates the iminodiacetate groups.

Catalysis

Ion exchangers can function as insoluble catalysts. If a solution of an acid-labile amide or ester is passed through a strong acid cation exchanger in the H^+ form, the amide C–N or ester C–O bond can be cleaved. The products emerge from the column (which might be heated to speed the process) uncontaminated by soluble catalyst.

23-4 MOLECULAR-EXCLUSION CHROMATOGRAPHY

Molecular-exclusion chromatography is also commonly called *gel-filtration* or *gel-permeation chromatography*. This type of chromatography, in which molecules are separated according to their size, is widely used in biochemistry for the resolution of large molecules, such as proteins and carbohydrates. It finds use in polymer chemistry in the separation and characterization of synthetic polymers.

Principles

A schematic representation of the principle of molecular-exclusion chromatography is given in Figure 22-9. The stationary phase contains small pores that can be penetrated by small molecules but not by large molecules. The volume available to small molecules is thus greater than that available to large molecules. Large molecules will therefore be eluted from the column before small molecules (Figure 23-15). Another approach is to consider that large molecules spend all of their time in the mobile phase, whereas small molecules spend only a fraction of their time in the mobile phase. Small molecules are therefore transported more slowly.

In molecular-exclusion chromatography, the volume of mobile phase (V_m) is usually called the **void volume**, V_0. In Chapter 22 (Equation 22-31), we derived the relationship

$$V_r = V_m + KV_s \qquad (22\text{-}31)$$

where V_r is the retention volume, V_s is the volume of stationary phase, and K is the partition coefficient (= concentration of solute in stationary phase/

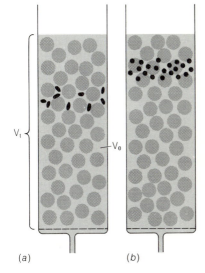

Figure 23-15
(a) Large molecules cannot penetrate the pores of the stationary phase. They are eluted by a volume of solvent equal to the volume of mobile phase. (b) Small molecules that can be found inside or outside the gel require a larger volume for elution. V_t is the total column volume occupied by gel plus solvent. V_0 ($= V_m$) is the volume of the mobile phase. $V_t - V_0$ is the volume occupied by the gel plus its internal liquid phase.

Large molecules pass through the column faster than small molecules.

598

23 / CHROMATOGRAPHIC METHODS

The void volume is the same as the volume of mobile phase: $V_0 \equiv V_m$.

concentration in mobile phase). Equation 22-31 can be rearranged to the form

$$K = \frac{V_r - V_m}{V_s} = \frac{V_r - V_0}{V_s} \qquad (23\text{-}12)$$

The volume of solvent inside the gel particles is V_s. If the gel matrix occupied no volume, then V_s would be $V_t - V_0$, where V_t is the total volume of the column (Figure 23-15). The gel matrix does occupy some space, so $V_t - V_0$ is greater than V_s. However, V_s is proportional to $V_t - V_0$, since the liquid inside the particles occupies a constant fraction of the particles' volume.

The quantity K_{av} (read "K average") is defined as

K_{av} is proportional to the partition coefficient, K.

$$K_{av} = \frac{V_r - V_0}{V_t - V_0} \qquad (23\text{-}13)$$

For a large molecule that does not penetrate the gel, $V_r = V_0$, and $K_{av} = 0$. For a small molecule that freely penetrates the gel, $V_r \approx V_t$, and $K_{av} \approx 1$. Molecules of intermediate size can penetrate some gel pores, but not others. For these molecules, K_{av} will be between 0 and 1.

The value of V_0 is measured by passing a large, inert molecule through the column. Its elution volume is defined as V_0. Blue Dextran 2 000, a blue dye of molecular weight 2×10^6, is most commonly used for this purpose. The total volume, V_t, is calculated from the column dimensions ($V_t = \pi r^2 \times$ length).

Ideally, gel penetration is the only mechanism by which molecules are retained in this type of chromatography. In fact, there is always some adsorption. Aromatic molecules, in particular, are known to be adsorbed by some Sephadex ion exchangers and can exhibit $K_{av} > 1$.

Types of Gels

Fractionation range is more properly defined in terms of molecular size than molecular weight.

The most widely used gels for molecular exclusion are of the Sephadex variety, whose structure is given in Figure 23-14. The Bio-Gel P gels are made of polyacrylamide crosslinked by N,N′-methylenebisacrylamide (Figure 23-16). The pore size of either gel is controlled by the extent of crosslinking in the manufacturing process. The approximate molecular-weight fractionation range of each gel is listed in Table 23-7. The fractionation range applies to "globular" molecules, which are roughly spherical. An elongated molecule, such as a polysaccharide, cannot penetrate the gel as well as a globular molecule of the same weight. The fractionation range for elongated molecules is therefore at a lower molecular weight than for globular molecules. Each gel is available in several particle sizes. The finer the particle size, the greater the resolution and the slower the flow rate of the column.

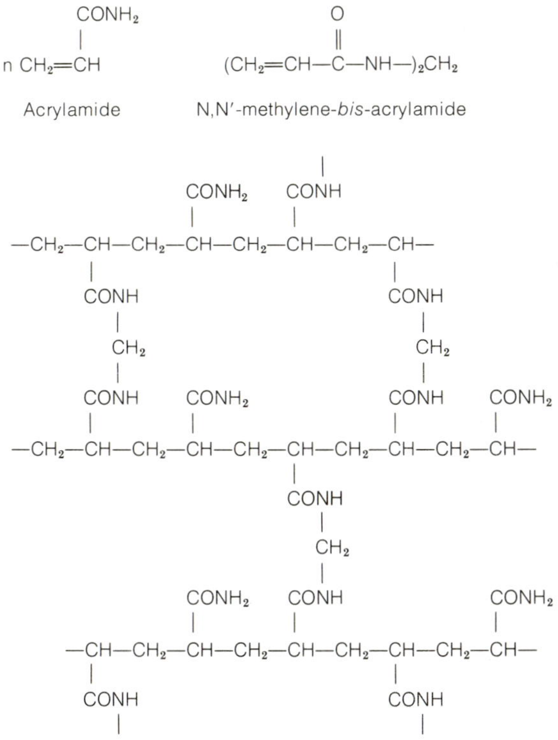

23-4 MOLECULAR-EXCLUSION CHROMATOGRAPHY

Acrylamide

N,N'-methylene-*bis*-acrylamide

Crosslinked polyacrylamide

Figure 23-16
Structure of polyacrylamide.

Table 23-7
Properties of some gel filtration media

Name	Fractionation range (M.W.) for globular proteins
Sephadex G-10	-700
Sephadex G-15	$-1\,500$
Sephadex G-25	$1\,000-5\,000$
Sephadex G-50	$1\,500-30\,000$
Sephadex G-75	$3\,000-80\,000$
Sephadex G-100	$4\,000-150\,000$
Sephadex G-150	$5\,000-300\,000$
Sephadex G-200	$5\,000-600\,000$
Sephacryl S-200	$5\,000-250\,000$
Sephacryl S-300	$10\,000-1\,500\,000$
Sepharose 2B	$70\,000-40\,000\,000$
Sepharose 4B	$60\,000-20\,000\,000$
Sepharose 6B	$10\,000-4\,000\,000$
Bio-Gel P-2	$100-1\,800$
Bio-Gel P-4	$800-4\,000$
Bio-Gel P-6	$1\,000-6\,000$
Bio-Gel P-10	$1\,500-20\,000$
Bio-Gel P-30	$2\,500-40\,000$
Bio-Gel P-60	$3\,000-60\,000$
Bio-Gel P-100	$5\,000-100\,000$
Bio-Gel P-150	$15\,000-150\,000$
Bio-Gel P-200	$30\,000-200\,000$
Bio-Gel P-300	$60\,000-400\,000$
Bio-Gel A-0.5 m	$<10\,000-500\,000$
Bio-Gel A-1.5 m	$<10\,000-1\,500\,000$
Bio-Gel A-5 m	$10\,000-5\,000\,000$
Bio-Gel A-15 m	$40\,000-15\,000\,000$
Bio-Gel A-50 m	$100\,000-50\,000\,000$
Bio-Gel A-150 m	$1\,000\,000-150\,000\,000$

Note: Sephadex and Sephacryl are manufactured by Pharmacia Fine Chemical Co., Piscataway, N.J. Bio-Gel is sold by Bio-Rad Laboratories, Richmond, Calif.

SOURCE: The information in this table was taken from the manufacturers' bulletins, which provide a great deal of useful information about these products.

Applications

Chromatography

Gel filtration is used mainly to separate mixtures of molecules of different molecular weight. Figure 23-17 illustrates the separation of the iron-storage protein ferritin (M.W. 450 000), the iron-transport protein transferrin (M.W. 80 000), and ferric citrate.

Molecular-weight determination

For each type of gel, there is a range over which K_{av} for a solute is roughly proportional to the logarithm of its molecular weight (Figure 23-18). By comparing K_{av} of an unknown with a series of standards, it is possible to estimate the molecular weight of the unknown. Caution is called for in interpreting molecular weights estimated by gel filtration, because molecules with the same weight but different shapes exhibit different elution characteristics. It is also important to use a high enough ionic strength ($\gtrsim 0.05$ M) to eliminate extraneous electrostatic interactions between solute and occasional charged sites on the gel.

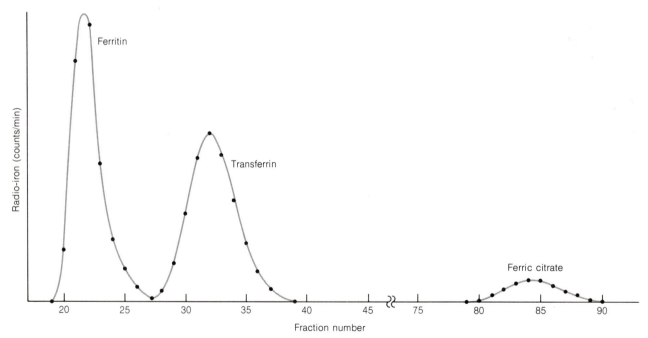

Figure 23-17
Separation of ferritin (M.W. 450 000), transferrin (M.W. 80 000), and ferric citrate on Bio-Gel P-300. 0.50 mL containing 70 μg of ferritin, 700 μg of transferrin, and 96 μg of citrate, each labeled with radioactive iron, was chromatographed on a 1.5 × 37 cm column eluted with 0.05 M N-2-hydroxyethylpiperazine-N'-2-ethanesulfonic acid (HEPES buffer, pH 8.0) plus 0.1 M KCl at a rate of 6.8 mL/hr. Each fraction contained 0.65 mL.

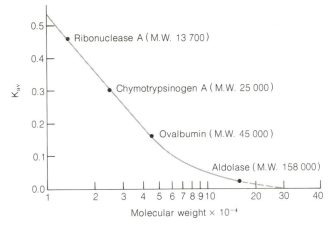

Figure 23-18
Relation of K_{av} and molecular weight for a mixture of proteins chromatographed on Sephadex G-75. Note that the semilogarithmic graph is equivalent to using linear graph paper and plotting log(M.W.) on the abscissa. [Courtesy Pharmacia Fine Chemical Co., Piscataway, N.J.]

A graph of K_{av} versus log(M.W.) is fairly linear.

Desalting

Salts of low molecular weight (or any small molecule) can be removed from solutions of large molecules by passage through a gel filtration column. This is useful for changing the buffer composition of a macromolecule solution or for separating small reactants and products from large molecules after a chemical reaction.

23-5 AFFINITY CHROMATOGRAPHY

Affinity chromatography is probably the most rational and powerful development in the field of separations. The principle is illustrated in Figure 22-9. A molecule that has a specific interaction with just one solute of a complex mixture is covalently attached to the stationary phase in a column. When the sample is passed through the column, just one solute binds to the stationary phase. After everything else has been washed through thoroughly, the one adhering solute is eluted by changing conditions to weaken the binding between the solute and the stationary phase.

This is the only form of chromatography in which the chemist *designs* the stationary phase to interact with a specific solute. The technique is especially applicable to biochemistry, where advantage can be taken of such well-known specific interactions as those involving enzymes and substrates or coenzymes, antibodies and antigens, or receptors and hormones. Affinity chromatography has even been used to separate one type of cell from others.

23 / CHROMATOGRAPHIC METHODS

Affinity chromatography has not yet had a large impact in chemistry, but it is a rapidly developing area limited mainly by the imagination and synthetic abilities of those who wish to use it.

The literature on affinity chromatography in biochemistry is growing explosively.[†] In this section, we mention one application of this technique to the separation of small molecules rather than macromolecules. Molecules with coplanar cis–diol groups ($- \overset{\displaystyle OH}{\underset{\displaystyle |}{C}} - \overset{\displaystyle OH}{\underset{\displaystyle |}{C}} -$) can be separated from a complex mixture by passage through a column to which is attached phenylboronic acid. Affi-Gel 601 is a commercially available[‡] form of Bio-Gel P-6 incorporating phenylboronic acid:

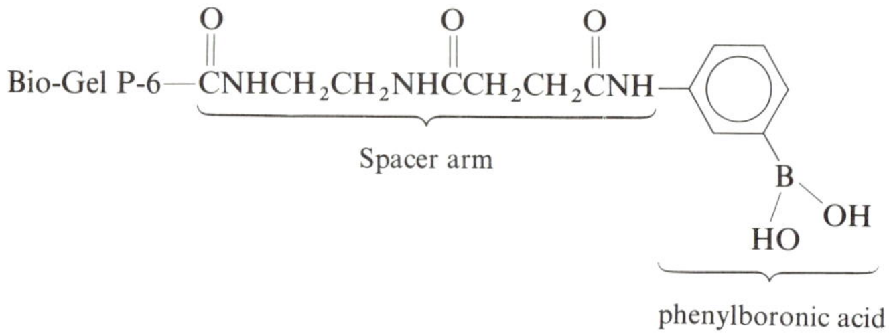

$$\text{Bio-Gel P-6}-\underbrace{\overset{O}{\overset{||}{C}}NHCH_2CH_2NH\overset{O}{\overset{||}{C}}CH_2CH_2\overset{O}{\overset{||}{C}}NH}_{\text{Spacer arm}}-\underbrace{\overset{\text{phenylboronic acid}}{\bigcirc\overset{B}{\underset{HO}{\diagdown}}OH}}$$

The boronic acid forms a covalent linkage with coplanar cis–diols. By this means, nucleotides (which have a cis–diol group) can be separated from deoxynucleotides or cyclic nucleotides, which do not have cis–diols. The nucleotide can be displaced from the affinity column by eluting the column with a citrate-containing buffer.

The spacer arm in the structure drawn above is a means of extending the boronic acid away from the bulk of the gel particle. Spacer arms are widely used for affinity chromatography of macromolecules that are sterically hindered from approaching the bulk gel particle closely enough to bind (Figure 23-19).

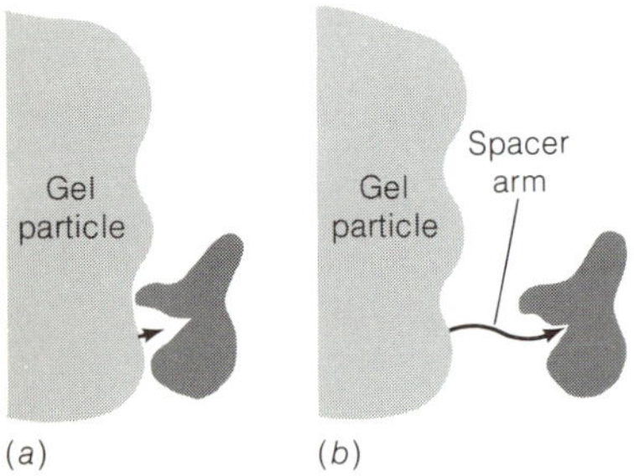

23-6 PRACTICAL NOTES

There is a certain degree of art to pouring uniform columns, applying samples evenly, and obtaining symmetric elution bands. In this section, we mention some of the practical aspects of liquid column chromatography.

Figure 23-19
The use of a spacer arm allows solute–stationary phase interactions that might otherwise be sterically forbidden.

[†] Practical, up-to-date information on affinity chromatography can be found in the following manufacturers' bulletins: *Affinity Chromatography: Principles and Methods* (Piscataway, N.J.: Pharmacia Fine Chemical Co.); and *Materials, Equipment and Systems for Chromatography, Electrophoresis, Immunochemistry and HPLC* Richmond, Calif.: Bio-Rad Laboratories). A few of the many excellent descriptions of applications of affinity chromatography can be found in T. G. Cooper, *The Tools of Biochemistry* (New York: Wiley, 1977), Chap. 7; M. Wilchek and W. B. Jakoby in W. B. Jacoby and M. Wilchek, eds., *Methods of Enzymology*, Vol. 34 (New York: Academic Press, 1974), p. 3; J. Porath and T. Kristiansen in H. Neurath and R. L. Hill, eds., *The Proteins*, Vol. I (New York: Academic Press, 1975), p. 95.

[‡] Bio-Rad Laboratories, Richmond, California.

Preparing the Stationary Phase

Before pouring a column, the stationary phase must be equilibrated with the solvent. For alumina and silica gel, this means simply making a slurry of the solid in the appropriate solvent. For ion-exchange and molecular-exclusion gels, this means following the manufacturer's instructions concerning time and temperature of equilibration, as well as what solutions might be needed to fully swell and hydrate the gel. When mixing a dry gel with solvent, the gel should be slowly sprinkled on top of the liquid and allowed to settle with gentle stirring, using a glass stirring rod. *Never* use magnetic stirring, which breaks and destroys gel particles.

After the gel has been equilibrated with solvent, it should be suspended in ~2–5 volumes of solvent and allowed to stand until 90–95% has settled. The smaller particles (called **fines**) that are still suspended should be removed by decanting or suction. If left in the gel, the fines significantly reduce the column flow rate. The procedure to remove fines should be repeated several times. Many gels are sold in a pre-equilibrated form from which the fines have already been removed.

Pouring the Column

After it has been equilibrated with solvent, the stationary phase should be suspended in enough solvent so that the solid would occupy from one-half to two-thirds of the total volume when it settles. The chromatography column should contain a few centimeters of liquid before pouring the gel.

The slurry of suspended phase is then poured gently down the wall of the column, preferably directed by a glass rod (Figure 23-20). Foaming and violent convection currents are undesirable. If most of the column is to be filled with stationary phase, additional volume must be available to hold the slurry (only about half of which is stationary phase). For this purpose, a wide-stem funnel held to the top of the column by a rubber stopper is adequate. Reservoirs that screw into the top of some columns are commercially available. It is *not* desirable to pour part of a column, allow it to settle, and then pour the rest of the column. This creates discontinuities in the bed.

When the entire slurry has been added to the column, it should be allowed to settle gently, like a fine snow. After a few centimeters have settled, begin a slow flow of solvent to pack the remainder. The hydrostatic pressure limit set by the manufacturer should not be exceeded during packing or any other phase of chromatography. (Hydrostatic pressure measurement is discussed below.)

Liquid should never be allowed to drain below the top of the gel in the column. Otherwise, air spaces might be introduced that could lead to irregular flow patterns. Solvent should be directed gently down the wall of the column or on top of a few centimeters of liquid above the packing. In *no* case should the solvent be allowed to dig a channel into the gel.

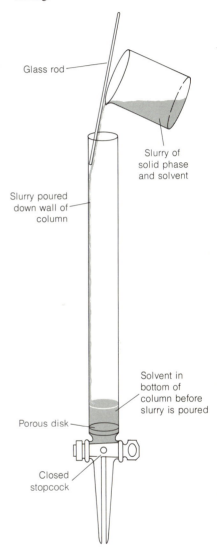

A slurry is a suspension of a solid in a liquid.

Add gel *to* solvent. *Never* use magnetic stirring.

Figure 23-20
A slurry of the stationary phase in starting solvent should be poured gently down the wall of the column.

Applying the Sample

To apply sample, the solvent is drained to the top of the gel, and flow is halted. Then sample is applied gently down the wall of the column by pipet. The sample should be applied evenly around the whole column. Flow is resumed, the sample is drained into the gel, and flow is halted again. Then *small* quantities of solvent are added down the walls by pipet to finish washing the sample into the gel. Finally, a layer of solvent is built up by pipet and maintained during column operation.

Running the Column

Flow is usually maintained by siphoning solvent from a reservoir into the column. The proper flow rate depends on the column diameter and the degree of separation between solutes. Maximum resolution usually demands a very slow flow rate. The manufacturer's bulletin for the particular stationary phase should be consulted to select a flow rate.

Pressure is measured from the inlet level to the outlet level.

The flow rate is governed by the hydrostatic pressure, measured as the distance between the level of liquid in the reservoir and the level of the column outlet (Figure 23-21a). If a **Mariotte flask** is used as a reservoir, the pressure is measured between the bottom of the air inlet and the level of the column outlet (Figure 23-21b). In an ordinary reservoir, the hydrostatic pressure pushing liquid into the outlet tube is equal to the height of liquid above the opening of the tube within the reservoir. This pressure decreases as liquid is drained from the reservoir. In a Mariotte flask, the pressure at the bottom

A Mariotte flask maintains a constant pressure and therefore a constant flow rate.

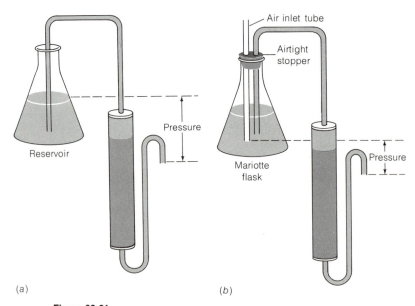

Figure 23-21
Measurement of the hydrostatic pressure with (a) an ordinary solvent reservoir and (b) a Mariotte reservoir.

of the air inlet tube must be 1 atm. This pressure remains constant unless the liquid level falls below the bottom of either tube. A Mariotte flask thus maintains a constant pressure for the duration of the chromatography.

Tubing that comes in contact with any organic solvent is a potential source of contamination because the plasticizers used in tubing manufacture can be leached out by some solvents. For work with nonaqueous solutions, common tubing such as Tygon should be avoided, and fluoroplastic or glass tubing should be used instead.

Gradients

In ion-exchange chromatography, a gradient of eluent ionic strength may be needed to displace the strongly retained solutes from the stationary phase. A device for making a linearly increasing gradient is shown in Figure 23-22. A solution of low ionic strength is placed in the right compartment, and an equal volume at high ionic strength is placed in the left compartment. The first drop of liquid leaving the right side has low ionic strength. As liquid from the left side flows into the right side to keep the levels equal, the ionic strength in the right compartment increases linearly. The last drop to be drained from this system has the ionic strength of the left-hand reservoir.

Accessories

Figure 23-23 shows a rather complete setup for liquid chromatography. Eluent is fed to the column by a peristaltic pump. The flow rate is set by the pump speed, so regulation of hydrostatic pressure is not necessary. The

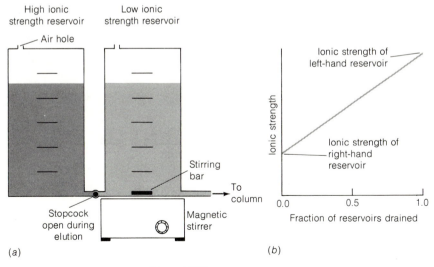

Figure 23-22
A device for gradient elution.

eluate passes through a spectrophotometric flow cell on its way to a fraction collector, which automatically changes test tubes when a preset volume has been collected in each one. The absorbance of the eluate is displayed on a recorder, which also marks each time the fraction collector changes tubes.

All chromatographic devices require a great deal of loving attention, but the automatic features shown in Figure 23-23 allow you to go home and sleep fitfully as the column runs through the night.

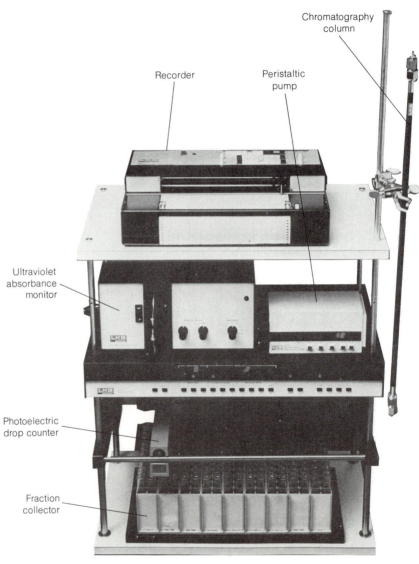

Figure 23-23
Automated apparatus for liquid chromatography.
[Courtesy LKB–Produkter AB, Pleasant Hill, Calif.]

Summary

In gas chromatography, a volatile liquid or gaseous solute is carried by a gaseous mobile phase over a stationary liquid coated on a solid support. A compound can be identified by its retention time on different columns and quantified by the area under its elution peak. An empirical response factor must be determined for each solute in quantitative analysis. Particularly difficult analytical separations may be effected by very long glass capillary columns, while preparative separations require wide columns with a high capacity. A given stationary phase most strongly retains compounds in its own polarity class. The most common detectors are based on thermal conductivity or flame ionization.

High-performance liquid chromatography (HPLC) is similar to gas chromatography, except that the solvent is a liquid forced through the column by a high-pressure pump. The most common stationary phase contains nonpolar groups covalently bound to $5-10$ μm silica particles. Isocratic (single-solvent) elution or gradient elution (with decreasing polarity) is employed in this form of liquid–liquid partition chromatography. Detectors commonly measure ultraviolet absorbance, refractive index, or polarographic current of elution bands. Classical liquid–solid adsorption chromatography employs columns of alumina or silica gel as adsorbents. A gradient of increasing eluent strength (increasing solvent polarity) is frequently used.

Ion-exchange resins and gels contain covalently bound charged groups that attract solute counterions (and that exclude solute ions having the same charge as the resin). Resins of the polystyrene type are useful for the separation of small ions. Greater crosslinking of the resin increases the capacity, selectivity, and time needed for equilibration. Ion-exchange gels based on cellulose and dextran have large pore sizes and low charge densities, suitable for the separation of macromolecules. Certain inorganic salts have ion-exchange properties useful at extremes of temperature and radiation. All ion exchangers operate by the principle of mass action, with a gradient of increasing ionic strength most commonly used to effect a separation.

Molecular-exclusion chromatography is based on the relative inability of large molecules to enter the pores in the stationary phase. Small molecules can enter these spaces and therefore exhibit longer elution times than large molecules. Exclusion chromatography is used for separations based on size and also for approximate molecular-weight determinations of macromolecules. In affinity chromatography, the chemist designs a stationary phase to interact with one particular solute in a complex mixture. After all other components have been eluted, the desired species is liberated by a change in conditions.

Terms to Understand

adjusted retention time
anion exchanger
bonded-phase particles
capacity factor
cation exchanger
chromatogram
cochromatography
crosslinking
deionization
desalting
Donnan equilibrium
eluent strength
elutropic series
fines
flame ionization detector
flow adaptor
gel
gradient elution
HPLC

internal standard
ion-exclusion chromatography
isocratic elution
Mariotte flask
microporous particles
pellicular particles
polarizability
precolumn
preconcentration
resin
response factor
retention index
retention time
retention volume
reverse-phase chromatography
septum
silanization
thermal conductivity detector
thin-layer chromatography

Exercises

23-A. When 1.06 mmol of 1-pentanol and 1.53 mmol of 1-hexanol were dissolved and chromatographed, they gave peak areas of 922 and 1570 units, respectively.
  (a) Calculate the response factor for hexanol relative to pentanol, the internal standard.
  (b) The reaction mixture in Figure 23-7 contained 0.57 mmol of pentanol. How much hexanol did it contain?
  (c) The solution in Figure 23-7 contained 0.44 mmol of thexyl alcohol. Calculate the response factor for this compound relative to that of 1-pentanol.

23-B. An unknown compound was known to be a member of the family $(CH_3)_2CH(CH_2)_nCH_2OSi(CH_3)_3$.
  (a) From the retention times below, estimate the value of n for the unknown.

  $n = 7$: 4.0 min   air: 1.1 min
  $n = 8$: 6.5 min   unknown: 42.5 min
  $n = 14$: 86.9 min

  (b) Calculate the capacity factor for the unknown.

23-C. Vanadyl sulfate ($VOSO_4$), as supplied commercially, is contaminated with H_2SO_4 and H_2O. A solution was prepared by dissolving 0.244 7 g of impure $VOSO_4$ in 50.0 mL of water. Spectrophotometric analysis indicated that the concentration of the blue VO^{2+} ion was 0.024 3 M. A 5.00 mL sample was passed through a cation-exchange column loaded with H^+ to bind VO^{2+} and release H^+, which required 13.03 mL of 0.022 74 M NaOH for titration. Find the weight percent of each component ($VOSO_4$, H_2SO_4, and H_2O) in the vanadyl sulfate.

23-D. Blue Dextran 2000 was eluted in a volume of 36.4 mL from a 2.0 × 40 cm (diameter × length) column of Sephadex G-50.
  (a) At what retention volume would hemoglobin (M.W. 64 000) be expected?
  (b) At what volume is $^{22}NaCl$ expected?
  (c) What would be the retention volume of a molecule with $K_{av} = 0.65$?

23-E. Make a graph showing the qualitative shape of the ionic-strength gradient that would be produced by each device below.

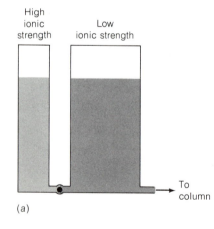
(a)

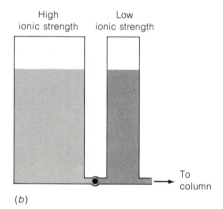
(b)

Problems

23-1. Why is gradient elution useful in liquid adsorption and partition chromatography?

23-2. Name two methods that can be used to select a suitable solvent for large-scale chromatographic separation before performing the large-scale separation.

23-3. Why is high pressure needed in HPLC?

23-4. What is the advantage of temperature programming in gas chromatography?

23-5. Why do glass capillary columns afford more theoretical plates than do ordinary gas chromatography columns?

23-6. Explain why the relative eluent strengths of solvents in adsorption chromatography are fairly independent of solute.

23-7. The exchange capacity of an ion-exchange resin can be defined as the number of moles of charged sites per gram of dry resin. Describe how you would measure the exchange capacity of an anion-exchange resin using standard NaOH, standard HCl, or any other reagent you wish.

23-8. Consider a protein with a net negative charge tightly adsorbed on an anion-exchange gel at pH 8.
  (a) How will a gradient of eluent pH (from pH 8 to some lower pH) be useful for eluting the protein? Assume that the ionic strength of the eluent is kept constant.
  (b) How would a gradient of ionic strength (at constant pH) be useful for eluting the protein?

23-9. How should Figure 23-22 be modified to produce a linear gradient of decreasing ionic strength?

23-10. Propose a scheme for separating trimethylamine, dimethylamine, methylamine, and ammonia from each other by ion-exchange chromatography.

23-11. An unknown compound was cochromatographed with heptane and decane. The adjusted retention times were: heptane (12.6 min), decane (22.9 min), unknown (20.0 min). Realizing that the scale of the Kovats index is logarithmic and that the indices for heptane and decane are 700 and 1 000, respectively, find the Kovats index for the unknown.

23-12. A compound is eluted from a gas chromatograph at $t_r' = 17.0$ min when $T = 140°C$ and 29.2 min at 120°C. Predict the retention time at 155°C.

23-13. What is the average diameter (in micrometers) of particles labeled 100/120 mesh? (Assume that the average particle size is 110 mesh and that the screen-wire thickness is negligible.) What mesh size corresponds to 10 μm?

23-14. Assuming that ferritin is eluted at the void volume of the column in Figure 23-17, calculate K_{av} for transferrin and for ferric citrate.

23-15. From Figure 23-18, estimate the molecular weight of a globular protein with $V_r = 80.0$ mL if $V_0 = 53.1$ mL and the column dimensions are 2.5 × 35 cm.

23-16. Suppose that an ion-exchange resin (R^-Na^+) is immersed in a solution of NaCl. Let the concentration of R^- in the resin be 3.0 M.
  (a) What will be the ratio $[Cl^-]_0/[Cl^-]_i$ if $[Cl^-]_0$ is 0.10 M?

  (b) What will be the ratio $[Cl^-]_0/[Cl^-]_i$ if $[Cl^-]_0$ is 1.0 M?
  (c) Will the fraction of electrolyte inside the resin increase or decrease as the outside concentration of electrolyte increases?

23-17. Compounds with
$$-\overset{\overset{\displaystyle OH}{|}}{C}-\overset{\overset{\displaystyle OH}{|}}{C}- \text{ or } -\overset{\overset{\displaystyle OH}{|}}{C}-\overset{\overset{\displaystyle NH_2}{|}}{C}-$$
linkages can be analyzed by cleavage with periodate. For example, one mole of 1,2-ethanediol consumes one mole of iodate:

$$\begin{matrix} CH_2OH \\ | \\ CH_2OH \end{matrix} \quad + \quad IO_4^- \quad \longrightarrow$$

1,2-ethanediol   periodate
M.W. 62.068

$$2CH_2{=}O + H_2O + IO_3^-$$

formaldehyde   iodate

To analyze 1,2-ethanediol, oxidation with excess IO_4^- is followed by passage of the whole reaction solution through an anion-exchange resin that binds both IO_4^- and IO_3^-.[†] The IO_3^- is then selectively and quantitatively removed from the resin by elution with NH_4Cl. The absorbance of the eluate is measured at 232 nm ($\varepsilon = 900$ M^{-1} cm^{-1}) to find the quantity of IO_3^- produced by the reaction. In one experiment, 0.213 9 g of aqueous 1,2-ethanediol was dissolved in 10.00 mL. Then 1.000 mL of the solution was treated with 3 mL of 0.15 M KIO_4 and subjected to ion-exchange separation of IO_3^- from unreacted IO_4^-. The eluate (diluted to 250.0 mL) gave $A_{232} = 0.521$ in a 1.000 cm cell, and a blank gave $A_{232} = 0.049$. Find the weight percent of 1,2-ethanediol in the original sample.

23-18. The substances below were chromatographed on a 1.5 × 30 cm (diameter × length) gel filtration column. Estimate the molecular weight of the unknown.

Compound	V_r (mL)	Molecular weight
Blue Dextran 2 000	17.7	2 × 10⁶
Aldolase	35.6	158 000
Catalase	32.3	210 000
Ferritin	28.6	440 000
Thyroglobulin	25.1	669 000
Unknown	30.3	?

[†] J. X. Khym and W. E. Cohn, *J. Amer. Chem. Soc.*, **82**, 6380 (1960).

24 / Experiments

The experiments described in this chapter are intended to illustrate the major analytical techniques described in the text. These procedures are organized roughly in the same order as the topics in the text.

Although these procedures are safe when carried out with reasonable care, *all chemical experiments are potentially hazardous.* Any solution that fumes (such as concentrated HCl), and all nonaqueous solvents, should be handled in a fume hood. Pipetting should never be done by mouth. Spills on your body should be flooded immediately with water and washed with soap and water; your instructor should be notified for possible further action. Spills on the benchtop should be cleaned immediately. Toxic chemicals should not be flushed down the drain.

24-1 CALIBRATION OF VOLUMETRIC GLASSWARE

An important trait of good analysts is the ability to extract the best possible data from their equipment. For this purpose, it is desirable to calibrate your own volumetric glassware (burets, pipets, flasks, etc.) to measure the exact volumes delivered or contained. This experiment also promotes improved technique in beginning students' handling of volumetric glassware.

Calibration is normally done by measuring the mass of distilled water delivered or contained by the apparatus. (For some small or odd-shaped vessels, mercury can be used instead of water. Mercury is easier to pour out of glass and weighs 13.5 times as much as an equal volume of water.) Sections 2-3 to 2-5 describe the proper techniques for using volumetric glassware.

Table 24-1 shows that pure water expands $\sim 0.02\%$ per degree near $20°C$. This has a practical implication for the calibration of glassware and for the temperature-dependence of reagent concentrations.

612

24 / EXPERIMENTS

Table 24-1
Density of water

Temperature (°C)	Density of water (g cm^{-3})	Volume of 1 g of water (cm^3)	
		At temperature shown[†]	Corrected to 20°C[‡]
0	0.999 842 5	—	—
4	975 0	—	—
5	966 8	—	—
10	702 6	1.001 4	1.001 5
11	608 1	1.001 5	1.001 6
12	500 4	1.001 6	1.001 7
13	380 1	1.001 7	1.001 8
14	247 4	1.001 8	1.001 9
15	102 6	1.002 0	1.002 0
16	0.998 946 0	1.002 1	1.002 1
17	777 9	1.002 3	1.002 3
18	598 6	1.002 5	1.002 5
19	408 2	1.002 7	1.002 7
20	207 1	1.002 9	1.002 9
21	0.997 995 5	1.003 1	1.003 1
22	773 5	1.003 3	1.003 3
23	541 5	1.003 5	1.003 5
24	299 5	1.003 8	1.003 8
25	047 9	1.004 0	1.004 0
26	0.996 786 7	1.004 3	1.004 2
27	516 2	1.004 6	1.004 5
28	236 5	1.004 8	1.004 7
29	0.995 947 8	1.005 1	1.005 0
30	650 2	1.005 4	1.005 3
35	0.994 034 9	—	—
37	0.993 331 6	—	—
40	0.992 218 7	—	—
100	0.958 366 5	—	—

[†] Corrected for buoyancy with Equation 2-2, using 0.001 2 g/mL air density and assuming stainless-steel weights.

[‡] Corrected for expansion of borosilicate glass (0.001 0% per degree).

EXAMPLE

A dilute aqueous solution with a molarity of 0.031 46 M was prepared in the winter when the lab temperature was 17°C. What will be the molarity of that solution on a warm spring day when the temperature is 25°C?

We assume that the thermal expansion of the solution is similar to the thermal expansion of pure water. Since the concentration of a solution is proportional to its density, we can write

$$\frac{c'}{d'} = \frac{c}{d} \qquad (24\text{-}1)$$

where c' and d' are the concentration and density at temperature T′, and c and d apply at temperature T. Using the densities in column 2 of Table 24-1, we can write

$$\frac{c \text{ at } 25°}{0.997\,05} = \frac{0.031\,46}{0.998\,78} \Rightarrow c = 0.031\,41 \text{ M}$$

The concentration has decreased by 0.16%.

Glass itself expands when it is heated. Pyrex and other borosilicate glasses, which are the most common types, expand by about 1.0×10^{-3} percent per degree near room temperature. This means that if a glass container is heated 10°C, its volume will increase by about $(10)(0.001\,0\%) = 0.01\%$. For all but the most accurate work, this expansion is insignificant. Soft glass expands approximately 2–3 times as much as borosilicate glass.

Calibration of 50 mL Buret

In this procedure, we will construct a graph (such as Figure 3-3) needed to convert the measured volume delivered by a buret to the true volume delivered at 20°C.

PROCEDURE

1. Fill the buret with distilled water and force any air bubbles out the tip. See that the buret drains without leaving drops on the walls. If drops are left, clean the buret with soap and water or soak it with chromic acid cleaning solution (footnote on page 19). Adjust the meniscus to be at or slightly below 0.00 mL, and touch the buret tip to a beaker to remove the suspended drop of water. Allow the buret to stand for 5 min while you weigh a 125 mL flask fitted with a rubber stopper. (Do not touch the flask with your hands, to avoid changing its mass with fingerprints.) If the level of the liquid in the buret has changed, tighten the stopcock and repeat the procedure.

2. Drain approximately 10 mL of water (at a rate of <20 mL/min) into the weighed flask, and cap it tightly to prevent evaporation. Allow about 30 s for the film of liquid on the walls to descend before you read the buret. Estimate all readings to the nearest 0.01 mL. Weigh the flask again to determine the mass of water delivered.

3. Now drain the buret from 10 to 20 mL, and measure the mass of water delivered. Repeat the procedure for 30, 40, and 50 mL. Then do the entire procedure (10, 20, 30, 40, 50 mL) a second time.

4. Use Table 24-1 to convert the mass of water to the volume delivered. Repeat any set of duplicate buret corrections that do not agree to within 0.04 mL. Prepare a calibration graph such as Figure 3-3, showing the correction factor at each 10 mL interval.

EXAMPLE

When draining the buret at 25°C, you observe the following values:

Final reading	10.01	10.08 mL
Initial reading	0.03	0.04
Difference	9.98	10.04 mL
Mass	9.984	10.056 g
Actual volume delivered	10.02	10.09 mL
Correction factor	+0.04	+0.05 mL
Average correction		+0.045 mL

To calculate the actual volume delivered when 9.984 g of water is delivered at 24°C, look at the column of Table 24-1 headed "Corrected to 20°C." Across from 24°C, you find that 1.000 0 g of water occupies 1.003 8 mL. Therefore, 9.984 g occupies (9.984 g)(1.003 8 mL/g) = 10.02 mL. The average correction factor for both sets of data is +0.045 mL.

To obtain the correction factor for a volume greater than 10 mL, add successive masses of water collected in the flask. Suppose that the following masses were measured:

Volume interval (mL)	Mass delivered (g)
0.03–10.01	9.984
10.01–19.90	9.835
19.90–30.06	10.071
Sum 30.03 mL	29.890 g

The total mass of water delivered corresponds to (29.890 g)(1.0038 mL/g) = 30.00 mL. Since the indicated volume is 30.03 mL, the buret correction at 30 mL is −0.03 mL.

What does this mean? Suppose that Figure 3-3 applies to your buret. If you begin a titration at 0.04 mL and end at 29.00 mL, you would deliver 28.96 mL if the buret were perfect. In fact, Figure 3-3 tells you that the buret delivers 0.03 mL less than the indicated amount, so only 28.93 mL was actually delivered. To use the calibration curve, either begin all titrations near 0.00 mL or correct both the initial and the final readings. Use the calibration curve whenever you use your buret.

Other Calibrations

Pipets can be calibrated by weighing the water delivered from them. A volumetric flask can be calibrated by weighing it empty and then weighing it filled to the mark with distilled water. Perform each procedure at least twice. Compare your results with the tolerances in Tables 2-1, 2-2, and 2-3.

24-2 GRAVIMETRIC DETERMINATION OF CALCIUM AS $CaC_2O_4 \cdot H_2O$[†]

Calcium ion can be analyzed by precipitation with oxalate in basic solution to form $CaC_2O_4 \cdot H_2O$ ($K_{sp} = 1.9 \times 10^{-9}$). The precipitate is soluble in acidic solution because the oxalate anion is a weak base (Reactions 8-2 and 8-3). Large, easily filtered, relatively pure crystals of product are obtained if the precipitation is carried out slowly. This can be done by dissolving Ca^{2+} and $C_2O_4^{2-}$ in acidic solution and gradually raising the pH by thermal decomposition of urea (Reaction 8-4).

REAGENTS

Ammonium oxalate solution: Each student needs 80 mL of solution containing 40 g of $(NH_4)_2C_2O_4$ plus 25 mL of 12 M HCl per liter.

Unknowns: Each student should receive 100 mL of solution containing 15–18 g of $CaCO_3$ plus 38 mL of 12 M HCl per liter. Alternatively, solid unknowns are available from Thorn Smith.[‡]

PROCEDURE

1. Dry three medium-porosity, sintered-glass funnels for 1–2 h at 105°C; cool them in a desiccator for 30 min and weigh them. Repeat the procedure with 30 min heating periods until successive weighings agree to within 0.3 mg. Use a paper towel or tongs, not your fingers, to handle the funnels.

2. Use a few small portions of unknown to rinse a 25 mL transfer pipet, and discard the washings. *Use a rubber bulb, not your mouth, to provide suction.* Transfer exactly 25 mL of unknown to each of three 250–400 mL beakers, and dilute each with $\sim$75 mL of 0.1 M HCl. Add 5 drops of methyl red indicator solution (Table 12-3) to each beaker. This indicator is red below pH 4.8 and yellow above pH 6.0.

3. Add $\sim$25 mL of ammonium oxalate solution to each beaker while stirring with a glass rod. Remove the rod and rinse it with the beaker. Add $\sim$15 g of solid urea to each sample, cover it with a watchglass, and boil gently for $\sim$30 min until the indicator turns yellow.

4. Filter each hot solution through a weighed funnel using suction (Figure 2-13). Add $\sim$3 mL of ice-cold water to the beaker, and use a rubber policeman to help transfer the remaining solid to the funnel. Repeat this procedure with small portions of ice-cold water until all of the precipitate has been transferred. Finally, use two 10 mL portions of ice-cold water to rinse each beaker, and pour the washings over the precipitate.

5. Dry the precipitate, first with aspirator suction for 1 min, then in an oven at 105°C for 1–2 h. Bring each filter to constant weight. The product is somewhat hygroscopic, so only one filter at a time should be removed from the desiccator, and weighings should be done rapidly.

6. Calculate the molarity of Ca^{2+} in the unknown solution or the weight percent of Ca in the unknown solid. Report the standard deviation and relative standard deviation ($s/\bar{x}$).

[†] C. H. Hendrickson and P. R. Robinson, *J. Chem. Ed.*, **56,** 341 (1979).

[‡] Thorn Smith Inc., 7755 Narrow Gauge Road, Beulah, MI 49617.

24-3 GRAVIMETRIC DETERMINATION OF IRON AS Fe_2O_3[†]

A sample containing iron can be analyzed by precipitation of the hydrous oxide from basic solution, followed by ignition to produce Fe_2O_3:

$$Fe^{3+} + (2 + x)H_2O \xrightarrow{\text{base}} FeOOH \cdot xH_2O(s) + 3H^+$$

$$FeOOH \cdot xH_2O \xrightarrow{900°C} Fe_2O_3(s)$$

The hydrous oxide is gelatinous and may occlude impurities. If this is suspected, the initial precipitate is dissolved in acid and reprecipitated. Since the concentration of impurities is lower during the second precipitation, occlusion is diminished. Solid unknowns may be prepared from reagent ferrous ammonium sulfate or purchased from Thorn Smith.

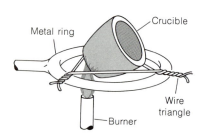

Figure 24-1
Positioning of a crucible above a burner.

PROCEDURE

1. Bring three porcelain crucibles and caps to constant weight by heating to redness for 15 min over a burner (Figure 24-1). Cool for 30 min in a desiccator and weigh each crucible. Repeat this procedure until successive weighings agree within ±0.3 mg. Be sure that all oxidizable substances on the entire surface of each crucible have burned off.

2. Accurately weigh three samples of unknown containing enough Fe to produce ~0.3 g of Fe_2O_3. Dissolve each sample in 10 mL of 3 M HCl (with heating, if necessary). If there are insoluble impurities, filter through qualitative filter paper and wash the filter very well with distilled water. Add 5 mL of 6 M HNO_3 to the filtrate, and boil for a few minutes to ensure that all iron is oxidized to Fe(III).

3. Dilute the sample to 200 mL with distilled water and add 3 M ammonia[‡] with constant stirring until the solution is basic (as determined with litmus paper or pH indicator paper). Digest the precipitate by boiling for 5 min and allow the precipitate to settle.

4. Decant the supernatant liquid through coarse, ashless filter paper (Whatman 41 or Schleicher and Schuell Black Ribbon, as in Figures 2-14 and 2-15). Do not pour liquid higher than 1 cm from the top of the funnel. Proceed to Step 5 if a reprecipitation is desired. Wash the precipitate repeatedly with hot 1% NH_4NO_3 until little or no Cl⁻ is detected in the filtered supernate. (Test for Cl⁻ by acidifying a few milliliters of filtrate with 1 mL of dilute HNO_3 and adding a few drops of 0.1 M $AgNO_3$.) Finally, transfer the solid to the filter with the aid of a rubber policeman and more hot liquid. Proceed to Step 6 if a reprecipitation is not used.

5. Wash the gelatinous mass twice with 30 mL of boiling 1% aqueous NH_4NO_3, decanting the supernate through the filter. Then put the filter paper back into the beaker with the precipitate, add 5 mL of 12 M HCl to dissolve the iron, and tear the filter paper into small pieces with a glass rod. Add ammonia with stirring and reprecipitate the iron. Decant through a funnel fitted with a fresh sheet of ashless filter paper. Wash the solid repeatedly with hot 1% NH_4NO_3 until little or no Cl⁻

[†] Adapted from D. A. Skoog and D. M. West, *Fundamentals of Analytical Chemistry*, 3rd ed. (New York: Holt, Rinehart and Winston, 1976).

[‡] Basic reagents should not be stored in glass bottles, because they slowly dissolve the glass. If ammonia from a glass bottle is used, it may contain silica particles and should be freshly filtered.

is detected in the filtered supernate. Then transfer all of the solid to the filter with the aid of a rubber policeman and more hot liquid.

6. Allow the filter to drain overnight—if possible, protected from dust. Carefully lift the paper out of the funnel, fold it as shown in Figure 24-2, and transfer it to a porcelain crucible that has been brought to constant weight.

7. Dry the crucible cautiously with a small flame, as shown in Figure 24-1. The flame should be directed at the top of the container, and the lid should be off. Avoid spattering. After it is dry, *char* the filter paper by increasing the flame temperature. The crucible should have free access to air to avoid reduction of iron by carbon. (The lid should be kept handy to smother the crucible if the paper inflames.) Any carbon left on the crucible or lid should be removed by directing the burner flame at it. Use tongs to manipulate the crucible. Finally, *ignite* the product for 15 min with the full heat of the burner.

8. Cool the crucible briefly in air and then in a desiccator for 30 min. Weigh the crucible and the lid, reignite, and bring to constant weight (± 0.3 mg) with repeated heatings.

9. Calculate the weight percent of iron in each sample, the average, the standard deviation, and the relative standard deviation ($s/\bar{x}$).

24-4 PREPARING STANDARD ACID AND BASE

Section 12-6 provides background information related to the procedures described below. Unknown samples of potassium acid phthalate or sodium carbonate (available from Thorn Smith) may be analyzed by the procedures described in this section.

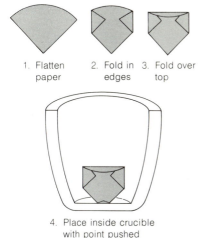

Figure 24-2
Folding filter paper and placing it inside a crucible for ignition. Continued folding of the paper so that the entire packet fits on the bottom of the crucible may be necessary. Be careful not to puncture the paper.

Standard NaOH

PROCEDURE

1. A 50% (wt/wt) aqueous NaOH solution must be prepared in advance and allowed to settle. Sodium carbonate is insoluble in this solution and precipitates. The solution is stored in a tightly sealed polyethylene bottle and handled gently to avoid stirring the precipitate when supernate is taken. The density is close to 1.50 g of solution/mL.

2. Primary-standard grade potassium acid phthalate should be dried for 1 h at 110°C and stored in a desiccator.

$$\text{(benzene ring with } CO_2K \text{ and } CO_2H) + NaOH \longrightarrow \text{(benzene ring with } CO_2K \text{ and } CO_2Na) + H_2O \quad (24\text{-}2)$$

F.W. 204.223

3. Boil 1 L of water for 5 min to expel CO_2. Pour the water into a polyethylene bottle, which should be tightly capped whenever possible. Calculate the volume of aqueous 50% NaOH needed to produce 1 L of ~ 0.1 M NaOH. Use a graduated cylinder to transfer this much concentrated NaOH to the bottle of water. Mix well and allow the solution to cool to room temperature (preferably overnight).

618

24 / EXPERIMENTS

4. Weigh four samples of solid potassium acid phthalate and dissolve each in ~ 25 mL of distilled water in a 125 mL flask. Each sample should contain enough solid to react with ~ 25 mL of 0.1 M NaOH. Add 3 drops of phenolphthalein indicator (Table 12-3) to each, and titrate one of them rapidly to find the approximate endpoint. The buret should have a loosely fitted cap to minimize entry of CO_2.

5. Calculate the volume of NaOH required for each of the other three samples and titrate them carefully. During each titration, you should periodically tilt and rotate the flask to wash all liquid from the walls into the bulk solution. When very near the end, you should deliver less than one drop of titrant at a time. To do this, carefully suspend a fraction of a drop from the buret tip, touch it to the inside wall of the flask, wash it into the bulk solution by careful tilting, and swirl the solution. The endpoint is the first appearance of faint pink color that persists for 15 s. (The color will slowly fade as CO_2 from the air dissolves in the solution.)

6. Calculate the average molarity, the standard deviation, and the relative standard deviation ($s/\bar{x}$). If you have used some care, the relative standard deviation should by $<0.2\%$.

Standard HCl

PROCEDURE

1. Use the information in the table on the inside cover of this text to calculate the volume of concentrated ($\sim 37\%$ wt/wt) HCl that should be added to 1 L of distilled water to produce ~ 0.1 M HCl, and prepare this solution.

2. Dry primary-standard-grade sodium carbonate for 1 h at 110°C and cool it in a desiccator.

3. Weigh four samples containing enough Na_2CO_3 to react with ~ 25 mL of 0.1 M HCl and place each in a 125 mL flask. As you are ready to titrate each one, dissolve it in ~ 25 mL of distilled water.

$$2HCl + Na_2CO_3 \rightarrow CO_2 + 2NaCl \qquad (24\text{-}3)$$

F.W. 105.989

Add 3 drops of bromocresol green indicator (Table 12-3) to each and titrate one rapidly (to a green color) to find the approximate endpoint.

4. Carefully titrate each sample until it just turns from blue to green. Then boil the solution to expel CO_2. The solution should return to a blue color.

5. Carefully add HCl from the buret until the solution turns green again. If desired, a blank titration can be performed, using 3 drops of indicator in 50 mL of 0.05 M NaCl. Subtract the volume of HCl needed for the blank titration from that required to titrate Na_2CO_3.

6. Calculate the mean HCl molarity, standard deviation, and relative standard deviation.

24-5 ANALYSIS OF A MIXTURE OF CARBONATE AND BICARBONATE

This procedure involves two titrations. First, total alkalinity $(= [HCO_3^-] + 2[CO_3^{2-}])$ is measured by titrating the mixture with standard HCl to a bromocresol green endpoint:

$$HCO_3^- + H^+ \rightarrow H_2CO_3$$

$$CO_3^{2-} + 2H^+ \rightarrow H_2CO_3$$

A separate aliquot of unknown is treated with excess standard NaOH to convert HCO_3^- to CO_3^{2-}:

$$HCO_3^- + OH^- \rightarrow CO_3^{2-} + H_2O$$

Then all of the carbonate is precipitated with $BaCl_2$:

$$Ba^{2+} + CO_3^{2-} \rightarrow BaCO_3$$

The excess NaOH is immediately titrated with standard HCl to determine how much HCO_3^- was present. From the total alkalinity and bicarbonate concentration, we can calculate the original carbonate concentration. Solid unknowns may be prepared from reagent-grade K_2CO_3 and $NaHCO_3$.

PROCEDURE

1. Dry 3 g of solid unknown for 2 h at 110°C, cool in a desiccator, and accurately weigh a 2.0–2.5 g sample into a 250 mL volumetric flask. This is conveniently done by weighing the sample in a capped weighing bottle, delivering some to a funnel in the volumetric flask, and reweighing the bottle. Continue this process until the desired mass of reagent has been transferred to the funnel. Rinse the funnel repeatedly with small portions of freshly boiled and cooled water to dissolve the sample. Remove the funnel, dilute to the mark, and mix well.

2. Pipet a 25.00 mL aliquot of the unknown solution into a 250 mL flask and titrate with standard 0.1 M HCl, using bromocresol green indicator as described in Section 24-4. Repeat this procedure with two more 25.00 mL samples of unknown.

3. Pipet a 25.00 mL aliquot of unknown and 50.00 mL of standard 0.1 M NaOH into a 250 mL flask. Swirl and add 10 mL of 10% (wt/wt) $BaCl_2$, using a graduated cylinder. Swirl again to precipitate $BaCO_3$, add 2 drops of phenolphthalein indicator (Table 12-3), and immediately titrate with standard 0.1 M HCl.

4. From the results of Step 2, calculate the total alkalinity and its standard deviation. From the results of Step 3, calculate the bicarbonate concentration and its standard deviation. Using the standard deviations as estimates of uncertainty, calculate the concentration (and uncertainty) of carbonate in the sample. Express the composition of the solid unknown as weight percent ($\pm$ uncertainty) for each component. For example, your final result might be written 63.4 ($\pm$0.5) wt % K_2CO_3 and 36.6 ($\pm$0.2) wt % $NaHCO_3$.

24-6 KJELDAHL NITROGEN ANALYSIS

Developed in 1883, this analytical technique remains one of the most accurate and widely used methods for determining nitrogen. The nitrogen-containing substance is first *digested* in boiling sulfuric acid, which converts the nitrogen to NH_4^+ and oxidizes other elements present:

$$\text{Organic C,H,N} \xrightarrow[H_2SO_4]{\text{boiling}} NH_4^+ + CO_2 + H_2O \qquad (24\text{-}4)$$

Mercury, copper, and selenium compounds catalyze the digestion process. To speed the rate of reaction, the boiling point of concentrated sulfuric acid (338°C) is raised by adding K_2SO_4. The digestion procedure given below is applicable to amines and amides (proteins). Modifications are necessary for nitro compounds, nitrites, azo compounds, cyanides, and derivatives of hydrazine.[†] (Kjeldahl digestion is also recommended for the destruction of highly toxic organic waste.[‡])

After digestion is complete, the solution containing NH_4^+ is made basic and the NH_3 formed is distilled into a receiver containing a known amount of HCl (see Equations 9-8 to 9-10). The unreacted HCl is titrated with NaOH to determine how much HCl was consumed by NH_3. Since the solution to be titrated contains both HCl and NH_4^+, we must choose an indicator that permits titration of HCl without beginning to titrate NH_4^+. Bromocresol green (transition range 3.8–5.4) fulfils this purpose. (See Problem 12-22.)

Biological samples such as powdered milk, dry cereal, or flour can be analyzed by the procedure described below. Alternatively, unknowns can be prepared from pure acetanilide, N-2-hydroxyethylpiperazine-N′-2-ethanesulfonic acid (HEPES buffer, Table 11-2) or *tris*(hydroxymethyl)aminomethane (TRIS buffer, Table 11-2).

Digestion

PROCEDURE

1. Dry your unknown at 105°C for 45 min and accurately weigh an amount that will produce 2–3 mmol of NH_3. Place the sample in a *dry* 500 mL Kjeldahl flask (Figure 24-3) so that as little as possible sticks to the walls. Add 10 g of K_2SO_4 and 3 selenium-coated boiling chips.[§] Pour in 25 mL of 98% (wt/wt) H_2SO_4,

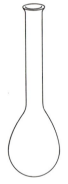

Figure 24-3
A Kjeldahl digestion flask. The long neck is designed to prevent loss of sample by spattering.

[†] H. C. Moore, *Ind. Eng. Chem.*, **12**, 669 (1920); *J. Assoc. Offic. Agr. Chemists*, **8**, 411 (1924–1925); I. K. Phelps and H. W. Daudt, *J. Assoc. Offic. Agr. Chemists*, **3**, 306 (1920); **4**, 72 (1921).

[‡] To destroy highly toxic organic compounds, place 10 g of K_2SO_4 and 100 mL of 98% (wt/wt) H_2SO_4 in a 500 mL flask fitted with a reflux condenser. Add up to 5 g of waste and a few glass beads and reflux until the solution is clear. If clarification does not occur in 1 h, cool the flask, slowly add 10 mL of 30% (wt/wt) H_2O_2 with stirring, add a fresh glass bead, and reflux again. Organic nitro, azo, and peroxide compounds should be reduced prior to this destruction procedure. The H_2SO_4 mixture can be reused until it solidifies when cooled.

[§] Hengar selenium-coated granules are available from Scientific Products, 1430 Waukegan Road, McGraw Park, IL 60085. Alternatively, the catalyst may be 0.1 g of Se, 0.2 g of $CuSeO_3$, or a crystal of $CuSO_4$.

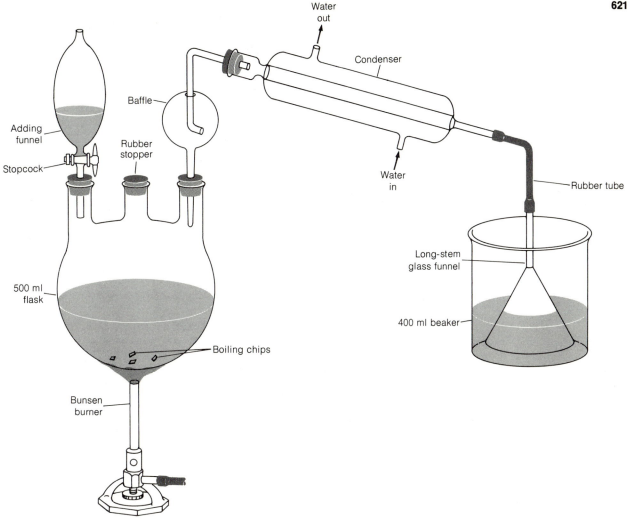

Figure 24-4
Apparatus for the Kjeldahl distillation.

washing down any solid from the walls. (*Caution:* Concentrated H_2SO_4 eats people. If you get any on your skin, flood it immediately with water, followed by soap and water.)

2. In a fume hood, clamp the flask at a 30° angle away from you. Heat gently with a burner until foaming ceases and the solution becomes homogeneous. Gentle boiling should then be continued for an additional 30 min.

3. Cool the flask for 30 min *in the air*, and then in an ice bath for 15 min. Slowly, and with constant stirring, add 50 mL of ice-cold water. Dissolve any solids that have crystallized. Transfer the liquid to a distillation flask (Figure 24-4), wash the Kjeldahl flask five times with 10 mL portions of water, and pour the washings into the distillation flask.

622

Distillation

24 / EXPERIMENTS

PROCEDURE

1. Apparatus such as that in Figure 24-4 should be set up and the connections made airtight. Pipet 50.00 mL of standard 0.1 M HCl into the receiving beaker and clamp the funnel in place below the liquid level.

2. Pour 50 mL of 6 M NaOH into the adding funnel and drip this into the distillation flask over a period of 1 min. Do not let the last 1 mL through the stopcock, so that gas cannot escape from the flask. Close the stopcock and heat the flask gently until two-thirds of the liquid has distilled.

3. Remove the funnel from the beaker *before* removing the burner from the flask (to avoid sucking distillate back into the condenser). Rinse the funnel well with distilled water and catch the rinses in the beaker. Add 6 drops of bromocresol green indicator solution (Table 12-3) to the beaker and carefully titrate to the blue endpoint with standard 0.1 M NaOH. You are looking for the first appearance of light blue color. (Several practice titrations with HCl and NaOH will familiarize you with the endpoint.)

4. Calculate the weight percent of nitrogen in the unknown.

24-7 ANALYSIS OF AN ACID–BASE TITRATION CURVE: THE GRAN PLOT

In this experiment, you will titrate a sample of pure potassium acid phthalate (Equation 24-2) with standard NaOH. The Gran plot (which you should read about in Section 12-5) will be used to find the equivalence point and K_A. Activity coefficients are used in the calculations of this experiment.

PROCEDURE (AN EASY MATTER)

1. Dry about 1.5 g of potassium acid phthalate at $105°$ for 1 h and cool it in a desiccator for 20 min. Accurately weigh out ~ 1.5 g and dissolve it in water in a 250 mL volumetric flask. Dilute to the mark and mix well.

2. Following the instructions for your particular pH meter, calibrate a meter and glass electrode using buffers with pH values near 7 and 4. Rinse the electrodes well with distilled water and blot them dry with a tissue before immersing in a solution.

3. Pipet 100.0 mL of phthalate solution into a 250 mL beaker containing a magnetic stirring bar. Position the electrode(s) in the liquid so that the stirring bar will not strike the electrode. If a combination electrode is used, the small hole near the bottom on the side must be immersed in the solution. This hole is the reference electrode salt bridge. Allow the electrodes to equilibrate for 1 min (with stirring) and record the pH.

4. Add 1 drop of phenolphthalein indicator (Table 12-3) and titrate the solution with standard ~ 0.1 M NaOH. Until you are within 4 mL of the theoretical equivalence point, add base in ~ 1.5 mL aliquots, recording the volume and pH 30 s after each addition. Thereafter, use 0.4 mL aliquots until you are within 1 mL of the equivalence point. After that, add base 1 drop at a time until you have passed the pink endpoint by a few tenths of a milliliter. (Record the volume at which the pink color is observed.) Then add five more 1 mL aliquots.

5. Construct a graph of pH versus V_b (volume of added base). Locate the equivalence volume (V_e) by eye or "feel" as described in Section 12-5. Compare this to the theoretical and phenolphthalein endpoints.

CALCULATIONS (THE WORK BEGINS!)

1. Construct a Gran plot (a graph of $V_b 10^{-pH}$ versus V_b) using the data collected between $0.9\ V_e$ and V_e. Draw a line through the linear portion of this curve and extrapolate it to the abscissa to find V_e. Use this value of V_e in the calculations below. Compare this value to those found with phenolphthalein and estimated from the graph of pH versus V_b.

2. Compute the slope of the Gran plot and use Equation 12-51 to find K_A for potassium acid phthalate as follows: The slope of the Gran plot is $-K_A \gamma_{HP^-}/\gamma_{P^{2-}}$. In this equation, P^{2-} is the phthalate anion, and HP^- is monohydrogen phthalate. Because the ionic strength changes slightly as the titration proceeds, so also do the activity coefficients. Calculate the ionic strength at $0.95\ V_e$ and use this "average" ionic strength to find the activity coefficients.

EXAMPLE

Find γ_{HP^-} and $\gamma_{P^{2-}}$ at $0.95\ V_e$ in the titration of 100.0 mL of 0.020 0 M potassium acid phthalate with 0.100 M NaOH.

The equivalence point is 20.0 mL, so $0.95\ V_e = 19.0$ mL. The concentrations of H^+ and OH^- are negligible compared to those of K^+, Na^+, HP^-, and P^{2-}, whose concentrations are

$$[K^+] = \left(\frac{100}{119}\right)(0.020\ 0) = 0.016\ 8\ \text{M}$$

$$[Na^+] = \left(\frac{19}{119}\right)(0.100) = 0.016\ 0\ \text{M}$$

$$[HP^-] = (0.050)\left(\frac{100}{119}\right)(0.020\ 0) = 0.000\ 84\ \text{M}$$

$$[P^{2-}] = (0.95)\left(\frac{100}{119}\right)(0.020\ 0) = 0.016\ 0\ \text{M}$$

The ionic strength is

$$\mu = \tfrac{1}{2}\sum c_i z_i^2$$
$$= \tfrac{1}{2}[(0.016\ 8) \cdot 1^2 + (0.016\ 0) \cdot 1^2 + (0.000\ 84) \cdot 1^2 + (0.016\ 0) \cdot 2^2]$$
$$= 0.048\ 8\ \text{M}$$

To estimate $\gamma_{P^{2-}}$ and γ_{HP^-} at $\mu = 0.048\ 8$ M, interpolate in Table 6-1. At the bottom of this table, we find that the hydrated radius of P^{2-}—phthalate, $C_6H_4(COO)_2^{2-}$—is 600 pm. The size of HP^- is not listed, but we will suppose that it is also 600 pm. An ion with charge ± 2 and a size of 600 pm has $\gamma = 0.485$ at $\mu = 0.05$ and $\gamma = 0.675$ at $\gamma = 0.01$. Interpolating between these values, we estimate $\gamma_{P^{2-}} = 0.49$ when $\mu = 0.048\ 8$ M. Similarly, we estimate $\gamma_{HP^-} = 0.84$ at this same ionic strength.

624

24 / EXPERIMENTS

3. From the measured slope of the Gran plot and the values of γ_{HP^-} and $\gamma_{P^{2-}}$, calculate pK_A. Then choose an experimental point near $\frac{1}{3}V_e$ and one near $\frac{2}{3}V_e$. Use Equation 11-56 to find pK_A with each point. (You will have to calculate $[P^{2-}]$, $[HP^-]$, $\gamma_{P^{2-}}$, and γ_{HP^-} at each point.) Compare the average value of pK_A from your experiment with pK_2 for phthalic acid listed in Appendix G.

24-8 EDTA TITRATION OF Ca^{2+} AND Mg^{2+} IN NATURAL WATERS

The most common multivalent metal ions in natural waters are Ca^{2+} and Mg^{2+}. In this experiment, we will find the total concentration of metal ions that can react with EDTA, and we will assume that this equals the concentration of Ca^{2+} and Mg^{2+}. In a second experiment, Ca^{2+} is analyzed separately by precipitating $Mg(OH)_2$ with strong base.

REAGENTS

Buffer (pH 10): Add 142 mL of 28% (wt/wt) aqueous NH_3 to 17.5 g of NH_4Cl and dilute to 250 mL with water.

Eriochrome Black T indicator: Dissolve 0.2 g of the solid indicator in 15 mL of triethanolamine plus 5 mL of absolute ethanol.

PROCEDURE

1. Dry $Na_2H_2EDTA \cdot 2H_2O$ (F.W. 372.25) at 80° for 1 h and cool in a desiccator. Accurately weigh out ~ 0.6 g and dissolve it with heating in 400 mL of water in a 500 mL volumetric flask. Cool to room temperature, dilute to the mark, and mix well.

2. Pipet a sample of unknown into a 250 mL flask. A 1.000 mL sample of sea water or a 50.00 mL sample of tap water is usually reasonable. If you use 1.000 mL of sea water, add 50 mL of distilled water. To each sample, add 3 mL of pH 10 buffer and 6 drops of Eriochrome Black T indicator. Titrate with EDTA from a 50 mL buret and note when the color changes from wine-red to blue. You may need to practice finding the endpoint several times by adding a little tap water and titrating with more EDTA. Save a solution at the endpoint to use as a color comparison for other titrations.

3. Repeat the titration with three samples to find an accurate value of the total Ca^{2+} and Mg^{2+} concentration. Perform a blank titration with 50 mL of distilled water and subtract the value of the blank from each result.

4. For the determination of Ca^{2+}, pipet four samples of unknown into clean flasks (adding 50 mL of distilled water if you use 1.000 mL of sea water). Add 30 drops of 50% (wt/wt) NaOH to each solution and swirl for 2 min to precipitate $Mg(OH)_2$ (which may not be visible). Add ~ 0.1 g of solid hydroxynaphthol blue to each flask. (This indicator is used because it remains blue at higher pH than does Eriochrome Black T.) Titrate one sample rapidly to find the endpoint; practice finding it several times, if necessary.

5. Titrate the other three samples carefully. After reaching the blue endpoint, allow each sample to sit for 5 min with occasional swirling so that any $Ca(OH)_2$ precipitate may redissolve. Then titrate back to the blue endpoint. (Repeat this procedure if the blue color turns to red upon standing.) Perform a blank titration with 50 mL of distilled water.

6. Calculate the total concentration of Ca^{2+} and Mg^{2+}, as well as the individual concentrations of each ion. Calculate the relative standard deviation of replicate titrations.

24-9 SYNTHESIS AND ANALYSIS OF AMMONIUM DECAVANADATE[†]

The balance of species in a solution of vanadium(V) is a delicate function of both pH and concentration.

$$VO_4^{3-} \rightleftharpoons HVO_4^{2-} \rightleftharpoons V_2O_7^{4-} \rightleftharpoons H_2V_2O_7^{2-}$$

In strong base

$$VO_2^+ \rightleftharpoons H_2V_{10}O_{28}^{4-} \rightleftharpoons \begin{bmatrix} V_4O_{12}^{4-} \\ V_3O_9^{3-} \end{bmatrix}$$

In strong acid

NH_4^+ / alcohol

$$(NH_4)_6V_{10}O_{28} \cdot 6H_2O$$
F.W. 1 173.7

The decavanadate ion ($V_{10}O_{28}^{6-}$), which we will isolate in this experiment as the ammonium salt, consists of ten VO_6 octahedra sharing edges with each other (Figure 24-5).

After preparing this salt, we will determine the vanadium content by a redox titration and NH_4^+ by the Kjeldahl method. In the redox titration, V(V) will first be reduced to V(IV) with sulfurous acid and then titrated with standard permanganate.

$$V_{10}O_{28}^{6-} + H_2SO_3 \rightarrow VO^{2+} + SO_2 \qquad (24\text{-}5)$$

$$VO^{2+} + MnO_4^- \rightarrow VO_2^+ + Mn^{2+} \qquad (24\text{-}6)$$

  blue purple yellow colorless

Figure 24-5
Structure of the $V_{10}O_{28}^{6-}$ anion, consisting of ten VO_6 octahedra sharing edges with each other.

Synthesis

PROCEDURE

1. Heat 3.0 g of ammonium metavanadate (NH_4VO_3) in 100 mL of water with constant stirring (but not boiling) until most or all of the solid has dissolved. Filter the solution and add 4 mL of 50% (vol/vol) aqueous acetic acid with stirring.

2. Add 150 mL of 95% ethanol with stirring and then cool the solution in a refrigerator or ice bath.

[†] G. G. Long, R. L. Stanfield, and F. C. Hentz, Jr., *J. Chem. Ed.*, **56**, 195 (1979).

626

24 / EXPERIMENTS

3. After maintaining a temperature of 0–10°C for 15 min, filter the orange product with suction and wash with two 15 mL portions of ice-cold 95% ethanol.

4. Dry the product in the air (protected from dust) for two days. Typical yield is 2.0–2.5 g.

Analysis of Vanadium with $KMnO_4$

Preparation and standardization of $KMnO_4$[†]

See Section 17-4.

PROCEDURE

1. Prepare a 0.02 M permanganate solution by dissolving 1.6 g of $KMnO_4$ in 500 mL of distilled water. Boil gently for 1 h, cover, and allow the solution to cool overnight. Filter through a clean, fine sintered glass funnel, discarding the first 20 mL of filtrate. Store the solution in a clean glass amber bottle. Do not let the solution touch the cap.

2. Dry sodium oxalate ($Na_2C_2O_4$) at 105°C for 1 h, cool in a desiccator, and weigh three $\sim$0.25 g samples into 500 mL flasks or 400 mL beakers. To each, add 250 mL of 0.9 M H_2SO_4 that has been recently boiled and cooled to room temperature. Stir with a thermometer to dissolve the sample, and add 90–95% of the theoretical amount of $KMnO_4$ solution needed for the titration. (This can be calculated from the mass of $KMnO_4$ used to prepare the permanganate solution. The chemical reaction is given by Equation 9-1.)

3. Leave the solution at room temperature until it is colorless. Then heat it to 55–60°C and complete the titration by adding $KMnO_4$ until the first pale pink color persists. Proceed slowly near the end, allowing 30 s for each drop to lose its color. As a blank, titrate 250 mL of 0.9 M H_2SO_4 to the same pale pink color.

Vanadium analysis

PROCEDURE

1. Accurately weigh two 0.3 g samples of ammonium decavanadate into 250 mL flasks and dissolve each in 40 mL of 1.5 M H_2SO_4 (with warming, if necessary).

2. In a fume hood, add 50 mL of water and 1 g of $NaHSO_3$ to each and dissolve with swirling. After 5 min, boil the solution gently for 15 min to remove SO_2.

3. Titrate the warm solution with standard 0.02 M $KMnO_4$ from a 50 mL buret. The endpoint is taken when the yellow color of VO_2^+ takes on a dark shade (from excess MnO_4^-) that persists for 15 s.

Analysis of Ammonium Ion

Ammonium ion is analyzed by a modification of the Kjeldahl procedure. Transfer 0.6 g of accurately weighed ammonium decavanadate to the distillation flask in Figure 24-4 and add 200 mL of water. Then proceed as described under "Distillation" in Section 24-6.

[†] R. M. Fowler and H. A. Bright, *J. Res. National Bureau of Standards*, **15**, 493 (1935).

24-10 IODIMETRIC TITRATION OF VITAMIN C[†]

Ascorbic acid (vitamin C) is a mild reducing agent that reacts rapidly with triiodide (Equation 17-70). In this experiment, we will generate a known excess of I_3^- by the reaction of iodate with iodide (Equation 17-65), allow the reaction with ascorbic acid to proceed, and then back-titrate the excess I_3^- with thiosulfate (Equation 17-66).

Preparation and Standardization of Thiosulfate Solution

PROCEDURE

1. Starch indicator is prepared by making a paste of 5 g of soluble starch and 5 mg of HgI_2 in 50 mL of water. Pour the paste into 500 mL of boiling water and boil until it is clear.

2. Prepare 0.07 M $Na_2S_2O_3$ by dissolving ~8.7 g of $Na_2S_2O_3 \cdot 5H_2O$ in 500 mL of freshly boiled water containing 0.05 g of Na_2CO_3. Store this solution in a tightly capped amber bottle. Prepare ~0.01 M KIO_3 by accurately weighing ~1 g of solid reagent and dissolving it in a 500 mL volumetric flask.

3. Standardize the thiosulfate solution as follows: Pipet 50.00 mL of KIO_3 solution into a flask. Add 2 g of solid KI and 10 mL of 0.5 M H_2SO_4. Immediately titrate with thiosulfate until the solution has lost almost all of its color (pale yellow). Then add 2 mL of starch indicator and complete the titration. Repeat the titration with two additional 50.00 mL volumes of KIO_3 solution.

Analysis of Vitamin C

Commercial vitamin C containing 100 mg per tablet may be used. Perform the following analysis three times, and find the mean value (and relative standard deviation) for the number of milligrams of vitamin C per tablet.

PROCEDURE

1. Dissolve two tablets in 60 mL of 0.3 M H_2SO_4, using a glass rod to help break the solid. (Some solid binding material will not dissolve.)

2. Add 2 g of solid KI and 50.00 mL of standard KIO_3. Then titrate with standard thiosulfate as above. Add 2 mL of starch indicator just before the endpoint.

24-11 POTENTIOMETRIC HALIDE TITRATION WITH Ag$^+$

Mixtures of halides may be titrated with $AgNO_3$ solution as described in Section 9-5. In this experiment, we will use the apparatus in Figure 9-5 to monitor the activity of Ag^+ as the titration proceeds. The theory of the potentiometric measurement is described in Section 16-2.

Each student is given a vial containing 0.22–0.44 g of KCl plus 0.50–

[†] D. N. Bailey, *J. Chem. Ed.*, **51**, 488 (1974).

628

24 / EXPERIMENTS

1.00 g of KI (both weighed accurately). The object is to determine the quantity of each salt in the mixture. A 0.4 M bisulfate buffer (pH 2) should be available in the lab. This is prepared by titrating 1 M H_2SO_4 with 1 M NaOH to a pH near 2.0.

PROCEDURE

1. Pour your unknown carefully into a 50 or 100 mL beaker. Dissolve the solid in ~20 mL of water and pour it into a 100 mL volumetric flask. Rinse the sample vial and beaker many times with small portions of H_2O and transfer the washings to the flask. Dilute to the mark and mix well.

2. Dry 1.2 g of $AgNO_3$ at 105° for 1 h and cool in a desiccator for 30 min with minimal exposure to light. Some discoloration is normal (and tolerable in this experiment) but should be minimized. Accurately weigh 1.2 g and dissolve it in a 100 mL volumetric flask.

3. The apparatus in Figure 9-5 should be set up. The silver electrode is simply a 3 cm length of silver wire connected to copper wire. (Fancier electrodes can be prepared by housing the connection in a glass tube sealed with epoxy at the lower end. Only the silver should protrude from the epoxy.) The copper wire is fitted with a jack that goes to the reference socket of a pH meter. The reference electrode for this titration is a glass pH electrode connected to its usual socket on the meter. If a combination pH electrode is employed, the reference jack of the combination electrode is not used. The silver electrode should be taped to the inside of the 100 mL beaker so that the Ag/Cu junction remains dry for the entire titration. The stirring bar should not hit either electrode.

4. Pipet 25.00 mL of unknown into the beaker, add 3 mL of bisulfate buffer, and begin magnetic stirring. Record the initial level of $AgNO_3$ in a 50 mL buret and add ~1 mL of titrant to the beaker. Turn the pH meter to the millivolt scale and record the volume and potential. It is convenient (but is not essential) to set the initial potential reading to +800 mV by adjusting the meter.

5. Titrate the solution with ~1 mL aliquots until 50 mL of titrant has been added or until you can see two clear potentiometric endpoints. You need not allow more than 15–30 s for each point. Record the volume and potential at each point. Make a graph of millivolts versus milliliters to find the approximate positions (± 1 mL) of the two endpoints.

6. Turn the pH meter to standby, remove the beaker, rinse the electrodes well with water, and blot them dry with a tissue.[†] Clean the beaker and set up the titration apparatus again. (The beaker need not be dry.)

7. Now perform an accurate titration, using 1 drop aliquots near the endpoints (and 1 mL aliquots elsewhere). You need not allow more than 30 s per point for equilibration.

8. Prepare a graph of millivolts versus milliliters and locate the endpoints as in Figure 9-4. The I^- endpoint is taken as the intersection of the two dashed lines in the inset of Figure 9-4. The Cl^- endpoint is the inflection point at the second break. Calculate the mg of KI and mg of KCl in your solid unknown.

[†] Silver halide adhering to the glass electrode may be removed by soaking in concentrated sodium thiosulfate solution. This thorough cleaning is not necessary between Steps 6 and 7 in this experiment. The silver halides in the titration beaker can be saved and converted back to pure $AgNO_3$ using the procedure of E. Thall, *J. Chem. Ed.*, **58**, 561 (1981).

24-12 ELECTROGRAVIMETRIC ANALYSIS OF COPPER

Most copper-containing compounds can be electrolyzed in acidic solution, with quantitative deposition of Cu at the cathode. Sections 18-1 to 18-3 discuss the theory of this technique.

Students may analyze a preparation of their own (such as copper acetylsalicylate[†]) or be given unknowns prepared from $CuSO_4 \cdot 5H_2O$ or metallic Cu. In the latter case, dissolve the metal in 8 M HNO_3, boil to remove HNO_2, neutralize with ammonia, and *barely* acidify the solution with dilute H_2SO_4. Samples must be free of chloride and nitrous acid. Copper oxide unknowns (soluble in acid) are available from Thorn Smith.

Commercial apparatus, such as that in Figure 18-7, is convenient. Alternatively, any 6–12 V direct-current power supply may be set up as in Figure 18-6. A tall-form 150 mL beaker is used as the reaction vessel.

PROCEDURE

1. Handle the Pt gauze cathode with a tissue, touching only the thick stem, not the wire gauze. Immerse the electrode in hot 8 M HNO_3 to remove previous deposits, rinse with water and alcohol, dry at 110°C for 5 min, cool for 5 min, and weigh accurately. If the electrode contains any grease, it can be heated to red heat over a burner after the treatment above.[‡]

2. The sample should contain 0.2–0.3 g of Cu in 100 mL. Add 3 mL of 98% (wt/wt) H_2SO_4 and 2 mL of freshly boiled 8 M HNO_3. Position the cathode so that the top 5 mm are above the liquid level after magnetic stirring is begun. Adjust the current to 2 A, which should require 3–4 V. When the blue color of Cu(II) has disappeared, add some distilled water so that new Pt surface is exposed to the solution. If no further deposition of Cu occurs on the fresh surface in 15 min at a current of 0.5 A, the electrolysis is complete. If deposition is observed, continue electrolysis and test the reaction for completeness again.

3. *Without* turning off the power, lower the beaker while washing the electrode with a squirt bottle. Then the current can be turned off. (If current is disconnected before removing the cathode from the liquid and rinsing off the acid, some Cu could redissolve.) Wash the cathode gently with water and alcohol, dry at 110°C for 3 min, cool in a desiccator for 5 min, and weigh.

24-13 POLAROGRAPHIC MEASUREMENT OF AN EQUILIBRIUM CONSTANT[§]

In this experiment, we will find the overall formation constant and stoichiometry for the reaction of oxalate with Pb^{2+}:

[†] E. Dudek, *J. Chem. Ed.*, **54**, 329 (1977).

[‡] Some metals, such as Zn, Ga, and Bi, form alloys with Pt and should not be deposited directly on the Pt surface. The electrode should be coated first with Cu, dried, and then used. Alternatively, Ag may be used in place of Pt for depositing these metals. Platinum anodes are attacked by Cl_2 formed by electrolysis of Cl^- solutions. To prevent this, 1–3 g of a hydrazinium salt (per 100 mL of solution) may be used as an *anodic depolarizer*, since hydrazine is more readily oxidized than Cl^-: $N_2H_4 \rightarrow N_2 + 4H^+ + 4e^-$.

[§] W. C. Hoyle and T. M. Thorpe, *J. Chem. Ed.*, **55**, A229 (1978).

$$Pb^{2+} + pC_2O_4^{2-} \rightleftharpoons Pb(C_2O_4)_p^{2-2p} \qquad (24\text{-}7)$$

$$\beta_p = \frac{[Pb(C_2O_4)_p^{2-2p}]}{[Pb^{2+}][C_2O_4^{2-}]^p} \qquad (24\text{-}8)$$

We will do this by measuring the polarographic half-wave potential for solutions containing Pb^{2+} and varying amounts of oxalate. According to Equation 19-26, the change of half-wave potential, $\Delta E_{1/2} [= E_{1/2}$ (observed) $- E_{1/2}$ (for Pb^{2+} without oxalate)] should obey the equation

$$\Delta E_{1/2} = -\frac{RT}{nF} \ln \beta_p - \frac{pRT}{nF} \ln[C_2O_4^{2-}] \qquad (24\text{-}9)$$

Here we have converted $0.059\,16 \log x$ in Equation 19-26 to $(RT/F)\ln x$ in Equation 24-9. R is the gas constant, F is the Faraday constant, and T is kelvins. You should measure the lab temperature at the time of the experiment or use a thermostatically controlled cell.

PROCEDURE

1. Pipet 1.00 mL of 0.020 M $Pb(NO_3)_2$ into each of five 50 mL volumetric flasks (A–E) and add 1 drop of 1% Triton X-100 to each. Then add the following solutions and dilute to the mark with water. The KNO_3 may be delivered carefully with a graduated cylinder. The oxalate should be pipetted.
   A. Add nothing else. Dilute to the mark with 1.20 M KNO_3.
   B. Add 5.00 mL of 1.00 M $K_2C_2O_4$ and 37.5 mL of 1.20 M KNO_3.
   C. Add 10.00 mL of 1.00 M $K_2C_2O_4$ and 25.0 mL of 1.20 M KNO_3.
   D. Add 15.00 mL of 1.00 M $K_2C_2O_4$ and 12.5 mL of 1.20 M KNO_3.
   E. Add 20.00 mL of 1.00 M $K_2C_2O_4$.

2. Transfer each solution to a polarographic cell, deoxygenate for 10 min, and record the polarogram from -0.20 to -0.95 V (versus S.C.E.). Measure the residual current using the same settings and a solution containing just 1.20 M KNO_3 (plus 1 drop of 1% Triton X-100). Record each polarogram on a scale sufficiently expanded to allow accurate measurements. (Use a sweep rate of 0.05 V/min and a chart speed of 2.5 cm/min, with a mercury drop interval of 4 s.)

3. For each polarogram, make a graph of E versus $\log[I/(I_d - I)]$, using 6–8 points for each graph. According to Equation 19-11, $E = E_{1/2}$ when $\log[I/(I_d - I)] = 0$. Use this condition to locate $E_{1/2}$ on each graph. When you measure currents on the polarograms, be sure to subtract the residual current at each potential. Current on the polarographic wave is measured at the top of each oscillation, which corresponds to the maximum size of each mercury drop. Residual current is measured as follows: The electrocapillary maximum is the potential at which the residual current shows no oscillations. If the potential is more *positive* than the electrocapillary maximum, use the top of the oscillation, which corresponds to the maximum mercury drop size. If the potential is more *negative* than the electrocapillary maximum, use the bottom of each oscillation, which also corresponds to the maximum mercury drop size.

4. Make a graph of $\Delta E_{1/2}$ versus $\ln[C_2O_4^{2-}]$. Use Equation 24-9 to find p, the stoichiometry coefficient, from the slope of the graph. Then use the intercept to find the

value of β_p. A worthwhile exercise is to use the method of least squares (Section 4-4) to find the standard deviations of the slope and intercept. From the standard deviations, find the uncertainties in p and β_p and express each with the correct number of significant figures.

24-14 COULOMETRIC TITRATION OF CYCLOHEXENE WITH BROMINE[†]

This experiment is described in Section 18-4, and the apparatus is shown in Figure 18-11. A conventional coulometric power supply can be employed, but we use the homemade circuits in Figure 24-6. A stopwatch is manually started as the generator switch is closed. Alternatively, a double-pole, double-throw switch may be used to simultaneously start the generator circuit and an electric clock.

PROCEDURE

1. The electrolyte is a 60:26:14 (vol/vol/vol) mixture of acetic acid, methanol, and water. The solution contains 0.15 M KBr and 0.1 g of mercuric acetate per 100 mL. (The latter catalyzes the reaction between Br_2 and cyclohexane.) The electrodes should be covered with electrolyte. Begin vigorous magnetic stirring (without spattering) and adjust the potential of the indicator circuit to 0.25 V.

2. Generate Br_2 with the generator circuit until the detector current is 20.0 μA. (The generator current is 5–10 mA.)

3. Pipet 2–5 mL of unknown (containing 1–5 mg of cyclohexene in methanol) into the flask and set the clock or coulometer to zero. The detector current should drop to near zero because the cyclohexene consumes the Br_2.

4. Turn the generator circuit on and simultaneously begin timing. While the reaction is in progress, measure the potential across the precision resistor (100.0 ± 0.1 Ω) to find the exact current flowing through the cell (I = E/R). Continue the electrolysis until the detector current rises to 20.0 μA. Then stop the coulometer and clock and record the time.

5. Repeat the procedure twice more and find the average molarity (and relative standard deviation) of cyclohexene.

6. When you are finished, be sure all switches are off. The generator electrodes should be soaked in 8 M HNO_3 to dissolve Hg that is deposited during the electrolysis.

24-15 SPECTROPHOTOMETRIC DETERMINATION OF IRON IN VITAMIN TABLETS[‡]

In this procedure, iron from a vitamin supplement tablet is dissolved in acid, reduced to Fe(II) with hydroquinone, and complexed with *o*-phenanthroline

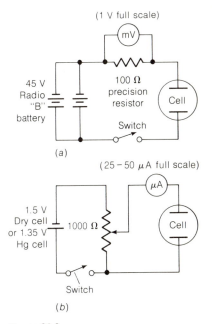

Figure 24-6
Simple circuits for coulometric titrations. (a) Generator circuit. (b) Detector circuit.

[†] D. H. Evans, *J. Chem. Ed.*, **45**, 88 (1968). A similar experiment in which vitamin C is analyzed has been described by D. G. Marsh, D. L. Jacobs, and H. Veening, *J. Chem. Ed.*, **50**, 626 (1973). Home-built circuits for constant-current or controlled-potential coulometry have been described by E. Grimsrud and J. Amend, *J. Chem. Ed.*, **56**, 131 (1979).

[‡] R. C. Atkins, *J. Chem. Ed.*, **52**, 550 (1975).

to form an intensely colored complex.

$$2Fe^{3+} + HO-\langle\bigcirc\rangle-OH \rightleftharpoons 2Fe^{2+} + O=\langle\bigcirc\rangle=O + 2H^+$$

hydroquinone   quinone

$$Fe^{2+} + 3 \;\; \langle\text{o-phenanthroline}\rangle \;\; \rightleftharpoons \;\; \left[\quad \right]_3 Fe^{2+}$$

o-phenanthroline

$\lambda_{max} = 508$ nm

REAGENTS

Hydroquinone: Freshly prepared solution containing 10 g/L in water. Store in an amber bottle.

Trisodium citrate: 25 g/L in water.

o-Phenanthroline: Dissolve 2.5 g in 100 mL of ethanol and add 900 mL of water. Store in an amber bottle.

Standard Fe (0.04 mg Fe/mL): Dissolve 0.281 g of reagent-grade $Fe(NH_4)_2(SO_4)_2 \cdot 6H_2O$ in water in a 1 L volumetric flask containing 1 mL of 98% (wt/wt) H_2SO_4.

PROCEDURE

1. Place one tablet of the iron-containing vitamin in a 125 mL flask or 100 mL beaker and boil gently (*in a fume hood*) with 25 mL of 6 M HCl for 15 min. Filter the solution directly into a 100 mL volumetric flask. Wash the beaker and filter several times with small portions of water to complete a quantitative transfer. Allow the solution to cool, dilute to the mark, and mix well.

2. Pipet 10 mL of standard Fe solution into a beaker and measure the pH (with pH paper or a glass electrode). Add sodium citrate solution one drop at a time until a pH of ~ 3.5 is reached. Count the drops needed. (It will require about 30 drops.)

3. Pipet a fresh 10.00 mL aliquot of Fe standard into a 100 mL volumetric flask and add the same number of drops of citrate solution as required in Step 2. Add 2.00 mL of hydroquinone solution, 3.00 mL of o-phenanthroline solution, dilute to the mark with water, and mix well.

4. Prepare three more solutions from 5.00, 2.00, and 1.00 mL of Fe standard and prepare a blank containing no Fe. Use sodium citrate solution in proportion to the volume of Fe solution. (If 10 mL of Fe requires 30 drops of citrate solution, 5 mL of Fe requires 15 drops of citrate solution.)

5. Find out how many drops of citrate solution are needed to bring 5.00 mL of the iron supplement tablet solution to pH 3.5. (Use a 10.00 mL aliquot if the tablet contains <15 mg Fe.)

6. Transfer 5.00 mL of the solution containing the dissolved tablet to a 100 mL volumetric flask. Add the required amount of citrate solution, 2.00 mL of hydroquinone solution, 3.00 mL of o-phenanthroline solution, dilute to the mark, and mix well.

7. Allow the solutions to stand for at least 10 min. Then measure the absorbance of each solution at 508 nm. (The color is stable, so all solutions may be prepared and all the absorbances measured at once.) Use distilled water in the reference cuvette and subtract the absorbance of the blank from the absorbance of the Fe standards.

8. Make a graph of absorbance versus micrograms of Fe in the standards. If desired, find the slope and intercept (and standard deviations) by the method of least squares, as described in Section 4-4. Calculate the molarity of Fe (o-phenanthroline)$_3^{2+}$ in each solution and find the average molar absorptivity (ε in Beer's law) from the four absorbances. (Remember that all of the iron has been converted to the phenanthroline complex.) If a Spectronic 20 is used for absorbance measurements, assume that the pathlength is 1.1 cm for the sake of this calculation.

9. Using the calibration curve (or its least-squares parameters), find the milligrams of Fe in the tablet.

24-16 SPECTROPHOTOMETRIC MEASUREMENT OF AN EQUILIBRIUM CONSTANT

In this experiment, we will use the Scatchard plot described in Section 20-6 to find the equilibrium constant for the formation of a complex between iodine and pyridine in cyclohexane:

$$I_2 + N\hexagon \longrightarrow I_2 \cdot N\hexagon \qquad (24\text{-}10)$$

Both I_2 and $I_2 \cdot$pyridine absorb visible radiation, but pyridine is colorless. Analysis of the spectral changes associated with variation of pyridine concentration (with a constant total concentration of iodine) will allow us to evaluate K for Reaction 24-10. The experiment is best performed with a recording spectrophotometer, but single-wavelength measurements can be used.

PROCEDURE

All operations described below should be carried out *in a fume hood*, including pouring solutions into and out of the spectrophotometer cell. Only a *capped* cuvette containing the solution whose spectrum is to be measured should be taken from the hood. Do not spill solvent on your hands or breathe the vapors. Used solutions should be discarded in a waste container *in the hood*, not down the drain.

1. The following stock solutions should be available in the lab:
    (a) 0.050–0.055 M pyridine in cyclohexane (40 mL for each student, concentration known accurately).
    (b) 0.012 0–0.012 5 M I_2 in cyclohexane (10 mL for each student, concentration known accurately).

2. Pipet the following volumes of stock solutions into six 25 mL volumetric flasks (A–F), dilute to the mark with cyclohexane, and mix well.

Flask	Pyridine stock solution (mL)	I_2 stock solution (mL)
A	0	1.00
B	1.00	1.00
C	2.00	1.00
D	4.00	1.00
E	5.00	1.00
F	10.00	1.00

3. Using glass or quartz cells, record a baseline between 350 and 600 nm with solvent in both the sample and the reference cells. Subtract the absorbance of the baseline from all future absorbances. If possible, record all spectra, including the baseline, on one sheet of chart paper. (If a fixed-wavelength instrument is used, first find the positions of the two absorbance maxima in solution E. Then make all measurements at these two wavelengths.)

4. Record the spectrum of each solution A–F or measure the absorbance at each maximum if a fixed-wavelength instrument is used.

DATA ANALYSIS

1. Measure the absorbance at the wavelengths of the two maxima in each spectrum. Be sure to subtract the absorbance of the blank from each.

2. The analysis of this problem follows that of Reaction 20-22, in which P is iodine and X is pyridine. As a first approximation, assume that the concentration of free pyridine equals the total concentration of pyridine in the solution (since [pyridine] $\gg [I_2]$). Prepare a graph of $\Delta A/[\text{free pyridine}]$ versus ΔA (a Scatchard plot), using the absorbance at the $I_2 \cdot$pyridine maximum.

3. From the slope of the graph, find the equilibrium constant using Equation 20-32. From the intercept, find $\Delta\varepsilon \ (= \varepsilon_{PX} - \varepsilon_X)$.

4. Now refine the values of K and $\Delta\varepsilon$. Use $\Delta\varepsilon$ to find ε_{PX}. Then use the absorbance at the wavelength of the $I_2 \cdot$pyridine maximum to find the concentration of bound and free pyridine in each solution. Make a new graph of $\Delta A/[\text{free pyridine}]$ versus ΔA using the new values of [free pyridine]. Find a new value of K and $\Delta\varepsilon$. If you feel it is justified, perform another cycle of refinement.

5. Using the values of free pyridine concentration from your last refinement and the values of absorbance at the I_2 maximum, prepare another Scatchard plot and see if you get the same value of K.

6. Explain why an isosbestic point is observed in this experiment.

24-17 PROPERTIES OF AN ION-EXCHANGE RESIN[†]

In this experiment, we explore the properties of a cation-exchange resin, which is an organic polymer containing many sulfonic acid groups ($-SO_3H$). When a cation, such as Cu^{2+}, flows into the resin, the cation is tightly bound by sulfonate groups, and one H^+ is released for each positive charge bound to the resin (Figure 24-7). The bound cation can be displaced from the resin

---

[†] Part of this experiment is taken from M. V. Olson and J. M. Crawford, *J. Chem. Ed.*, **52**, 546 (1975).

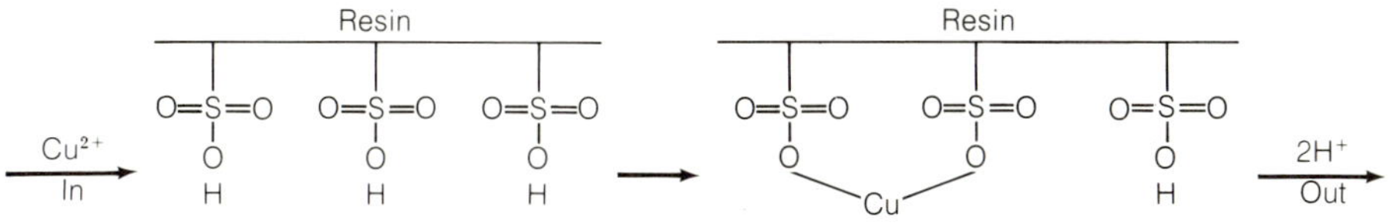

24-17 PROPERTIES OF AN ION-EXCHANGE RESIN

Figure 24-7
Stoichiometry of ion exchange.

by a large excess of H^+ or by an excess of any other cation for which the resin has some affinity.

First, known quantities of NaCl, $Fe(NO_3)_3$, and NaOH will be passed through the resin in the H^+ form. The H^+ released by each cation will be measured by titration with NaOH.

In the second part of the experiment, we will analyze a sample of impure vanadyl sulfate ($VOSO_4 \cdot 2H_2O$). As supplied commercially, this salt contains $VOSO_4$, H_2SO_4, and H_2O. A solution will be prepared from a known mass of reagent. The VO^{2+} content can be assayed spectrophotometrically, and the total cation (VO^{2+} and H^+) content can be assayed by ion exchange. Together, these measurements enable us to establish the quantities of $VOSO_4$, H_2SO_4, and H_2O in the sample.

REAGENTS

0.3 M NaCl: A bottle containing 5–10 mL per student, with an accurately known concentration.

0.1 M Fe(NO$_3$)$_3$ · 6H$_2$O: A bottle containing 5–10 mL per student, with an accurately known concentration.

VOSO$_4$: The commonly available grade (usually designated "purified") is used for this experiment. Students may make their own solutions and measure the absorbance at 750 nm, or a bottle of stock solution (25 mL per student) can be supplied. The stock should contain 8 g/L (accurately weighed) and state the absorbance.

0.02 M NaOH: Each student should prepare an accurate 1/5 dilution of standard 0.1 M NaOH.

PROCEDURE

1. Prepare a chromatography column from a 0.7 cm diameter, 15 cm length of glass tubing, fitted at the bottom with a cork having a small hole that serves as the outlet. Place a small ball of glass wool above the cork to retain the resin. Use a small glass rod to plug the outlet and shut off the column. (Alternatively, an inexpensive column such as 0.7 × 15 cm Econo-Column from Bio-Rad Laboratories[†] works well in this experiment.) Fill the column with water, close it off, and test for leaks. Then drain the water until 2 cm remains and close the column again.

2. Make a slurry of 1.1 g of Bio-Rad Dowex 50W-X2 (100/200 mesh) cation-exchange resin in 5 mL of water and pour it into the column. If the resin cannot be poured

---

[†] Bio-Rad Laboratories, 2200 Wright Avenue, Richmond, CA 94804.

636

24 / EXPERIMENTS

all at once, allow some to settle, remove the supernatant liquid with a pipet, and pour in the rest of the resin. If the column is stored between laboratory periods, it should be upright, capped, and contain water above the level of the resin.[†]

3. The general procedure for analysis of a sample is:
   (a) Generate the H^+-saturated resin by passing ~ 10 mL of 1 M HCl through the column. Apply the liquid sample to the glass wall so as not to disturb the resin.
   (b) Wash the column with ~ 15 mL of water. Use the first few milliliters to wash the glass walls and allow the water to soak into the resin before continuing the washing.[‡]
   (c) Place a clean 125 mL flask under the outlet and pipet the sample onto the column.
   (d) After the reagent has soaked in, wash it through with 10 mL of H_2O, collecting all eluate.
   (e) Add 3 drops of phenolphthalein indicator (Table 12-3) to the flask and titrate with standard 0.02 M NaOH.

4. Analyze 2.000 mL aliquots of 0.3 M NaCl and 0.1 M $Fe(NO_3)_3$, following the procedures in Step 3. Calculate the theoretical volume of NaOH needed for each titration. If you do not come within 2% of this volume, repeat the analysis.

5. Pass 10.0 mL of your 0.02 M NaOH through the column as in Step 3, and analyze the eluate. Explain what you observe.

6. Analyze 10.00 mL of $VOSO_4$ solution as described in Step 3.

7. Using the molar absorptivity of vanadyl ion ($\varepsilon = 18.0$ M^{-1} cm^{-1} at 750 nm) and the results of Step 6, express the composition of the vanadyl sulfate in the form $(VOSO_4)_{1.00}(H_2SO_4)_x(H_2O)_y$.

24-18 QUANTITATIVE ANALYSIS BY GAS CHROMATOGRAPHY OR HPLC

We now describe an experiment illustrating the use of internal standards for quantitative analysis. The directions are purposely general, since the specific mixtures and analytical conditions depend on the equipment and columns employed. The procedure can be modified for gas or liquid chromatography.

You will receive an unknown solution containing two or three compounds from a specified list of possibilities.[§] Pure samples of each possible component of the unknown should also be available.

[†] When the experiment is finished, the resin can be collected, washed with 1 M HCl and water, and reused.

[‡] Unlike most other chromatography resins, the one used in this experiment retains water when allowed to run "dry." Ordinarily, you must not let liquid fall below the top of the solid phase in a chromatography column.

[§] For example, we have done this experiment with Carle student gas chromatographs having columns packed with Carbowax or dinonyl phthalate. We used dichloromethane, chloroform, ethyl acetate, 1-propanol, and toluene as unknowns. We have also done this experiment with a DuPont 841 HPLC instrument containing a 1 m column with an octadecyl bonded phase eluted with 65% (vol/vol) methanol/water. The unknowns included toluene, biphenyl, naphthalene, 2-naphthol, and 9-fluorenone.

PROCEDURE

1. Identify the components of your mixture. Run a chromatogram of the pure unknown. Then mix pure samples of suspected components with samples of the unknown and run a new chromatogram of each. By observing which peaks grow or whether new peaks appear, you should be able to identify each species in your unknown.

2. Perform a quantitative analysis of one component (designated A) of the unknown, which is well separated from all other components. Find a compound (designated S) that is not a component of the unknown and is separated from all other peaks in the chromatogram. Prepare a mixture of unknown plus S such that the peak areas (or heights, if the peaks are sharp) of A and S are within a factor of two of each other. This is done by trial and error. The mass (or volume) of S and the mass of unknown (not A, but total unknown) in the mixture must be accurately known.

3. Prepare known mixtures of pure A and pure S to establish a calibration curve. If a solvent is used, the mixtures should have the same concentration of S required in the previous paragraph, plus variable quantities of unknown. If pure liquids are used without a solvent, the concentrations of both S and A will necessarily vary. Prepare a calibration curve in which you plot the signal ratio (peak area of A/peak area of S) versus the known concentration ratio (concentration of A/concentration of S). The graph should span the range including the signal ratio measured for the unknown mixture plus standard.

4. From your graph, calculate the concentration of species A in the unknown. Calculate the response factor (Equation 23-6) for A relative to S.

Appendixes

A/Logarithms and Exponents 641
B/Graphs of Straight Lines 644
C/A Detailed Look at Propagation of Error 646
D/Oxidation Numbers and Balancing Redox Equations 649
E/Normality 656
F/Solubility Products 659
G/Acid Dissociation Constants 663
H/Standard Reduction Potentials 673

A / Logarithms and Exponents

The base 10 logarithm of a number, n, is the exponent to which 10 must be raised to produce n:

$$\text{If } n = 10^a, \qquad \log n = a. \tag{A-1}$$

On a calculator, you can get the log of n by merely pressing the "log" function. If you know $a = \log n$ and you wish to find n, you can calculate the antilog of a or you can raise 10 to the ath power:

$$a = \log n \tag{A-2}$$

$$10^a = 10^{\log n} = n \tag{A-3}$$

Natural logarithms (ln) work the same way, but are based on the number e ($= 2.718\,281\ldots$) instead of the number 10. The ln of a number, n, is the exponent to which e must be raised to produce n:

$$b = \ln n \tag{A-4}$$

$$e^b = e^{\ln n} = n \tag{A-5}$$

On a calculator, you can find the ln of n with a button marked "ln." To find n when you know $b = \ln n$, use the e^x key.

Logarithms have certain useful properties:

$$\log(a \cdot b) = \log a + \log b \tag{A-6}$$

$$\log(a/b) = \log a - \log b \tag{A-7}$$

$$\log(a^b) = b \log a \tag{A-8}$$

$$\log 10^a = a \tag{A-9}$$

Some useful properties of exponents are

$$a^b \cdot a^c = a^{(b+c)} \qquad \text{(A-10)}$$

$$a^b/a^c = a^{(b-c)} \qquad \text{(A-11)}$$

Problems

Here are some problems to help you practice using logs and exponents. Simplify each expression as much as possible. Answers are given at the end of this appendix.

(a) $e^{\ln a}$

(b) $10^{\log a}$

(c) $\log 10^a$

(d) $10^{-\log a}$

(e) $e^{-\ln a^3}$

(f) $e^{\ln a^{-3}}$

(g) $\log 10^{1/a^3}$

(h) $\log 10^{-a^2}$

(i) $\log(10^{a^2 - b})$

(j) $\log(2a^3 10^{b^2})$

(k) $e^{a + \ln b}$

(l) $10^{(\log 3) - (4 \log 2)}$

In working with the Nernst equation or the Henderson–Hasselbalch equation, the case often arises in which an equation of the following form must be solved for the unknown, x:

$$a = b - c \log \frac{d}{gx} \qquad \text{(A-12)}$$

In this equation a, b, c, d, and g are constants. We can isolate x by first bringing the log term to the left and bringing a to the right:

$$c \log \frac{d}{gx} = b - a$$

Dividing by c gives

$$\log \frac{d}{gx} = \frac{b - a}{c}$$

Now raise 10 to the value of each side of the equation:

$$10^{\log(d/gx)} = 10^{(b-a)/c}$$

But $10^{\log(d/gx)}$ is just d/gx, so

$$\frac{d}{gx} = 10^{(b-a)/c}$$

Finally, we rearrange to solve for x:

$$\frac{d}{g\,10^{(b-a)/c}} = x \tag{A-13}$$

To test yourself on the algebra, solve for the numerical value of x in the equation:

$$-0.317 = 0.111 - \frac{0.059\,16}{2}\log\frac{(x^2)}{0.01}$$

Answer: $x = 1.72 \times 10^6$

Sometimes it is convenient to be able to convert between $\ln x$ and $\log x$. The relation between these is derived by writing

$$x = 10^{\log x}$$

and finding the ln of both sides:

$$\ln x = \ln(10^{\log x}) \tag{A-14}$$

$$\ln x = (\log x)(\ln 10) \tag{A-15}$$

Equation A-15 follows from Equation A-14 and the fact that $\ln a^b = b \ln a$.

Answers

(a) a (e) $1/a^3$ (i) $a^2 - b$
(b) a (f) $1/a^3$ (j) $b^2 \log(2a^3)$
(c) a (g) $1/a^3$ (k) be^a
(d) $1/a$ (h) $-a^2$ (l) $3/16$

B / Graphs of Straight Lines

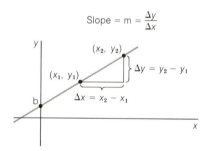

Figure B-1
Parameters of a straight line.

The general form of the equation of a straight line is

$$y = mx + b \tag{B-1}$$

where $m = \text{slope} = \dfrac{\Delta y}{\Delta x} = \dfrac{y_2 - y_1}{x_2 - x_1}$

$b = \text{intercept on } y \text{ axis}$

The meanings of slope and intercept are illustrated in Figure B-1.

If you know two points $[(x_1, y_1) \text{ and } (x_2, y_2)]$ that lie on the line, you can generate the equation of the line by noting that the slope is the same for every pair of points on the line. Calling some general point on the line (x, y), we can write

$$\frac{y - y_1}{x - x_1} = \frac{y_2 - y_1}{x_2 - x_1} \equiv m \tag{B-2}$$

which can be rearranged to the form

$$y - y_1 = \left(\frac{y_2 - y_1}{x_2 - x_1}\right)(x - x_1)$$

$$y = \underbrace{\left(\frac{y_2 - y_1}{x_2 - x_1}\right)}_{m} x + \underbrace{y_1 - \left(\frac{y_2 - y_1}{x_2 - x_1}\right) x_1}_{b} \tag{B-3}$$

When you have a series of experimental points that should lie on a line, the best line is generally obtained by the method of least squares, described in Chapter 4. This method gives the slope and the intercept directly.

If, instead, you wish to draw the "best" line by eye, you can derive the equation of the line by selecting two points *that lie on the line* and applying Equation B-2. In general, your experimental points probably will not lie exactly on the best line. To apply Equation B-2, select two points that do lie on the line in your graph.

Sometimes you are presented with a linear plot in which x and/or y are nonlinear functions. An example is shown in Figure B-2, in which the potential of an electrode is expressed as a function of the activity of analyte. Given that the slope is 29.6 mV and the line passes through the point ($\mathscr{A} = 10^{-4}$, $E = -10.2$), find the equation of the line. To do this, first note that the y axis is linear, but that the x axis is *logarithmic*. That is, the function E versus $\mathscr{A}$ is *not* linear, but E versus log $\mathscr{A}$ is linear. The form of the straight line should therefore be

$$E = \underbrace{(29.6)}_{m} \underbrace{\log \mathscr{A}}_{x} + b \qquad \text{(B-4)}$$
$\uparrow \ y$

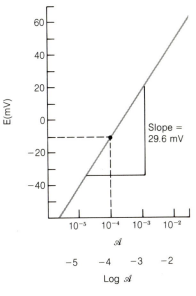

Figure B-2
A linear graph in which one axis is a logarithmic function.

To find b, we can use the coordinates of the one known point in Equation B-2:

$$\frac{y - y_1}{x - x_1} = \frac{E - E_1}{\log \mathscr{A} - \log \mathscr{A}_1} = \frac{E - (-10.2)}{\log \mathscr{A} - \log(10^{-4})} = m = 29.6$$

or

$$E + 10.2 = 29.6 \log \mathscr{A} + (29.6)(4)$$

$$E \text{ (mV)} = 29.6 \text{ (mV)} \log \mathscr{A} + 108.2 \text{ (mV)} \qquad \text{(B-5)}$$

C / A Detailed Look at Propagation of Error

The rules given for propagation of error in Equations 3-6 and 3-9 are special cases of a general formula. Suppose you wish to calculate the function, F, of several experimental quantities, x, y, z, ... If the errors in x, y, z, ..., are small, random, and independent of each other, then the error in F is given approximately by

$$e_F = \sqrt{\left(\frac{\partial F}{\partial x}\right)^2 e_x^2 + \left(\frac{\partial F}{\partial y}\right)^2 e_y^2 + \left(\frac{\partial F}{\partial z}\right)^2 e_z^2 + \ldots} \qquad \text{(C-1)}$$

Here e_F is the error in F and each e_i is the error in the measured quantities (x, y, z, etc.). The quantities in parentheses are partial derivatives. A partial derivative is calculated using the same rules for ordinary derivatives, except that all but one variable are treated as constants. For example, if $F = 3xy^2$, $\partial F/\partial x = 3y^2$ and $\partial F/\partial y = (3x)(2y) = 6xy$.

Let's apply Equation C-1 to the function $F = x - y$. For this function,

$$\frac{\partial F}{\partial x} = 1 \qquad \frac{\partial F}{\partial y} = -1$$

$$e_F = \sqrt{(1)^2 e_x^2 + (-1)^2 e_y^2} = \sqrt{e_x^2 + e_y^2}$$

which is just the same as the rule given for addition and subtraction in Chapter 3.

POWERS AND ROOTS

Suppose $F = x^a$, where the exponent a is a constant such as 3 or $-2/3$:

$$\frac{\partial F}{\partial x} = ax^{a-1} = a\frac{x^a}{x} = \left(\frac{a}{x}\right)F$$

$$e_F = \sqrt{\left(\frac{a}{x}\right)^2 F^2 e_x^2} = \left(\frac{a}{x}\right)F e_x$$

The relative error will be

$$\frac{e_F}{F} = \frac{a}{x} e_x = a\left(\frac{e_x}{x}\right) \tag{C-2}$$

That is, the relative error in F will be a times the relative error in x. For example, if $F = \sqrt{x} = x^{1/2}$, a 2% error in x will yield a $(\frac{1}{2})(2\%) = 1\%$ error in F. If $F = x^3$, a 2% error in x leads to a $(3)(2\%) = 6\%$ error in F.

LOGS AND ANTILOGS

Now consider the function $F = \log x$. This means that

$$x = 10^F$$

But 10 can be rewritten as $10 = e^{\ln 10}$. Putting this into the expression above gives

$$x = 10^F = (e^{\ln 10})^F = e^{F \ln 10}$$

or

$$\ln x = F \ln 10$$

or

$$F = \frac{\ln x}{\ln 10} \approx 0.434\,29 \ln x$$

Recalling that $d(\ln x)/dx = 1/x$ allows us to write

$$\frac{\partial F}{\partial x} = \left(\frac{1}{\ln 10}\right) \frac{\partial \ln x}{\partial x} = \frac{1}{\ln 10} \frac{1}{x}$$

Inserting this derivative into Equation C-1 shows that if $F = \log x$,

$$e_F = \sqrt{\left(\frac{1}{\ln 10}\right)^2 \left(\frac{1}{x}\right)^2 e_x^2} = \left(\frac{1}{\ln 10}\right) \frac{e_x}{x} = (0.434\,29) \frac{e_x}{x} \tag{C-3}$$

That is, the absolute error in F is proportional to the relative error in x.

Finally, consider $F = \text{antilog } x$, which is the same as saying $F = 10^x$. Noting that $d(a^x)/dx = a^x \ln a$, we can write

$$\frac{\partial F}{\partial x} = \frac{\partial(10^x)}{\partial x} = 10^x \ln 10$$

$$e_F = \sqrt{(10^x \ln 10)^2 e_x^2} = (10^x \ln 10)e_x$$

648

C / A DETAILED LOOK AT PROPAGATION OF ERROR

or

$$\frac{e_F}{F} = (\ln 10)e_x = (2.302\ 6)e_x \qquad\qquad (\text{C-4})$$

Here the relative error in F is proportional to the absolute error in x.

Problems

C-1. Verify the following calculations.
    (a) $\sqrt{2.36\ (\pm 0.06)} = 1.53_6 \pm 0.02_0$
    (b) $2.36^{4.39\ (\pm 0.08)} = 43._4 \pm 3._0$
    (c) $2.36\ (\pm 0.06)^{4.39\ (\pm 0.08)} = 43._4 \pm 5._7$
    (d) $\log[3.141\ 5\ (\pm 0.001\ 1)] = 0.497\ 1_4 \pm 0.00\ 1_5$
    (e) $\text{antilog}[2.08\ (\pm 0.06)] = 1.2_0\ (\pm 0.1_7) \times 10^2$
    (f) $\ln[3.141\ 5\ (\pm 0.001\ 1)] = 1.144\ 7_0 \pm 0.000\ 3_5$
    (g) $2.08\ (\pm 0.06)^{3.141\ 5\ (\pm 0.001\ 1)}\ -$
        $10.4\ (\pm 0.1)\log\sqrt{9.28\ (\pm 0.11)} = 4.9_5 \pm 0.9_1$

C-2. Consider the definition of pH to be $\text{pH} = -\log[\text{H}^+]$.
If a pH measurement has an uncertainty of ± 0.02,
Show that the relative error in $[\text{H}^+]$ is 4.6%.

D / Oxidation Numbers and Balancing Redox Equations

The *oxidation number*, or *oxidation state*, is a bookkeeping device used to keep track of the number of electrons formally associated with a particular element. The oxidation number is meant to tell how many electrons have been lost or gained by a neutral atom when it forms a compound. Because oxidation numbers have no real physical meaning, they are somewhat arbitrary and not all chemists will assign the same oxidation number to a given element in an unusual compound. However, there are some ground rules that provide a useful start:

1. The oxidation number of an element by itself—e.g., $Cu(s)$ or $Cl_2(g)$—is zero.

2. The oxidation number of H is almost always $+1$, except in metal hydrides—e.g., NaH—in which H is -1.

3. The oxidation number of oxygen is almost always -2. The only common exceptions are peroxides, in which two oxygen atoms are connected and each has an oxidation number of -1. Two examples are hydrogen peroxide ($H—O—O—H$) and its anion ($H—O—O^-$). The oxidation number of oxygen in gaseous O_2 is, of course, zero.

4. The alkali metals (Li, Na, K, Rb, Cs, Fr) almost always have an oxidation number $+1$. The alkaline earth metals (Be, Mg, Ca, Sr, Ba, Ra) are almost always in the $+2$ oxidation state.

5. The halogens (F, Cl, Br, I) are usually in the -1 oxidation state. Exceptions occur when two different halogens are bound to each other or when a halogen is bound to more than one atom. When different halogens are bound to each other, we assign the oxidation number -1 to the more electronegative halogen.

The sum of the oxidation numbers of each atom in a molecule must equal the charge of the molecule. In H_2O, for example, we have

$$
\begin{array}{lrr}
2 \text{ Hydrogen} = 2(+1) = & +2 \\
\text{Oxygen} & = & -2 \\
\hline
\text{Net charge} & & 0
\end{array}
$$

In SO_4^{2-}, sulfur must have an oxidation number of $+6$ in order that the sum of the oxidation numbers be -2:

$$
\begin{array}{lrr}
\text{Oxygen} = 4(-2) = & -8 \\
\text{Sulfur} & = & +6 \\
\hline
\text{Net charge} & & -2
\end{array}
$$

650

D / OXIDATION NUMBERS AND BALANCING REDOX EQUATIONS

In benzene (C_6H_6), the oxidation number of carbon must be -1 if hydrogen is assigned the number $+1$. In cyclohexane (C_6H_{12}), the oxidation number of carbon must be -2, for the same reason. Benzene therefore represents a higher oxidation state of carbon than does cyclohexane.

The oxidation number of iodine in ICl_2^- is $+1$. This is unusual since halogens are usually -1. However, since chlorine is more electronegative than iodine, we assign Cl as -1, forcing I to be $+1$.

The oxidation number of As in As_2S_3 is $+3$, and the value for S is -2. This is arbitrary, but reasonable. Since S is more electronegative than As, we make S negative and As positive. Because sulfur is in the same family as oxygen, which is usually -2, we assign S as -2, leaving As as $+3$.

The oxidation number of S in $S_4O_6^{2-}$ (tetrathionate) is $+2.5$. The *fractional oxidation state* comes about since six O atoms contribute -12. Since the charge is -2, the four S atoms must contribute $+10$. The average oxidation number of S must be $+10/4 = 2.5$.

The oxidation number of Fe in $K_3Fe(CN)_6$ is $+3$. To make this assignment, we first recognize cyanide (CN^-) as a common ion that carries a charge of -1. Six cyanide ions give -6, and three potassium ions (K^+) give $+3$. Therefore, Fe should have an oxidation number of $+3$ for the whole formula to be neutral. In this approach, it is not necessary to assign individual oxidation numbers to carbon and nitrogen, so long as we recognize that the charge of CN is -1.

Problems

Answers are given at the end of this appendix.

D-1. Write the oxidation state of the boldfaced atom in each species below.

(a) **Ag**Br
(b) **S**$_2$O$_3^{2-}$
(c) **Se**F$_6$
(d) H**S**$_2$O$_3^-$
(e) H**O**$_2$
(f) **N**O
(g) **Cr**$^{3+}$
(h) **Mn**O$_2$
(i) **Pb**(OH)$_3^-$
(j) **Fe**(OH)$_3$
(k) **Cl**O$^-$
(l) K$_4$**Fe**(CN)$_6$

(m) **Cl**O$_2$
(n) **Cl**O$_2^-$
(o) **Mn**(CN)$_6^{4-}$
(p) **N**$_2$
(q) **N**H$_4^+$
(r) **N**$_2$H$_5^+$
(s) H**As**O$_3^{2-}$
(t) **Co**$_2$(CO)$_8$ (CO group is neutral)
(u) (CH$_3$)$_4$**Li**$_4$
(v) **P**$_4$O$_{10}$
(w) **C**$_2$H$_6$O (ethanol, CH$_3$CH$_2$OH)
(x) **V**O(SO$_4$)

(y) **Fe**$_3$O$_4$
(z) **C**$_3$H$_3^+$

Structure: $H-C\overset{\oplus}{\underset{}{\langle}}\begin{matrix} C-H \\ C-H \end{matrix}$

D-2. Identify the oxidizing agent and the reducing agent on the left side of each reaction below.

(a) $Cr_2O_7^{2-} + 3Sn^{2+} + 14H^+ \rightarrow 2Cr^3 + 3Sn^{4+} + 7H_2O$

(b) $4I^- + O_2 + 4H^+ \rightarrow 2I_2 + 2H_2O$

(c) $5CH_3\overset{O}{\overset{\|}{C}}H + 2MnO_4^- + 6H^+ \rightarrow 5CH_3\overset{O}{\overset{\|}{C}}OH + 2Mn^{2+} + 3H_2O$

(d) $HOCH_2CHOHCH_2OH + 2IO_4^- \rightarrow 2H_2C{=}O + HCO_2H + 2IO_3^- + H_2O$

        glycerol   formaldehyde   formic
                                                        acid

(e) $C_8H_8 + 2Na \rightarrow C_8H_8^{2-} + 2Na^+$

C_8H_8 is cyclooctatetraene with the structure

(f) $I_2 + OH^- \rightarrow HOI + I^-$

                   hypoiodite

(g) ascorbic acid $+ I_3^- \rightarrow$ dehydroascorbate $+ 3I^- + 2H^+$

(You should be able to do this one without knowing the formula for ascorbic acid.)

BALANCING REDOX REACTIONS

To balance a reaction involving oxidation and reduction, we must first identify which element is oxidized and which is reduced. We then break the net reaction into two imaginary *half-reactions*, one of which involves only oxidation and the other only reduction. Although free electrons never appear in a balanced net reaction, they do appear in balanced half-reactions. If we are dealing with aqueous solutions, we proceed to balance each half-reaction, using H_2O and either H^+ or OH^-, as necessary. *A reaction is balanced when the number of atoms of each element is the same on both sides, and the net charge is the same on both sides.*[†]

Acidic Solutions

Here are the steps we will follow:

1. Assign oxidation numbers to the elements that are oxidized or reduced.
2. Break the reaction into two half-reactions, one involving oxidation and the other reduction.
3. For each half-reaction, balance the number of atoms that are oxidized or reduced.
4. Balance the electrons to account for the change in oxidation number by adding electrons to one side of each half-reaction.
5. Balance oxygen atoms by adding H_2O to one side of each half-reaction.
6. Balance the H atoms by adding H^+ to one side of each half-reaction.
7. Multiply each half-reaction by the number of electrons in the other half-reaction so that the number of electrons on each side of the total reaction will cancel. Then add the two half-reactions and simplify to the smallest integral coefficients.

[†] A completely different method for balancing complex redox equations by inspection has been described by D. Kolb, *J. Chem. Ed.*, **58**, 642 (1981).

D / OXIDATION NUMBERS AND BALANCING REDOX EQUATIONS

EXAMPLE

Balance the following equation using H^+, but not OH^-:

$$Fe^{2+} + MnO_4^- \rightleftharpoons Fe^{3+} + Mn^{2+}$$
$$+2 +7 \phantom{\rightleftharpoons Fe^{3+}}+3 +2$$

permanganate

The steps listed above apply to this example as shown below:

1. *Assign oxidation numbers.* These are assigned for Fe and Mn in each species in the above reaction.

2. *Break the reaction into two half-reactions.*

   Oxidation half-reaction: $Fe^{2+} \rightleftharpoons Fe^{3+}$
   $$+2 +3$$

   Reduction half-reaction: $MnO_4^- \rightleftharpoons Mn^{2+}$
   $$+7 +2$$

3. *Balance the atoms that are oxidized or reduced.* Since there is only one Fe or Mn in each species on each side of the equation, the atoms of Fe and Mn are already balanced.

4. *Balance electrons.* Electrons are added to account for the change in each oxidation state:

$$Fe^{2+} \rightleftharpoons Fe^{3+} + e^-$$
$$MnO_4^- + 5e^- \rightleftharpoons Mn^{2+}$$

   In the second case, we need $5e^-$ on the left side to take Mn from $+7$ to $+2$.

5. *Balance oxygen atoms.* There are no oxygen atoms in the Fe half-reaction. There are four oxygen atoms on the left side of the Mn reaction, so we add four molecules of H_2O to the right side:

$$MnO_4^- + 5e^- \rightleftharpoons Mn^{2+} + 4H_2O$$

6. *Balance hydrogen atoms.* The Fe equation is already balanced. The Mn equation needs $8H^+$ on the left.

$$MnO_4^- + 5e^- + 8H^+ \rightleftharpoons Mn^{2+} + 4H_2O$$

   At this point, each half-reaction must be completely balanced (the same number of atoms and charge on each side), or you have made a mistake.

7. *Multiply and add the reactions.* We multiply the Fe equation by 5 and the Mn equation by 1 and add:

$$5Fe^{2+} \rightleftharpoons 5Fe^{3+} + 5e^-$$
$$\underline{MnO_4^- + 5e^- + 8H^+ \rightleftharpoons Mn^{2+} + 4H_2O}$$
$$5Fe^{2+} + MnO_4^- + 8H^+ \rightleftharpoons 5Fe^{3+} + Mn^{2+} + 4H_2O$$

The total charge on each side is $+17$, and we find the same number of each atom on each side. The equation is balanced.

EXAMPLE

Now try the next reaction, which represents the reverse of a *disproportionation*. (In a disproportionation, an element in one oxidation state reacts to give the same element in higher and lower oxidation states.)

$$I_2 \;+\; IO_3^- + Cl^- \rightleftharpoons ICl_2^-$$

$$\quad 0 \qquad +5 \qquad -1 \qquad +1 - 1$$

iodine   iodate

1. The oxidation numbers are assigned above. Note that chlorine has an oxidation number of -1 on both sides of the equation. Only iodine is involved in electron transfer.

2. Oxidation half-reaction: $I_2 \rightleftharpoons ICl_2^-$

$$\qquad\qquad 0 \qquad +1$$

Reduction half-reaction: $IO_3^- \rightleftharpoons ICl_2^-$

$$\qquad\qquad +5 \qquad +1$$

3. We need to balance I atoms in the first reaction and add Cl^- to each reaction to balance Cl.

$$I_2 + 4Cl^- \rightleftharpoons 2ICl_2^-$$

$$IO_3^- + 2Cl^- \rightleftharpoons ICl_2^-$$

4. Now add electrons to each.

$$I_2 + 4Cl^- \rightleftharpoons 2ICl_2^- + 2e^-$$

$$IO_3^- + 2Cl^- + 4e^- \rightleftharpoons ICl_2^-$$

The first reaction needs $2e^-$ because there are two I atoms, each of which changes from 0 to $+1$.

5. The second reaction needs $3H_2O$ on the right side to balance oxygen atoms.

$$IO_3^- + 2Cl^- + 4e^- \rightleftharpoons ICl_2^- + 3H_2O$$

6. The first reaction is balanced, but the second needs $6H^+$ on the left.

$$IO_3^- + 2Cl^- + 4e^- + 6H^+ \rightarrow ICl_2^- + 3H_2O$$

As a check, the charge on each side of this half-reaction is -1, and all atoms are balanced.

7. Multiply and add.

$$2(I_2 + 4Cl^- \rightleftharpoons 2ICl_2^- + 2e^-)$$

$$\underline{IO_3^- + 2Cl^- + 4e^- + 6H^+ \rightleftharpoons ICl_2^- + 3H_2O}$$

$$2I_2 + IO_3^- + 10Cl^- + 6H^+ \rightleftharpoons 5ICl_2^- + 3H_2O \qquad \text{(D-1)}$$

We multiplied the first reaction by 2 so there would be the same number of electrons in each half-reaction. You could have multiplied the first reaction by 4 and the second by 2, but then all coefficients would simply be doubled. We customarily write the smallest coefficients.

654

D / OXIDATION NUMBERS AND BALANCING
REDOX EQUATIONS

Basic Solutions

The method many people prefer for basic solutions is to balance the equation first with H^+. The answer can then be converted to one in which OH^- is used instead. This is done by adding to each side of the equation a number of hydroxide ions equal to the number of H^+ ions appearing in the equation. For example, to balance Eq. D-1 with OH^- instead of H^+, proceed as follows:

$$2I_2 + IO_3^- + 10Cl^- + 6H^+ \rightleftharpoons 5ICl_2 + 3H_2O$$

$$+ 6OH^- \qquad\qquad + 6OH^-$$

$$2I_2 + IO_3^- + 10Cl^- + \underbrace{6H^+ + 6OH^-}_{\substack{6H_2O \\ \Downarrow \\ 3H_2O}} \rightleftharpoons 5ICl_2 + 3H_2O + 6OH^-$$

Realizing that $6H^+ + 6OH^- = 6H_2O$ and canceling $3H_2O$ on each side gives the final result:

$$2I_2 + IO_3^- + 10Cl^- + 3H_2O \rightleftharpoons 5ICl_2 + 6OH^-$$

Problems

D-3. Balance the following reactions using H^+, but not OH^-.
(a) $Fe^{3+} + Hg_2^{2+} \rightleftharpoons Fe^{2+} + Hg^{2+}$
(b) $Ag + NO_3^- \rightleftharpoons Ag^+ + NO$
(c) $VO^{2+} + Sn^{2+} \rightleftharpoons V^{3+} + Sn^{4+}$
(d) $SeO_4^{2-} + Hg + Cl^- \rightleftharpoons SeO_3^{2-} + Hg_2Cl_2$
(e) $CuS + NO_3^- \rightleftharpoons Cu^{2+} + SO_4^{2-} + NO$
(f) $S_2O_3^{2-} + I_2 \rightleftharpoons I^- + S_4O_6^{2-}$
(g) $ClO_3^- + As_2S_3 \rightleftharpoons Cl^- + H_2AsO_4^- + SO_4^{2-}$

(h) $Cr_2O_7^{2-} + CH_3\overset{\text{O}}{\overset{\|}{C}}H \rightleftharpoons CH_3\overset{\text{O}}{\overset{\|}{C}}OH + Cr^{3+}$
(i) $MnO_4^{2-} \rightleftharpoons MnO_2 + MnO_4^-$
(j) $Hg_2SO_4 + Ca^{2+} + S_8 \rightleftharpoons Hg_2^{2+} + CaS_2O_3$
(k) $ClO_3^- \rightleftharpoons Cl_2 + O_2$

D-4. Balance the following equations using OH^-, but not H^+.
(a) $PbO_2 + Cl^- \rightleftharpoons ClO^- + Pb(OH)_3^-$
(b) $HNO_2 + SbO^+ \rightleftharpoons NO + Sb_2O_5$
(c) $Ag_2S + CN^- + O_2 \rightleftharpoons S + Ag(CN)_2^- + OH^-$
(d) $HO_2^- + Cr(OH)_3^- \rightleftharpoons CrO_4^{2-} + OH^-$
(e) $ClO_2 + OH^- \rightleftharpoons ClO_2^- + ClO_3^-$
(f) $WO_3^- + O_2 \rightleftharpoons HW_6O_{21}^{5-} + OH^-$
(g) $Mn_2O_3 + CN^- \rightleftharpoons Mn(CN)_6^{4-} + (CN)_2$
(h) $Cu^{2+} + H_2 \rightleftharpoons Cu + H_2O$
(i) $BH_4^- + H_2O \rightleftharpoons H_3BO_3 + H_2$
(j) $Mn_2O_3 + Hg + CN^- \rightleftharpoons Mn(CN)_6^{4-} + Hg(CN)_2$

Answers

D-1. (a) $+1$ (g) $+3$ (m) $+4$ (s) $+3$ (y) $+8/3$
(b) $+2$ (h) $+4$ (n) $+3$ (t) 0 (z) $-2/3$
(c) $+6$ (i) $+2$ (o) $+2$ (u) -4
(d) $+2$ (j) $+3$ (p) 0 (v) $+5$
(e) $-\frac{1}{2}$ (k) $+1$ (q) -3 (w) -2
(f) $+2$ (l) $+2$ (r) -2 (x) $+4$

D-2.

	Oxidizing agent	Reducing agent
(a)	$Cr_2O_7^{2-}$	Sn^{2+}
(b)	O_2	I^-
(c)	MnO_4^-	CH_3CHO
(d)	IO_4^-	Glycerol
(e)	C_8H_8	Na
(f)	I_2	I_2
(g)	I_3^-	Ascorbic acid

Reaction f is called a *disproportionation*, since an element in one oxidation state is transformed into two different oxidation states—one higher and one lower than the original oxidation state. In Reaction g, I_3^- is the oxidizing agent, since its oxidation number changes from $-1/3$ to -1. Therefore, ascorbic acid *must* be the reducing agent, since one reactant is oxidized and one is reduced.

D-3. (a) $2Fe^{3+} + Hg_2^{2+} \rightleftharpoons 2Fe^{2+} + 2Hg^{2+}$
(b) $3Ag + NO_3^- + 4H^+ \rightleftharpoons 3Ag^+ + NO + 2H_2O$
(c) $4H^+ + 2VO^{2+} + Sn^{2+} \rightleftharpoons 2V^{3+} + Sn^{4+} + 2H_2O$
(d) $2Hg + 2Cl^- + SeO_4^{2-} + 2H^+ \rightleftharpoons$
$\qquad Hg_2Cl_2 + SeO_3^{2-} + H_2O$
(e) $3CuS + 8NO_3^- + 8H^+ \rightleftharpoons$
$\qquad 3Cu^{2+} + 3SO_4^{2-} + 8NO + 4H_2O$
(f) $2S_2O_3^{2-} + I_2 \rightleftharpoons S_4O_6^{2-} + 2I^-$
(g) $14ClO_3^- + 3As_2S_3 + 18H_2O \rightleftharpoons$
$\qquad 14Cl^- + 6H_2AsO_4^- + 9SO_4^{2-} + 24H^+$
(h) $Cr_2O_7^{2-} + 3CH_3CHO + 8H^+ \rightleftharpoons$
$\qquad 2Cr^{3+} + 3CH_3CO_2H + 4H_2O$

(i) $4H^+ + 3MnO_4^{2-} \rightleftharpoons MnO_2 + 2MnO_4^- + 2H_2O$
(j) $2Hg_2SO_4 + 3Ca^{2+} + \frac{1}{2}S_8 + H_2O \rightleftharpoons$
$\qquad 2Hg_2^{2+} + 3CaS_2O_3 + 2H^+$
(k) $2H^+ + 2ClO_3^- \rightleftharpoons Cl_2 + \frac{5}{2}O_2 + H_2O$

The balanced half-reaction for As_2S_3 in part g is

$$As_2S_3 + 20H_2O \rightleftharpoons$$
$$+3 \quad -2$$
$$2H_2AsO_4^- + 3SO_4^{2-} + 28e^- + 36H^+$$
$$+5 \qquad +6$$

Since As_2S_3 is a single compound, we must consider the $As_2S_3 \rightarrow H_2AsO_4^-$ and $As_2S_3 \rightarrow SO_4^{2-}$ reactions together. The net change in oxidation number for the *two* As atoms is $2(5 - 3) = +4$. The net change in oxidation number for the *three* S atoms is $3[6 - (-2)] = +24$. Therefore, $24 + 4 = 28e^-$ are involved in the half-reaction.

D-4. (a) $H_2O + OH^- + PbO_2 + Cl^- \rightleftharpoons$
$\qquad Pb(OH)_3^- + ClO^-$
(b) $4HNO_2 + 2SbO^+ + 2OH^- \rightleftharpoons$
$\qquad 4NO + Sb_2O_5 + 3H_2O$
(c) $Ag_2S + 4CN^- + \frac{1}{2}O_2 + H_2O \rightleftharpoons$
$\qquad S + 2Ag(CN)_2^- + 2OH^-$
(d) $2OH_2^- + Cr(OH)_3^- \rightleftharpoons CrO_4^{2-} + OH^- + 2H_2O$
(e) $2ClO_2 + 2OH^- \rightleftharpoons ClO_2^- + ClO_3^- + H_2O$
(f) $12WO_3^- + 3O_2 + 2H_2O \rightleftharpoons$
$\qquad 2HW_6O_{21}^{5-} + 2OH^-$
(g) $Mn_2O_3 + 14CN^- + 3H_2O \rightleftharpoons$
$\qquad 2Mn(CN)_6^{4-} + (CN)_2 + 6OH^-$
(h) $Cu^{2+} + H_2 + 2OH^- \rightleftharpoons Cu + 2H_2O$
(i) $BH_4^- + 4H_2O \rightleftharpoons H_3BO_3 + 4H_2 + OH^-$
(j) $3H_2O + Mn_2O_3 + Hg + 14CN^- \rightleftharpoons$
$\qquad 2Mn(CN)_6^{4-} + Hg(CN)_2 + 6OH^-$

E / Normality

The *normality* (N) of a redox reagent is n times the molarity, where n is the number of electrons donated or accepted by that species in a chemical reaction.

$$N = n M \qquad (E\text{-}1)$$

For example, in the half-reaction

$$MnO_4^- + 8H^+ + 5e^- \rightleftharpoons Mn^{2+} + 4H_2O \qquad (E\text{-}2)$$

the normality of permangante ion is five times its molarity, since each MnO_4^- accepts $5e^-$. If the molarity of permanganate is 0.1 M, the normality for the reaction

$$MnO_4^- + 5Fe^{2+} + 8H^+ \rightleftharpoons Mn^{2+} + 5Fe^{3+} + 4H_2O \qquad (E\text{-}3)$$

is $5 \times 0.1 = 0.5$ N (read "0.5 normal"). In this reaction, each Fe^{2+} ion donates one electron. The normality of ferrous ion *equals* the molarity of ferrous ion, even though it takes five ferrous ions to balance the reaction.

In the half-reaction

$$MnO_4^- + 4H^+ + 3e^- \rightleftharpoons MnO_2 + 2H_2O \qquad (E\text{-}4)$$

each MnO_4^- ion accepts only *three* electrons. The normality of permanagante for this reaction is equal to three times the molarity of permanganate. A 0.06 N permanganate solution for this reaction contains 0.02 M MnO_4^-.

The normality of a solution is a statement of the moles of "reacting units" per liter. One mole of reacting units is called one *equivalent*. Therefore, the units of normality are equivalents per liter (equiv/L). For redox reagents, *one equivalent is the amount of substance that can donate or accept one electron.* It is possible to speak of equivalents only with respect to a particular half-reaction. For example, in Reaction E-2 there are five equivalents per mole of MnO_4^-; but in Reaction E-4, there are only three equivalents per mole of MnO_4^-. The mass of substance containing one equivalent is called the *equivalent weight.* The formula weight of $KMnO_4$ is 158.033 9. The equivalent weight of $KMnO_4$ for Reaction E-2 is $158.033\,9/5 = 31.606\,8$ g/equiv. The equivalent weight of $KMnO_4$ for Reaction E-4 is $158.033\,9/3 = 52.678\,0$ g/equiv.

EXAMPLE

Find the normality of a solution containing 6.34 g of ascorbic acid in 250.0 mL if the relevant half-reaction is

ascorbic acid
(vitamin C)

dehydroascorbic acid

The formula weight of ascorbic acid $(C_6H_8O_6)$ is 176.126. In 6.34 g, there are $(6.34\ \text{g})/(176.126\ \text{g/mol}) = 3.60 \times 10^{-2}$ mol. Since each mole contains 2 equivalents in this example, $6.34\ \text{g} = (2\ \text{equiv/mol})(3.60 \times 10^{-2}\ \text{mol}) = 7.20 \times 10^{-2}$ equivalents. The normality is $(7.20 \times 10^{-2}\ \text{equiv})/(0.250\,0\ \text{L}) = 0.288$ N.

EXAMPLE

How many grams of potassium oxalate should be dissolved in 500.0 mL to make a 0.100 N solution for titration of MnO_4^-?

$$5H_2C_2O_4 + 2MnO_4^- + 6H^+ \rightleftharpoons 2Mn^{2+} + 10CO_2 + 8H_2O \qquad \text{(E-5)}$$

It is first necessary to write the oxalic acid half-reaction:

$$H_2C_2O_4 \rightleftharpoons 2CO_2 + 2H^+ + 2e^-$$

It is apparent that there are two equivalents per mole of oxalic acid. Hence a 0.100 N solution will be 0.050 0 M:

$$\frac{0.100\ \text{equiv/L}}{2\ \text{equiv/mol}} = 0.050\,0\ \text{mol/L} = 0.050\,0\ \text{M}$$

Therefore, we must dissolve $(0.050\,0\ \text{mol/L})(0.500\,0\ \text{L}) = 0.025\,0$ mol in 500.0 mL. Since the formula weight of $K_2C_2O_4$ is 166.216, we should use $(0.025\,0\ \text{mol}) \times (166.216\ \text{g/mol}) = 4.15$ g of potassium oxalate.

The utility of normality in volumetric analysis lies in the equation

$$N_1V_1 = N_2V_2 \qquad \text{(E-6)}$$

where N_1 is the normality of reagent 1, V_1 is the volume of reagent 1, N_2 is the normality of reagent 2, and V_2 is the volume of reagent 2. V_1 and V_2 may be expressed in any units, so long as the same units are used for both.

E / NORMALITY

EXAMPLE

A solution containing 25.0 mL of oxalic acid required 13.78 mL of 0.041 62 N $KMnO_4$ for titration, according to Reaction E-5. Find the normality and molarity of the oxalic acid.

Setting up Equation E-6, we write

$$N_1(25.0 \text{ mL}) = (0.041\ 62\ \text{N})(13.78 \text{ mL})$$

$$N_1 = 0.022\ 94 \text{ equiv/L}$$

Since there are two equivalents per mole of oxalic acid in Reaction E-5,

$$M = N/n = 0.022\ 94/2 = 0.011\ 47 \text{ M}$$

Normality is sometimes used in acid–base or ion-exchange chemistry. With respect to acids and bases, the equivalent weight of a reagent is the amount that can donate or accept one mole of H^+. With respect to ion exchange, the equivalent weight is the mass of reagent containing one mole of charge.

F / Solubility Products[†]

Formula	pK_{sp}	K_{sp}	Temperature (°C)	Ionic strength (M)
Azides: L = N_3^-				
CuL	8.31	4.9×10^{-9}	25	0
AgL	8.56	2.8×10^{-9}	25	0
Hg_2L_2	9.15	7.1×10^{-10}	25	0
TlL	3.66	2.2×10^{-4}	25	0
$PdL_2(\alpha)$	8.57	2.7×10^{-9}	25	0
Bromates: L = BrO_3^-				
$BaL \cdot H_2O$	5.11	7.8×10^{-6}	25	0.5
AgL	4.26	5.5×10^{-5}	25	0
TlL	3.78	1.7×10^{-4}	25	0
PbL_2	5.10	7.9×10^{-6}	25	0
Bromides: L = Br^-				
CuL	8.3	5×10^{-9}	25	0
AgL	12.30	5.0×10^{-13}	25	0
Hg_2L_2	22.25	5.6×10^{-23}	25	0
TlL	5.44	3.6×10^{-6}	25	0
HgL_2	18.9	1.3×10^{-19}	25	0.5
PbL_2	5.68	2.1×10^{-6}	25	0
Carbonates: L = CO_3^{2-}				
MgL	7.46	3.5×10^{-8}	25	0
CaL(calcite)	8.35	4.5×10^{-9}	25	0
CaL(aragonite)	8.22	6.0×10^{-9}	25	0
SrL	9.03	9.3×10^{-10}	25	0
BaL	8.30	5.0×10^{-9}	25	0
Y_2L_3	30.6	2.5×10^{-31}	25	0
La_2L_3	33.4	4.0×10^{-34}	25	0
MnL	9.30	5.0×10^{-10}	25	0
FeL	10.68	2.1×10^{-11}	25	0
CoL	9.98	1.0×10^{-10}	25	0
NiL	6.87	1.3×10^{-7}	25	0
CuL	9.63	2.3×10^{-10}	25	0
Ag_2L	11.09	8.1×10^{-12}	25	0
Hg_2L	16.05	8.9×10^{-17}	25	0
ZnL	10.00	1.0×10^{-10}	25	0
CdL	13.74	1.8×10^{-14}	25	0
PbL	13.13	7.4×10^{-14}	25	0

[†]Data for all salts except oxalates are taken from A. E. Martell and R. M. Smith, *Critical Stability Constants*, Vol. 4 (New York: Plenum Press, 1976). Data for oxalates are from L. G. Sillén and A. E. Martell, *Stability Constants of Metal-Ion Complexes*, Supplement No. 1 (London: The Chemical Society, Special Publication No. 25, 1971).

Formula	pK_{sp}	K_{sp}	Temperature (°C)	Ionic strength (M)
Chlorides: L = Cl^-				
CuL	6.73	1.9×10^{-7}	25	0
AgL	9.74	1.8×10^{-10}	25	0
Hg_2L_2	17.91	1.2×10^{-18}	25	0
TlL	3.74	1.8×10^{-4}	25	0
PbL_2	4.78	1.7×10^{-5}	25	0
Chromates: L = CrO_4^{2-}				
BaL	9.67	2.1×10^{-10}	25	0
CuL	5.44	3.6×10^{-6}	25	0
Ag_2L	11.92	1.2×10^{-12}	25	0
Hg_2L	8.70	2.0×10^{-9}	25	0
Tl_2L	12.01	9.8×10^{-13}	25	0
Cobalticyanides: L = $Co(CN)_6^{3-}$				
Ag_3L	25.41	3.9×10^{-26}	25	0
$(Hg_2)_3L_2$	36.72	1.9×10^{-37}	25	0
Cyanides: L = CN^-				
AgL	15.66	2.2×10^{-16}	25	0
Hg_2L_2	39.3	5×10^{-40}	25	0
ZnL_2	15.5	3×10^{-16}	25	3
Ferrocyanides: L = $Fe(CN)_6^{4-}$				
Ag_4L	44.07	8.5×10^{-45}	25	0
Zn_2L	15.68	2.1×10^{-16}	25	0
Cd_2L	17.38	4.2×10^{-18}	25	0
Pb_2L	18.02	9.5×10^{-19}	25	0
Fluorides: L = F^-				
LiL	2.77	1.7×10^{-3}	25	0
MgL_2	8.18	6.6×10^{-9}	25	0
CaL_2	10.41	3.9×10^{-11}	25	0
SrL_2	8.54	2.9×10^{-9}	25	0
BaL_2	5.76	1.7×10^{-6}	25	0
ThL_4	28.3	5×10^{-29}	25	3
PbL_2	7.44	3.6×10^{-8}	25	0
Hydroxides: L = OH^-				
MgL_2	11.15	7.1×10^{-12}	25	0
CaL_2	5.19	6.5×10^{-6}	25	0
$BaL_2 \cdot 8H_2O$	3.6	3×10^{-4}	25	0
YL_3	23.2	6×10^{-24}	25	0
LaL_3	20.7	2×10^{-21}	25	0
CeL_3	21.2	6×10^{-22}	25	0
$UO_2 (\rightleftharpoons U^{4+} + 4OH^-)$	56.2	6×10^{-57}	25	0
$UO_2L_2 (\rightleftharpoons UO_2^{2+} + 2OH^-)$	22.4	4×10^{-23}	25	0
MnL_2	12.8	1.6×10^{-13}	25	0
FeL_2	15.1	7.9×10^{-16}	25	0
CoL_2	14.9	1.3×10^{-15}	25	0
NiL_2	15.2	6×10^{-16}	25	0
CuL_2	19.32	4.8×10^{-20}	25	0
VL_3	34.4	4.0×10^{-35}	25	0

Formula	pK_{sp}	K_{sp}	Temperature (°C)	Ionic strength (M)
CrL_3	29.8	1.6×10^{-30}	25	0.1
FeL_3	38.8	1.6×10^{-39}	25	0
CoL_3	44.5	3×10^{-45}	19	0
$VOL_2 \ (\rightleftharpoons VO^{2+} + 2OH^-)$	23.5	3×10^{-24}	25	0
PdL_2	28.5	3×10^{-28}	25	0
$ZnL_2(amorphous)$	15.52	3.0×10^{-16}	25	0
$CdL_2(\beta)$	14.35	4.5×10^{-15}	25	0
$HgO \ (red) \ (\rightleftharpoons Hg^{2+} + 2OH^-)$	25.44	3.6×10^{-26}	25	0
$Cu_2O \ (\rightleftharpoons 2Cu^+ + 2OH^-)$	29.4	4×10^{-30}	25	0
$Ag_2O \ (\rightleftharpoons 2Ag^+ + 2OH^-)$	15.42	3.8×10^{-16}	25	0
AuL_3	5.5	3×10^{-6}	25	0
$AlL_3(\alpha)$	33.5	3×10^{-34}	25	0
$GaL_3(amorphous)$	37	10^{-37}	25	0
InL_3	36.9	1.3×10^{-37}	25	0
$SnO \ (\rightleftharpoons Sn^{2+} + 2OH^-)$	26.2	6×10^{-27}	25	0
$PbO \ (yellow) \ (\rightleftharpoons Pb^{2+} + 2OH^-)$	15.1	8×10^{-16}	25	0
$PbO \ (red) \ (\rightleftharpoons Pb^{2+} + 2OH^-)$	15.3	5×10^{-16}	25	0
Iodates: L = IO_3^-)				
CaL_2	6.15	7.1×10^{-7}	25	0
SrL_2	6.48	3.3×10^{-7}	25	0
BaL_2	8.81	1.5×10^{-9}	25	0
YL_3	10.15	7.1×10^{-11}	25	0
LaL_3	10.99	1.0×10^{-11}	25	0
CeL_3	10.86	1.4×10^{-11}	25	0.5
ThL_4	14.62	2.4×10^{-15}	25	0.2
$UO_2L_2 \ (\rightleftharpoons UO_2^{2+} + 2IO_3^-)$	7.01	9.8×10^{-8}	25	0.5
CrL_3	5.3	5×10^{-6}	25	0
AgL	7.51	3.1×10^{-8}	25	0
Hg_2L_2	17.89	1.3×10^{-18}	25	0
TlL	5.51	3.1×10^{-6}	25	0
ZnL_2	5.41	3.9×10^{-6}	25	0
CdL_2	7.64	2.3×10^{-8}	25	0
PbL_2	12.61	2.5×10^{-13}	25	0
Iodides: L = I^-				
CuL	12.0	1×10^{-12}	25	0
AgL	16.08	8.3×10^{-17}	25	0
$CH_3HgL \ (\rightleftharpoons CH_3Hg^+ + I^-)$	11.46	3.5×10^{-12}	20	1
$CH_3CH_2HgL \ (\rightleftharpoons CH_3CH_2Hg^+ + I^-)$	4.11	7.8×10^{-5}	25	1
TlL	7.23	5.9×10^{-8}	25	0
Hg_2L_2	27.95	1.1×10^{-28}	25	0.5
SnL_2	5.08	8.3×10^{-6}	25	4
PbL_2	8.10	7.9×10^{-9}	25	0
Oxalates: L = $C_2O_4^{2-}$				
CaL	7.9	1.3×10^{-8}	20	0.1
SrL	6.4	4×10^{-7}	20	0.1
BaL	6.0	1×10^{-6}	20	0.1
La_2L_3	25.0	1×10^{-25}	20	0.1
ThL_2	21.38	4.2×10^{-22}	25	1
$UO_2L \ (\rightleftharpoons UO_2^{2+} + C_2O_4^{2-})$	8.66	2.2×10^{-9}	20	0.1

Formula	pK_{sp}	K_{sp}	Temperature (°C)	Ionic strength (M)
Phosphates: L = PO_4^{3-}				
$MgHL \cdot 3H_2O$ ($\rightleftharpoons Mg^{2+} + HL^{2-}$)	5.82	1.5×10^{-6}	25	0
$CaHL \cdot 2H_2O$ ($\rightleftharpoons Ca^{2+} + HL^{2-}$)	6.58	2.6×10^{-7}	25	0
$SrHL$ ($\rightleftharpoons Sr^{2+} + HL^{2-}$)	6.92	1.2×10^{-7}	20	0
$BaHL$ ($\rightleftharpoons Ba^{2+} + HL^{2-}$)	7.40	4.0×10^{-8}	20	0
LaL	22.43	3.7×10^{-23}	25	0.5
$Fe_3L_2 \cdot 8H_2O$	36.0	1×10^{-36}	25	0
$FeL \cdot 2H_2O$	26.4	4×10^{-27}	25	0
$(VO)_3L_2$ ($\rightleftharpoons 3VO^{2+} + 2L^{3-}$)	25.1	8×10^{-26}	25	0
Ag_3L	17.55	2.8×10^{-18}	25	0
Hg_2HL ($\rightleftharpoons Hg_2^{2+} + HL^{2-}$)	12.40	4.0×10^{-12}	25	0
$Zn_3L_2 \cdot 4H_2O$	35.3	5×10^{-36}	25	0
Pb_3L_2	43.53	3.0×10^{-44}	38	0
GaL	21.0	1×10^{-21}	25	1
InL	21.63	2.3×10^{-22}	25	1
Sulfates: L = SO_4^{2-}				
CaL	4.62	2.4×10^{-5}	25	0
SrL	6.50	3.2×10^{-7}	25	0
BaL	9.96	1.1×10^{-10}	25	0
RaL	10.37	4.3×10^{-11}	20	0
Ag_2L	4.83	1.5×10^{-5}	25	0
Hg_2L	6.13	7.4×10^{-7}	25	0
PbL	6.20	6.3×10^{-7}	25	0
Sulfides: L = S^{2-}				
MnL (pink)	10.5	3×10^{-11}	25	0
MnL (green)	13.5	3×10^{-14}	25	0
FeL	18.1	8×10^{-19}	25	0
$CoL(\alpha)$	21.3	5×10^{-22}	25	0
$CoL(\beta)$	25.6	3×10^{-26}	25	0
$NiL(\alpha)$	19.4	4×10^{-20}	25	0
$NiL(\beta)$	24.9	1.3×10^{-25}	25	0
$NiL(\gamma)$	26.6	3×10^{-27}	25	0
CuL	36.1	8×10^{-37}	25	0
Cu_2L	48.5	3×10^{-49}	25	0
Ag_2L	50.1	8×10^{-51}	25	0
Tl_2L	21.2	6×10^{-22}	25	0
$ZnL(\alpha)$	24.7	2×10^{-25}	25	0
$ZnL(\beta)$	22.5	3×10^{-23}	25	0
CdL	27.0	1×10^{-27}	25	0
HgL (black)	52.7	2×10^{-53}	25	0
HgL (red)	53.3	5×10^{-54}	25	0
SnL	25.9	1.3×10^{-26}	25	0
PbL	27.5	3×10^{-28}	25	0
In_2L_3	69.4	4×10^{-70}	25	0
Thiocyanates: L = SCN^-				
CuL	13.40	4.0×10^{-14}	25	5
AgL	11.97	1.1×10^{-12}	25	0
Hg_2L_2	19.52	3.0×10^{-20}	25	0
TlL	3.79	1.6×10^{-4}	25	0
HgL_2	19.56	2.8×10^{-20}	25	1

G / Acid Dissociation Constants

Name	Structure[†]	pK_A[‡]	K_A
Acetic acid (ethanoic acid)	CH_3CO_2H	4.757	1.75×10^{-5}
Alanine	NH₃⁺ | CHCH₃ | CO₂H	2.348 (CO_2H) 9.867 (NH_3)	4.49×10^{-3} 1.36×10^{-10}
Aminobenzene (aniline)	$C_6H_5-NH_3^+$	4.601	2.51×10^{-5}
4-Aminobenzenesulfonic acid (sulfanilic acid)	$^-O_3S-C_6H_4-\overset{+}{N}H_3$	3.232	5.86×10^{-4}
2-Aminobenzoic acid (anthranilic acid)	$C_6H_4(NH_3^+)(CO_2H)$	2.08 (CO_2H) 4.96 (NH_3)	8.3×10^{-3} 1.10×10^{-5}
2-Aminoethanethiol (2-mercaptoethylamine)	$HSCH_2CH_2NH_3^+$	8.21 (SH) ($\mu = 0.1$) 10.71 (NH_3) ($\mu = 0.1$)	6.2×10^{-9} 1.95×10^{-11}
2-Aminoethanol (ethanolamine)	$HOCH_2CH_2NH_3^+$	9.498	3.18×10^{-10}
2-Aminophenol	$C_6H_4(OH)(NH_3^+)$	4.78 (NH_3) (20°) 9.97 (OH) (20°)	1.66×10^{-5} 1.05×10^{-10}
Ammonia	NH_4^+	9.244	5.70×10^{-10}
Arginine	NH₃⁺ | CHCH₂CH₂CH₂NHC(=NH₂⁺)NH₂ | CO₂H	1.823 (CO_2H) 8.991 (NH_3) (12.48) (NH_2)	1.50×10^{-2} 1.02×10^{-9} 3.3×10^{-13}
Arsenic acid (hydrogen arsenate)	O ‖ HO—As—OH | OH	2.24 6.96 11.50	5.8×10^{-3} 1.10×10^{-7} 3.2×10^{-12}

[†] Each acid is written in its protonated form. The acidic protons are indicated in bold type.
[‡] pK_A values refer to 25°C and zero ionic strength unless otherwise indicated. Values in parentheses are considered to be less reliable. Data are from A. E. Martell and R. M. Smith, *Critical Stability Constants* (New York: Plenum Press, 1974).

Name	Structure	pK_A	K_A
Arsenious acid (hydrogen arsenite)	$As(OH)_3$	9.29	5.1×10^{-10}
Asparagine	$\overset{NH_3^+}{\underset{CO_2H}{CHCH_2}}\overset{O}{CNH_2}$	2.14 (CO_2H) ($\mu = 0.1$) 8.72 (NH_3) ($\mu = 0.1$)	7.2×10^{-3} 1.9×10^{-9}
Aspartic acid	$\overset{NH_3^+}{\underset{CO_2H}{CHCH_2CO_2H}}$	1.990 (α-CO_2H) 3.900 (β-CO_2H) 10.002 (NH_3)	1.02×10^{-2} 1.26×10^{-4} 9.95×10^{-11}
Aziridine (dimethyleneimine)	$\triangleright NH_2^+$	8.04	9.1×10^{-9}
Benzene-1,2,3-tricarboxylic acid (hemimellitic acid)	(benzene ring with three CO_2H groups)	2.88 4.75 7.13	1.32×10^{-3} 1.78×10^{-5} 7.4×10^{-8}
Benzoic acid	$C_6H_5-CO_2H$	4.202	6.28×10^{-5}
Benzylamine	$C_6H_5-CH_2NH_3^+$	9.35	4.5×10^{-10}
2,2'-Bipyridine	(bipyridine structure, $\overset{}{\underset{H}{N}}_+ \; N$)	4.35	4.5×10^{-5}
Boric acid (hydrogen borate)	$B(OH)_3$	9.236 (12.74) (20°) (13.80) (20°)	5.81×10^{-10} 1.82×10^{-13} 1.58×10^{-14}
Bromoacetic acid	$BrCH_2CO_2H$	2.902	1.25×10^{-3}
Butane-2,3-dione dioxime (dimethylglyoxime)	$\overset{HON \quad NOH}{\underset{CH_3 \quad CH_3}{}}$	10.66 12.0	2.2×10^{-11} 1×10^{-12}
Butanoic acid	$CH_3CH_2CH_2CO_2H$	4.819	1.52×10^{-5}
cis-Butenedioc acid (maleic acid)	(cis structure with two CO_2H)	1.910 6.332	1.23×10^{-2} 4.66×10^{-7}
trans-Butenedioc acid (fumaric acid)	(trans structure, CO_2H and HO_2C)	3.053 4.494	8.85×10^{-4} 3.21×10^{-5}
Butylamine	$CH_3CH_2CH_2CH_2NH_3^+$	10.640	2.29×10^{-11}
Carbonic acid[†] (hydrogen carbonate)	$HO-\overset{O}{\underset{}{C}}-OH$	6.352 10.329	4.45×10^{-7} 4.69×10^{-11}

[†] The concentration of "carbonic acid" is considered to be the sum $[H_2CO_3] + [CO_2(aq)]$. See Box 10-2.

Name	Structure	pK_A	K_A
Chloroacetic acid	$ClCH_2CO_2H$	2.865	1.36×10^{-3}
3-Chloropropanoic acid	$ClCH_2CH_2CO_2H$	4.11	7.8×10^{-5}
Chlorous acid (hydrogen chlorite)	$HOCl{=}O$	1.95	1.12×10^{-2}
Chromic acid (hydrogen chromate)	HO—Cr—OH (with two O double bonds)	−0.2 (20°) 6.51	1.6 3.1×10^{-7}
Citric acid (2-hydroxypropane-1,2,3-tricarboxylic acid)	$HO_2CCH_2\underset{OH}{\overset{CO_2H}{C}}CH_2CO_2H$	3.128 4.761 6.396	7.44×10^{-4} 1.73×10^{-5} 4.02×10^{-7}
Cyanoacetic acid	$NCCH_2CO_2H$	2.472	3.37×10^{-3}
Cyclohexylamine	cyclohexyl—NH_3^+	10.64	2.3×10^{-11}
Cysteine	$\underset{CO_2H}{\overset{NH_3^+}{CHCH_2SH}}$	(1.71) (CO_2H) 8.36 (SH) 10.77 (NH_3)	1.95×10^{-2} 4.4×10^{-9} 1.70×10^{-11}
Dichloroacetic acid	Cl_2CHCO_2H	1.30	5.0×10^{-2}
Diethylamine	$(CH_3CH_2)_2NH_2^+$	10.933	4.7×10^{-10}
1,2-Dihydroxybenzene (catechol)	benzene with OH, OH	9.40 12.8	4.0×10^{-10} 1.6×10^{-13}
1,3-Dihydroxybenzene (resorcinol)	benzene with OH, OH	9.30 11.06	5.0×10^{-10} 8.7×10^{-12}
D-2,3-Dihydroxybutanedioc acid (D-tartaric acid)	$HO_2C\underset{OH}{\overset{OH}{CCHCHCO_2H}}$	3.036 4.366	9.20×10^{-4} 4.31×10^{-5}
2,3-Dimercaptopropanol	$HOCH_2\underset{SH}{CHCH_2SH}$	8.58 ($\mu = 0.1$) 10.68 ($\mu = 0.1$)	2.6×10^{-9} 2.1×10^{-11}
Dimethylamine	$(CH_3)_2NH_2^+$	10.774	1.68×10^{-11}
Ethane-1,2-dithiol	$HSCH_2CH_2SH$	8.85 (30°, $\mu = 0.1$) 10.43 (30°, $\mu = 0.1$)	1.4×10^{-9} 3.7×10^{-11}
Ethylamine	$CH_3CH_2NH_3^+$	10.636	2.31×10^{-11}
Ethylenediamine (1,2-diaminoethane)	$H_3\overset{+}{N}CH_2CH_2\overset{+}{N}H_3$	6.848 9.928	1.42×10^{-7} 1.18×10^{-10}

Name	Structure	pK_A	K_A
Ethylenedinitrilotetraacetic acid (EDTA)	$(HO_2CCH_2)_2\overset{+}{N}HCH_2CH_2\overset{+}{N}H(CH_2CO_2H)_2$	0.0 (CO_2H) ($\mu = 1.0$) 1.5 (CO_2H) ($\mu = 0.1$) 2.0 (CO_2H) ($\mu = 0.1$) 2.68 (CO_2H) ($\mu = 0.1$) 6.11 (NH) ($\mu = 0.1$) 10.17 (NH) ($\mu = 0.1$)	1.0 0.032 0.010 0.0021 7.8×10^{-7} 6.8×10^{-11}
Formic acid (methanoic acid)	HCO_2H	3.745	1.80×10^{-4}
Glutamic acid	$\overset{NH_3^+}{\underset{CO_2H}{\overset{|}{\underset{|}{CHCH_2CH_2CO_2H}}}}$	2.23 (α-CO_2H) 4.42 (γ-CO_2H) 9.95 (NH_3)	5.9×10^{-3} 3.8×10^{-5} 1.12×10^{-10}
Glutamine	$\overset{NH_3^+}{\underset{CO_2H}{\overset{|}{\underset{|}{CHCH_2CH_2\overset{O}{\overset{\|}{C}}NH_2}}}}$	2.17 (CO_2H) ($\mu = 0.1$) 9.01 (NH_3) ($\mu = 0.1$)	6.8×10^{-3} 9.8×10^{-10}
Glycine (aminoacetic acid)	$\overset{NH_3^+}{\underset{CO_2H}{\overset{|}{\underset{|}{CH_2}}}}$	2.350 (CO_2H) 9.778 (NH_3)	4.47×10^{-3} 1.67×10^{-10}
Guanidine	$\overset{+NH_2}{\underset{H_2N-C-NH_2}{\overset{\|}{}}}$	13.54 (27°, $\mu = 1.0$)	2.9×10^{-14}
1,6-Hexanedioic acid (adipic acid)	$HO_2CCH_2CH_2CH_2CH_2CO_2H$	4.42 5.42	3.8×10^{-5} 3.8×10^{-6}
Hexane-2,4-dione	$CH_3\overset{O}{\overset{\|}{C}}CH_2\overset{O}{\overset{\|}{C}}CH_2CH_3$	9.38	4.2×10^{-10}
Histidine	$\overset{NH_3^+}{\underset{CO_2H}{\overset{|}{\underset{|}{CHCH_2}}}}$ imidazole	1.7 (CO_2H) ($\mu = 0.1$) 6.02 (NH) ($\mu = 0.1$) 9.08 (NH_3) ($\mu = 0.1$)	2×10^{-2} 9.5×10^{-7} 8.3×10^{-10}
Hydrazoic acid (hydrogen azide)	$H\overset{+}{N}=\overset{}{N}=\overset{-}{N}$	4.65	2.2×10^{-5}
Hydrogen cyanate	$HOC\equiv N$	3.48	3.3×10^{-4}
Hydrogen cyanide	$HC\equiv N$	9.21	6.2×10^{-10}
Hydrogen fluoride	HF	3.17	6.8×10^{-4}
Hydrogen peroxide	$HOOH$	11.65	2.2×10^{-12}
Hydrogen sulfide	H_2S	7.02 13.9	9.5×10^{-8} 1.3×10^{-14}
Hydrogen thiocyanate	$HSC\equiv N$	0.9	0.13
Hydroxyacetic acid (glycolic acid)	$HOCH_2CO_2H$	3.831	1.48×10^{-4}

Name	Structure	pK_A	K_A
Hydroxybenzene (phenol)	C₆H₅—OH	9.98	1.05×10^{-10}
2-Hydroxybenzoic acid (salicylic acid)	(benzene ring with CO_2H and OH)	2.97 (CO_2H) 13.74 (OH)	1.07×10^{-3} 1.82×10^{-14}
L-Hydroxybutanedioic acid (malic acid)	$HO_2CCH_2CHCO_2H$ (with OH)	3.459 5.097	3.48×10^{-4} 8.00×10^{-6}
Hydroxylamine	$HONH_3^+$	5.96	1.10×10^{-6}
8-Hydroxyquinoline (oxine)	(quinoline with HO and NH⁺)	4.91 (NH) 9.81 (OH)	1.23×10^{-5} 1.55×10^{-10}
Hypobromous acid (hydrogen hypobromite)	HOBr	8.63	2.3×10^{-9}
Hypochlorous acid (hydrogen hypochlorite)	HOCl	7.53	3.0×10^{-8}
Hypoiodous acid (hydrogen hypoiodite)	HOI	10.64	2.3×10^{-11}
Hypophosphorous acid (hydrogen hypophosphite)	H_2POH (with O double bond)	1.23	5.9×10^{-2}
Imidazole (1,3-diazole)	(imidazole ring with NH⁺ and N—H)	6.993	1.02×10^{-7}
Iminodiacetic acid	$H_2\overset{+}{N}(CH_2CO_2H)_2$	1.82 (CO_2H) ($\mu = 0.1$) 2.84 (CO_2H) 9.79 (NH_2)	1.51×10^{-2} 1.45×10^{-3} 1.62×10^{-10}
Iodic acid (hydrogen iodate)	$HOI{=}O$ (with O double bond)	0.77	0.17
Iodoacetic acid	ICH_2CO_2H	3.175	6.68×10^{-4}
Isoleucine	$CHCH(CH_3)CH_2CH_3$ (with NH_3^+ and CO_2H)	2.319 (CO_2H) 9.754 (NH_3)	4.80×10^{-3} 1.76×10^{-10}
Leucine	$CHCH_2CH(CH_3)_2$ (with NH_3^+ and CO_2H)	2.329 (CO_2) 9.747 (NH_3)	4.69×10^{-3} 1.79×10^{-10}
Lysine	$CHCH_2CH_2CH_2CH_2NH_3^+$ (with NH_3^+ and CO_2H)	2.04 (CO_2H) ($\mu = 0.1$) 9.08 (α-NH_3) ($\mu = 0.1$) 10.69 (ε-NH_3) ($\mu = 0.1$)	9.1×10^{-3} 8.3×10^{-10} 2.0×10^{-11}

Name	Structure	pK_A	K_A
Malonic acid (propanedioic acid)	$HO_2CCH_2CO_2H$	2.847 / 5.696	1.42×10^{-3} / 2.01×10^{-10}
Mercaptoacetic acid (thioglycolic acid)	$HSCH_2CO_2H$	(3.60) (CO_2H) / 10.55 (SH)	2.5×10^{-4} / 2.82×10^{-11}
2-Mercaptoethanol	$HSCH_2CH_2OH$	9.72	1.91×10^{-10}
Methionine	$\overset{\displaystyle NH_3^+}{\underset{\displaystyle CO_2H}{CHCH_2CH_2SCH_3}}$	2.20 ($\mu = 0.1$) / 9.05 ($\mu = 0.1$)	6.3×10^{-3} / 8.9×10^{-10}
2-Methoxyaniline (o-anisidine)		4.527	2.97×10^{-5}
4-Methoxyaniline (p-anisidine)	CH_3O—⬡—$\overset{+}{N}H_3$	5.357	4.40×10^{-6}
Methylamine	$CH_3\overset{+}{N}H_3$	10.64	2.3×10^{-11}
2-Methylaniline (o-toluidine)		4.447	3.57×10^{-5}
4-Methylaniline (p-toluidine)	CH_3—⬡—$\overset{+}{N}H_3$	5.084	8.24×10^{-6}
2-Methylphenol (o-cresol)		10.09	8.1×10^{-11}
4-Methylphenol (p-cresol)	CH_3—⬡—OH	10.26	5.5×10^{-11}
Morpholine (perhydro-1,4-oxazine)		8.492	3.22×10^{-9}
1-Naphthoic acid		3.70	2.0×10^{-4}
2-Naphthoic acid		4.16	6.9×10^{-5}
1-Naphthol		9.34	4.6×10^{-10}

Name	Structure	pK_A	K_A
2-Naphthol		9.51	3.1×10^{-10}
Nitrilotriacetic acid	$H\overset{+}{N}(CH_2CO_2H)_3$	1.1 (CO_2H) (20°, $\mu = 1.0$) 1.650 (CO_2H) (20°) 2.940 (CO_2H) (20°) 10.334 (NH) (20°)	8×10^{-2} 2.24×10^{-2} 1.15×10^{-3} 4.63×10^{-11}
2-Nitrobenzoic acid		2.179	6.62×10^{-3}
3-Nitrobenzoic acid		3.449	3.56×10^{-4}
4-Nitrobenzoic acid	$O_2N-\!\!\!\bigcirc\!\!\!-CO_2H$	3.442	3.61×10^{-4}
Nitroethane	$CH_3CH_2NO_2$	8.57	2.7×10^{-9}
2-Nitrophenol		7.21	6.2×10^{-8}
3-Nitrophenol		8.39	4.1×10^{-9}
4-Nitrophenol	$O_2N-\!\!\!\bigcirc\!\!\!-OH$	7.15	7.1×10^{-8}
2,4-Dinitrophenol		4.11	7.8×10^{-5}
N-Nitrosophenylhydroxylamine (cupferron)		4.16 ($\mu = 0.1$)	6.9×10^{-5}
Nitrous acid	$HON\!=\!O$	3.15	7.1×10^{-4}
Oxalic acid (ethanedioic acid)	HO_2CCO_2H	1.252 4.266	5.60×10^{-2} 5.42×10^{-5}
Oxoacetic acid (glyoxylic acid)	$H\overset{\displaystyle O}{\overset{\|}{C}}CO_2H$	3.46	3.5×10^{-4}
Oxobutanedioic acid (oxaloacetic acid)	$HO_2CCH_2\overset{\displaystyle O}{\overset{\|}{C}}CO_2H$	2.56 4.37	2.8×10^{-3} 4.3×10^{-5}

Name	Structure	pK_A	K_A
2-Oxopentanedioic (α-ketoglutaric acid)	$HO_2CCH_2CH_2\overset{O}{\overset{\|}{C}}CO_2H$	1.85 ($\mu = 0.5$) 4.44 ($\mu = 0.5$)	1.41×10^{-2} 3.6×10^{-2}
2-Oxopropanoic acid (pyruvic acid)	$CH_3\overset{O}{\overset{\|}{C}}CO_2H$	2.55	2.8×10^{-3}
1,5-Pentanedioic acid (glutaric acid)	$HO_2CCH_2CH_2CH_2CO_2H$	4.34 5.43	4.6×10^{-5} 3.7×10^{-6}
Pentanoic acid (valeric acid)	$CH_3CH_2CH_2CH_2CO_2H$	4.843	1.44×10^{-5}
1,10-Phenanthroline		4.86	1.38×10^{-5}
Phenylacetic acid	$\langle\ \rangle\!-\!CH_2CO_2H$	4.310	4.90×10^{-5}
Phenylalanine	$\overset{NH_3^+}{\underset{CO_2H}{CHCH_2}}\!-\!\langle\ \rangle$	2.20 (CO_2H) 9.31 (NH_3)	6.3×10^{-3} 4.9×10^{-10}
Phosphoric acid (hydrogen phosphate)	$HO\!-\!\overset{O}{\overset{\|}{P}}\!-\!OH$ $\underset{OH}{}$	2.148 7.199 12.35	7.11×10^{-3} 6.32×10^{-8} 4.5×10^{-13}
Phosphorous acid (hydrogen phosphate)	$HP\!-\!OH$	1.5 6.79	3×10^{-2} 1.62×10^{-7}
Phthalic acid (benzene-1,2-dicarboxylic acid)		2.950 5.408	1.12×10^{-3} 3.90×10^{-6}
Piperazine (perhydro-1,4-diazine)	$H_2^+N\quad NH_2^+$	5.333 9.731	4.65×10^{-6} 1.86×10^{-10}
Piperidine	NH_2^+	11.123	7.53×10^{-12}
Proline	$\overset{}{\underset{\overset{N}{+}H_2}{}}\!-\!CO_2H$	1.952 (CO_2H) 10.640 (NH_2)	1.12×10^{-2} 2.29×10^{-11}
Propanoic acid	$CH_3CH_2CO_2H$	4.874	1.34×10^{-5}
Propenoic acid (acrylic acid)	$H_2C{=}CHCO_2H$	4.258	5.52×10^{-5}

Name	Structure	pK_A	K_A
Propylamine	$CH_3CH_2CH_2NH_3^+$	10.566	2.72×10^{-11}
Pyridine (azine)	(pyridinium, NH^+)	5.229	5.90×10^{-6}
Pyridine-2-carboxylic acid (picolinic acid)	(pyridinium with 2-CO_2H)	1.01 5.39	9.8×10^{-2} 4.1×10^{-6}
Pyridine-3-carboxylic acid (nicotinic acid)	(pyridinium with 3-HO_2C)	2.05 4.81	8.9×10^{-3} 1.55×10^{-5}
Pyridoxal-5-phosphate	(pyridinium ring: 4-$O{=}CH$, 3-OH, 5-$CH_2OPO_3H_2$ written as $HO{-}PO$ / HO, 2-CH_3, N-H^+)	1.4 (POH) ($\mu = 0.1$) 3.44 (POH) ($\mu = 0.1$) 6.01 (NH) ($\mu = 0.1$) 8.45 (OH) ($\mu = 0.1$)	0.04 3.6×10^{-4} 9.8×10^{-7} 3.5×10^{-9}
Pyrophosphoric acid (hydrogen diphosphate)	$(HO)_2POP(OH)_2$ (with two $P{=}O$)	0.8 2.2 6.70 9.40	0.16 6×10^{-3} 2.0×10^{-7} 4.0×10^{-10}
Pyrrolidine	(pyrrolidinium, NH_2^+)	11.305	4.95×10^{-12}
Serine	NH_3^+ / $CHCH_2OH$ / CO_2H	2.187 9.209	6.50×10^{-3} 6.18×10^{-10}
Succinic acid (butanedioic acid)	$HO_2CCH_2CH_2CO_2H$	4.207 5.636	6.21×10^{-5} 2.31×10^{-6}
Sulfuric acid (hydrogen sulfate)	$HO{-}S{-}OH$ (with two $S{=}O$)	1.99 (pK_2)	1.02×10^{-2}
Sulfurous acid (hydrogen sulfite)	$HOSOH$ (with $S{=}O$)	1.91 7.18	1.23×10^{-2} 6.6×10^{-8}
Thiosulfuric acid (hydrogen thiosulfate)	$HOSOH$ (with $S{=}O$ and $S{=}S$)	0.6 1.6	0.3 3×10^{-2}
Threonine	NH_3^+ / $CHCHOHCH_3$ / CO_2H	2.088 (CO_2H) 9.100 (NH_3)	8.17×10^{-3} 7.94×10^{-10}

Name	Structure	pK_A	K_A
Trichloroacetic acid	Cl_3CCO_2H	0.66 ($\mu = 0.1$)	0.22
Triethanolamine	$(HOCH_2CH_2)_3NH^+$	7.762	1.73×10^{-8}
Triethylamine	$(CH_3CH_2)_3NH^+$	10.715	1.93×10^{-11}
1,2,3-Trihydroxybenzene (pyrogallol)	(structure)	8.94 / 11.08 / (14)	1.15×10^{-9} / 8.3×10^{-12} / 10^{-14}
Trimethylamine	$(CH_3)_3NH^+$	9.800	1.58×10^{-10}
$tris$(hydroxymethyl)aminomethane (TRIS or THAM)	$(HOCH_2)_3CNH_3^+$	8.075	8.41×10^{-9}
Tryptophan	(structure)	2.35 (CO_2H) ($\mu = 0.1$) / 9.33 (NH_3) ($\mu = 0.1$)	4.5×10^{-3} / 4.7×10^{-10}
Tyrosine	(structure)	2.17 (CO_2H) ($\mu = 0.1$) / 9.19 (NH_3) / 10.47 (OH)	6.8×10^{-3} / 6.5×10^{-10} / 3.4×10^{-11}
Valine	(structure)	2.286 (CO_2H) / 9.718 (NH_3)	5.18×10^{-3} / 1.91×10^{-10}
Water	H_2O	13.997	1.01×10^{-14}

H / Standard Reduction Potentials[†]

Reaction	E^0 (volts)
Aluminum	
$Al^{3+} + 3e^- \rightleftharpoons Al(s)$	-1.66
$AlCl^{2+} + 3e^- \rightleftharpoons Al(s) + Cl^-$	-1.802
$AlF_6^{3-} + 3e^- \rightleftharpoons Al(s) + 6F^-$	-2.069
$Al(OH)_4^- + 3e^- \rightleftharpoons Al(s) + 4OH^-$	-2.33
Antimony	
$SbO^+ + 2H^+ + 3e^- \rightleftharpoons Sb(s) + H_2O$	0.212
$Sb_2O_3(s) + 6H^+ + 6e^- \rightleftharpoons 2Sb(s) + 3H_2O$	0.152
$Sb(s) + 3H^+ + 3e^- \rightleftharpoons SbH_3(g)$	-0.51
Arsenic	
$H_3AsO_4 + 2H^+ + 2e^- \rightleftharpoons H_3AsO_3 + H_2O$	0.559
$H_3AsO_3 + 3H^+ + 3e^- \rightleftharpoons As(s) + 3H_2O$	0.248
$As(s) + 3H^+ + 3e^- \rightleftharpoons AsH_3(g)$	-0.60
Barium	
$Ba^{2+} + 2e^- \rightleftharpoons Ba(s)$	-2.912
Beryllium	
$Be^{2+} + 2e^- \rightleftharpoons Be(s)$	-1.85
Bismuth	
$Bi^{3+} + 3e^- \rightleftharpoons Bi(s)$	0.200
$BiCl_4^- + 3e^- \rightleftharpoons Bi(s) + 4Cl^-$	0.16
$BiOCl(s) + 2H^+ + 3e^- \rightleftharpoons Bi(s) + H_2O + Cl^-$	0.160
Boron	
$2B(s) + 6H^+ + 6e^- \rightleftharpoons B_2H_6(g)$	-0.143
$B_4O_7^{2-} + 14H^+ + 12e^- \rightleftharpoons 4B(s) + 7H_2O$	-0.792
$B(OH)_3 + 3H^+ + 3e^- \rightleftharpoons B(s) + 3H_2O$	-0.869
Bromine	
$BrO_3^- + 6H^+ + 5e^- \rightleftharpoons \frac{1}{2}Br_2(l) + 3H_2O$	1.52
$Br_2(aq) + 2e^- \rightleftharpoons 2Br^-$	1.087
$Br_2(l) + 2e^- \rightleftharpoons 2Br^-$	1.065
$Br_3^- + 2e^- \rightleftharpoons 3Br^-$	1.051
$BrO^- + H_2O + 2e^- \rightleftharpoons Br^- + 2OH^-$	0.76
$BrO_3^- + 3H_2O + 6e^- \rightleftharpoons Br^- + 6OH^-$	0.61

[†] All species are aqueous unless otherwise indicated.

SOURCE: L. G. Sillén and A. E. Martell, *Stability Constants of Metal-Ion Complexex*, London: The Chemical Society, Special Publications No. 17 and 25, 1964 and 1971; G. Milazzo and S. Caroli, *Tables of Standard Electrode Potentials* (New York: Wiley, 1978).

Reaction	E^0 (volts)

Cadmium

$Cd^{2+} + 2e \rightleftharpoons Cd(\text{in Hg, saturated})$ -0.352

$Cd^{2+} + 2e^- \rightleftharpoons Cd(s)$ -0.402

$Cd(C_2O_4)(s) + 2e^- \rightleftharpoons Cd(s) + C_2O_4^{2-}$ -0.522

$Cd(C_2O_4)_2^{2-} + 2e^- \rightleftharpoons Cd(s) + 2C_2O_4^{2-}$ -0.572

$Cd(NH_3)_4^{2+} + 2e^- \rightleftharpoons Cd(s) + 4NH_3$ -0.613

$CdS(s) + 2e^- \rightleftharpoons Cd(s) + S^{2-}$ -1.175

Calcium

$Ca(s) + 2H^+ + 2e^- \rightleftharpoons CaH_2(s)$ 0.776

$Ca^{2+} + 2e^- \rightleftharpoons Ca(s)$ -2.868

$Ca(\text{acetate})^+ + 2e^- \rightleftharpoons Ca(s) + \text{acetate}^-$ -2.891

$CaSO_4(s) + 2e^- \rightleftharpoons Ca(s) + SO_4^{2-}$ -2.936

$Ca(\text{malonate})(s) + 2e^- \rightleftharpoons Ca(s) + \text{malonate}^{2-}$ -3.608

Carbon

$C_2H_2(g) + 2H^+ + 2e^- \rightleftharpoons C_2H_4(g)$ 0.731

$O{=}\langle\rangle{=}O + 2H^+ + 2e^- \rightleftharpoons HO{-}\langle\rangle{-}OH$ 0.700

$CH_3OH + 2H^+ + 2e^- \rightleftharpoons CH_4(g) + H_2O$ 0.588

Dehydroascorbic acid $+ 2H^+ + 2e^- \rightleftharpoons$ ascorbic acid 0.390

$(CN)_2 + 2H^+ + 2e^- \rightleftharpoons 2HCN$ 0.373

$H_2CO + 2H^+ + 2e^- \rightleftharpoons CH_3OH$ 0.232

$HCO_2H + 2H^+ + 2e^- \rightleftharpoons H_2CO + H_2O$ 0.056

$CO_2(g) + 2H^+ + 2e^- \rightleftharpoons CO(g) + H_2O$ -0.116

$C(s) + 4H^+ + 4e^- \rightleftharpoons CH_4(g)$ -0.132

$CO_2(g) + 2H^+ + 2e^- \rightleftharpoons HCO_2H$ -0.196

$2CO_2(g) + 2H^+ + 2e^- \rightleftharpoons H_2C_2O_4$ -0.49

Cerium

$Ce^{4+} + e^- \rightleftharpoons Ce^{3+}$

   $\begin{cases} 1.70 \ 1\text{F } HClO_4 \\ 1.44 \ 1\text{F } H_2SO_4 \\ 1.61 \ 1\text{F } HNO_3 \\ 1.28 \ 1\text{F } HCl \end{cases}$

$Ce^{3+} + 3e^- \rightleftharpoons Ce(s)$ -2.335

Cesium

$Cs^+ + e^- \rightleftharpoons Cs(s)$ -2.923

Chlorine

$HClO_2 + 2H^+ + 2e^- \rightleftharpoons HOCl + H_2O$ 1.64

$HClO + H^+ + e^- \rightleftharpoons \frac{1}{2}Cl_2(g) + H_2O$ 1.63

$ClO_3^- + 6H^+ + 5e^- \rightleftharpoons \frac{1}{2}Cl_2(g) + 3H_2O$ 1.47

$Cl_2(aq) + 2e^- \rightleftharpoons 2Cl^-$ 1.395

$Cl_2(g) + 2e^- \rightleftharpoons 2Cl^-$ 1.358

$ClO_3^- + 3H^+ + 2e^- \rightleftharpoons HClO_2 + H_2O$ 1.21

$ClO_4^- + 2H^+ + 2e^- \rightleftharpoons ClO_3^- + H_2O$ 1.19

$ClO_3^- + 2H^+ + e^- \rightleftharpoons ClO_2(g) + H_2O$ 1.15

$ClO_2(g) + e^- \rightleftharpoons ClO_2^-$ 0.954

Chromium

$Cr_2O_7^{2-} + 14H^+ + 6e^- \rightleftharpoons 2Cr^{3+} + 7H_2O$ 1.33

$CrO_4^{2-} + 4H_2O + 3e^- \rightleftharpoons Cr(OH)_3 \ (s, \text{hydrated}) + 5OH^-$ -0.13

$Cr^{3+} + e^- \rightleftharpoons Cr^{2+}$ -0.41

$Cr^{3+} + 3e^- \rightleftharpoons Cr(s)$ -0.74

$Cr^{2+} + 2e^- \rightleftharpoons Cr(s)$ -0.91

Reaction	E^0 (volts)

Cobalt

$$Co^{3+} + e^- \rightleftharpoons Co^{2+}$$
$\begin{cases} 1.95 \\ 1.817 \text{ 8F } H_2SO_4 \\ 1.850 \text{ 4F } HNO_3 \end{cases}$

$$Co(NH_3)_5(H_2O)^{3+} + e^- \rightleftharpoons Co(NH_3)_5(H_2O)^{2+} \qquad 0.37 \text{ 1F } NH_4NO_3$$
$$Co(NH_3)_6^{3+} + e^- \rightleftharpoons Co(NH_3)_6^{2+} \qquad 0.1$$
$$Co^{2+} + 2e^- \rightleftharpoons Co(s) \qquad -0.277$$
$$Co(OH)_2(s) + 2e^- \rightleftharpoons Co(s) + 2OH^- \qquad -0.73$$

Copper

$$Cu^+ + e^- \rightleftharpoons Cu(s) \qquad 0.518$$
$$Cu^{2+} + 2e^- \rightleftharpoons Cu(s) \qquad 0.337$$
$$Cu^{2+} + e^- \rightleftharpoons Cu^+ \qquad 0.159$$
$$CuCl(s) + e^- \rightleftharpoons Cu(s) + Cl^- \qquad 0.137$$
$$Cu(IO_3)_2(s) + 2e^- \rightleftharpoons Cu(s) + 2IO_3^- \qquad -0.079$$
$$Cu(ethylenediamine)_2^+ + e^- \rightleftharpoons Cu(s) + 2 \text{ ethylenediamine} \qquad -0.119$$
$$CuI(s) + e^- \rightleftharpoons Cu(s) + I^- \qquad -0.185$$
$$Cu(EDTA)^{2-} + 2e^- \rightleftharpoons Cu(s) + EDTA^{4-} \qquad -0.216$$
$$Cu(OH)_2(s) + 2e^- \rightleftharpoons Cu(s) + 2OH^- \qquad -0.222$$
$$Cu(CN)_2^- + e^- \rightleftharpoons Cu(s) + 2CN^- \qquad -0.429$$
$$CuCN(s) + e^- \rightleftharpoons Cu(s) + CN^- \qquad -0.639$$

Dysprosium

$$Dy^{3+} + 3e^- \rightleftharpoons Dy(s) \qquad -2.35$$

Erbium

$$Er^{3+} + 3e^- \rightleftharpoons Er(s) \qquad -2.30$$

Europium

$$Eu^{3+} + 3e^- \rightleftharpoons Eu(s) \qquad -2.41$$
$$Eu^{3+} + e^- \rightleftharpoons Eu^{2+} \qquad -0.43 \text{ 1F } KCl$$

Fluorine

$$F_2(g) + 2e^- \rightleftharpoons 2F^- \qquad 2.87$$
$$F_2O(g) + 2H^+ + 4e^- \rightleftharpoons 2F^- + H_2O \qquad 2.153$$

Gadolinium

$$Gd^{3+} + 3e^- \rightleftharpoons Gd(s) \qquad -2.40$$

Gallium

$$Ga^{3+} + 3e^- \rightleftharpoons Ga(s) \qquad -0.560$$
$$Ga(OH)_3(s) + 3H^+ + 3e^- \rightleftharpoons Ga(s) + 3H_2O \qquad -0.419$$

Germanium

$$Ge^{2+} + 2e^- \rightleftharpoons Ge(s) \qquad 0.231$$
$$H_2GeO_3 + 4H^+ + 4e^- \rightleftharpoons Ge(s) + 3H_2O \qquad -0.131$$

Gold

$$Au^+ + e^- \rightleftharpoons Au(s) \qquad 1.692$$
$$Au^{3+} + 2e^- \rightleftharpoons Au^+ \qquad 1.401$$
$$AuCl_2^- + e^- \rightleftharpoons Au(s) + 2Cl^- \qquad 1.154$$
$$AuCl_4^- + 2e^- \rightleftharpoons AuCl_2^- + 2Cl^- \qquad 0.926$$

Hafnium

$$Hf^{4+} + 4e^- \rightleftharpoons Hf(s) \qquad -1.700$$

Holmium

$$Ho^{3+} + 3e^- \rightleftharpoons Ho(s) \qquad -2.32$$

Reaction	E^0 (volts)

Hydrogen

$2H^+ + 2e^- \rightleftharpoons H_2(g)$ — 0.000

$H_2O + e^- \rightleftharpoons \frac{1}{2}H_2(g) + OH^-$ — -0.828

Indium

$In^{3+} + 3e^- \rightleftharpoons In(s)$ — -0.336

$In(OH)_3(s) + 3e^- \rightleftharpoons In(s) + 3OH^-$ — -1.00

Iodine

$H_5IO_6 + H^+ + 2e^- \rightleftharpoons IO_3^- + 3H_2O$ — 1.6

$HOI + H^+ + e^- \rightleftharpoons \frac{1}{2}I_2(s) + H_2O$ — 1.45

$ICl_3(s) + 3e^- \rightleftharpoons \frac{1}{2}I_2(s) + 3Cl^-$ — 1.28

$ICl(s) + e^- \rightleftharpoons \frac{1}{2}I_2(s) + Cl^-$ — 1.22

$IO_3^- + 6H^+ + 5e^- \rightleftharpoons \frac{1}{2}I_2(s) + 3H_2O$ — 1.195

$I_2(aq) + 2e^- \rightleftharpoons 2I^-$ — 0.622

$I_2(s) + 2e^- \rightleftharpoons 2I^-$ — 0.536

$I_3^- + 2e^- \rightleftharpoons 3I^-$ — 0.536

$IO_3^- + 3H_2O + 6e^- \rightleftharpoons I^- + 6OH^-$ — 0.26

Iridium

$IrCl_6^{2-} + e^- \rightleftharpoons IrCl_6^{3-}$ — 1.026 1F HCl

$IrBr_6^{2-} + e^- \rightleftharpoons IrBr_6^{3-}$ — 0.947 2F NaBr

$IrO_2(s) + 4H^+ + 4e^- \rightleftharpoons Ir(s) + 2H_2O$ — 0.93

$IrCl_6^{2-} + 4e^- \rightleftharpoons Ir(s) + 6Cl^-$ — 0.835

$IrI_6^{2-} + e^- \rightleftharpoons IrI_6^{3-}$ — 0.485 1F KI

Iron

$Fe(phenanthroline)_3^{3+} + e^- \rightleftharpoons Fe(phenanthroline)_3^{2+}$ — 1.147

$Fe(bipyridyl)_3^{2+} + e^- \rightleftharpoons Fe(bipyridyl)_3^{2+}$ — 1.120

$FeOOH(s) + 3H^+ + e^- \rightleftharpoons Fe^{2+} + 2H_2O$ — 0.908

$Fe^{3+} + e^- \rightleftharpoons Fe^{2+}$ —
$\begin{cases} 0.770 \\ 0.732 \text{ 1F HCl} \\ 0.767 \text{ 1F HClO}_4 \\ 0.746 \text{ 1F HNO}_3 \end{cases}$

$FeO_4^{2-} + 3H_2O + 3e^- \rightleftharpoons FeOOH(s) + 5OH^-$ — 0.71

$ferricinium^+ + e^- \rightleftharpoons ferrocene$ — 0.400

$Fe(CN)_6^{3-} + e^- \rightleftharpoons Fe(CN)_6^{4-}$ — 0.356

$Fe(glutamate)^{3+} + e^- \rightleftharpoons Fe(glutamate)^{2+}$ — 0.240

$Fe^{2+} + 2e^- \rightleftharpoons Fe(s)$ — -0.440

Lanthanum

$La^{3+} + 3e^- \rightleftharpoons La(s)$ — -2.52

$La(succinate)^+ + 3e^- \rightleftharpoons La(s) + succinate^{2-}$ — -2.601

Lead

$Pb^{4+} + 2e^- \rightleftharpoons Pb^{2+}$ — 1.69 1F HNO$_3$

$PbO_2(s) + 4H^+ + SO_4^{2-} + 2e^- \rightleftharpoons PbSO_4(s) + 2H_2O$ — 1.685

$PbO_2(s) + 4H^+ + 2e^- \rightleftharpoons Pb^{2+} + 2H_2O$ — 1.455

$3PbO_2(s) + 2H_2O + 4e^- \rightleftharpoons Pb_3O_4(s) + 4OH^-$ — 0.295

$Pb_3O_4(s) + H_2O + 2e^- \rightleftharpoons 3PbO(s) + 2OH^-$ — 0.249

$Pb^{2+} + 2e^- \rightleftharpoons Pb(s)$ — -0.126

$PbF_2(s) + 2e^- \rightleftharpoons Pb(s) + 2F^-$ — -0.350

$PbSO_4(s) + 2e^- \rightleftharpoons Pb(s) + SO_4^{2-}$ — -0.355

Lithium

$Li^+ + e^- \rightleftharpoons Li(s)$ — -3.045

Reaction	E^0 (volts)

Lutetium
$$Lu^{3+} + 3e^- \rightleftharpoons Lu(s) \qquad -2.25$$

Magnesium
$$Mg^{2+} + 2e^- \rightleftharpoons Mg(s) \qquad -2.375$$
$$Mg(OH)^+ + 2e^- \rightleftharpoons Mg(s) + OH^- \qquad -2.440$$
$$Mg(C_2O_4)(s) + 2e^- \rightleftharpoons Mg(s) + C_2O_4^{2-} \qquad -2.493$$
$$Mg(OH)_2(s) + 2e^- \rightleftharpoons Mg(s) + 2OH^- \qquad -2.690$$

Manganese
$$MnO_4^- + 4H^+ + 3e^- \rightleftharpoons MnO_2(s) + 2H_2O \qquad 1.695$$
$$Mn^{3+} + e^- \rightleftharpoons Mn^{2+} \qquad 1.542$$
$$MnO_4^- + 8H^+ + 5e^- \rightleftharpoons Mn^{2+} + 4H_2O \qquad 1.51$$
$$MnO_2(s) + 4H^+ + 2e^- \rightleftharpoons Mn^{2+} + 2H_2O \qquad 1.229$$
$$MnO_4^- + e^- \rightleftharpoons MnO_4^{2-} \qquad 0.558$$
$$Mn^{2+} + 2e^- \rightleftharpoons Mn(s) \qquad -1.182$$
$$Mn(OH)_2(s) + 2e^- \rightleftharpoons Mn(s) + 2OH^- \qquad -1.55$$
$$Mn(EDTA)^- + e^- \rightleftharpoons Mn(EDTA)^{2-} \qquad 0.825$$

Mercury
$$2Hg^{2+} + 2e^- \rightleftharpoons Hg_2^{2+} \qquad 0.908$$
$$Hg^{2+} + 2e^- \rightleftharpoons Hg(l) \qquad 0.854$$
$$Hg_2^{2+} + 2e^- \rightleftharpoons 2Hg(l) \qquad 0.792$$
$$Hg_2SO_4(s) + 2e^- \rightleftharpoons 2Hg(l) + SO_4^{2+} \qquad 0.614$$
$$Hg_2Cl_2(s) + 2e^- \rightleftharpoons 2Hg(l) + 2Cl^- \qquad \begin{cases} 0.268 \\ 0.241 \text{(saturated calomel} \\ \qquad \text{electrode)} \end{cases}$$
$$Hg_2Br_2(s) + 2e^- \rightleftharpoons 2Hg(l) + 2Br^- \qquad 0.140$$

Molybdenum
$$MoO_4^{2-} + 4H_2O + 6e^- \rightleftharpoons Mo(s) + 8OH^- \qquad -1.05$$

Neodymium
$$Nd^{3+} + 3e^- \rightleftharpoons Nd(s) \qquad -2.246$$

Neptunium
$$NpO_2^{2+} + 2e^- \rightleftharpoons NpO_2^+ \qquad 1.14 \text{ 1F HCl}$$
$$NpO_2^+ + 4H^+ + e^- \rightleftharpoons Np^{4+} + 2H_2O \qquad 0.74 \text{ 1F HCl}$$
$$Np^{4+} + e^- \rightleftharpoons Np^{3+} \qquad 0.14 \text{ 1F HCl}$$
$$Np^{3+} + 3e^- \rightleftharpoons Np(s) \qquad -1.86 \text{ 1F HCl}$$

Nickel
$$Ni(OH)_3(s) + 3H^+ + e^- \rightleftharpoons Ni^{2+} + 3H_2O \qquad 2.08$$
$$Ni^{2+} + 2e^- \rightleftharpoons Ni(s) \qquad -0.231$$
$$Ni(CN)_4^{2-} + e^- \rightleftharpoons Ni(CN)_3^{2-} + CN^- \qquad -0.401$$
$$Ni(OH)_2(s) + 2e^- \rightleftharpoons Ni(s) + 2OH^- \qquad -0.72$$

Niobium
$$\tfrac{1}{2}Nb_2O_5(s) + 5H^+ + 5e^- \rightleftharpoons Nb(s) + \tfrac{5}{2}H_2O \qquad -0.65$$

Nitrogen
$$N_2O(g) + 2H^+ + 2e^- \rightleftharpoons N_2(g) + H_2O \qquad 1.77$$
$$2NO + 2H^+ + 2e^- \rightleftharpoons N_2O(g) + H_2O \qquad 1.59$$
$$NO^+ + e^- \rightleftharpoons NO(g) \qquad 1.46$$
$$2NH_3OH^+ + H^+ + 2e^- \rightleftharpoons N_2H_5^+ + 2H_2O \qquad 1.42$$
$$NH_3OH^+ + 2H^+ + 2e^- \rightleftharpoons NH_4^+ + H_2O \qquad 1.35$$

Reaction	E^0 (volts)

$$N_2H_5^+ + 3H^+ + 2e^- \rightleftharpoons 2NH_4^+ \qquad 1.275$$
$$HNO_2 + H^+ + e^- \rightleftharpoons NO(g) + H_2O \qquad 1.00$$
$$NO_3^- + 4H^+ + 3e^- \rightleftharpoons NO(g) + 2H_2O \qquad 0.96$$
$$NO_3^- + 3H^+ + 2e^- \rightleftharpoons HNO_2 + H_2O \qquad 0.94$$
$$NO_3^- + 2H^+ + e^- \rightleftharpoons \tfrac{1}{2}N_2O_4(g) + H_2O \qquad 0.80$$
$$N_2(g) + 5H^+ + 4e^- \rightleftharpoons N_2H_5^+ \qquad -0.23$$
$$N_2(g) + 2H_2O + 4H^+ + 2e^- \rightleftharpoons 2NH_3OH^+ \qquad -1.87$$
$$\tfrac{3}{2}N_2(g) + H^+ + e^- \rightleftharpoons HN_3 \qquad -3.1$$

Osmium
$$OsO_4(s) + 8H^+ + 8e^- \rightleftharpoons Os(s) + 4H_2O \qquad 0.85$$
$$OsCl_6^{2-} + e^- \rightleftharpoons OsCl_6^{3-} \qquad \begin{cases} 0.452 \\ 0.85 \text{ 1F HCl} \end{cases}$$

Oxygen
$$O_3(g) + 2H^+ + 2e^- \rightleftharpoons O_2(g) + H_2O \qquad 2.07$$
$$H_2O_2 + 2H^+ + 2e^- \rightleftharpoons 2H_2O \qquad 1.776$$
$$\tfrac{1}{2}O_2(g) + 2H^+ + 2e^- \rightleftharpoons H_2O \qquad 1.229$$
$$O_2(g) + 2H^+ + 2e^- \rightleftharpoons H_2O_2 \qquad 0.682$$

Palladium
$$Pd^{2+} + 2e^- \rightleftharpoons Pd(s) \qquad 0.915$$
$$PdCl_6^{4-} + 2e^- \rightleftharpoons Pd(s) + 6Cl^- \qquad 0.615$$
$$PdO_2(s) + H_2O + 2e^- \rightleftharpoons PdO(s) + 2OH^- \qquad 0.73$$

Phosphorus
$$\tfrac{1}{4}P_4(s) + 3H^+ + 3e^- \rightleftharpoons PH_3(g) \qquad 0.06$$
$$H_3PO_4 + 2H^+ + 2e^- \rightleftharpoons H_3PO_3 + H_2O \qquad -0.276$$
$$H_3PO_3 + 2H^+ + 2e^- \rightleftharpoons H_3PO_2 + H_2O \qquad -0.50$$
$$H_3PO_2 + H^+ + e^- \rightleftharpoons \tfrac{1}{4}P_4(s) + 2H_2O \qquad -0.51$$

Platinum
$$Pt^{2+} + 2e^- \rightleftharpoons Pt(s) \qquad 1.188$$
$$PtCl_4^{2-} + 2e^- \rightleftharpoons Pt(s) + 4Cl^- \qquad 0.755$$
$$PtCl_6^{2-} + 2e^- \rightleftharpoons PtCl_4^{2-} + 2Cl^- \qquad 0.68$$

Plutonium
$$PuO_2^{2+} + 4H^+ + 2e^- \rightleftharpoons Pu^{4+} + 2H_2O \qquad 1.04$$
$$Pu^{4+} + e^- \rightleftharpoons Pu^{3+} \qquad 0.97$$
$$PuO_2^{2+} + e^- \rightleftharpoons PuO_2^+ \qquad 0.93$$
$$Pu^{3+} + 3e^- \rightleftharpoons Pu(s) \qquad -2.03$$

Potassium
$$K^+ + e^- \rightleftharpoons K(s) \qquad -2.925$$

Praseodymium
$$Pr^{3+} + 3e^- \rightleftharpoons Pr(s) \qquad -2.47$$

Promethium
$$Pm^3 + 3e^- \rightleftharpoons Pm(s) \qquad -2.42$$

Radium
$$Ra^{2+} + 2e^- \rightleftharpoons Ra(s) \qquad -2.92$$

Rhenium
$$ReO_4^- + 2H^+ + e^- \rightleftharpoons ReO_3(s) + H_2O \qquad 0.768$$
$$ReO_4^- + 2H_2O + 3e^- \rightleftharpoons ReO_2(s) + 4OH^- \qquad -0.594$$

Reaction	E^0 (volts)

Rhodium

$Rh^{6+} + 3e^- \rightleftharpoons Rh^{3+}$	1.48 1F $HClO_4$
$Rh^{4+} + e^- \rightleftharpoons Rh^{3+}$	1.44 3F H_2SO_4
$Rh^{3+} + e^- \rightleftharpoons Rh^{2+}$	1.2
$RhCl_6^{2-} + e^- \rightleftharpoons RhCl_6^{3-}$	1.2
$Rh^{2+} + e^- \rightleftharpoons Rh^+$	0.6
$Rh^+ + e^- \rightleftharpoons Rh(s)$	0.6
$RhCl_6^{3-} + 3e^- \rightleftharpoons Rh(s) + 6Cl^-$	0.44

Rubidium

$Rb^+ + e^- \rightleftharpoons Rb(s)$	-2.924

Ruthenium

$RuO_4^{2-} + 8H^+ + 4e^- \rightleftharpoons Ru^{2+} + 4H_2O$	1.563
$RuO_4^- + 8H^+ + 5e^- \rightleftharpoons Ru^{2+} + 4H_2O$	1.368
$Ru(dipyridyl)_3^{3+} + e^- \rightleftharpoons Ru(dipyridyl)_3^{2+}$	1.29
$Ru^{2+} + 2e^- \rightleftharpoons Ru(s)$	0.455
$Ru^{3+} + e^- \rightleftharpoons Ru^{2+}$	0.249
$Ru(NH_3)_6^{3+} + e^- \rightleftharpoons Ru(NH_3)_6^{2+}$	0.214

Samarium

$Sm^{3+} + 3e^- \rightleftharpoons Sm(s)$	-2.41

Scandium

$Sc^{3+} + 3e^- \rightleftharpoons Sc(s)$	-2.08

Selenium

$SeO_4^{2-} + 4H^+ + 2e^- \rightleftharpoons H_2SeO_3 + H_2O$	1.15
$H_2SeO_3 + 4H^+ + 4e^- \rightleftharpoons Se(s) + 3H_2O$	0.740
$Se(s) + 2H^+ + 2e^- \rightleftharpoons H_2Se(g)$	-0.37
$Se(s) + 2e^- \rightleftharpoons Se^{2-}$	-0.92

Silicon

$Si(s) + 4H^+ + 4e^- \rightleftharpoons SiH_4(g)$	0.102
$SiO_2(s, \text{hydrated}) + 4H^+ + 4e^- \rightleftharpoons Si(s)$	-0.807
$SiO_2(s, \text{quartz}) + 4H^+ + 4e^- \rightleftharpoons Si(s) + 2H_2O$	-0.857
$SiF_6^{2-} + 4e^- \rightleftharpoons Si(s) + 6F^-$	-1.24

Silver

$Ag^{2+} + e^- \rightleftharpoons Ag^+$	$\begin{cases} 2.000 \text{ 4F } HClO_4 \\ 1.929 \text{ 4F } HNO_3 \end{cases}$
$AgO(s) + H^+ + e^- \rightleftharpoons \frac{1}{2}Ag_2O(s) + \frac{1}{2}H_2O$	1.40
$Ag^+ + e^- \rightleftharpoons Ag(s)$	0.799
$Ag_2C_2O_4(s) + 2e^- \rightleftharpoons 2Ag(s) + C_2O_4^{2-}$	0.465
$AgN_3(s) + e^- \rightleftharpoons Ag(s) + N_3^-$	0.293
$AgCl(s) + e^- \rightleftharpoons Ag(s) + Cl^-$	$\begin{cases} 0.222 \\ 0.197 \text{ saturated KCl} \end{cases}$
$AgBr(s) + e^- \rightleftharpoons Ag(s) + Br^-$	0.071
$Ag(S_2O_3)_2^{3-} + e^- \rightleftharpoons Ag(s) + 2S_2O_3^{2-}$	0.017
$AgI(s) + e^- \rightleftharpoons Ag(s) + I^-$	-0.152
$Ag_2S(s) + H^+ + 2e^- \rightleftharpoons 2Ag(s) + SH^-$	-0.272

Sodium

$Na^+ + e^- \rightleftharpoons Na(s)$	-2.713

Reaction	E^0 (volts)

Strontium

$Sr^{2+} + 2e^- \rightleftharpoons Sr(s)$ — -2.886

Sulfur

$S_2O_8^{2-} + 2e^- \rightleftharpoons 2SO_4^{2-}$ — 2.01
$S_2O_6^{2-} + 4H^+ + 2e^- \rightleftharpoons 2H_2SO_3$ — 0.57
$2H_2SO_3 + 2H^+ + 4e^- \rightleftharpoons S_2O_3^{2-} + 3H_2O$ — 0.40
$S(s) + 2H^+ + 2e^- \rightleftharpoons H_2S(g)$ — 0.171
$S(s) + 2H^+ + 2e^- \rightleftharpoons H_2S(aq)$ — 0.141
$S_4O_6^{2-} + 2e^- \rightleftharpoons 2S_2O_3^{2-}$ — 0.08
$5S(s) + 2e^- \rightleftharpoons S_5^{2-}$ — -0.340
$2SO_3^{2-} + 3H_2O + 4e^- \rightleftharpoons S_2O_3^{2-} + 6OH^-$ — -0.58
$SO_3^{2-} + 3H_2O + 4e^- \rightleftharpoons S(s) + 6OH^-$ — -0.66
$SO_4^{2-} + H_2O + 2e^- \rightleftharpoons SO_3^{2-} + 2OH^-$ — -0.93
$2SO_3^{2-} + 2H_2O + 2e^- \rightleftharpoons S_2O_4^{2-} + 4OH^-$ — -1.12

Tantalum

$Ta_2O_5(s) + 10H^+ + 10e^- \rightleftharpoons 2Ta(s) + 5H_2O$ — -0.750

Technetium

$TcO_3(s) + 2H^+ + 2e^- \rightleftharpoons TcO_2(s) + H_2O$ — 0.8
$TcO_4^- + 2H_2O + 3e^- \rightleftharpoons TcO_2(s) + 4OH^-$ — -0.311

Tellurium

$Te(s) + 2e^- \rightleftharpoons Te^{2-}$ — -1.14

Terbium

$Tb^{3+} + 3e^- \rightleftharpoons Tb(s)$ — -2.39

Thallium

$Tl^{3+} + 2e^- \rightleftharpoons Tl^+$
$\begin{cases} 1.28 \\ 0.77 \ 1\text{F HCl} \\ 1.22 \ 1\text{F H}_2\text{SO}_4 \\ 1.23 \ 1\text{F HNO}_3 \\ 1.26 \ 1\text{F HClO}_4 \end{cases}$

$Tl^+ + e^- \rightleftharpoons Tl$ (in Hg, saturated) — -0.334
$Tl^+ + e^- \rightleftharpoons Tl(s)$ — -0.336
$TlCl(s) + e^- \rightleftharpoons Tl(s) + Cl^-$ — -0.557

Thorium

$Th^{4+} + 4e^- \rightleftharpoons Th(s)$ — -1.90

Thullium

$Tm^{3+} + 3e^- \rightleftharpoons Tm(s)$ — -2.28

Tin

$Sn(OH)_3^- + 3H^+ + 2e^- \rightleftharpoons Sn(s) + 3H_2O$ — 0.333

$Sn^{4+} + 2e^- \rightleftharpoons Sn^{2+}$ — $\begin{cases} 0.154 \\ 0.139 \ 1\text{F HCl} \end{cases}$

$Sn(OH)_4(s) + 4H^+ + 4e^- \rightleftharpoons Sn(s) + 4H_2O$ — -0.008
$Sn(OH)_2(s) + 2H^+ + 2e^- \rightleftharpoons Sn(s) + 2H_2O$ — -0.091
$SnF_6^{2-} + 4e^- \rightleftharpoons Sn(s) + 6F^-$ — -0.25
$Sn^{2+} + 2e^- \rightleftharpoons Sn(s)$ — -0.136
$Sn(OH)_4(s) + H^+ + 2e^- \rightleftharpoons Sn(OH)_3^- + H_2O$ — -0.349
$Sn(OH)_6^{2-} + 2e^- \rightleftharpoons Sn(OH)_3^- + 3OH^-$ — -0.93

Reaction	E^0 (volts)

Titanium
$$TiO^{2+} + 2H^+ + e^- \rightleftharpoons Ti^{3+} + H_2O \qquad 0.100$$
$$Ti^{3+} + e^- \rightleftharpoons Ti^{2+} \qquad -0.368$$
$$TiO_2(s, hydrated) + 4H^+ + 4e^- \rightleftharpoons Ti(s) + 2H_2O \qquad -0.86$$
$$TiF_6^{2-} + 4e^- \rightleftharpoons Ti(s) + 6F^- \qquad -1.191$$
$$Ti^{2+} + 2e^- \rightleftharpoons Ti(s) \qquad -1.630$$

Tungsten
$$W(CN)_8^{3-} + e^- \rightleftharpoons W(CN)_8^{4-} \qquad 0.457$$
$$W^{6+} + e^- \rightleftharpoons W^{5+} \qquad 0.26 \; 12\text{F HCl}$$
$$WO_3(s) + 6H^+ + 6e^- \rightleftharpoons W(s) + 3H_2O \qquad -0.09$$
$$W^{5+} + e^- \rightleftharpoons W^{4+} \qquad -0.3 \; 12\text{F HCl}$$
$$WO_4^{2-} + 4H_2O + 6e^- \rightleftharpoons W(s) + 8OH^- \qquad -1.05$$

Uranium
$$UO_2^+ + 4H^+ + e^- \rightleftharpoons U^{4+} + 2H_2O \qquad 0.55$$
$$UO_2^{2+} + 4H^+ + 2e^- \rightleftharpoons U^{4+} + 2H_2O \qquad 0.334$$
$$UO_2^{2+} + e^- \rightleftharpoons UO_2^+ \qquad 0.052$$
$$U^{4+} + e^- \rightleftharpoons U^{3+} \qquad -0.61$$
$$U^{3+} + 3e^- \rightleftharpoons U(s) \qquad -1.80$$

Vanadium
$$VO_2^+ + 2H^+ + e^- \rightleftharpoons VO^{2+} + H_2O \qquad 1.000$$
$$VO^{2+} + 2H^+ + e^- \rightleftharpoons V^{3+} + H_2O \qquad 0.337$$
$$V^{3+} + e^- \rightleftharpoons V^{2+} \qquad -0.255$$
$$V^{2+} + 2e^- \rightleftharpoons V(s) \qquad -1.18$$

Xenon
$$H_4XeO_6 + 2H^+ + 2e^- \rightleftharpoons XeO_3 + 3H_2O \qquad 3.0$$
$$XeF_2 + 2H^+ + 2e^- \rightleftharpoons Xe(s) + 2HF \qquad 2.2$$
$$XeO_3 + 6H^+ + 6e^- \rightleftharpoons Xe(g) + 3H_2O \qquad 1.8$$

Ytterbium
$$Yb^{3+} + 3e^- \rightleftharpoons Yb(s) \qquad -2.27$$

Yttrium
$$Y^{3+} + 3e^- \rightleftharpoons Y(s) \qquad -2.37$$

Zinc
$$ZnO(s) + 2H^+ + 2e^- \rightleftharpoons Zn(s) + H_2O \qquad -0.439$$
$$Zn^{2+} + 2e^- \rightleftharpoons Zn(s) \qquad -0.764$$
$$Zn(NH_3)_4^{2+} + 2e^- \rightleftharpoons Zn(s) + 4NH_3 \qquad -1.04$$
$$ZnCO_3(s) + 2e^- \rightleftharpoons Zn(s) + CO_3^{2-} \qquad -1.06$$
$$Zn(OH)_3^- + 2e^- \rightleftharpoons Zn(s) + 3OH^- \qquad -1.183$$
$$Zn(OH)_4^{2-} + 2e^- \rightleftharpoons Zn(s) + 4OH^- \qquad -1.214$$
$$Zn(OH)_2(s) + 2e^- \rightleftharpoons Zn(s) + 2OH^- \qquad -1.245$$
$$ZnS(s) + 2e^- \rightleftharpoons Zn(s) + S^{2-} \qquad -1.405$$

Zirconium
$$ZrO_2(s) + 4H^+ + 4e^- \rightleftharpoons Zr(s) + 2H_2O \qquad -1.456$$
$$Zr^{4+} + 4e^- \rightleftharpoons Zr(s) \qquad -1.539$$

Glossary

Absolute error An expression of the margin of uncertainty associated with a measurement. It could also refer to the difference between a measured value and the "true" value.

Absorbance, A, or **Optical Density, OD** Defined as $A = \log(P_0/P)$, where P_0 is the radiant power of light striking the sample on one side and P is the radiant power emerging from the other side.

Absorption Occurs when a substance is taken up *inside* another. See also **Adsorption.**

Absorption Spectrum A graph of absorbance or transmittance of light versus wavelength, frequency, or wavenumber.

Accuracy A measure of how close a measured value is to the "true" value.

Acid A substance that increases the concentration of H^+ when added to water.

Acid–Base Titration One in which the reaction between analyte and titrant is an acid–base reaction.

Acid Dissociation Constant, K_A The equilibrium constant for the reaction of an acid, HA, with H_2O:

$$HA + H_2O \rightleftharpoons A^- + H_3O^+ \qquad K_A = \frac{\mathscr{A}_{A^-}\mathscr{A}_{H_3O^+}}{\mathscr{A}_{HA}}$$

Acid Error Occurs in strongly acidic solutions, where glass electrodes tend to indicate a value of pH that is too high.

Activity, $\mathscr{A}$ The value that replaces concentration in a thermodynamically correct equilibrium expression. The activity of X is given by $\mathscr{A}_X = [X]\gamma_X$, where γ_X is the activity coefficient and $[X]$ is the concentration.

Activity Coefficient, γ The number by which the concentration must be multiplied to give activity.

Adjusted Retention Time, t_r' In chromatography, this is given by $t_r' = t_r - t_m$, where t_r is the retention time of a solute and t_m is the time needed for mobile phase to travel the length of the column.

Adsorption Occurs when a substance becomes attached to the *surface* of another substance. See also **Absorption.**

Adsorption Chromatography A technique in which the solute equilibrates between the mobile phase and adsorption sites on the stationary phase.

Adsorption Indicator Used for precipitation titrations, it becomes attached to a precipitate and changes color when the surface charge of the precipitate changes sign at the equivalence point.

Affinity Chromatography A technique in which a particular solute is retained by a column by virtue of a specific interaction with a molecule covalently bound to the stationary phase.

Aliquot Portion.

Alkalimetric Titration With reference to EDTA titrations, this involves titration of the protons liberated from EDTA upon binding to a metal.

Alkaline Error Occurs when a glass pH electrode is placed in a strongly basic solution containing very little H^+ and a high concentration of Na^+. The electrode begins to respond to Na^+ as if it were H^+, so that the pH reading is lower than the actual pH.

Amalgam A solution of anything in mercury.

Amine A compound with the general formula RNH_2, R_2NH, or R_3N, where R is any group of atoms.

Amino Acid One of twenty building blocks of proteins, having the general structure

$$\overset{\displaystyle R}{\underset{\displaystyle ^+H_3NCHCO_2^-}{|}}$$

where R is a different substituent for each acid.

Ammonium Ion *The* ammonium ion is NH_4^+. *An* ammonium ion is any ion of the type RNH_3^+, $R_2NH_2^+$, R_3NH^+, or R_4N^+.

Ampere, A One ampere is the current that will produce a force of exactly 2×10^{-7} N/m when that current flows through two "infinitely" long, parallel conductors

GLOSSARY

of negligible cross section, with a spacing of 1 m, in a vacuum.

Amperometric Titration One in which the endpoint is determined by monitoring the current passing between two electrodes immersed in the sample solution and maintained at a constant potential difference.

Amperometry The measurement of electric current for analytical purposes.

Amphiprotic Molecule One that can act as both a proton donor and a proton acceptor. The intermediate species of polyprotic acids are amphiprotic.

Analyte The substance being analyzed.

Analytical Concentration See **Formal Concentration.**

Anion Exchanger An ion exchanger with positively charged groups covalently attached to the support. It can reversibly bind anions.

Anode The electrode at which oxidation occurs.

Anodic Depolarizer A molecule that is easily oxidized, thereby preventing the anode potential of an electrochemical cell from becoming too large.

Anodic Wave In polarography, a flow of current due to oxidation of analyte.

Anolyte The solution present in the anode chamber of an electrochemical cell.

Antilogarithm The antilogarithm of a is b if $10^a = b$.

Aquo Ion The species $M(H_2O)_n^{m+}$, containing just the cation M and its tightly bound water ligands.

Argentometric Titration One using Ag^+ ion.

Ashless Filter Paper Specially treated paper that leaves a negligible residue after ignition. It is used for gravimetric analysis.

Asymmetry Potential When the activity of analyte is the same on the inside and outside of an ion-selective electrode, there should be no potential across the membrane. In fact, the two surfaces are never identical, and some potential (called the asymmetry potential) is usually observed. The asymmetry potential changes with time and leads to electrode drift.

Atmosphere, atm One atm is defined as a pressure of $101\ 325\ N/m^2$. It is also equal to the pressure exerted by a column of Hg 760 mm in height at the earth's surface.

Atomic Absorption Spectroscopy A technique in which the absorption of light by free gaseous atoms in a flame or furnace is used to measure the concentration of atoms.

Atomic Emission Spectroscopy A technique in which the emission of light by thermally excited atoms in a flame or furnace is used to measure the concentration of atoms.

Atomic Fluorescence Spectroscopy A technique in which atomic electronic transitions are excited by light, and fluorescence is observed at a right angle to the incident beam.

Atomic Weight The number of grams of an element containing Avogadro's number of atoms.

Atomization The process in which a compound is decomposed into its atoms at high temperature.

Autoprotolysis The reaction of a neutral solvent, in which two of the same molecules transfer a proton between each other; e.g., $CH_3OH + CH_3OH \rightleftharpoons CH_3OH_2^+ + CH_3O^-$.

Auxiliary Complexing Agent A species, such as ammonia, that is added to a solution to stabilize another species and keep that other species in solution. It binds loosely enough to be displaced by a titrant.

Auxiliary Electrode The current-carrying partner of the working electrode in an electrolysis.

Azeotrope The distillate produced by two liquids. It is of constant composition, containing both substances.

Back Titration One in which an excess of reagent is added to react with analyte. Then the excess reagent is titrated with a second reagent or with a standard solution of analyte.

Bandwidth Usually, the range of wavelengths or frequencies of an absorption or emission band at a height equal to half of the peak height. It also refers to the width of radiation emerging from the exit slit of a monochromator.

Base A substance that decreases the concentration of H^+ when added to water.

Base "Dissociation" Constant, K_B The equilibrium constant for the reaction of a base, B, with H_2O:

$$B + H_2O \rightleftharpoons BH^+ + OH^- \qquad K_B = \frac{[BH^+][OH^-]}{[B]}$$

Beam Chopper A rotating mirror that directs light alternately through the sample and reference cells of a double-beam spectrophotometer. In atomic absorption, periodic blocking of the beam allows a distinction to be made between light from the source and light from the flame.

Beer's Law Relates the absorbance (A) of a sample to its concentration (c), pathlength (b), and molar absorptivity (ε): $A = \varepsilon bc$.

Biamperometric Titration An amperometric titration conducted with two polarizable electrodes held at a constant potential difference.

Bipotentiometric Titration A potentiometric titration in which a constant current is passed between two polarizable electrodes immersed in the sample solution.

An abrupt change in potential characterizes the endpoint

Blank Titration One in which a solution containing all of the reagents except analyte is titrated. The volume of titrant needed in the blank titration should be subtracted from the volume needed to titrate an unknown.

Blocking Occurs when metal ion binds tightly to a metal ion indicator. A blocked indicator is unsuitable for a titration because no color change is observed at the endpoint.

Boltzmann Distribution The relative population of two states at thermal equilibrium:

$$\frac{N_2}{N_1} = \frac{g_2}{g_1} e^{-(E_2 - E_1)/kT}$$

where N_i is the population of the state, g_i is the degeneracy of the state, E_i is the energy of the state, k is Boltzmann's constant, and T is kelvins; degeneracy refers to the number of states with the same energy.

Bonded Phase In HPLC, a stationary liquid phase covalently attached to the solid support.

Brønsted Acid A proton donor.

Brønsted Base A proton acceptor.

Buffer A mixture of an acid and its conjugate base. A buffered solution is one that resists changes in pH when acids or bases are added.

Buffer Capacity or **Buffer Intensity** A measure of the ability of a buffer to resist changes in pH. The larger the buffer capacity, the greater the resistance to pH change. The definition of buffer capacity (β) is $\beta = dC_B/dpH = -dC_A/dpH$, where C_A and C_B are the number of moles of strong acid or base per liter needed to produce a unit change in pH.

Buoyancy Occurs when an object is weighed in air and the observed mass is less than the true mass because the object has displaced an equal volume of air from the balance pan.

Buret A calibrated glass tube with a stopcock at the bottom. Used to deliver known volumes of liquid.

Calibration Curve A graph showing the value of some property versus concentration of analyte. When the corresponding property of an unknown is measured, its concentration can be determined from the graph.

Calomel Electrode A common reference electrode based on the half-reaction $Hg_2Cl_2(s) + 2e^- \rightleftharpoons 2Hg(l) + 2Cl^-$.

Candela, cd The basic SI unit of luminous intensity. It is the intensity emitted by $1/60\ cm^2$ of Pt at 2045 K.

Capacity Factor, k′ In chromatography, the adjusted retention time of a peak divided by the time for mobile phase to travel through the column.

Capillary Constant The quantity $m^{2/3}t^{1/6}$ characteristic of each dropping Hg electrode. The rate of flow is m (mg s^{-1}) and t is the drop interval (s). The capillary constant is proportional to the square root of the Hg height.

Carboxylate Anion The conjugate base (RCO_2^-) of a carboxylic acid.

Carboxylic Acid A molecule with the general structure RCO_2H, where R is any group of atoms.

Catalytic Wave One that results when the product of a polarographic reaction is rapidly regenerated by reaction with another species and the polarographic wave height increases.

Cathode The electrode at which reduction occurs.

Cathodic Depolarizer A molecule that is easily reduced, thereby preventing the cathode potential of an electrochemical cell from becoming very low.

Catholyte The solution present in the cathode chamber of an electrochemical cell.

Cation Exchanger An ion exchanger with negatively charged groups covalently attached to the support. It can reversibly bind cations.

Character The part to the left of the decimal point in a logarithm.

Charge Balance A statement that the sum of all positive charge in solution equals the magnitude of the sum of all negative charge in solution.

Charge Effect With respect to the strength of acids and bases, the repulsion between H^+ and a positive charge in the same molecule. It could also refer to the attraction between H^+ and a negative charge. A positive charge within a molecule increases the acidity, and a negative charge decreases the acidity.

Charring In a gravimetric analysis, the precipitate (and filter paper) are first dried gently. Then the filter paper is *charred* at intermediate temperature to destroy the paper without letting it inflame. Finally, the precipitate is ignited at high temperature to convert it to its analytical form.

Chelate Effect The observation that a single multidentate ligand forms metal complexes that are more stable than those formed by several individual ligands with the same ligand atoms.

Chelator A ligand that binds to a metal through more than one atom.

Chemical Coulometer A device that measures the yield of an electrolysis reaction in order to determine how much electricity has flowed through a circuit.

Chemical Interference In atomic spectroscopy, any

GLOSSARY

chemical reaction that decreases the efficiency of atomization.

Chromatogram A graph showing the concentration of solutes emerging from a chromatography column as a function of elution time or volume.

Chromatograph A machine used to perform chromatography.

Chromatography A technique in which molecules in a mobile phase are separated because of their different affinities for a stationary phase. The greater the affinity for the stationary phase, the longer the molecule is retained.

Chromophore The part of a molecule responsible for absorption of light of a particular frequency.

Chronoamperometry A technique in which the potential applied between a pair of electrodes in an unstirred solution is varied rapidly while the current is measured. Suppose that the analyte is reducible and that the potential is varied from positive to negative. Initially, no reduction occurs. At a certain potential, the analyte begins to be reduced and the current increases. As the potential becomes more negative, the current increases further until the concentration of analyte at the surface of the electrode is sufficiently depleted. Then the current decreases, even though the potential is still being raised. The maximum current is proportional to the concentration of analyte in bulk solution.

Chronopotentiometry A technique in which a constant current is forced to flow between two electrodes. The potential remains fairly steady until the concentration of an electroactive species becomes depleted. Then the potential changes rapidly as a new redox reaction assumes the burden of current flow. The elapsed time when the potential suddenly changes is proportional to the concentration of the initial electroactive species in bulk solution.

Clark Electrode One that measures the activity of dissolved oxygen by amperometry.

Coagulation With respect to gravimetric analysis, small crystallites coming together to form larger crystals.

Cochromatography Simultaneous chromatography of known compounds with an unknown. If a known and an unknown have the same retention time on several columns, they are probably identical.

Colloid A dissolved particle with a diameter in the approximate range 1–100 nm. It is too large to be considered one molecule but too small to simply precipitate.

Combination Electrode Consists of a glass electrode with a concentric reference electrode built on the same body.

Combustion Analysis A technique in which a sample is heated in an atmosphere of O_2 to oxidize it to CO_2 and H_2O, which are collected and weighed. Modifications permit the simultaneous analysis of N, S, and halogens.

Common Ion Effect Occurs when a salt is dissolved in a solution already containing one of the ions of the salt. The salt is less soluble than it would be in a solution without that ion. An application of Le Châtelier's principle.

Complex Ion Historical name for any ion containing two or more ions or molecules that are each stable by themselves; e.g., $CuCl_2^-$ contains $Cu^+ + 2Cl^-$.

Complexometric Titration One in which the reaction between analyte and titrant involves complex formation.

Compound Electrode An ion-selective electrode consisting of a conventional electrode surrounded by a barrier that is selectively permeable to the analyte of interest. Alternatively, the barrier region might convert external analyte into a different species, to which the inner electrode is sensitive.

Concentration An expression of the quantity per unit volume or unit mass of a substance. Common measures of concentration are molarity (mol/L) and molality (mol/kg of solvent).

Concentration Cell A galvanic cell in which both half-reactions are the same, but the concentrations in each half-cell are not identical. The cell reaction increases the concentration of species in one half-cell and decreases the concentration in the other.

Concentration Polarization Occurs when an electrode reaction occurs so rapidly that the concentration of solute near the surface of the electrode is not the same as the concentration in bulk solution.

Conditional Formation Constant See **Effective Formation Constant.**

Confidence Interval The range of values within which there is a specified probability that the true value will occur.

Conjugate Acid–Base Pair An acid and a base that differ only through the gain or loss of a single proton.

Constant Mass In gravimetric analysis, the product is heated and cooled to room temperature in a desiccator until successive weighings are "constant." There is not a standard definition of "constant mass"; but for ordinary work, it is usually taken as about ± 0.3 mg. Constancy is usually limited by the irreproducible regain of moisture picked up by the sample during cooling in the desiccator and during weighing.

Controlled Potential Electrolysis A technique for selective reduction (or oxidation), in which the potential of the working electrode is held at a selected value with respect to a reference electrode.

GLOSSARY

Coprecipitation Occurs when a substance whose solubility is not exceeded precipitates along with one whose solubility is exceeded.

Coulomb The amount of charge per second that flows past any point in a circuit when the current is one ampere. There are 96 484.56 coulombs in a mole of electrons.

Coulometric Titration One conducted with a constant current for a measured time.

Coulometry A technique in which the quantity of analyte is determined by measuring the number of coulombs needed for complete electrolysis.

Countercurrent Distribution A technique in which a series of solvent extractions is used to separate solutes from each other.

Counterion Most ionic substances contain two kinds of ions. Often one is of particular interest to us, and the other is said to be the counterion. It is necessary for electrical neutrality, but its nature is unimportant for the purpose at hand.

Crosslinking The covalent linkage between different strands of a polymer.

Crystallization Occurs when a substance comes out of solution slowly to form a solid with a regular arrangement of atoms.

Current, I Tells how much charge flows through a circuit per unit time.

Cuvette A cell used to hold samples for spectrophotometric measurements.

Cyclic Voltammetry A polarographic technique in which a triangular wave form is applied with a period of a few seconds. Both cathodic and anodic currents are observed for reversible reactions.

Dead Stop Endpoint The endpoint of a biamperometric titration.

Debye–Hückel Equation Gives the activity coefficient (γ) as a function of ionic strength (μ). The extended Debye–Hückel equation, applicable to ionic strengths up to about 0.1 M, is $\log \gamma = [-0.51 \, z^2 \sqrt{\mu}]/[1 + (\alpha \sqrt{\mu}/305)]$, where z is the ionic charge and α is the effective hydrated radius in pm.

Decant To pour liquid off a solid or, perhaps, a denser liquid. The denser phase is left behind.

Decomposition Potential In an electrolysis, that voltage at which rapid reaction first begins.

Deionized Water Water that has been passed through a cation exchanger (in the H^+ form) and an anion exchanger (in the OH^- form) to remove ions from the solution.

Deliquescent Substance Like a hygroscopic substance, one that spontaneously picks up water from the air. It can eventually absorb so much water that the substance completely dissolves.

Demasking The removal of a masking agent from the species protected by the masking agent.

Density The mass per unit volume of a substance.

Depolarizer A molecule that is oxidized or reduced at a modest potential: It is added to an electrolytic cell to prevent the cathode or anode potential from becoming too extreme.

Desalting The removal of salts (or any small molecules) from a solution of macromolecules. Gel filtration is conveniently used for desalting.

Desiccant A drying agent.

Desiccator A sealed chamber in which samples can be dried in the presence of a desiccant and/or vacuum pumping.

Detection Limit That concentration of an element which gives a signal equal to twice the peak-to-peak noise level of the baseline.

Determinant The value of the two-dimensional determinant $\begin{vmatrix} a & b \\ c & d \end{vmatrix}$ is the difference $ad - bc$.

Determinate Error See Systematic Error.

Dialysis A technique in which solutions are placed on either side of a semipermeable membrane that allows small molecules, but not large molecules, to cross. The small molecules in the two solutions diffuse across and equilibrate with each other. The large molecules are retained on their original side.

Dielectric Constant The electrostatic force, F, between two charged particles, given by $F = kq_1q_2/\varepsilon r^2$, where k is a constant, q_1 and q_2 are the charges, r is the separation between particles, and ε is the dielectric constant of the medium. The higher the dielectric constant, the less force is exerted by one charged particle on another.

Differential Pulse Polarography A technique in which current is measured before and at the end of pulses of potential superimposed on the ordinary wave form. It is more sensitive than ordinary polarography, and the signal closely approximates the derivative of a polarographic wave.

Diffraction Occurs when electromagnetic radiation passes through slits with a spacing comparable to the wavelength. Interference of waves from adjacent slits produces a spectrum of radiation, with each wavelength at a different angle.

Diffusion Coefficient, D Defined by Fick's first law of diffusion: $J = -D(dc/dx)$, where J is the rate at which molecules diffuse across a plane of unit area

GLOSSARY

and dc/dx is the concentration gradient in the direction of diffusion.

Diffusion Current In polarography, the current observed when the rate of reaction is limited by the rate of diffusion of analyte to the electrode. See also **Limiting Current.**

Digestion The process in which a precipitate is left (usually warm) in the presence of mother liquor to promote particle recrystallization and growth. Purer, more easily filterable crystals result. Also used to describe any chemical treatment in which a substance is decomposed to transform the analyte into a form suitable for analysis.

Dimer A molecule made from two identical units.

Diprotic Acid One that can donate two protons.

Direct Current Polarography The classical form of polarography, in which a linear voltage ramp is applied to the working electrode.

Direct Titration One in which the analyte is treated with titrant, and the volume required for complete reaction is measured.

Displacement Titration An EDTA titration procedure in which analyte is treated with excess $MgEDTA^{2-}$ to displace Mg^{2+}: $M^{n+} + MgEDTA^{2-} \rightleftharpoons MEDTA^{n-4} + Mg^{2+}$. The liberated Mg^{2+} is then titrated with EDTA. This procedure is useful if there is not a suitable indicator for direct titration of M^{n+}.

Disproportionation A reaction in which an element in one oxidation state gives products containing that element in higher and lower oxidation states; e.g., $Cu^+ \rightleftharpoons Cu^{2+} + Cu(s)$.

Distribution Coefficient Describes the distribution of a solute partitioned between two phases. The distribution coefficient is defined as the total concentration of all forms of solute in phase 2 divided by the total concentration in phase 1.

Donnan Equilibrium The phenomenon that ions of the same charge as those fixed on an exchange resin are repelled from the resin. Thus, anions do not readily penetrate a cation-exchange resin, and cations are repelled from an anion-exchange resin.

Dopant When small amounts of substance B are added to substance A, we call B a dopant and say that A is doped with B. Doping is done to alter the properties of A.

Doppler Effect The phenomenon that a molecule moving toward a source of radiation experiences a higher frequency than one moving away from the source.

Dropping-Mercury Electrode One that delivers fresh drops of Hg to a polarographic cell.

E^0 The standard reduction potential.

$E^{0\prime}$ The effective standard reduction potential at pH 7 (or at some other specified conditions).

Eddy Diffusion One factor that broadens the zone of solute in chromatography is the random nature of paths followed by solute molecules traveling through the column. In addition to following paths of different length, some solute molecules become trapped in little pools (eddies) and must diffuse out into the mainstream of eluent to continue passing through the column.

EDTA (Ethylenediaminetetraacetic Acid) $(HO_2CCH_2)_2$-$NCH_2CH_2N(CH_2CO_2H)_2$—the most widely used reagent for complexometric titrations. It forms 1:1 complexes with virtually all cations with a charge of 2 or more.

Effective Formation Constant or **Conditional Formation Constant** The equilibrium constant for formation of a complex under a particular stated set of conditions, such as ionic strength and concentration of auxiliary complexing species.

Effluent See **Eluate.**

Einstein A mole of photons.

Electric Discharge Emission Spectroscopy A technique in which atomization and excitation are stimulated by an electric arc, a spark, or a microwave discharge.

Electric Double Layer The region comprising the charged surface of a particle plus the oppositely charged ionic atmosphere immediately surrounding the particle in solution.

Electric Potential The potential difference (in volts) between two points is the energy (in joules) needed to transport one coulomb of positive charge from the negative point to the positive point.

Electroactive Species Any species that can be oxidized or reduced at an electrode.

Electrocapillary Maximum The potential at which the net charge on a mercury drop from a dropping mercury electrode is zero (and the surface tension of the drop is maximal).

Electrode A device at which or through which electrons flow into or out of chemical species involved in a redox reaction.

Electrogravimetric Analysis A technique in which the mass of an electrolytic deposit is used to quantify the analyte.

Electrolysis The process in which the passage of electric current causes a chemical reaction to occur.

Electrolyte A substance that produces ions when dissolved.

Electromagnetic Spectrum The spectrum of all electro-

GLOSSARY

magnetic radiation (visible light, radio waves, x-rays, etc.).

Electronic Transition One in which an electron is promoted from one energy level to another.

Eluate or **Effluent** What comes out of a chromatography column.

Eluent The solvent applied to the beginning of a chromatography column.

Eluent Strength A measure of the absorption energy of a solvent on the stationary phase in chromatography. The greater the eluent strength, the more rapidly will the solvent elute solutes from the column.

Elution The process of passing a liquid or a gas through a chromatography column.

Elutropic Series Ranks solvents according to their ability to displace solutes from the stationary phase in adsorption chromatography.

Emission Spectrum A graph of luminescence intensity versus luminescence wavelength (or frequency or wavenumber), using a fixed excitation wavelength.

Endergonic Reaction One for which ΔG is positive; it is not spontaneous.

Endothermic Reaction One for which ΔH is positive; heat must be supplied to reactants for them to react.

Endpoint The point in a titration at which there is a sudden change in a physical property, such as indicator color, pH, conductivity, or absorbance. Used as a measure of the equivalence point.

Enthalpy Change, ΔH The heat absorbed when the reaction occurs at constant pressure.

Enthalpy of Hydration The heat liberated when a gaseous species is transferred to water.

Entropy A measure of the "disorder" of a substance.

Equilibrium The state in which the forward and reverse rates of all reactions are equal, so the concentrations of all species remain constant.

Equilibrium Constant, K For the reaction $a\text{A} + b\text{B} \rightleftharpoons c\text{C} + d\text{D}$, $\text{K} = \mathscr{A}_\text{C}^\text{c} \mathscr{A}_\text{D}^\text{d} / \mathscr{A}_\text{A}^\text{a} \mathscr{A}_\text{B}^\text{b}$, where $\mathscr{A}_i$ is the activity of the ith species.

Equimolar Mixture of Compounds One that contains an equal number of moles of each compound.

Equivalence Point The point in a titration at which the quantity of titrant is exactly sufficient for stoichiometric reaction with the analyte.

Equivalent For a redox reaction, the amount of reagent that can donate or accept one mole of electrons. For an acid–base reaction, the amount of reagent that can donate or accept one mole of protons.

Equivalent Weight The mass of substance containing one equivalent.

Excitation Spectrum A graph of luminescence (measured at a fixed wavelength) versus excitation frequency or wavelength. It closely corresponds to an absorption spectrum because the luminescence is generally proportional to the absorbance.

Excited State Any state of an atom or a molecule having more than the minimum possible energy.

Exergonic Reaction One for which ΔG is negative; it is spontaneous.

Exothermic Reaction One for which ΔH is negative; heat is liberated when products are formed.

Extensive Property A property of a system or chemical reaction, such as entropy, that depends on the amount of matter in the system; e.g., ΔG, which is twice as large if two moles of product are formed than if one mole is formed. See also **Intensive Property.**

Extinction Coefficient, ε See **Molar Absorptivity.**

Extraction The process in which a solute is allowed to equilibrate between two phases, usually for the purpose of separating solutes from each other.

Fajans Titration A precipitation titration in which the endpoint is signaled by adsorption of a colored indicator on the precipitate.

Faradaic Current That component of current in an electrochemical cell due to oxidation and reduction reactions.

Faraday Constant $9.648\,456 \times 10^4$ C/mol of charge.

Faraday's Laws These two laws state that the extent of an electrochemical reaction is directly proportional to the quantity of electricity that has passed through the cell. The mass of substance that reacts is proportional to its formula weight and inversely proportional to the number of electrons required in its half-reaction.

Filtrate The liquid that passes through a filter.

Fines The smallest particles of stationary phase used for chromatography. It is desirable to remove the fines before packing a column because they retard solvent flow.

Fischer Titration See **Kart Fischer Titration.**

Flame Ionization Detector A gas chromatography detector in which solute is burned in a H_2–O_2 flame to produce CHO^+ ions. The current carried through the flame by these ions is proportional to the concentration of susceptible species in the eluate.

Flame Photometer A device that uses flame atomic emission and a filter photometer to quantify Li, Na, K, and Ca in liquid samples. It is widely used in clinical laboratories.

Flow Adaptor An adjustable plungerlike device that may be used on either side of a chromatographic bed to support the bed and to minimize the dead space through which liquid can flow outside of the column bed.

Fluorescence The process in which a molecule emits a photon shortly (10^{-9}–10^{-4} s) after absorbing a photon. It results from a transition between states of the same spin multiplicity.

Flux An agent used as the medium for a fusion.

Formal Concentration or **Analytical Concentration** The molarity of a substance if it did not change its chemical form upon being dissolved. It represents the total number of moles of substance dissolved in a liter of solution, regardless of any reactions that take place when the solute is dissolved.

Formal Potential The potential of a half-reaction (relative to a standard hydrogen electrode) when the formal concentrations of reactants and products are unity. Any other conditions (such as pH, ionic strength, and concentrations of ligands) must also be specified.

Formation Constant or **Stability Constant** The equilibrium constant for the reaction of a metal with its ligands to form a metal–ligand complex.

Formula Weight The mass containing one mole of the indicated chemical formula of a substance. For example, the formula weight of $CuSO_4 \cdot 5H_2O$ is the sum of the masses of copper, sulfate, and five water molecules.

Fraction of Association, α For the reaction of a base (B) with H_2O, the fraction of base in the form BH^+.

Fraction of Dissociation, α For the dissociation of an acid (HA), the fraction of acid in the form A^-.

Frequency The number of oscillations of a wave per second.

Fugacity The activity of a gas. The activity coefficient for a gas is called the **Fugacity Coefficient.**

Fusion The process in which an otherwise insoluble substance is dissolved in a molten salt such as Na_2CO_3, Na_2O_2, or KOH. Once the substance has dissolved, the melt is cooled, dissolved in aqueous solution, and analyzed.

Galvanic Cell One that produces electricity by means of a spontaneous chemical reaction.

Gathering A process in which a trace constituent of a solution is intentionally coprecipitated with a major constituent.

Gaussian Curve or **Normal Error Curve** This function describes the theoretical bell-shaped distribution of measurements when all error is random. The center of the curve is the mean, and the width is characterized by the standard deviation.

Gel Chromatographic stationary-phase particles, such as Sephadex or polyacrylamide, which are soft and pliable.

Gel Filtration or **Gel-Permeation Chromatography** See **Molecular Exclusion Chromatography.**

Geometric Mean For a series of n measurements with the values x_i, $[\Pi_i(x_i)]^{1/n}$.

Gibbs Free Energy The change in Gibbs free energy (ΔG) for any process at constant temperature is related to the change in enthalpy (ΔH) and entropy (ΔS) by the equation $\Delta G = \Delta H - T\Delta S$, where T is kelvins. A process is spontaneous (thermodynamically favorable) if ΔG is negative.

Glass Electrode One that has a thin glass membrane across which a pH-dependent electric potential develops. The potential (and hence pH) is measured by a pair of reference electrodes on either side of the membrane.

Gooch Crucible A short, cup-shaped container with holes at the bottom, used for filtration and ignition of precipitates. For ignition, the crucible is made of porcelain or platinum and lined with a mat of purified asbestos to retain the precipitate. For precipitates that do not need ignition, the crucible is made of glass and has a porous glass disk instead of holes at the bottom.

Gradient Elution Chromatography in which the composition of the mobile phase is progressively changed to increase the eluent strength of the solvent.

Graduated Cylinder or **Graduate** A tube with volume calibrations along its length.

Gram-Atom The amount of an element containing Avogadro's number of atoms; it is the same as a mole of the element.

Gran Plot A graph such as ($V_b 10^{-pH}$ versus volume) used to find the endpoint of a titration.

Graphite Furnace A hollow graphite rod that can be heated electrically to about 2 500 K to decompose and atomize a sample for atomic spectroscopy.

Grating Either a reflective or a transmitting surface etched with closely spaced lines; used to disperse light into its component wavelengths.

Gravimetric Analysis Any analytical method that relies on measuring the mass of a substance (such as a precipitate) to complete the analysis.

Ground State The state of an atom or a molecule with the minimum possible energy.

Half-Reaction Any redox reaction can be conceptually broken into two half-reactions, one involving only oxidation and one involving only reduction.

Half-Wave Potential The potential at the midpoint of the rise in the current of a polarographic wave.

Hammett Acidity Function, H_0 Used to measure the acidity of nonaqueous solutions or concentrated aqueous solutions of strong acids.

Hanging-Drop Electrode One with a stationary drop of Hg that is used for stripping analysis.

Henderson–Hasselbalch Equation A logarithmic rearranged form of the acid dissociation equilibrium equation:

$$pH = pK_A + \log \frac{[A^-]}{[HA]}$$

Hertz, Hz The unit of frequency, s^{-1}.

HETP (Height Equivalent to a Theoretical Plate) The length of a chromatography column divided by the number of theoretical plates in the column.

Hexadentate Ligand One that binds to a metal atom through six ligand atoms.

Hollow-Cathode Lamp One that emits sharp atomic lines characteristic of the element from which the cathode is made.

Homogeneous Precipitation A technique in which a precipitating agent is generated slowly by a reaction in homogeneous solution, effecting a slow crystallization instead of a rapid precipitation of product.

HPLC (High Performance Liquid Chromatography) A chromatographic technique using very small stationary-phase particles and high pressure to force solvent through the column.

Hydrated Radius The effective size of an ion or a molecule plus its associated water molecules in solution.

Hydrolysis "Reaction with water." The reaction $B + H_2O \rightleftharpoons BH^+ + OH^-$ is often called hydrolysis of a base.

Hydronium Ion, H_3O^+ What we really mean when we write $H^+(aq)$.

Hydrophilic Substance One that is soluble in water or attracts water to its surface.

Hydrophobic Substance One that is insoluble in water or repels water from its surface.

Hygroscopic Substance One that readily picks up water from the atmosphere.

Ignition The heating to high temperature of some gravimetric precipitates to convert them to a known, constant composition that can be weighed.

Inclusion An impurity that occupies random sites in a crystal lattice.

Indeterminate Error See **Random Error.**

Indicator A compound having a physical property (usually color) that changes abruptly near the equivalence point of a chemical reaction.

Indicator Electrode One that develops a potential whose magnitude depends on the activity of one or more species in contact with the electrode.

Indicator Error The difference between the indicator endpoint of a titration and the true equivalence point.

Indirect Titration One that is used when the analyte cannot be directly titrated. For example, analyte A may be precipitated with excess reagent R. The product is filtered, and the excess R washed away. Then AR is dissolved in a new solution, and R can be titrated.

Inductive Effect The attraction of electrons by an electronegative element through the sigma-bonding framework of a molecule.

Inductively Coupled Plasma A high-temperature plasma that derives its energy from an oscillating radio-frequency field. It is used to atomize a sample for atomic emission spectroscopy.

Inflection Point One at which the derivative of the slope is zero: $d^2y/dx^2 = 0$. That is, the slope reaches a maximum or minimum value.

Intensity See **Radiant Power.**

Intensive Property A property of a system or chemical reaction that does not depend on the amount of matter in the system; e.g., temperature and electric potential. See also **Extensive Property.**

Intercept For a straight line whose equation is $y = mx + b$, the value of b is the intercept. It is the value of y when $x = 0$.

Interference The effect when the presence of one substance changes the signal in the analysis of another substance.

Internal Conversion A radiationless isoenergetic electronic transition between states of the same electron-spin multiplicity.

Internal Standard A known quantity of a compound added to a solution containing an unknown quantity of analyte. The concentration of analyte is then measured relative to that of the internal standard.

Interpolation The estimation of the value of a quantity that lies between two known values.

Intersystem Crossing A radiationless isoenergetic electronic transition between states of different electron-spin multiplicity.

Iodimetry The use of triiodide (or iodine) as a titrant.

Iodometry A technique in which an oxidant is treated with I^- to produce I_3^-, which is then titrated (usually with thiosulfate).

Ion-Exchange Chromatography A technique in which

GLOSSARY

solute ions are retained by oppositely charged sites in the stationary phase.

Ion-Exclusion Chromatography A technique in which electrolytes are separated from nonelectrolytes by means of an ion-exchange resin.

Ionic Atmosphere The region of solution around an ion or a charged particle. It contains an excess of oppositely charged ions.

Ionic Radius The effective size of an ion in a crystal.

Ionic Strength, μ Given by $\mu = \frac{1}{2}\sum_i c_i z_i^2$, where c_i is the concentration of the ith ion in solution and z_i is the charge on that ion. The sum extends over all ions in solution, including the ions whose activity coefficients are being calculated.

Ionization Interference In atomic spectroscopy, a lowering of signal intensity due to ionization of analyte atoms.

Ionization Suppressor An element used in atomic spectroscopy to decrease the extent of ionization of the analyte.

Ionophore A molecule with a hydrophobic outside and a polar inside that can engulf an ion and carry the ion through a hydrophobic phase (such as a cell membrane).

Ion Pair A closely associated anion and cation, held together by electrostatic attraction. In solvents less polar than water, ions are usually found as ion pairs.

Ion-Selective Electrode One that produces a potential selectively dependent on the concentration of one particular ion in solution.

Isocratic Elution Chromatography using a single solvent for the mobile phase.

Isoelectric Focusing A technique in which a sample containing polyprotic molecules is subjected to a strong electric field in a medium with a pH gradient. Each species migrates until it reaches the region of its isoelectric pH. In that region, the molecule has no net charge, ceases to migrate, and remains focused in a narrow band.

Isoelectric pH or Isoelectric Point That pH at which the average charge of a polyprotic species is zero.

Isoionic pH or Isoionic Point The pH of a pure solution of a neutral, polyprotic molecule. The only ions present are H^+, OH^-, and those derived from the polyprotic species.

Isosbestic Point A wavelength at which the absorbance spectra of two species cross each other. The appearance of isosbestic points in a solution in which a chemical reaction is occurring is evidence that there are only two components present, with a constant total concentration.

Jones Reductor A column packed with zinc amalgam. An oxidized analyte is passed through to reduce the analyte, which is then titrated with an oxidizing agent.

Joule, J The SI unit of energy. One joule is required to heat one milliliter of water by 0.24°C, to lift a one-kilogram mass 0.98 m at the earth's surface, or to move a charge of one coulomb through a potential difference of one volt.

Junction Potential An electric potential that exists at the junction between two different electrolyte solutions or substances. It arises in solutions from unequal rates of diffusion of different ions.

Karl Fischer Titration A sensitive technique for determining water, based on the reaction of H_2O with pyridine, I_2, SO_2, and methanol.

Kelvin, K The absolute unit of temperature defined such that the temperature of water at its triple point (where water, ice, and water vapor are at equilibrium) is 273.16 K and the absolute zero of temperature is zero kelvin.

Kieselguhr The German term for diatomaceous earth, which is used as a solid support in gas chromatography.

Kilogram, kg The mass of a particular Pt–Ir cylinder kept at the International Bureau of Weights and Measures, Sèvres, France.

Kinetic Current The wave that occurs when the height of a polarographic wave is affected by the rate of a chemical reaction involving the analyte and some species in the solution.

Kinetic Polarization Occurs whenever an overpotential is associated with an electrode process.

Kovats Index In chromatography, a retention index comparing the adjusted retention time of an unknown to those of linear alkanes eluted before and after the unknown.

Latimer Diagram One that shows the reduction potentials connecting a series of species containing an element in different oxidation states.

Law of Mass Action States that for the chemical reaction $aA + bB \rightleftharpoons cC + dD$, the condition at equilibrium is $K = \mathscr{A}_C^c \mathscr{A}_D^d / \mathscr{A}_A^a \mathscr{A}_B^b$, where $\mathscr{A}_i$ is the activity of the ith species. The law is usually used in approximate form, in which the activities are replaced by concentrations.

Le Châtelier's Principle States that if a system at equilibrium is disturbed, the direction in which it proceeds back to equilibrium is such that the disturbance is partially offset.

Leveling Effect The strongest acid that can exist in solution is the protonated form of the solvent. Any acid stronger than this will donate its proton to the solvent and be leveled to the acid strength of the protonated solvent. Similarly, the strongest base that can exist in a solvent is the deprotonated form of the solvent.

Lewis Acid One that can form a chemical bond by sharing a pair of electrons donated by another species.

Lewis Base One that can form a chemical bond by sharing a pair of its electrons with another species.

Ligand An atom or a group attached to a central atom in a molecule. The term is often used to mean any group attached to anything else of interest.

Limiting Current In a polarographic experiment, the current that is reached at the plateau of a polarographic wave. See also **Diffusion Current.**

Linear Interpolation A form of interpolation in which it is assumed that the variation in some quantity is linear. For example, to find the value of b when $a = 32.4$ in the table below

a	32	32.4	33
b	12.85	x	17.96

you can set up the proportion

$$\frac{32.4 - 32}{33 - 32} = \frac{x - 12.85}{17.96 - 12.85}$$

which gives $x = 14.89$.

Linear Voltage Ramp The linearly increasing potential that is applied to the working electrode in polarography.

Liquid-Based Ion-Selective Electrode One that has a hydrophobic membrane separating an inner reference electrode from the analyte solution. The membrane is saturated with a liquid ion exchanger dissolved in a nonpolar solvent. The ion-exchange equilibrium of analyte between the liquid ion exchanger and the aqueous solution gives rise to the electrode potential.

Liter, L Defined in 1964 as exactly $1\,000$ cm^3.

Logarithm The logarithm of a is b if $10^b = a$.

Longitudinal Diffusion Diffusion of solute molecules parallel to the direction of travel through a chromatography column.

Luminescence Any emission of light by a molecule.

Mantissa The part of a logarithm to the right of the decimal point.

Mariotte Flask A reservoir that maintains a constant hydrostatic pressure for liquid chromatography.

Masking Agent A reagent that selectively reacts with one (or more) component(s) of a solution to prevent the component(s) from interfering in a chemical analysis.

Mass Balance A statement that the sum of the moles of any element in all of its forms in a solution must equal the moles of that element delivered to the solution.

Mass Spectrograph An apparatus in which a sample is bombarded with electrons to produce charged molecular fragments that are then separated according to their mass in a magnetic field.

Matrix The medium containing analyte. For many analyses, it is important that standards be prepared in the same matrix as the unknown.

Maximum Suppressor A surface-active agent (such as the detergent Triton X-100) used to eliminate current maxima in polarography.

Mean The average of a set of all results.

Mean Activity Coefficient For the salt (cation)$_m$(anion)$_n$, the mean activity coefficient, $\gamma_\pm$, is related to the individual ion activity coefficients (γ_+ and γ_-) by the equation $\gamma_\pm = (\gamma_+^m \gamma_-^n)^{1/(m+n)}$.

Median That value above and below which there is an equal number of data points.

Mediator With reference to electrolysis, a molecule added to a solution to carry electrons between the electrode and a dissolved species. Used when the target species cannot react directly at the electrode or when the target concentration is so low that other reagents react instead.

Meniscus The curved surface of a liquid.

Mesh Size The number of spacings per linear inch in a standard screen used to sort particles.

Metal Ion Buffer Consists of a metal–ligand complex plus excess free ligand. The two serve to fix the concentration of free metal ion through the reaction $M + nL \rightleftharpoons ML_n$.

Metal Ion Indicator A compound whose color changes when it binds to a metal ion.

Meter, m Defined as the length equal to $1\,650\,763.73$ wavelengths (in vacuum) of the radiation corresponding to the $2p^{10}$–$5d^5$ transition of ^{86}Kr.

Microporous Particles A type of stationary phase used in HPLC consisting of 5–10 μm diameter porous particles with high efficiency and high capacity for solute.

Miscible Liquids Two liquids that form a single phase when mixed in any ratio.

Mobile Phase In chromatography, the phase that travels through the column.

Mobility The terminal velocity that an ion reaches in a field of 1 V m^{-1}. Velocity = mobility $\times$ field.

Modulation Amplitude   In pulsed polarography, the magnitude of the voltage pulse.

Molality   The number of moles of solute per kilogram of solvent.

Molar Absorptivity, ε, or Extinction Coefficient   The constant of proportionality in Beer's law: $A = \varepsilon bc$, where A is absorbance, b is pathlength, and c is the molarity of the absorbing species.

Molarity   The number of moles of solute per liter of solution.

Mole, mol   The amount of substance that contains as many molecules as there are atoms in 12 g of ^{12}C. There are approximately $6.022\,045 \times 10^{23}$ molecules per mole.

Molecular Exclusion Chromatography or Gel Filtration or Gel-Permeation Chromatography   A technique in which the stationary phase has a porous structure into which small molecules can enter but large molecules cannot. Molecules are separated by size, with larger molecules moving faster than smaller ones.

Molecular Orbital   Describes the distribution of an electron within a molecule.

Molecular Weight, M.W.   The number of grams of a substance that contains Avogadro's number of molecules.

Mole Fraction   The number of moles of a substance in a mixture divided by the total number of moles of all components present.

Monochromatic Light   Light of a single wavelength (color).

Monochromator   A device (usually a prism, grating, or filter) for selecting a single wavelength of light.

Monodentate Ligand   One that binds to a metal ion through only one atom.

Mother Liquor   The solution from which a substance has crystallized.

Multidentate Ligand   One that binds to a metal ion through more than one atom.

Nebulizer   In atomic spectroscopy, this breaks the liquid sample into a mist of fine droplets.

Nephelometry   A technique in which the intensity of light scattered by a suspension is measured to determine the concentration of suspended particles.

Nernst Equation   Relates the potential of a cell to the activities of reactants and products:

$$E = E^0 - \frac{RT}{nF} \ln Q$$

where R is gas constant, T is kelvins, F is Faraday constant, Q is reaction quotient, and n is number of electrons transferred in the balanced reaction.

Neutralization   The process in which a stoichiometric equivalent of acid (or base) is added to a base (or acid).

Neutron-Activation Analysis   A technique in which radiation is observed from a sample bombarded by slow neutrons. The radiation gives both qualitative and quantitative information about the sample composition.

Nonelectrolyte   A substance that does not dissociate into ions when dissolved.

Nonpolarizable Electrode   One whose potential remains nearly constant, even when current flows; e.g., a saturated calomel electrode.

Normal Error Curve   See **Gaussian Curve.**

Normal Hydrogen Electrode (N.H.E.)   See **Standard Hydrogen Electrode (S.H.E.)**

Normality n times the molarity of a redox reagent, where n is the number of electrons donated or accepted by that species in a particular chemical reaction. For acids and bases, it is also n times the molarity, but n is the number of protons donated or accepted by the species.

Normal Phase Chromatography   A chromatographic separation utilizing a polar stationary phase and a less polar mobile phase.

Nucleation   The process whereby molecules in solution come together randomly to form small aggregates.

Occlusion   An impurity that becomes trapped (sometimes with solvent) in a pocket within a growing crystal.

Ohm's Law   States that the current (I) in a circuit is proportional to voltage (E) and inversely proportional to resistance (R): $I = E/R$.

Optical Density, OD   See **Absorbance.**

Osmolarity   An expression of concentration that gives the total number of particles (ions and molecules) per liter of solution.

Overall Formation Constant, β_n   The equilibrium constant for a reaction of the type $M + nX \rightleftharpoons MX_n$.

Overpotential   The potential above that expected from the equilibrium potential, concentration polarization, and ohmic potential needed to cause an electrolytic reaction to occur at a given rate. It is zero for a reversible reaction.

Oxidant   See **Oxidizing Agent.**

Oxidation   A loss of electrons or a raising of the oxidation state.

Oxidation State (Number)   A bookkeeping device used to tell how many electrons have been gained or lost by a neutral atom when it forms a compound.

Oxidizing Agent or Oxidant   A substance that takes electrons in a chemical reaction.

Parallax The apparent displacement of an object when the observer changes position. Occurs when the scale of an instrument is viewed from a position that is not perpendicular to the scale, so that the apparent reading is not the true reading.

Particle Growth The process in which molecules become attached to a crystal to form a larger crystal.

Partition Chromatography A technique in which separation is achieved by equilibration of solute between two phases.

Partition Coefficient The equilibrium constant for the reaction in which a solute is partitioned between two phases: solute (in phase 1) $\rightleftharpoons$ solute (in phase 2).

Pascal, Pa A unit of pressure equal to $1\ N\ m^{-1}$. There are 101 325 Pa in 1 atm.

Pellicular Particles The type of stationary phase used in liquid chromatography. Contains a thin layer of liquid coated on a large spherical bead. It has high efficiency (low HETP), but low capacity.

Peptization Occurs when washing some ionic precipitates with distilled water causes the ions that neutralize the charges of individual particles, and thereby help to hold the particles together, to be washed away. The particles then disintegrate and pass through the filter with the wash liquid.

p Function The negative logarithm (base 10) of a quantity: $pX = -\log X$.

pH Defined as $pH = -\log \mathscr{A}_{H^+}$, where $\mathscr{A}_{H^+}$ is the activity of H^+. In most approximate applications, the pH is taken as $-\log[H^+]$.

Phase-Transfer Catalysis A technique in which a compound such as a crown ether is used to extract a reactant from one phase into another in which a chemical reaction can occur.

pH Meter A very sensitive potentiometer used in conjunction with a glass electrode to measure pH.

Phosphorescence The emission of light during a transition between states of different spin multiplicity (e.g., triplet $\rightarrow$ singlet).

Photomultiplier Tube One in which the cathode emits electrons when struck by light. The electrons then strike a series of dynodes (plates that are positive with respect to the cathode), and more electrons are released each time a dynode is struck. As a result, more than 10^6 electrons may reach the anode for every photon striking the cathode.

Phototube A vacuum tube with a photoemissive cathode. The electric current flowing between the cathode and the anode is proportional to the intensity of light striking the cathode.

pH-Stat A device that maintains a constant pH in a solution by continually injecting (or electrochemically generating) acid or base to counteract pH changes.

Pilot Ion In polarography, an internal standard.

Pipet A glass tube calibrated to deliver a fixed or variable volume of liquid.

pK The negative logarithm (base 10) of an equilibrium constant: $pK = -\log K$.

Polarizability The proportionality constant relating the induced dipole to the strength of the electric field. When a molecule is placed in an electric field, a dipole is induced in the molecule by attraction of the electrons toward the positive pole and attraction of the nuclei toward the negative pole.

Polarizable Electrode One whose potential can change readily when a small current flows. Examples are Pt or Ag wires used as indicator electrodes.

Polarogram A graph showing the relation between current and potential during a polaragraphic experiment.

Polarograph An instrument used to obtain and record a polarogram.

Polarographic Wave The S-shaped increase in current during a redox reaction in polarography.

Polarography A technique in which the current flowing into an electrolysis cell is measured as a function of the applied potential.

Polyprotic Acids and Bases Compounds that can donate or accept more than one proton.

Postprecipitation The adsorption of otherwise soluble impurities on the surface of a precipitate after the precipitation is over.

Potentiometer A device that measures electric potential by balancing it with a known potential of the opposite sign.

Potentiometry Any analytical method in which electric potential is measured.

Potentiostat An electronic device that maintains a constant potential between a pair of electrodes.

Power The amount of energy per unit time ($J\ s^{-1}$) being expended.

ppb (parts per billion) An expression of concentration that refers to nanograms (10^{-9} g) of solute per gram of solution.

ppm (parts per million) An expression of concentration that refers to micrograms (10^{-6} g) of solute per gram of solution.

ppt (parts per thousand) An expression of concentration that refers to milligrams (10^{-3} g) of solute per gram of solution.

Precipitant A substance that precipitates a species from solution.

Precipitation Occurs when a substance leaves solution

GLOSSARY

rapidly (to form either microcrystalline or amorphous solid).

Precipitation Titration One in which the analyte forms a precipitate with the titrant.

Precision A measure of the reproducibility of a measurement.

Precolumn In HPLC, a short precolumn packed with the same material as the main column may be used to purify solvent prior to entering the main column. It removes impurities that are irreversibly bound to the column and that would slowly ruin the main column.

Preconcentration The process of concentrating trace components of a mixture prior to their analysis.

Premix Burner In atomic spectroscopy, one in which the sample is nebulized and simultaneously mixed with fuel and oxidant before being fed into the flame.

Preoxidation In some redox titrations, adjustment of the analyte oxidation state to a higher value so that it can be titrated with a reducing agent.

Prereduction The process of reducing an analyte to a lower oxidation state prior to performing a titration with an oxidizing agent.

Pressure Force per unit area, commonly measured in pascals (N/m^{-1}) or atmospheres.

Pressure Broadening In spectroscopy, line broadening due to collisions between molecules.

Primary Standard A reagent that is pure enough and stable enough to be used directly after weighing. The entire mass is considered to be pure reagent.

Prism A transparent, triangular solid. Each wavelength of light passing through the prism is bent at a different angle. Therefore, light is dispersed into its component wavelengths by the prism.

Protic Solvent One with an acidic hydrogen atom.

Q Test Used to decide whether to discard a datum that appears discrepant.

Quaternary Ammonium Ion A cation containing four substituents attached to a nitrogen atom; e.g., $(CH_3CH_2)_4N^+$, the tetraethylammonium ion.

Radiant Power or **Intensity** The energy per unit time per unit area carried by a beam of light.

Random Error or **Indeterminate Error** A type of error, which can be either positive or negative and cannot be eliminated, based on the ultimate limitations on a physical measurement.

Range or **Spread** The difference between the highest and the lowest value in a set of data.

Reactant The species that is consumed in a chemical reaction. It appears on the left side of a chemical equation.

Reaction Quotient, Q Has the same form as the equilibrium constant for a reaction. However, the reaction quotient is evaluated for a particular set of existing activities (concentrations), which are generally not the equilibrium values. At equilibrium, $Q = K$.

Reagent Blank A solution prepared from all of the reagents, but no analyte. The blank measures the response of the analytical method to impurities in the reagents or any other effects caused by any component other than the analyte.

Redox Couple A pair of reagents related by electron transfer (e.g., $Fe^{3+}|Fe^{2+}$ or $MnO_4^-|Mn^{2+}$).

Redox Indicator A compound whose different oxidation states have different colors, and which is used to find the endpoint of a redox titration. The potential of the indicator must be such that its color changes near the equivalence point potential of the titration reaction.

Redox Reaction A chemical reaction involving transfer of electrons from one element to another.

Redox Titration One in which the reaction between analyte and titrant is an oxidation–reduction reaction.

Reduced Plate Height In chromatography, the quotient $HETP/d$, where the numerator is the height equivalent to a theoretical plate and the denominator is the diameter of stationary-phase particles.

Reducing Agent or **Reductant** A substance that donates electrons in a chemical reaction.

Reduction A gain of electrons or a lowering of the oxidation state.

Reference Electrode One that maintains a constant potential against which the potential of a half-cell may be measured.

Refractive Index The speed of light in any medium is c/n, where c is the speed of light in vacuum and n is the refractive index of the medium. The refractive index also measures the angle at which a light ray will be bent when it passes from one medium into another. Snell's law states that $n_1 \sin \theta_1 = n_2 \sin \theta_2$, where n_i is the refractive index for each medium and θ_i is the angle of the ray with respect to a normal between the two media.

Relative Error or **Relative Uncertainty** The uncertainty of a quantity divided by the value of the quantity. It is usually expressed as a percent of the measured quantity.

Relative Supersaturation Defined as $(Q - S)/S$, where S is the concentration of solute in a saturated solution and Q is the concentration in a particular supersaturated solution.

Releasing Agent In atomic spectroscopy, a substance that prevents chemical interference.

Reprecipitation Sometimes a gravimetric precipitate can be freed of impurities only by redissolving it and reprecipitating it. The impurities are present at lower concentration during the second precipitation and are less likely to coprecipitate.

Residual Current The small current that is observed prior to the decomposition potential in an electrolysis.

Resin An ion exchanger, such as polystyrene, which exists as small, hard particles.

Resistance, R A measure of the retarding force opposing the flow of electric current.

Resolution How close two bands in a spectrum or a chromatogram can be to each other and still be seen as two peaks. In chromatography, it is defined as the difference in retention time between adjacent peaks divided by their width.

Resonance Effect Contribution to a physical property by delocalization of electrons through the pi orbitals of a molecule.

Response Factor An empirically determined factor measuring the response of a detector to a given compound. It is usually used in gas chromatography to determine the amount of unknown relative to an internal standard.

Retention Index In gas chromatography, the Kovats retention index is a logarithmic scale that relates the retention time of a compound to those of linear alkanes.

Retention Ratio In chromatography, the time required for solvent to pass through the column divided by the time required for solute to pass through the column.

Retention Time The time, measured from injection, needed for a solute to be eluted from a chromatography column.

Retention Volume The volume of solvent needed to elute a solute from a chromatography column.

Reverse Phase Chromatography A technique in which the stationary phase is less polar than the mobile phase.

Rotating Platinum Electrode One that consists of a Pt wire projecting from a rotating shaft. It is particularly useful for studying anodic processes, for which a mercury electrode would be too easily oxidized.

Rotational Transition Occurs when a molecule changes its rotation energy.

Rubber Policeman A glass rod with a flattened piece of rubber on the tip. The rubber is used to scrape solid particles from glass surfaces in gravimetric analysis.

Salt An ionic solid.

Salt Bridge A conducting ionic medium in contact with two electrolyte solutions. It allows an ionic current to flow without allowing immediate diffusion of one electrolyte solution into the other.

Saturated Solution One that contains the maximum amount of a compound that can dissolve at equilibrium.

Scatchard Plot A graph used to find the equilibrium constant for a reaction such as $X + P \rightleftharpoons PX$. It is a graph of $[PX]/[X]$ versus $[PX]$, or any functions proportional to these quantities. The magnitude of the slope of the graph is the equilibrium constant.

S.C.E. (Saturated Calomel Electrode) A calomel electrode saturated with KCl. The electrode half-reaction is $Hg_2Cl_2(s) + 2e^- \rightleftharpoons 2Hg(l) + 2Cl^-$.

Second, s The duration of 9 192 631 770 periods of the radiation corresponding to the transition between two hyperfine levels of the ground state of ^{133}Cs.

Selectivity Coefficient With respect to an ion-selective electrode, a measure of the relative response of the electrode to two different ions. In ion-exchange chromatography, this is the equilibrium constant for displacement of one ion by another from the resin.

Self-Absorption In flame emission atomic spectroscopy, there is a lower concentration of excited-state atoms in the cool, outer part of the flame than in the hot, inner flame. The cool atoms can absorb emission from the hot ones and thereby decrease the observed signal.

Sensitivity In a spectrophotometric analysis, the concentration of analyte necessary to produce 99% T (or an absorbance of 0.004 4).

Septum A disk, usually made of silicone rubber, covering the injection port of a gas chromatograph. The sample is injected by syringe through the septum.

Significant Figure The number of significant figures in a quantity is the minimum number of figures needed to express the quantity in scientific notation. In experimental data, the first uncertain figure is the last significant figure.

Silanization The treatment of a chromatographic solid support or glass column with silicon compounds that bind to the most reactive Si–OH groups. It reduces irreversible adsorption and tailing of polar solutes.

Silver–Silver Chloride Electrode A common reference electrode containing a silver wire coated with AgCl paste and dipped in a solution saturated with AgCl and (usually) KCl. The half-reaction is $AgCl(s) + e^- \rightleftharpoons Ag(s) + Cl^-$.

Single-Electrode Potential The potential measured when the electrode of interest is connected to a standard hydrogen electrode anode.

Singlet State One in which all electron spins are paired.

Slope For a straight line whose equation is $y = mx + b$, the value of m is the slope. It is the ratio $\Delta y/\Delta x$ for any segment of the line.

Slurry A suspension of a solid in a solvent.

Solid-State Electrode A type of ion-selective electrode that has a solid membrane made of an inorganic salt crystal. Ion-exchange equilibria between the solution and the surface of the crystal account for the cell potential.

Solubility Product, K_{sp} The equilibrium constant for the dissolution of a solid salt to give its ions in solution. For the reaction $M_mN_n(s) \rightleftharpoons mM^{n+} + nN^{m-}$, $K_{sp} = \mathscr{A}_{M^{n+}}^m \mathscr{A}_{N^{m-}}^n$ where $\mathscr{A}$ is the activity of each species.

Solute A minor component of a solution.

Solvation The interaction of solvent molecules with solute. In general, solvent molecules will orient themselves around a solute to minimize the energy of the solution through dipole and van der Waals forces.

Solvent The major constituent of a solution.

Solvent Extraction A method in which a chemical species is transferred from one liquid phase to another. It is used to separate components of a mixture.

Specific Gravity A dimensionless quantity equal to the mass of a substance divided by the mass of an equal volume of water at $4°C$. Since the density of water at $4°C$ is 1.0000 g/mL, density and specific gravity are synonymous.

Spectral Interference In atomic spectroscopy, any physical process that affects the light intensity at the analytical wavelength. Created by substances that absorb, scatter, or emit light of the analytical wavelength.

Spectrophotometer A device used to measure absorption of light. It includes a source of light, a wavelength selector (monochromator), and an electrical means of detecting light.

Spectrophotometric Analysis Any method in which light absorption, emission, or scattering is used to measure chemical concentrations.

Spectrophotometric Titration One in which absorption of light is used to monitor the progress of the chemical reaction.

Spectrophotometry In a broad sense, any method using light to measure chemical concentrations.

Spontaneous Process One that is energetically favorable. It will eventually occur, but thermodynamics makes no prediction as to how long it will take.

Spread See **Range**.

Stability Constant See **Formation Constant**.

Standard Addition Method A technique in which an analytical signal due to an unknown is first measured. Then a known quantity of analyte is added, and the increase in signal is recorded. Assuming linear response, it is possible to calculate what quantity of analyte must have been present in the unknown.

Standard Curve A graph showing the response of an analytical technique to known quantities of analyte.

Standard Deviation Measures how closely data are clustered about the mean value. For a finite set of data, the standard deviation, s, is computed from the formula

$$s = \sqrt{\frac{\sum(x_i - \bar{x})^2}{n - 1}}$$

where n is the number of results, x_i is an individual result, and $\bar{x}$ is the mean result.

Standard Hydrogen Electrode (S.H.E.) or **Normal Hydrogen Electrode (N.H.E.)** One that contains $H_2(g)$ bubbling over a catalytic Pt surface in contact with aqueous H^+. The activities of H_2 and H^+ are both unity in the hypothetical standard electrode. The cell reaction is $H^+ + e^- \rightleftharpoons \frac{1}{2}H_2(g)$.

Standardization The process whereby the concentration of a reagent is determined by reaction with a known quantity of a second reagent.

Standard Reduction Potential, E^0 The potential that would be measured when a hypothetical cell containing the desired half-reaction (with all species present at unit activity) is connected to a standard hydrogen electrode anode.

Standard State When writing equilibrium constants, the standard state of a solute is 1 M and the standard state of a gas is 1 atm. Pure solids and liquids are considered to be in their standard states.

Stationary Phase In chromatography, the phase that does not move through the column.

Stepwise Formation Constant, K_n The equilibrium constant for a reaction of the type $ML_{n-1} + L \rightleftharpoons ML_n$.

Stripping Analysis A very sensitive polarographic technique in which analyte is concentrated from a dilute solution by reduction into a single drop of Hg. It is then analyzed polarographically during an anodic redissolution process.

Strong Acids and Bases Those that are completely dissociated (to H^+ or OH^-) in water.

Strong Electrolyte One that dissociates completely into its ions when dissolved.

Supersaturated Solution One that contains more dissolved solute than would be present at equilibrium.

Supporting Electrolyte An unreactive salt added in high concentration to most solutions for voltammetric measurements (such as polarography). The supporting electrolyte carries most of the ion-migration current and therefore decreases the coulombic migration of electroactive species to a negligible level. The electrolyte also decrease the resistance of the solution.

Syringe A device having a calibrated barrel into which liquid is sucked by a plunger. The liquid is expelled through a needle by pushing on the plunger.

Systematic Error or **Determinate Error**   A type of error due to procedural or instrumental factors that cause a measurement to be systematically too large or too small. The error can, in principle, be discovered and corrected.

Systematic Treatment of Equilibrium   A method that uses the charge balance, mass balance(s), and equilibria to completely specify the system's composition.

Tailing   An asymmetric chromatographic elution band in which the later part of the band is drawn out. It often results from adsorption of a solute to a few active adsorption sites on the stationary phase.

Tare   The mass of an empty vessel used to receive a substance to be weighed. Many balances can be tared. That is, with the empty receiver in place, the balance can be set to read zero grams.

Thermal Conductivity Detector   A device that detects bands eluted from a gas chromatography column by measuring changes in the thermal conductivity of the gas stream.

Thermogravimetric Analysis   A technique in which the mass of a substance is measured as the substance is heated. Changes in mass reflect decomposition of the substance, often to well-defined products.

Thermometric Titration   One in which the temperature is measured to determine the endpoint. Most titration reactions are exothermic, so the temperature rises during the reaction and suddenly stops rising when the equivalence point is reached.

Thin-Layer Chromatography   A technique in which the stationary phase is coated on a flat glass or plastic plate. Solute is spotted near the bottom of the plate. The bottom edge of the plate is placed in contact with solvent, which is allowed to creep up the plate by capillary action.

Titer   A measure of concentration, usually defined as how many milligrams of reagent B will react with 1 mL of reagent A. Consider a $AgNO_3$ solution with a titer of 1.28 mg of NaCl per milliliter of $AgNO_3$. The reaction is $Ag^+ + Cl^- \rightarrow AgCl(s)$. Since 1.28 mg of NaCl = 2.19×10^{-5} mol, the concentration of Ag^+ is 2.19×10^{-5} mol/mL = 0.021 9 M. The same solution of $AgNO_3$ has a titer of 0.993 mg of KH_2PO_4, because three moles of Ag^+ react with one mole of PO_4^{3-} (to precipitate Ag_3PO_4) and 0.993 mg of KH_2PO_4 equals $\frac{1}{3}(2.19 \times 10^{-5}$ mol).

Titrant   The substance added to the analyte in a titration.

Titration   A procedure in which one substance (titrant) is carefully added to another (analyte) until complete reaction has occurred. The quantity of titrant required for complete reaction tells how much analyte is present.

Titration Error   Caused by the difference between the observed endpoint and the true equivalence point of the reaction.

Transmittance, T   Defined as $T = P/P_0$, where P_0 is the radiant power of light striking the sample on one side and P is the radiant power of light emerging from the other side of the sample.

Triple Point   The one temperature and pressure at which the solid, liquid, and gaseous forms of a substance are in equilibrium with each other.

Triplet State   An electronic state in which there are two unpaired electrons.

t Test   Used to decide whether the results of two experiments are within experimental uncertainty of each other. The uncertainty must be specified to within a certain probability.

Turbidimetry   A technique in which the decrease in radiant power of light traveling through a turbid solution is measured.

Turbidity   The light-scattering property associated with suspended particles in a liquid. A turbid solution appears cloudy.

Turbidity Coefficient   The transmittance of a turbid solution, given by $P/P_0 = e^{-\tau b}$, where P is the transmitted radiant power, P_0 is the incident radiant power, b is the pathlength, and τ is the turbity coefficient.

van Deemter Equation   Describes the dependence of chromatographic plate height on the velocity of elution: $HETP = A + Bv + C/v$.

Variance   The square of the standard deviation.

Vibrational Transition   Occurs when a molecule changes its vibrational energy.

Void Volume, V_0   The volume of the mobile phase, V_m.

Volatile   Easily vaporized.

Volatilization   The selective removal of a component from a mixture by transforming the component into a volatile (low-boiling) species and removing it by heating, pumping, or bubbling a gas through the mixture.

Volhard Titration   That of Ag^+ with SCN^-, in which the formation of the red complex $Fe(SCN)^{2+}$ marks the endpoint.

Voltammetry   An analytical method in which the relationship between current and voltage is observed during an electrochemical reaction.

700

GLOSSARY

Volume Percent Defined as (Volume of solute/Volume of solution) $\times$ 100.

Volumetric Analysis A technique in which the volume of material needed to react with the analyte is measured.

Volumetric Flask One having a tall, thin neck with a calibration mark. When the liquid level is at the calibration mark, the flask contains its specified volume of liquid.

Von Weimarn Ratio The quotient $(Q - S)/S$, where Q is the concentration of a solute and S is the concentration at equilibrium. A large value of this ratio means that the solution is highly supersaturated.

Walden Reductor A column packed with silver and eluted with HCl. An oxidized analyte is reduced upon passage through the column. The reduced product is titrated with an oxidizing agent.

Watt, W The SI unit of power, equal to an energy flow of one joule per second. When an electric current of one ampere flows through a potential difference of one volt, the power is one watt.

Wavelength The distance between consecutive crests of a wave.

Wavenumber, $\bar{\nu}$ The reciprocal of the wavelength, λ.

Weak Acids and Bases Those whose dissociation constants are not large.

Weak Electrolyte One that only partially dissociates into ions when it dissolves.

Weighing Paper Used as a base on which to place a solid reagent on a balance. The paper has a very smooth surface, from which solids fall easily for transfer to a vessel.

Weight Percent Defined as (Mass of solute/Mass of solution) $\times$ 100.

Weight/Volume Percent Define as (Mass of solute/Volume of solution) $\times$ 100.

Weston Cell An extremely stable source of electric potential, based on the reaction $Cd(s) + HgSO_4(aq) \rightleftharpoons CdSO_4(aq) + Hg(l)$. It is often used to standardize a potentiometer.

Wet Ashing The destruction of organic matter in a sample by a liquid reagent (such as boiling aqueous $HClO_4$) prior to analysis of an inorganic component.

Working Electrode The one at which the reaction of analytical interest in coulometry or polarography occurs.

Zwitterion A molecule with a positive charge localized in one position and a negative charge localized at another position.

Solutions to Exercises

Chapter 1

1-A. (a) $\dfrac{(25.00 \text{ mL})(0.791\ 4 \text{ g/mL})/(32.042 \text{ g/mol})}{0.500\ 0 \text{ L}}$

$= 1.235 \text{ M}$

(b) 500.0 mL of solution weighs (1.454 g/mL) $(500.0 \text{ mL}) = 727.0$ g and contains 25.00 mL $(= 19.78$ g) of methanol. The mass of chloroform in 500 mL must be $727.0 - 19.78 = 707.2$ g. The molality of methanol is

$$\text{molality} = \frac{\text{mol methanol}}{\text{kg chloroform}}$$

$$= \frac{(19.78 \text{ g})/(32.042 \text{ g/mol})}{0.707\ 2 \text{ kg}}$$

$$= 0.872\ 9 \text{ m}$$

1-B. (a) $\left(\dfrac{48.0 \text{ g HBr}}{100 \text{ g solution}}\right)\left(1.50 \dfrac{\text{g solution}}{\text{mL solution}}\right)$

$$= \frac{0.720 \text{ g HBr}}{\text{mL solution}}$$

$$= \frac{720 \text{ g HBr}}{\text{L solution}}$$

$$= 8.90 \text{ F}$$

(b) $\dfrac{36.0 \text{ g HBr}}{0.480 \text{ g HBr/g solution}} = 75.0 \text{ g solution}$

(c) $233 \text{ mmol} = 0.233 \text{ mol}$

$$\frac{0.233 \text{ mol}}{8.90 \text{ mol/L}} = 26.2 \text{ mL}$$

(d) $V = 250 \text{ mL}\left(\dfrac{0.160 \text{ M}}{8.90 \text{ M}}\right) = 4.49 \text{ mL}$

1-C. $\dfrac{\text{wt Cl}}{\text{wt MgCl}_2} = \dfrac{(2)(35.453)}{24.305 + 2(35.453)} = 0.745$

If $MgCl_2 = 12.6$ ppt,

$$Cl = (0.745)(12.6) = 9.38 \text{ ppt}$$

Chapter 3

3-A. (a) $21.0_9\ (\pm 0.1_6)$ or $21.1\ (\pm 0.2)$; relative error $= 0.8\%$
(b) $11.55\ (\pm 0.13)$; relative error $= 1.1\%$
(c) $14._9 \pm 1._3$ or 15 ± 1; relative error $= 9\%$

3-B. (a) 2.000 L of 0.169 M NaOH (F.W. $= 39.997\ 1$) requires 0.338 mol $= 13.52$ g NaOH

$$\frac{13.52 \text{ g NaOH}}{0.534 \text{ g NaOH/g solution}} = 25.32 \text{ g solution}$$

$$\frac{25.32 \text{ g solution}}{1.52 \text{ g solution/mL solution}} = 16.66 \text{ mL}$$

(b) Molarity

$$= \frac{[16.66\ (\pm 0.10) \text{ mL}]\left[1.52\ (\pm 0.01)\ \dfrac{\text{g solution}}{\text{mL}}\right]}{\left(39.997\ 1\ \dfrac{\text{g NaOH}}{\text{mol}}\right)(2.000 \text{ L})} \times \left[0.534(\pm 0.004)\ \dfrac{\text{g NaOH}}{\text{g solution}}\right]$$

Since the relative errors in molecular weight and final volume are negligible (≈ 0), we can write

Relative error in molarity $= \sqrt{\left(\dfrac{0.10}{16.66}\right)^2 + \left(\dfrac{0.01}{1.52}\right)^2 + \left(\dfrac{0.004}{0.534}\right)^2}$

$$= 1.16\%$$

Molarity $= 0.169\ (\pm 0.002)$

3-C. $0.050\,0\,(\pm 2\%)$ mol

$$= \frac{[4.18\,(\pm x)\,\text{mL}]\left[1.18\,(\pm 0.01)\dfrac{\text{g solution}}{\text{mL}}\right]}{36.461\dfrac{\text{g HCl}}{\text{mol}}} \times \left[0.370\,(\pm 0.005)\dfrac{\text{g HCl}}{\text{g solution}}\right]$$

Error analysis:

$$(0.02)^2 = \left(\frac{x}{4.18}\right)^2 + \left(\frac{0.01}{1.18}\right)^2 + \left(\frac{0.005}{0.370}\right)^2$$

$$x = 0.05 \text{ mL}$$

Chapter 4

4-A. Mean $= \frac{1}{5}(116.0 + 97.9 + 114.2 + 106.8 + 108.3)$

$$= 108._4$$

Standard deviation

$$= \sqrt{\frac{(116.0 - 108.6_4)^2 + \cdots + (108.3 - 108.6_4)^2}{5 - 1}}$$

$$= 7.1_4$$

Median $= 108.3$ (the middle value)

Geometric mean

$$= [(116.0)(97.9)(114.2)(106.8)(108.3)]^{1/5} = 108.4_5$$

Range $= 116.0 - 97.9 = 18.1$

90% Confidence interval

$$= 108.6_4 \pm \frac{(2.132)(7.1_4)}{\sqrt{5}} = 108.6_4 \pm 6.8_1$$

$$Q = \frac{106.8 - 97.9}{116.0 - 97.9} = 0.49 < [Q(\text{Table 4-4}) = 0.64]$$

Therefore, 97.9 should be retained.

4-B. (a) The mean, $\bar{x}$, is greater than 45 800 miles by 16 900 miles, and 16 900 is 1.625 standard deviations from $\bar{x}$. In Table 4-1, we see that the area listed for $x = 1.6$ is 0.445 2, and the area for $x = 1.7$ is 0.455 4. By linear interpolation, the area for 1.625 is

$$0.445\,2 + \left(\frac{1.625 - 1.6}{1.7 - 1.6}\right)(0.455\,4 - 0.445\,2)$$

$$= 0.447\,8$$

The area *beyond* 1.625 standard deviations must be $0.500\,0 - 0.447\,8 = 0.052\,2$. The fraction of brakes expected to wear out in less than 45 800 miles is 0.052 2 (or 5.22%).

(b) 60 000 miles lies 0.259 6 standard deviations below $\bar{x}$. 70 000 miles lies 0.701 9 standard deviations above $\bar{x}$. By linear interpolation, the area from $\bar{x}$ to $0.259\,6s = 0.102\,3$. The area from $\bar{x}$ to $0.701\,9s = 0.258\,6$. Total area $= 0.102\,3 + 0.258\,6 = 0.360\,9$.

4-C. Confidence 90%: $\mu = \bar{x} \pm \dfrac{ts}{\sqrt{n}}$

$$= 116._4 \pm \frac{(2.132)(3._{58})}{\sqrt{5}}$$

$$= 112._9 \text{ to } 119._8$$

Confidence 99%: $\quad = 116._4 \pm \dfrac{(4.604)(3._{58})}{\sqrt{5}}$

$$= 109._0 \text{ to } 123._8$$

You can be 90% sure, but not 99% sure, that the result is high.

4-D. Using Equations. 4-6 and 4-7, we find

$$t = \frac{0.027\,5_6 - 0.026\,9_0}{0.000\,4_5}\sqrt{\frac{5 \cdot 5}{5 + 5}} = 2._{32}$$

Since $t(\text{calculated}) = 2.32 > t(\text{Table 4-2}) = 1.86$, the difference is significant.

4-E. (a)

x_i	y_i	$x_i y_i$	x_i^2	d_i	d_i^2
0.00	.466	0	0	$-0.004\,6$	2.12×10^{-5}
9.36	.676	6.327	87.61	$+0.001\,6$	2.58×10^{-6}
18.72	.883	16.530	350.44	$+0.004\,8$	2.31×10^{-5}
28.08	1.086	30.495	788.49	$+0.004\,0$	1.61×10^{-5}
37.44	1.280	47.923	1 401.75	$-0.005\,8$	3.34×10^{-5}
Sum: 93.60	4.391	101.275	2 628.29		9.64×10^{-5}

$$D = \begin{vmatrix} \Sigma(x_i^2) & \Sigma x_i \\ \Sigma x_i & n \end{vmatrix}$$

$$= (2\,628.29)(5) - (93.60)(93.60) = 4\,380.5$$

$$m = \begin{vmatrix} \Sigma x_i y_i & \Sigma x_i \\ \Sigma y_i & n \end{vmatrix} \div D$$

$$= \frac{(101.275)(5) - (93.60)(4.391)}{D}$$

$$= 95.377 \div 4\,380.5 = 0.021\,773$$

$$b = \begin{vmatrix} \sum(x_i^2) & \sum x_i y_i \\ \sum x_i & \sum y_i \end{vmatrix} \div D$$

$$= \frac{(2\,628.29)(4.391) - (101.275)(93.60)}{D}$$

$$= 2\,061.48 \div 4\,380.5 = 0.470\,60$$

$$\sigma_y^2 \approx s_y^2 = \frac{\sum(d_i^2)}{n-2} = \frac{9.64 \times 10^{-5}}{3}$$

$$= 3.21 \times 10^{-5}; \ \sigma_y = 0.005\,7$$

$$\sigma_m = \sqrt{\frac{\sigma_y^2 n}{D}} = \sqrt{\frac{(3.21 \times 10^{-5})5}{4\,380.5}} = 0.000\,191$$

$$\sigma_b = \sqrt{\frac{\sigma_y^2 \sum(x_i^2)}{D}} = \sqrt{\frac{(3.21 \times 10^{-5})(2\,628.29)}{4\,380.5}}$$

$$= 0.004\,39$$

Equation of the best line:

$$y = [0.021\,8\,(\pm 0.000\,2)]x + [0.471\,(\pm 0.004)]$$

(c) $$x = \frac{y(\pm \sigma_y) - b(\pm \sigma_b)}{m(\pm \sigma_m)}$$

$$= \frac{0.973\,(\pm 0.005_7) - 0.471\,(\pm 0.004_4)}{0.021\,8\,(\pm 0.000\,1_9)}$$

$$= 23.0 \pm 0.4\ \mu g$$

Chapter 5

5-A. (a) $Ag^+ + \cancel{Cl^-} \rightleftharpoons AgCl(aq)$

$$\underline{AgCl(s) \rightleftharpoons Ag^+ + \cancel{Cl^-}}$$
$$AgCl(s) \rightleftharpoons AgCl(aq)$$

$$K_1 = 2.0 \times 10^3$$
$$K_2 = 1.8 \times 10^{-10}$$
$$\overline{K_3 = K_1 K_2 = 3.6 \times 10^{-7}}$$

(b) The answer to part a tells us $[AgCl(aq)]$
$= 3.6 \times 10^{-7}$ M.

(c) $\quad AgCl_2^- \rightleftharpoons Ag\cancel{Cl}(aq) + Cl^-$

$\quad Ag^+ + \cancel{Cl^-} \rightleftharpoons AgCl(s)$

$\quad \underline{Ag\cancel{Cl}(aq) \rightleftharpoons Ag^+ + \cancel{Cl^-}}$

$\quad AgCl_2^- \rightleftharpoons AgCl(s) + Cl^-$

$$K_1 = 1/(9.3 \times 10^1)$$
$$K_2 = 1/(1.8 \times 10^{-10})$$
$$K_3 = 1/(2.0 \times 10^3)$$
$$\overline{K_4 = K_1 K_2 K_3 = 3.0 \times 10^4}$$

5-B. (a)

	BrO_3^-	Cr^{3+}	Br^-	$Cr_2O_7^{2-}$	H^+
Initial concentration:	0.010 0	0.010 0	0	0	1.00
Final concentration:	$0.010\,0 - x$	$0.010\,0 - 2x$	x	x	$1.00 + 8x$

$$\frac{(x)(x)(1.00 + 8x)^8}{(0.010\,0 - x)(0.010\,0 - 2x)^2} = 1 \times 10^{11}$$

(b) $[Br^-]$ and $[Cr_2O_7^{2-}]$ will both be 0.005 00 M because Cr^{3+} is the *limiting reagent*. Reaction 5-12 requires two moles of Cr^{3+} per mole of BrO_3^-. The Cr^{3+} will be used up first, making one mole of Br^- and one mole of $Cr_2O_7^{2-}$ per two moles of Cr^{3+} consumed. To solve the equation above, we set $x = 0.005\,00$ M in all terms except $[Cr^{3+}]$. The concentration of $[Cr^{3+}]$ will be a small, unknown quantity.

$$\frac{(0.005\,00)(0.005\,00)[1.00 + 8(0.005\,00)]^8}{(0.010\,0 - 0.005\,00)[Cr^{3+}]^2}$$

$$= 1 \times 10^{11}$$

$$[Cr^{3+}] = 2.6 \times 10^{-7}\ M$$

$$[BrO_3^-] = 0.010\,0 - 0.005\,00 = 0.005\,00\ M$$

5-C. (a) $[La^{3+}][IO_3^-]^3 = x(3x)^3 = 1.0 \times 10^{-11}$

$$\Rightarrow x = 7.8 \times 10^{-4}\ M$$
$$= 0.13\ g\ La(IO_3)_3/250\ mL$$

(b) $[La^{3+}][IO_3^-]^3 = x(3x + 0.050)^3$

$$\approx x(0.050)^3 = 1.0 \times 10^{-11}$$
$$\Rightarrow x = 8 \times 10^{-8} = 1.3 \times 10^{-5}\ g$$

5-D. (a) $Ca(IO_3)_2$ (since it has a larger K_{sp})

(b) The two salts do not have the same stoichiometry. Therefore, K_{sp} values cannot be compared. For $TlIO_3$, $x^2 = K_{sp} \Rightarrow x = 1.8 \times 10^{-3}$. For $Sr(IO_3)_2$, $x(2x)^2 = K_{sp} \Rightarrow 4.4 \times 10^{-3}$. $Sr(IO_3)_2$ is more soluble.

5-E. $[Fe^{3-}][OH^-]^3 = (10^{-10})[OH^-]^3 = 1.6 \times 10^{-39}$

$$\Rightarrow [OH^-] = 2.5 \times 10^{-10}$$

$[Fe^{2+}][OH^-]^2 = (10^{-10})[OH^-]^2 = 7.9 \times 10^{-16}$

$$\Rightarrow [OH^-] = 2.8 \times 10^{-3}$$

5-F. First we need to find which salt will precipitate at the lowest $[C_2O_4^{2-}]$ concentration:

$$[Ca^{2+}][C_2O_4^{2-}] = x^2 = 1.3 \times 10^{-8}$$
$$\Rightarrow [C_2O_4^{2-}] = x$$
$$= 1.1 \times 10^{-4}\ M$$

$$[Ce^{3+}]^2[C_2O_4^{2-}]^3 = (2x)^2(3x)^3 = 3 \times 10^{-29}$$
$$\Rightarrow [C_2O_4^{2-}] = 3x$$
$$= 7.7 \times 10^{-7}\ M$$

$Ce_2(C_2O_4)_3$ is less soluble than CaC_2O_4. The concentration of $C_2O_4^{2-}$ needed to reduce Ce^{3+} to 1% of 0.010 M is

$$[C_2O_4^{2-}] = [K_{sp}/(0.000\ 10)^2]^{1/3} = 1.4 \times 10^{-7}$$

This concentration of $C_2O_4^{2-}$ will not precipitate Ca^{2+} because

$$Q = [Ca^{2+}][C_2O_4^{2-}] = (0.010)(1.4 \times 10^{-7})$$
$$= 1.4 \times 10^{-9} < K_{sp}$$

The separation is feasible.

5-G. Assuming that all of the Ni is in the form $Ni(en)_3^{2+}$, $[Ni(en)_3^{2+}] = 1.00 \times 10^{-5}$ M. This uses up just 3×10^{-5} mol of en, which leaves the en concentration at 0.100 M. Adding the three equations gives

$$Ni^{2+} + 3en \rightleftharpoons Ni(en)_3^{2+}$$

$$K = K_1K_2K_3 = 2.1_4 \times 10^{18}$$

$$[Ni^{2+}] = \frac{[Ni(en)_3^{2+}]}{K[en]^3}$$

$$= \frac{(1.00 \times 10^{-5})}{(2.1_4 \times 10^{18})(0.100)^3} = 4.7 \times 10^{-21}\ \text{M}$$

Chapter 6

6-A. (a) $\mu = \frac{1}{2}([K^+] \cdot 1^2 + [Br^-] \cdot (-1)^2) = 0.02$ M

(b) $\mu = \frac{1}{2}([Cs^+] \cdot 1^2 + [CrO_4^{2-}] \cdot (-2)^2)$
$= \frac{1}{2}([0.04] \cdot 1 + [0.02] \cdot 4) = 0.06$ M

(c) $\mu = \frac{1}{2}([Mg^{2+}] \cdot 2^2 + [Cl^-] \cdot (-1)^2$
$+ [Al^{3+}] \cdot 3^2)$
$= \frac{1}{2}([0.02] \cdot 4 +$
$[\quad 0.04 \quad + \quad 0.09 \quad] \cdot 1 + [0.03] \cdot 9)$
$\qquad\uparrow \qquad\qquad \uparrow$
$\text{from } MgCl_2 \quad \text{from } AlCl_3$
$= 0.24$ M

6-B. For $0.005\ 0$ M $(CH_3CH_2CH_2)_4N^+Br^-$ plus $0.005\ 0$ M $(CH_3)_4N^+Cl^-$, $\mu = 0.010$ M. The ion $(CH_3CH_2\text{-}CH_2)_4N^+$ has a size of 800 pm. [It is listed at the bottom of Table 6-1 as $(C_3H_7)_4N^+$.] At $\mu = 0.01$ M, $\gamma = 0.912$ for an ion of charge ± 1 with $\alpha = 800$ pm. $\mathscr{A} = (0.005\ 0)(0.912) = 0.004\ 6$.

6-C. (a) $\mu = 0.060$ M from KNO_3 (assuming that AgSCN has negligible solubility)

$$[Ag^+]\gamma_{Ag^+}[SCN^-]\gamma_{SCN^-} = K_{sp}$$
$$[x](0.79)[x](0.80) = 1.1 \times 10^{-12}$$
$$\Rightarrow x = [Ag^+] = 1.3 \times 10^{-6}\ \text{M}$$

(b) $\mu = 0.060$ M from KSCN

$$[Ag^+]\gamma_{Ag^+}[SCN^-]\gamma_{SCN^-} = K_{sp}$$
$$[x](0.79)\underbrace{[x + 0.060]}_{\approx 0.060}(0.80) = 1.1 \times 10^{-12}$$
$$\Rightarrow x = [Ag^+]$$
$$= 2.9 \times 10^{-11}\ \text{M}$$

6-D. Assuming that $Mn(OH)_2$ gives a negligible concentration of ions, $\mu = 0.075$ from $CaCl_2$.

$$[Mn^{2+}]\gamma_{Mn^{2+}}[OH^-]^2\gamma_{OH^-}^2 = K_{sp}$$
$$[x](0.445)[2x]^2(0.785)^2 = 1.6 \times 10^{-13} \Rightarrow 2x$$
$$= [OH^-] = 1.1 \times 10^{-4}\ \text{M}$$

6-E. For 0.02 M $MgCl_2$ (which dissociates into 0.020 M Mg^{2+} and 0.040 M Cl^-), $\mu = 0.06$ M. At this ionic strength, interpolation in Table 6-1 gives $\gamma_{Mg^{2+}} = 0.50_6$ and $\gamma_{Cl^-} = 0.795$.

$$\gamma_{\pm} = (\gamma_{Mg^{2+}}\gamma_{Cl^-}^2)^{1/(1+2)} = 0.68_4$$

Chapter 7

7-A. $[H^+] + 2[Ca^{2+}] + [CaF^+] = [OH^-] + [F^-]$
Remember that H^+ and OH^- are present in every aqueous solution.

7-B. (a) $[Cl^-] = 2[Ca^{2+}]$
(b) $\underbrace{[Cl^-] + [CaCl^+]}_{\text{moles of Cl}} = 2(\underbrace{[Ca^{2+}] + [CaCl^+]}_{\text{moles of Ca}})$

7-C. (a) $[F^-] + [HF] = 2[Ca^{2+}]$
(b) $\underbrace{[F^-] + [HF] + 2[HF_2^-]}_{\text{moles of F}} = 2[Ca^{2+}]$
(One mole of HF_2^- contains two moles of fluorine.)

7-D. $2[Ca^{2+}] = 3\{[PO_4^{3-}] + [HPO_4^{2-}]$
$+ [H_2PO_4^-] + [H_3PO_4]\}$

7-E. (a) Charge balance: Invalid because pH is fixed.

Mass balance: $[Ag^+] = [CN^-] + [HCN]$ (1)

Equilibria: $K_B = \dfrac{[HCN][OH^-]}{[CN^-]}$ (2)

$K_{sp} = [Ag^+][CN^-]$ (3)

$K_w = [H^+][OH^-]$ (4)

Since $[H^+] = 10^{-9.00}$ M, $[OH^-] = 10^{-5.00}$ M. Putting this value $[OH^-]$ into Equation 2 gives

$$[HCN] = \frac{K_B}{[OH^-]}[CN^-] = 1.6[CN^-]$$

Substituting into Equation 1 gives

$$[Ag^+] = [CN^-] + 1.6[CN^-] = 2.6[CN^-]$$

Substituting into Equation 3 gives

$$[Ag^+]\left(\frac{[Ag^+]}{2.6}\right) = K_{sp} \Rightarrow [Ag^+] = 2.4 \times 10^{-8} \text{ M}$$

$$[CN^-] = [Ag^+]/2.6 = 9.2 \times 10^{-9} \text{ M}$$

$$[HCN] = 1.6[CN^-] = 1.5 \times 10^{-8} \text{ M}$$

(b) With activities:

$$[Ag^+] = [CN^-] + [HCN] \tag{1'}$$

$$K_B = \frac{[HCN]\gamma_{HCN}[OH^-]\gamma_{OH^-}}{[CN^-]\gamma_{CN^-}} \tag{2'}$$

$$K_{sp} = [Ag^+]\gamma_{Ag^+}[CN^-]\gamma_{CN^-} \tag{3'}$$

$$K_w = [H^+]\gamma_{H^+}[OH^-]\gamma_{OH^-} \tag{4'}$$

Since pH $= 9.00$, $[OH^-]\gamma_{OH^-} = K_w/[H^+]\gamma_{H^+} = 10^{-5.00}$. Putting this value into Equation 2' gives

$$[HCN] = \frac{K_B\gamma_{CN^-}[CN^-]}{\gamma_{HCN}[OH^-]\gamma_{OH^-}}$$

$$= \frac{(1.6 \times 10^{-5})(0.755)[CN^-]}{1 \cdot 10^{-5.00}}$$

$$= 1.208[CN^-]$$

Here we have assumed $\gamma_{HCN} = 1$. Using the above relation in the mass balance (Equation 1') gives

$$[Ag^+] = 2.208[CN^-]$$

Substituting into Equation 3' gives

$$K_{sp} = [Ag^+](0.75)\left(\frac{Ag^+}{2.208}\right)(0.755) \Rightarrow [Ag^+]$$

$$= 2.9 \times 10^{-8} \text{ M}$$

$$[CN^-] = [Ag^+]/2.208 = 1.3 \times 10^{-8} \text{ M}$$

$$[HCN] = 1.208[CN^-] = 1.6 \times 10^{-8} \text{ M}$$

7-F.   Charge balance: Invalid because pH is fixed.

Mass balance:

$$[Zn^{2+}] = [C_2O_4^{2-}] + [HC_2O_4^-] \tag{1}$$
$$+ [H_2C_2O_4]$$

Equilibria:

$$K_{sp} = [Zn^{2+}][C_2O_4^{2-}] \tag{2}$$

$$K_{B1} = \frac{[HC_2O_4^-][OH^-]}{[C_2O_4^{2-}]} \tag{3}$$

$$K_{B2} = \frac{[H_2C_2O_4][OH^-]}{[HC_2O_4^-]} \tag{4}$$

$$K_w = [H^+][OH^-] \tag{5}$$

If pH $= 3.0$, $[OH^-] = 1.0 \times 10^{-11}$. Putting this into Equation 3 gives

$$[HC_2O_4^-] = \frac{K_{B1}}{[OH^-]}[C_2O_4^{2-}] = 18[C_2O_4^{2-}]$$

Using this result in Equation 4 gives

$$[H_2C_2O_4] = \frac{K_{B2}}{[OH^-]}[HC_2O_4^-]$$

$$= \frac{K_{B2}}{[OH^-]}\frac{K_{B1}}{[OH^-]}[C_2O_4^{2-}]$$

$$= 0.324[C_2O_4^{2-}]$$

Using these values of $[H_2C_2O_4]$ and $[HC_2O_4^-]$ in Equation 1 gives

$$[Zn^{2+}] = [C_2O_4^{2-}](1 + 18 + 0.324)$$

$$= [C_2O_4^{2-}](19.324)$$

Putting this last equality into Equation 2 produces the result

$$[Zn^{2+}]\left(\frac{[Zn^{2+}]}{19.324}\right) = K_{sp} \Rightarrow [Zn^{2+}] = 3.8 \times 10^{-4} \text{ M}$$

If $[Zn^{2+}] = 3.8 \times 10^{-4}$ M, this many moles of ZnC_2O_4 ($= 0.058$ g) must be dissolved in each liter.

Chapter 8

8-A.   One mole of ethoxyl groups produces one mole of AgI. 29.03 mg of AgI $= 0.12365$ mmol. The amount of compound analyzed is 25.42 mg/(417 g/mol) $= 0.06096$ mmol. There are

$$\frac{0.12365 \text{ mmol ethoxyl groups}}{0.06096 \text{ mmol compound}}$$

$$= 2.03 (= 2) \text{ ethoxyl groups/molecule.}$$

8-B.   There is one mole of SO_4^{2-} in each mole of each reactant and of the product. Let $x = $ g of K_2SO_4

and $y =$ g of $(NH_4)_2SO_4$.

$$x + y = 0.649 \text{ g} \quad (1)$$

$$\underbrace{\frac{x}{174.27}}_{\substack{\text{Moles of} \\ K_2SO_4}} + \underbrace{\frac{y}{132.14}}_{\substack{\text{Moles of} \\ (NH_4)_2SO_4}} = \underbrace{\frac{0.977}{233.40}}_{\substack{\text{Moles of} \\ BaSO_4}} \quad (2)$$

Making the substitution $y = 0.649 - x$ in Equation 2 gives $x = 0.397$ g $= 61.1\%$ of the sample.

8-C. Formula and atomic weights: Ba(137.34), Cl(35.453), K(39.102), H_2O(18.015), KCl(74.555), $BaCl_2 \cdot 2H_2O$(244.28). H_2O lost $= 1.7839 - 1.5623 = 0.2216$ g $= 1.2301 \times 10^{-2}$ mol of H_2O. For every two moles of H_2O lost, one mole of $BaCl_2 \cdot 2H_2O$ must have been present. 1.2301×10^{-2} mol of H_2O implies that 6.1504×10^{-3} mol of $BaCl_2 \cdot 2H_2O$ must have been present. This much $BaCl_2 \cdot 2H_2O$ equals 1.5024 g. The Ba and Cl contents of the $BaCl_2 \cdot 2H_2O$ are

$$Ba = \left(\frac{137.34}{244.28}\right)(1.5024 \text{ g}) = 0.84468 \text{ g}$$

$$Cl = \left(\frac{2(35.453)}{244.28}\right)(1.5024 \text{ g}) = 0.43609 \text{ g}$$

Since the total sample weighs 1.7839 g and contains 1.5024 g of $BaCl_2 \cdot 2H_2O$, the sample must contain $1.7839 - 1.5024 = 0.2815$ g of KCl, which contains

$$K = \left(\frac{39.102}{74.555}\right)(0.2815) = 0.14764 \text{ g}$$

$$Cl = \left(\frac{35.453}{74.555}\right)(0.2815) = 0.13386 \text{ g}$$

Weight percent of each element:

$$Ba = \frac{0.84468}{1.7839} = 47.35\%$$

$$K = \frac{0.14764}{1.7839} = 8.276\%$$

$$Cl = \frac{0.43609 + 0.13386}{1.7839} = 31.95\%$$

8-D. Let $x =$ weight of $Al(BF_4)_3$ and $y =$ weight of $Mg(NO_3)_3$. We can say that $x + y = 0.2828$ g. We also know that

$$\underset{\substack{\text{Moles of nitron} \\ \text{tetrafluoroborate}}}{} = 3(\text{moles of } Al(BF_4)_3) = \frac{3x}{287.39}$$

$$\underset{\substack{\text{Moles of nitron} \\ \text{nitrate}}}{} = 2(\text{moles of } Mg(NO_3)_2) = \frac{2y}{148.31}$$

Equating the mass of product to the mass of nitron tetrafluoroborate plus the mass of nitron nitrate, we can write

$$1.322 = \underbrace{\left(\frac{3x}{287.39}\right)(400.18)}_{\substack{\text{Mass of} \\ \text{product}}} + \underbrace{\left(\frac{2y}{148.31}\right)(375.39)}_{\substack{\text{Mass of nitron} \\ \text{nitrate}}}$$

$$\underset{\substack{\text{Mass of} \\ \text{product}}}{} \quad \underset{\substack{\text{Mass of nitron} \\ \text{tetrafluoroborate}}}{} \quad \underset{\substack{\text{Mass of nitron} \\ \text{nitrate}}}{}$$

Making the substitution $x = 0.2828 - y$ in the above equation allows us to find $y = 0.1589$ g of $Mg(NO_3)_2 = 1.072$ mmol of $Mg = 0.02605$ g of $Mg = 9.210\%$ of the original solid sample.

Chapter 9

9-A. (a) Formula weight of ascorbic acid $= 176.126$
0.1970 g of ascorbic acid $= 1.1185$ mmol
Molarity of $I_3^- = 1.1185$ mmol/29.41 mL $= 0.03803$ M
(b) 31.63 mL of $I_3^- = 1.203$ mmol of $I_3^- = 1.203$ mmol of ascorbic acid $= 0.2119$ g $= 49.94\%$ of the tablet

9-B. 34.02 mL of 0.08771 M NaOH $= 2.9839$ mmol of OH^-. Let x be the mass of malonic acid and y be the mass of anilinium chloride. Then $x + y = 0.2376$ g and

(Moles of anilinium chloride)

$$+ 2(\text{moles of malonic acid}) = 0.0029839$$

$$\frac{y}{129.59} + 2\left(\frac{x}{104.06}\right) = 0.0029839$$

Substituting $y = 0.2376 - x$ gives $x = 0.09997$ g $= 42.07\%$ malonic acid. Anilinium chloride $= 57.93\%$.

9-C. The reaction is $SCN^- + Cu^+ \rightarrow CuSCN(s)$. The equivalence point occurs when moles of $Cu^+ =$ Moles of $SCN^- \Rightarrow V_e = 100.0$ mL. Before the equivalence point, there is excess SCN^- remaining in the solution. We calculate the molarity of SCN^- and then find $[Cu^+]$ from the relation $[Cu^+] = K_{sp}/[SCN^-]$. For example, when 0.10 mL of Cu^+ has been added,

$$[SCN^-] = \left(\frac{100.0 - 0.10}{100.0}\right)(0.0800)\left(\frac{50.0}{50.1}\right)$$

$$= 7.98 \times 10^{-2} \text{ M}$$

$[Cu^+] = 4.8 \times 10^{-15}/7.98 \times 10^{-2}$
$= 6.0 \times 10^{-14}$
$pCu^+ = 13.22$

At the equivalence point, $[Cu^+][SCN^-] = x^2 = K_{sp} \Rightarrow x = [Cu^+] = 6.9 \times 10^{-8} \Rightarrow pCu^+ = 7.16$.
Past the equivalence point, there is excess Cu^+. For example, when $V = 101.0$ mL,

$$[Cu^+] = (0.0400)\left(\frac{101.0 - 100.0}{151.0}\right) = 2.6 \times 10^{-4} \text{ M}$$

$pCu^+ = 3.58$

mL	pCu	mL	pCu	mL	pCu
0.10	13.22	75.0	12.22	100.0	7.16
10.0	13.10	95.0	11.46	100.1	4.57
25.0	12.92	99.0	10.75	101.0	3.58
50.0	12.62	99.9	9.75	110.0	2.60

9-D. (a) Moles of $I^- = 2$(moles of Hg_2^{2+})
$(V_e)(0.100 \text{ M}) = 2(40.0 \text{ mL})(0.0400 \text{ M}) \Rightarrow V_e = 32.0$ mL

(b) Virtually all of the Hg_2^{2+} has precipitated, along with 3.20 mmol of I^-. The ions remaining in solution are

$$[NO_3^-] = \frac{3.20 \text{ mmol}}{1.00 \text{ mL}} = 0.0320 \text{ M}$$

$$[I^-] = \frac{2.80 \text{ mmol}}{100.0 \text{ mL}} = 0.0280 \text{ M}$$

$$[K^+] = \frac{6.00 \text{ mmol}}{100.0 \text{ mL}} = 0.0600 \text{ M}$$

$$\mu = \tfrac{1}{2}\sum c_i z_i^2 = 0.0600 \text{ M}$$

(c) $\mathscr{A}_{Hg_2^{2+}} = K_{sp}/\mathscr{A}_{I^-}^2$
$= 4.5 \times 10^{-29}/(0.0280)^2(0.795)^2$
$= 9.1 \times 10^{-26} \Rightarrow pHg_2^{2+} = 25.04$

9-E. $V_e = 23.66$ mL for AgBr. At 2.00, 10.00, 22.00, and 23.00 mL, AgBr is partially precipitated and excess Br^- remains.

2.00 mL: $[Ag^+] = \dfrac{K_{sp} \text{ (for AgBr)}}{[Br^-]}$

$$= \frac{5.0 \times 10^{-13}}{\underbrace{\left(\dfrac{23.66 - 2.00}{23.66}\right)}_{\substack{\text{Fraction} \\ \text{remaining}}} \underbrace{(0.05000)}_{\substack{\text{Original} \\ \text{molarity} \\ \text{of } Br^-}} \underbrace{\left(\dfrac{40.00}{42.00}\right)}_{\substack{\text{Dilution} \\ \text{factor}}}}$$

$[Ag^+] = 1.15 \times 10^{-11}$ M $\Rightarrow pAg^+ = 10.94$

By similar reasoning, we find

10.00 mL: $pAg^+ = 10.66$
22.00 mL: $pAg^+ = 9.66$
23.00 mL: $pAg^+ = 9.25$

At 24.00, 30.00 and 40.00 mL, AgCl is precipitating and excess Cl^- remains in solution.

24.00 mL:

$$[Ag^+] = \frac{K_{sp}(\text{for AgCl})}{[Cl^-]}$$

$$= \frac{1.8 \times 10^{-10}}{\left(\dfrac{47.32 - 24.00}{23.66}\right)(0.05000)\left(\dfrac{40.00}{64.00}\right)}$$

$[Ag^+] = 5.8 \times 10^{-9}$ M $\Rightarrow pAg^+ = 8.23$

By similar reasoning, we find

30.00 mL: $pAg^+ = 8.07$
40.00 mL: $pAg^+ = 7.63$

At the second equivalence point (47.32 mL), $[Ag^+] = [Cl^-]$, and we can write

$[Ag^+][Cl^-] = x^2 = K_{sp}(\text{for AgCl})$
$\Rightarrow [Ag^+] = 1.34 \times 10^{-5}$ M
$pAg^+ = 4.87$

At 50.00 mL, there is an excess of $(50.00 - 47.32) = 2.68$ mL of Ag^+.

$$[Ag^+] = \left(\frac{2.68}{90.00}\right)(0.08454 \text{ M}) = 2.5 \times 10^{-3} \text{ M}$$

$pAg^+ = 2.60$

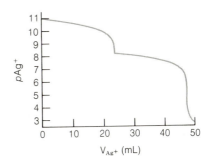

9-F. (a) 12.6 mL of Ag^+ is required to precipitate I^-. $(27.7 - 12.6) = 15.1$ mL is required to precipitate SCN^-.

SCN^-

$= \dfrac{\text{Moles of } Ag^+ \text{ needed to react with } SCN^-}{\text{Original volume of } SCN^-}$

$= \dfrac{\left[\begin{array}{c} 27.7\,(\pm 0.3) \\ -12.6\,(\pm 0.4) \end{array}\right][0.068\,3\,(\pm 0.000\,1)]}{50.00\,(\pm 0.05)}$

$= \dfrac{[15.1\,(\pm 0.5)][0.068\,3\,(\pm 0.000\,1)]}{50.00\,(\pm 0.05)}$

$= \dfrac{[15.1\,(\pm 3.31\%)][0.068\,3\,(\pm 0.146\%)]}{50.00\,(\pm 0.100\%)}$

$= 0.020\,6\,(\pm 0.000\,7)$ M

(b) SCN^- $(\pm 4.0\%)$

$= \dfrac{[27.7\,(\pm 0.3) - 12.6\,(\pm\,?)][0.068\,3\,(\pm 0.000\,1)]}{50.00\,(\pm 0.05)}$

Let the error in 15.1 mL be $y\%$:

$(4.0\%)^2 = (y\%)^2 + (0.146\%)^2 + (0.100\%)^2$

$\Rightarrow y = 4.00\% = 0.603$ mL

$27.7\,(\pm 0.3) - 12.6\,(\pm\,?) = 15.1\,(\pm 0.603)$

$\Rightarrow 0.3^2 + ?^2 = 0.603^2 \Rightarrow\,? = 0.5$ mL

9-G. The transmittance should decrease, and the scattering should increase until the equivalence point is reached. Note that the transmittance is related logarithmically to the absorbance, and hence logarithmically to the particle concentration.

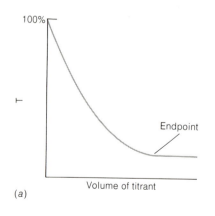

(a)

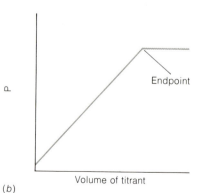

(b)

Chapter 10

10-A. (a) Neutral—neither Na^+ nor Br^- has any acidic or basic properties.
(b) Basic—$CH_3CO_2^-$ is the conjugate base of acetic acid, and Na^+ is neither acidic nor basic.
(c) Acidic—NH_4^+ is the conjugate acid of NH_3, and Cl^- is neither acidic nor basic.
(d) Basic—PO_4^{3-} is a base, and K^+ is neither acidic nor basic.
(e) Neutral—Neither ion is acidic or basic.
(f) Basic—The quaternary ammonium ion is neither acidic nor basic, and the $C_6H_5CO_2^-$ anion is the conjugate base of benzoic acid.

10-B. $K_{B1} = K_w/K_{A2} = 4.4 \times 10^{-9}$
$K_{B2} = K_w/K_{A1} = 1.6 \times 10^{-10}$

10-C. $K = K_{B2} = K_w/K_{A2} = 1.0 \times 10^{-8}$

10-D. $[H^+][OH^-] = x^2 = K_w \Rightarrow x = \sqrt{K_w} \Rightarrow pH = -\log\sqrt{K_w} = 7.471\,8$ at 0°C, 7.083 4 at 20°C and 6.767 4 at 40°C. Since $\mu \approx 0$, $\gamma \approx 1$; thus, the answers would hardly be affected by the inclusion of activity coefficients.

Chapter 11

11-A. $pH = -\log \mathscr{A}_{H^+}$. But $\mathscr{A}_{H^+}\mathscr{A}_{OH^-} = K_w \Rightarrow \mathscr{A}_{H^+} = K_w/\mathscr{A}_{OH^-}$. For 1.0×10^{-2} M NaOH, $[OH^-] = 1.0 \times 10^{-2}$ M and $\gamma_{OH^-} = 0.900$ (using Table 6-1, with ionic strength = 0.010 M).

$\mathscr{A}_{H^+} = K_w/[OH^-]\gamma_{OH^-}$
$= 1.0 \times 10^{-14}/(1.0 \times 10^{-2})(0.900)$
$= 1.11 \times 10^{-12} \Rightarrow pH$
$= -\log \mathscr{A}_{H^+} = 11.95$.

11-B. (a) Charge balance: $[H^+] = [OH^-] + [Br^-]$
Mass balance: $[Br^-] = 1.0 \times 10^{-8}$ M
Equilibrium: $[H^+][OH^-] = K_w$
Setting $[H^+] = x$ and $[Br^-] = 1.0 \times 10^{-8}$ M, the charge balance tells us that $[OH^-] = x - 1.0 \times 10^{-8}$. Putting this into the K_w equilibrium gives

$$(x)(x - 1.0 \times 10^{-8}) = 1.0 \times 10^{-14} \Rightarrow x$$
$$= 1.0_5 \times 10^{-7} \text{ M} \Rightarrow pH = 6.98$$

(b) Charge balance: $[H^+] = [OH^-] + 2[SO_4^{2-}]$
Mass balance: $[SO_4^{2-}] = 1.0 \times 10^{-8}$ M
Equilibrium: $[H^+][OH^-] = K_w$
As above, writing $[H^+] = x$ and $[SO_4^{2-}] = 1.0 \times 10^{-8}$ M gives $[OH^-] = x - 2.0 \times 10^{-8}$ and $[H^+][OH^-] = (x)[x-(2.0 \times 10^{-8})] = 1.0 \times 10^{-14} \Rightarrow x = 1.10 \times 10^{-7}$ M $\Rightarrow$ pH = 6.96

11-C.

2-nitrophenol   F.W. = 139.110

$C_6H_5NO_3$ $K_A = 6.2 \times 10^{-8}$

$$F_{HA}(\text{formal concentration}) = \frac{1.23 \text{ g}/(139.110 \text{ g/mol})}{0.250 \text{ L}}$$

$$= 0.035 \, 4 \text{ M}$$

$$HA \rightleftharpoons H^+ + A^-$$
$$F - x \qquad x \qquad x$$

$$\frac{x^2}{0.035 \, 4 - x} = 6.2 \times 10^{-8} \Rightarrow$$

$$x = 4.7 \times 10^{-5} \text{ M} \Rightarrow$$
$$pH = -\log x = 4.33$$

11-D.

$$\rightleftharpoons H^+ +$$
$$F - x \qquad x \qquad x$$

But $[H^+] = 10^{-pH} = 8.9 \times 10^{-7}$ M $\Rightarrow [A^-] = 8.9 \times 10^{-7}$ M and $[HA] = 0.010 - [H^+] = 0.010$.

$$K_A = \frac{[H^+][A^-]}{[HA]} = \frac{(8.9 \times 10^{-7})^2}{0.010}$$

$$= 7.9 \times 10^{-11} \Rightarrow pK_A = 10.10$$

11-E. As $[HA] \to 0$, pH $\to 7$. If pH = 7,

$$\frac{[H^+][A^-]}{[HA]} = K_A \Rightarrow [A^-] = \frac{K_A}{[H^+]}[HA]$$

$$= \frac{10^{-5.00}}{10^{-7.00}} = 100[HA]$$

$$\alpha = \frac{[A^-]}{[HA] + [A^-]} = \frac{100[HA]}{[HA] + 100[HA]}$$

$$= \frac{100}{101} = 99\%$$

If $pK_A = 9.00$, we find $\alpha = 0.99\%$.

11-F. $CH_3CH_2CH_2CO_2^- + H_2O \rightleftharpoons$
$\quad F - x$

$$CH_3CH_2CH_2CO_2H + OH^-$$
$$\qquad x \qquad\qquad\qquad x$$

$$K_B = \frac{K_w}{K_A} = 6.58 \times 10^{-10}$$

$$\frac{x^2}{F - x} = K_B \Rightarrow x = 5.7_4 \times 10^{-6} \text{ M}$$

$$pH = -\log(K_w/x) = 8.76$$

11-G. (a) $CH_3CH_2NH_2 + H_2O \rightleftharpoons$
$\qquad\quad F - x$

$$CH_3CH_2NH_3^+ + OH^-$$
$$\qquad x \qquad\qquad x$$

Since pH = 11.80, $[OH^-] = K_w/10^{-pH} = 6.3 \times 10^{-3}$ M = $[BH^+]$. $[B] = F - x = 0.094$ M.

$$K_B = \frac{[BH^+][OH^-]}{[B]} = \frac{(6.3 \times 10^{-3})^2}{0.094}$$

$$= 4.2 \times 10^{-4}$$

(b) $CH_3CH_2NH_3^+ \rightleftharpoons CH_3CH_2NH_2 + H^+$
$\quad F - x \qquad\qquad x \qquad\qquad x$

$$K_A = \frac{K_w}{K_B} = 2.4 \times 10^{-11}$$

$$\frac{x^2}{F - x} = K_A \Rightarrow x = 1.5_5 \times 10^{-6} \text{ M}$$

$$\Rightarrow pH = 5.81$$

11-H.

Compound	pK_A (for conjugate acid)	
ammonia	9.24	← Most suitable, since pK_A is closest to pH.
aniline	4.60	
hydrazine	8.48	
pyridine	5.23	

11-I. pH = $4.25 + \log 0.75 = 4.13$

11-J. (a) $\quad pH = pK_2 \text{ (for } H_2CO_3) + \log \dfrac{[CO_3^{2-}]}{[HCO_3^-]}$

$$10.80 = 10.329 + \log \dfrac{(4.00/138.206)}{(x/84.007)}$$

$$\Rightarrow x = 0.822 \text{ g}$$

(b)
$$CO_3^{2-} \ + \ H^+ \ \rightarrow HCO_3^-$$

Initial moles:	$0.028\,9_4$	$0.010\,0$	$0.009\,78$
Final moles:	$0.018\,9_4$	—	$0.019\,7_8$

$$pH = 10.329 + \log \dfrac{0.018\,9_4}{0.019\,7_8} = 10.31$$

(c)
$$CO_3^{2-} \ + H^+ \rightarrow HCO_3^-$$

Initial moles:	$0.028\,9_4$	x	—
Final moles:	$0.028\,9_4 - x$	—	x

$$10.00 = 10.329 + \log \dfrac{0.028\,9_4 - x}{x}$$

$$\Rightarrow x = 0.019\,7 \text{ mol}$$

$$\Rightarrow \text{Volume} = \dfrac{0.019\,7 \text{ mol}}{0.320 \text{ M}} = 61.6 \text{ mL}$$

11-K. $H_2C_2O_4$ — F.W. = 90.035. 3.38 g = 0.037 5 mol

$$H_2Ox \xrightarrow{OH^-} HOx^- \xrightarrow{OH^-} Ox^{2-}$$

The pH of HOx^- is approximately $\frac{1}{2}(pK_1 + pK_2) = 2.76$. At pH = 2.40, the predominant species will be H_2Ox and HOx^-.

$$H_2Ox \ + OH^- \rightarrow HOx^- + H_2O$$

Initial moles:	$0.037\,5$	x	—
Final moles:	$0.037\,5 - x$	—	x

$$pH = pK_1 + \log \dfrac{[HOx^-]}{[H_2Ox]}$$

$$2.40 = 1.252 + \log \dfrac{x}{0.037\,5 - x}$$

$$x = 0.035\,0 \text{ mol} \Rightarrow \text{Volume} = \dfrac{0.035\,0 \text{ mol}}{0.800 \text{ M}}$$

$$= 43.8 \text{ mL}$$

(Note that this problem involves a rather low pH, at which Equations 11-63 and 11-64 would be more appropriate to use.)

11-L. (a) $\quad H_2SO_3 \ \rightleftharpoons HSO_3^- + H^+$

$\qquad\qquad 0.050 - x \qquad x \qquad x$

$$\dfrac{x^2}{0.050 - x} = K_1 = 1.23 \times 10^{-2} \Rightarrow x = 1.94 \times 10^{-2}$$

$$[HSO_3^-] = [H^+] = 1.94 \times 10^{-2} \text{ M} \Rightarrow pH = 1.71$$

$$[H_2SO_3] = 0.050 - x = 0.031 \text{ M}$$

$$[SO_3^{2-}] = \dfrac{K_2[HSO_3^-]}{[H^+]} = K_2 = 6.6 \times 10^{-8} \text{ M}$$

(b) $\quad [H^+] = \sqrt{\dfrac{K_1 K_2(.050) + K_1 K_w}{K_1 + (.050)}}$

$$= 2.55 \times 10^{-5} \Rightarrow pH = 4.59$$

$$[H_2SO_3] = \dfrac{[H^+][HSO_3^-]}{K_1}$$

$$= \dfrac{(2.55 \times 10^{-5})(0.050)}{1.23 \times 10^{-2}} = 1.0 \times 10^{-4} \text{ M}$$

$$[SO_3^{2-}] = \dfrac{K_2[HSO_3^-]}{[H^+]} = 1.3 \times 10^{-4} \text{ M}$$

$$[HSO_3^-] = 0.050 \text{ M}$$

(c) $\quad SO_3^{2-} \ + H_2O \rightleftharpoons HSO_3^- + OH^-$

$\qquad\quad 0.050 - x \qquad\qquad x \qquad x$

$$\dfrac{x^2}{0.050 - x} = K_{B1} = \dfrac{K_w}{K_{A2}} = 1.52 \times 10^{-7}$$

$$[HSO_3^-] = x = 8.7 \times 10^{-5} \text{ M}$$

$$[H^+] = K_w/x = 1.15 \times 10^{-10} \text{ M} \Rightarrow pH = 9.94$$

$$[SO_3^{2-}] = 0.050 - x = 0.050 \text{ M}$$

$$[H_2SO_3] = \dfrac{[H^+][HSO_3^-]}{K_1} = 8.1 \times 10^{-13}$$

11-M. (a) Calling the three forms of glutamine H_2G^+, HG, and G^-, the form shown is HG.

$$[H^+] = \sqrt{\dfrac{K_1 K_2(0.010) + K_1 K_w}{K_1 + 0.010}}$$

$$= 1.9_9 \times 10^{-6} \Rightarrow pH = 5.70$$

(b) Calling the four forms of cysteine H_3C^+, H_2C, HC^-, and C^{2-}, the form shown is HC^-.

$$[H^+] = \sqrt{\dfrac{K_2 K_3(0.010) + K_2 K_w}{K_2 + 0.010}}$$

$$= 2.8_1 \times 10^{-10} \Rightarrow pH = 9.55$$

(c) Calling the four forms of arginine H_3A^{2+}, H_2A^+, HA, and A^-, the form shown is HA.

$$[H^+] = \sqrt{\frac{K_2K_3(0.010) + K_2K_w}{K_2 + 0.010}}$$

$$= 3.6_8 \times 10^{-11} \Rightarrow pH = 10.43$$

11-N. The reaction of phenylhydrazine with water is

$$B + H_2O \rightleftharpoons BH^+ + OH^- \qquad K_B$$

We know that pH = 8.13, so we can find $[OH^-]$.

$$[OH^-] = \frac{\mathscr{A}_{OH^-}}{\gamma_{OH^-}} = \frac{K_w/10^{-pH}}{\gamma_{OH^-}} = 1.78 \times 10^{-6} \text{ M}$$

(using $\gamma_{OH^-} = 0.76$ for $\mu = 0.10$ M).

$$K_B = \frac{[BH^+]\gamma_{BH^+}[OH^-]\gamma_{OH^-}}{[B]\gamma_B}$$

$$= \frac{(1.78 \times 10^{-6})(0.80)(1.78 \times 10^{-6})(0.76)}{[0.010 - (1.78 \times 10^{-6})](1.00)}$$

$$= 1.93 \times 10^{-10}.$$

$$K_A = K_w/K_B = 5.19 \times 10^{-5} \Rightarrow pK_A = 4.28$$

To find K_B, we made use of the equality $[BH^+] = [OH^-]$.

Chapter 12

12-A. The titration reaction is $H^+ + OH^- \rightarrow H_2O$ and $V_e = 5.00$ mL. Three representative calculations are given:

$$1.00 \text{ mL}: [OH^-] = \left(\frac{4.00}{5.00}\right)(0.010\,0)\left(\frac{50.00}{51.00}\right)$$

$$= 0.007\,84 \text{ M}$$

$$pH = -\log(K_w/[OH^-]) = 11.89$$

$$5.00 \text{ mL}: H_2O \rightleftharpoons H^+ + OH^-$$
$$\phantom{5.00 \text{ mL}: H_2O \rightleftharpoons} x \quad\;\; x$$

$$x^2 = K_w \Rightarrow x = 1.0 \times 10^{-7} \text{ M}$$

$$pH = -\log x = 7.00$$

$$5.01 \text{ mL}: [H^+] = \left(\frac{0.01}{55.01}\right)(0.100)$$

$$= 1.82 \times 10^{-5} \text{ M} \Rightarrow pH = 4.74$$

V_a(mL)	pH	V_a	pH	V_a	pH
0.00	12.00	4.50	10.96	5.10	3.74
1.00	11.89	4.90	10.26	5.50	3.05
2.00	11.76	4.99	9.26	6.00	2.75
3.00	11.58	5.00	7.00	8.00	2.29
4.00	11.27	5.01	4.74	10.00	2.08

12-B. The titration reaction is $HCO_2H + OH^- \rightarrow HCO_2^-$ $+ H_2O$ and $V_e = 50.0$ mL. For formic acid, $K_A = 1.80 \times 10^{-4}$. Four representative calculations are given:

$$0.0 \text{ mL}: HA \rightleftharpoons H^+ + A^-$$
$$0.050\,0 - x \qquad x \qquad x$$

$$\frac{x^2}{0.050\,0 - x} = K_A \Rightarrow x = 2.91 \times 10^{-3}$$

$$\Rightarrow pH = 2.54$$

$$48.0 \text{ mL}: \qquad HA + OH^- \rightarrow A^- + H_2O$$

Initial:	50	48	—	—
Final:	2	—	48	48

$$pH = pK_A + \log\frac{[A^-]}{[HA]} = 3.745 + \log\frac{48.0}{2.0} = 5.13$$

$$50.0 \text{ mL}: \; A^- + H_2O \xrightarrow{K_B} HA + OH^-$$
$$\phantom{50.0 \text{ mL}: \;} F - x \qquad\qquad x \quad\;\; x$$

$$K_B = \frac{K_w}{K_A} \quad \text{and} \quad F = \left(\frac{50}{100}\right)(0.05)$$

$$\frac{x^2}{0.025\,0 - x} = 5.56 \times 10^{-11} \Rightarrow x = 1.18 \times 10^{-6} \text{ M}$$

$$pH = -\log(K_w/x) = 8.07$$

$$60.0 \text{ mL}: [OH^-] = \left(\frac{10.0}{110.0}\right)(0.050\,0)$$

$$= 4.55 \times 10^{-3} \text{ M} \Rightarrow pH = 11.66$$

V_b(mL)	pH	V_b	pH	V_b	pH
0.0	2.54	45.0	4.70	50.5	10.40
10.0	3.14	48.0	5.13	51.0	10.69
20.0	3.57	49.0	5.44	52.0	10.99
25.0	3.74	49.5	5.74	55.0	11.38
30.0	3.92	50.0	8.07	60.0	11.66
40.0	4.35				

12-C. The titration reaction is $B + H^+ \rightarrow BH^+$ and $V_e = 50.00$ mL. Representative calculations:

$V_a = 0.0$ mL:
$$B + H_2O \rightleftharpoons BH^+ + OH^-$$
$$0.100 - x \qquad\qquad x \qquad x$$

$$\frac{x^2}{0.100 - x} = 2.6 \times 10^{-6} \Rightarrow x = 5.09 \times 10^{-4}$$

$$pH = -\log(K_w/x) = 10.71$$

$V_a = 20.0$ mL:
$$B + H^+ \rightarrow BH^+$$

Initial:	50.00	20.0	—
Final:	30.0	—	20.0

$$pH = pK_A \text{ (for } BH^+\text{)} + \log \frac{[B]}{[BH^+]}$$

$$= 8.41 + \log \frac{30.0}{20.0} = 8.59$$

$V_a = V_e = 50.0$ mL: All B has been converted to the conjugate acid, BH^+. The formal concentration of BH^+ is $\left(\frac{100}{150}\right)(0.100) = 0.066\,7$ M. The pH is determined by the reaction

$$BH^+ \rightleftharpoons B + H^+$$
$$0.066\,7 - x \qquad x \qquad x$$

$$\frac{x^2}{0.066\,7 - x} = K_A = \frac{K_w}{K_B} \Rightarrow x$$

$$= 1.60 \times 10^{-5} \Rightarrow pH = 4.80$$

$V_a = 51.0$ mL: There is excess H^+:

$$[H^+] = \left(\frac{1.0}{151.0}\right)(0.200) = 1.32 \times 10^{-3} \Rightarrow pH = 2.88$$

V_a(mL)	pH	V_a	pH	V_a	pH
0.0	10.71	30.0	8.23	50.0	4.80
10.0	9.01	40.0	7.81	50.1	3.88
20.0	8.59	49.0	6.72	51.0	2.88
25.0	8.41	49.9	5.71	60.0	1.90

12-D. The titration reactions are

$$HO_2CCH_2CO_2H + OH^- \rightarrow$$
$$^-O_2CCH_2CO_2H + H_2O$$
$$^-O_2CCH_2CO_2H + OH^- \rightarrow$$
$$^-O_2CCH_2CO_2^- + H_2O$$

and the equivalence points occur at 25.0 and 50.0 mL. We will designate malonic acid as H_2M.

0.0 mL:
$$H_2M \rightleftharpoons H^+ + HM^-$$
$$0.050\,0 - x \qquad x \qquad x$$

$$\frac{x^2}{0.050\,0 - x} = K_1 \Rightarrow x = 7.75 \times 10^{-3} \Rightarrow pH = 2.11$$

8.0 mL:
$$H_2M + OH^- \rightarrow HM^- + H_2O$$

Initial:	25	8	—
Final:	20	—	8

$$pH = pK_1 + \log \frac{[HM^-]}{[H_2M]} = 2.847 + \log \frac{8}{17} = 2.52$$

12.5 mL: $V_b = \frac{1}{4}V_e \Rightarrow pH = pK_1 = 2.85$

19.3 mL:
$$H_2M + OH^- \rightarrow HM^- + H_2O$$

Initial:	25	19.3	—
Final:	5.7	—	19.3

$$pH = pK_1 + \log \frac{19.3}{5.7} = 3.38$$

25.0 mL: At the first equivalence point, H_2M has been converted to HM^-.

$$[H^+] = \sqrt{\frac{K_1 K_2 F + K_1 K_w}{K_1 + F}},$$

$$\text{where } F = \left(\frac{50}{75}\right)(0.050\,0)$$

$$[H^+] = 5.23 \times 10^{-5} \text{ M} \Rightarrow pH = 4.28$$

37.5 mL: $V_b = \frac{3}{2}V_e \Rightarrow pH = pK_2 = 5.70$

50.0 mL: At the second equivalence point H_2M has been converted to M^{2-}.

$$M^{2-} + H_2O \rightleftharpoons HM^- + OH^-$$
$$\left(\frac{50}{100}\right)(0.050\,0) - x \qquad x \qquad x$$

$$\frac{x^2}{0.025\,0 - x} = K_{B1} = \frac{K_w}{K_{A2}}$$

$$\Rightarrow x = 1.12 \times 10^{-5} \text{ M}$$

$$\Rightarrow pH = -\log(K_w/x) = 9.05$$

56.3 mL: There is 6.3 mL of excess NaOH.

$$[OH^-] = \left(\frac{6.3}{106.3}\right)(0.100) = 5.93 \times 10^{-3} \text{ M}$$

$$\Rightarrow pH = 11.77$$

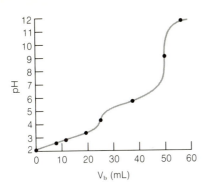

12-E.

$$\underset{HHis}{\overset{NH_3^+}{\underset{CO_2^-}{CHCH_2}}\diagup\text{imidazole}} \xrightarrow{H^+}$$

$$\underset{H_2His^+}{\overset{NH_3^+}{\underset{CO_2^-}{CHCH_2}}\diagup\text{imidazole-}^+H} \xrightarrow{H^+} \underset{H_3His^{2+}}{\overset{NH_3^+}{\underset{CO_2H}{CHCH_2}}\diagup\text{imidazole-}^+H}$$

The equivalence points occur at 25.0 and 50.0 mL.

0 mL: HHis is the second intermediate form derived from the triprotic acid, H_3His^{2+}.

$$[H^+] = \sqrt{\frac{K_2K_3(0.050\,0) + K_2K_w}{K_2 + (0.050\,0)}}$$

$$= 2.81 \times 10^{-8} \text{ M} \Rightarrow pH = 7.55$$

4.0 mL: $\qquad HHis + H^+ \rightarrow H_2His^+$

Initial:	25	4	—
Final:	21	—	4

$$pH = pK_2 + \log\frac{21}{4} = 6.74$$

12.5 mL: $pH = pK_2 = 6.02$

25.0 mL: The histidine has been converted to H_2His^+ at the formal concentration $F = (\frac{25}{50}) \times (0.050\,0) = 0.025\,0$ M.

$$[H^+] = \sqrt{\frac{K_1K_2F + K_1K_w}{K_1 + F}}$$

$$= 1.03 \times 10^{-4} \Rightarrow pH = 3.99$$

26.0 mL: $\qquad H_2His^+ + H^+ \rightarrow H_3His^{2+}$

Initial:	25	1	—
Final:	24	—	1

$$pH = pK_1 + \log\frac{24}{1} = 3.08$$

50.0 mL: The histidine has been converted to H_3His at the formal concentration $F = (\frac{25}{75})(0.050\,0) = 0.016\,7$ M.

$$H_3His^{2+} \rightleftharpoons H_2His^+ + H^+$$
$$0.016\,7 - x \qquad x \qquad x$$

$$\frac{x^2}{0.016\,7 - x} = K_1 \Rightarrow x = 0.010\,8 \text{ M} \Rightarrow pH = 1.97$$

12-F. Figure 12-2: bromothymol blue: blue → yellow
Figure 12-3: thymol blue: yellow → blue
Figure 12-4: thymolphthalein: colorless → blue

12-G. (a) $A = 2\,080[HIn] + 14\,200[In^-]$
(b) $[HIn] = x; [In^-] = 1.84 \times 10^{-4} - x$

$$A = 0.868 = 2\,080x + 14\,200(1.84 \times 10^{-4} - x)$$
$$\Rightarrow x = 1.44 \times 10^{-4} \text{ M}$$

$$pK_A = pH - \log\frac{[In^-]}{[HIn]}$$

$$= 6.23 - \log\frac{(1.84 \times 10^{-4}) - (1.44 \times 10^{-4})}{1.44 \times 10^{-4}}$$

$$pK_A = 6.79$$

12-H. The titration reaction is $HA + OH^- \rightarrow A^- + H_2O$. It requires one mole of NaOH to react with one mole of HA. Therefore, the formal concentration of A^- at the equivalence point is

$$\underbrace{\left(\frac{27.63}{127.63}\right)}_{\substack{\text{Dilution factor}\\\text{for NaOH}}} \times \underbrace{(0.093\,81)}_{\substack{\text{Initial concentration}\\\text{of NaOH}}} = 0.020\,31 \text{ M}$$

Since the pH is 10.99, $[OH^-] = 9.77 \times 10^{-4}$ and we can write

$$A^- + H_2O \rightleftharpoons HA + OH^-$$

$$K_B = \frac{[HA][OH^-]}{[A^-]} = \frac{(9.77 \times 10^{-4})^2}{0.020\,31 - 9.77 \times 10^{-4}}$$

$$= 4.94 \times 10^{-5}$$

$$K_A = K_w/K_B = 2.03 \times 10^{-10} \Rightarrow pK_A = 9.69$$

714

For the 19.47 mL point, we have

$$HA + OH^- \rightarrow A^- + H_2O$$

Initial:	27.63	19.47	—
Final:	8.16	—	19.47

$$pH = pK_A + \log \frac{[A^-]}{[HA]} = 9.69 + \log \frac{19.47}{8.16} = 10.07$$

12-I. When $V_b = \frac{1}{2}V_e$, $[HA] = [A^-] = 0.033\,3$ M (using a correction for dilution by NaOH). $[Na^+] = 0.033\,3$ M, as well. Ionic strength $= 0.033\,3$ M.

$$pK_A = pH - \log \frac{[A^-]\gamma_{A^-}}{[HA]\gamma_{HA}} \quad \text{(from Equation 11-56)}$$

$$pK_A = 4.62 - \log \frac{(0.033\,3)(0.854)}{(0.033\,3)(1.00)} = 4.69$$

The activity coefficient of A^- was found by interpolation in Table 6-1.

Chapter 13

13-A.

pH 9.00	pH 11.00

Principal species:

Secondary species:

Percent in major form: 66.5%   52.9%

The percent in the major form was calculated with the formulas for α_0 (Equation 13-19 at pH 9.00) and α_1 (Equation 13-20 at pH 11.00).

13-B.

pH 9.0	pH 10.0

Predominant form:

$$\underset{CO_2^-}{\overset{NH_3^+}{CHCH_2CH_2CO_2^-}} \qquad \underset{CO_2^-}{\overset{NH_2}{CHCH_2CH_2CO_2^-}}$$

Secondary form:

$$\underset{CO_2^-}{\overset{NH_2}{CHCH_2CH_2CO_2^-}} \qquad \underset{CO_2^-}{\overset{NH_3^+}{CHCH_2CH_2CO_2^-}}$$

13-C. The isoionic pH is the pH of a solution of pure neutral lysine, which is

$$\underset{CO_2^-}{\overset{NH_2}{CHCH_2CH_2CH_2CH_2NH_3^+}}$$

$$[H^+] = \sqrt{\frac{K_2 K_3 F + K_2 K_w}{K_2 + F}} \Rightarrow pH = 9.88$$

13-D. We know that the isoelectric point will be near $\frac{1}{2}(pK_2 + pK_3) \approx 9.88$. At this pH, the fraction of lysine in the form H_3L^{2+} is negligible. Therefore, the electroneutrality condition reduces to $[H_2L^+] = [L^-]$, for which the expression, Isoelectric pH $= \frac{1}{2}(pK_2 + pK_3) = 9.88$, applies.

13-E.

$$HF + NH_3 \overset{K}{\rightleftharpoons} F^- + NH_4^+$$

$$K = \frac{K_A K_B}{K_w} = \frac{(6.8 \times 10^{-4})(K_w/5.70 \times 10^{-10})}{K_w}$$

$$= 1.2 \times 10^6$$

Since $K \gg 1$, this falls under Case 1 of Section 13-4.

(a)

$$HF + NH_3 \rightleftharpoons F^- + NH_4^+$$

Relative initial amount:	20.0	14.0	—	—
Relative final amount:	6.0	—	14.0	14.0

$$pH = pK_A + \log \frac{[F^-]}{[HF]} = 3.17 + \log \frac{14.0}{6.0} = 3.54$$

(b)

$$HF + NH_3 \rightleftharpoons F^- + NH_4^+$$

Relative initial amount:	14.0	20.0	—	—
Relative final amount:	—	6.0	14.0	14.0

$$pH = pK_{BH^+} + \log \frac{[NH_3]}{[NH_4^+]}$$

$$= 9.244 + \log \frac{6.0}{14.0} = 8.88$$

13-F. (a) $[(CH_3CH_2)_4N^+][HCO_2^-]$ is simply a monoprotic weak base.

$$HCO_2^- + H_2O \rightleftharpoons HCO_2H + OH^-$$
$$0.100 - x \qquad\qquad x \qquad\quad x$$

$$\frac{x^2}{0.100 - x} = K_B = \frac{K_w}{K_A} \Rightarrow x = 2.36 \times 10^{-6} \text{ M}$$

$$\Rightarrow pH = 8.37$$

(b) $[(CH_3CH_2)_3NH^+][HCO_2^-]$ behaves as the intermediate form of a diprotic acid with $pK_1 = 3.745$ (HCO_2H) and $pK_2 = 10.715$ ($(CH_3CH_2)_3NH^+$).

$$[H^+] = \sqrt{\frac{K_1 K_2(0.100) + K_1 K_w}{K_1 + 0.100}} \Rightarrow pH = 7.23$$

13-G.
$$\underset{\substack{|\\CH-CH_2-\bigcirc-OH + NH_3 \overset{K}{\rightleftharpoons}\\|\\\overset{+}{N}H_3}}{}$$
$$\overset{|}{CO_2^-}$$

$$\text{HA} \qquad\qquad\qquad \text{B}$$

$$\underset{\substack{|\\CH-CH_2-\bigcirc-OH + NH_4^+\\|\\NH_2}}{}$$
$$\overset{|}{CO_2^-}$$

$$\text{A}^- \qquad\qquad\qquad \text{BH}^+$$

$$K = \frac{K_A K_B}{K_w} = \frac{K_2 \text{(for tyrosine)} \cdot K_B \text{(for NH}_3)}{K_w} = 1.14$$

Since K is not large, this falls under Case 2 of Section 13-4.

$$\text{HA} \quad + \quad \text{B} \quad \rightleftharpoons \text{A}^- + \text{BH}^+$$

Initial
millimoles: 0.250   0.225   —   —
Final
millimoles: $0.250 - x$ $0.225 - x$ x x

$$\frac{x^2}{(0.250 - x)(0.225 - x)} = 1.14 \Rightarrow x = 0.122$$

$$pH = pK_A + \log\frac{[A^-]}{[HA]} = 9.19 + \log\frac{.122}{.250 - .122} = 9.17$$

13-H. The answer is K_2HPO_4, since $\frac{1}{2}(pK_2 + pK_3) = 9.77$ for phosphoric acid. The pH values of the other solutions are estimated to be: potassium acid phthalate—$\frac{1}{2}(pK_1 + pK_2) = 4.18$; monosodium malonate—$\frac{1}{2}(pK_1 + pK_2) = 4.27$; glycine—

$\frac{1}{2}(pK_1 + pK_2) = 6.06$; ammonium bicarbonate—$\frac{1}{2}[K_1 \text{ (of } H_2CO_3) + K_A \text{ (for } NH_4^+)] = 7.80$.

13-I. The pH of the solution is 7.50, and the total concentration of indicator is 5.00×10^{-5} M. At pH 7.50, there is a negligible amount of H_2In, since $pK_1 = 1.00$. We can write

$$[HIn^-] + [In^{2-}] = 5.0 \times 10^{-5}$$

$$pH = pK_2 + \log\frac{[In^{2-}]}{[HIn^-]}$$

$$7.50 = 7.95 + \log\frac{[In^{2-}]}{5.00 \times 10^{-5} - [In^{2-}]}$$

$$\Rightarrow [In^{2-}] = 1.31 \times 10^{-5} \text{ M}$$
$$[HIn^-] = 3.69 \times 10^{-5} \text{ M}$$

$$A_{435} = \varepsilon_{435}[HIn^-] + \varepsilon_{435}[In^{2-}]$$
$$A_{435} = (1.80 \times 10^4)(3.69 \times 10^5)$$
$$+ (1.15 \times 10^4)(1.31 \times 10^{-5}) = 0.815$$

Chapter 14

14-A. For every mole of K^+ entering the first reaction, four moles of EDTA are produced in the second reaction.

Moles of EDTA = Moles of Zn^{2+} used in titration

$$[K^+] = (\tfrac{1}{4})(\text{moles of Zn}^{2+})/(\text{volume of original sample})$$

$$[K^+] = \frac{(\tfrac{1}{4})[28.73\,(\pm 0.03)][0.043\,7\,(\pm 0.000\,1)]}{250.0\,(\pm 0.1)}$$

$$= \frac{[\tfrac{1}{4}(\pm 0\%)][28.73(\pm 0.104\%)][0.043\,7(\pm 0.229\%)]}{250.0(\pm 0.040\,0\%)}$$

$$[K^+] = 1.256\,(\pm 0.255\%) \times 10^{-3} \text{ M}$$
$$= 1.256\,(\pm 0.003) \text{ mM}$$

14-B. Total $Fe^{3+} + Cu^{2+}$ in 25.00 mL = (16.06 mL) $\times$ (0.050 83 M) = 0.816 3 mmol.

Second titration:

Millimoles EDTA used: (25.00)(0.050 83) = 1.270 8
Millimoles Pb^{2+} needed: (19.77)(0.018 83) = 0.372 3
Millimoles Fe^{3+} present
(difference):   0.898 5
Since 50.00 mL of unknown was used in the second titration, the millimoles of Fe^{3+} in 25.00 mL are 0.449 2. The millimoles of Cu^{2+} in 25.00 mL are $0.816 3 - 0.449 2 = 0.367 1$ mmol/25.00 mL = 0.014 68 M.

14-C. Designating the total concentration of free EDTA as [EDTA], we can write

$$\frac{[Ga^{3+}][EDTA]}{[GaY^-]} = \frac{1}{\alpha_{Y^{4-}} \cdot K_f} = 1.32 \times 10^{-12}$$

Some representative calculations are shown below:

0.1 mL: $[EDTA] = \left(\dfrac{25.0-0.1}{25.0}\right)(0.040\,0)\left(\dfrac{50.0}{50.1}\right)$

$$= 0.039\,8 \text{ M}$$

$$[GaY^-] = \left(\frac{0.1}{50.1}\right)(0.080\,0)$$

$$= 1.60 \times 10^{-4} \text{ M}$$

$$[Ga^{3+}] = \frac{(1.32 \times 10^{-12})[GaY^-]}{[EDTA]}$$

$$= 5.31 \times 10^{-15} \Rightarrow pGa^{3+} = 14.28$$

25.0 mL: Formal concentration of GaY^-

$$= \left(\frac{25.0}{75.0}\right)(0.0800) = 0.026\,7 \text{ M}$$

	GaY^-	$\rightleftharpoons$	Ga^{3+}	+ EDTA
Initial concentration:	0.026 7		—	—
Final concentration:	0.026 7 − x		x	x

$$\frac{x^2}{0.0267-x} = 1.32 \times 10^{-12} \Rightarrow [Ga^{3+}] = 1.88 \times 10^{-7} \text{ M}$$

$$\Rightarrow pGa^{3+} = 6.73$$

26.0 mL: $Ga^{3+} = \left(\dfrac{1.0}{76.0}\right)(0.080\,0)$

$$= 1.05 \times 10^{-3} \text{ M} \Rightarrow pGa^{3+} = 2.98$$

Summary:

Volume (mL)	pGa^{3+}	Volume	pGa^{3+}
0.1	14.28	24.0	10.50
5.0	12.48	25.0	6.73
10.0	12.06	26.0	2.98
15.0	11.70	30.0	2.30
20.0	11.28		

14-D. $HY^{3-} \rightleftharpoons H^+ + Y^{4-}$ K_6

$H_2Y^{2-} \rightleftharpoons H^+ + HY^{3-}$ K_5

$H_2Y^{2-} \rightleftharpoons 2H^+ + Y^{4-}$ $K = K_5K_6 = \dfrac{[H^+]^2[Y^{4-}]}{[H_2Y^{2-}]}$

$$[H_2Y^{2-}] = \frac{[H^+]^2[Y^{4-}]}{K_5K_6} = \frac{[H^+]^2\alpha_{Y^{4-}}[EDTA]}{K_5K_6}$$

Using the values $[H^+] = 10^{-4.00}$, $\alpha_{Y^{4-}} = 3.8 \times 10^{-9}$, and $[EDTA] = 1.88 \times 10^{-7}$ gives $[H_2Y^{2-}] = 1.8 \times 10^{-7}$ M.

14-E. (a) One volume of Fe^{3+} will require two volumes of EDTA to reach the equivalence point. The formal concentration of FeY^- at the equivalence point is $(\frac{1}{3})(0.010\,0) = 0.003\,33$ M.

	FeY^-	$\rightleftharpoons$	Fe^{3+}	+ EDTA
	0.003 33 − x		x	x

$$\frac{x^2}{0.003\,33 - x} = \frac{1}{\alpha_{Y^{4-}} \cdot K_f} \Rightarrow x = [Fe^{3+}]$$

$$= 8.8 \times 10^{-8} \text{ M}$$

(b) Since the pH is 2.00, the *ratio* $[H_3Y^-]/[H_2Y^{2-}]$ is constant throughout the *entire* titration.

$$\frac{[H_2Y^{2-}][H^+]}{[H_3Y^-]} = K_4 \Rightarrow \frac{[H_3Y^-]}{[H_2Y^{2-}]} = \frac{[H^+]}{K_4}$$

$$= 4.6$$

14-F. The titration reaction is $Cu(NH_3)_4^{2+} + EDTA \rightleftharpoons CuY^{2+} + 4NH_3$ and $V_e = 50.00$ mL. The equilibrium constant for the titration reaction can be found as follows:

$$\frac{K_f}{\beta_4} = \frac{[CuY^{2-}][NH_3]^4}{[Cu(NH_3)_4^{2+}][Y^{4-}]}$$

$$= \frac{[CuY^{2-}][NH_3]^4}{[Cu(NH_3)_4^{2+}]\alpha_{Y^{4-}} \cdot EDTA}$$

$$K(\text{titration}) = \frac{[CuY^{2-}][NH_3]^4}{[Cu(NH_3)_4^{2+}][EDTA]} = \frac{\alpha_{Y^{4-}} \cdot K_f}{\beta_4}$$

$$= 0.85 \, K_f/\beta_4$$

$$K(\text{titration}) = 2.6 \times 10^5$$

0 mL: We need only consider the dissociation of $Cu(NH_3)_4^{2+}$ at this point. We assume that almost all of the Cu is present in the form $Cu(NH_3)_4^{2+}$.

$$[Cu^{2+}] = \frac{[Cu(NH_3)_4^{2+}]}{\beta_4[NH_3]^4} = \frac{(0.001\,00)}{(1/4.8 \times 10^{-14})(0.100)^4}$$

$$= 4.8 \times 10^{-13} \text{ M} \Rightarrow pCu^{2+} = 12.32$$

1.00 mL: Now 1/50 of the $Cu(NH_3)_4^{2+}$ has been converted to CuY^{2-}.

$[Cu(NH_3)_4^{2+}]$

$$= \underbrace{\left(\frac{50.00 - 1.00}{50.00}\right)}_{\substack{\text{Fraction} \\ \text{remaining}}} \underbrace{(0.001\ 00)}_{\substack{\text{Initial} \\ \text{concentration}}} \underbrace{\left(\frac{50.00}{51.00}\right)}_{\substack{\text{Dilution} \\ \text{factor}}}$$

$$= 9.61 \times 10^{-4}\ \text{M}$$

$$[Cu^{2+}] = \frac{[Cu(NH_3)_4^{2+}]}{\beta_4[NH_3]^4} = 4.6 \times 10^{-13}\ \text{M}$$

$$\Rightarrow pCu^{2+} = 12.34$$

45.00 mL: At this point, 9/10 of the $Cu(NH_3)_4^{2+}$ has been converted to CuY^{2-}.

$$[Cu(NH_3)_4^{2+}] = \left(\frac{50.00 - 45.00}{50.00}\right)(0.001\ 00)\left(\frac{50.00}{95.00}\right)$$

$$= 5.26 \times 10^{-5}\ \text{M}$$

$$[Cu^{2+}] = \frac{[Cu(NH_3)_4^{2+}]}{\beta_4[NH_3]^4} = 2.5 \times 10^{-14}\ \text{M}$$

$$\Rightarrow pCu^{2+} = 13.60$$

50.00 mL: The formal concentration of CuY^{2-} is

$$[CuY^{2-}] = \left(\frac{50.00}{100.00}\right)(0.001\ 00) = 0.000\ 500\ \text{M}$$

We can write

$$CuY^{2-} \quad + \quad 4NH_3 \rightleftharpoons Cu(NH_3)_4^{2+} \quad + \quad Y^{4-}$$

Initial:

0.000 500	0.100	—	—

Final:

$0.000\ 500 - x$	0.100	x	$\alpha_{Y^{4-}} \cdot x$

$$\frac{(0.85)x^2}{(.000\ 500 - x)(0.100)^4} = \frac{1}{3.0 \times 10^5}$$

$$\Rightarrow x = 4.43 \times 10^{-7}\ \text{M}$$

$$[Cu^{2+}] = \frac{[Cu(NH_3)_4^{2+}]}{\beta_4[NH_3]^4} = \frac{4.43 \times 10^{-7}}{(1/4.8 \times 10^{-14})(0.100)^4}$$

$$= 2.13 \times 10^{-16}\ \text{M}$$

$$pCu^{2+} = 15.67$$

55.00 mL: There is an excess of 5.00 mL of EDTA.

$$[EDTA] = \left(\frac{5.00}{105.00}\right)(0.001\ 00) = 4.76 \times 10^{-5}\ \text{M}$$

$$[CuY^{2-}] = \left(\frac{50.00}{105.00}\right)(0.001\ 00) = 4.76 \times 10^{-4}\ \text{M}$$

$$[Cu^{2+}] = \frac{[CuY^{2-}]}{\alpha_{Y^{4-}} \cdot K_f[EDTA]} = 1.87 \times 10^{-18}$$

$$\Rightarrow pCu^{2+} = 17.73$$

14-G. 25.0 mL of 0.120 M iminodiacetic acid $= 3.00$ mmol
25.0 mL of 0.050 0 M $Cu^{2+} = 1.25$ mmol

$$Cu^{2+} \quad + \quad \begin{matrix} 2\ \text{iminodiacetic} \\ \text{acid} \end{matrix} \rightleftharpoons CuX_2^{2+}$$

Initial millimoles:	1.25	3.00	—
Final millimoles:	—	0.50	1.25

$$\frac{[CuX_2^{2-}]}{[Cu^{2+}][X^{2-}]^2} = K_f$$

$$\frac{[1.25/50.0]}{[Cu^{2+}][(0.50/50.0)(4.6 \times 10^{-3})]^2} = 3.5 \times 10^{16}$$

$$\Rightarrow [Cu^{2+}] = 3.4 \times 10^{-10}\ \text{M}$$

Chapter 15

15-A. The cell potential will be 1.35 V because all activities are unity.

$$I = P/E = 0.010\ 0\ \text{W}/1.35\ \text{V} = 7.41 \times 10^{-3}\ \text{C/s}$$
$$= 7.68 \times 10^{-8}\ \text{mol e}^-/\text{s} = 2.42\ \text{mol e}^-/365\ \text{days}$$
$$= 1.21\ \text{mol HgO}/365\ \text{days} = 0.262\ \text{kg HgO}$$
$$= 0.578\ \text{lb}$$

15-B. (a) $I_2(s) + 6H_2O \rightleftharpoons$

$$\begin{array}{ll} 2IO_3^- + 12H^+ + 10e^- & E^0 = 1.195\ \text{V} \\ 5Br_2(aq) + 10e^- \rightleftharpoons 10Br^- & E^0 = 1.087\ \text{V} \\ \hline I_2(s) + 5Br_2(aq) + 6H_2O \rightleftharpoons & E^0 = 1.087 - 1.195 \\ 2IO_3^- + 10Br^- + 12H^+ & = -0.108\ \text{V} \end{array}$$

$$K = 10^{10(-0.108)/0.059\ 16} = 6 \times 10^{-19}$$

(b)
$$\begin{array}{ll} Cr^{2+} + 2e^- \rightleftharpoons Cr(s) & E^0 = -0.91\ \text{V} \\ Fe(s) \rightleftharpoons Fe^{2+} + 2e^- & E^0 = -0.440\ \text{V} \\ \hline Cr^{2+} + Fe(s) \rightleftharpoons & E^0 = -0.91 - (-0.440) \\ Cr(s) + Fe^{2+} & = -0.47\ \text{V} \end{array}$$

$$K = 10^{2(-0.47)/0.059\ 16} = 10^{-16}$$

(c)
$$\begin{array}{ll} Mg(s) \rightleftharpoons Mg^{2+} + 2e^- & E^0 = -2.375\ \text{V} \\ Cl_2(g) + 2e^- \rightleftharpoons 2Cl^- & E^0 = 1.358\ \text{V} \\ \hline Mg(s) + Cl_2(g) \rightleftharpoons & E^0 = 1.358 - \\ Mg^{2+} + 2Cl^- & (-2.375) \\ & = 3.733\ \text{V} \end{array}$$

$$K = 10^{2(3.733)/0.059\ 16} = 2 \times 10^{-126}$$

(d)
$$2[MnO_2(s)+2H_2O \rightleftharpoons$$
$$MnO_4^- +4H^+ +3e^-] \quad E^0 = 1.695 \text{ V}$$
$$3[MnO_2(s)+4H^+ +2e^- \rightleftharpoons$$
$$\underline{Mn^{2+} +2H_2O]} \qquad E^0 = 1.229 \text{ V}$$
$$5MnO_2(s)+4H^+ \rightleftharpoons \qquad E^0 = 1.229 - 1.695$$
$$2MnO_4^- +3Mn^{2+} \qquad = -0.466 \text{ V}$$
$$+2H_2O$$

$$K = 10^{6(-0.466)/0.059\,16} = 5 \times 10^{-48}$$

An alternative way to answer d is the following:

$$5[MnO_2(s)+4H^+ +2e^- \rightleftharpoons$$
$$Mn^{2+} +2H_2O] \qquad E^0 = 1.229 \text{ V}$$
$$2[Mn^{2+} +4H_2O \rightleftharpoons$$
$$\underline{MnO_4^- +8H^+ +5e^-]} \qquad E^0 = 1.51 \text{ V}$$
$$5MnO_2(s)+4H^+ \rightleftharpoons \qquad E^0 = 1.229 - 1.51$$
$$2MnO_4^- +3Mn^{2+} +2H_2O \qquad = -0.281 \text{ V}$$

$$K = 10^{10(-0.281)/0.059\,16} = 3 \times 10^{-48}$$

(e)
$$Ag^+ +e^- \rightleftharpoons Ag(s) \qquad E^0 = 0.799 \text{ V}$$
$$Ag(s)+2S_2O_3^{2-} \rightleftharpoons$$
$$\underline{Ag(S_2O_3)_2^{3-} +e^-} \qquad E^0 = 0.017 \text{ V}$$
$$Ag^+ +2S_2O_3^{2-} \rightleftharpoons \qquad E^0 = 0.799 - 0.017$$
$$Ag(S_2O_3)_2^{3-} \qquad = 0.782 \text{ V}$$

$$K = 10^{0.782/0.059\,16} = 2 \times 10^{13}$$

(f)
$$Cu(s) \rightleftharpoons Cu^+ +e^- \qquad E^0 = 0.518 \text{ V}$$
$$\underline{CuI(s)+e^- \rightleftharpoons Cu(s)+I^-} \quad E^0 = -0.185 \text{ V}$$
$$CuI \rightleftharpoons Cu^+ +I^- \qquad E^0 = -0.185 - 0.518$$
$$\qquad = -0.703 \text{ V}$$

$$K = 10^{-0.703/0.059\,16} = 1 \times 10^{-12}$$

15-C. (a) Anode: $\quad Fe(s) \rightleftharpoons Fe^{2+} + 2e^- \quad E^0 = -0.440$
Cathode: $Br_2(l) + 2e^- \rightleftharpoons 2Br^- \quad E^0 = 1.065$

$$E = 1.505 - \frac{0.059\,16}{2}\log(0.010)(0.050)^2 = 1.641 \text{ V}$$

(b) Anode: $\quad Cu(s) \rightleftharpoons Cu^{2+} + 2e^- \quad E^0 = 0.337$
Cathode: $Fe^{2+} + 2e^- \rightleftharpoons Fe(s) \quad E^0 = -0.440$

$$E = -0.777 - \frac{0.059\,16}{2}\log\frac{0.020}{0.050} = -0.765 \text{ V}$$

(c) Anode: $\quad 2Hg(l)+2Cl^- \rightleftharpoons \qquad E^0 = 0.268$
$$Hg_2Cl_2(s)+2e^-$$
Cathode: $Cl_2(g)+2e^- \rightleftharpoons 2Cl^- \qquad E^0 = 1.358$

$$E = 1.090 - \frac{0.059\,16}{2}\log\frac{(0.040)^2}{(0.060)^2}(0.50) = 1.092 \text{ V}$$

15-D. (a)
$$E(\text{cell}) = -0.799 - 0.059\,16\,\log\frac{[Ag^+]P_{H_2}^{1/2}}{[H^+]}$$

(b)
$$[Ag^+] = K_{sp}/[I^-] = 8.3 \times 10^{-17}/0.10$$
$$= 8.3 \times 10^{-16} \text{ M}$$

$$E(\text{cell}) = -0.799 - 0.059\,16$$

$$\times \log\frac{(8.3 \times 10^{-16})\sqrt{0.20}}{(0.10)} = 0.055 \text{ V}$$

(c) $E^0(\text{cell}) = -E^0$ (for Reaction 1)

$$E(\text{cell}) = E^0(\text{cell}) - 0.059\,16\,\log\frac{P_{H_2}^{1/2}}{[H^+][I^-]}$$

$$0.055 = -E^0 \text{ (for Reaction 1)} - 0.059\,16$$

$$\times \log\frac{\sqrt{0.20}}{(0.10)(0.10)}$$

$$E^0 \text{ (for Reaction 1)} = -0.153 \text{ V}$$

15-E. Anode: $\quad Cu(s) \rightleftharpoons Cu^{2+} +2e^- \quad E^0 = 0.337 \text{ V}$
Cathode: $2Ag(CN)_2^- +2e^- \rightleftharpoons$
$$2Ag(s)+4CN^- \qquad E^0 = -0.310 \text{ V}$$

$$E(\text{cell}) = -0.647 - \frac{0.059\,16}{2}\log\frac{[Cu^{2+}][CN^-]^4}{[Ag(CN)_2^-]^2}$$

At pH 8.21, $\dfrac{[CN^-]}{[HCN]} = \dfrac{K_A}{[H^+]} = 0.10 \Rightarrow$

$$[CN^-] = 0.10\,[HCN]$$

But since $[CN^-] + [HCN] = 0.10 \text{ M}$, $[CN^-] = 0.009\,1 \text{ M}$

$$E(\text{cell}) = -0.647 - \frac{0.059\,16}{2}\log\frac{(0.030)(0.009\,1)^4}{(0.010)^2}$$

$$= -0.479 \text{ V}$$

15-F. (a) $PuO_2^+ + e^- + 4H^+ \rightleftharpoons Pu^{4+} + 2H_2O$

$$\begin{array}{ll} PuO_2^{2+} \rightarrow PuO_2^+ & \Delta G = -1F\,(0.93) \\ PuO_2^+ \rightarrow Pu^{4+} & \Delta G = -1F\,E^0 \\ \underline{Pu^{4+} \rightarrow Pu^{3+}} & \underline{\Delta G = -1F\,(0.97)} \\ PuO_2^{2+} \rightarrow Pu^{3+} & \Delta G = -3F\,(1.02) \end{array}$$

$$-3F\,(1.02) = -1F\,(0.93) - 1F\,E^0$$
$$-1F\,(0.97) \Rightarrow E^0 = 1.16 \text{ V}$$

(b)
$$2PuO_2^{2+} +2e^- \rightleftharpoons 2PuO_2^+ \qquad E^0 = 0.93 \text{ V}$$
$$\underline{H_2O \rightleftharpoons \tfrac{1}{2}O_2(g)+2H^+ +2e^-} \quad E^0 = 1.229 \text{ V}$$
$$2PuO_2^{2+} +H_2O \rightleftharpoons$$
$$2PuO_2^+ +\tfrac{1}{2}O_2(g)+2H^+ \qquad E^0 = -0.30 \text{ V}$$

$$E = -0.30 - \frac{0.059\,16}{2} \log \frac{[PuO_2^+]^2 P_{O_2}^{1/2}[H^+]^2}{[PuO_2^{2+}]^2}$$

$$(\text{since } [PuO_2^+] = [PuO_2^{2+}])$$

$$= -0.30 - \frac{0.059\,16}{2} \log \sqrt{0.20}\,(0.010)^2$$

$$= -0.17 \text{ V}$$

Water will not be oxidized at pH 2.00.
At pH 7.00, $E = +0.124$ V. Water will be oxidized.

15-G. Anode: $2Hg(l) + 2Cl^- \rightleftharpoons$
$Hg_2Cl_2(s) + 2e^- \qquad E^0 = 0.268$
Cathode: $2H^+ + 2e^- \rightleftharpoons H_2(g) \qquad E^0 = 0$

$$E(\text{cell}) = -0.268 - \frac{0.059\,16}{2} \log \frac{P_{H_2}}{[H^+]^2[Cl^-]^2}$$

We find $[H^+]$ in the right half-cell by considering the acid–base chemistry of KHP, the intermediate form of a diprotic acid.

$$[H^+] = \sqrt{\frac{K_1 K_2(0.050) + K_1 K_w}{K_1 + 0.050}} = 6.5 \times 10^{-5} \text{ M}$$

$$E(\text{cell}) = -0.268 - \frac{0.059\,16}{2}$$

$$\times \log \frac{1}{(6.5 \times 10^{-5})^2 (0.10)^2} = -0.575 \text{ V}$$

15-H. Anode: $Hg(l) \rightleftharpoons Hg^{2+} + 2e^- \qquad E^0 = 0.854$ V
Cathode: $2H^+ + 2e^- \rightleftharpoons H_2(g) \qquad E^0 = 0$ V

$$E(\text{cell}) = -0.854 - \frac{0.059\,16}{2} \log \frac{[Hg^{2+}]P_{H_2}}{[H^+]^2}$$

$$-0.321 = -0.854 - \frac{0.059\,16}{2} \log \frac{[Hg^{2+}](1)}{(1)^2}$$

$$\Rightarrow [Hg^{2+}] = 9.6 \times 10^{-19} \text{ M}$$

Since $[Hg^{2+}]$ is so small, $[HgI_4^{2-}] = 0.0010$ M. To make this much HgI_4^{2-}, the concentration of I^- must have been reduced from 0.010 M to 0.0060 M, since one Hg^{2+} ion reacts with four I^- ions.

$$K = \frac{[HgI_4^{2-}]}{[Hg^{2+}][I^-]^4} = \frac{(0.001\,0)}{(9.6 \times 10^{-19})(0.006\,0)^4}$$

$$= 8.0 \times 10^{23}$$

15-I. $CuY^{2-} + 2e^- \rightleftharpoons Cu(s) + Y^{4-} \qquad E_1^0$
$Cu(s) \rightleftharpoons Cu^{2+} + 2e^- \qquad E_2^0 = 0.337$

———————————————

$CuY^{2-} \xrightarrow{1/K_f} Cu^{2+} + Y^{4-} \qquad E_3^0$

$$E_3^0 = \frac{0.059\,16}{2} \log \frac{1}{K_f} = -0.556 \text{ V}$$

$$E_1^0 = E_3^0 + E_2^0 = -0.219 \text{ V}$$

15-J. To compare glucose and H_2 at pH $= 0$, we need to know E^0 for each. For H_2, $E^0 = 0$ V. For glucose, we find E^0 from $E^{0'}$:

$$\underset{\substack{\text{gluconic}\\\text{acid}}}{HA} + 2H^+ + 2e^- \rightleftharpoons \underset{\text{glucose}}{G} + H_2O$$

$$E = E^0 - \frac{0.059\,16}{2} \log \frac{[G]}{[HA][H^+]^2}$$

But $F_G = [G]$ and $[HA] = \dfrac{[H^+]F_{HA}}{[H^+] + K_A}$. Putting these into the Nernst equation gives

$$E = E^0 - \frac{0.059\,16}{2} \log \frac{F_G}{\left(\dfrac{[H^+]F_{HA}}{[H^+] + K_A}\right)[H^+]^2}$$

$$E = E^0 - \underbrace{\frac{0.059\,16}{2} \log \frac{[H^+] + K_A}{[H^+]^3}}_{\text{This is } E^{0'} = -0.45 \text{ V}} - \frac{0.059\,16}{2} \log \frac{F_G}{F_{HA}}$$

$$-0.45 \text{ V} = E^0 - \frac{0.059\,16}{2} \log \frac{10^{-7.00} + 10^{-3.56}}{(10^{-7.00})^3}$$

$$\Rightarrow E^0 = +0.066 \text{ V for glucose.}$$

Since E^0 for H_2 is more negative than E^0 for glucose, H_2 is the stronger reducing agent at pH 0.00.

15-K. (a) Each H^+ must provide $\frac{1}{2}(34.5$ kJ) when it passes from outside to inside.

$$\Delta G = -(\tfrac{1}{2})(34.5 \times 10^3 \text{ J}) = -RT \ln \frac{\mathscr{A}_{\text{high}}}{\mathscr{A}_{\text{low}}}$$

$$\frac{\mathscr{A}_{\text{high}}}{\mathscr{A}_{\text{low}}} = 1.05 \times 10^3 \Rightarrow \Delta pH = \log(1.05 \times 10^3)$$

$$= 3.02 \text{ pH units}$$

(b) $\Delta G = -nFE$ (where $n =$ charge of $H^+ = 1$)

$$-(\tfrac{1}{2})(34.5 \times 10^3 \text{ J}) = -1 \cdot FE \Rightarrow E = 0.179 \text{ V}$$

(c) If $\Delta pH = 1.00$, $\mathscr{A}_{\text{high}}/\mathscr{A}_{\text{low}} = 10$.

$$\Delta G(\text{pH}) = -RT \ln 10 = -5.7 \times 10^3 \text{ J}$$

$$\Delta G(\text{electric}) = [\tfrac{1}{2}(34.5) - 5.7] \text{ kJ} = 11.5 \text{ kJ}$$

$$E = \Delta G(\text{electric})/F = 0.120 \text{ V}$$

720   **Chapter 16**

16-A. The reaction at the silver electrode (written as a reduction) is $Ag^+ + e^- \rightleftharpoons Ag(s)$, and the cell potential is written as

$$E(\text{cell}) = E(\text{cathode}) - E(\text{anode})$$

$$E(\text{cell}) = (+0.200) - \left(0.799 - 0.059\,16\,\log\frac{1}{[Ag^+]}\right)$$

$$E(\text{cell}) = -0.599 - 0.059\,16\,\log[Ag^+]$$

Titration reactions:

$$Br^- + Ag^+ \rightarrow AgBr(s) \qquad (K_{sp} = 5.0 \times 10^{-13})$$

$$Br^- + Tl^+ \rightarrow TlBr(s) \qquad (K_{sp} = 3.6 \times 10^{-6})$$

The two equivalence points are at 25.0 and 50.0 mL. Between 0 and 25 mL, there is unreacted Ag^+ in the solution.

1.0 mL: $[Ag^+] = \left(\dfrac{24.0}{25.0}\right) \underbrace{(0.050\,0)}_{\substack{\text{Initial concentration} \\ \text{of } Ag^+}} \left(\dfrac{100.0}{101.0}\right)$

$$= 0.047\,5 \text{ M} \Rightarrow E(\text{cell}) = -0.521$$

15.0 mL: $[Ag^+] = \left(\dfrac{10.0}{25.0}\right)(0.050\,0)\left(\dfrac{100.0}{115.0}\right)$

$$= 0.017\,4 \text{ M} \Rightarrow E = -0.495$$

24.0 mL: $[Ag^+] = \left(\dfrac{1.0}{25.0}\right)(0.050\,0)\left(\dfrac{100.0}{124.0}\right)$

$$= 0.001\,61 \text{ M} \Rightarrow E = -0.434$$

24.9 mL: $[Ag^+] = \left(\dfrac{0.10}{25.0}\right)(0.050\,0)\left(\dfrac{100.0}{124.9}\right)$

$$= 1.60 \times 10^{-4} \text{ M} \Rightarrow E = -0.374$$

Between 25 mL and 50 mL, all AgBr has precipitated and TlBr is in the process of precipitating. There is some unreacted Tl^+ left in solution in this region.

25.2 mL:

$$[Tl^+] = \left(\dfrac{24.8}{25.0}\right)(0.050\,0)\left(\dfrac{100.0}{125.2}\right)$$

$$= 3.96 \times 10^{-2} \text{ M}$$

$$[Br^-] = K_{sp} \text{ (for TlBr)}/[Tl^+] = 9.0_9 \times 10^{-5} \text{ M}$$

$$[Ag^+] = K_{sp} \text{ (for AgBr)}/[Br^-] = 5.5 \times 10^{-9} \text{ M}$$

$$E(\text{cell}) = -0.599 - 0.059\,16\,\log[Ag^+] = -0.110 \text{ V}$$

35.0 mL: $[Tl^+] = \left(\dfrac{15.0}{25.0}\right)(0.050\,0)\left(\dfrac{100.0}{135.0}\right)$

$$= 0.022\,2 \text{ M} \Rightarrow$$

$$[Br^-] = 1.62 \times 10^{-4} \text{ M} \Rightarrow$$

$$[Ag^+] = 3.08 \times 10^{-9} \text{ M} \Rightarrow$$

$$E = -0.095 \text{ V}$$

50.0 mL is the second equivalence point, at which $[Tl^+] = [Br^-]$.

50.0 mL:

$$[Tl^+][Br^-] = K_{sp} \text{ (for TlBr)} \Rightarrow [Tl^+] = \sqrt{K_{sp}}$$

$$= 1.90 \times 10^{-3} \text{ M} \Rightarrow$$

$$[Br^-] = 1.90 \times 10^{-3} \text{ M} \Rightarrow$$

$$[Ag^+] = 2.64 \times 10^{-10} \text{ M} \Rightarrow E = -0.032 \text{ V}$$

At 60.0 mL, there is excess Br^- in the solution.

60.0 mL:

$$[Br^-] = \left(\dfrac{10.0}{160.0}\right)(0.200) = 0.012\,5 \text{ M} \Rightarrow$$

$$[Ag^+] = 4.00 \times 10^{-11} \text{ M} \Rightarrow E = +0.016 \text{ V}$$

16-B. The cell potential is given by Equation C, in which K_f is the formation constant for $Hg(EDTA)^{2-}$ ($= 5.0 \times 10^{21}$). To find the potential, we must calculate $[HgY^{2-}]$ and $[Y^{4-}]$ at each point. The concentration of HgY^{2-} is 1.0×10^{-4} M when $V = 0$, and is thereafter affected only by dilution because $K_f(HgY^{2-}) \gg K_f(MgY^{2-})$. The concentration of Y^{4-} is found from the Mg–EDTA equilibrium at all but the first point. At $V = 0$ mL, the Hg–EDTA equilibrium determines $[Y^{4-}]$.

0 mL:

$$\frac{[Hg^{2+}][EDTA]}{[HgY^{2-}]} = \frac{1}{\alpha_{Y^{4-}} K_f \text{ (for } HgY^{4-})}$$

$$\frac{(x)(x)}{1.0 \times 10^{-4} - x} = 5.56 \times 10^{-22} \Rightarrow x = [EDTA]$$

$$= 2.36 \times 10^{-13} \text{ M}$$

$$[Y^{4-}] = \alpha_{Y^{4-}}[EDTA] = 8.49 \times 10^{-14} \text{ M}$$

Using Equation C, we write

$$E(\text{cell}) = 0.854 - 0.241 - \frac{0.059\,16}{2} \log \frac{5.0 \times 10^{21}}{1.0 \times 10^{-4}}$$

$$- \frac{0.059\,16}{2} \log 8.49 \times 10^{-14}$$

$$= 0.239 \text{ V}$$

10.0 mL: Since $V_e = 25.0$ mL, 10/25 of the Mg^{2+} is in the form MgY^{2-}, and 15/25 is in the form Mg^{2+}.

$$[Y^{4-}] = \frac{[MgY^{2-}]}{[Mg^{2+}]} \bigg/ K_f \text{ (for } MgY^{2-})$$

$$= \left(\frac{10}{15}\right) \bigg/ 6.2 \times 10^8 = 1.08 \times 10^{-9} \text{ M}$$

$$[HgY^{2-}] = \left(\frac{50.0}{60.0}\right)(1.0 \times 10^{-4}) = 8.33 \times 10^{-5} \text{ M}$$

Dilution factor

$$E(\text{cell}) = 0.854 - 0.241 - \frac{0.059\,16}{2} \log \frac{5.0 \times 10^{21}}{8.33 \times 10^{-5}}$$

$$- \frac{0.059\,16}{2} \log 1.08 \times 10^{-9}$$

$$= 0.116 \text{ V}$$

20.0 mL:

$$[Y^{4-}] = \left(\frac{20}{5}\right) \bigg/ 6.2 \times 10^8 = 6.45 \times 10^{-9} \text{ M}$$

$$[HgY^{2-}] = \left(\frac{50.0}{70.0}\right)(1.0 \times 10^{-4}) = 7.14 \times 10^{-5} \text{ M} \Rightarrow$$

$$E(\text{cell}) = 0.091 \text{ V}$$

24.9 mL: $[Y^{4-}] = \left(\dfrac{24.9}{0.1}\right) \bigg/ 6.2 \times 10^8$

$$= 4.02 \times 10^{-7} \text{ M}$$

$$[HgY^{2-}] = \left(\frac{50.0}{74.9}\right)(1.0 \times 10^{-4})$$

$$= 6.68 \times 10^{-5} \text{ M} \Rightarrow E(\text{cell}) = 0.037 \text{ V}$$

25.0 mL: This is the equivalence point, at which $[Mg^{2+}] = [EDTA]$.

$$\frac{[Mg^{2+}][EDTA]}{[MgY^{2-}]} = \frac{1}{\alpha_{Y^{4-}} \cdot K_f \text{ (for } MgY^{2-})}$$

$$\frac{x^2}{\left(\dfrac{50.0}{75.0}\right)(0.0100) - x} = \frac{1}{2.22 \times 10^8} \Rightarrow$$

$$x = 5.48 \times 10^{-6} \text{ M}$$

$$[Y^{4-}] = \alpha_{Y^{4-}} (5.48 \times 10^{-6})$$

$$= 1.97 \times 10^{-6} \text{ M}$$

$$[HgY^{2-}] = \left(\frac{50.0}{75.0}\right)(1.0 \times 10^{-4})$$

$$= 6.67 \times 10^{-5} \text{ M}$$

$$\Rightarrow E(\text{cell}) = 0.016 \text{ V}$$

26.0 mL: Now there is excess EDTA in the solution:

$$[Y^{4-}] = \alpha_{Y^{4-}} \cdot [EDTA] = (0.36)\left[\left(\frac{1.0}{76.0}\right)(0.020\,0)\right]$$

$$= 9.47 \times 10^{-5} \text{ M}$$

$$[HgY^{2-}] = \left(\frac{50.0}{76.0}\right)(1.0 \times 10^{-4}) = 6.58 \times 10^{-5} \text{ M}$$

$$\Rightarrow E(\text{cell}) = -0.034 \text{ V}$$

16-C. At intermediate pH, the potential will be constant at 100 mV. When $[OH^-] \approx [F^-]/10 = 10^{-6}$ M (pH = 8), the electrode begins to respond to OH^- and the potential will decrease (i.e., the potential will change in the same direction as if more F^- were being added). Near pH = 3.17 (= pK_A for HF), F^- reacts with H^+ and the concentration of free F^- decreases. At pH = 1.17, $[F^-] \approx 1\%$ of 10^{-5} M = 10^{-7} M, and $E \approx 100 + 2(59) = 218$ mV. A qualitative sketch of this behavior is shown below. The slope at high pH is less than 59 mV/pH unit, because the response of the electrode to OH^- is less than the response to F^-.

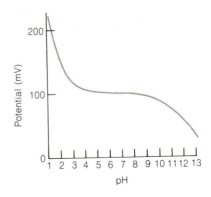

16-D. (a) For 1.00 mM Na^+ at pH 8.00, we can write

$$E = \text{Constant} + 0.059\,16$$
$$\times \log([Na^+] + 36[H^+])$$
$$-0.038 = \text{Constant} + 0.059\,16$$
$$\times \log[(1.00 \times 10^{-3}) + (36 \times 10^{-8})]$$
$$\Rightarrow \text{Constant} = +0.139 \text{ V}$$

For 5.00 mM Na^+ at pH 8.00, we have

$$E = +0.139 + 0.059\,16$$
$$\times \log[(5.00 \times 10^{-3}) + (36 \times 10^{-8})] = 0.003 \text{ V}$$

(b) For 1.00 mM Na^+ at pH 3.87, we have

$$E = +0.139 + 0.059\,16$$
$$\times \log[(1.00 \times 10^{-3}) + (36 \times 10^{-3.87})]$$
$$= 0.007 \text{ V}$$

16-E. A graph of $E(mV)$ versus $\log[NH_{3(M)}]$ gives a straight line whose equation is $E = 563.4 + 59.05 \log[NH_3]$. For $E = 339.3$ mV, $[NH_3] = 1.60 \times 10^{-4}$ M. The sample analyzed contains $(100 \text{ mL}) (1.60 \times 10^{-4} \text{ M}) = 0.016\,0$ mmol of nitrogen. But this sample represents just 2.00% (20.0 mL/1.00 L) of the food sample. Therefore, the food contains $0.016/0.020\,0 = 0.800$ mmol of nitrogen = 11.2 mg of N = 3.59% nitrogen.

16-F. (a) For the original solution, we can write

$$[Ag^+] = \frac{[Ag(CN)_2^-]}{[CN^-]^2(7.1 \times 10^{19})}$$
$$= \frac{(1.00 \times 10^{-5})}{(8.0 \times 10^{-6})^2(7.1 \times 10^{19})}$$
$$= 2.2 \times 10^{-15} \text{ M}$$
$$E = \text{Constant} + 0.059\,16$$
$$\times \log(2.2 \times 10^{-15}) = 206.3 \text{ mV}$$
$$\Rightarrow \text{Constant} = 1.073\,4 \text{ V}$$

After addition of CN^-, we can say that

$$[Ag^+] = \frac{(1.00 \times 10^{-5})}{(12.0 \times 10^{-6})^2(7.1 \times 10^{19})}$$
$$= 9.8 \times 10^{-16} \text{ M}$$
$$E = 1.073\,4 + 0.059\,16$$
$$\times \log(9.8 \times 10^{-16}) = 185.4 \text{ mV}$$

(b) Let there be x mol of CN^- in 50.0 mL of unknown. After the standard addition, the unknown contains

$$x + (1.00 \times 10^{-3} \text{ L})(2.50 \times 10^{-4} \text{ M})$$
$$= (x + 2.50 \times 10^{-7}) \text{ mol } CN^-$$
$$\Rightarrow [CN^-] = \frac{x + 2.50 \times 10^{-7} \text{ mol}}{0.051\,0 \text{ L}}$$
$$[Ag^+]_2 = \frac{(1.00 \times 10^{-5})}{\left(\dfrac{x + 2.50 \times 10^{-7}}{0.051\,0}\right)^2 (7.1 \times 10^{19})}$$

Before the addition, we have

$$E = \text{Constant} + 0.059\,16 \log[Ag^+]$$
$$0.134\,8 = \text{Constant} + 0.059\,16 \log[Ag^+]_1$$

After the addition, we can write

$$0.118\,6 = \text{Constant} + 0.059\,16 \log[Ag^+]_2$$

Subtracting the second equation from the first gives

$$0.134\,8 - 0.118\,6 = 0.059\,16 \log \frac{[Ag^+]_1}{[Ag^+]_2} \Rightarrow$$
$$\frac{[Ag^+]_1}{[Ag^+]_2} = 1.879$$

But we can also write that

$$1.879 = \frac{[Ag^+]_1}{[Ag^+]_2} = \frac{\dfrac{1.00 \times 10^{-5}}{\left(\dfrac{x}{0.0500}\right)^2 (7.1 \times 10^{19})}}{\dfrac{(1.00 \times 10^{-5})(50.0/51.0)}{\left(\dfrac{x + 2.50 \times 10^{-7}}{0.051\,0}\right)^2 7.1 \times 10^{19}}}$$
$$\Rightarrow x = 6.50 \times 10^{-7} \text{ mol}$$
$$[CN^-] = 6.50 \times 10^{-7} \text{ mol}/0.050\,0 \text{ L}$$
$$= 1.30 \times 10^{-5} \text{ M}$$

Chapter 17

17-A. Titration reaction: $6Fe^{2+} + Cr_2O_7^{2-} + 14H^+ \rightarrow 6Fe^{3+} + 2Cr^{3+} + 7H_2O$ $V_e = 10.00$ mL

Representative calculations:

0.100 mL:

$$E(\text{cathode}) = 0.770 - 0.059\,16 \log \frac{[Fe^{2+}]}{[Fe^{3+}]}$$
$$E(\text{cathode}) = 0.770 - 0.059\,16 \log \frac{9.90}{0.100} = 0.652 \text{ V}$$

10.0 mL:

$$E(\text{cathode}) = +0.770 - 0.05916 \log \frac{[\text{Fe}^{2+}]}{[\text{Fe}^{3+}]}$$

$$6E(\text{cathode}) = 6(1.33) - (6)\frac{0.05916}{6} \log \frac{[\text{Cr}^{3+}]^2}{[\text{Cr}_2\text{O}_7^{2-}][\text{H}^+]^{14}}$$

$$7E(\text{cathode}) = 8.75 - 0.05916 \log \frac{[\text{Fe}^{2+}][\text{Cr}^{3+}]^2}{[\text{Fe}^{3+}][\text{Cr}_2\text{O}_7^{2-}][\text{H}^+]^{14}}$$

At the equivalence point, $[\text{Fe}^{3+}] = 3[\text{Cr}^{3+}]$ and $[\text{Fe}^{2+}] = 6[\text{Cr}_2\text{O}_7^{2-}]$. Putting in these values gives

$$7E(\text{cathode}) = 8.75 - 0.05916$$
$$\times \log \frac{6[\cancel{\text{Cr}_2\text{O}_7^{2-}}][\text{Cr}^{3+}]^2}{3[\cancel{\text{Cr}^{3+}}][\cancel{\text{Cr}_2\text{O}_7^{2-}}][\text{H}^+]^{14}}$$

At the equivalence point, all of the $\text{Cr}_2\text{O}_7^{2-}$ has been converted to Cr^{3+}. Therefore,

$$[\text{Cr}^{3+}] = 2(0.0200)\left(\frac{10.00}{130.00}\right) = 0.00308 \text{ M}$$

Using this value in the log term gives

$$7E(\text{cathode}) = 8.75 - 0.05916 \log \frac{2(0.00308)}{(0.10)^{14}}$$

$$E(\text{cathode}) = 1.15 \text{ V} \Rightarrow E(\text{cell}) = 0.953 \text{ V}$$

10.10 mL: $E(\text{cathode}) = 1.33 - \frac{0.05916}{6}$
$$\times \log \frac{[\text{Cr}^{3+}]^2}{[\text{Cr}_2\text{O}_7^{2-}][\text{H}^+]^{14}}$$

$$[\text{Cr}^{3+}] = 2(0.0200)\left(\frac{10.00}{130.10}\right) = 0.00307 \text{ M}$$

$$[\text{Cr}_2\text{O}_7^{2-}] = (0.0200)\left(\frac{0.10}{130.10}\right) = 1.54 \times 10^{-5} \text{ M}$$

$$E(\text{cathode}) = 1.33 - \frac{0.05916}{6} \times \log \frac{(0.00307)^2}{(1.54 \times 10^{-5})(0.10)^{14}}$$

$$= 1.19_4 \text{ V}$$

The final results are

mL	E(cell) (V)	mL	E(cell) (V)
0.100	0.455	9.90	0.691
2.00	0.537	10.00	0.953
4.00	0.563	10.10	0.99₇
6.00	0.583	11.00	1.00₇
8.00	0.609	12.00	1.01₀
9.00	0.629		

17-B. Titration reactions:

$$\text{V}^{2+} + \text{Ce}^{4+} \rightarrow \text{V}^{3+} + \text{Ce}^{3+}$$
$$\text{V}^{3+} + \text{Ce}^{4+} + \text{H}_2\text{O} \rightarrow \text{VO}^{2+} + \text{Ce}^{3+} + 2\text{H}^+$$
$$\text{VO}^{2+} + \text{Ce}^{4+} + \text{H}_2\text{O} \rightarrow \text{VO}_2^+ + \text{Ce}^{3+} + 2\text{H}^+$$

5.0 mL:

$$E(\text{cathode}) = -0.255 - 0.05916 \log \frac{[\text{V}^{2+}]}{[\text{V}^{3+}]}$$

$$\Rightarrow E(\text{cell}) = -0.496 \text{ V}$$

15.0 mL:

$$E(\text{cathode}) = 0.337 - 0.05916 \log \frac{[\text{V}^{3+}]}{[\text{VO}^{2+}](1.00)^2}$$

$$\Rightarrow E(\text{cell}) = 0.096 \text{ V}$$

25.0 mL:

$$E(\text{cathode}) = 1.000 - 0.05916 \log \frac{[\text{VO}^{2+}]}{[\text{VO}_2^+](1.00)^2}$$

$$\Rightarrow E(\text{cell}) = 0.759 \text{ V}$$

35.0 mL:

$$E(\text{cathode}) = 1.70 - 0.05916 \log \frac{[\text{Ce}^{3+}]}{[\text{Ce}^{4+}]}$$

$$E(\text{cathode}) = 1.70 - 0.05916 \log \left(\frac{30.0}{5.0}\right)$$

$$\Rightarrow E(\text{cell}) = 1.41 \text{ V}$$

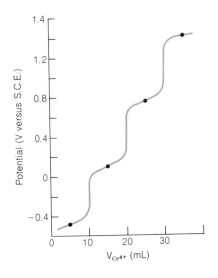

724

17-C. (a) The second endpoint is halfway between 0.096 and 0.759 = 0.43 V versus S.C.E. (= 0.67 V versus S.H.E.). Diphenylamine would change from colorless to violet at this endpoint.

(b) The third endpoint is approximately midway between 1.41 and 0.759 = 1.08 V versus S.C.E. (= 1.32 V versus S.H.E.). Tris(2,2'-bipyridine)-ruthenium would change from yellow to pale blue at this endpoint.

17-D. (a)

$$\begin{array}{c} R \\ | \\ HO-CH \\ | \\ H_3^+N-CH \\ | \\ CO_2^- \end{array} + IO_4^- \rightarrow \begin{array}{c} R \\ | \\ O=CH \\ \\ + \\ O=CHCO_2^- \end{array} + NH_4^+ + IO_3^-$$

(For serine, $R = H$, and for threonine, $R = CH_3$.) One mole of IO_4^- consumes one mole of serine or one mole of threonine.

(b) $I_3^- + 2S_2O_3^{2-} \rightarrow 3I^- + S_4O_6^{2-}$

823 μL of 0.098 8 M $S_2O_3^{2-}$ = 81.3_1 μmol of $S_2O_3^{2-}$ = 40.6_6 μmol of I_3^-. But one mole of unreacted IO_4^- gives one mole of I_3^- in Reaction 17-72. Therefore, 40.6_6 μmol of IO_4^- was left from the periodate oxidation of the amino acids. The original amount of IO_4^- was 2.000 mL of 0.048 7 M IO_4^- = 97.4_0 μmol. The difference (97.40 − 40.66 = 56.74) is the number of micromoles of serine + threonine in 128.6 mg of protein. But with M.W. = 58 600, 128.6 mg of protein = 2.195 μmol. (Serine + threonine)/protein = 56.74 μmol/2.195 μmol = 25.85 ≈ 26 residues/molecule.

(c) 40.66 μmol of I_3^- will react with 40.66 μmol of H_3AsO_3 (Reaction 17-64), which is produced by $\frac{1}{4}(40.66)$ = 10.16 μmol of As_4O_6 = 4.02 mg.

Chapter 18

18-A. The anode reaction is $Zn(s) \rightarrow Zn^{2+} + 2e^-$.

$$5.0 \text{ g Zn} = 7.65 \times 10^{-2} \text{ mol Zn}$$
$$= 1.52 \times 10^{-1} \text{ mol e}^-$$

$$(0.152 \text{ mol e}^-)(96\,485 \text{ C/mol}) = 1.48 \times 10^4 \text{ C}$$

The current flowing through the circuit is $I = E/R = 1.02 \text{ V}/2.8 \text{ }\Omega = 0.364 \text{ A} = 0.364 \text{ C/s}$.

$$1.48 \times 10^4 \text{ C}/(0.364 \text{ C/s}) = 4.06 \times 10^4 \text{ s} = 11.3 \text{ h}$$

18-B. Anode: $H_2O \rightleftharpoons \frac{1}{2}O_2(g) + 2H^+ + 2e^-$ $E^0 = 1.229$

Cathode: $2H^+ + 2e^- \rightleftharpoons H_2(g)$ $E^0 = 0.00$

$H_2O \rightleftharpoons \frac{1}{2}O_2(g) + H_2(g)$ $E^0 = -1.229$

$$E_{equilibrium} = -1.229 - \frac{0.059\,16}{2} \log P_{O_2}^{1/2} P_{H_2}$$

$$= -1.229 \text{ V}$$

$$E_{applied} = E_{equilibrium} - IR - \text{Overpotential}$$
$$= -1.229 - (0.100 \text{ A})(2.00 \text{ }\Omega)$$
$$\underbrace{-0.85 \text{ V}}_{\substack{\text{Anode} \\ \text{overpotential}}} - \underbrace{0.068 \text{ V}}_{\substack{\text{Cathode} \\ \text{overpotential}}} = -2.35 \text{ V}$$

From Table 18-1

For Au electrodes, $E_{applied} = -2.78 \text{ V}$.

18-C. (a) To electrolyze 0.010 M SbO^+ requires a potential of

$$E(\text{cathode})$$

$$= 0.212 - \frac{0.059\,16}{3} \log \frac{1}{[SbO^+][H^+]^2}$$

$$= 0.212 - \frac{0.059\,16}{3} \log \frac{1}{(0.010)(1.0)^2}$$

$$= 0.173 \text{ V}$$

The concentration of $[Cu^{2+}]$ that would be in equilibrium with $Cu(s)$ at this potential is found as follows:

$$Cu^{2+} + 2e^- \rightleftharpoons Cu(s) \qquad E^0 = 0.337$$

$$E(\text{cathode}) = 0.337 - \frac{0.059\,16}{2} \log \frac{1}{[Cu^{2+}]}$$

$$0.173 = 0.337 - \frac{0.059\,16}{2} \log \frac{1}{[Cu^{2+}]}$$

$$\Rightarrow [Cu^{2+}] = 2.86 \times 10^{-6} \text{ M}$$

Percent of Cu^{2+} not reduced

$$= \frac{2.86 \times 10^{-6}}{0.10} \times 100 = 2.86 \times 10^{-3}\%$$

Percent of Cu^{2+} reduced = 99.997%

(b) In part a, E(cathode versus S.H.E.) = 0.173 V.

E(cathode versus Ag|AgCl)

$$= E(\text{versus S.H.E.}) - E(Ag|AgCl)$$

$$= 0.173 - 0.197 = -0.024 \text{ V}$$

18-D. (a) $Co^{2+} + 2e^- \rightleftharpoons Co(s)$ $E^0 = -0.277$ V

E(cathode versus S.H.E.)

$$= -0.277 - \frac{0.059\,16}{2} \log \frac{1}{[Co^{2+}]}$$

Putting in $[Co^{2+}] = 1.0 \times 10^{-6}$ M gives E = -0.454 V and

E(cathode versus S.C.E.)

$$= -0.454 - \underline{0.241} = -0.695 \text{ V}$$
$$\quad\quad\quad\quad\quad E(\text{S.C.E.})$$

(b) $Co(C_2O_4)_2^{2-} + 2e^- \rightleftharpoons$
$\quad Co(s) + 2C_2O_4^{2-}$ $E^0 = -0.474$ V

E(cathode versus S.C.E.)

$$= -0.474 - \frac{0.059\,16}{2}$$

$$\times \log \frac{[C_2O_4^{2-}]^2}{[Co(C_2O_4)_2^{2-}]} - 0.241$$

Putting in $[C_2O_4^{2-}] = 0.10$ M and $[Co(C_2O_4)_2^{2-}]$ = 1.0×10^{-6} M gives E = -0.833 V.

(c) We can think of the reduction as $Co^{2+} + 2e^- \rightleftharpoons$ Co(s), for which $E^0 = -0.277$ V. But the concentration of Co^{2+} is that tiny amount in equilibrium with 0.10 M EDTA plus 1.0×10^{-6} M $Co(EDTA)^{2-}$. In Table 14-2, we find that the formation constant for $Co(EDTA)^{2-}$ is $10^{16.31} = 2.0 \times 10^{16}$.

$$K_f = \frac{[Co(EDTA)^{2-}]}{[Co^{2+}][EDTA^{4-}]} = \frac{[Co(EDTA)^{2-}]}{[Co^{2+}]\,\alpha_{Y^{4-}}\cdot F}$$

where F is the formal concentration of EDTA (=0.10 M) and $\alpha_{Y^{4-}} = 5.0 \times 10^{-4}$ at pH 7.00 (Table 14-1). Putting in $[Co(EDTA)^{2-}] = 1.0 \times 10^{-6}$ M and solving for $[Co^{2+}]$ gives $[Co^{2+}] = 1.0 \times 10^{-18}$ M.

E(cathode versus S.C.E.)

$$= -0.277 - \frac{0.059\,16}{2} \log \frac{1}{1.0 \times 10^{-18}} - 0.241$$

$$= -1.050 \text{ V}$$

18-E. (a) 75.00 mL of 0.023 80 M KSCN = 1.785 mmol of SCN^-, which gives 1.785 mmol of AgSCN, containing 0.103 7 g of SCN. Final mass = 12.463 8 + 0.103 7 = 12.567 5 g.

(b) The concentration of Ag^+ in equilibrium with 0.10 M Br^- is $[Ag^+] = K_{sp}/[Br^-] = (5.0 \times 10^{-13})(0.10) = 5.0 \times 10^{-12}$ M. Writing both cell reactions as reductions

Anode: $Ag^+ + e^- \rightleftharpoons Ag(s)$

$$E^0 = 0.799 \text{ V}$$

Cathode: $Hg_2Cl_2(s) + 2e^- \rightleftharpoons 2Hg(l) + 2Cl^-$

$$E(\text{S.C.E.}) = 0.241 \text{ V}$$

we can say

E(cell) = E(cathode) − E(anode)

$$= 0.241 - \left(0.799 - 0.059\,16 \log \frac{1}{[Ag^+]}\right)$$

Putting in $[Ag^+] = 5.0 \times 10^{-12}$ M gives E(cell) = 0.111 V.

(c) To remove 99.99% of 0.10 M KI will leave $[I^-] = 1.0 \times 10^{-5}$ M. The concentration of Ag^+ in equilibrium with this much I^- is $[Ag^+] = K_{sp}/[I^-] = 8.3 \times 10^{-17}/1.0 \times 10^{-5}$ $= 8.3 \times 10^{-12}$ M. In part b, we found that it takes only 5.0×10^{-12} M Ag^+ to precipitate 0.10 M Br^-. Therefore, the separation is not possible.

18-F. 1.00 ppt corresponds to 30.0/1 000 = 0.030 0 mL of O_2/min = 5.00×10^{-4} mL of O_2/s. The moles of oxygen in this volume are

$$n = \frac{PV}{RT} = \frac{(1.00 \text{ atm})(5.00 \times 10^{-7} \text{ L})}{(0.082\,06 \text{ L atm K}^{-1} \text{ mol}^{-1})(293 \text{ K})}$$

$$= 2.080 \times 10^{-8} \text{ mol}$$

For each mole of O_2, four moles of e^- flow through the circuit, so $e^- = 8.320 \times 10^{-8}$ mol/s = 8.03 × 10^{-3} C/s = 8.03 mA. An oxygen content of 1.00 ppm would give a current of 8.03 μA instead.

18-G. The Zn^{2+} reacts first with PDTA freed by the reduction of $Hg(PDTA)^{2-}$ in the region BC. Then additional Zn^{2+} goes on to liberate Hg^{2+} from $Hg(PDTA)^{2-}$. This additional Hg^{2+} is reduced in the region DE. The total Hg^{2+}, equivalent to the added Zn^{2+}, equals one-half the coulombs measured in regions BC and DE (since $2e^-$ reacts with $1Hg^{2+}$). Coulombs = 3.89 + 14.47 + 18.36. Moles of Hg^{2+} reduced = $0.5(18.36\ C)/(96\ 485\ C/mol)$ = 9.514×10^{-5} mol. $[Zn^{2+}] = 9.514 \times 10^{-5}$ mol$/2.00 \times 10^{-3}$ L = 0.047 57 M

Chapter 19

19-A. The standard curve is moderately linear with slope = 0.004 19 μA/ppb and intercept of 0.019 8. The concentration of Ni(II) when 54.0 μL of 10.0 ppm solution is added to 5.00 mL is

$$\left(\frac{0.0540\ mL}{5.0540\ mL}\right)(10.0\ ppm) = 0.107\ ppm = 107\ ppb$$

The expected current is

$$I = m[Ni(II)] + b$$
$$= (0.004\ 19)(107) + 0.019\ 8 = 0.468\ \mu A$$

A careful examination of the standard curve shows that a better fit might be obtained if the slope and intercept of just the first seven points are calculated. The curve appears to be starting to level off at the higher concentrations in this experiment.

19-B. If the conditions were perfectly reproducible, the diffusion current for Tl^+ in experiment B would be $(1.21/1.15)(6.38) = 6.71\ \mu A$. The observed current in experiment B is only $6.11/6.71 = 91.1\%$ of the expected value. That is, in experiment B the response is only 91.1% as great as in experiment A. Therefore, the responses to Cd^{2+} and Zn^{2+} in experiment B are expected to be only 91.1% as great as they are in experiment A.

$$[Cd^{2+}] = \frac{(4.76/6.48)}{0.911}(1.02) = 0.82\ mM$$

$$[Zn^{2+}] = \frac{(8.54/6.93)}{0.911}(1.23) = 1.66\ mM$$

19-C. Sample height (mm) = 26.8 − 2.4 = 24.4.
Sample + 1 ppm Cu = 42.2 − 5.6 = 36.6.
Sample + 2 ppm Cu = 57.8 − 8.7 = 49.1.

The average response to added Cu is

$$\frac{(36.6 - 24.4) + (49.1 - 24.4)}{3} = 12.3\ \frac{mm}{ppm\ Cu}$$

The initial sample must have contained

$$\frac{24.4}{12.3} = 1.98\ ppm\ Cu$$

19-D. A graph of $E_{1/2}$ versus log $[OH^-]$ is shown below.

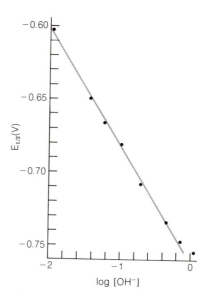

All but the lowest two points appear to lie on a line whose equation is

$$E_{1/2} = -0.080\ 6\ \log[OH^-] - 0.763$$

According to Equation 19-26, the slope of the graph is $-0.059\ 16\ p/n$. Assuming that $n = 2$, we calculate p as follows:

$$p = \frac{(n)(\text{slope})}{-0.059\ 16} = 2.72 \approx 3$$

The intercept of Equation 19-26 is given by

$$\text{Intercept} = E_{1/2}\ (\text{for free Pb}^{2+}) - \frac{0.059\ 16}{n}\log \beta_3$$

$$-0.763 = -0.41 - \frac{0.059\ 16}{2}\log \beta_3$$

$$\Rightarrow \beta_3 = 9 \times 10^{11}$$

19-E. We see two consecutive reductions. From the value of $E_a - E_c$, we find that one electron is involved in each reduction (using Equation 19-30). A possible sequence of reactions is

$$Co(III)(B_9C_2H_{11})_2^- \rightarrow Co(II)(B_9C_2H_{11})_2^{2-}$$
$$\rightarrow Co(I)(B_9C_2H_{11})_2^{3-}$$

The equality of the anodic and cathodic peak heights suggests that the reactions are reversible.

The expected DC (a) and DPP (b) polarograms are sketched below.

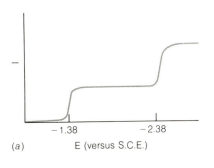
(a)   E (versus S.C.E.)

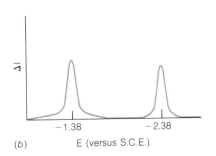
(b)   E (versus S.C.E.)

19-F. In curve a, the current decreases prior to the endpoint because Pb^{2+} is reduced to Pb(Hg) at -0.8 V. Beyond the endpoint, excess $Cr_2O_7^{2-}$ can be reduced and the current increases again. Curve b is level near zero current prior to the equivalence point because Pb^{2+} is not reduced at 0 V (versus S.C.E.). Beyond the endpoint, excess $Cr_2O_7^{2-}$ is reduced, even at 0 V (versus S.C.E.).

19-G. Initially there is no redox couple to carry current, so the potential will be high. As Ce(IV) is added, Fe(II) is converted to Fe(III), the mixture of which can support current flow by the reactions

Anode: $Fe(II) \rightleftharpoons Fe(III) + e^-$

Cathode: $Fe(III) + e^- \rightleftharpoons Fe(II)$

The potential will therefore decrease. At the equivalence point, all of the Fe(II) and all of the Ce(IV) are consumed, so the potential is very high. Beyond the equivalence point, the redox couple Ce(IV)|Ce(III) can support a current and the potential will be low again. The expected curve is shown below.

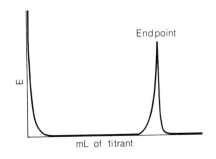

19-H. 34.61 mL of methanol with 4.163 mg of H_2O/mL contains 144.08 mg H_2O = 7.997 8 mmol of H_2O. The titration of "dry" methanol tells us that 25.00 mL of methanol reacts with 3.18 mL of reagent. Therefore, 34.61 mL of methanol will react with $(34.61/25.00)(3.18) = 4.40$ mL of Karl Fischer reagent. The titer of the reagent is

$$\frac{7.997\ 8 \text{ mmol } H_2O}{(25.00 - 4.40) \text{ mL reagent}} = 0.388\ 24 \frac{\text{mmol } H_2O}{\text{mL reagent}}$$

Reagent needed to react with 1.000 g of salt in 25.00 mL of methanol $= (38.12 - 3.18) = 34.94$ mL. H_2O in 1.000 g of salt $= (0.388\ 24)(34.94) = 13.565$ mmol $= 244.38$ mg of $H_2O = 24.44\%$ (wt/wt) of the crystal.

Chapter 20

20-A. (a) $A = -\log P/P_0 = -\log T = -\log(0.45) = 0.347$

(b) The absorbance will double to 0.694, giving $T = 10^{-A} = 10^{-0.694} = 0.202 \Rightarrow \%T = 20.2\%$.

20-B. (a) 1.00×10^{-2} g of NH_4Cl in 1.00 L $= 1.869 \times 10^{-4}$ M. In the colored solution, the concentration is $(\frac{10}{50})(1.869 \times 10^{-4} \text{ M}) = 3.739 \times 10^{-5}$ M. $\varepsilon = A/bc = (0.308 - 0.140)/(1.00)(3.739 \times 10^{-5}) = 4.49_3 \times 10^3$ M^{-1} cm^{-1}.

(b) $\dfrac{\text{Absorbance of unknown}}{\text{Absorbance of reference}}$

$= \dfrac{0.592 - 0.140}{0.308 - 0.140}$

$= \dfrac{\text{Concentration of unknown}}{\text{Concentration of reference}}$

$\Rightarrow$ Concentration of NH_3 in unknown

$$= \left(\frac{0.452}{0.168}\right)(1.869 \times 10^{-4})$$

$$= 5.028 \times 10^{-4} \text{ M}.$$

100.0 mL of unknown

$$= 5.028 \times 10^{-5} \text{ mol of N}$$

$$= 7.043 \times 10^{-4} \text{ g of N}$$

$$\Rightarrow \text{weight } \% \text{ of N}$$

$$= (7.043 \times 10^{-4} \text{ g})/(4.37 \times 10^{-3} \text{ g})$$

$$= 16.1\%.$$

20-C. (a) Milligrams of Cu in flask C $= (1.00)(\frac{10}{250})(\frac{15}{30}) = 0.020\,0$ mg. This entire quantity is in the isoamyl alcohol (20.00 mL), so the concentration is $(2.00 \times 10^{-5} \text{ g})/(0.020\,0 \text{ L})(63.546 \text{ g/mol}) = 1.57 \times 10^{-5}$ M.

(b) Observed absorbance

$$= \text{Absorbance due to Cu in rock}$$
$$\quad + \text{ Blank absorbance}$$

$$= \varepsilon bc + 0.056$$

$$= (7.90 \times 10^3)(1.00)(1.574 \times 10^{-5}) + 0.056$$

$$= 0.180$$

Note that the observed absorbance is equal to the absorbance from Cu in the rock *plus* the blank absorbance. In the lab we measure the observed absorbance and subtract the blank absorbance from it to find the absorbance due to copper.

(c) $\dfrac{\text{Cu in unknown}}{\text{Cu in known}} = \dfrac{A \text{ of unknown}}{A \text{ of known}}$

$$\frac{x \text{ mg}}{1.00 \text{ mg}} = \frac{0.874 - 0.056}{0.180 - 0.056} \Rightarrow x = 6.60 \text{ mg Cu}$$

20-D. (a) $c = A/\varepsilon b = 0.463/(4\,170)(1.00) = 1.110 \times 10^{-4}$ M $= 8.99$ g/L $= 8.99$ mg of transferrin/mL. The Fe concentration is 2.220×10^{-4} M $= 0.012\,4$ g/L $= 12.4$ μg/mL.

(b) $A_\lambda = \sum \varepsilon bc$

$$0.424 = 4\,170[\text{T}] + 2\,290[\text{D}]$$

$$0.401 = 3\,540[\text{T}] + 2\,730[\text{D}]$$

where [T] and [D] are the concentrations of transferrin and desferrioxamine, respectively. Solving for [T] and [D] gives [T] $= 7.30 \times 10^{-5}$M and [D] $= 5.22 \times 10^{-5}$ M. The fraction

of iron in transferrin (which binds two ferric ions) is $2[\text{T}]/(2[\text{T}] + [\text{D}]) = 73.7\%$.

20-E. The absorbance must be corrected by multiplying each observed absorbance by (total volume/initial volume). For example, at 36.0 μL, A(corrected) $= (0.399)[(2\,025 + 36)/2\,025] = 0.406$. A graph of corrected absorbance versus volume of Pb^{2+} (μL) is similar to Figure 20-22, with the endpoint at 46.7 μL. The moles of Pb^{2+} in this volume are $(46.7 \times 10^{-6} \text{ L})(7.515 \times 10^{-4} \text{ M}) = 3.510 \times 10^{-8}$ mol. The concentration of semi-xylenol orange is $(3.510 \times 10^{-8} \text{ mol})/(2.025 \times 10^{-3} \text{ L}) = 1.73 \times 10^{-5}$ M.

20-F. The appropriate Scatchard plot is a graph of $\Delta A/[\text{X}]$ versus ΔA (Equation 20-32).

Experiment	ΔA	$\Delta A/[\text{X}]$
1	0.090	20 360
2	0.181	19 890
3	0.271	16 940
4	0.361	14 620
5	0.450	12 610
6	0.539	9 764
7	0.627	7 646
8	0.713	5 021
9	0.793	2 948
10	0.853	1 453
11	0.904	93.6

Points 2–10 lie on a reasonably straight line whose slope is -2.72×10^4 M^{-1}, giving K $= 2.72 \times 10^4$ M^{-1}.

20-G. If self-absorption can be neglected, Equation 20-40 reduces to

$$\text{I} = k'\text{P}_0(1 - 10^{-\varepsilon_{\text{ex}}b_2c}) \qquad \text{(a)}$$

At low concentrations, this expression reduces to

$$\text{I} = k'\text{P}_0(\varepsilon_{\text{ex}}b_2c \ln 10) \qquad \text{(b)}$$

(using the first term of a power series expansion). As the concentration is increased, Expression b becomes greater than a. When Expression a is 5% below b, we can say

$$k'\text{P}_0(1 - 10^{-\varepsilon_{\text{ex}}b_2c}) = 0.95k'\text{P}_0\varepsilon_{\text{ex}}b_2c \ln 10$$

$$1 - 10^{-A} = 0.95A \ln 10$$

By trial and error, this equation can be solved to find that when $A = 0.045$, $1 - 10^{-A} = 0.95A \ln 10$. (Alternatively, you could make a graph of $1 - 10^{-A}$ versus A and $0.95A \ln 10$ versus A. The solution is the intersection of the two curves.)

Chapter 21

21-A.

Emission intensity	Concentration of added standard (μg/mL)
309	0
452	0.081
600	0.162
765	0.243
906	0.324

A graph of intensity versus concentration of added standard intercepts the x axis at -0.164 μg/mL. Since the sample was diluted by a factor of 10, the original sample concentration is 1.64 μg/mL.

21-B. The concentration of Mn in the unknown mixture is $(13.5)(1.00/6.00) = 2.25$ μg/mL.

$$\frac{\text{Signal ratio in standard}}{\text{Concentration ratio in standard}}$$

$$= \frac{\text{Signal ratio in unknown}}{\text{Concentration ratio in unknown}}$$

$$\frac{1.05}{2.50/2.00} = \frac{0.185/0.128}{[\text{Fe}]/2.25}$$

$$[\text{Fe}] = 3.87 \ \mu\text{g/mL}$$

The original concentration of Fe must have been

$$\frac{6.00}{5.00}(3.87) = 4.65 \ \mu\text{g/mL} = 8.33 \times 10^{-5} \text{ M}.$$

21-C. The ratio of signal to peak-to-peak noise level is measured to be 17 in the figure. The concentration of Fe needed to give a signal-to-noise ratio of 2 is $(2/17)(0.048\ 5 \ \mu\text{g/mL}) = 0.005\ 7 \ \mu\text{g/mL} (= 5.7 \text{ ppb})$.

21-D. In the excitation spectrum, we are looking at emission over a band of wavelengths 1.6 nm wide, while exciting the sample with different narrow bands (0.03 nm) of laser light. The sample can absorb the laser light only when the laser frequency coincides with the atomic frequency. Therefore, emission is observed only when the narrow laser line is in resonance with the atomic levels. In the emission spectrum, the sample is excited by a fixed laser frequency and then emits radiation. The monochromator bandwidth is not narrow enough to discriminate between emission at different wavelengths, so a broad envelope is observed.

Chapter 22

22-A. (a) Fraction remaining $= q = \dfrac{V_1}{V_1 + KV_2}$

$$0.01 = \frac{10}{10 + 4.0V_2} \Rightarrow V_2 = 248 \text{ mL}$$

(b) $q^3 = 0.01 = \left(\dfrac{10}{10 + 4.0V_2}\right)^3 \Rightarrow V_2 = 9.1$ mL,

and total volume $= 27.3$ mL

22-B. (a) For A, $p = D/(D + 1) = 0.5$. For B, $p = 0.375$. For A, $r_{\max} = 100p = 50$, and for B, $r_{\max} = 37.5$. $\sigma = \sqrt{npq} = \sqrt{100(0.5)(0.5)} = 5$ for A and $\sqrt{100(0.375)(0.625)} = 4.84$ for B.

(b) The center of A is at tube 50, and $\sigma = 5$ tubes. Tube 28 lies $(50 - 28)/5 = 4.4\sigma$ from the center, and tube 44 lies $(50 - 44)/5 = 1.2\sigma$ from the center. In Table 4-1, we find that the fraction of the area of A to the left of tube 44 is $0.5 - 0.384\ 9 = 0.115\ 1$. A negligible fraction of A lies to the left of tube 28. Therefore, the fraction of A in tubes 28–44 is 0.115 1. For component B, centered at tube 37.5 and with a standard deviation of 4.84, tube 28 lies $(37.5 - 28)/4.84 = 1.96\sigma$ to the left of center, and tube 44 lies 1.34σ to the right of center. The area between these limits is found to be $0.474\ 9 + 0.409\ 6 = 0.884\ 5$ (by interpolation in Table 4-1). Percent recovery of B $= 88.45\%$, and purity of B $= 0.884\ 5/(0.115\ 1 + 0.884\ 5) = 88.49\%$. Similarly, tubes 44–60 contain 86.22% of A with a purity of $0.862\ 2/(0.090\ 4 + 0.862\ 2) = 90.51\%$.

22-C. (a) For ethyl acetate, we measure $t_r = 11.3$ and $w = 1.5$ millimeters. Therefore, $N = 16\ t_r^2/w^2 = 900$ plates. For toluene, the figures are $t_r = 36.2$, $w = 4.2$, and $N = 1200$ plates.

(b) From Figure 22-10, we expect $w_{1/2} = (2.35/4)w$. The measured value of $w_{1/2}$ is in good agreement with the calculated value.

22-D. The column is overloaded, causing a gradual rise and an abrupt fall of the peak. As the sample size is decreased, the overloading decreases and the peak becomes more symmetric.

Chapter 23

23-A. (a) $\dfrac{[\text{Solute}]}{[\text{Standard}]} = F\dfrac{\text{Solute area}}{\text{Standard area}} \Rightarrow$

$$\frac{1.53}{1.06} = F\frac{1\ 570}{922} \Rightarrow F = 0.848$$

(b) $\dfrac{\text{mmol hexanol}}{\text{mmol pentanol}} = F \dfrac{\text{Hexanol area}}{\text{Pentanol area}}$

$\Rightarrow \dfrac{\text{mmol hexanol}}{0.57 \text{ mmol}} = 0.848 \dfrac{1\,320}{1\,331}$

$\Rightarrow$ mmol hexanol = 0.48

(c) $\dfrac{[\text{Solute}]}{[\text{Standard}]} = F \dfrac{\text{Solute area}}{\text{Standard area}} \Rightarrow$

$\dfrac{0.44}{0.57} = F \dfrac{1\,353}{1\,331} \Rightarrow F = 0.759$

23-B. (a) A plot of log t'_r versus (Number of carbon atoms) should be a fairly straight line for a homologous series of compound.

Peak	t'_r	log t'_r
$n = 7$	2.9	0.46
$n = 8$	5.4	0.73
$n = 14$	85.8	1.93
unknown	41.4	1.62

From a graph of log t'_r versus n, it appears that $n = 12$ for the unknown.

(b) $k' = t'_r/t_m = 41.4/1.1 = 38$

23-C. 13.03 mL of 0.022 74 M NaOH = 0.296 3 mmol of OH$^-$, which must equal the total cation charge ($= 2[\text{VO}^{2+}] + 2[\text{H}_2\text{SO}_4]$) in the 5.00 mL aliquot. 50.0 mL therefore contains 2.963 mmol of cation charge. The VO^{2+} content is (50.0 mL)(0.024 3 M) = 1.215 mmol = 2.43 mmol of charge. The H$_2$SO$_4$ must therefore be (2.963 − 2.43)/2 = 0.267 mmol. 1.215 mmol VOSO$_4$ = 0.198 g VOSO$_4$ in 0.244 7 g sample = 80.9%
0.267 mmol H$_2$SO$_4$ = 0.026 2 g H$_2$SO$_4$ in 0.244 7 g sample = 10.7%
H$_2$O (by difference) = 8.4%

23-D. (a) Since the fractionation range of Sephadex G-50 is 1 500–30 000, hemoglobin should not be retained and ought to be eluted in a volume of 36.4 mL.

(b) ^{22}NaCl ought to require one column volume to pass through. Neglecting the volume occupied by gel, we expect NaCl to require $\pi r^2 \times$ length $= \pi(1.0 \text{ cm})^2(40 \text{ cm}) = 126$ mL of solvent.

(c) $K_{av} = \dfrac{V_r - V_0}{V_t - V_0} \Rightarrow V_r = K_{av}(V_t - V_0) + V_0$

$= 0.65(126 - 36.4) + 36.4 = 95$ mL

23-E. In a, the low ionic strength reservoir drains more rapidly than the high ionic strength reservoir, since the levels must remain equal. In b, the high ionic strength reservoir drains faster than the low ionic strength reservoir.

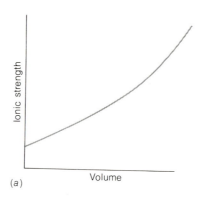
(a)

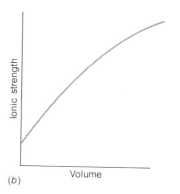
(b)

Answers to Problems

Chapter 1

5. 0.445 F   6. 0.138 M
7. 0.0548 ppm, 54.8 ppb   8. (a) 55.6 mL
(b) 1.81 g/mL   9. 0.119 osmolar   10. 1.51 m
11. 1.52 g/mL

Chapter 3

1. (a) 5   (b) 4   (c) 3   2. (a) 1.237   (b) 1.238
(c) 0.135   (d) 2.1   (e) 2.00   3. (a) 0.217   (b) 0.216
(c) 0.217   4. (a) 12.3   (b) 75.5   (c) 5.520×10^3
(d) 3.04   (e) 3.04×10^{-10} (f) 11.9   (g) 4.600
(h) 4.9×10^{-7} 5. (a) 12.6251   (b) 6.0×10^{-4}
(c) 242   6. (a) 208.24   (b) 560.602
7. $3.124(\pm0.005)$, $3.124(\pm0.2\%)$ 8. (a) $10.18(\pm0.07$
or $\pm0.7\%)$ (b) $174(\pm3$ or $\pm2\%)$ (c) $0.147(\pm0.003$ or
$\pm2\%)$ 9. (a) c   (b) b   (c) d   (d) a
10. $1.054589(\pm0.000006)$ 11. 0.000037

Chapter 4

1. (a) 0.05004   (b) 0.4037   2. 104.7   3. (a) no
(b) 0.14111   4. 90%: $0.14_8 \pm 0.02_8$, 99%: $0.14_8 \pm 0.05_6$
5. yes   6. yes   7. no   8. no, yes   9. no
10. 88   11. $y = [-0.7_5(\pm0.1_4)]x + [3.9_2(\pm0.4_9)]$
12. $5.6(\pm0.8) \times 10^3$

Chapter 5

2. 1.2×10^{10} 3. true   4. 3.0×10^{-6}
5. 2×10^{-9} 6. 9.4×10^6 8. 1.0×10^{-56}
9. 5×10^{-11} 10. (a) endergonic   (b) endothermic
11. (a) 4.7×10^{-4} atm   (b) 153°C
12. (a) 7.82 kJ/mol   13. (a) right   (b) right
(c) neither   (d) right   (e) left   14. unchanged
15. (a) right   (b) $H_2 = 1366$ Pa, $Br_2 = 3306$ Pa,
$HBr = 57.0$ Pa   (c) neither   (d) formed

Chapter 6 (right column continued)

16. (a) 0.29 g   (b) 4.5×10^{-4} g   17. 0.018
18. AgCl: 1400 ppb, AgBr: 76 ppb, AgI: 0.98 ppb
19. $I^- < Br^- < Cl^- < CrO_4^{2-}$ 20. no, 0.0014 M
21. (a) 1.3 mg/L   (b) 1.3×10^{-13} M   (c) 8.4×10^{-4} M
22. $[Zn^{2+}] = 2.9 \times 10^{-3}$ M, $[Zn(OH)^+] = 2.3 \times 10^{-5}$ M,
$[Zn(OH)_3^-] = 6.9 \times 10^{-7}$ M, $[Zn(OH)_4^{2-}] = 8.6 \times 10^{-14}$ M
23. 5.2×10^{-7} M   24. (a) 1.5×10^{-4} M
(b) 3.3×10^{-3} M   (c) 0.059 M   25. 0.22 g
26. (a) 12   (b) $[Pb^{2+}] = 2.0 \times 10^{-3}$ M

Chapter 6

1. (a) true   (b) true   (c) true   3. (a) 0.0012 M
(b) 0.08 M   (c) 0.30 M   4. (a) 0.660   (b) 0.54
(c) 0.18   (d) 0.83   5. (a) 0.42_2 (b) 0.43_2 6. 0.929
7. increase   8. (a) 2.6×10^{-8} M   (b) 3.0×10^{-8} M
(c) 4.1×10^{-8} M   9. 1.5×10^{-6} M

Chapter 7

4. $[H^+] + 2[Ca^{2+}] + [Ca(HCO_3)^+] + [Ca(OH)^+] +$
$[K^+] = [OH^-] + [HCO_3^-] + 2[CO_3^{2-}] + [ClO_4^-]$
5. $[H^+] = [OH^-] + [HSO_4^-] + 2[SO_4^{2-}]$
6. $[CH_3CO_2^-] + [CH_3CO_2H] = 0.1$ M
7. $[OH^-] = 1.5 \times 10^{-7}$ M   8. $[Ca^{2+}] = 1.4 \times 10^{-3}$ M,
$[F^-] = 1.7 \times 10^{-4}$ M, $[HF] = 2.5 \times 10^{-3}$ M
9. 4.0×10^{-5} M   10. 5.8×10^{-4} M
11. (a) 5.0×10^{-9} mol   (b) 2.1×10^{-6} mol
(c) 1.1×10^{-8} mol   12. 4.3×10^{-3} M
13. 5.2×10^5 lb

Chapter 8

7. 11.69 mg CO_2, 2.051 mg H_2O 8. 10.5%
9. 7.22 mL   10. 0.08660 g   11. 0.19140 g, 0.10730 g
12. 0.339 g   13. 14.5% K_2CO_3, 14.6% NH_4Cl
14. 40.4%   15. 22.65%   16. 13.99%

731

732 **Chapter 9**

3. 43.2 mL $KMnO_4$, 270.0 mL $H_2C_2O_4$ 4. 32.0 mL
5. 1.72 mg   6. 12.4%   7. 15.1%   8. 0.149 M
9. (a) 0.020 34 M   (b) 0.125 7 g   (c) 0.019 82 M
10. 0.052 12 M   11. 0.092 54 M   12. 56.28%
13. 13.08, 8.04, 2.53   14. 947 mg   15. 0.037 3 M
17. negative   18. 90% T $\Rightarrow \tau = 0.105$ cm^{-1},
10% T $\Rightarrow \tau = 2.30$ cm^{-1}, A $= 1 \Rightarrow \tau = 2.30$ cm^{-1}
19. (a) 20.17 mL   (b) 9.57   20. (a) 19.00, 18.85,
18.65, 17.76, 14.17, 13.81, 7.83, 1.95   (b) no
21. (a) 25.0 mL   (b) 0.090 9 M   (c) 24.07
22. upper $= 9.6 \times 10^3$ M (which means there is no
upper limit), lower $= 6.2 \times 10^{-4}$ M

Chapter 10

9. pH $= 6.99$, $[H^+] = 1.2 \times 10^{-7}$ M
10. (a) 0.012   (b) increase   11. 15%   17. a, c
18. 4.79   20. $K_{A1} = 7.0 \times 10^{-3}$, $K_{A2} = 6.3 \times 10^{-8}$,
$K_{A3} = 4.3 \times 10^{-13}$

Chapter 11

2. (a) 3.00   (b) 12.0   3. 7.38, 7.12
5. pH $= 11.28$, $[B] = 0.058$ M, $[BH^+] = 1.9 \times 10^{-3}$ M
6. pH $= 5.51$, $[B] = 3.1 \times 10^{-6}$ M, $[BH^+] = 0.060$ M
7. pH $= 10.95$ 8. (a) 3.03, 9.4%   (b) 7.00, 99.9%
(c) 8.00, 0.010%   9. pH $= 3.15$ 10. $pK_A = 4.19$
11. 3.6×10^{-9} 12. 3.55
13. 0.007 56%, 0.023 9%, 0.568%
14. 5°—endothermic, 45°—exothermic
15. 0.180, 1.00, 1.80   16. 4-aminobenzenesulfonic acid
19. 13.9 mL   20. (a) 8.19   (b) 8.25
21. $[A^-] = 0.004 01$ M   22. (a) 14   (b) 1.4×10^{-7}
23. 3.27 mL   24. (b) 7.18   (c) 7.00   (d) 6.86 mL
25. (a) 2.56   (b) 2.61   (c) 2.86   26. 11.70, 11.48
27. 4.37×10^{-4}, 8.95×10^{-13} 28. H_3O^+, 0, 14
29. pH $= 11.60$, $[B] = 0.296$ M, $[BH^+] = 3.99 \times 10^{-3}$ M,
$[BH_2^{2+}] = 2.15 \times 10^{-9}$ M   30. (a) pH $= 1.95$,
$[H_2M] = 0.089$ M   (b) pH $= 4.28$, $[H_2M] = 3.7 \times 10^{-3}$ M
(c) pH $= 9.35$, $[H_2M] = 7.04 \times 10^{-12}$ M
32. pH $= 5.70$ 33. 2.8×10^{-3}, 2.9×10^{-8}
34. (a) 1.37   (b) 12.61   35. 0.005 40%
36. (a) 5.88   (b) 5.59   37. pH $= 3.69$ 38. (b) 3.96

Chapter 12

3. 2.2×10^9 4. 10.92, 9.57, 9.35, 8.15, 5.53, 2.74
5. $\frac{1}{11}V_e$, $\frac{10}{11}V_e$, $V_e/11$—pH $= 3.60$, $V_e/2$—pH $= 4.60$,
$10V_e/11$—pH $= 5.60$, V_e—pH $= 8.65$, $1.2V_e$—pH $= 12.07$
6. 8.74, 5.35, 4.87, 4.40, 3.22, 2.57   7. 8.18   8. 5.09
9. 11.36, 10.21, 9.73, 9.25, 7.53, 5.81, 5.33, 4.85, 3.41,

2.11, 1.85   10. (a) 1.99   11. 7.18   12. 9.81
13. 2.66   14. (a) 9.56   (b) 7.4×10^{-10}
15. 2.97 mL   16. no   18. red, orange, yellow
19. red, orange, yellow, red   20. 2.47   21. violet,
blue, yellow   22. 5.62   23. 0.100 0 M   24. 0.30 g
25. 7.1×10^7 26. (a) 9.45   (b) 2.55   (c) 5.15
27. 0.091 78 M   28. 6.28 g   29. 0.139 M
30. $V_e = 23.40$ mL   31. 0.063 56 M
32. (a) 20.254%   (b) 17.985 g   33. 9.72
34. (a) acetic acid   (b) pyridine

Chapter 13

2. (b) 8.62×10^{-6}, 0.613, 0.387, 1.74×10^{-6}
3. 5.59, 5.72   5. $\alpha_1 = 0.123$, 0.694   6. $\alpha_1 = 0.110$,
0.500, 0.682, 0.500, 2.15×10^{-4} 7. 1.06×10^5
8. 9.44   9. 4.16   10. 4.48   11. 4.82
12. 4.44   13. 7.73   14. 3.62   15. 18%
16. (a) 8.34   (b) 7.00   17. 1.83

Chapter 14

4. (a) 3.4×10^{-10} (b) 0.64   5. 0.995 mg
6. 21.45 mL   7. 49.9—4.87, 50.0—6.90, 50.1—8.92
8. 49.9—7.55, 50.0—6.21, 50.1—4.88
9. (a) 2.93   (b) 6.77   (c) 10.49   10. 2.7×10^{-11} M
11. $[Ni^{2+}] = 0.012 4$ M, $[Zn^{2+}] = 0.007 18$ M
12. 0.092 6 M   13. 0.024 30 M   14. Mn—69.64,
Mg—5.150, Zn—20.89   15. 2 600

Chapter 15

1. (a) $6.241 460 \times 10^{18}$ (b) 96 484.56   2. 0.054 20,
0.061 54   3. (a) 1.87×10^{16} e^-/s
(b) 9.63×10^{-19} J/e^- (c) 5.59×10^{-5} mol   (d) 447 V
4. 1.32 kg   5. Cl_2 6. Cr^{2+}
7. stronger—$Cr_2O_7^{2-}$, MnO_4^-, IO_3^-, unchanged—Cl_2,
Fe^{3+} 8. (a) K $= 10^{48}$ (b) K $= 8.5 \times 10^{-7}$
(c) K $= 1.9 \times 10^{-6}$ (d) K $= 2.4 \times 10^5$
9. (a) $Pt(s)|Cr^{2+}, Cr^{3+}\|Tl^+|Tl(s)$ (b) 0.07_4 V   (d) Pt
10. (b) K $= 1._1 \times 10^{16}$ (c) -0.02_3 V   (d) 11 kJ,
(e) 0.24   12. (a) Fe^{3+} (b) Fe^{2+}
13. (a) $Fe(s)|FeO(s)|KOH(aq)|Ag_2O(s)|Ag(s)$;
$Fe(s) + 2OH^- \rightleftharpoons FeO(s) + H_2O + 2e^-$;
$Ag_2O(s) + H_2O + 2e^- \rightleftharpoons 2Ag(s) + 2OH^-$
(b) $Pb(s)|PbSO_4(s)|K_2SO_4(aq)\|H_2SO_4(aq)|PbSO_4(s)|$
$PbO_2(s)|Pb(s)$; $Pb(s) + SO_4^{2-} \rightleftharpoons PbSO_4(s) + 2e^-$;
$PbO_2(s) + 4H^+ + SO_4^{2-} + 2e^- \rightleftharpoons PbSO_4(s) + 2H_2O$
14. (a) $Fe(CN)_6^{4-} \rightleftharpoons Fe(CN)_6^{3-} + e^-$;
$Ag(CN)_2^- + e^- \rightleftharpoons Ag(s) + 2CN^-$
(b) $2Hg(l) + 2Cl^- \rightleftharpoons Hg_2Cl_2(s) + 2e^-$; $Zn^{2+} +$
$2e^- \rightleftharpoons Zn(s)$ 15. (a) 0.053 V   (b) 2.5×10^{-13} M
(c) 1.6×10^{13} 16. (b) -2.82 V   (c) Br_2

(d) 1.31 kJ   (e) 2.69×10^{-8} g/s   17. 0.573 V
18. (a) 0.572 V   (b) 0.569 V   19. (a) A = -0.414 V,
B = 0.059 16 V   (b) Hg $\rightarrow$ Pt   20. 9.6×10^{-7}
21. 7.5×10^{-8} 22. 0.101 V   23. 1.065 V
24. K = 1.0×10^{-9} 25. 1.7×10^{-4}
26. -0.447 V   27. 0.117 V   28. 0.76 V
29. 1.003 V   30. $E_2^0 > E_1^0$ 31. -0.398 V
32. -0.184 V   33. -0.036 V   34. 7.2×10^{-4}
35. 5.4×10^{14} 36. (a) 1.95×10^{-4} M
(b) 2.7×10^{-8} 37. 0.76
38. (a) $[Ox] = 3.82 \times 10^{-5}$ M, $[Red] = 1.88 \times 10^{-5}$ M
(b) $[S^-] = [Ox], [S] = [Red]$ (c) -0.092 V

Chapter 16

1. 0.683 V   2. 0.243 V   3. 0.627   5. H^+—42.4 s,
NO_3^-—208 s   6. 0.10 pH units   7. (a) $3._2 \times 10^{13}$
(b) 8%   (c) 49.0, 8%   9. (a) 274 mV   (b) 285 mV
11. 10.67   13. (a) K_I/0.033 3   (b) K_{Cl}/0.020 0
(d) 2.2×10^6 14. $5._2 \times 10^{21}$ 15. 1.366 V
17. (b) 0.951 3   18. 0.301 2 V, Zn^{2+}
19. (b) $2.43 (\pm 0.09) \times 10^{-3}$ M   20. (a) 8.9×10^{-8} M
(b) 1.90 mmol   21. 1.49×10^{-8} M, 1.25×10^{-13} M
22. (a) 2.32×10^{-4} M   (c) Divide each ΔE by 2
23. (a) 1.13×10^{-4} (b) 4.8×10^4

Chapter 17

2. $-0.71, -0.65, -0.06, 0.529$ V   3. (c) 0.095 V
(d) 0.236 V   4. 0.493, 0.519, 0.529, 0.539, 0.565, 0.585,
0.647, 1.07, 1.15, 1.16, 1.17 V   5. (a) -0.031 V
(b) 0.047 V   6. $-0.102, 0.53, 0.96, 1.24$ V
7. (b) 0.025 1 M UO_2^+, 0.036 5 M Fe^{3+} (c) -0.189,
0.526, 1.15, 1.25 V   8. $-0.043, 0.438, 0.851$ V
9. no   10. no   11. 0.390, 0.450, 0.526, 0.602, 0.662,
0.850, 0.906, 0.962, 1.14, 1.32, 1.38
12. -0.207 V   13. (a) 7.0×10^2 (b) 0.9
(c) 0.33_4 g/L   14. 0.011 29 M   15. (a) 0.029 14 M
(b) no   16. 41.9%   17. 40.3%   18. 5.730 mg
19. 78.67%   20. 3.826×10^{-3} M   21. (a) $5IO_4^-$ +
$C_6H_{12}O_6 \rightarrow 5HCO_2H + H_2CO + 5IO_3^-$
(b) $C_6H_{13}O_5N + H_2O + 5IO_4^- \rightarrow 5HCO_2H +$
$H_2CO + NH_3 + 5IO_3^-$ (c) $C_3H_6O_3 + IO_4^- \rightarrow$
$H_2CO + HO_2CCH_2OH + IO_3^-$ 22. 9.107 mg

Chapter 18

5. 2.68 h   6. (a) anode   (b) 52.0 g/mol
(c) 0.039 6 M   7. (a) 6.64×10^3 J   (b) 0.012 4 g/h
8. (a) 1.946 mmol   (b) 0.041 09 M   (c) 4.75 h
9. (a) 0.84 V   (b) -1.06 V   10. 0.53 V
11. 54.77%   12. -0.614 V, negative

13. -2.193 V   14. (a) -1.23 V
15. (b) 0.000 26 mL   16. 26.3% trichloroacetic acid,
49.5% dichloroacetic acid   17. 1.51×10^2 μg/mL

Chapter 19

3. (a) 0.000 35 $\min^{-1}$ (b) 3.4 min   (c) 0.118%
5. $E_{3/4} - E_{1/4} = 0.056 5/n$ V   6. 1.03×10^{-9} m^2 s^{-1}
7. 0.007 6 9 $(\pm 0.003 2)$ mM   8. 2.371 (± 0.024) mM
9. 0.760 ppm   10. similar to Exercise 19-F, lower curve
11. similar to Demonstration 19-1   12. 2.95 mM
13. (c) $E' = 0.853$ V (versus S.H.E.)   (d) $n \approx 2$

Chapter 20

1. double, halve, double   2. (a) 184 kJ/mol
(b) 299 kJ/mol   3. 5.33×10^{14} Hz, 1.78×10^4 cm^{-1},
213 kJ/mol   4. 3.56×10^4 $M^{-1}cm^{-1}$ 5. violet blue
6. Wavenumber = 1.697 834 5 and $1.696 114 4 \times 10^4$ cm^{-1}
7. 1.02 mg   8. 0.289   9. (a) 4.97×10^4 $M^{-1}cm^{-1}$
(b) 4.69 μg   10. (a) $2.42_5 \times 10^4$ M^{-1} cm^{-1}
(b) 1.26%   11. yes   12. 2.19×10^{-4} M
13. $[A] = 9.11 \times 10^{-3}$ M   14. 78 (± 3) μL
18. (a) 2.38×10^3 (b) 143 lines/cm   19. (a) 228 μM
(b) 357 μM   20. 88.2

Chapter 21

3. 589.3 nm   4. (a) 283.0 kJ/mol   (b) 3.67×10^{-6}
(c) 8.4%   (d) 1.03×10^{-2} 5. 17.4 μg/mL
6. Na—3.3 GHz, Hg—2.6 GHz   7. (b) 1.84 μg/mL
8. 0.079 μg/mL

Chapter 22

6. (a) 0.158 M in benzene   (b) 2×10^{-6} M in benzene
7. 2   8. (a) 2.6×10^{-4} at pH 1 and 260 at pH 4
(b) 0.037   9. fraction of A in each tube: 0.000 012 8,
0.000 358, 0.004 30, 0.028 7, 0.115, 0.275, 0.367, 0.210;
fraction of B in each tube: 0.007 81, 0.054 7, 0.164,
0.273, 0.273, 0.164, 0.054 7, 0.007 81
10. (b) recovery = 53.4%, purity = 99.1%
11. (a) 19.6 mL   (b) 13.9 mL   12. (a) 360   (b) 9 096
14. 10.4 mL   15. 405 000 plates   16. 4.69
17. (a) 44 000   (b) 12 000   (c) 11 000

Chapter 23

11. 932   12. 11.7 min   13. 230 μm, 2 540 mesh
14. transferrin—0.127, ferric citrate—0.789   15. 35 000
16. (a) 30   (b) 3.3   (c) decrease   17. 38.0%
18. 320 000

Index

A page number not followed by a letter indicates that the subject appears in the main body of the text or in an example. Otherwise, the following abbreviations are used:

b = box
d = demonstration
e = experiment
f = footnote
i = illustration
m = marginal comment
p = problem or exercise

A (ampere), 1
Å (angstrom), 4t
Absolute error, 33
Absorbance, 470, 471b
  light scattering, 137b
  Spectronic 20 scale, 28
Absorption, 115
Absorption quenching, 503b
Absorption spectrum, 470, 472d
Absorptivity, 470
Accuracy, 33
ACES (N-2-acetamido-2-aminoethanesulfonic acid), 197t
Acetic acid, 197t
  ionic-strength effect on K_A, 88
  nonaqueous titrations, 249
  temperature-dependence of pK_A, 218t
Acetoacetic acid, 321t
Acetoxyquinoline, 115t
Acetylene flame, 516
Acetylsalicylic acid, 178
Acid, 155
  amphiprotic, 205–210
  Brønsted–Lowry, 155
  carboxylic, 165
  conjugate, 157, 169–170, 185, 187–188
  diprotic, 199–212
  dissociation constant, 165, 169–171
  intermediate form, 205–210
  leveling, 162–164
  Lewis, 156, 271
  nature of H^+, 157–158
  pH scale, 161
  pK, 167
  polyprotic, 166, 213–215
  solvent extraction, 537
  strong, 161t
  strong acid pH calculation, 175–178
  titration of dibasic compound with strong acid, 233–237

titration of strong acid with strong base, 221–225
titration of weak acid with strong base, 225–231
titration of weak base with strong acid, 231–232
water-dissociation effect, 178, 180
weak, 164–167, 178–182
Acid–base equilibria, 175–215
Acid–base titration, 221–250, 617–618e, 619e, 622–624e
Acid dissociation constants, 663–672t
  how to use Appendix G, 170
Acid error, 348
Acidic pH, 161
Acidity of strong acids, 242–243b
Activation energy, 404i
Activity, 83
  equilibrium constant, 84
  pH calculation, 175, 189, 196, 219p
  relation of gradient to free energy difference, 326–327p, 345
Activity calculations, 89–91
Activity coefficient, 83, 623–624
  acid dissociation constant, 165
  extended Debye–Hückel equation, 85
  ionic-size effect, 85–87
  K_w, 159
  mass balance, 95m
  nonionic compounds, 88
  table, 85–86t
Activity gradient free energy, 326–327p, 345
ADA (N-2-acetamidoiminodiacetic acid), 197t
Addition:
  propagation of error, 34–35
  significant figures, 28–30
Adenosine triphosphate (ATP), 271, 272i, 326p
Adjusted retention time, 570
Adsorption, 115
Adsorption chromatography, 551, 552i

Affi-Gel, 602
Affinity chromatography, 552i, 553, 601–602
Agar, 301
Alanine, 201t
  isoelectric pH, 261–262
  isoionic pH, 261–262
Alcohol oxidation, 383, 388–389
Aldehyde, 386
Aldehyde oxidation, 383, 388–389
Aldolase, 601
Alizarin garnet R, 505
Alizarin red S, 149t
Alizarin yellow, 241t
Alkaline error, 348
α (fraction of association), 185, 187–188
α (fraction of dissociation), 181–182, 187–188, 258–259
$\alpha_{Y^{4-}}$ (fraction of EDTA in form Y^{4-}), 275–276
Alumina, 583m–584
  drying agent, 24t
Aluminum:
  EDTA titration, 287
  gravimetric analysis, 120t, 124
Aluminum chloride, 6
Amalgam, 378, 436
Amberlite, 589t
Amine, 166i
  Malaprade reaction, 389
Amino acid, 200, 201–202t, 575b
  isoelectric pH, 260–261
  isoionic pH, 260–261
  pH calculations, 200–211
  titration in ribonuclease, 221, 222i
Aminocarboxylic acids, 271, 272i
Ammonia, 9
  auxiliary complexing agent, 283, 286–287
  buffer preparation, 624
  copper complex formation, 290p
  electrode, 354
  Kjeldahl titration, 134

735

736 Ammonia (*continued*)
  pH, 186
  principal species, 256
  reaction with water, 63
  spectrophotometric analysis, 506p
  temperature-dependence of K_A, 77p
Ammonium bicarbonate, 268
Ammonium bifluoride, 348
Ammonium decavanadate, 625–626e
Ammonium hexanitratocerate(IV), 382
Ammonium hydroxide, 9
Ammonium ion, 166
  gravimetric analysis, 120t
  Kjeldahl analysis, 626
Ampere, 1, 2t, 294
Amperometric titration, 456–461
  biamperometric titration, 457–459
  bipotentiometric titration, 461
  one polarizable electrode, 456–457
  rotating platinum electrode, 456–457
  two polarizable electrodes, 457–461
Amperometry, 414–415, 456
Amphiprotic species, 205
  weak acid–weak base equilibrium, 266–268
Ampholyte, 262b
Amylose, 376
Analyte, 129
Analytical concentration, 6
Angstrom, 4t
Anhydrone, 24t
Anion exchanger, 587
Anode, 299
Anodic wave, 452
Anthraquinone-2,6-disulfonate, 331p
Antibonding orbital, 474–475
Antilogarithm, 30
  significant figures, 30–31
Antimony:
  iodimetric titration, 387t
  iodometric titration, 388t
  permanganate titration, 381t
Antimony pentachloride, 156
Apiezon, 576t
Apotransferrin, 495
Approximations, 70b
Aragonite, 332
Area measurement, 580
Argentometric titration, 146, 456i, 627–628e
Arginine, 201t, 256
Arsenate, 149t
Arsenic:
  iodimetric titration, 387t
  iodometric titration, 388t
  permanganate titration, 381t
Arsenious oxide, 380, 384, 385i
Arsenite, 377t
Ascarite, 110, 248
  drying agent, 24t
Ascorbic acid, 274b, 377t, 460d, 489, 627e
  formal potential, 321t, 322–323
  iodimetric titration, 386
Ashless filter paper, 23, 616
Asparagine, 201t
Aspartic acid, 201t
Aspirator trap, 22i
Aspirin, 178
Asymmetric bend, 476i
Asymmetric stretch, 476i
Asymmetry potential, 346
Atmosphere, 4, 4t
Atomic absorption, 513–529
Atomic emission, 513, 514, 518, 529

Atomic fluorescence, 514
Atomic spectroscopy, 513–529
  absorption, 514
  analytical methods, 524–528
  burner, 516–517
  detection limits, 524i
  emission, 514, 515
  flame photometer, 525b
  fluorescence, 514
  graphite furnace, 517–518
  hollow-cathode lamp, 514i, 522
  inductively coupled plasma, 518–519, 529
  interference, 528–529
  linewidth, 521–522
  spectrometer, 515i, 521–524
  temperature effect, 519–521
  viscosity effect, 528
Atomic weight, 5
Atomization, 514, 516–519, 528
ATP synthesis, 326p
Autoprotolysis, 158–159
Auxiliary complexing agent, 283, 286–287
Auxiliary electrode, 412
Average, 43
Avogadro's number, 5
Azobenzene, 390
Azo compound, 390m
Azo indicators, 286

Background correction (atomic spectroscopy), 523
Back titration, 287, 390
Balance, 14–17
Balance point, 15
Balancing redox reactions, 651–655
Ballistic galvanometer, 431f
Band shape (chromatography), 560–561
Band spreading, 560
Bandwidth, 484, 486–487
  chromatography, 555
  countercurrent distribution, 548–551
Barbituric acid, 217p
Barium:
  gravimetric analysis, 120t
  permanganate titration, 381t
Barium oxide (drying agent), 24t
Barium perchlorate (drying agent), 24t
Base, 155
  alkali metal equilibria, 173p
  alkaline metal equilibria, 162t
  amines, 168
  amphiprotic, 205–210
  Brønsted–Lowry, 155
  conjugate, 157, 169–170, 185, 187–188
  diprotic, 199–212
  "dissociation" constant, 165, 169–171
  intermediate form, 205–210
  leveling, 162–163
  Lewis, 156, 271
  nature of OH^-, 158
  pH scale, 161
  pK, 167
  polyprotic, 166–167, 213–215
  solvent extraction, 536
  storage, 616f
  strong, 161t, 175–178
  titration of dibasic compounds, 233–237
  titration of strong acid with strong base, 221–225
  titration of weak acid with strong base, 225–231

  titration of weak base with strong acid, 231–232
  water-dissociation effect, 178
  weak, 164–167, 183–186
Base association constant, 100
Basic pH, 161
Beam chopper, 523
Beer's law, 470, 471b, 491
Belousov–Zhabotinskii reaction, 337d
Benzoic acid, 255
β (buffer capacity), 193–196
β (cumulative formation constant), 73
Biacetyl, 503b
Biamperometric titration, 457–459, 631e
Bicarbonate analysis, 619e
Bicarbonate reference buffer, 347t
BICINE (N,N-*bis*(2-hydroxyethyl)glycine), 198t
Binomial expansion, 547
Bio-Gel, 598, 599t
Bipotentiometric titration, 460d, 461
bis(aminoethyl)glycolether-N,N,N',N'-tetra-acetic acid, 272i
Bismuth, 381t
Bismuthate, 377t, 378
1,3-*bis*[*tris*(hydroxymethyl)amino]propane, 197t
BIS–TRIS propane, 197t
Bisulfite, 194d
Blank, 489
Bleach, 189
Blocking, 286, 287–288
Blood, 8, 489
Blue dextran, 598
Boltzmann distribution, 519–520
Bonded-phase particle, 587
Borax reference buffer, 347t
Boric acid, 198t
Bradford protein analysis, 59p
British thermal unit, 4t
Bromate, 377t
  iodometric titration, 388t
Bromide:
  gravimetric analysis, 120t
  ion-selective electrode, 352t
  permanganate titration, 381t
Bromine, 377t
  generation by coulometry, 631e
  iodometric titration, 388t
  reaction with cyclohexene, 414
Bromocresol green, 241t, 618, 619, 620
Bromocresol purple, 241t
Bromophenol blue spectrum, 472d
Bromothymol blue, 241t
Brønsted–Lowry acids and bases, 155–157
Btu (British thermal unit), 4t
Buffer, 186–193, 194d, 195–199
  acid–base titration, 226–228, 232, 234, 235
  activity effects, 196
  diprotic, 211–212
  electrochemical cell, 317–318
  Henderson–Hasselbalch equation, 188–189
  hydrogen electrode, 304m
  limitations, 196
  metal ion, 356–357
  method of operation, 192, 194d
  National Bureau of Standards pH references, 347t
  pH range, 195
  preparation, 193
  temperature-dependence, 196
  weak acid–weak base equilibrium, 266

Buffer capacity, 193–196, 230
Buffer intensity, 193–196
Bulb (for pipet), 21f
Buoyancy, 16–17
Buret, 17–19
   calibration, 611–614
   calibration curve, 32
   cleaning, 18–19
   drainage rate, 19
   method of reading, 18
   reading scale, 28
   tolerance, 18t, 32
   use, 18–19
Burner, 516–517

c (centi), 5t
C (coulomb), 3t
c (speed of light), 468
Cacodylic acid, 197t, 270p
Cadmium:
   electrode, 342
   amine complexes, 272–273
   iodimetric titration, 387t
   ion-selective electrode, 351
Cadmium reductor, 379
cal (calorie), 4t
Calcein, 505
Calcite, 332
Calcium:
   buffer, 356–357
   EDTA titration, 624–625e
   fluorescence analysis, 505
   gravimetric analysis, 114, 120t
   gravimetric determination, 615e
   permanganate titration, 381t
   water hardness, 289
Calcium chloride (drying agent), 24t
Calcium fluoride (solubility), 100
Calcium oxide (drying agent), 24t
Calcium salicylate (ignition), 118
Calcium sulfate (drying agent), 24t
Calculator, 67
Calibration:
   glass electrode, 346–348, 349
   glassware, 611–614
Calmagite, 285t
Calomel electrode, 336
Calorie, 4t
Candela, 2t
Capacitance, 3t
Capacity factor, 570
Capillary constant, 432, 446
CAPS (3-(cyclohexylamino)propanesulfonic
   acid), 198t
Carbon analysis, 110
Carbonate, 619e
   EDTA titration, 288
   gravimetric analysis, 120t
Carbonate reference buffer, 347t
Carbon dioxide:
   acidity, 239d
   electrode, 354
   exclusion from basic solutions, 247–
   248
Carbowax 20M, 577t
Carboxylate anion, 165
Carboxylic acid, 165, 383
Catalysis, 597
Catalytic wave, 446
Cathode, 299
Cathodic depolarizer, 411

Cathodic wave, 452
Cation exchanger, 587
cd (candela), 2t
Cell positioning error, 488
Cellulose, 584
centi, 5t
Ceric ammonium sulfate, 382
Cerium, 377t, 382–383
   gravimetric analysis, 120t
   iodometric titration, 388t
   oxidation of malonic acid, 337d
   permanganate titration, 381t
   potentiometric titration calculations, 364–
   369
   standardization, 382
Cesium hydroxide, 161t
Character, of logarithm, 30
Charge, 294
   polyprotic system, 260–261
Charge balance, 93–95, 102b
Charging current, 434
Charring, 518
Chelate effect, 272–273
Chelate extraction, 537–538, 540–541d
Chelating ligand, 271
Chelation therapy, 274b
Chelator, 271
Chelex 100, 596–597
Chemical coulometer, 421–422
Chemical interference, 528
Chemiluminescence, 479
Chemiosmotic hypothesis, 326p
CHES (cyclohexylaminoethanesulfonic acid),
   198t
Chloramine T, 377t
Chlorate, 383
Chlordiazepoxide, 448i
Chloride:
   gravimetric analysis, 109, 120t
   ion-selective electrode, 352t
   titration, 627–628e
Chlorine, 388t
Chlorophenol red, 241t
Chlorophyll, 455b
Chopper, 481, 523
Chromate:
   EDTA titration, 288
   homogeneous precipitant, 115t
Chromatogram, 570
Chromatography, 551–561, 567–606, 634–636e,
   636–637e
   adjusted retention time, 570
   adsorption, 551, 552i
   affinity, 552i, 553, 601–602
   anion exchanger, 587
   applications of gel filtration, 600
   applications of ion exchange, 596–597
   automated apparatus, 605–606
   band shape, 560–561
   band spreading, 560
   bandwidth, 555i
   bonded-phase particles, 587
   capacity factor, 570
   capillary column, 574–575b
   cation exchanger, 587
   chromatogram, 570
   cochromatography, 580
   column, 573, 583, 586, 595, 603, 606i
   column-length effect, 559
   detector, 578–580, 584, 585i
   discovery, 551m
   Donnan equilibrium, 593–594

eddy diffusion, 556, 557i, 557f
eluate, 553
eluent, 553
eluent strength, 584
elutropic series, 584
fines, 603
flow adaptor, 583
gas, 567–582
gel, 598–599
gel filtration, 551, 597–601
gel permeation, 551
gradient elution, 586, 595, 605
H.E.T.P., 554, 557, 558
HPLC (high performance liquid chromatography), 585–587
hydrostatic pressure, 604–605
inorganic ion exchangers, 590, 594–595
instrumentation, 567–569, 586, 606i
internal standard, 582
ion exchange, 551, 552i, 587–597
ion-exchange gels, 590, 591i, 591t, 594–595
ion-exchange resins, 588–590, 594–595
ion-exchange selectivity, 592–593
isocratic elution, 586
K_{av}, 598
Kovats index, 571
liquid, 583–587
longitudinal diffusion, 556, 557i
Mariotte flask, 604–605
microporous particles, 587
molecular-exclusion, 551, 552i, 597–601
normal phase, 587
overloading, 560–561, 563p
particle size, 558
partition, 551, 552i
partition coefficient, 554, 560
pellicular particles, 587
plate theory, 554–555
pouring columns, 603
precolumn, 586
quantitative analysis, 580–582
rate theory, 556–558
reduced plate height, 557
resolution, 558–559
response factor, 581–582
retention index, 571
retention ratio, 553
retention time, 553, 555i, 570
retention volume, 554, 555i
reverse phase, 587
sample application, 604
silanization, 561
solid support, 573
solvent, 584
stationary phase, 573–578, 583–584, 587,
   588–591, 594–595, 598–599, 602,
   603
tailing, 560–561
techniques, 602–606
temperature programming, 571–572
terminology, 553–554
theoretical plate, 554–555
thin-layer, 584–585
van Deemter equation, 556–558
velocity, 558
void volume, 597
Chromic acid cleaning solution, 19f
Chromium, 120t
Chromium(II), 377t, 378, 390
Chromophore, 470
Chromous chloride, 378
Chymotrypsin, 186i, 187

737

738

Chymotrypsinogen, 601i
Circuit, 297
Citrate, 286
Citric acid, 197t
Clark electrode, 458b
Cleaning solution (chromic acid), 19f
Clock reaction, 194d
Coagulation, 115
Cobalt:
  gravimetric analysis, 120t
  permanganate titration, 381t
Cobaltinitrate, 381t
Cocaine, 184
Cochromatography, 580
Colloid, 112d
Colloidal suspension, 112
Color, 470, 472d, 473t
Colorimetric analysis, 488m
Column (chromatography), 573–578, 583, 595, 603, 606
Combination electrode, 343, 344i
Combusion analysis, 110–111, 125
Commas, in writing numbers, 4
Common ion effect, 68–69
Comparison of means, 48–51
Complementary color, 470, 473t
Complex formation, 72–74, 271–272
Complex ion, 72, 271
Composition of precipitates, 118
Compound electrode (ion-selective), 354
Concentration, 5
  changes in galvanic cell, 312b
Concentration gradient, 326–327p, 345
Concentration polarization, 402–403, 430
Condenser current, 434, 447i, 449–450
Conditional formation constant, 277–279
Conductance, 3t
Conduction band, 400b
Conductivity, 183d
Confidence interval, 46–48, 52
Congo red, 241t
Conjugate acids and bases, 157, 169–170, 185–186, 187–188
Constant-applied-potential electrolysis, 410–411
Constant-current coulometry, 416
Constant-current electrolysis, 412
Continuous-flow analysis, 492b
Controlled-potential coulometry, 416
Controlled-potential electrolysis, 412–413
Controlled precipitation, 114
Conversion factors, 4t
Coordination number, 277
Copper:
  EDTA titration, 286
  electrode, 342
  electrogravimetric analysis, 629e
  gravimetric analysis, 120t
  iodometric titration, 388t
  ion-selective electrode, 351
Copper acetylsalicylate preparation, 629
Coprecipitation, 72, 117
Coulomb, 3t, 294, 295, 397
Coulomb's law, 108p
Coulometric titration, 416
Coulometry, 413–422, 631e
  apparatus, 414i, 420
  chemical coulometer, 421
  constant-current, 416
  controlled-potential, 416
  endpoint, 416

mediators, 417–420
  practical circuits, 631m
Countercurrent distribution, 542–551
  band spreading, 550–551
  bandwidth and resolution, 548–551
  extraction scheme, 542–545
  purity, 545, 550
  theoretical distribution, 546–548
Cresol purple, 241t
Cresol red, 241t
Crosslinking, 590
Crown ether, 540–541b
Crystallization, 112–114
Cumulative formation constant, 73
Cupferron, 121t, 127p, 537
Current, 1, 2t, 294, 296
  galvanic cell, 312b
  relation to coulombs, 397
Current maxima, 437
Current–voltage relation (electrolysis), 397–405, 408–412
Cuvette, 479, 484
Cyanide:
  demasking, 289
  EDTA titration, 291p
  gravimetric analysis, 120t
  ion-selective electrode, 352t
  masking agent, 288–289
Cyanogen flame, 516t, 517
Cyclic voltammetry, 452–454
Cyclohexene, 414, 631
Cyclohexylaminoethanesulfonic acid, 198t
3-(cyclohexylamino)propanesulfonic acid, 198t
Cysteine, 201t, 387t
Cystine, 321t
Cytochrome, 263b, 321t
Czerney–Turner grating monochromator, 482i

d (deci), 5t
da (deca), 5t
Dark current noise, 487–488
DCTA (diaminohexanetetraacetic acid), 272
Debye–Hückel equation, 85
deca, 5t
Decavanadate, 625
deci, 5t
Decomposition potential, 408
Degrees of freedom, 43, 49, 56, 475
Dehydration, of glass electrode, 346–347, 348
Deionization, 596
Deionized water, 596
Demasking, 289
Demonstrations:
  absorption spectra, 472
  colloids and dialysis, 112
  conductivity of weak electrolytes, 183
  effect of ionic strength on ion dissociation, 84
  electrochemical writing, 398
  extraction with dithizone, 540–541
  Fajans titration, 149d
  HCl fountain, 162–163
  how buffers work, 194
  human salt bridge, 302
  indicators and acidity of CO_2, 239
  Karl Fischer jacks of pH meter, 460
  metal ion indicator color changes, 284
  potentiometric titration of Fe^{2+} with MnO_4^-, 373
  potentiometry with oscillating reaction, 337–339

Density, 7
  effect on weighing, 17
  water, 612t
Derived units, 3t
Desalting, 601
Desferrioxamine, 274b, 507p
Desiccant, 24
Desiccator, 24
Detection limit, 524
Detector:
  chromatography, 578–580, 584, 585i, 586i, 606i
  spectrophotometer, 484–485
Determinant, 54
Determinate error, 32
Deuterium lamp, 482, 523
Dextran, 590, 591i
Dialysis, 112d, 261
Diaminohexanetetraacetic acid, 272
Diatomite, 561
Dibenzo-30-crown-10, 540b
Dibutyl tetrachlorophthalate, 576t
2,6-dichlorophenolindophenol, 321t
Dichromate, 377t, 383
  iodometric titration, 388t
  redox titrations, 383
  spectrum, 472d
Dielectric constant, 250
Diethylenetriaminepentaacetic acid, 272i
Differential pulse polarography, 446, 447–451
Diffraction, 483–484
Diffraction grating, 472d, 482–484
Diffusion coefficient, 432f
Diffusion current, 430, 431–433
Digestion, 116, 620
2,3-dimercaptopropanol, 289
Dimethyldichlorosilane, 561
Dimethylglyoxime, 121t, 122–123
Dimethyl oxalate, 115t
Dimethyl sulfate, 115t
Dinonyl phthalate, 576t
Dioctylsulfoxide, 539
Diol, 602
Diphenylamine, 364, 375t
Diphenylamine sulfonic acid, 375t, 383
Diphenylbenzidine sulfonic acid, 375t, 383m
Diphenyldisulfide, 321t
Diphenylthiocarbazone, 537
Dipole attraction, 85
Diprotic acids and bases, 199–212
Diprotic buffer, 211–212
Diprotic titration, 233–237
Diprotic fractional composition equations, 258–259
Direct titration, 286
Disorder (entropy), 64
Disproportionation, 68, 378
Distribution coefficient, 536
Dithionite, 377t
Dithizone, 537, 539i, 540–541d
Divinylbenzene, 588
Division:
  propagation of error, 35–36
  significant figures, 30, 37
Donnan equilibrium, 593–594
Dopant, 478b
Doppler effect, 521–522
Double-beam spectrophotometer, 480–481
Double monochromator, 484
Dowex, 589t

Drierite, 24t
Drop counter, 606i
Dropping-mercury electrode, 427–429
  relation of height to limiting current, 432, 446
Dry ice, 239d
Drying, 23–24
Drying agent, 24
DTPA (diethylenetriaminepentaacetic acid), 272i
Dumas method, 111
Dyne, 4t
Dynode, 485

E^0 (standard reduction potential), 303
$E^{0'}$ (formal potential), 320–323, 321t
Eddy diffusion, 556, 557i, 557f, 574b, 585
EDTA, 624–625e
  acid–base properties, 273–276
  back titration, 287–288
  calcium complex, 277
  conditional formation constant, 277–279
  direct titration, 286–287
  displacement titration, 288
  drying instructions, 275f
  effective formation constant, 278
  formation constant from cell potential, 318–320
  formation constants, 276t
  fraction of Y^{4-} ($\alpha_{Y^{4-}}$), 276t
  indirect titration, 288
  iron complex, 277i
  literature, 286f
  manganese complex, 277i
  masking, 288–289
  metal complexes, 276–277
  metal ion indicators, 283–286, 284d, 285t
  pH effect on titrations, 279
  primary standard, 275f
  releasing agent, 528
  solubility of complexes, 537
  solution preparation, 624
  stability constants, 276t
  structure, 272i
  titration curve, 279i, 280–283
Effective formation constant, 278
EGTA (bis(aminoethyl)glycolether-N,N,N',N'-tetraacetic acid), 272i
Eight-coordination, 277
Electrical power, 297
Electrical resistance, 296
Electrical work, 295, 296, 297, 307
Electric charge, 3t, 294
Electric circuit, 297
Electric current, 1, 2t, 294, 296, 312b
Electric double layer, 115
Electric potential, 3t, 295, 296, 307
  concentration gradient, 345
  free energy, 296, 308b
  sign in galvanic cell, 299
Electroactive species, 333, 407m
Electrocapillary maximum, 434
Electrochemical cell (chemical probe), 316–320
Electrochemical writing, 398d
Electrochemistry:
  amperometric titrations, 456–461
  basic electrical quantities, 294–297
  biamperometric titration, 457–459
  bipotentiometric titration, 460b, 461
  coulometry, 413–422
  current-flow effect, 397–405
  cyclic voltammetry, 452–454

differential pulse polarography, 447–451
electrogravimetric analysis, 405–413
electrolysis, 395–405
equilibrium constant and E^0, 314
formal potential ($E^{0'}$), 320–323
galvanic cells, 297–303
galvanic cells as chemical probes, 316–320
glass electrode, 343–349
indicator electrodes, 336–342
ion-selective electrodes, 349–356
junction potential, 342–343
Nernst equation, 307–314
optically transparent thin-layer electrode, 455b
oxidation-reduction reactions, 293
pH electrode, 343–349
polarography, 427–446
reference electrodes, 333–336
standard potentials, 303–307
stripping analysis, 446, 451–452
thin-layer electrode, 453b
voltammetry, 427–461
Electrode, 295, 299
  indicator, 365
  nonaqueous acid–base titration, 249i
  nonpolarizable, 456
  optically transparent, 455b
  polarizable, 456
  potential, 313–314
  reference, 365
  rotating platinum, 456–457
  thin-layer, 453b, 455b
Electrogravimetric analysis, 395, 405–413, 629e
  apparatus, 406i, 407i
  constant-applied-potential, 410–411
  constant-current, 412
  controlled-potential, 412–413
  current–voltage relation, 406–410
Electrolysis, 395–405
  photo-assisted, 400–401b
Electrolyte:
  charge type, 82
  conductivity, 183b
  gravimetric analysis, 115–116
  strong, 6, 156, 161t
  volatile, 117
  weak, 6, 182, 183d
Electromagnetic radiation, 467, 468, 469i
Electromotive efficiency, 346
Electromotive force, 3t
Electronic transition, 474–475
Electron volt, 4t
Electrophoresis, 262b
Eluate, 553
Eluent, 553
Eluent strength, 584
Elutropic series, 584t
Emission:
  atomic, 513, 514, 529
  temperature effect, 520–521
Emission intensity, 504
Emission monochromator, 501i
Emission spectrum, 478b, 499–504
Endergonic, 65
Endothermic, 63
Endpoint:
  acid–base titrations, 223i, 224, 225, 228–229, 230i, 231i, 235–236, 237–246
  amperometry, 456–461
  coulometry, 416–417

EDTA titrations, 283–286
indistinct, 236–237
precipitation titrations, 136, 137b, 146–148, 149d
redox reaction, 374–375
redox titrations, 374–376
spectrophotometric titration, 496i
Energy, 3t, 4t
  electrical, 297
  electromagnetic radiation, 469i
Enthalpy, 63, 273
Entropy, 64, 273
Enzyme, 186i
Eppendorf pipet, 21
ε (molar absorptivity), 470
Equilibrium:
  acid–base, 161–171
  cell potential, 314–316
  complex formation, 73b
  EDTA complex formation, 276–279
  electrochemical cell, 316–317
  strong plus weak acids and bases, 191b
  systematic treatment, 93–106
  weak acid–weak base, 261–268
Equilibrium constant, 61, 629–631e, 633–634e
  cell potential, 314–316
  chelate effect, 272–273
  consecutive reactions, 62–63
  determination, 496f, 496–499
  Le Châtelier's principle, 65–66
  literature sources, 74
  polarography, 443–445
  reverse reaction, 62–63
  Scatchard plot, 496–499, 633–634
  temperature effect, 66
Equivalence point:
  acid–base titrations, 223i, 224, 225, 228–229, 230i, 231i, 235–236, 237–246
  amperometry, 456–461
  coulometry, 416–417
  EDTA titrations, 283–286
  indistinct, 236–237
  precipitation titrations, 136, 137b, 146–148, 149d
  redox reaction, 374–375
  redox titrations, 374–376
  spectrophotometric titration, 496i
Erg, 4t
Eriochrome black T, 284, 285t, 286, 624
Error:
  buret, 18
  determinate, 32
  experimental, 32–34
  indeterminate, 32
  pH measurement, 348–349
  random, 32–33
  spectrophotometry, 485–488
  systematic, 32
  weighing, 17
Erythrosine, 241t
Ethoxide, 163
4'-ethoxy-2,4-diaminoazobenzene, 375t
Ethoxyl group gravimetric analysis, 126p
Ethylenediamine, 272
Ethyl orange, 241t
eV (electron volt), 4t
Excitation monochromator, 501i
Excitation spectrum, 500, 503b
Excited state, 468, 469i, 473–479, 520
Exergonic, 65

Exothermic, 63
Expansion:
    glass, 613
    water, 8, 611–613
Experiments, 611–637
    analysis of acid–base titration curve (Gran plot), 622–624
    analysis of mixture of carbonate and bicarbonate, 619
    calibration of volumetric glassware, 611–614
    coulometric titration of cyclohexene with bromine, 631
    EDTA titration of calcium and magnesium, 624
    electrogravimetric analysis of copper, 629
    gravimetric determination of calcium with oxalate, 615
    gravimetric determination of iron as ferric oxide, 616
    iodimetric titration of vitamin C, 627
    Kjeldahl nitrogen analysis, 620–622
    polarographic measurement of equilibrium constant, 629–631
    potentiometric halide titration with silver, 627–628
    preparing standard acid and base, 617–618
    properties of ion-exchange resin, 634–636
    quantitative analysis by gas chromatography or HPLC, 636–637
    spectrophotometric determination of iron in vitamin tablets, 631–633
    spectrophotometric measurement of equilibrium constant, 633–634
    synthesis and analysis of ammonium decavanadate, 625–626
Exponent, 641–643
Exponential power series, 502m
Extended Debye–Hückel equation, 85
Extinction coefficient, 470

F (farad), 3t
F (Faraday constant), 294
f (femto), 5t
F (formal concentration), 6
Factorial, 547m
FAD (flavin adenine dinucleotide), 321t
Fajans titration, 149d
Farad, 3t
Faradaic current, 434, 447i, 449–450
Faraday constant, 294, 418–419b
femto, 5t
Ferric citrate, 600i
Ferric nitrilotriacetate, 496m
Ferric oxide, 616
Ferricyanide, 388t
Ferrioxamine B, 274b
Ferritin, 600i
Ferrocyanide, 377t
Ferroin, 374–375
Ferrous ammonium sulfate, 380
Ferrous ethylenediammonium sulfate, 380
Ferrous ion 377t
Ferrozine, 487i
Fick's law, 432f
Filterability of precipitate, 112
Filter paper:
    ashless, 23
    folding, 23, 23i, 616, 617i
Filtrate, 22
Filtration, 22–23
Fines, 603

Flame, 516–517
Flame ionization detector, 579–580
Flame photometer, 525b
Flame temperature, 516t, 519i
Flavin adenine dinucleotide, 321t
Florisil, 584
Flow adaptor, 583
Flow analysis, 492b
Flow cell, 484i
Fluorapatite, 103b
Fluorescence, 472d, 477–478, 478b, 499–505
    analytical applications, 504–505
    atomic, 514
    emission spectrum, 532p
    excitation spectrum, 531p, 532p
    lifetime, 505
    sensitivity, 504
    spectrophotometer, 501i
Fluorescent lamp, 478b
Fluoride:
    fluorescence analysis, 505
    gravimetric analysis, 120t
    hydrogen bonding, 164b
    ion-selective electrode, 350–351, 352t
    masking agent, 288
Fluorometer, 50li, 505
Fluorosulfuric acid, 243b
Force, 3t, 4t
Formal concentration, 6
Formaldehyde:
    clock reaction with sulfite, 194d
    demasking agent, 289
    geometry, 473i
    iodimetric titration, 387t
    molecular orbitals, 473–474
    vibrations, 475–476
Formality, 6
Formal potential, 320, 382
Formation constant:
    cell potential, 316, 318–320
    EDTA complexes, 276–279, 276t
    notation, 73
Formic acid, 197t
Formula weight, 6
Fraction:
    association, 185, 187–188
    dissociation, 181–182, 187–188, 258–259
    saturation, 499
Fractional composition:
    $E^{0\prime}$ calculations, 322–323
    equations, 257–259
Fractionation range, 598, 599t
Fraction collector, 606i
Free energy, 273
    absolute potential for cell, 304f
    concentration gradient, 326–327p, 345
    electrochemical reactions, 296, 299
    enthalpy and entropy, 64
    equilibrium constant, 64
    half-reactions, 308b
Frequency, 3t, 467
Fritted-glass funnel, 22
Fugacity, 88
Fugacity coefficient, 88
Fulcrum, 15
Fumaric acid:
    formal potential, 321t
    fractional composition, 259i
Functional groups (polarographic behavior), 440t

Funnel, 23
Furnace, 517–518
F.W. (formula weight), 6

G (giga), 5t
Galvanic cell, 297–303
γ (activity coefficient), 83
$\gamma_\pm$ (mean activity coefficient), 89
Gamma ray, 469i
Gap, 51
Gas:
    activity coefficient, 88
    standard state, 61
Gas chromatography, 567–582, 636–637e
    adjusted retention time, 570
    capacity factor, 570
    capillary column, 574–575b
    carrier gas, 568, 578, 580
    chromatogram, 570
    collecting fractions, 568
    column, 573–578
    detector, 578–580
    flame ionization detector, 579–580
    homologous series, 571
    instrumentation, 567–569
    integrator, 581
    internal standard, 582
    Kovats index, 571
    liquid phases, 576–577t, 578t
    polarity of stationary phase, 578t
    preparation of packing, 578
    quantitative analysis, 580–582
    response factor, 581
    retention index, 571
    retention time, 570
    silanization, 573
    solid support, 573
    splitter, 574b
    stationary phase, 573–578
    stationary-phase loading effect, 578
    temperature-dependence of retention, 572
    temperature programming, 571–572
    thermal conductivity detector, 578–579, 580
    volatile derivative, 575
Gas–liquid partition chromatography, 567
Gas–solid adsorption chromatography, 567
Gathering, 117
Gaussian error curve, 42–46
    area, 45t
    ordinate value, 45t
Gel, 590, 594–595, 598, 599t
Gel electrophoresis, 262b
Gel-filtration chromatography, 597–601
Gel-permeation chromatography, 597–601
Geometric mean, 44
Gibbs free energy, 64
giga, 5t
Glass:
    adsorption of ions, 356m
    attack by base, 248, 275f
    silanization, 561
    storage of base, 616f
    structure, 343, 344i
Glass electrode, 343–349
    acid–base titrations, 240, 244i
    calibration, 346–348
    compound electrode, 354
    errors, 348–349
    reconditioning, 346–348
    reference electrode, 340
    sodium electrode, 349

theory of operation, 343–346
Glassware calibration, 611–614
Globar, 482
Globular molecule, 598
Gluconic acid, 321t
Glucose, 386
Glutamic acid, 201t
Glutamine, 201t
Glutathione:
  formal potential, 321t
  iodimetric titration, 387t
Glycerol, 389
Glycine, 201t
Glycine amide, 198t
Glycylglycine, 198t
Glyoxalic acid, 321t
Gold electrode, 336
Gooch filter crucible, 22i
Gradient elution, 584, 595, 605, 608p
Gradient maker, 605i
Gramicidin, 541b
Gran plot, 244f–246, 254p, 622–624
Graph:
  significant figures, 31
  straight line, 644–645
Graphite furnace, 517–518
Grating, 472d, 482–484
Gravimetric analysis, 22, 109–126
  atomic weights, 110
  calculations, 119–125
  combustion analysis, 110–111
  precipitation process, 111–119
  scope, 119, 120t, 121t
Ground state, 468, 469i

h (hecto), 5t
H (henry), 3t
Half-reaction, 298, 308b
Half-wave potential, 434–437
Halfwidth, 555
Halide electrode, 341–342
Halide titration, 627–628e
Hamilton syringe, 22
Hammett acidity function, 242–243b
Hardness of water, 289
Heat, 3t
  equilibrium-constant effect, 66
  solution, 66f
hecto, 5t
Helium, 578
Hemoglobin, 274b
Henderson–Hasselbalch equation, 188–189
Hengar granules, 620
Henry, 3t
HEPES (N-2-hydroxyethylpiperazine-N′-2-ethanesulfonic acid), 197t
HEPPS (N-2-hydroxyethylpiperazine-N′-3-propanesulfonic acid), 198t
Hertz, 3t, 467
H.E.T.P. (height equivalent to a theoretical plate), 554, 557, 558, 573, 574b, 578
Hexachloroantimonate, 156
Hexamethyldisilazine, 561
High performance liquid chromatography, 585–587, 636–637e
Histidine, 201t
  pH calculations, 214
Hollow-cathode lamp, 522
Homogeneous precipitation, 114, 115t
Horsepower, 4t
HPLC, 585–587, 636–637e

Hydrated gel (glass electrode), 344i, 345
Hydrated radius, 85, 85–86t, 87
  inorganic ions, 85–86t
  organic ions, 86t
Hydrazine, 377t
  iodimetric titration, 387t
Hydrochloric acid, 161t
  fountain, 162–163d
  preparation, 618
  primary standard, 247t, 254p
  solubility in water, 162d
  strength in acetic acid solvent, 164
  titrant in acetic acid, 249
Hydrofluoric acid, 164b
Hydrogen:
  analysis, 110
  formal potential, 321t
  generation by photo-assisted electrolysis, 400–401b
  overpotential, 405t
Hydrogen bond, 158, 164b
  pK_A effect, 179
Hydrogen bromide, 161t
Hydrogen cyanide, 387t
Hydrogen electrode, 303–304
Hydrogen iodide, 161t
Hydrogen ion, 155
  hydrogen bonding to fluoride, 164b
  ion exchange with glass surface, 345
  measurement of activity, 343–349
  structure, 157–158
Hydrogen–oxygen flame, 516t
Hydrogen peroxide, 377t, 378
  continuous-flow analysis, 492b
  demasking agent, 289
  iodometric titration, 388t
  permanganate titration, 381t
  polarographic wave, 438
Hydrogen sulfide, 378
  electrode, 354
  iodimetric titration, 387t
Hydrolysis, 204m
Hydronium ion, 155
  hydrogen bonding to fluoride, 164b
  ion exchange with glass surface, 345
  measurement of activity, 343–349
  structure, 157–158
Hydroquinone, 329–330p, 363, 377t, 632
Hydrostatic pressure, 604–605
Hydroxamate group, 274b
Hydroxide:
  alkali metal equilibria, 173p
  alkaline metal equilibria, 162t
  homogeneous precipitant, 115t
  masking agent, 289
  structure, 158
Hydroxyapatite, 103b
Hydroxylamine, 377t, 489
Hydroxynaphthol blue, 624
8-hydroxyquinoline, 121t, 124, 537
  homogeneous precipitant, 115t
  releasing agent, 528
Hygroscopic, 15, 118
Hypochlorite, 377t
Hypochlorous acid, 388t
Hypoiodous acid, 385m
Hz (hertz), 3t

Ignition, 23, 118, 120t, 616
Ilkovič equation, 431–432
Illuminance, 3t

Imidazole hydrochloride, 197t
Iminodiacetic acid, 291p
Impurities in precipitates, 117
Inch, 4t
Inclusion, 117
Indeterminate error, 32
Indicator:
  acid–base, 238–240, 239d, 241t
  error, 239
  metal ion, 283–286, 284d, 285t
  precipitation, 146–148, 149d
  preparation of acid–base indicator solutions, 241t
  redox, 374–375, 375t
Indicator electrode, 333, 336, 337–339d, 340–342, 365
Indigo tetrasulfonate, 375t
Indirect titration, 288, 383
Indium, 533
Inductance, 3t
Inductively coupled plasma, 518–519, 529
Inflection point, 223i
Infrared radiation, 469i
Infrared sample cell, 484
in. (inch), 4t
Inorganic ion exchanger, 590, 594–595
Instrument errors, 487–488
Insulator, 400b
Integrator, 581
Intercept, 645
  least-squares calculation, 54
Interference:
  atomic spectroscopy, 528–529
  EDTA titrations, 288–289
  gravimetric analysis, 119
  ion-selective electrodes, 349–350, 352t
  light rays, 483i
  removal, 533
Intermediate form of diprotic acid, 205–210
Internal conversion, 477
Internal standard, 442–443, 527–528, 582, 636–637e
Interpolation, 46, 87
Intersystem crossing, 477
Iodate, 377t, 385
  iodometric titration, 388t
  redox titration, 370–374
Iodide:
  gravimetric analysis, 120t
  ion-selective electrode, 352t
  titration, 627–628e
Iodimetric titration of vitamin C, 627e
Iodimetry, 383, 387t, 460d
Iodine, 377t
  pyridine complex, 633
  redox titrations, 383–386, 387t, 388t
  standardization, 384–385
  starch indicator, 376–384, 376, 384
Iodometry, 383, 388t
Ion exchange:
  glass surface, 345
  liquid ion exchanger, 351–353
  water purification, 161m, 596
Ion-exchange chromatography, 551, 552i, 587–597, 634–636e
  applications, 596–597
  catalysis, 597
  choice of exchanger, 594–595
  column dimensions, 595
  concentration of trace species, 596–597
  deionization, 596

742 Ion-exchange chromatography (*continued*)
  Donnan equilibrium, 593–594
  functional groups, 588–590, 591t
  gels, 590, 591i, 594–595
  gradient elution, 595
  interconversion of salts, 596
  ion exchangers, 588–591, 594–595
  ion exclusion, 594
  resins, 588–590, 594–595
  selectivity coefficient, 592–593
Ion-exclusion chromatography, 594
Ionic atmosphere, 83, 115, 116i
Ionic radius, 85
Ionic strength, 81
Ionic-strength effects:
  activity coefficient, 84–89
  ion dissociation, 84d
  pH, 159–160
  solubility, 82–83, 89–91
Ionization interference, 528
Ionization suppressor, 529
Ion mobility, 342t
Ionophore, 540–541b
Ion pair, 164b, 250, 538–539
Ion-selective electrode, 349–356
  advantages, 355
  calcium, 351–353, 352t
  carbon dioxide, 354
  compound, 354
  fluoride, 350–351, 352t
  gas-sensing, 354
  liquid-based, 351–353, 352t
  microelectrode, 353b
  operating tips, 355
  relative error, 355m
  selectivity coefficient, 349–350
  solid-state, 350–351, 352t
Iron:
  binding to transferrin, 495–496
  blood, 489–491
  ferrozine complex spectrum, 487i
  gravimetric analysis, 114, 120t, 127p
  gravimetric determination, 616–617e
  hydrolysis, 112d
  Jones reductor treatment, 379
  overload, 274b
  permanganate standardization, 380
  permanganate titration, 381t
  reaction with thiocyanate, 84
  redox titration, 364–369, 373d
  spectrophotometric determination, 631–633e
Isocratic elution, 586
Isocyanate, 504–505
Isoelectric focussing, 262b
Isoelectric pH, 260–261, 262b
Isoelectric point, 222i, 260–261, 262b
Isoionic pH, 260–261
Isoionic point, 222i, 260–261
Isolation of electrode, 420
Isoleucine, 201t
Isosbestic point, 494–495
Isotherm, 560

J (joule), 3t
Jones reductor, 378–379
Joule, 3t, 295
Junction potential, 332f, 342–343, 343t
  glass electrode, 348

K (kelvin), 1
k (kilo), 5t

K_A (acid dissociation constant), 164–165, 166–167
  relation to K_B, 169–170
Karl Fischer jacks of pH meter, 460d
Karl Fischer titration, 549–461
K_B (base association constant), 100
  relation to K_A, 169–170
Kelvin, 1, 2t
Ketone oxidation, 383, 388–389
kg (kilogram), 1
KHP (potassium acid phthalate), 209, 211, 247t
Kidney dialysis, 113d
kilo, 5t
Kilogram, 1, 2t
Kinetic polarization, 405t
Kinetic prediction, 66
Kinetic wave, 446
Kjeldahl flask, 620i
Kjeldahl nitrogen analysis, 134, 620–622e, 626e
Knife edges, of balance, 17
Kovats index, 571
K_w, 159, 159t

L (liter), 4t
Laboratory notebook, 13–14
Lactic acid, 103b
Lactoglobulin, 263b
λ (wavelength), 467, 468
Laminar flow, 560
Lanthanide elements, 86f
Lanthanum:
  permanganate titration, 381t
  releasing agent, 528
Lanthanum fluoride, 350
Latimer diagram, 308b
Law of mass action, 61
1b (pound), 4t
Lead:
  EDTA titration, 286–287
  gravimetric analysis, 120t
  iodide complexes, 72p
  iodimetric titration, 387t
  ion-selective electrode, 351
  oxalate formation constant, 629–631e
  permanganate titration, 381t
Lead(IV) acetate, 377t
Lean flame, 517
Least squares, 53–58
  error analysis, 55–58
  intercept, 54
  slope, 54
Le Châtelier's principle, 65–66
  acid–base dissociation equilibrium, 187–188
  Nernst equation, 307m
Length, 1, 2t, 4t
Leucine, 202t
  pH calculations, 200–211
Leveling effect, 162–164, 249
Lewis acid, 156, 271
Lewis base, 156, 271
Librium, 448i
Lifetime, 478
Ligand, 73, 271
Light, 467–473
  absorbance, 470
  absorption, 468–473
  absorption spectrum, 472d
  Beer's law, 470, 471b, 491
  diffraction, 482–484
  Doppler effect, 521–522
  electric field, 467

electromagnetic spectrum, 468, 469i, 472d, 473t
  emission, 477–478, 499–505
  fluorescence, 477, 478b, 499–505
  frequency, 467
  interference, 483
  luminescence, 478, 499–505
  magnetic field, 467
  molecular excitation, 473–479
  phosphorescence, 477
  plane-polarized, 467
  properties, 467–468
  quantitative analysis, 488–499, 504–505
  refractive index, 468
  sources, 482
  spectrophotometer, 479–485
  speed, 468
  transmittance, 470
  wavelength, 467
  wavenumber, 468
Limiting current, 431, 432t
Linear graph, 644–645
Linear interpolation, 46, 87
Linear regression, 53–58
Line notation, 302–303
Linewidth, 521–522
Liquid, standard state, 61
Liquid chromatography, 583–587, 636–637e
  column, 583, 606
  detector, 584, 585i, 586
  elutropic series, 584t
  flow adaptor, 583
  gradient elution, 584
  HPLC, 585–587
  instrumentation, 586i
  normal-phase, 587
  particle-size effect, 585
  reverse-phase, 587
  selecting conditions, 584–585
  stationary phase, 583–584, 587
  thin-layer, 584–585
Liquid junction, 332f
Liter, 4t
Lithium, 525b
Lithium hydroxide, 161t, 173p
Lithium perchlorate (drying agent), 24t
Litmus, 241t
1m (lumen), 3t
Logarithm, 30, 641–643
  significant figures, 30–31
Longitudinal diffusion, 556, 557i
Lowry protein determination, 52
Luciferin, 511–512p
Lumen, 3t
Luminescence, 478, 499–505
  absorption and emission, 499–500
  absorption quenching, 503b
  analytical applications, 504–505
  emission intensity, 500–504
  sensitivity, 504
Luminous flux, 3t
Luminous intensity, 2t
1x (lux), 3t
Lycine, 202t

M (mega), 5t
m (meter), 1
m (milli), 5t
m (molality), 8
M (molarity), 5, 6

Magnesia, 584
Magnesium;
  EDTA titration, 286, 624–625e
  gravimetric analysis, 120t, 124
  permanganate titration, 381t
  water hardness, 289
Magnesium oxide (drying agent), 24t
Magnesium perchlorate (drying agent), 24t
Magnetic flux, 3t
Magnetic flux density, 3t
Malaprade reaction, 389
Malic acid, 208b
Malonic acid, 337d
Manganate, 380
Manganese, 120t
Manganese dioxide, 379, 388t
Mantissa, 30
Mariotte flask, 604–605
Masking, 288–289, 490
Mass, 1, 2t, 4t
Mass action, 61
Mass balance, 95–96, 257, 259
Matrix, 355
Maximum suppressor, 437
Mean, 43
  comparison, 48–51
Mean activity coefficient, 89
Measuring pipet, 20
Median, 44
Mediator, 417–420
mega, 5t
Membrane potential, 326p
Meniscus, 18, 20, 21
Mercaptoethanol, 387t
Mercuric sulfide solubility, 102
Mercurous ion, 68, 377t, 390
Mercury, 387t
Mercury electrode, 279i, 342, 358–359p,
  427–429
Mesh size, 573
MES (2-(N-morpholino)ethanesulfonic acid),
  197t
  titration with NaOH, 225–230
Metal ion buffer, 356–357
Metal ion extraction, 537–538, 540–541d
Metal ion indicator, 283, 284d, 285t, 286
Metaperiodic acid, 388
Meter, 1, 2t
Methionine, 202t
Methoxide, 158
Methylammonium chloride, 166
Methylene blue, 321t, 375t
Methyl orange, 241t
Methyl red, 241t, 284d, 437, 494–495
Methyl violet, 241t
Methyl viologen, 420
micro, 5t
Microelectrode, 353b
Micropipet, 21–22
Microporous particle, 587
Microwave radiation, 469i
milli, 5t
Milliliter, 4t
Mitochondrion, 326p
Mixing (gravimetric analysis), 114
Mixture, analysis by spectrophotometry,
  491–494
mL (milliliter), 4t
Mobile phase, 545
Mobility, 342t
Modulation amplitude, 447, 449, 450i

Mohr pipet, 20
Moisture removal, 24t
mol (mole), 1
Molality, 8
Molar absorptivity, 470
Molarity, 5
Mole, 1, 2t, 5
Molecular-exclusion chromatography, 551,
  552i, 597–601
  applications, 600–601
  desalting, 601
  fractionation range, 598, 599t
  gels, 598, 599t
  K_{av}, 598
  molecular weight determination, 600–601
  void volume, 597
Molecular orbital, 473–474
Molecular weight, 5
  determination, 600–601
  significant figures, 29
Molybdenum, 381t
Monochromatic radiation, 479m, 482
Monochromator, 469, 482–484, 487, 501i
Monodentate ligand, 271
Monoprotic system, fractional composition
  equations, 257–258
MOPS (3-(N-morpholine)propanesulfonic
  acid), 197t
Morphine, 270p
Mother liquor, 22
μ (ionic strength), 81
μ (micro), 5t
Multidentate ligand, 271
Multiplication:
  propagation of error, 35–36
  significant figures, 30
Murexide, 285t
M.W. (molecular weight), 5
Myoglobin, 263b

n (nano), 5t
N (newton), 3t
N-2-acetamido-2-aminoethanesulfonic acid,
  197t
N-2-acetamidoiminodiacetic acid, 197t
NADH (nicotine adenine dinucleotide), 321t
NADPH (nicotine adenine dinucleotide
  phosphate), 321t
nano, 5t
Nanometer, 469i
Naphtholphthalein, 241t
1-naphthyl-methylamine, 504–505
National Bureau of Standards pH reference
  buffers, 347
Nebulization, 516
Neocuproine, 490
Nernst equation, 307–314, 335
  glass electrode, 346
  ion-selective electrodes, 346, 349, 350, 353
Neutralization, 156
Neutral pH, 161
Neutral red, 241t
Newton, 3t
N-2-hydroxyethylpiperazine-N′-2-ethanesul-
  fonic acid, 197t
N-2-hydroxyethylpiperazine-N′-3-propanesul-
  fonic acid, 198t
Nickel:
  EDTA titration, 287
  gravimetric analysis, 120t, 122–123
Nicotine adenine dinucleotide, 321t

Nicotine adenine dinucleotide phosphate, 321t
Nicotine titration, 233i, 236–237
Nigericin, 541b
Nitramine, 241t
Nitrate:
  cathodic depolarizer, 411
  gravimetric analysis, 120t
  indirect titration, 383
  ion-selective electrode, 352t
Nitric acid, 161t
Nitrilotriacetic acid, 272i, 356, 496m
Nitrite:
  iodometric titration, 388t
  spectrophotometric analysis, 509p
Nitro compound, 390m
Nitrogen, Kjeldahl analysis, 111, 620–621e
Nitrogen oxide electrode, 354
Nitrogen oxides, 379
Nitron, 121t, 126p
1-nitroso-2-naphthol, 121t
Nitroso compound, 390m
Nitrous acid, 381t
Nitrous oxide flame, 516t
N-methylcarbazole, 499i
3-(N-morpholine)propanesulfonic acid, 197t
2-(N-morpholino)ethanesulfonic acid, 197t,
  225–230
N,N-bis(2-hydroxyethyl)glycine,N′-methylene
  bisacrylamide, 198t, 598, 599i
Nonactin, 541b
Nonaqueous titrations, 248–250, 250f
Nonbonding orbital, 473–475
Nonpolarizable electrode, 456
Normal error curve, 42–46
Normality, 656–658
Normalization, 45
Normal-phase chromatography, 587
Notebook, 13–14
NTA (nitrilotriacetic acid), 272i
N-tris(hydroxymethyl)methyl-2-aminoethane-
  sulfonic acid, 197t
N-tris(hydroxymethyl)methylglycine, 198t
ν (frequency), 467, 468
Nucleation, 112

Occlusion, 117
Ohm, 3t
Ohmic potential, 399
Ohm's law, 296–297
o-hydroxybenzoic acid, 178–179, 182i, 185–186
Oleum, 243b
Ω (ohm), 3t
o-phenanthroline solution preparation, 632
Optical isomer separation, 574–575b
Optically transparent thin-layer electrode,
  455b
Optical rotation, 424p
Oscillating reaction, 337–339d
Osmolarity, 8
Ostwald–Folin pipet, 20i, 21
o-tolidine, 455b
OTTLE (optically transparent thin-layer
  electrode), 455b
OV-17, 576t
Ovalbumin, 601
Overall formation constant, 73
Overloading (chromatography), 560–561, 563p
Overpotential, 403–405, 405t
Ovotransferrin, 263b
Oxalate:
  homogeneous precipitant, 115t

743

744

Oxalate (*continued*)
  lead formation constant, 629–631e
  permanganate standardization, 380
Oxalate oxidase, 492b
Oxalic acid, 212
  continuous-flow analysis, 492b
  permanganate titration, 381t
  principal species, 256
Oxaloacetic acid, 321t
Oxidant, 293
Oxidation, 293
Oxidation number, 649–651, 655
Oxidizing agent, 293
Oxidizing strength, 306t
Oxine, 121t
Oxygen:
  coulometry, 423p
  electrode, 458b
  formal potential, 321t
  iodometric titration, 388t
  overpotential, 405t
  polarographic waves, 438
Ozone, 388t

p (pico), 5t
Pa (pascal), 3t
PAN (pyridylazonaphthol), 285t, 288
Parallax, 18
Paraperiodic acid, 388
Particle growth (gravimetric precipitate), 112
Particle size, in chromatography, 588
Partition chromatography, 551, 552i
Partition coefficient, 534, 554, 560, 570
Parts per billion, 8
Parts per million, 8, 513m
Parts per thousand, 8
Pascal, 3t, 4e
Peak area measurement, 580
Pellicular particle, 587
Peptization, 117
Percent composition, 6–7
Percent transmittance, 470
Perchlorate:
  gravimetric analysis, 120t
  ion-selective electrode, 352t
Perchloric acid, 157–158, 161t, 377t, 378
  strength in acetic acid solvent, 164
  titrant in acetic acid, 249–250
Periodate, 377t, 388t
Periodic acid, 388–389
Permanganate, 377t, 379–381
  indirect titration, 383
  iodometric titration, 388t
  preparation and standardization, 626e
  standardization, 380
  titration 373d, 381t
Peroxides, 383
Peroxydisulfate, 377t, 377–378
  iodometric titration, 388t
  permanganate titration, 381t
Persulfate, 377–378
pH, 98, 160
  acidity scale, 98
  activity coefficient use, 98–99
  ATP synthesis, 326p
  from cell potential, 317–318
  electrochemical cell calculations, 311
  glass electrode, 343–349
  measurement errors, 348–349
  negative, 242–243b
  pOH, 176

principal species, 255–257
scale, 161
Phase transfer catalyst, 540b
pH effects:
  activity coefficient, 159–160
  cell potential, 320–323
  EDTA complex formation, 279
  enzyme reaction rate, 186i
  extractions, 535–538
  formal potential, 323i
  solubility, 100
  tooth decay, 103b
Phenol, 249i
Phenolphthalein, 241t, 618, 622
  spectrum, 472d
Phenol red, 241t
Phenosafranine, 375t
Phenylalanine, 202t
Phenylboronic acid, 602
Phenylhydrazine, 217p
pH meter:
  calibration, 346, 349
  as potentiometer, 302d, 337d
  resistance, 312b
Phosphate:
  base association constants, 167
  gravimetric analysis, 120t
  permanganate titration, 381t
  reference buffers, 347t
  spectrophotometric analysis, 509p
Phosphor, 478b
Phosphorescence, 477–478, 499
Phosphoric acid, 197t
Phosphorous acid, 387t
Phosphorus pentoxide, 24t, 110
Phosphorylation, 326p
Photo-assisted electrolysis, 400–401b
Photochemistry, 479
Photomultiplier, 485
Photon, 468
Phototube, 485
pH scale, 161
p-hydroxybenzoic acid, 178–179, 182i, 185–186
pico, 5t
Pi orbital, 473–474
Piperazine, 119
Piperazine-N,N′-*bis*(2-ethanesulfonic acid). 197t
PIPES (piperazine-N,N′-*bis*(2-ethanesulfonic acid), 197t
Pipet, 20–22
  bulb, 21f
  calibration, 611–614
  tolerance, 21t
pK, 167
  strong acids, 242b
  titration curve, 230
pK_A, 167
  information from protein titration, 221, 222i
  principal species, 255–257
pK_B, 167
pK_w, 159t, 167
Planck's constant, 468
Plane angle, 3t
Plane-polarized light, 467
Planimeter, 580
Plasma, 489
Plate height, 573, 574b, 578, 585
Plate theory of chromatography, 554–555
Platinum electrode, 336, 337–339d, 407i
p-nitroaniline, 242b

p-nitrophenol, 241t
pOH, 176
Polarity, 578t
Polarizability, 592m
Polarizable electrode, 456
Polarized light, 467
Polarogram, 430
Polarographic wave, 430, 434–435
Polarography, 427–452, 453b, 629–631e
  apparatus, 428i, 451i
  applications, 438–446
  catalytic wave, 446
  chemical kinetics, 445–446
  current maxima, 437
  differential pulse polarography, 446, 447–451
  diffusion current, 431–433
  dropping-mercury electrode, 427–429
  functional groups, 440t
  half-wave potential, 435–437
  internal standard, 442–443
  kinetic wave, 446
  measurement of equilibrium constants, 443–445
  organic functional groups, 440t
  oxygen waves, 438
  polarogram, 430–438
  polarographic wave, 430, 434–435
  qualitative analysis, 438–439
  quantitative analysis, 439–443
  relation of wave height to mercury height, 432, 446
  residual current, 433–434
  standard addition, 441–442
  standard curve, 439–441
  stripping analysis, 446, 451–452
  supporting electrolyte, 432–433
  thin-layer cell, 453b
Polyacrylamide, 598, 599i
Polybasic species, 167, 170, 199
Polychromatic light, 482m
Polyprotic acid, 166–167, 170, 199
  pH calculations, 213–215
  weak acid–weak base analogy, 266–268
Polyprotic base, 167, 170, 199
Polystyrene, 588
Polyurethane, 504
Pooled standard deviation, 49
Population standard deviation, 55
Postprecipitation, 117
Potassium:
  crown ether complex, 540b
  EDTA titration, 290p
  flame photometer, 525b
  gravimetric analysis, 120t
  ionophore complex, 541b
  permanganate titration, 381t
Potassium acid phthalate, 209, 211, 247t, 617, 622
Potassium chloride:
  junction potential of salt bridge, 343
  salt bridge, 301
Potassium cyanate, 115t
Potassium dichromate:
  redox titrations, 383
  spectrum, 472d
Potassium dihydrogen citrate reference buffer, 347t
Potassium hydrogen phthalate reference buffer, 347t
Potassium hydrogen tartrate reference buffer, 347t

Potassium hydroxide, 24t, 161t
  solution preparation, 246–248
Potassium iodate, generation of iodine, 385, 627
Potassium permanganate, 379–381
  preparation and standardization, 626e
  standardization, 380
  titrations, 381t
  water purification, 161m
Potential, 295, 296, 307
  absolute potential for cell, 304f
  ATP synthesis, 326p
  ohmic, 399
  sign in galvanic cell, 299
Potential energy, 404i
Potentiometer, 337d
  convention in cells, 299
  pH meter as, 302d
Potentiometric titration, 364–372, 373d, 374, 622–624e, 627–628e
Potentiostat, 412
Pound, 4t
  conversion to newton, 108p
Pouring columns, 603
Power, 3t, 4t, 297
Precipitant, 114m
Precipitation, 111–119
  controlled, 114
  crystal growth, 112–114
  digestion, 116
  EDTA titrations, 287
  electrolyte role, 115–116
  filterability of product, 112–116
  homogeneous, 114, 115t
  impurities, 117
  product composition, 118
  separations, 70
  solubility control, 111
Precision, 33
Precolumn, 586
Prefix (multiplier), 5, 5t
Premix burner, 516–517
Preoxidation, 376–378
Preparative scale electrolysis, 439, 439f
Preparing solutions, 9–10
Prereduction, 378–379
Pressure, 3t, 4, 4t
  equilibrium constant, 62
Pressure broadening, 521–522
Primary amine, 166
Primary-standard acids and bases, 246, 246f, 247t
Principal species, in acid–base chemistry, 255–257
Principle of Le Châtelier, 65–66
Probe, electrochemical cell, 316–320
Proline, 202t
Propagation of error, 34–37, 646–648
  addition and subtraction, 34–35
  general formula, 646
  logs and antilogs, 647–648
  multiplication and division, 35–36
  powers and roots, 646–647
Protein, 489
  analysis, 52, 59p, 620–621
  isoelectric focussing, 262b
Protic solvent, 159
psi (pounds per square inch), 4t
Pyridine, 232
Pyridoxal phosphate, 171
Pyridylazonaphthol, 285t, 288

4-(2-pyridylazo)-resorcinol, 533
Pyrocatechol violet, 285t
Pyruvic acid, 321t

QF-1, 576t
Q test, 51–52, 51t
Quadratic equation, 177, 180
Qualitative analysis (polarography), 438–439
Quartz, 484
Quaternary ammonium hydroxide, 161t
Quinhydrone electrode, 329–330p
Quinone, 329–330p, 363

rad (radian), 3t
Radiant flux, 3t
Radiant power, 469, 479, 501–504
Radio wave, 469i
Random error, 32–33
Range, 44, 51
Rate theory of chromatography, 556–558
Reaction quotient, 65
  Nernst equation, 309
Reagent blank, 489
Redox indicators, 374–375, 375t
Redox reaction, 293
  balancing, 651–655
Redox reagents, 377t
Redox titration, 363–390
  calculations of curve shape, 364–374
  cerium(IV), 382–383
  dichromate, 383
  indicators, 374–375
  iodine, 383–386, 387t, 388t
  periodic acid, 388–389
  permanganate, 379–380, 381t
  preoxidation, 376–378
  prereduction, 378–379
  reagents, 376–390, 377t
  reducing titrants, 389–390
  titration of mixture, 373–374
Reduced plate height, 557, 558
Reduced velocity, 557, 558
Reducing agent, 293
Reducing strength, 306t
Reducing sugar, 386
Reducing titrations, 389–390
Reductant, 293
Reduction, 293
Reduction potential:
  equilibrium constant, 314–316
  Latimer diagram, 308b
  use of tables, 305–307
Reductor, 378–379
Reference cuvette, 479
Reference electrode, 333–336, 365, 412
  calomel, 336
  glass electrode, 340
  silver–silver chloride, 336
Reflection grating, 483i
Refractive index, 468, 486f, 586
Refractory element, 517m
Rejection of data, 51–52
Relative error, 33–34
Relative supersaturation, 113
Releasing agent, 528
Reprecipitation, 117, 616
Residual current, 408, 430, 433–434
Resin, 588–590
Resistance, 3t, 296

glass electrode, 346
pH meter, 312b
Resolution, 484, 548–551, 558–559
Response factor, 581
Retention index, 571
Retention ratio, 553
Retention time, 553, 555i, 570
Retention volume, 554, 555i
Reverse-phase chromatography, 587
Riboflavin, 321t, 504m
Ribonuclease, 221, 222i, 601i
Rich flame, 517, 528
Roots (finding with calculator), 67
Rotational energy, 476
Rounding-off rules, 29
Rubber policeman, 23
Rubidium hydroxide, 161t

s (second), 1
S (siemens), 3t
s (standard deviation), 43
Safranine T, 321t
Salicylic acid, 249i
Salt, 156
Salt bridge, 300–301, 302d, 303m
Salting out, 92p
Saturated calomel electrode, 336
Saturated solution, 67
Scatchard plot, 60p, 496–499, 633–634e
S.C.E. (saturated calomel electrode), 336
Schematic circuit diagram, 297
SE-30, 576T
Second, 1, 2t
Secondary amine, 166
Selectivity coefficient:
  ion exchange, 592–593
  ion-selective electrode, 349–350
Selenium, 620
Self-absorption, 503b, 529
Self-ionization, 158–159
Self-quenching, 503b
Semiconductor, 400b
Semipermeable membrane, 112d
Semi-xylenol orange, 508p
Sensitivity, 523–524
Separations, 70–72, 412–413, 533–561, 567–606
  affinity chromatography, 601–602
  chromatography, 551–561, 636–637e
  controlled-potential electrolysis, 412–413
  countercurrent distribution, 542–551
  gas chromatography, 567–582
  ion-exchange chromatography, 587–597
  liquid chromatography, 583–587
  molecular-exclusion chromatography, 597–601
  precipitation, 70–72
  solvent extraction, 534–541
Sephacryl, 599t
Sephadex, 590, 591i, 598, 599t
Sepharose, 599t
Septum, 567–568
Serine, 202t
Serological pipet, 20i, 21
Serum, 489
Serum iron determination, 489
Seven-coordination, 277
Short-circuited cell, 300
Siemens, 3t
σ (standard deviation), 44, 56

746

Sign, of cell potential, 299
Significant figures, 27–31, 37
   addition and subtraction, 28–30
   exponent, 315
   graphs, 31
   least-squares parameters, 57
   logarithms and antilogarithms, 30–31, 315
   multiplication and division, 30, 37
   retention in calculations, 35m
   zeros, 28
Silanization, 561, 573
Silica gel, 24t, 583
Silver:
   argentometric titration, 627–628e
   chloride complexes, 75p, 79p
   coulometric measurement of Faraday
      constant, 418–419b
   ion-selective electrode, 351
   permanganate titration, 381t
   Walden reductor, 379
Silver electrode, 340–342, 627–628e
Silver halide removal, 628f
Silver(II), 377–378
Silver(II) oxide, 378
Silver–silver chloride electrode, 336, 458b
   pH electrode, 343, 344i
Silver sulfide, 351
Single-electrode potential, 313–314
Singlet state, 475, 477i
SI units, 1–5
Size, activity-coefficient effect, 85–87
Slope, 644
   least-squares calculation, 54
Sodium:
   flame photometer, 525b
   ion-selective electrode, 349
   mobility in glass, 343
   permanganate titration, 381t
Sodium bismuthate, 378
Sodium carbonate, 618
Sodium error, 348
Sodium hydroxide, 161t, 173p
   drying agent, 24t
   preparation, 246–248, 617–618
Sodium hypochlorite, 189
Sodium o-hydroxybenzoate, 185
Sodium oxalate:
   permanganate standardization, 380
   standardization of permanganate, 626e
Sodium tetraphenylborate, 121t, 127p
Sodium thiosulfate, 385–386, 627
Solar energy, 400–401b
Solid, standard state, 61
Solid angle, 3t
Solid support (chromatography), 573
Solubility:
   cell potential, 315
   gravimetric analysis, 111
   ionic-strength effect, 82–83, 89–91
   ligand-concentration effect, 73i
   metal complexes, 537
   pH-dependence, 100
   salts with basic anions, 100–105
   separations, 70
   temperature-dependence, 66f
   tables, 659–662t
Solubility product, 67–68, 315
   tables, 659–662t
Solute, standard state, 61–62
Solvation, 499–500
Solvent, standard state, 62

Solvent extraction, 533, 534–542
   countercurrent distribution, 542–551
   crown ether, 540–541b
   distribution coefficient, 536
   dithizone, 537, 539i, 540–541d
   efficiency, 535
   ionophore, 540–541b
   metal chelator, 537–538
   partition coefficient, 534
   pH effect, 535–538
   strategies, 538–539
Source, spectrophotometer, 482
Soybean trypsin inhibitor, 263b
Spacer arm, 602
Spaces, in writing numbers, 4
Specific gravity, 7
Spectral interference, 528
Spectronic 20 spectrophotometer, 28, 479, 480i
Spectrophotometer, 469, 479–485, 487–488
   atomic spectroscopy, 523–524
   cells, 479, 484
   cuvette, 479, 484
   detector, 484–485
   double-beam, 480–481
   errors, 487–488
   monochromator, 482–484
   sources, 482
   Spectronic 20, 28, 479, 480i
Spectrophotometric titration, 495–496
Spectrophotometry, 631–633e, 633e, 634e
   absorption of light, 468–472
   analysis of mixture, 491–494
   Beer's law, 470, 486, 491
   continuous-flow, 492b
   detectors, 484–485
   double-beam spectrophotometer, 480–481
   emission, 477–478, 499–505
   equilibrium constant measurement, 496–499
   errors, 485–488
   excitation of molecules, 473–479
   fluorescence, 477, 478b, 499–505
   isosbestic point, 494–495
   luminescence, 478, 499–505
   monochromator, 482–484
   phosphorescence, 477
   properties of light, 467–468
   sample cells, 484
   Scatchard plot, 496–499
   serum iron determination, 489–491
   sources of light, 482
   Spectronic 20, 479, 480i
   spectrophotometer, 479–485
   titration, 495–496
Spectrum, 468, 469i, 472d
Speed of light, 468
Spread, 44
Sputtering, 522
Squalane, 576t
sr (steradian), 3t
SrTiO$_3$, 400–401b
Stability constant, 496f
   EDTA complexes, 276–279, 276t
   cell potential, 316, 318–320
   literature sources, 74
   polarography, 444–445
Standard addition, 355–356, 355m, 361t, 441–
   442, 526f–527
Standard curve, 439–441, 490i, 525–526,
   631–633e
Standard deviation, 43–46
   least-squares parameters, 55–57

measure of experimental error, 34
   population, 55
Standard enthalpy change, 63
Standard HCl, 618e
Standard hydrogen electrode, 303–304
Standard NaOH, 617–618e
Standard reduction potential, 673–681t
   equilibrium constant, 314–316
   Latimer diagram, 308b
   use of tables, 305–307
Standard state, 61–62, 63f
Stannous chloride, 378
Stannous ion, 377t, 378
Starch indicator, 376, 384, 627
Static dropping-mercury electrode, 451i
Stationary phase, 545, 603
Statistics, 41–58
   comparison of means, 48–51
   confidence interval, 46–48
   Gaussian error curve, 42–46
   least squares, 53–58
   linear regression, 53–58
   Q test, 51–52
   rejection of data, 51–52
   standard deviation, 43–46
   Student's t, 46–51
Stepwise formation constant, 73
Steradian, 3t
Storage, of dilute solutions, 356m
Straight-line graph, 644–645
   least-squares calculation, 53–58
Strength of redox reagents, 306t
Stripping analysis, 446, 451–452, 453b
Strong acid, 161–164
   pH calculation, 175–178
   reaction with weak base, 191b
   titration of strong base, 221–225
   titration of weak base, 231–232
Strong base, 161–164
   pH calculation, 175–178
   reaction with weak acid, 191b
   titration of strong acid, 221–225
   titration of weak acid, 225–231
Strontium, permanganate titration, 381t
Student's t, 46–51f
   comparison of means, 48
   confidence interval, 46–48
   table, 47t
Styrene, 588
Subtraction:
   propagation of error, 34–35
   significant figures, 28–30
Successive approximations, 180–181, 208–209b
Succinic acid, 197t
Sulfamic acid, 115t
Sulfate:
   EDTA titration, 288
   formal potential, 321t
   gravimetric analysis, 120t
   homogeneous precipitant, 115t
Sulfide:
   EDTA titration, 288, 292p
   homogeneous precipitant, 115t
   ion-selective electrode, 351, 352t
Sulfite, 194d, 321t, 377t
Sulfur dioxide, 377t, 378
   electrode, 354
   iodimetric titration, 387t
Sulfuric acid, 161t
   acidity, 243b
   drying agent, 24

"Super acid," 243b
Superoxide, 417
Supersaturation, 113
Supporting electrolyte, 432
Symmetric bend, 476i
Symmetric stretch, 476i
Synthesis of ammonium decavanadate, 625–626e
Syringe, 21–22
Systematic error, 32
Systematic treatment of equilibrium, 93, 96–97
   amphiprotic species, 205–206
   weak acids, 179

T (tera), 5t
T (tesla), 3t
Tailing, 561
Tare, 15
Tartarate, 286
TC (to contain), 20
TD (to deliver), 20
Teeth, 103b
Teflon, 78p, 573
Temperature, 1, 2t
Temperature effects:
   atomic spectroscopy, 519–521
   buffer, 196
   equilibrium constant, 66
   gas chromatography, 571–572
   gravimetric analysis, 113
   molarity of solution, 8
   pH measurements, 349
   pK_A of acetic acid, 218t
   volume of glassware, 613
   volume of water, 8, 611–613
   weighing, 17
Temperature programming, 571–572
tera, 5t
Tertiary amine, 166
Tesla, 3t
TES (N-tris(hydroxymethyl)methyl-2-amino-ethanesulfonic acid), 197t
Tetrabutylammonium cation, 539
Tetrabutylammonium hydroxide, 249–250
Tetracyanoethyl pentaerythritol, 577t
Tetradentate ligand, 271
Tetrafluoroborate electrode, 352t
Tetrahydroxyethylethylenediamine, 577t
Tetraphenylarsonium chloride, 121t
Tetraphenylborate, 121t, 127p, 290p
Tetrathionate, 385
Thalassemia, 274b
Thallium, redox titration, 370–374
Theoretical plate, 554–555, 557, 558
Thermal cell, 484i
Thermal conductivity detector, 578–579, 580
Thermal expansion, 611–613
Thermodynamic prediction, 66
Thermogravimetric analysis, 118
Thin-layer chromatography, 584–585
Thin-layer differential pulse polarography, 453b
Thioacetamide, 115t
Thiocyanate:
   gravimetric analysis, 120t
   ion-selective electrode, 352t
Thiocyanatoiron(III), 84d
Thioglycolic acid, 489
   iodimetric titration, 387t
Thiosulfate, 377t, 385–386, 627e

Thiourea, 289
Thorium, 381t
Threonine, 202t
Thymol blue, 238–239, 241t
Thymolphthalein, 241t
Time, 1, 2t
Tin:
   gravimetric analysis, 120t
   iodimetric titration, 387t
   redox titration, 373–374
Titanium:
   gravimetric analysis, 120t
   permanganate titration, 381t
   Ti(III) reductant, 390
Titration:
   acid–base, 221–250
   argentometric, 146–150
   back, 287
   blurred endpoint, 236–237
   calculation of curve for redox titration, 364–374
   cerium(IV), 382–383
   dichromate, 383
   diprotic system, 233–237
   direct, 286
   displacement, 288
   EDTA, 271–289
   endpoint, 136, 141–143, 144–146, 146–150, 237–246, 282–286, 374–376, 496i
   Fajans, 149d
   Gran plot, 622–624e
   indirect, 288, 383
   iodine, 383–386, 387t, 388t
   nonaqueous, 248–250
   periodic acid, 388–389
   permanganate, 379–380, 381t
   pK effect, 230, 231i
   potentiometric, 364–374
   precipitation, 136–150
   primary-standard acids and bases, 246–248
   principles, 129–131
   protein, 221, 222i
   redox, 363–390
   redox indicators, 374–375
   redox reagents, 376–390
   redox titration of mixture, 373–374
   reducing agents, 389–390
   spectrophotometric, 495
   stoichiometry calculations, 131–136
   strong acid–strong base, 221–225
   strong base plus diprotic acid, 265i
   techniques with EDTA, 286–289
   weak acid–strong base, 225–231
   weak acid–weak base, 265i
   weak base–strong acid, 231–232
Tolerance:
   buret, 18t
   pipet, 21t
   volumetric flask, 19t
Tooth decay, 103b
Torr, 4t
Toxic chemical destruction, 620f
Transfer pipet, 20, 21
Transferrin, 489, 495–496, 507p, 600i
Transition range:
   acid–base indicators, 238–240, 241t
   redox indicators, 374–375
Transmittance, 470
   Spectronic 20 scale, 28
Transport of electrolyte, 432
Trichloroacetic acid, 489

TRICINE (N-tris(hydroxymethyl)methyl-glycine), 198t
Triethanolamine, 286, 289
Trimethylammonium chloride, 182
Trimethyl phosphate, 115t
Trioctylphosphine, 539
Triodide, 383–384
Triplet state, 475, 477i
tris(2,2′-bipyridine)iron, 375t
tris(2,2′-bipyridine)ruthenium, 375t
tris(hydroxymethyl)aminomethane, 190, 193, 198t
tris(5-nitro-1,10-phenanthroline)iron, 375t
tris(1,10-phenanthroline)iron, 375t
TRIS (tris(hydroxymethyl)aminomethane), 198t
Tritium, 346
Triton X-100, 437
Tropaeolin 0, 241t
Tryptophan, 202t
Tungsten, 381t
Tungsten lamp, 482
Tyrosine, 202t, 221

Ultraviolet light, 469i
Units of measurement, 1–5
Uranium, 381t
Urea, 114, 115t

V (volt), 3t
Vacuum desiccator, 24i
Valence band, 400b
Valinomycin, 541b
Valine, 202t
Vanadate, 120t, 625
Vanadium equilibria, 625
Vanadyl sulfate analysis, 634–636e
van Deemter equation, 556–557
Variance, 43
Versamid 900, 577t
Vibrational relaxation, 477
Vibrational states, 475, 476i, 477
Vibronic structure, 499i, 500
Visible light, 469i
Vitamin C, 274b, 460d
   analysis, 627e
   formal potential, 321t, 322–323
Void volume, 597
Volatilization, 533
Volt, 3t
Voltage, 295
Voltage ramp, 447
Voltammetry, 427–461
Voltmeter, 299
Volume, 4t, 5
   column, 598
   gravimetric analysis, 114
Volume percent, 7
Volumetric analysis, redox reagents, 376f–390
Volumetric flask, 19–20
   tolerance, 19t

W (watt), 3t
Walden reductor, 379
Waste chemical destruction, 620f
Watchglass, 23i
Water:
   acid–base effect on dissociation, 178, 180
   carbon dioxide content, 161
   density, 612t

748 Water (*continued*)
electrolysis, 400–401b
hardness, 289
ionization, 97–98
Karl Fischer titration, 459–461
pH, 159–161
purification, 161m
standard buffer preparation, 347t
Watt, 3t, 297
Wavelength, 467, 486–487
color of light, 473t
Wavenumber, 468
Wb (weber), 3t
Weak acid:
amphiprotic, 205–210
conjugate, 185, 187–188
diprotic, 199–212
equilibria, 164–171, 262
fraction of dissociation, 181–182, 187–188
intermediate form, 205–210
pH calculation, 178–182
polyprotic, 213–215
reaction with strong base, 191b

reaction with weak base 261–268
titration with strong base, 225–231
Weak base:
amphiprotic, 205–210
conjugate, 185, 187–188
diprotic, 199–212
equilibria, 165–171, 262
fraction of association, 185, 187–188
intermediate form, 205–210
pH calculation, 183–186
polyprotic, 213–215
reaction with strong acid, 191b
reaction with weak acid, 261–268
titration with strong acid, 231–232
Weak electrolyte, 182, 183d
Weber, 3t
Weighing bottle, 23i
Weighing procedure, 15
Weight, 5
Weight percent, 6–7
Weight/volume percent, 7
Weston cell, 425p
Wet ashing, 378

Work, 3t, 295, 297, 307
Working electrode, 412

$\bar{x}$ (mean), 43
Xanthine, 321t
XE-60, 577t
X-ray, 469i
Xylenol orange, 285t, 286, 511p
titration with NaOH, 244, 246i

Zeeman effect, 523f
Zeros, significant figures, 28
Zinc:
EDTA titrations, 287
electrode, 342
gravimetric analysis, 120t
iodimetric titration, 387t
Jones reductor, 378–379
permanganate titration, 381t
Zinc–hydroxide complexes, 78p
Zonyl E-7, 577t
Zwitterion, 200

Atomic Weights, Based on ^{12}C

ELEMENT	SYMBOL	ATOMIC WEIGHT	ELEMENT	SYMBOL	ATOMIC WEIGHT
Actinium	Ac	(227)	Mercury	Hg	200.5_9
Aluminum	Al	26.981 54	Molybdenum	Mo	95.94
Americium	Am	(243)	Neodymium	Nd	144.24
Antimony	Sb	121.7_5	Neon	Ne	20.17_9
Argon	Ar	39.948	Neptunium	Np	237.048 2
Arsenic	As	74.921 6	Nickel	Ni	58.7_1
Astatine	At	($\sim$210)	Niobium	Nb	92.906 4
Barium	Ba	137.33	Nitrogen	N	14.006 7
Berkelium	Bk	(247)	Nobelium	No	(259)
Beryllium	Be	9.012 18	Osmium	Os	190.2
Bismuth	Bi	208.980 8	Oxygen	O	15.999_4
Boron	B	10.81	Palladium	Pd	106.4
Bromine	Br	79.904	Phosphorus	P	30.973 76
Cadmium	Cd	112.41	Platinum	Pt	195.0_9
Calcium	Ca	40.08	Plutonium	Pu	(244)
Californium	Cf	(251)	Polonium	Po	($\sim$210)
Carbon	C	12.011	Potassium	K	39.098_3
Cerium	Ce	140.12	Praseodymium	Pr	140.907_7
Cesium	Cs	132.905 4	Promethium	Pm	(145)
Chlorine	Cl	35.453	Protactinium	Pa	231.035 9
Chromium	Cr	51.996	Radium	Ra	226.025 4
Cobalt	Co	58.933 2	Radon	Rn	($\sim$222)
Copper	Cu	63.546	Rhenium	Re	186.2
Curium	Cm	(247)	Rhodium	Rh	102.905 5
Dysprosium	Dy	162.50	Rubidium	Rb	85.467_8
Einsteinium	Es	(254)	Ruthenium	Ru	101.0_7
Erbium	Er	167.26	Samarium	Sm	150.4
Europium	Eu	151.96	Scandium	Sc	44.955 9